The
ISO 9000
Handbook

The
ISO 9000
Handbook

Third Edition

Edited by
Robert W. Peach

McGraw-Hill

New York San Francisco Washington, D.C. Auckland Bogotá
Caracas Lisbon London Madrid Mexico City Milan
Montreal New Delhi San Juan Singapore
Sydney Tokyo Toronto

McGraw-Hill

A Division of The **McGraw·Hill** *Companies*

© 1992, 1995, 1997 by the McGraw-Hill Companies, Inc.

This publication is designed to provide accurate and
authoritative information in regard to the subject matter
covered. It is sold with the understanding that neither the
author nor the publisher is engaged in rendering legal, accounting,
or other professional service. If legal advice or other expert
assistance is required, the services of a competent professional
person should be sought.

*From a Declaration of Principles jointly adopted by a Committee
of the American Bar Association and a Committee of Publishers.*

Library of Congress Cataloging in Publication Data

The IS0 9000 handbook / edited by Robert W. Peach.—3rd ed.
   p. cm.
  Includes index.
  ISBN 0-7863-0786-2
   1. Quality control—Standards. 2. Quality assurance—Standards.
3. ISO 9000 Series Standards. 4. Manufactures—Quality control—
Standards.   I. Peach, Robert W.
TS 156.I838   1997
658.5'62'0218—dc20   96–23785

Printed in the United States of America
  5 6 7 8 9 0 DOC/DOC 9 0 2 1 0 9 8

Contents

Preface to the Third Edition   xi

How to Use This Handbook   xii

About the Editor   xv

SECTION I   Introduction to ISO 9000

Chapter 1
ISO 9000—Ten Years of Marketplace
Development   3

Chapter 2
Background and Development of the ISO 9000
Standards   9
   Introduction   9
   Need for the ISO 9000 Standards   11
   Background of the ISO 9000 Standards   12
   Standardization: Standards and Their
   Implementation   14
   The Role of Regional Blocks of Countries   14
   The ISO 9000 Family of Standards   21
   Quality Management and Quality
   Assurance—Similarities and Differences   22
   Roles of the ISO 9000 Standards   24
   The Growth of Third-Party Registration   25
   Other ISO 9000 Considerations   26
   Concerns About Registration   27
   Outlook for the Future   30

SECTION II   The ISO 9000 Series Standard

Chapter 3
Overview of the ISO 9000 Series Standard   33
   Introduction   34
   Uses of the Standards   34
   Definition of Terms   35
   Types of Standards in the ISO 9000 Series   39
   Other Useful Standards in the ISO 9000
   Family   53

Chapter 4
The ISO 9001 Standard   59
   Introduction   60
   Clause 4: Quality System Requirements   62
   4.1 Management Responsibility   62
   4.1.1 Quality Policy   63
   4.1.2 Organization   64
   4.1.2.2 Resources   65
   4.1.2.3 Management Representative   66
   4.1.3 Management Review   68
   4.1.4 Business Plan   70
   4.1.5 Analysis and Use of Company-Level
   Data   70
   4.1.6 Customer Satisfaction   70
   4.2. Quality System   70
   4.2.1 General   71
   4.2.2 Quality System Procedures   72
   4.2.3 Quality Planning   72
   4.3 Contract Review   75
   4.3.2 Review   76

4.3.3 Amendment to Contract   76
4.3.4 Records   76
4.4 Design Control   78
4.4.2 Design And Development Planning   79
4.4.3 Organizational and Technical Interfaces   80
4.4.4 Design Input   80
4.4.5 Design Output   81
4.4.6 Design Review   82
4.4.7 Design Verification   82
4.4.8 Design Validation   83
4.4.9 Design Changes   83
4.5 Document and Data Control   85
4.51 General   87
4.5.2 Document and Data Approval
and Issue   90
4.5.3 Document and Data Changes   91
4.6.1 Purchasing General   91
4.6.2 Evaluation of Subcontractors   97
4.6.3 Purchasing Data   98
4.6.4 Verification of Purchased Product   99
4.7 Control of Customer-Supplied Product   105
4.8 Product Identification and Traceability   105
4.9 Process Control   107
4.9.1 Process Monitoring and Operation
Instructions   118
4.10 Inspection and Testing   118
4.10.1 General   118
4.10.2 Receiving Inspection and
Testing   118
4.10.22.3 Release for Urgent Production
Purposes   119
4.10.3 In-Process Inspection and Testing   120
4.10.4 Final Inspection and Testing   120
4.10.5 Inspection and Test Records   121
4.11 Control of Inspection, Measuring, and Test
Equipment   125
4.11.1 General   125
4.11.2 Control Procedure   126
4.12 Inspection and Test Status   132
4.13 Control of Nonconforming Product   133
4.13.1 General   133
4.13.2 Review and Disposition of Nonconforming
Product   135
4.13.3 Control of Reworked Product   135
4.13.4 Engineering Approved Product
Authorization   136
4.14 Corrective and Preventive Action   137
4.14.1 General   137
4.14.2 Corrective Action   137
4.14.3 Preventive Action   138
4.15 Handling, Storage, Packaging, Preservation,
and Delivery   138
4.16 Control of Quality Records   144

4.17 Internal Quality Audits   148
4.18 Training   152
4.19 Servicing   157
4.20 Statistical Techniques   158

SECTION III   The Registration and Audit
Process

Chapter 5
Registration and Development of ISO 9000
Standards   165

How to Select a Registrar   166
Introduction   166
Selecting an Accredited Registrar   169
Evaluating Preassessment   175
Assessment and Registration   176
Auditor Qualification   177
Other Considerations   179

Steps in the Registration Process   180
Introduction   180
Integrated Management Systems   180
ISO 9000 Registration   181
Time and Costs of Registration   185

Chapter 6
The Audit Process   191

Internal Quality Audits   192
What Is an Audit?   192
The Role of the Auditor   193
Phases of the Audit: PERC   196

Interview, Not Inquisition: Successful
Communication Techniques   208
Introduction   208
Putting the Auditee at Ease   208
Interview/Communication Techniques   209
General Considerations   210
Dealing With Unusual Situations or Conflicts
Ethics   211
Conclusion   211

SECTION IV   Implementing ISO 9000

Chapter 7
A Basic Guide to Implementing ISO 9000   215
Why Do Companies Take the ISO 9000
Journey?   216

ISO as a Milestone and a Foundation   221
The Journey: A Proven Implementation
Process   233
The Formal Registration Process   296
Conclusion   302

Chapter 8

Quality System Documentation   303

Introduction   304
Writing Documentation   304
Defining the Scope of the System   305
System Structure   313
The Quality Manual: Structure and
Content   319
Procedure Planning and Development   322
Writing Procedures   325
Document Control   333
Computerized Documentation   336
Implementing the System   337
Conclusion   337
Appendix 1: Sample Quality Manual   341
Appendix 2: Sample Procedures   355
Appendix 3: ISO/DIS 10013: Guidelines for
Developing Quality Manuals   368

Beyond Compliance: Managing Records for Increased Protection   371

ISO 9000 Records Requirements and Legal
Issues   372
Statue of Limitations and Product Liability
Records   376
Development of a Records Retention
Schedule   376
Access to Records   379
Interview Key Personnel   379
Research Records Retention Periods   380
Appraise the Records   380
Prepare a Draft Schedule   380
Discuss the Schedule with Affected Departmental
Managers and Legal Counsel   381
Obtain the Necessary Signatures   381
Duplicate and Distribute the Records Retention
Schedule   381
Implementing the Retention
Schedule   382
Records on Developing a Retention
Schedule   383
Suspension of Destruction When Litigation Is
Imminent   384
Personal Records Versus Organization
Records   385
Terminology   385

Chapter 9

A Quality System Checklist   391

Introduction   391
4.1   Management Responsibility   392
4.2   Quality System   393
4.3   Contract Review   394
4.4   Design Control   395
4.5   Document and Data Control   396
4.6   Purchasing   397
4.7   Control of Customer-Supplied
Product   398
4.8   Product Identification and Traceability   398
4.9   Process Control   399
4.10 Inspection and Testing   399
4.11 Control of Inspection, Measuring, and Test
Equipment   400
4.12 Inspection and Test Status   401
4.13 Control of Nonconforming Product   401
4.14 Corrective and Preventive Action   402
4.15 Handling, Storage, Packaging, Preservation,
and Delivery   403
4.16 Control of Quality Records   404
4.17 Internal Quality Audits   404
4.18 Training   405
4.19 Servicing   406
4.20 Statistical Techniques   406

Chapter 10

Using ISO 9000 in Service Organizations   409

What Is a Service Organization?   410
How Does a Service Organization Use ISO 9001
or ISO 9002?   416
Comparisons with Other Service Industry
Standards   441
A Proven Implementation Process for an ISO
9000 Quality System   443
   *A Case Study: ISO 9002 at a Staffing Services
   Company   445*
   *A Case Study in Progress: An Educational
   Institution   447*
Appendix 1: Quality System Model for the
Construction Industry   449
Appendix 2: Quality System Model or
Education   451
Appendix 3: Quality System Model for
Healthcare   453
Appendix 4: Quality System Model for the
Staffing Industry   455

Application of ISO 9000 to the Construction Industry   457

The Dorma Experience   457

Industry Changes   458
How the Standard Relates to Construction   459

Section V   The ISO 9000 Family and Related Standards

Chapter 11
The Future of the ISO 9000 Standards   465

Introduction   465
The Importance of Third-Party Registration   467
Developments in the ISO 9000 Family   470
Marketplace Forces at Work Today   473
Summary   475

Chapter 12
ISO 14000—The International Environmental Management Standard   477

An Introduction to ISO 14000   478
Reasons Behind Development   478
ISO 14000, Part of an International Trend   480
How ISO 14000 Was Developed   480
Structure of the Standard   481
Structure of the ISO 14001 Standard   485
Developing an EMS System   488
Formulation Strategy   490
How to Participate in the ISO 14000 Process   495

Integrating ISO 9001 and 14001   495
Differences Between ISO 9001 and ISO 14001   496
Implementing ISO 14001   497
Critical EMS Elements   497
Comparison to ISO 9000   498

Integrating ISO 14001 With the Chemical Manufacturers Association's Responsible Care   499
Comparing ISO 14001 and Responsible Care   499
Conclusion   502

Important Legal Considerations in Implementing ISO 14001   504
Reasons to Comply With ISO 14001   505
Collecting Sensitive Corporate Information   505
Identifying Environmental Priorities   507
ISO 14001 and Current Legal Requirements   507
Industrywide Standard of Care   508
Conclusion   509

Chapter 13
Comparing ISO 9000, Malcolm Baldrige, and Total Quality Management   511

Introduction   511
Overview of the Two Systems   513
ISO 9000 Compared to MBNQA   526
ISO 9000, MBNQA, and Total Quality in Summary   530

Chapter 14
Other Standards, Guidelines, and Business Process Initiatives based on ISO 9000   537

An ISO Occupational Health and Safety Management System   537
Introduction   537
OHSMS Overview   538
Rationale for an OHSMS   538
Expected Benefits of an OHSMS   541
Current Status of Development   543
AN OHSMS and Its Relation to Existing Standards/Policies   546
Conclusion   552

ISO 9000 in U.S. Government Agencies
Department of Agriculture (USDA)   556
Department Commerce (DOC)   556
International Trade Administration (ITA)   556
National Institute of Standards and Technology (NIST)   557
National Voluntary Conformity Assessment System Evaluation (NVCASE)   557
Office of Standards Services' Weights and Measures Program   558
National Oceanic and Atmosphere Administration (NOAA)   558
Department of Defense (DOD)   558
Department of Education (DOED)   559
Department of Energy (DOE)   559
Department of Health and Human Services (DHHS)   559
Department of Interior (DOI)   560
Department of Labor (DOL)   561
Department of State   561
Federal Trade Commission (FTC)   562
Genereal Services Administration (GSA)   562
International Trade Commission (USITC)   562
National Aeronautics and Space Administration (NASA)   562
Nuclear Regulatory Commission (NRC)   563
Office of Management and Budget   563

U.S. Postal Service   564

**Guide 25 and Laboratory
Accreditation   565**
Introduction   565
Similarities and Differences   566
Other Considerations   567
Laboratory Product   568
Complementary Functions   569
Scope of Accreditation/Registration   569
The Special Role of Accredited Calibration
Laboratories   570
European Position   570
Conclusion   570

The Supplier Audit Confirmation   572
Basic Principles of SAC   572

Chapter 15
QS-9000 Quality System Requirements   575

An Introduction to QS-9000   575
Companies Affected   576
Registration Process   476
The QS-9000 Audit   578
Quality System Assessment (QSA)   578
Additional QS-9000 Requirements   578

The Benefits of QS-9000   587

QS-9000 Implementation   589
Building Support   589
Positioning for QS-9000 Change   590
Teams and QS-9000 Implementation   592
Path to Implementation   593
Registration and Beyond   597

Section VI   Conformity Assessment

Chapter 16
**European Union and Conformity Assessment
Requirements   605**

**The European Union and Conformity
Assessment   606**
Introduction   606
The EU's Single Internal Market   607
Goals of the New System   607
The European Union and U.S. Conformity
Assessment   626
The Critical Role of Conformity Assessment
Conclusion   627

Chapter 17
Registrar Accreditation   629
Registrar Accreditation   630
Accreditation Bodies in Europe   630
Accreditation and Registration   630
Accreditation in the United States   631
Designating US Notified Bodies   633
Accreditation Bodies in Canada   634
Criteria for Registrar Accreditation   635
Interpreting EN 45000   637
Recognition of Registration Certificates   639
The European Accreditation of Certification
(EAC)   643
The European Organization for testing and
Certification (EOTC)   643
Auditor Certification Programs   645
Mutual Recognition of Auditor
Certification   647
ISO 14000 Series Standard for Environmental
Management Systems   648
QS-9000 Quality System Requirements   648
ISO 9001 Interpretations   469
QS-9000 Interpretations   650
Conclusion   650

Chapter 18
Challenges Facing the ISO 9000 Industry   659
ISO 9000 as an Industry   659
The Challenge of Registration Credibility   660
Responsibilities of the Organizations
Involved   663
The Challenge of Continuous
Improvement   664
The Challenge of Statistical Techniques   666
The Challenge of Standards Interpretation   667
The Challenge of Alternate Routes to
Registration   667
The Challenge of Industry-Specific Adoption and
Extension of ISO 9000 Standards   668
The Challenge of Computer Software   670
The Challenge of Environmental Management
Systems   672
Appendix 1—Part 1: A Registrar Accreditation
Board (RAB) White Paper Defining the Scope of
the Certification/Registration of a Supplier's
Quality System   673
Appendix 1—Part 2: Registrar Accreditation
Board (RAB)—Policy, Principles, and Implemen-
tation Regarding Conflict of Interest by RAB-
Accredited Registrars   683
Appendix 3—Part 3: Code of Conduct—Registrar
Accreditation Board (RAB)   682

SECTION VII   ISO 9000 Around the World and in Industry

Chapter 19
ISO 9000 Registration Growth Around the World   689

Registration in the United States and Around the World   689
Registration Around the World   691
Benefits for Individual Companies—The U.S. Experience   691

Status of the Canadian ISO 9000 Registration System   696
Analysis of Canadian Registrars   697
Trends in Canadian ISO 9000 Registration   697

Chapter 20
ISO 9000 in Various Industry Sectors   699

The Application of ISO 9001 to Software Development   700
Introduction   700
Applying ISO 9001 to Software   700
The Guidance in ISO 9000-3   701
Terms and Definitions   701
Mapping ISO 9000-3:1991 to
ISO 9001: 1994   706
Beyond ISO 9000-3   708
Summary: ISO 9001 and Configuration Management   711
ISO 9000-3 and the IEEE Software Engineering Standards   715
Other Programs   715
Sources of Standards and Information   719

Chemical Industry Harmonization of process Management Initiatives Using ISO 9000   724
The Heart of the System   725

ISO 9000 and QS-9000 for the Chemical Industry   730
Introduction   730
ISO 9000 and the CPI   730
Selecting a Registrar   737
QS-9000 Compared to ISO 9001   739
QS-9000 and the CPI   740
ISO 14001 and the CPI   744
Conclusion   744

ISO 9000 and QS-9000 in the Metals Industry   745
Introduction   745
The Metals Industry Continues to Undergo Change   745
ISO 9000 and the Metals Industry   746
Who's Getting ISO 9000 Registered?   747
Internal Improvements Are Recognized   747
ISO 90000—A Good Foundation for Total Quality Management (TQM)   747
The Impact of the Automotive Industry and QS-9000   748
What's Ahead for the Metals Industry?   748

Appendix A
Contributors   749

Appendix B
Standards and Directives   760

Appendix C
Consultants and Training Services   781

Appendix D
ISO 9000/QS-9000/ISO 14000 Registrars   837

Appendix E
Additional Resources   849

Appendix F
Vision 2000: The Strategy for the ISO 9000 Series Standards in the '90s   881

Appendix G
E-mail and Home Page Addresses   897

Appendix H
Acronyms and Glossary   908

Appendix I
ANSI/ISO/ASQC Q9000 Series Standards   916

Preface to the Third Edition

As the 21st century draws near, American business faces expanding international competition. The economic superpowers, the United States and the Far East, will soon be joined by the unified market of the European Union.

The key to economic success in this global marketplace will be higher-quality products and services. This emphasis on increased quality is demonstrated by the growing acceptance of international quality standards such as the ISO 9000 series standard. Meeting and exceeding the requirements of ISO 9000 quality assurance and quality management standards is fast becoming essential to succeed in an ever more competitive marketplace.

The primary objective of this third edition is to fully update and expand the second edition of *The ISO 9000 Handbook.* In a single, comprehensive source, this book contains all the information that an organization needs to understand the ISO 9000 series and initiate the process of implementing the standards. Included in this third edition is information about QS-9000, the American auto industry Big Three producers' publication that places additional quality system requirements on their suppliers.

This book also describes recent developments in the European Union in the "bigger picture" of product standards, product certification, and conformity assessment.

Whatever the motivation—whether to protect sales to the European Union, respond to the requirements of large customers, or adopt the standard on the basis of good quality practice—there is a need to understand not only the content and use of the standard, but also the marketplace factors that are influencing adoption of the ISO 9000 series standard worldwide.

The publishers of *Quality Systems Update (QSU)* are well qualified to provide the material for such a handbook. (Subscribers to *QSU* will recognize that some material contained in the *Handbook* has appeared in previous *QSU* issues. This material has been fully updated.) My responsibility as editor has been to work with the Irwin Professional Publishing staff to ensure that the contents are accurate and unbiased.

In carrying out this process, input has been solicited from a variety of knowledgeable contributors. While many of the conditions affecting registration to the standard are constantly changing, each contributor has made every effort to be as current and as accurate as possible in discussing the subject matter of this *Handbook*.

The contributors to this *Handbook* come from a variety of backgrounds and thus reflect different points of view regarding quality system standards issues. Most of these differences are not substantive, however, and we believe they will contribute positively to the reader's appreciation of the broad spectrum of factors that influence the application of ISO 9000.

In a few cases, however, the opinions of the contributors conflict, or may seem to conflict. As long as the facts used by the contributors are correct, no attempt has been made to resolve them. Rather, such debates reflect the diverse judgments and perspectives of people throughout American industry and should contribute to the comprehensiveness of the *Handbook*.

Underlying the increasing level of ISO 9000 registration activity is the fact that the ISO 9000 standards describe a technically sound quality system for use by manufacturing and service organizations. The standards are proving to be a valuable foundation for expanded quality practice to which principles of Total Quality Management (TQM) are applied. Many companies initially make use of the standards in response to external demands—customer requirements, regulatory compliance, or market competition. They soon find that meeting all requirements of the standard results in significant internal benefits and that the rewards are well worth the necessary cost and effort.

I trust that this *Handbook* will provide the information readers need to apply the standard successfully in their own organizations and to achieve the benefits of an improved quality system, and that readers will discover that this is only the beginning of an era of continual improvement in the quality capability of all segments of commerce and industry.

Robert Peach

How to Use This Handbook

The third edition of *The ISO 9000 Handbook* is fully updated and expanded from the 1994 second edition. The 20 chapters in this edition include sections on European Union history and its complex conformity assessment framework, emerging uses of the standard by other industry segments, and sections on QS-9000 and ISO 14000. New, expanded chapters on documenting and implementing an ISO 9000 system, along with important articles on how the service industry is using ISO 9000, make the third edition an even more useful industry guide.

Like the second edition, the third edition also includes the ANSI/ISO/ASQC Q9000:1994 series standards, verbatim, including Q9000-1, Q9001, Q9002, Q9003, and Q9004-1.

The third edition's expanded resource section provides readers with information on obtaining standards and directives; profiles of consultants and training services; and a list of ISO 9000, QS-9000, and ISO 14000 registrars. Other resources include: publications, software packages, networks, and databases. The third edition of the *Handbook* even includes an e-mail and home page directory drawn from all electronic addresses located throughout the book.

USER-FRIENDLINESS

Implementing international standards can be complex. The editors have confronted this challenge in Chapter 4 by paraphrasing the requirements of the standard in simple, practical language. Each paraphrased clause is noted with a corresponding icon. Guidance from other standards in the series immediately follows each requirement and is distinguished by a different icon.

As an additional feature, Chapter 4 now includes paraphrased QS-9000 requirements in addition to the ISO 9000 requirements.* This feature should be useful for the thousands of companies implementing QS-9000. Another popular feature of the second edition, interpretations of the ISO 9000 standard, is continued here and can be found throughout Chapter 4 as boxes.

We hope that readers of this book will find *The ISO 9000 Handbook,* third edition, a useful tool in implementing any ISO 9000-based quality management system.

Mark Morrow
Executive Editor, Irwin Professional Publishing

---

*The paraphrased QS-9000 requirements in Chapter 4 are taken from *The Memory Jogger™ 9000* published by GOAL/QPC, 13 Branch Street, Methuen, MA 01844, (800) 685-3900, Robert W. Peach and Diane Ritter authors.

About the Editor

Robert W. Peach is principal of Robert Peach and Associated, Inc., Quality Management Consultants, Cary, North Carolina. As a member of the U.S. Delegation to the ISO TC176 Committee on Quality Assurance, he served as Convener of the Working Group that developed ISO Quality System Standard 9004-1 (ANSI/ASQC Standard Q9004-1). He is immediate past chair of the Registrar Accreditation Board, an affiliate of the American Society of Quality Control (ASQC), which he helped form and where he continues as a board member. He is a member of the Executive Committee of the American National Standards Institute (ANSI) Z-1 Accredited Standards Committee on Quality Assurance and serves on the Electronics Components Certification Board (ECCB). At intervals, Mr. Peach evaluates and plans quality management training in developing nations for the World Bank.

Mr. Peach established the quality assurance activity at Sears Roebuck and Company and managed it for over 25 years. In this capacity, he and his staff worked with quality systems in the plants of hundreds of Sears' suppliers.

Mr. Peach has spoken over 300 times to organizations on the subject of quality management. He is editor of *The ISO 9000 Handbook* published by Irwin Professional Publishing, coauthors the *Memory Jogger™ 9000*, published by GOAL/QPC, and is author of the ASQC home-study course, "Successfully Managing the Quality Function." He has received the Edwards Medal of the ASQC for leadership in the application of modern Quality Control methods. Past service in the ASQC includes: vice president, publication; technical editor, quality progress; chairman, standards council; deputy chair, general technical council; chairman, textile and needle trades division; and chairman, awards board. Through ASQC he aided in the evaluation of contractor quality programs for NASA's Excellence Award for Quality and Productivity.

For its initial three years, Mr. Peach served as project manager of the Malcolm Baldrige National Quality Award Consortium, which administered the awards program managed by the National Institute of Standards and Technology. He chaired the writing of the

ANSI/ASQC Z1.15 Standard "Generic Guidelines for Quality Systems." He is a member of ASTM Committee E-11 on Statistical Methods. He is a delegate to the International Laboratory Accreditation Conference (ISAC) and has served on the board of the American Association for Laboratory Accreditation.

Mr. Peach has received degrees from Massachusetts Institute of Technology and the University of Chicago. He is a Fellow of the ASQC, a certified quality engineer and registered professional engineer in quality engineering. He also has been an instructor in quality control in the Graduate School of the Illinois Institute of Technology and has taught courses in Quality Engineering for the ASQC Professional and Technical Development Division.

INTRODUCTION TO ISO 9000

Chapter 1   ISO 9000—Ten Years of Marketplace Development

Chapter 2   Background and Development of ISO 9000 Standards

ISO 9000—Ten Years of Marketplace Development

1

by Mark Morrow

This chapter provides an overview of the influence ISO 9000 is beginning to have on the international marketplace. It suggests directions for the ISO 9000 series of standards and other industry guidelines along with the development of standards associated with ISO 9000, including ISO 14000 and QS-9000.

> Strict adherence to 'made in the USA' policy does not make economic or competitive sense, a CEO of major U.S. manufacturer told an interviewer recently. Most products, the CEO said, cannot ethically make this claim in the global marketplace.

The ISO 9000 Series of Standards will celebrate its 10-year publication anniversary in 1997. In that time, the standard has been attacked and praised in equal measure. Its doom has been predicted and its importance inflated. Yet, despite the misunderstanding and misinterpretation of the purpose and promise of the standard, registration continues to grow by an average of about 250 certificates a month. (See Figure 1–1 and Table 1–1.) Entire industries have adopted the standard including the chemical processing, semiconductors, and much of the electronic components manufacturing industry.

Pinpointing a single reason for interest in ISO 9000 is difficult. Companies seeking registration in the early 1990s, were reacting to either customer demand or a mistaken impression that the developing European Community (now European Union) would soon demand ISO 9000 registration as a prerequisite to do business in Europe. Today, companies still seek registration for marketplace and competitive reasons; however, a growing number of companies now pursue ISO 9000 registration as a valuable business and process management tool.

Yet neither inflated promises, effective marketing campaigns, or a growing sense of business management utility can sustain any movement for long. The real value, and ultimately the reason behind the continued marketplace success of the ISO 9000 standard, is its ability to deliver an organization-specific road map that helps organizations build a continual improving quality management system. In short, ISO 9000, effectively applied, has demonstrated itself as a foundation for a quality management program (see Chapter 13 for a comparison of ISO 9000 to Malcolm Baldrige National Quality Award criteria and to Total Quality Management).

One of the principal developers of the ISO 9000 standard, Donald Marquardt, argues in Chapter 2 of this *Handbook,* "Background and Development of the ISO 9000 Standards," that the importance of the ISO 9000 standards extends well beyond simply providing a

Figure 1–1

ISO 9000 Registrations Have Skyrocketed over a Four-Year Period in Both the United States and Canada

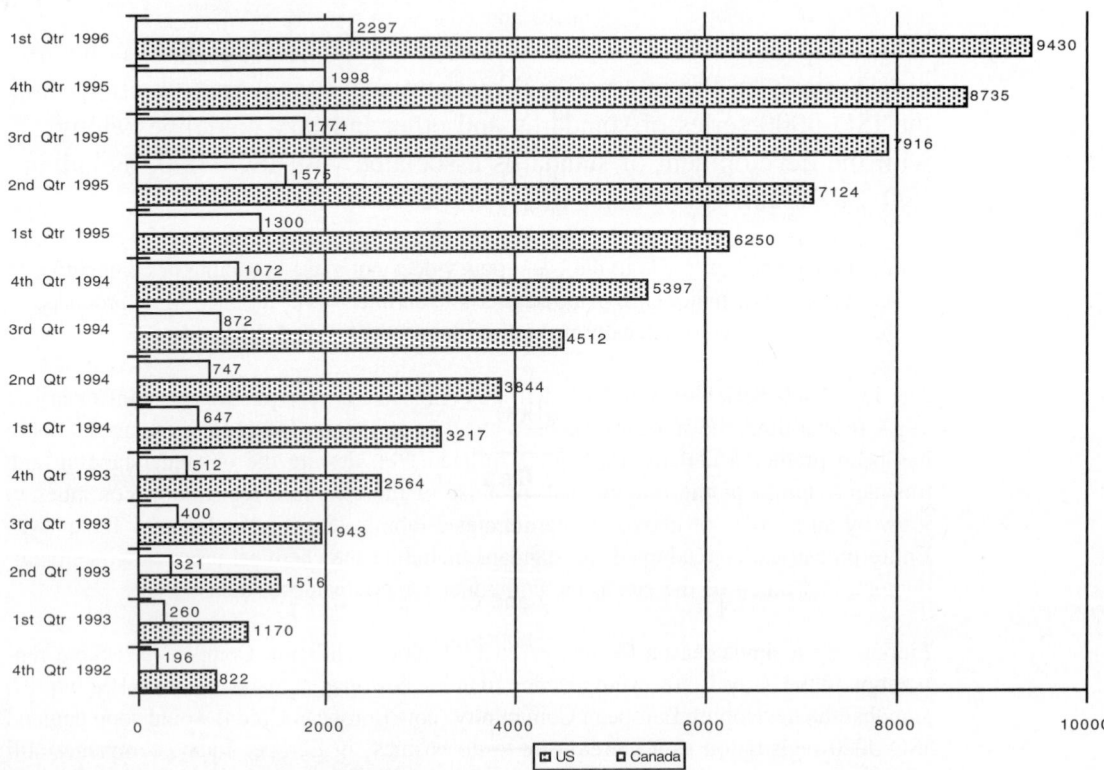

Source: *ISO 9000 Registered Company Directory,* North America, Irwin Professional Publishing

universally understood measure of process quality consistency. Marquardt says that ISO 9000 is part of a much larger fabric of international trade and the rapidly developing technology that supports the production and delivery of goods and services around the world. "The ISO 9000 standards are a natural and necessary result of the global economy," Marquardt notes.

Ian Durand, who has been a key figure in the development of the ISO 9000 standard points out in Chapter 11, "The Future of the ISO 9000 Standards," coauthored by April Cormaci, that integration of ISO 9000 with other emerging standards such as ISO 14000 will further solidify the importance of ISO 9000 as the baseline international process management standard.

ISO 9000 has become a pervasive catch phrase throughout the U.S. economy and dozens of government agencies are either studying the standard for use within their agencies, or have adopted it outright. The Department of Defense (DoD) and the National Aeronautics and Space Administration (NASA) have both adopted ISO 9000 and dropped long-standing quality standards such as MIL-Q-9858. A 1995 legislative act even requires federal agencies that do not use voluntary consensus standards such as ISO 9000 and ISO 14000 to provide a written explanation. Experts predict that the act is likely to push more government agencies toward use of these standards (see Chapter 14 for more information on other government agencies adopting or using ISO 9000).

In addition to government and private industry use, the Big Three automakers (Ford, Chrysler, and General Motors) and certain truck manufacturers have created common quality system requirements for its suppliers known as Quality Systems requirements (QS-9000). QS-9000 incorporates all 20 elements of ISO 9001, plus industry and manufacturer-specific requirements. QS-9000 will ultimately affect thousands of suppliers in the auto and truck manufacturing industry, was developed to reduce overlapping and redundant quality system requirements. Both the manufacturers and the suppliers global marketplace (see Chapter 15 for more on QS-9000 and the history of its development).

Table 1–1

Industries Registered to ISO 9000

Industry SIC*	Description	Registered†
3600	Electronics Industry	1697
2800	Chemical	1423
3500	Machinery and Computer Equipment	1309
3400	Fabricated Metals	927
3800	Measuring Equipment	866
5000	Durable Goods	404

*Standard Industrial Classification
†Registered Sites as of July 1996

The success of ISO 9000 as an internationally accepted process-management standard has also aided the enthusiasm for the ISO 14000 series of environmental management standards. ISO 14001, the environmental process-management standard, is at least a first cousin to ISO 9000. While important differences do exist between the standards (as explained in Chapter 12 of the *Handbook*), the general consensus among those involved in international standards is that the ISO 14000 series would not have attracted so much international interest without the success of ISO 9000 in the world marketplace.

If fact, the success of ISO 9000, ISO 14000, and plans for other standards, such as an occupational health and safety management system standard, have driven debate toward a single process-management standard that would encompass all three areas—quality management, environmental management, and the management of worker health and safety. While that goal seems lofty, some registrars already plan to offer integrated audits for ISO 9000 and ISO 14000 and emphasize the use of the term "management systems" rather than specific standards (see Chapter 14 for more information on the developing occupational health and safety standard).

The reason for this move to integrate standards is simple—implementing three or four unrelated or marginally related management systems does not add business value, especially for those companies that may have to meet ISO 9000, QS-9000, ISO 14000, and perhaps someday a separate standard for occupational health and safety. Certainly, an integrated approach to these standards is part of the natural evolution process toward an efficient global marketplace.

ISO 9000 and the future

Most agree that ISO 9000 is not just another "flavor of the month" quality system standard. ISO 9000 and its related standards are fast becoming part of the "glue" that holds the world economy together. It is just as vital as negotiated trade pacts and agreements between countries. Yet, challenges to the viability of international voluntary consensus standards do exist.

Donald Marquardt points out some of these challenges in Chapter 18 of this *Handbook*, including the credibility of third-party registration, interpretation of the standard, and industry-specific editions of the ISO 9000 standard. The long-term viability of these voluntary standards are also challenged by those who serve the "ISO 9000 industry"—trainers, course providers, consultants, accreditation bodies, registrars, publishers, and others. Inconsistency among any of these players, whether in the provision of consultant services, registration services, or other services, slows the process that will allow ISO 9000 and related standards to have positive impact on the global marketplace.

Despite these and other challenges to the system, a new global trade language has developed that will change the way all companies plan their business strategies and allocate resources. By the time the fourth edition of the *ISO 9000 Handbook* is printed, more consolidation will have taken place among registrars. At the same time, a global auditor and training certification program will be in place, as well as agreements to recognize ISO 9000 registrations in any marketplace. Thousands of suppliers around the world will have met the QS-9000 requirements, and perhaps, tens of thousands of companies around the world will have an ISO 14000 registration certificate. Stopping this train will require more road blocks than anyone can dare to imagine at the time of this *Handbook's* printing.

Taking Care of Business with ISO 9000

by Greg Hutchins

Many organizations are chasing fads such as downsizing, re-engineering, outsourcing, and customer-supplier partnering. Each of these initiatives has its place to achieve certain goals yet, another marketplace trend is also emerging. Some of these same companies are discovering they must go back to basics—simple but effective organizational, managerial and operational processes, and techniques and methods embodied in the ISO 9000 standard.

Contrary to many naysayers, ISO 9000 is not disappearing. Many companies that have been striving for "world class" quality, now realized that they must first ensure that they have the foundation of an effective quality system. Since the publication of the ISO 9000 standard, focus has shifted from looking at ISO 9000 as a family of compliance documents to now recognizing it as a set of common sense business systems and processes.

Benefits can be accrued through ISO systems implementation as well as formal ISO 9000 registration. ISO 9000 benefits can be categorized in terms of three broad areas: (1) customer-marketing benefits; (2) internal benefits; and (3) customer-supplier partnering benefits

Customer/Marketing Benefits

- Assists in developing products.
- Provides access to markets.
- Conveys commitment to quality and partnering and allows for promotional credibility.

(Continueud on next page)

(Continued from page 7)

Internal Benefits

- Guarantees that new and existing products and services satisfy customers.
- Facilitates business and quality planning.
- Provides a universal approach to quality and business.
- Assists in establishing operational baselines and operationalizes and proceduralizes quality.
- Provides insights on organizational interrelationships, encourages internal focus, facilitates internal operational control, and assists employees in understanding and improving operations.
- Encourages self-assessment and maintains internal consistency.
- Controls process and systems and establishes operational controls.
- Makes internal operations more efficient and effective.
- Ensures product development and design changes are controlled.
- Creates awareness of the need for training and encourages operational problem solving.

Customer-Supplier Partnering Benefits

- Forms the basis for a common language of quality.
- Ensures a minimum level of quality.
- Facilitates development of seamless operations.
- Reduces supplier base and assists in selecting suppliers.
- Facilitates just-in-time delivery.
- Assists in monitoring suppliers.

Background and Development of ISO 9000 Standards

2

by Donald W. Marquardt

This chapter provides an overview of the background and development of the ISO 9000 standards. The chapter includes:

- The development of the ISO 9000 standards.

- The development of third-party registration systems.

- A brief description of the European Union and the EU Conformity Assessment System.

- Reasons for the marketplace success of ISO 9000, and some related concerns about the implementation of the standards.

INTRODUCTION

International Standards and the Development of the Global Economy

During the second half of the 20th century profound changes have occurred in the way companies do business. Although importing, exporting, and international trade have gone on for hundreds of years, such trade was peripheral to the everyday business activities of most companies.

As the 20th century ends, products offered for sale typically involve raw materials, parts, design know-how, assembly operations, software, services, and other inputs from multiple countries. Most products now incorporate added value from various countries—very few products are produced entirely in a single country of origin. "Domestic content" percentages more accurately reflect where and how a product is produced. Automobiles and computers are good examples of products produced for the global marketplace.

Yet these changes in trade and commerce are not confined to large companies. Small companies, with only a handful of employees, often do business in multiple countries; they form joint ventures, produce products, market, and implement competitive strategies in the global marketplace.

The Technological Basis of the Global Economy

The global economy could not have developed without two dramatic technological advances:

- Information technology—the ability to move large amounts of information rapidly, accurately, and inexpensively to any part of the world.

- Transportation technology—the ability to move people and goods rapidly, safely, and inexpensively to any part of the world.

The Impact on Standards and Regulations

People in any economic system need standards for designing the things they make and use and for the ways people work together; otherwise the system becomes chaotic. Figure 2–1 illustrates how this reality has impacted the world of standards and regulations during the second half of the 20th century.

In the 1950s most business activities were focused within a local economy and/or within a specific company. In such a setting, the most useful standards were those devised locally, often specifically for use within the company. Such standards often dealt with both technical aspects of products and management procedures for the company's activ-

Figure 2–1

What's Happened in the World of Standards and Regulations

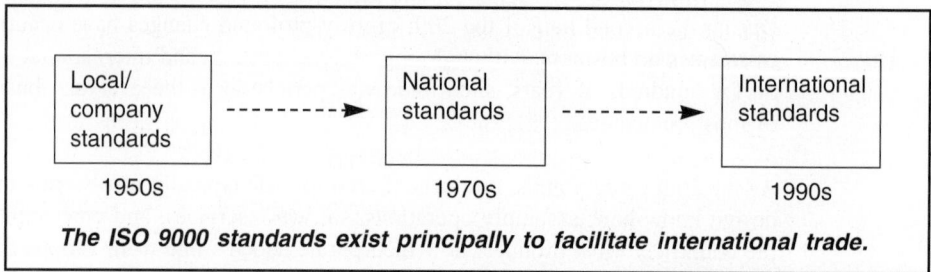

The ISO 9000 standards exist principally to facilitate international trade.

ities. Standards were viewed as proprietary information because they formed part of a company's unique competitive position.

By the 1970s many company-specific standards were being replaced by national standards, as a vast number of companies expanded their operations beyond local borders. Many U.S. companies began to sell U.S.-designed and produced products in the international market. U.S. standards also gained stature internationally and were exported and sold; this enabled other countries to use effectively the products of U.S. industry.

In the 1990s, many countries have the technology and economic infrastructure to compete effectively in the global marketplace. Company-specific or country-specific standards have become nontariff barriers to trade. Countries can no longer succeed economically by insisting that their country's standard is better than another country's standard. Such differences make contract negotiations more difficult and create barriers beyond the inevitable language translation problems. These problems can lead to costly, non-value-adding activities in the conduct of trade.

NEED FOR THE ISO 9000 STANDARDS

The ISO 9000 Standards and the Global Economy

The ISO 9000 standards are a natural and necessary result of the global economy. The ISO 9000 standards, with their internationally harmonized requirements and guidelines, remove the nontariff trade barriers that arise from differences and inadequacies among national, local, or company standards. There are two basic tenets of the ISO 9000 standards:

The primary purpose of the ISO 9000 standards is to facilitate international trade. All activities within any nation's economy take place in the context of the global economy. Consequently, the use of harmonized standards both within individual countries and among these countries produces a valuable economic benefit. Companies around the world representing a wide variety of industry and economic sectors are using the ISO 9000 standards as a fundamental basis for their own operations as well as their trading-partner relationships.

The roles of the ISO 9000 standards and product technical standards are "separate and complementary." The founding principle of the ISO 9000 standards is the concept that the assurance of consistent product quality is best achieved by the simultaneous application of two kinds of standards:

- Product standards (technical specifications).

- Quality system (management system) standards.

I call this the "separate and complementary" concept. The two kinds of standards, when implemented together, can provide confidence that products will meet consistently their requirements for quality.

Product standards provide the technical specifications for the design of the products and often of the process by which the product is produced. Product standards are specific to the particular product; both its intended functionality and the end use situations it may encounter.

The ISO 9000 standards are management system standards exclusively and are not related to any product's technical specifications. Organizations use the ISO 9000 standards to define and implement the management systems by which they design, produce, deliver, and support their products. While the ultimate purpose of the ISO 9000 standards is to achieve and assure the quality of products, the ISO 9000 standards focus directly on the management system of an organization.

The ISO 9000 Standards Apply to All Industry/Economic Sectors

The ISO 9000 standards are being applied in a wide range of industry/economic sectors and in government regulatory areas. One of the characteristics that makes the standards so widely useful is that they apply to all generic product categories (see Vision 2000, Appendix F), namely hardware, software, processed materials, and services.

The ISO 9000 standards provide guidelines or requirements on what features are to be present in the management system of an organization but do not prescribe how these features should be implemented. This nonprescriptive characteristic enables the ISO 9000 standards to have wide applicability for various products and situations of use. Each organization is free to determine how individual ISO 9000 requirements or guidelines will be incorporated in its own management system.

BACKGROUND OF THE ISO 9000 STANDARDS
The Initial (1987) Series of ISO 9000 Standards

The ISO 9000 standards are prepared and maintained by Technical Committee 176 of the International Organization for Standardization (ISO/TC176). The committee had its first meeting in 1980. The vocabulary standard for the ISO 9000 family was first published in 1986. The initial ISO 9000 series of standards was published in 1987. The original ISO 9000 family consisted of:

- ISO 8402, the vocabulary standard.

- ISO 9000, the fundamental concepts and roadmap guideline standard.

- ISO 9001, ISO 9002, ISO 9003, three alternative quality system requirements standards to be used for quality assurance.

- ISO 9004, the guideline standard to be used for quality management purposes.

These six initial standards were quickly adopted as national standards by many countries and quickly began to have enormous impact on international trade (see Figure 2–2).

National Adoption of International Standards

ISO standards are published in English, French, and sometimes in Russian, the three official languages of ISO. Adoption by a country as a national standard involves three steps. First, the standard must be translated into the national language. Second, the formal procedure for adoption of a national standard is followed. Third, the standard is given a national identification number and is published. (Further detail on this subject is provided in Chapter 18.)

Figure 2–2

101 Countries Have Adopted the ISO 9000 Standards

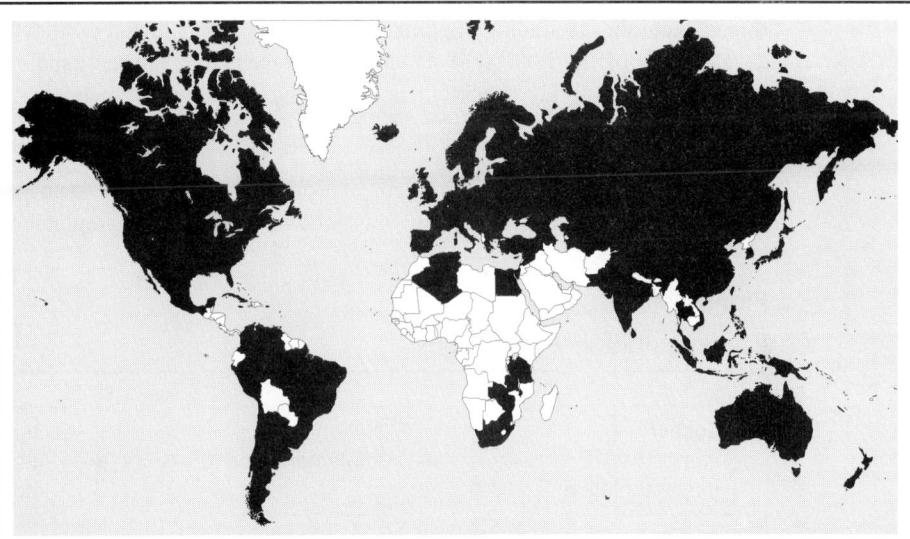

Algeria, Argentina, Armenia, Australia, Austria, Azerbaijan, Barbados, Belarus, Belgium, Brazil. Bulgaria, Canada, Chile, China, Columbia, Croatia, Cuba, Cyprus, Czech Republic, Denmark, Egypt, Finland. France, Germany, Greece, Hungary, Iceland, India, Indonesia, Ireland, Israel, Italy, Jamaica, Japan, Kazakhstan, Kyrgyzstan, Malawi, Malaysia, Mexico, Moldova, Mongolia, Netherlands, New Zealand, Norway, Pakistan, Papua New Guinea, Peru, Phillipines, Poland, Portugal, Romania, Russia, Singapore, Slovakia, Slovenia, South Africa, South Korea, Spain, Sri Lanka, Sweden, Switzerland, Syria, Taiwan, Tajikistan, Tanzania, Thailand, Trinidad/Tobago, Tunisia, Turkey, Turkmenistan, United Kingdom, Ukraine, Uruguay, Uzbekistan, United States, Venezuela, Vietnam, Yugoslavia, Zambia, and Zimbabwe. The standards have also been adopted the the European Union.

Courtesy of Robert Peach and Associates, Inc.

STANDARDIZATION: STANDARDS AND THEIR IMPLEMENTATION

Standardization Defined

Standardization encompasses activities in two interrelated areas:

- The conception, planning, production, periodic revision, promotion, and selling of standards.

- The conception, planning, establishment, control, and maintenance of standards implementation,

Both areas of standardization generate challenges and opportunities. In practice, the implementation activities of standardization often give rise to more challenges than the standards themselves (see Chapter 18).

THE ROLE OF REGIONAL BLOCKS OF COUNTRIES

International and Regional Approaches

Developing and implementing international standards requires the interplay of many forces, including technology approaches, economic issues, and political alignments. The development of the ISO 9000 standards has taken this "international" approach, allowing each member body to participate according to individual technology, economic, and political circumstances.

As shown in Figure 2–3, the implementation of the ISO 9000 standards is following a mix of these two approaches, an *international* approach to implementation occurring

Figure 2–3

Implementation Approaches

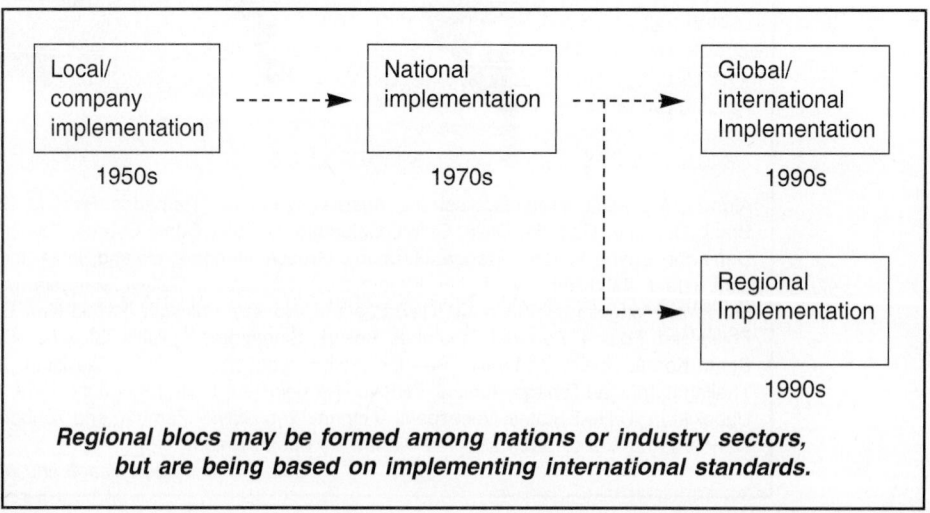

Regional blocs may be formed among nations or industry sectors, but are being based on implementing international standards.

What Is ISO?

ISO is the acronym for the International Organization for Standardization, founded in 1946 to develop a common set of manufacturing, trade, and communications standards. According to ISO officials, the organization borrowed the acronym from the Greek word, *isos,* meaning equal. Isos also is the root of the prefix *iso,* as in isometric (of equal measure or dimensions) and isonomy (equality of laws or of people before the law). Its selection was based on the conceptual path that leads from "equal" to "uniform" to "standard."

The Geneva, Switzerland-based organization is composed of approximately 100 member countries. In ISO each country is represented by its *member body,* the national organization that coordinates national standards. Each member body has a single vote, irrespective of the size of the country. The American National Standards Institute (ANSI) is the United States member body to ISO.

All standards developed by the International Organization for Standardization are voluntary; no legal requirements force countries to adopt them. However, countries and industries often adopt ISO standards as national standards. In some instances countries attach legal requirements to ISO standards they have adopted, thereby making the standards mandatory in that country. The International Organization for Standardization develops standards in all industries except those related to electrical and electronic engineering. Standards in these areas are made by the Geneva-based International Electrotechnical Commission (IEC), which has more than 40 member countries, including the United States. In practice ISO and IEC cooperate closely in their activities. They publish a common set of directives governing standards development.

ISO is structured into approximately 200 technical committees that draft standards. Member nations form national committees—in the United States these are called technical advisory groups (TAGs)—that develop national negotiating positions and strategies and select delegates that provide input into the ISO standards development process. Through this mechanism ISO receives wide input and establishes consensus from industry, government, and other interested constituencies before promulgating a standard.

simultaneously with a European, *regional* approach. In a regional approach, the countries that comprise a regional bloc act in a relatively unified, regional manner.

It is not clear what the future will bring. In recent years other regional blocs of countries have been formed in various parts of the world. The NAFTA treaty is one example. The "mixed" approach to standards implementation is likely to expand to some degree in these other regional blocs.

The Early Role of the European Union in Worldwide Implementation of the ISO 9000 Standards

The European Union (EU, formally, the European Community, see Chapter 16), played a decisive role in the early, rapid implementation of the ISO 9000 standards. Perhaps the single most visible factor driving the early acceptance of the ISO 9000 series has been the effort to unify the 15 major European nations that form the EU into a single internal market. The full members are Austria, Belgium, Denmark, Finland, France, Germany, Greece, Ireland, Italy, Luxembourg, the Netherlands, Portugal, Spain, Sweden, and the United Kingdom.

The European Union originated with the 1957 Treaty of Rome, which was established to abolish tariffs and quotas among its six member states and to stimulate economic growth in Europe. (The original members were France, West Germany, Italy, Luxembourg, the Netherlands, and Belgium).

Economic growth slowed during the 1970s and early 1980s, and Europe began to fear that the U.S., Japanese, and Pacific Rim economies would dominate the world economy of the 21st century. The European nations were concerned that they would fall behind partly due to the differences among their technical standards and requirements.

Differing national product certification requirements made selling products in multiple national markets in the European Union a costly undertaking, requiring duplication of tests and documentation and separate approvals from national or local regulatory authorities.

In response, the European Union called for a greater push toward a unified market and for the removal of physical, technical, and fiscal barriers to trade. In 1985 the EU Commission presented a program for establishing a single internal market. The goal was to create a single set of procedures for conformity assessment that is simpler and less costly for manufacturers.

The single internal market program was based in part on the 1979 Casis de Dijon decision of the European Court of Justice, which established the principle of mutual recognition. This principle states that products that meet the requirements of one EU member state could freely circulate in other member states, a concept similar to the interstate commerce clause of the U.S. Constitution.

The 1985 program, presented in a white paper, drew upon the rationale of the Casis de Dijon decision. The move to a single internal market was further expedited by the Single European Act, adopted in February 1986. This act amends the 1957 Treaty of Rome. The combined intent of the white paper and the Single European Act was to abolish barriers to trade among the (then) 12 member states and to complete an internal European market by the end of 1992.

The single market—known as EC 92—became effective at midnight on December 31, 1992. Its goal is to encourage trade and to increase confidence in the safety and reliability of products marketed in the European Union. To better understand this framework of standards and product certification—and ISO 9000's role in that framework—it is important to understand the European Union's conformity assessment program.

European countries have extensive trade with countries worldwide. The EU single market approach was introduced into international trade before any other regional blocs had anything comparable. As a result, the European conformity assessment model has had a major effect on all international trade, whether or not any of the trading partners are European.

European Union Conformity Assessment

In preparation for EC 92, the European Union began developing a comprehensive framework for conformity assessment activities. Conformity assessment refers to all processes—product testing and certification, quality system registration, standards, and laboratory accreditation—that may be used to ensure that a product conforms to requirements. The conformity assessment system, when successfully completed, will give customers confidence that products conform to all requirements.

This system has the following four major components:

- EU-wide directives.
- Harmonized standards.
- Consistent conformity assessment procedures.
- Competent certification and testing bodies.

EU-Wide Directives and Harmonized Standards

In the European system, products are classified into two categories: regulated products and nonregulated products.

Most products sold in the European Union are nonregulated products—they are not covered by EU legislation. The European Union's strategy for removing technical barriers to nonregulated products is to rely on the principle of mutual recognition and product certification by a third party.

In effect, a U.S. exporter of a nonregulated product can certify to a U.S. standard for technical specifications, and if these standards are accepted in at least one EU country, they will be accepted in the European Union through the principle of mutual recognition.

The EU Regulatory Hierarchy

The 15 nations that make up the European Union (formerly the European Community) are bound by its regulations. The regulatory process begins with the European Commission. A proposal is drafted by the Commission and is sent to the European Parliament. The Parliament, with its 626 members, votes on the proposal. However, this vote is not binding on the European Council. The European Council comprises many working groups, each of which consists of civil servants from the member states. These groups examine the proposals and, if necessary, make changes. The proposal is then forwarded to the Committee of Permanent Representatives, made up of civil servants from the member states. When this Committee reaches agreement, the proposal is forwarded to the Council of Ministers. The makeup of the Council depends on the subject matter of the proposal. The Council of Ministers decides whether to adopt the legislative proposal. It acts by majority vote. As a result of the process, regulations and directives are created.

Regulations are directly binding on the member states; they do not require any action on the part of each state. Directives do not directly create a new law but instruct member states to amend their national legislation within a prescribed period.

The European Court of Justice has judicial oversight. It interprets and applies European Union law.

Regulated products are those that have important health, safety, or environmental implications. Their requirements are spelled out in directives—official EU legislation—that are binding on all member states of the European Union.

Each directive deals with a class of regulated products and spells out the essential requirements for compliance. However, directives do not list specific technical requirements; they reference appropriate technical standards that are being developed by the major European or international standards organizations. (These organizations are developing "harmonized" standards to eliminate the jumble of standards of the individual 15 member states.)

Consistent Conformity Assessment Procedures

Depending upon the requirements of a particular directive, which takes into account the health, safety, and environmental risks of the product involved, conformity assessment can require one or more of the following:

- Type-testing of the product.
- Third-party audit of the quality system.
- Testing of regular production.
- A manufacturer's self-declaration of conformity.

Accreditation, Certification, and Registration

Terms such as accreditation, certification, and registration are often used inter-changeably, creating some confusion. To clarify the meanings, the Conformity Assessment Committee of the International Organization for Standardization, in its *ISO/IEC Guide 2: General terms and their definitions concerning standardization and certification*, defines the terms as follows:

Accreditation: Procedure by which an authoritative body gives formal recognition that a body* or person is competent to carry out specific tasks.

Certification: Procedure by which a third party gives written assurance that a product, process, or service conforms to specific requirements.

Registration: Procedure by which a body indicates relevant characteristics of a product, process, or service, or particulars of a body or person, and then includes or registers the product, process, or service in an appropriate publicly available list.

Editor's note: Although certification and registration are slightly different steps in the same process, they are used interchangeably. In Europe the term *quality systems certification* is used widely; the term *quality systems registration* is the preferred U.S. terminology.

* In particular, a registration body. —(Editor's clarification)

This approach, known as the modular approach to conformity assessment, provides manufacturers with options from which to choose in order to demonstrate compliance with a directive. (Please refer to Chapter 16 for an in-depth discussion of conformity assessment procedures.)

The Quality Assurance Route to Conformity Assessment

Some EU directives require quality system registration. For other directives, quality system registration is not an absolute requirement. However, to ensure confidence in the quality of products circulating throughout Europe, the European Union has strongly emphasized quality assurance.

The EU has adopted the ISO 9000 series as part of its conformity assessment plan to establish uniform systems for product certification and quality systems registration. Registration involves the audit and approval of a quality system against ISO 9001, ISO 9002, or ISO 9003 by a third-party, independent organization known as a certification body in the European Union and a registrar in the United States.

For some businesses, achieving ISO 9000 registration satisfies an immediate customer requirement. However, for many companies ISO 9000 registration is only an important first step for doing business internationally. Additional regional, national, or trading bloc requirements may exist with respect to product technical requirements which are outside the scope of the ISO 9000 standards, in accordance with the "separate and complementary" concept.

Competent Certification and Testing Bodies

The European Union has recognized that confidence in the conformity assessment system and in the products sold throughout the European Union depends on confidence in the competence of certification and testing bodies. Thus, the European Union has encouraged the development of standards such as the EN 45000 series. The EN 45000 standards establish requirements for testing, certification, and accreditation bodies (see Chapter 17 for more on EN 45000).

The European Union has also encouraged the formation of organizations such as the European Organization for Testing and Certification (EOTC) to promote consistent practices in testing and certification and to promote the mutual recognition of test results.

Implications for the United States

The economic implications of the European Union's efforts are far-reaching. The European Economic Area (EEA) Treaty became effective January 1, 1994, extending the European single market to include four of the six EFTA countries, Finland, Iceland, Norway, and Sweden (the two EFTA countries not included are Switzerland, which rejected membership in a referendum, and Liechtenstein, which is redefining its relationship with Switzerland to become part of the EEA).

This expanded European single market is now the largest free trade zone in the world, comprising 372 million people and 17 countries. The treaty reinforces the free flow of goods, services, people, and capital throughout the EEA.[1]

Interest in European integration has also extended to other countries with the signing of Association Agreements with Poland, Hungary, and the Czech and Slovak Federal Republic. With the inclusion of other Eastern European countries and the former Soviet republics, the European Union could eventually develop into a multitrillion dollar market of 500 to 800 million people.[2]

The United States is the European Union's biggest foreign supplier. In 1994, the United States exported more than $112 billion in goods to the countries of the European Union. About half of all sales of U.S. subsidiaries overseas are in Europe—about $600 billion annually.[3]

U.S. companies are understandably eager to gain and maintain an economic foothold in this market. Manufacturers can anticipate millions of new customers for goods and ser-

---

Origins of ISO 9000

In the past several decades, quality has emerged as an important aspect in international trade. Various national and multinational standards were developed in the quality systems arena for military or nuclear power industry needs. Some standards were guidance documents, others were quality system requirements to be used in contracts between purchaser and supplier organizations.

In 1959 the U.S. Department of Defense (DoD) established the MIL-Q9858 quality assurance program. In 1968 the North Atlantic Treaty Organization (NATO) essentially adopted the tenets of the DoD program in the NATO AQAP1, AQAP4, and AQAP9 series of standards. In 1979 the United Kingdom's British Standards Institution (BSI) developed from the predecessors the first quality assurance system standards intended for commercial and industry use. These standards were designated the BS 5750 series, Parts 1, 2, and 3.

Despite the commonality among these predecessors to the ISO 9000 standards, there was no real consistency until Technical Committee 176 (TC176) of ISO issued the ISO 9000 series standards in 1987.

---

vices. The challenge is to meet both product and quality system standards and conformity assessment procedures necessary for unrestricted trade within this market.

THE ISO 9000 FAMILY OF STANDARDS

Table 2–1 lists the standards in the ISO 9000 family as of the beginning of 1996. The five standards ISO 9000 through ISO 9004, together with the vocabulary standard, ISO 8402, are often referred to as the ISO 9000 series. Revisions of the ISO 9000 series were published in 1994. Since the initial series was published, a number of additional standards have been published by ISO/TC176. Some are numbered in the 10000 range. Some are currently numbered as parts of ISO 9000 or ISO 9004. Part numbers are shown in abbreviated form by a dash before the part number. For example ISO 9000–1:1994 describes the 1994 revision of the former ISO 9000:1987. Part numbers –2, –3, etc., designate other standards numbered as part numbers to ISO 9000.

The term *ISO 9000 family* refers to all the standards published by ISO/TC176. As shown in Table 2–1, there are a variety of additional standards. However, only ISO 9001, ISO 9002, and ISO 9003 are quality assurance requirements standards. The remainder are all guideline standards. During the next few years ISO/TC176 plans to renumber into the 10000 range all the part numbers 2 and higher now associated with ISO 9000 and ISO 9004. The renumbering will occur at the time of the next revision of those documents. Many of the part numbers may be absorbed into revisions of the basic series. All ISO standards are reviewed at approximately five-year intervals and are reaffirmed, revised, or withdrawn.

Table 2–1

The ISO 9000 Family of International Standards

ISO 8402	Quality vocabulary (1994)
ISO 9000	Quality management and quality assurance standards
	Part 1: Guidelines for selection and use (1994)
	Part 2: Generic guidelines for the application of ISO 9001, ISO 9002, and ISO 9003 (1993)
	Part 3: Guidelines for the application of ISO 9001 to the development, supply, and maintenance of software (1991; reissue 1993)
	Part 4: Application for dependability management (1993)
ISO 9001	Quality systems—Model for quality assurance in design, development, production, installation, and servicing (1994)
ISO 9002	Quality systems—Model for quality assurance in production, installation, and servicing (1994)
ISO 9003	Quality systems—Model for quality assurance in final inspection and test (1994)
ISO 9004	Quality management and quality system elements
	Part 1: Guidelines (1994)
	Part 2: Guidelines for services (1991; reissue 1993)
	Part 3: Guidelines for processed materials (1993)
	Part 4: Guidelines for quality improvement (1993)
ISO 10005	Quality management—Guidelines for quality plans (1995)
ISO 10007	Guidelines for configuration management (1994)
ISO 10011	Guidelines for auditing quality systems
	Part 1: Auditing (1990; reissue 1993)
	Part 2: Qualification criteria for quality systems auditors (1991; reissue 1993)
	Part 3: Management of audit programs (1991; reissue 1993)
ISO 10012	Quality assurance requirements for measuring equipment
	Part 1: Management of measuring equipment (1992)
ISO 10013	Guidelines for developing quality manuals (1994)

QUALITY MANAGEMENT AND QUALITY ASSURANCE—SIMILARITIES AND DIFFERENCES

In the early years of ISO/TC176 one of the most pressing needs was to harmonize internationally the meanings of terms such as quality control and quality assurance. These two terms, in particular, were used with diametrically different meanings among various nations and even within nations.

As convener of the working group that wrote the ISO 9000:1987 standard, I proposed early in the 1980s that the term *quality management* be introduced into the ISO 9000 standards as the umbrella term for quality control and quality assurance. The term quality management was defined, included in ISO 8402, adopted internationally, and is now used worldwide. This, in turn, enabled international agreement on harmonized definitions of the meanings of the terms quality control and quality assurance.

The commonalities and distinctions between quality management and quality assurance are still not universally understood, despite the existence of internationally agreed definitions. This may be due to the expansion of ISO 9000 standards use to many more countries than participated in the early 1980s, or lack of widespread reference to ISO 8402, or deficiencies in the ISO 8402 definitions. Undoubtedly, all these reasons have contributed.

Quality management is defined in ISO 8402:1994 as:

"All activities of the overall management function that determine the quality policy, objectives and responsibilities, and implement them by means such as quality planning, quality control, quality assurance and quality improvement . . ." (See Appendix I; specifically ISO 9000-1:1994, Annex A, clause A.3.)

These four quality management activities have important relationships to the classic Plan-Do-Check-Act management cycle (see Table 18–1). Quality planning focuses on the Plan step; quality control focuses on the Do step; quality assurance focuses on the Check step; and quality improvement focuses on the Act step. Many quality assurance activities are internal to the organization and therefore fall within quality management; internal quality audits are an example. Other quality assurance activities involve parties external to the organization; third-party audits are an example.

For simplicity, it is common today to use the term quality management when referring to internal activities (i.e., quality planning, quality control, quality improvement, and/or quality assurance for internal purposes), and the term quality assurance when referring to activities related to second- or third-party audits.

Table 2–2 describes the prime focus of quality management and quality assurance. The quality control aspects of the umbrella term quality management are focused on the word *achieving,* but all bullet points in the left-hand column of Table 2–2 relate at least indirectly to quality control. The right-hand column of Table 2–2 shows that the quality assurance aspects of the umbrella term quality management focus primarily on the notions of *demonstrating* and providing confidence through objective evidence.

These terminology distinctions need to be understood to use the ISO 9000 standards effectively.

Three terms currently in use all have the same essential meaning.

- *Quality system* is the formal term currently defined internationally in ISO 8402, the ISO/TC176 vocabulary standard.

- *Management system* is the term frequently used in the daily language of business.

- *Quality management system* is the term coming into increasing use for discussing an organization's management system, when the focus is upon the organization's overall performance and products in relation to the organization's objectives for quality.

A benefit of the term quality management system is its effectiveness in emphasizing both (a) the commonalties in management system features, and (b) the differences in the objectives for the results of an organization's management system for various areas of application. For example, quality management system and environmental management system describe two such areas of application.

ROLES OF THE ISO 9000 STANDARDS

The ISO 9000 standards have two primary roles:

- **Quality management**—ISO 9004 and related guideline standards provide guidance for suppliers of all types of products who want to implement effective quality systems in their organizations or improve their existing quality systems.

- **Quality assurance**—The standards ISO 9001, ISO 9002, or ISO 9003 provide quality system requirements against which a customer, or a third party acting on behalf of customers, can evaluate the adequacy of a supplier's quality system.

These two roles are complementary. In the context of programs for quality system registration the quality assurance role is more visible, but the business value of both roles is important.

Table 2–2

The Prime Focus of Quality Management and Quality Assurance

Quality Management	Quality Assurance
Achieving results that satisfy the requirements for quality	*Demonstrating* that the requirements for quality have been (and can be) achieved
• Motivated by stakeholders internal to the organization, especially the organization's management	• Motivated by stakeholders, especially customers, external to the organization
• Goal is to satisfy all stakeholders	• Goal is to satisfy all customers
• Effective, efficient, and continually improving overall quality-related performance is the intended result	• Confidence in the organization's products is the intended result
• Scope covers all activities that impact the total quality-related business results of the organization	• Scope of demonstration covers activities that directly impact quality-related process and product results

THE GROWTH OF THIRD-PARTY REGISTRATION

Origins

The earliest users of quality assurance requirements standards were large customer organizations such as electric power providers and military organizations. These customers often purchase complex products to specific functional design.

In these situations the quality assurance requirements are specified in a two-party contract, where the providing organization (i.e., the supplier) is referred to as the first party and the customer organization is referred to as the second party. Two-party contract quality assurance requirements typically include provisions for the supplier organization to have internal audits ("first-party audits") sponsored by its own management to verify that its quality system meets the contract requirements.

Two-party contracts typically also include provisions for external audits ("second-party audits") sponsored by the management of the customer organization to verify that the supplier's quality system meets the contract requirements. Within a contractual arrangement between two parties it is possible to tailor the requirements as appropriate and maintain an on-going dialog between customer and supplier.

Unfortunately, two-party quality assurance arrangements become burdensome once the practice is widespread throughout an economy. Soon each organization in the supply chain is subject to periodic management system audits by many customers, and is itself subjecting many of its subsuppliers to such audits. The supply chain is burdened by redundant audits for essentially the same requirements. The conduct of audits becomes a significant cost element for both the auditor organizations and auditee organizations.

Certification/Registration Level Activities

The development of quality system registration/certification is a means to reduce the redundant, non-value-adding effort of multiple audits. A third-party organization—a certification body or a registrar—conducts a formal audit(s) of a supplier organization to assess conformance to the appropriate quality system standard, say, ISO 9001 or ISO 9002.

When the supplier organization is judged to be in complete conformance, the third party issues a certificate to the supplier organization and registers the organization's quality system in a publicly available register. The terms certification and registration carry the same marketplace meaning because they are two successive steps signaling a successful conclusion to the same process. To maintain its registered status the supplier organization must pass periodic surveillance audits by the registrar.

Hundreds of registrars operate worldwide. Most of them are private, for-profit companies. It is critical that the registrars operate competently and objectively and that all registrars meet standard requirements for their business activities. The registrars are, in fact, supplier organizations that provide a needed service in the economy. As long as these registration services add value in the supply chain, audits will be valued by both the supplier and the suppliers' customer.

Accreditation Level Activities

To assure competence and objectivity of the registrars, systems of registrar accreditation have been set up worldwide. Accreditation bodies audit the registrars for conformity to standard international guides for the operation of certification bodies.

The quality system of the registrar comes under scrutiny by the accreditation body through audits that cover the registrar's documented quality management system, the qualifications and certification of auditors used by the registrar, record keeping, and other features of the registrar's office operations. In addition, the accreditation body witnesses selected audits done by the registrar's auditors at a client supplier organization's facility.

In the United States, registrar accreditation is carried out by the Registrar Accreditation Board (RAB) under a joint program with the American National Standards Institute (ANSI) (see Chapter 17). This joint program is called the ANSI-RAB National Accreditation Program.

OTHER ISO 9000 IMPLEMENTATION CONSIDERATIONS

Customer Requirements

The role of ISO 9000 standards in the global market is well established. Customer expectation of ISO 9000 registration by a supplier is now commonplace. Companies are being asked by customers to become registered to ISO 9001, ISO 9002, or ISO 9003 as a precondition to placing a purchase order. It should be noted that ISO 9003 is rarely used because the ISO 9001/ISO 9002 model is appropriate and more effective economically and technically in virtually all situations.

Legal Requirements

For companies whose products are subject to EU directives, registration to ISO 9001, ISO 9002, or ISO 9003 is a legal requirement to enter the regulated EU market. Registration might also help a company meet a domestic regulatory mandate.

Liability Concerns

Liability concerns are also driving registration. Some companies register a quality system, at least in part, for the role ISO 9000 registration may play in product liability defense. Companies that sell regulated products in Western Europe may be subject to increasingly stringent product liability and safety requirements that are moving toward the strict liability concepts prevalent in the United States.

An EU product liability directive, for example, holds a manufacturer liable, regardless of fault or negligence, if a person is harmed or an object is damaged by a faulty product. In addition, an EU product safety directive requires manufacturers to monitor product safety. A possible consequence of these directives would be the necessity for companies to document that they have adequate quality systems for their production processes. These procedures would demonstrate more thoroughly that products meet specified requirements, thus minimizing liability claims.

Registration of Subcontractors

The ISO 9001 and ISO 9002 standards require the supplier to ensure that materials purchased from subcontractors conform to specified requirements. As a consequence, increasingly many companies are requiring that their subcontractors become registered, even though the ISO 9001 and ISO 9002 standards do not specifically require quality system registration of subcontractors.

Internal Improvement

Although external market pressure has stimulated many companies to seek ISO 9000 registration, other companies have implemented the ISO 9000 standards to gain internal benefits. Companies that have implemented the standards have often discovered that internal improvements in facility performance and quality have lasting value at least equal to the market value of ISO 9000 registration. A well established quality system can increase productivity and reduce costs associated with inefficiencies.

The ISO 9000 standards can also be used as a foundation or building block for implementing broader quality systems, such as Total Quality Management (TQM), and for meeting more stringent quality goals, such as the criteria of the Malcolm Baldrige National Quality Award in the United States (see Chapter 13).

Marketplace Competition

Marketplace competition is a great impetus for ISO 9000 standards implementation. Companies are implementing ISO 9000 standards to keep up with registered competitors and distinguish themselves from nonregistered competitors.

CONCERNS ABOUT REGISTRATION

Despite the worldwide acceptance of the ISO 9000 standards, the issue of ISO 9000 registration has raised some concerns.

Are Standards and Product Certification Barriers to Trade?

Standards facilitate a common international industrial language, provide consumer confidence, and promote product safety. Standards can also facilitate and encourage trade. Used improperly, however, standards can hinder worldwide trade.

One argument is that standards generally lag behind the development of the latest technology and thus become nontariff barriers to trade. Although standards may not reflect the latest technical innovations, periodic review and revision prevents them from becoming nontariff barriers to trade.

A related argument states that, like standards, product certification systems adopted to facilitate trade within an area can consequently act as trade barriers to trading partners outside that area.

In the early 1970s, for example, Europe developed a regional certification system for electronic components. This system, in effect, became a non-tariff barrier to American and Japanese manufacturers. The groups adversely affected petitioned the international electronics standards body, the IEC, to develop an international system to replace the regional one.

A major misunderstanding about the ISO 9000 standards is revealed, however, when examples like this are cited. The ISO 9000 standards deal only with quality systems, not with a product's technical or performance specifications. (See the discussion of the "separate and complementary" concept earlier in this chapter.) The above example deals with product certification to regional technical specifications. These can become non-tariff trade barriers for suppliers in other countries that have different technical specification standards for the same class of products.

Time and Costs

Achieving registration to ISO 9001, ISO 9002, or ISO 9003 requires money and time. It takes companies an average of somewhat more than a year to prepare for their first registration. The average cost for registrar fees alone was $19,800, according to a 1996 Irwin Professional Publishing/Dun and Bradstreet Survey of registered companies (see Chapter 19 for more details on the survey).

Some companies have established quality systems considered to be above the level provided for in ISO 9001. In such instances, the ISO 9000 standards may appear to add cost without adding real value. Many companies that believed their quality system exceeded the requirements for ISO 9001 registration, however, subsequently discovered that this was not true for all elements of their systems.

A Level Playing Field

A related concern is whether a level playing field will hurt companies that already produce high-quality products. According to this argument, all products manufactured by

ISO 9000 registered companies may be viewed favorably. This recognition may benefit the manufacturer that previously produced to lower product technical standards, for now its products are viewed on a par with all others.

Another misunderstanding of the ISO 9000 standards is revealed by this example. While ISO 9000 does represent a universal standard, no expectation exists that the two companies' products are the same just because both companies are registered.

Registration means that in both companies:

- The quality system has elements that at least meet the scope of the stated standard (ISO 9001, ISO 9002, or ISO 9003).

- Each quality system element meets the requirements of the standard and is consistently deployed.

Ample opportunity remains for suppliers to succeed in the marketplace if they offer products that conform to better technical specifications and they are better at meeting customer expectations than their competitors.

Differences between the U.S. and the EU Systems

The EU regional product certification system has also created a good deal of uncertainty, making it difficult for some U.S. companies to plan. The regulatory framework is still developing. The status of directives and which products they cover is complex, as are the conformity assessment requirements for specific products.

These are genuine problems affecting both European and non-European suppliers; however, they concern the procedures for implementing the testing, accreditation, and certification systems and are not related to the content or structure of the ISO 9000 standards themselves.

U.S. companies are not as familiar as European companies with a government-driven standards system such as the one being implemented in the European Union. Unfamiliarity with the system, coupled with the uncertainties about the developing EU system for conformity assessment, may leave some U.S. manufacturers with the perception that Europeans have an advantage. However, ingenuity and flexibility have always been a cultural advantage for the United States. Many U.S. firms have proven their ability to become registered and compete effectively.

Some perceive the European Union to be exploiting the differences between the United States and European standard-setting systems and government/business relationships to its advantage. The European system is government-oriented with an emphasis on third-party verification, while the United States system is driven by the private sector and has often relied on manufacturers' self-declaration of conformity.

OUTLOOK FOR THE FUTURE

Both the European system and the U.S. system have unique advantages and drawbacks. The world scene is likely to be a paradoxical, simultaneous mix of approaches to standards for the foreseeable future. Despite the challenges, companies around the world will continue to benefit from the increasing use of international standards and to compete and prosper in the global marketplace.

Endnotes

[1] "European Single Market Expands," *European Community Quarterly Review*, Vol. II, Issue 2 (Spring 1994): 2.

[2] Timothy J. Hauser, "The European Community Single Market and U.S. Trade Relations," *Business America*, 8 March 1993.

[3] Ibid.

THE ISO 9000 SERIES STANDARD

Chapter 3   Overview of the ISO 9000 Series Standard

Chapter 4   The ISO 9001 Standard

Overview of the ISO 9000 Series Standard

3

by Robert. W. Peach

This chapter is an overview of the ISO 9000 series standards. It explains ISO 9000-1 and ISO 9004-1 in detail, while introducing ISO 9001, 9002, and 9003. (Please refer to Chapter 4 for a detailed explanation of ISO 9001.) Chapter 3 covers the following topics:

- Uses of the Standards
- Definition of Terms
- The ISO 9000 Series
- Conformance Standards
  - —ISO 9001
  - —ISO 9002
  - —ISO 9003
- Guidance Standards
  - —ISO 9000-1
  - —ISO 9004-1
- Other Useful Standards in the ISO 900 Quality Family

INTRODUCTION

The ISO 9000 series is a set of generic standards that provide quality management guidance and identify generic quality system elements necessary to achieve quality assurance. ISO 9000 standards are independent of any specific industry or economic sector. An individual company determines how to implement these standards to meet its specific needs and the needs of its customers.

The ISO 9000 series covers a broad scope of basic and uncomplicated quality system elements. A company that has achieved ISO 9000 registration can attest that it has a documented quality system that is fully deployed and consistently followed. This does not necessarily imply, however, that the company produces better quality products than those of its competitors.

Standards in the ISO 9000 series are not product standards. They do not include any technical requirements. According to Clause 4.3 of ISO 9000-1, the quality system requirements in the ISO 9000 series are complementary to but not substitutes for distinct product technical requirements.

Basically, the ISO 9001, ISO 9002, and ISO 9003 standards require a company to document what it does, do what it documents, review the process, and change it when necessary. The objective of ISO documentation can be stated as follows: If a company suddenly replaced all its personnel, their replacements, properly trained, could use the documentation to continue making the product or providing the service as before.

ISO 9001, ISO 9002, and ISO 9003 requirements do not constitute a full-fledged, total quality management system; rather, they provide many of the basic building blocks for such a system. The ISO 9000 standards tell companies what to do, but not how to do it. The choice of methods is left to the management of the organization.

USES OF THE STANDARDS

According to Section 6 of ISO 9000-1, there are four primary uses for the ISO 9000 standards:

- Guidance for quality management.
- Contractual agreements.
- Second-party approval or registration.
- Third-party certification or registration.

In both *contractual* and *noncontractual* situations, an organization—referred to in the standard as the "supplier"—wants to install and maintain a quality system to strengthen its competitiveness and achieve the needed product quality in a cost-effective way. Thus, the ISO 9000 standards offer valuable guidance for internal quality management.

Additionally, in a *contractual* situation, the customer wants to know whether its supplier can produce products or services that consistently meet necessary requirements. According to ISO 9000, in a contractual situation both supplier and customer must agree on what is acceptable.

In a *second-party* approval or registration situation, the customer assesses its supplier's quality system and grants formal recognition of conformance with the standard.

In a *third-party* situation—evaluation by a registration body—the supplier agrees to maintain its quality system for all customers, unless otherwise specified in an individual contract.

A particular organization can be involved in one or more of the above situations. For example, a supplier may purchase some materials *without* contractual quality system requirements and purchase others *with* contractual requirements.

DEFINITION OF TERMS

Before discussing the ISO 9000 series further, it is necessary to define common terms that are used in the standard. One purpose of the ISO 9000 standard is to create a consistent, international "language of quality." Most of the definitions quoted below are from ISO 8402, *Quality management and quality assurance—Vocabulary*, 2d Edition, 1994.

What Is an Organization?

An organization, for the purposes of the standard, is "a company, corporation, firm, enterprise or institution or part thereof, whether incorporated or not, public or private, that has its own functions and administration." This is a broad definition; the quality system elements in ISO 9000 apply to almost any type of organization.

Supplier-Chain Terminology

ISO 9000-1, Clause 3.8, describes the supplier-chain terminology and is illustrated in Table 3–1. It improves terminology in the 1987 version of ISO 9000 which was not consistent.

The requirements of ISO 9001, ISO 9002, and ISO 9003 are addressed to the supplier. The guidance in ISO 9004-1 addresses the organization.

The supplier provides products/services to its customer(s). In a contractual situation, the customer is referred to as the purchaser and the supplier as the contractor. The supplier receives goods and services, if necessary, from either subsuppliers or subcontractors. (Please refer to the beginning of Chapter 4 for a more detailed chart of the supply chain.)

What Is a Product?

A product is defined as the "result of activities or processes." A product can be tangible, such as assemblies or processed materials; intangible, such as information; or a combination of both, such as a service. In Clause 4.4 of ISO 9000-1, the standard classifies

Table 3–1

Supplier-Chain Terminology

Standard	Relationships of Organizations in the Supplier-Chain		
ISO 9000-1	Subsupplier	Supplier	Customer
ISO 9001	Subcontractor	Supplier	Customer
ISO 9004-1	Subcontractor	Organization	Customer

products into four generic product categories: hardware, software, processed materials, and services.

Hardware refers to a "tangible, discrete product with distinctive form." Thus, hardware products normally consist of "manufactured, constructed or fabricated pieces, parts and/or assemblies" (ISO 9000-1, 3.1).

Software is an "intellectual creation consisting of information expressed through supporting medium." Software can be in the form of "concepts, transactions or procedures." Some examples are computer programs and the content of books and procedures (ISO 9000-1, 3.2).

Processed Material

A processed material is "a tangible product generated by transforming raw material into a desired state." This state can be "liquid, gas, particulate, material, ingot, filament or sheet" (ISO 9000-1, 3.3).

A service is a "result generated by activities at the interface between the supplier and the customer, and by supplier internal activities to meet the customer needs" (ISO 8402:1994). Note from this definition that there are services provided within an organization; thus, an organization can have internal suppliers and customers.

According to the standard, many companies include more than one generic product category. Often, hardware, software, and services will all be part of the organization's offering to its customers.

What Is Quality?

Quality has various meanings—many of which are subjective, such as the concept of "excellence." In the quality management field, however, the meaning is more specific. According to ISO 8402:1994, quality is "the totality of characteristics of an entity that bear on its ability to satisfy stated and implied needs."

In a contractual situation, stated needs are specified in contract requirements and translated into product features and characteristics with specified criteria. In other situations, implied needs are identified and defined by the company, based on knowledge of its marketplace. The needs of the customer, of course, change with time, so companies should review quality requirements periodically.

Quality in a product or service generally refers to "fitness for use" or "fitness for purpose." Most organizations meet specific production criteria such as technical specifications. As ISO 9000-1 notes, however, "specifications may not in themselves guarantee that a customer's requirements will be met consistently…"

Clause 4.5 of ISO 9000-1 looks at the following four facets of quality:

- Quality due to definition of needs for the product.
- Quality due to product design.
- Quality due to conformance to product design.
- Quality due to product support.

An effective quality system will address all four facets of quality.

What Is a Quality System?

A quality system is the "organizational structure, procedures, processes and resources needed to implement quality management" (ISO 8402:1994, Clause 3.6). It should only "be as comprehensive as needed to meet the quality objectives."

Earlier in the industrial era, product quality was associated only with inspection after the fact. To improve quality control and prevent problems from occurring, manufacturers developed tools such as statistical process control and installed quality control departments. Quality standards such as ISO 9000 are based on the idea of building quality into every aspect of the enterprise with an integrated quality management system.

Clause 5, Quality system elements of ISO 9004-1 states that the quality system involves all processes in the life cycle of a product that affect quality, from initial identification of market needs to final satisfaction of requirements. Figure 3–1 illustrates the typical life cycle phases of the product.

What Is Quality Management?

Quality management refers to "all activities of the overall management function that determine the quality policy, objectives and responsibilities and implement them by means such as quality planning, quality control, quality assurance and quality improvement within the quality system" (ISO 8402:1994). According to quality systems consultant Ian Durand, "Quality management is not separate from general management. When

Figure 3–1

Typical Life Cycle Phases of a Project

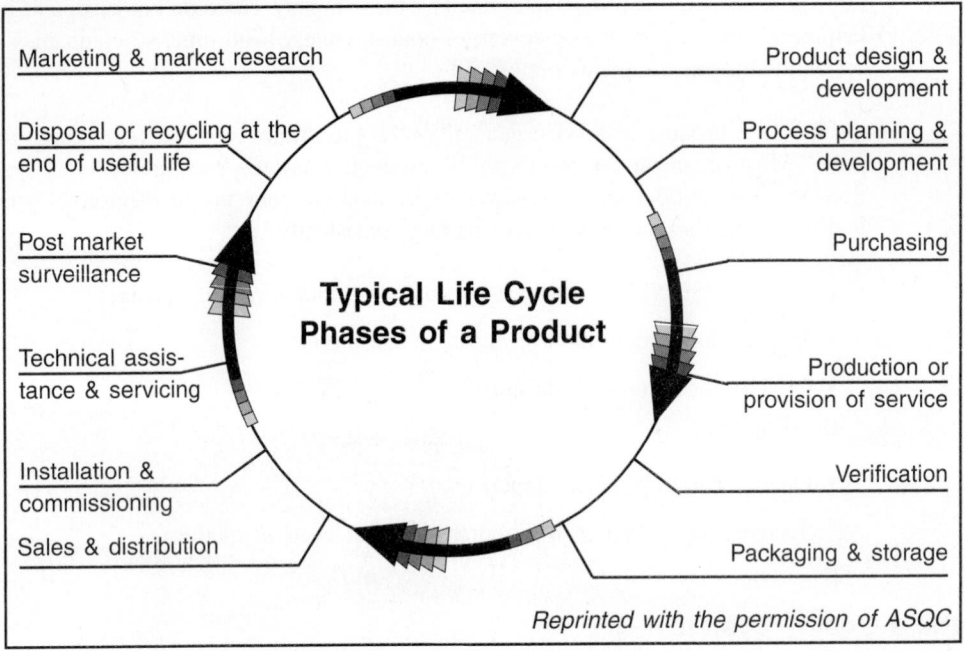

Reprinted with the permission of ASQC

used effectively, quality management should be an integral part of an organization's over-all management approach."

What Is Quality Assurance?

Quality assurance includes "all the planned and systematic activities implemented within the quality system, and demonstrated as needed, to provide adequate confidence that an entity will fulfill requirements for quality" (ISO 8402). An entity is anything that can be "individually described and considered." For example, an entity can be a process, a product, an organization, a system, person, or any combination of these.

The purpose of a quality assurance system is to prevent problems from occurring, detect them when they do, *identify the cause, remedy the cause, and prevent recurrence.* A more succinct summary is offered by Ian Durand: "The basis of a quality system," he says, "is to say what you do, do what you say, record what you did, check the results, and act on the difference" (see Figure 3–2). A summary of this process is offered by Donald W. Marquardt.

PLAN your objectives for quality, and the processes to achieve them.

Figure 3–2

Relationship of Concepts

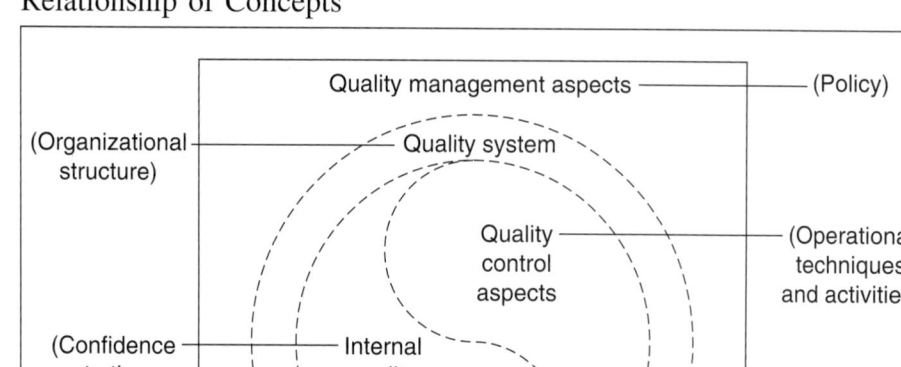

Reprinted with the permission of ASQC

DO the appropriate resource allocation, implementation, training and documentation
CHECK to see if:

- You are implementing as planned

- Your quality system is effective

- You are meeting your objectives for quality

ACT to improve the system as needed*

Figure 3–2 illustrates the relationship of these concepts.

TYPES OF STANDARDS IN THE ISO 9000 SERIES

The basic ISO 9000 series consists of five standards: ISO 9000-1, ISO 9001, ISO 9002, ISO 9003, and ISO 9004-1. The standards fall into two categories: contractual situations and noncontractual situations (see Figure 3–3).

*Editor's Note: excerpted from the *The Memory Jogger*™ *9000*, Published by GOAL/QPC, Methuen, MA, pp. 143–144.

Figure 3–3

Structure of the ISO 9000 Standards

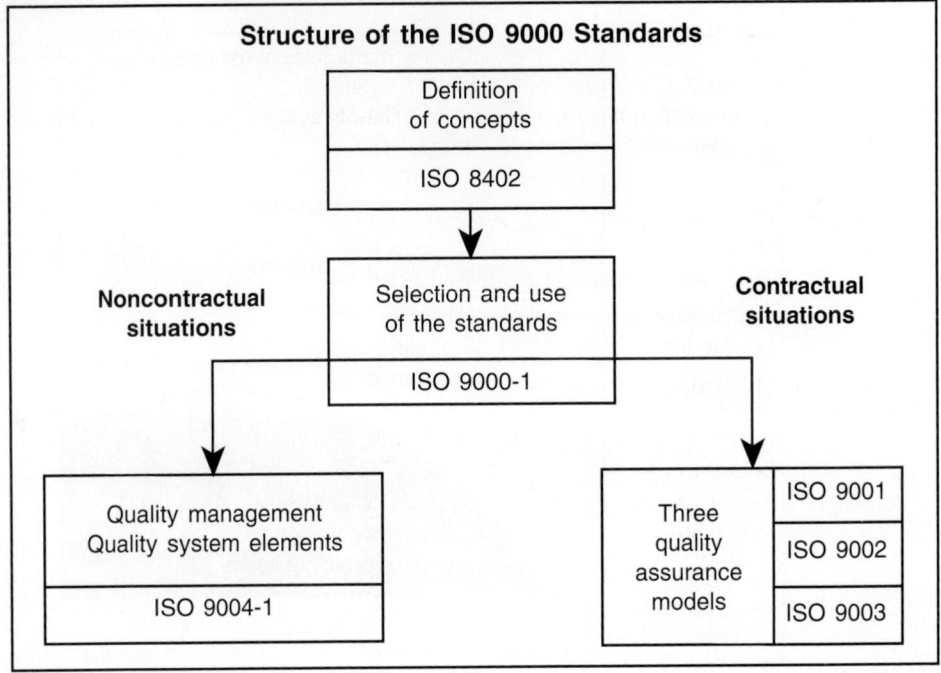

Conformance Standards

ISO 9001, ISO 9002, and ISO 9003, which are conformance standards, are used for *external quality assurance* to provide confidence to the customer that the company's quality system is capable of providing a satisfactory product or service.

The three conformance standards are not levels of quality; they differ only in comprehensiveness so that they can be adapted to different types of organizations. For example, ISO 9002 does not include design control as a quality system element.

ISO 9001

ISO 9001, Quality systems—Model for quality assurance in design/development, production, installation and servicing is the most comprehensive of the conformance standards. It includes all elements listed in ISO 9002 and ISO 9003. In addition, it addresses the design, development, and servicing capabilities not addressed in the other models.

ISO 9001 is used when the supplier must ensure product conformance to specified needs throughout the entire product cycle. It is also used when the contract specifically requires a design effort. ISO 9001 commonly applies to manufacturing or processing industries, but it can also apply to services such as construction, architecture, and engineering. (The requirements of ISO 9001 are discussed in detail in Chapter 4.)

ISO 9002

ISO 9002, Quality systems—Model for quality assurance in production, installation and servicing addresses production and installation. In the 1994 revised standard, the only distinction between ISO 9001 and ISO 9002 is that ISO 9002 does not include the design function.

ISO 9002 applies to a wide range of industries based on technical designs and specifications provided by customers. It is relevant for products that do not involve a design aspect and is used when the specified product requirements are stated in terms of an already established design or specification.

ISO 9003

ISO 9003, Quality systems—Model for quality assurance in final inspection and test is the least comprehensive standard. It addresses only the requirements for detection and control of problems during final inspection and testing. ISO 9003 applies to organizations whose products or services can be adequately assessed by testing and inspection. Generally, this refers to less complex products or services.

Guidance Standards

ISO 9000-1 and ISO 9004-1 are *guidance standards*. This means they are descriptive documents, not prescriptive requirements. ISO 9000-1 and ISO 9004-1 provide guidance to "all organizations for quality management purposes." These documents are used for internal quality assurance, "activities aimed at providing confidence to the management of an organization that the intended quality is being achieved."

ISO 9000-1

ISO 9000-1, Quality management and quality assurance standards—Guidelines for selection and use introduces the ISO 9000 series and explains fundamental quality concepts. It defines key terms and provides guidance on selecting, using, and tailoring ISO 9001, ISO 9002, and ISO 9003 for external quality assurance purposes. It also provides guidance on using ISO 9004-1 for internal quality management. It is the roadmap for use of the entire series.

Editor's Note: The 1994 revisions of ISO 9000 and 9004 are designated ISO 9000-1 and ISO 9004-1, with titles unchanged; this permits use of part numbers, such as ISO 9004-2 and ISO 9004-3, to accommodate additional guidance standards addressing individual subjects.

The following are some of the key points discussed in ISO 9000-1.

The quality objectives of an organization.   In *Clause 4.0, Principal concepts*, ISO 9000-1 describes an organization's basic quality objectives (see Table 3–2). Each organization should do the following:

- Achieve, maintain, and continually improve the quality of its products.

- Improve the quality of its own operations to meet the needs of its customers and other stakeholders.

- Provide confidence internally that quality is being fulfilled, maintained, and improved.

Table 3-2

Contents of ISO 9000-1: 1994

0 Introduction
1 Scope
2 Normative references
3 Definitions
4 Principal concepts
4.1 Key objectives and responsibilities for quality
4.2 Stakeholders and their expectations
4.3 Distinguishing between quality system requirements and product requirements
4.4 Generic product categories
4.5 Facets of quality
4.6 Concept of a process
4.7 Network of processes in an organization
4.8 Quality system in relation to the network of processes
4.9 Evaluating quality systems
5 Roles of documentation
6 Quality system situations
7 Selection and use of international standards on quality
8 Selection and use of international standards for external quality assurance
Annexes
A Terms and definitions taken from ISO 8402
B Product and process factors
C Proliferation of standards
D Cross-reference list of clause numbers for corresponding topics
E Bibliography

- Provide confidence to the customers and other stakeholders that requirements for quality will be achieved in the delivered product (ISO 9000-1).

- Provide confidence that quality-system requirements are fulfilled.

The standard emphasizes that every organization has five groups of stakeholders: customers, employees, owners, subcontractors, and society. In addition, there may be government requirements, environmental regulations, and other stakeholders. The quality expectations vary with each group. An effective quality system addresses the requirements of all groups of stakeholders. This concept of quality goes well beyond simply meeting customer specifications—it encourages companies to anticipate all stakeholder expectations and exceed them.

All work as a process.   ISO 9000 emphasizes a process-based view of the organization. Clauses 4.6 and 4.7 of 9000-1 emphasize that all work is accomplished through processes. A process is any transformation that adds value. Outputs of processes can be either tangible or intangible; that is, they can be either product or information related.

The work of organizations is accomplished through a network of processes. To achieve quality, an organization needs to identify, organize, and manage its network of processes and the interfaces between those processes. The final result should be an integrated quality system.

When evaluating a quality system, an organization should ask three key questions:

- Are the processes defined and are procedures appropriately documented?

- Are the processes fully deployed and implemented as documented?

- Are the processes effective in providing the expected results?

Management review of the quality system.   Subclause 4.9.2 describes the importance of systematic management review of the quality system. *Subclause 4.9.3, Quality system audits*, emphasizes the importance of audits, whether conducted by the organization itself, the customer, or an independent body. The goal of management reviews and internal audits is a more effective and efficient quality system. (Chapter 6 describes the internal audit process).

The role of documentation.   Section 5 of 9000-1 discusses the role of documentation. The purpose of ISO 9000 documentation is not to create a bureaucracy, nor to generate a paper factory, but to help the organization:

- Achieve required (product) quality.

- Evaluate quality systems.

- Achieve quality improvement.

- Maintain the improvements.

Documentation also plays a role in auditing, by providing objective evidence that a process is defined, procedures are approved, and changes to procedures are controlled. Documentation allows organizations to measure current performance and thus measure the effect of changes—both positive and negative.

Tailoring the standard to a contract. *Clause 8.4, Additional considerations in contractual situations*, notes that in certain situations the standards can be tailored to the contract; certain quality system elements or subelements called for in the standard may be deleted, or other elements may be added, such as statistical process control. When tailoring is required, it should be specified in the contract.

Both parties should review the proposed contract to make sure they understand its requirements. In second-party situations, the customer assesses the quality system prior to awarding the contract and continually audits the quality system after awarding the contract.

ISO 9004-1

ISO 9004-1, Quality management and quality system elements—Guidelines provides guidance to all organizations for internal quality management purposes, without regard to external contractual requirements. ISO 9004-1 examines most of the quality system elements contained in ISO 9001, ISO 9002, and ISO 9003 in greater detail (see Table 3–3). It can help organizations determine the extent that each quality system is applicable to them.

The Role of ISO 9004-1

Misunderstandings are common regarding ISO 9004-1's relationship to other standards in the ISO 9000 family and its own potential value. Although ISO 9004-1 is clearly identified as a guideline standard and has no specific requirements, suppliers proceeding through the registration process will find the guidance standard to be helpful. For example, ISO 9004-1 clauses—such as those on design review, statistical methods, and corrective action—contain elements that may lay the groundwork for fulfilling ISO 9001 requirements.

ISO 9004-1 also includes subjects not addressed explicitly in ISO 9001, such as quality economics (quality cost approaches) and continuous quality improvement. Perhaps it receives the most attention from the legal profession, since 9004-1 also addresses product safety, a subject not appearing in ISO 9001, 9002, and 9003.

ISO 9004-1 may also help suppliers that anticipate registering to ISO 9001 or ISO 9002 but are under no pressure to do so and have not yet set a registration timetable. Such a supplier may want to consider tailoring a quality management plan to its own present

Table 3–3

Contents of ISO 9004-1

```
0  Introduction
1  Scope
2  Normative references
3  Definitions
4  Management responsibility
5  Quality system elements
6  Financial considerations of quality systems
7  Quality in marketing
8  Quality in specification and design
9  Quality in purchasing
10  Quality in processes
11  Control of processes
12  Product verification
13  Control of inspection, measuring and test equipment
14  Control of nonconforming product
15  Corrective action
16  Post production activities
17  Quality records
18  Personnel
19  Product safety
20  Use of statistical methods
Annex A: Bibliography
```

and future needs based on a combination of ISO 9004-1, Malcolm Baldrige National Quality Award criteria, and relevant standards in its own industry. Later, if customers require evidence of an effective quality system, the organization can implement the requirements of ISO 9001, ISO 9002, or ISO 9003 to demonstrate the effectiveness of its quality system.

Another use of ISO 9004-1 is by suppliers already registered to ISO 9001 or ISO 9002. These suppliers recognize that the contents of the quality assurance standards are only a foundation and want to experience the benefits of improved quality management practice. In such cases, the discipline established by registration provides an excellent base to ensure that capabilities, once adopted, remain effective. The supplier can use the guidance of ISO 9004-1 (or other quality management standards) as it continues to improve its capabilities.

Topics Covered in ISO 9004-1

A brief introduction to some of the topics covered in ISO 9004-1 follows. (Guidance offered by 9004-1, cross-referenced to the corresponding clauses in ISO 9001, is also described in Chapter 4.)

Organizational Goals 0.2

ISO 9004-1 recommends each company "should ensure that the technical, administrative and human factors affecting the quality of its products will be under control, whether hardware, software, processed material or services." The aim is to reduce, eliminate, and prevent quality deficiencies.

To achieve its objectives, companies should develop a quality system that is appropriate to their activities and the product being offered. The goal is to meet desired customer needs for quality at an optimum cost to the organization. The standard encourages companies to consider the costs, benefits, and risks inherent in most products.

The benefits for the customer lie in product or service satisfaction; for the company in increased profitability and market share. A key issue is the cost of poor quality in marketing and design deficiencies, rework, repair, replacement, reprocessing, and other costs. An effective quality system satisfies customer needs while protecting the company's interests. It also addresses the needs of both society and the environment.

Management Responsibility (4.0)

ISO 9004-1 emphasizes that the responsibility for defining and documenting the organization's quality policy belongs to management. A company's quality policy is the "overall intentions and direction of an organization with regard to quality, as formally expressed by top management" (ISO 8402, 1994). It is one element of the corporate policy and is authorized by top management.

Management is ultimately responsible for establishing a quality policy and all decisions concerning the quality system. It is up to management to do the following:

- Define general and specific quality responsibilities.
- Delegate appropriate responsibility and authority.
- Clearly establish the organizational structure.
- Identify quality problems and initiate preventive measures.
- Provide sufficient and appropriate resources to achieve quality objectives.
- Determine training needs.
- Identify quality factors that affect new products, processes, or services.

- Exercise adequate and continuous control over all activities affecting quality.

- Emphasize preventive actions to avoid problems.

- Write procedures simply and clearly.

Quality System Elements (5.0)

This comprehensive section of ISO 9004-1 covers the key elements comprising the structure of a quality system. A quality system applies to all activities that affect quality and involves all phases in the life cycle of a product, from initial market research through post-market surveillance and disposal or recycling at the end of the product's useful life.

Configuration management. ISO 9004-1 adds configuration management to the structure of the quality system. According to ISO 9004-7, *Guidelines for configuration management*, Clause 3.1, a configuration consists of the "functional and physical characteristics of a product as set forth in technical documentation and achieved in the product." Configuration management is defined in ISO 9004-7 as the technical and organizational activities of configuration identification, control, status accounting, and audit. ISO 9004-1 emphasizes that the quality system should include documented procedures for configuration management to the extent appropriate.

Documentation of the quality system. The quality system of an organization should be documented in a systematic, orderly, and understandable manner in the form of policies and procedures. Not everything must be documented; only procedures that affect the quality of the product or service.

Typically, the quality system is documented in a quality manual. The purpose of a quality manual is to outline the quality system and to serve as a reference. Supporting the quality manual are documented quality system procedures and work instructions.

Quality plans. When there is a new product, process, or significant change to an existing product or process, the organization should prepare and maintain a documented *quality plan*, a "document setting out the specific quality practices, resources and sequence of activities relevant to a particular product, project or contract" (ISO 8402, 1994). The quality plan may be part of the overall quality system. A quality plan for a specific project usually references the applicable parts of the quality manual. (See Chapter 7 for a discussion of quality plans.)

Quality records. Documentation also includes maintaining quality records such as design charts, inspection and testing records, audit results, and so on. These provide evidence of conformance to specified requirements and the effective operation of the quality system.

Auditing the quality system   An effective quality system includes provisions for regular auditing to determine if the system is achieving its objectives. Section 5.4 outlines the necessary elements for an effective internal auditing program. These include:

- Planning and scheduling the activities and areas to audit.
- Gathering an audit team whose members are independent of those directly responsible for the specific activities being audited.
- Documenting the audit procedures, including recording its results and agreeing on timely corrective actions.
- Reporting the audit conclusions to management responsible for the area audited.
- Assessing and documenting the effectiveness of corrective actions.

(See Chapter 6 for more information on internal audits.)

Review and evaluation of the quality system.   ISO 9004-1 recommends that management regularly review and evaluate the quality system. The review should include results from internal audits, assessments of whether the objectives were fulfilled, and possible considerations for updating the quality system.

Quality improvement.   ISO 9004-1 adds a clause concerning quality improvement. The quality system should facilitate and promote continuous quality improvement. Management can create an environment for continuous improvement by doing the following:

- Encouraging and sustaining a supportive style of management.
- Promoting values, attitudes, and behavior that foster improvement.
- Setting clear quality improvement goals.
- Encouraging effective communication and teamwork.
- Recognizing successes.
- Providing training and education geared to improvement.

Financial considerations.   Clause 6 emphasizes the importance of measuring a quality system's effectiveness. The results can be used to identify inefficient activities and initiate internal improvement. Clause 6 describes the following three methods, among others, for analyzing the financial data.

The *quality-costing* approach addresses quality-related costs, arising both from internal operations and external activities. Using this approach, prevention and appraisal cost are considered investments, while failure costs, both internal and external, are considered losses.

The *process-cost* approach analyzes the costs of conformity and the costs of nonconformity for any process. The costs of the former are incurred to fulfill customer expectations. The costs of the latter result when processes and products fail to meet requirements.

The *quality loss* approach focuses on losses due to poor quality. Losses can be intangible, such as loss of future sales due to customer dissatisfaction or lower work efficiency, or they can be tangible, such as internal and external failure costs.

ISO 9004-1 recommends that the financial reporting of quality activities—whatever method is used—should be related to other conventional business measures, such as sales or added value. Note that no reference to financial considerations is contained in ISO 9001, 9002, or 9003, and is included in ISO 9004-1 only as guidance.

Quality in Marketing (7.0)

The standard notes that the marketing function should take the lead in establishing adequately defined and documented requirements for product or service quality. This involves determining the need for a product, defining the market demand, determining customer requirements, communicating these requirements, and ensuring that all relevant parts of the organization are capable of meeting them. Clause 7.0 emphasizes the importance of defining product specifications thoroughly and establishing a method of obtaining and analyzing customer feedback.

Quality in Specification and Design (8.0)

Clause 8 discusses ways to translate customer needs into technical specifications for materials, products, and processes. This clause discusses in detail the following:

- Design planning and objectives.
- Product testing and measurement.
- Design review.
- Design verification.
- Design qualification and validation.
- Final design review and production release.
- Market readiness review.
- Design change control.
- Design requalification.
- Configuration management in design.

Quality in Purchasing (9.0)

Since purchases become part of the organization's product, they directly affect quality. Clause 9.0 recommends that all purchasing be planned and controlled by documented procedures. This includes purchasing services such as testing, calibration, and subcontracted processing. Clause 9.0 covers the following elements of quality in purchasing:

- Requirements for specifications, drawings, and purchase documents.
- Selection of acceptable subcontractors.
- Agreement on quality assurance.
- Agreement on verification methods.
- Provisions for settlement of disputes.
- Procedures for receiving inspection, planning, and control.
- Maintenance of quality records related to purchasing.

Quality of Processes (10.0)

Clause 10 discusses the quality of processes. Planning of processes should ensure that they proceed under controlled conditions in the specified manner and sequence. Clause 10 covers the following management elements:

- Planning for process control.
- Verifying the capability of processes to produce in accordance with product specifications.
- Controlling and verifying supplies, utilities, and the environment insofar as these affect product quality characteristics.
- Proper planning, control, and documentation of product handling.

Control of Processes (11.0)

Clause 11 details the control of all processes in the life cycle of the product or service. Specifically, it covers the following:

- Material control, traceability, and identification.
- Equipment control and maintenance.
- Process control management.
- Documentation.
- Process change control.
- Control of verification status.
- Control of nonconforming product.

Product Verification (12.0)

This clause looks at the process of verifying products, including incoming materials and parts, in-process verification, and finished product verification.

Control of Inspection, Measuring, and Test Equipment (13.0)

Clause 13 emphasizes the importance of well-controlled measuring systems to provide confidence in decisions or actions that are taken based on measurement data. Topics discussed include the following:

- Elements of measurement control, including calibration and traceability to reference standards.
- Subcontractor measurement controls.
- Corrective action.
- Outside testing.

Control of Nonconforming Product (14.0)

Clause 14 recommends establishing procedures for dealing with nonconforming product to prevent the customer from inadvertently receiving the nonconforming product and to avoid further processing of nonconforming product. The steps outlined in the clause include:

- Identification
- Segregation
- Review
- Disposition
- Action
- Avoiding recurrences

Corrective Action (15.0)

Corrective action is necessary to eliminate quality problems or minimize their occurrence. Clause 15 covers the following:

- Assigning responsibility for instituting corrective action.
- Evaluating the importance of the problem.
- Investigating possible causes.
- Analyzing the problem.
- Eliminating the causes.

- Implementing process controls to avoid recurrence.
- Recording permanent changes resulting from corrective action.

Post-Production Activities (16.0)

The responsibility for quality continues beyond the production phase. Clause 16 covers the following:

- Storage methods to increase shelf life and avoid deterioration.
- Protection of product quality during delivery.
- Installation procedures.
- Servicing.
- Post-marketing surveillance to establish a warning system for reporting product failures or shortcomings.
- Market feedback regarding performance to monitor quality.

Quality Records (17.0)

Clause 17 focuses on establishing and maintaining documented procedures for quality records. Records can include inspection reports, test data, qualification reports, calibration data, and so on. The clause discusses documents that must be controlled, including drawings, specifications, inspection procedures and instructions, work instructions, and quality plans.

Personnel (18.0)

Clause 18 discusses the need for personnel training. Appropriate training should be provided to all levels of personnel that perform activities affecting quality. This includes executive and management personnel, technical personnel, process supervisors, and operating personnel. The clause covers the following:

- Qualification requirements for personnel.
- Motivating all personnel toward quality performance.

Product Safety (19.0)

This clause focuses on identifying safety aspects of the products and processes to enhance safety. Steps in the process include:

- Identifying relevant safety standards.
- Evaluating and testing for safety.
- Analyzing instructions and warnings for the user.

- Developing a product recall method.

- Developing an emergency plan in case recall becomes necessary.

Use of Statistical Methods (20.0)

The final clause in ISO 9004-1 recognizes the appropriate use of statistical methods. Documented procedures should be established and maintained for applying statistical methods to, among others, market analysis, product design, reliability specification, process control, and process improvement. The clause lists specific statistical methods that can be used, including design of experiments, statistical sampling, and other methods.

OTHER USEFUL STANDARDS IN THE ISO 9000 FAMILY

ISO 9000-1 includes references to other useful guidance standards in the ISO 9000 series (see Tables 3–4 and 3–5). Some of these standards have already been published. Others are in draft form. They are briefly introduced below. Those that have been published are discussed in greater detail in other chapters of this *Handbook*.

ISO 9000-2

ISO 9000-2, Quality management and quality assurance standards—Part 2: Generic guidelines for the application of ISO 9001, ISO 9002 and ISO 9003 offers application guidance in implementing the standards. Guidance from ISO 9000-2 that corresponds to each clause in ISO 9001 is described in Chapter 4.

ISO 9000-3

Software is discussed in *ISO 9000-3, Quality management and quality assurance standards—Part 3: Guidelines for the application of ISO 9001 to the development, supply and maintenance of software.* This standard provides guidance to supplier organizations that produce software or products that include a software element. The primary rationale for this standard is that software development, supply, and maintenance—unlike other manufacturing processes—does not have a distinct manufacturing phase. The key process is the design phase. ISO 9000-3 offers suggestions regarding appropriate controls and methods that apply to the design phase. (ISO 9000-3 and software applications are discussed in detail in Chapter 20.)

ISO 9000-4

ISO 9000-4, Quality management and quality assurance standards—Part 4: Guide to dependability program management provides guidance on dependability program management. It focuses on the reliability, maintainability, and availability characteristics of

Table 3–4

Cross Reference of ISO 9000 Quality Assurance Requirements

Clause and Title	Quality Assurance Requirements		
	ISO 9001	ISO 9002	ISO 9003
4.1 Management Responsibility	◆	◆	●
4.2 Quality System	◆	◆	●
4.3 Contract Review	◆	◆	◆
4.4 Design Control	◆	○	○
4.5 Document and Data Control	◆	◆	◆
4.6 Purchasing	◆	◆	○
4.7 Control of Customer-Supplied Product	◆	◆	◆
4.8 Product Identification and Traceability	◆	◆	●
4.9 Process Control	◆	◆	○
4.10 Inspection and Testing	◆	◆	●
4.11 Control of Inspection, Measuring and Test Equipment	◆	◆	◆
4.12 Inspection and Test Status	◆	◆	◆
4.13 Control of Nonconforming Product	◆	◆	●
4.14 Corrective and Preventive Action	◆	◆	●
4.15 Handling, Storage, Packaging, Preservation and Delivery	◆	◆	◆
4.16 Control of Quality Records	◆	◆	●
4.17 Internal Quality Audits	◆	◆	●
4.18 Training	◆	◆	●
4.19 Servicing	◆	◆	○
4.20 Statistical Techniques	◆	◆	●

Key:

◆ Comprehensive requirement

● Less comprehensive requirement than ISO 9001 and 9002

○ Element not present.

products such as transportation, electricity, telecommunications, and information services. It covers the essential features of a comprehensive dependability program. In Clause 1, *Scope,* the standard emphasizes that its requirements " are aimed primarily at controlling influences on dependability at all product lifecycle phases from product planning to operation."

Table 3-5

Cross-reference List of Clauses in ISO 9000 and ISO 9004-1

ISO 9001 Clause and Title	Quality Management Guidance ISO 9004-1
4.1 Management Responsibility	4
4.2 Quality System	5
4.3 Contract Review	Not Present
4.4 Design Control	8
4.5 Document and Data Control	5.3, 11.5
4.6 Purchasing	9
4.7 Control of Customer-Supplied Product	Not Present
4.8 Product Identification and Traceability	11.2
4.9 Process Control	10, 11
4.10 Inspection and Testing	12
4.11 Control of Inspection, Measuring and Test Equipment	13
4.12 Inspection and Test Status	11.7
4.13 Control of Nonconforming Product	14
4.14 Corrective and Preventive Action	15
4.15 Handling, Storage, Packaging, Preservation and Delivery	10.4, 16.1, 16.2
4.16 Control of Quality Records	5.3, 17.2, 17.3
4.17 Internal Quality Audits	5.4
4.18 Training	18.1
4.19 Servicing	16.4
4.20 Statistical Techniques	20
Quality Economics	6
Product Safety	19
Marketing	7

ISO 9004-2

ISO 9004-2, Quality management and quality system elements—Part 2: Guidelines for services is geared to organizations that provide services or whose products include a service component. It takes into account factors that may differ from a product offering, such as customer interaction and customer assessment. (ISO 9004-2 is discussed in greater detail in Chapter 10.)

ISO 9004-3

ISO 9004-3, Quality management and quality system elements—Part 3: Guidelines for processed materials applies to organizations whose products consist of processed materials such as solids, liquids, or gases that are delivered in pipelines, drums, tanks, or cans. Clause 7.11 points out the importance of statistical sampling and evaluation procedures and their application to in-process controls and final product specifications. ISO 9004-3 includes guidelines on process control, process capability, equipment control, and maintenance and documentation. (ISO 9004-3 is discussed in greater detail in Chapter 20.)

ISO 9004-4

All companies should strive for continuous quality improvement. *ISO 9004-4, Quality management and quality system elements—Part 4: Guidelines for quality improvement* describes fundamental concepts and methods for quality improvement. In the Introduction, the standard notes that "the motivation for quality improvement comes from the need to provide increased value and satisfaction to customers." Everyone in the organization should be alert as to how each process can be performed more effectively and more efficiently with less waste and resource consumption. A fundamental concept of quality improvement is that companies should "seek opportunities for improvement, rather than waiting for a problem to reveal opportunities" (Clause 3.1.4).

ISO 9004-7

ISO 9004-7, Quality management and quality systems elements—Part 7: Guidelines for configuration management provides guidance in adopting the management discipline of configuration management. It is applied over the lifecycle of a product to provide visibility and control of functional and physical characteristics. Configuration management describes the technical organizational activities of configuration identification, control, status accounting, and audit. Configuration management provides a rigid discipline for identification of product and part status, applicable for products with large numbers of component parts, where identification of design status is critical. This process is particularly applicable in industries such as aerospace.

ISO 10011

ISO 10011, Guidelines for auditing quality systems focuses on auditing. It is designed for all auditing situations: first-, second-, and third party. ISO 10011-1 looks at the overall process of establishing, planning, performing, and documenting quality system audits. 10011-2 provides guidance on the education, training, and experience needed to carry out an audit. And 10011-3 examines the process of managing an audit, from initial planning to the closing meeting. (See Chapter 6 for more information on the 10011 series.)

ISO 10012

In many products or processes, quality depends upon accurate measurements. *ISO 10012-1, Quality assurance requirements for measuring equipment—Part 1: Metrological confirmation system for measuring equipment* includes detailed guidance for a supplier's measurement system to ensure accurate and consistent measurement.

ISO 10013

ISO 10013, Guidelines for developing quality manuals, describes the development, preparation, and control of quality manuals, tailored to the specific user. Quality manuals developed in accordance with ISO 10013 will reflect documented quality system procedures required by ISO 9001 and ISO 9002. The standard does not cover detailed work instructions or quality plans. The standard may be used to develop quality manuals relating to quality system standards other than the ISO 9000 series.

The ISO 9001 Standard

4

ISO 9001 is the most comprehensive conformance model in the ISO 9000 series and includes all clauses contained in ISO 9002 and ISO 9003. This chapter discusses the requirements of ISO 9001:1994 in detail. It also discusses the changes made to ISO 9001:1987, where appropriate.

The chapter includes guidelines taken from ISO 9000-2, *Quality management and quality assurance—Part 2: Generic guidelines for the application of ISO 9001, ISO 9002 and ISO 9003*. Like ISO 9000-1 and ISO 9004-1, ISO 9000-2 is a guidance document; companies do not seek to become registered to any of these documents but rather use them for assistance when implementing ISO 9001, ISO 9002, or ISO 9003.

In addition, this chapter includes references taken from QS-9000 Quality System Requirements, a document prepared by the U.S. Big Three Automakers (Chrysler, Ford, and General Motors)*. For many suppliers in the American automotive industry, QS-9000 is a requirement to be met. For background on these requirements see Chapter 15.

Finally, this chapter includes some information from ISO 9004-1 and, where appropriate, interpretation of ISO 9001 elements by experts in the field. (See Appendix 1 for contributor biographies.) This interpretation takes two forms: comments by The Victoria Group, an ISO 9000, QS-9000, and ISO 14000 consulting, auditing, and training company, and question-and-answer interpretations that previously appeared in *Quality Systems Update* newsletter.

INTRODUCTION

The 1994 version of the standards has clarified the three key terms—supplier, purchaser, and subcontractor. Table 4–1 charts the various ways in which these terms are used, not only in the ISO 9000 series, but also in the U.S. DoD's military standards, the U.S. Food and Drug Administration's Good Manufacturing Practices (GMP) requirements, and in the Malcolm Baldrige National Quality Award Criteria. (The U.S. Department of Defense (DoD) standards have been discontinued; reference is included here because of their widespread use.)

The introduction to ISO 9001 introduces the three standards that can be used for external quality assurance purposes. The 1987 reference to quality systems as "suitable for two-party contractual purposes" has been replaced by the broader statement, "quality system requirements suitable for the purpose of a supplier demonstrating its capability, and for the assessment of the capability of a supplier by external parties." This statement recognizes the growth and importance of the third-party registration system.

ISO 9001 also emphasizes the generic quality of the standards. The standards are independent of any specific industry or economic sector. The introduction notes that "it is not the purpose of these International Standards to enforce uniformity of quality systems."

The Scope section states that ISO 9001 contains quality system requirements for use where a supplier's capability to design and supply conforming product needs to be

Table 4–1

Relationships of Organizations in the Supply Chain

	Companies Supplying Products to You	Your Organization	Companies to Whom You Provide Products
ISO 9001, 9002, 9003 (Contractual External)	Subcontractor	Supplier	Customer ~~Purchaser~~
ISO9004 Guidelines, Internal)	Subcontractor ~~Supplier~~	Organization ~~Company~~	Customer
Military Standards MIL-Q-9858A MIL-I-45208	Subcontractor	Contractor	Procurer
Medical Device Good Manufacturing Practices (GMPs)	Component Supplier	Manufacturer	User
Malcolm Baldrige National Quality Award Criteria	Supplier	Company	Customer

From chart developed by Dale Thanig, OHMEDA Medical Systems Division, Louisville, Co (Revisions to 1987 Issue Noted.)

demonstrated. ISO 9001:1987 emphasized that the requirements specified are "aimed primarily at preventing nonconformity at all stages from design through to servicing" (emphasis added). Customer satisfaction was implied throughout the 1987 standard. ISO 9001: 1994, however, adds an explicit statement regarding customer satisfaction by stating that the requirements in the standards are "aimed primarily at achieving customer satisfaction by preventing nonconformity at all stages from design through to servicing."

Thus, the expanded wording of the standard identifies customer satisfaction as an ultimate purpose of nonconformity prevention. This is reinforced in Subclause 4.1.1, Quality Policy, where the standard notes that the "quality policy shall be relevant to the supplier's organizational goals and the expectations and needs of its customers".

ISO 9001: Key Points

The introduction to ISO 9001 gives an overview of the structure of the ISO 9000 series, explains the applicability of each section, and emphasizes significant aspects of the ISO 9001 standard. Several key points include the following:

Three Distinct Models

First, the Introduction states that "the alternative quality assurance models . . . represent three distinct forms of quality-system requirements . . ." In other words, ISO 9001, ISO 9002, and ISO 9003 do not represent three steps on a ladder of excellence but rather are separate, independent standards. Equally important is to ensure that the scope of registration applies to the goods or services which a company produces or wishes to procure. It is not uncommon for a company to hold registration for only part of its commercial operations.

Complementary Requirements

The introduction further explains that requirements specified in the standard are always complementary (not alternative) to the technical (product) specified requirements. There are many situations where compliance with a published standard specific to a particular product or range of products will require more of the supplier than does the ISO 9000 standard. The individual product or service requirements remain supreme.

The ISO Standards are not about making everyone's systems the same. A company will do what is appropriate for its activities. ISO 9000 series registration will not solve all a company's problems, nor is registration a substitute for complying with government regulatory requirements such as the U.S. Environmental Protection Agency (EPA).

(Continued on next page)

(Continued from page 61)

Prevention of Nonconformity

Section 1.0, *Scope,* states that "the requirements specified in this International Standard are aimed primarily at achieving customer satisfaction by *preventing* nonconformity at all stages from design through to servicing" (emphasis added). The prevention of nonconformity is explicit and prominent. The standard reminds the user that the ultimate purpose is to achieve customer satisfaction. Using the ISO 9000 Standards as a model should create a system that is prevention oriented and monitored through all stages of design, production, and servicing. The language of ISO 9001 and ISO 9002 thus mirrors the tenets of Total Quality Management (TQM).

Related Documents

Sections 2.0 and 3.0 stress the importance of ensuring that related documents or special definitions are made clear to all readers. The terminology in International Standard ISO 8402:1994 should be used to ensure wide reader understanding of the standard. The definitions in ISO 8402:1994 constitute provisions of the standard through reference in the text.

The Victoria Group

CLAUSE 4: QUALITY SYSTEM REQUIREMENTS

The main body of ISO 9001 is contained in the quality system requirements of Section 4.0. There are 20 clauses in all. The first quality system requirement is management responsibility.

4.1 MANAGEMENT RESPONSIBILITY

Introduction

This clause describes the responsibility of management for developing a quality system. The major responsibilities include the following:

- Establish a quality policy.
- Organize personnel.
- Verify quality.
- Review the quality system.

ISO 9004-1 Guidance

 ISO 9004-1 adds a general introductory clause that defines quality management as encompassing "all activities of the overall management function that determine the quality policy, objectives and responsibilities, and implement them by means such as quality planning, quality control, quality assurance and quality improvement within the quality system" (Clause 4.1). The clause emphasizes that the responsibility for and commitment to a quality policy belongs to the highest level of management.

4.1.1 Quality Policy

ISO 8402, Quality Management and Quality Assurance—Vocabulary, defines a quality policy as "the overall intentions and direction of an organization with regard to quality, as formally expressed by top management . . . The quality policy forms one element of the corporate policy and is authorized by top management."

ISO 9001 Requirements

 According to ISO 9001, the requirements of the organization regarding the quality policy are the following:

- Define quality policy, quality objectives, and quality commitment.

- Document the quality policy.

- Make sure everyone in the organization understands, implements, and maintains the policy.

This is made clear with the following statement: "The quality policy shall be relevant to the supplier's organizational goals and the expectations and needs of its customers." This enhances the role of the quality policy with respect to the customers' expectations and needs and to the supplier's internal policy needs.

The standard also refers to "management with executive responsibility for quality" to define and document its policy. This change re-emphasizes the importance of high-level management involvement in the quality system development process. The phrase "management with executive responsibility" is repeated throughout the standard.

Finally, the clause emphasizes not only taking action to prevent the occurrence of product nonconformity but "any nonconformities relating to product, process and quality system."

ISO 9000-2 Guidance

 ISO 9000-2 recommends that management ensure that the quality policy is:

- Easy to understand.

- Relevant to the organization.

- Ambitious yet achievable.

Since commitment to a quality policy starts at the top of an organization, management should demonstrate its commitment visibly, actively, and continually.

ISO 9004-1 Guidance

In Clause 4.0, Management Responsibility, ISO 9004-1 counsels that management should do the following:

- Define objectives pertaining to key elements of quality, such as fitness for use, performance, safety, and reliability.

- Consider the costs associated with all quality elements to minimize quality losses.

- Ensure that appropriate levels of management define specialized quality objectives.

- Provide sufficient resources to achieve its objectives.

- Determine the level of competence, experience, and training necessary.

- Control all activities affecting quality.

- Emphasize preventive actions to avoid occurrence of problems.

- State written procedures simply, unambiguously, and clearly.

- Indicate methods to be used and criteria to be satisfied.

4.1.2 ORGANIZATION—

4.1.2.1 Responsibility and Authority

ISO 9001 Requirements

The organization is required to define the responsibility, authority, and interrelation of all personnel affecting the quality of product and service to customers. This refers to personnel who must prevent the occurrence of product nonconformity, identify and record any product quality problems, recommend solutions, verify their implementation, and control further processing, delivery, or installation of nonconforming products until the problem has been corrected.

ISO 9000-2 Guidance

Individuals in the organization should do the following:

- Be aware of the scope, responsibility, and authority of their functions.

- Be aware of their impact on product and service quality.

- Have adequate authority to carry out their responsibilities.

- Understand clearly their defined authority.

- Accept responsibility for achieving quality objectives.

QS-9000 Requirements

4.1.2.3 Organizational Interfaces

- Systems are to be in place during advanced planning stages. (See Advanced Product Quality and Control Plan Reference Manual.)

- Supplier shall use multidiscipline approach for decision making.

- Information and data is to be communicated in customer-prescribed format.

4.1.2.2 RESOURCES

ISO 9001 Requirements

Management shall make sure the company has adequate resources and trained personnel to carry out any verification work. Verification activities include:

- Inspection, testing, and monitoring.

- Design reviews.

- Internal quality system audits.

This clause expands the requirement to provide adequate resources and training personnel for management, work performance, and verification activities.

ISO 9000-2 Guidance

Effective verification requires objectivity and cooperation among those involved. Adequate verification resources and personnel can involve the following elements:

- Awareness of standards.

- Adequate training.

- Production schedules that allow time for inspection, testing, and verification.

- Appropriate equipment.

- Documented procedures.

- Access to quality records.

4.1.2.3 MANAGEMENT REPRESENTATIVE

ISO 9001 Requirements

The basic requirement is to appoint a management representative who has authority to implement and maintain the quality system.

This clause establishes the requirement that the management representative report "on the performance of the quality system to the supplier's management for review and as a basis for improvement of the quality system."

The clause also notes that the management representative's responsibility may also include liaison with external bodies on matters that relate to the supplier's quality system. This may include regulatory and product standards-setting bodies that set requirements affecting the supplier.

ISO 9000-2 Guidance

If the management representative has other functions to perform, there should be no conflict of interest with those functions.

Subclause 4.1.2.3: The Management Representative

ISO 9001 Subclause 4.1.2.3 states: "The supplier's management with executive responsibility for quality shall appoint a member of the supplier's own management who, irrespective of other responsibilities, shall have defined authority for ensuring that a quality system is established . . ." What does the term "irrespective of other responsibilities" mean? What background or qualifications should this person possess, and where should the position fall in an organizational structure?

A panel of experts disagreed on whether the management representative may oversee ISO 9000 series registration as a secondary responsibility. Elizabeth Potts, president, ABS Quality Evaluations, Inc. maintained that the term "irrespective of other responsibilities" does not preclude the representative from having other duties and from those other duties taking precedence.

"This phrase allows companies the flexibility and opportunity to structure their organizations to meet their needs and not be forced to comply with an imposed requirement to have a person assigned only to the quality system," she said. "This further enhances a wider application of the standard to all types of organizations and organizational structures."

(Continued on next page)

(Continued from page 66)

While agreeing that the management representative may have other responsibilities, the three other panelists said the quality system had to be the first priority.

As a Primary Responsibility

Stephen Gousie, operations director, Information Mapping, Inc. said the term *irrespective of other responsibilities* means that the management representative must assign equal or higher priority to ISO 9000 than to other responsibilities. The management representative's primary responsibilities are to ensure that the requirements of the ISO 9000 series standard are implemented and maintained, Gousie said.

This would include primary responsibility for ensuring that the organization is prepared for the initial registration audit and for the periodic, ongoing audits to maintain registration," he said. That person should be familiar with the requirements of ISO 9000 and the appropriate external quality assurance standard. He or she should also be familiar with the company and be able to show how the organization meets each of the requirements of the standard, Gousie said.

"Most organizations have their management representative attend one of the lead assessor courses," he said. "While this will help the representative to better understand the registration audit process, it is not an absolute requirement. Knowledge of the requirements of the standard can be obtained by attending one of the many ISO 9000 overview seminars or working closely with an ISO 9000 consultant."

The position should be considered a management position, and the person in it given direct access to the most senior person in the organization, Gousie said. In most cases, he added, the management representative is also the quality assurance manager.

"There is no specific requirement that the management representative be the quality assurance manager," Gousie said. "Where the quality function is closely meshed with production, the production manager might fill this responsibility."

Must Have Defined Authority and Responsibility

Robert Hammil, of Excelsior Consulting, said the management representative must have "defined authority and responsibility" for ensuring that the requirements of the ISO 9000 series are implemented and maintained. "Ultimately, the management representative is the facility's quality system overseer, sponsor, and champion," Hammil said. "He or she is the person whom others, both within the facility and outside of it, consult on any matter pertaining to the quality management system."

Hammil said the management representative shoulders a heavy burden. "Though the management representative may have duties apart from the quality system, the person

(Continued on next page)

(Continued from page 67)

assuming this role must, in order to be effective, make a serious and long-term commitment to the activity." According to Hammil, candidates should posses a number of qualifications, including the following:

- Be knowledgeable about traditional quality assurance and quality control technologies.

- Have an understanding of the series standards and the strategic role ISO 9000 plays in the company.

- Have a commitment to the importance of ISO 9000.

- Be an authority figure, ideally with seniority and experience that cuts across departmental lines, though not necessarily from senior management.

- Have superior communication skills.

- Carry the backing of the chief executive officer or general manager.

"He or she must have the trust, confidence, and backing of the highest authority in the facility," Hammil said. "This could and often does mean that the management representative reports directly to the chief executive officer, at least insofar as quality management system responsibilities are concerned."

(Continued on next page)

4.1.3 MANAGEMENT REVIEW

ISO 9001 Requirements

Continual review of the quality system is necessary to maintain its effectiveness. Subclause 4.1.3 calls on management to do the following:

- Conduct regular management reviews of the system to make sure it remains suitable and effective.

- Keep records of its reviews.

ISO 9000-2 Guidance

The scope of management reviews should encompass the following elements:

- Organizational structure.

- Implementation of the quality system.

(Continued from page 68)

Prioritizing Job Responsibilities

John Cachat, president of IQS, Inc. agrees that "the highest priority for the individual's time will always be to ensure that ISO requirements are implemented and maintained. The larger the organization the more full-time the job becomes." Cachat said the management representative should have a broad understanding of the organization and products or services offered. That person should also have excellent communication, interpersonal, and sales skills. "Knowledge of the ISO specs can be gained over time," he explained. The position's standing within the company hierarchy may vary, according to Cachat. "If a positive, quality-, and customer-oriented environment exists, the position in the structure is less important," he said. "If ISO, organization, structure, and discipline are lacking, the position should be a high—if not the highest—ranking member of management."

Selecting the Management Representative

Selecting the management representative is entirely up to the company, according to Potts. The background and qualifications of the person are up to the company. "For organizations with a distinct quality department it is typically the quality manager, but management representatives have ranged from lab supervisors to presidents of corporations," she said.

The decision regarding where to place the position on the corporate hierarchy is also left to the company, Potts said. "What is most important is that the individual be able to demonstrate that he or she has the authority and responsibility for implementation and maintenance of the ISO 9000 quality system and has access to all appropriate personnel—including top management—to raise issues and ensure resolution relative to that implementation and maintenance."

- The achieved quality of the product or service.

- Information based on customer feedback, internal audits, process, and product performance.

The frequency of reviews is not specified but depends on individual circumstances. In terms of follow-up, problems should be documented, analyzed, and resolved. Required changes to the quality system should be implemented in a timely manner.

QS-9000 Requirements

Review shall include all elements of the quality system.

4.1.4 BUSINESS PLAN

The plan should cover short term (one to two years) and longer term (three or more years) and include:

- Competitive analysis and benchmarking.
- How to determine customer expectations.
- The use of a valid information collection process.
- Documented procedures to assure use of plan.
- Data-driven process improvement.
- How employees will be empowered.

Note: Contents of the business plan are not subject to third-party audit.

4.1.5 ANALYSIS AND USE OF COMPANY-LEVEL DATA

The supplier shall:

- Document quality and performance trends.
- Compare your data to competitors and/or appropriate benchmarks.
- Take action to solve customer-related problems.
- Support status review, decision making, and long-term planning.

4.1.6 CUSTOMER SATISFACTION

The supplier shall:

- Document the process for determining the frequency, objectivity, and validity of customer satisfaction.
- Document trends and key indicators of dissatisfaction.
- Compare trends to competition and benchmarks.
- Review trends (task of senior management).
- Consider both immediate and final customers.

4.2 QUALITY SYSTEM

ISO 9001 Requirements

Clause 4.2 requires companies to prepare a documented quality system. This involves preparing documented quality procedures and instructions and effectively implementing them.

Subclause 4.1.3: Management Review

The management representative is an individual who, "irrespective of other responsibilities," has the duty and authority to review the quality system to "ensure its continuing suitability and effectiveness" in satisfying the requirements of the standard. The management representative must belong to the supplier's management team and not be an external person (i.e., a visiting consultant), and apart from ensuring that the system is effectively maintained, he or she must also report on the performance of the system to the management, not only for review purposes, but as a basis for improvement of the system.

The management review requirement is often mishandled. It is now required to be performed by executive management. Management must be concerned with the overall process, not just with audit results. The following questions can serve as guidelines:

Is the system working effectively?

What are the quality metrics, both internal and external?

Will changes be made within the operation of the company that will radically affect the system?

Is technology changing in a way that necessitates rewriting documentation?

Is the company achieving its stated quality policy and objectives?

Are opportunities for improvement being recognized and actioned?

The Victoria Group

The elements contained in Clause 4.2 were originally contained in a note to ISO 9001:1987. Notes are advisory language, not requirements. Now the same material is in the body of the standard. This is a significant change, since what was advisory now becomes a requirement. Clause 4.2 is divided into three subclauses, as described below.

4.2.1 GENERAL

ISO 9001 Requirements

Subclause 4.2.1 requires a quality manual that defines the documentation structure of the quality system: "The supplier shall prepare a quality manual covering the requirements of this International Standard. The quality manual shall include or make reference to the quality-system procedures and outline the structure of the documentation used in the quality system."

The subclause also references guidance on quality manuals, contained in ISO 10013. Remember, this is only advisory. The guidance contained in ISO 10013 is not to be interpreted as part of the requirements of Subclause 4.2.1.

4.2.2 QUALITY SYSTEM PROCEDURES

ISO 9001 Requirements

Subclause 4.2.2 states the basic purpose of the clause—to establish and implement a documented quality system. It notes that the degree of documentation required for procedures depends on the methods used, the skills needed, and the training acquired by the personnel responsible for carrying out the particular activities.

4.2.3 QUALITY PLANNING

ISO 9001 Requirements

Subclause 4.2.3 lists the elements required in quality planning:

- Preparing a quality plan and a quality manual.
- Identifying the controls, resources, and skills necessary to achieve required quality.
- Updating quality control, inspection, and testing techniques as necessary.
- Identifying extraordinary measurement requirements.
- Clarifying standards of acceptability.
- Ensuring the compatibility of the design, production, installation, inspection, and test procedures.
- Identifying suitable verification methods.
- Identifying and preparing quality records.

ISO 9001 adds the statement concerning quality plans: "the quality plans . . . may be in the form of a reference to the appropriate documented procedures that form an integral part of the supplier's quality system."

ISO 9000-2 Guidance

The quality manual can be structured as a tiered set of documents, with each tier becoming more detailed. Quality policy would be the top tier, while detailed procedures, work instructions, and record-keeping forms would be in lower tiers. (See Chapter 8 for more information about structuring the quality manual.)

Quality plans define how quality system requirements will be met in a specific contract or for a specific class of products. An example might include a detailed sequence of inspections for a particular product, types of inspection equipment, and quality record requirements.

ISO 9004-1 Guidance

 In Subclause 4.4.4, ISO 9004-1 stresses that the goal of the quality system is to provide confidence that:

- The system is understood and effective.

- Products and services actually satisfy requirements and customer expectations.

- The needs of both society and the environment have been addressed.

- Emphasis is on problem prevention rather than detection after occurrence.

The quality system applies to all activities related to the quality of a product or service. These activities range from initial market research and design through to installation, servicing, and disposal after use. Marketing and design are especially important for determining and defining customer needs and product requirements.

QS-9000 Requirements

 The advanced *Product Quality Planning and Control Plan* Reference Manual shall be used for:

- Planning.

- Product design and development.

- Process design and development.

- Product and process validation.

- Production.

- Feedback assessment and corrective action.

Special Characteristics

Appropriate process controls shall be established for special characteristics.

Use of Cross-Functional Teams

Internal cross-functional teams are to use techniques in the *Advanced Product Quality Planning* Reference Manual.

Feasibility Reviews

Manufacturing feasibility is to be confirmed prior to contracting.

Process FMEAs

Efforts are to be taken to prevent rather than detect defects, using techniques in the *Potential Failure Modes and Effects* Reference Manual.

The Control Plan

Control Plans— the output of the advanced quality planning process—shall be developed for:

- Prototype (may not be required from all suppliers).
- Prelaunch.
- Production.

Subclause 4.2.3: Quality Planning ISO 9001

Clause 4.2.3, Quality Planning of ISO 9001 external quality assurance standard, states: "The supplier shall give consideration to the following activities, as appropriate, in meeting the specified requirements for products, projects or contracts," and then lists eight activities. What is meant by "as appropriate?" How is "project" defined? Are the eight activities mandatory?

QSU's panel of experts agreed that the meaning of the term "as appropriate" may vary from company to company. Moderator Ira J. Epstein, VP of Government Services, STAT-A-MATRIX, said that the meaning of the term depends on the relationship of the activity to the quality of the company's product or services.

"The beauty of ISO 9000 in general is that it is full of subjective words," said Epstein. "Many people look at these terms as a weakness but in actuality these subjective words are the basis for its strength. The subjective term 'as appropriate' and other subjective terms are intended to permit the supplier to design his [or her] own quality system to make it efficient and effective."

Epstein said that only the company can determine what is appropriate. "No outside experts can tell a supplier what is appropriate for him [or her] in these types of situations," he said. "It is the supplier's responsibility to interpret the term 'as appropriate' and to make it less subjective in the written quality plan. Subjective terms should not be used in the supplier's quality plan as he [or she] develops it."

Tom Bair, CEO, Tom Bair & Associates, also agreed that the term "as appropriate" varies. "What is appropriate for one industry may not be appropriate for another," he said. "It really depends upon the application of what we're talking about, with the first consideration being safety."

(Continued on next page)

4.3 CONTRACT REVIEW

ISO 9001 Requirements

It is important for the supplier to thoroughly understand the customer's needs. Clause 4.3, Subclause 4.3.1, requires that the supplier establish and maintain documented procedures for contract review and coordinating contract review activities.

Clause 4.3 is divided into the following three additional subclauses.

(Continued from page 74)

What Is a Project?

According to Epstein, "project" refers to a specific program or activity as defined by the company. "Quality plans are usually written to be applicable to a specific project or program or contract, whereas the quality manual is intended to apply across the entire supplier's organization or operation."

Quality manuals are now mandatory in the revised ISO 9001 series standard. In those cases where a supplier provides a single product or single service, all of the quality system procedures may be included in the quality manual and the quality plan would be unnecessary. Quality plans, on the other hand, are not mandatory. "It certainly takes on a stronger emphasis since it has been moved from a note to the normal text," he said. "However, the subjective terms 'timely consideration' and 'as appropriate' provide the leeway and flexibility for the suppliers."

Blair said that the term "project" may refer to an internal project relating to the quality system. "It could very well mean a project for determining how a total cost-of-quality system would be implemented in every one of the standards of ISO or one of the elements of a company's quality assurance manual."

Mandatory Planning Activities

Epstein said that the eight activities outlined by the clause only become requirements if they are applicable to the company. "If they are applicable then they are mandatory," Epstein said. "If they don't apply, obviously the supplier does not have to include that element in his [or her] quality plan."

"If it does not apply because of the nature of the company's product then it needn't be addressed," he said. "Where it's obvious that any of these elements do apply then it becomes a prescriptive requirement." The experts said that most companies complying with the ISO 9001 series standard probably already are complying with the provisions of the clause, but these companies may not have been documenting how the requirements for quality have been met. Companies literally have to document how any of these activities have been achieved or integrated into the development process, the experts noted.

Bair said that, in his opinion, the eight activities are not mandatory: "They are illustrative in nature, samples to basically get your thought train going."

4.3.2 REVIEW

ISO 9001 Requirements

ISO 9001 includes precontract tender arrangements, as well as contract and ordering requirements.

In Clause 3, Definitions, the standard defines the terms "tender" and "contract; accepted order." A tender is an "offer made by a supplier in response to an invitation to satisfy a contract award to provide product." A contract or accepted order are "agreed requirements between a supplier and customer transmitted by any means."

The subclause and the definition thus take into account verbal orders. In these situations, the "supplier shall ensure that the order requirements are agreed before their acceptance."

The basic requirements of the review process are the following:

- Contract requirements are adequately defined and documented.

- Any requirements that differ from those in the contract or tender are resolved.

- The supplier is capable of meeting the contract requirements.

QS-9000 Requirements

Tender, contract, or order shall be reviewed before submission, to assure that all customers' requirements, including QS-9000 Section III Customer-specific Requirements, can be met.

4.3.3 AMENDMENT TO CONTRACT

ISO 9001 Requirements

Subclause 4.3.3 focuses specifically on contract amendments and calls on the supplier to identify how amendments to contracts are made and "correctly transferred to the functions concerned" within the organization.

4.3.4 RECORDS

ISO 9001 Requirements

Subclause 4.3.4 requires suppliers to maintain records of contract review.

ISO 9000-2 Guidance

The contract review process steps are:

- Review the contract.

- Achieve agreement.

- Discuss results of contract review.

- Discuss draft quality plan (if existing).

The contract review procedure should have the following features:

- An opportunity for all interested parties to review the contract.

- A verification checklist.

- A method for questioning the contract requirements and addressing the questions.

- A draft quality plan.

- Provision for changing the contract.

ISO 9004-1 Guidance

ISO 9004-1, Clause 7.0, Quality in Marketing, discusses the role of the marketing function in establishing quality requirements for the product. Marketing should determine the need for a product or service, define the market demand accurately, determine customer requirements, and communicate these requirements clearly within the company.

ISO 9004-1 emphasizes the importance of considering requirements for all elements of the total product, whether hardware, software, processed materials, or services. The marketing function should also ensure that all relevant organizational functions be capable of meeting the requirements.

These product requirements can be detailed in a statement or outline of product requirements, which are translated into "a preliminary set of specifications as the basis for subsequent design work." This is also known as a product brief. A product brief may include the following elements:

- Performance characteristics (e.g., environmental and usage conditions and reliability).

- Sensory characteristics (e.g., style, color, taste, smell).

- Installation configuration or fit.

- Applicable standards and statutory regulations.

- Packaging.

- Quality assurance/verification.

The marketing function should establish an "information monitoring and feedback system . . ." to analyze the quality of a product or service.

Clause 4.3: Contract Review

Contract review is another area that is often poorly handled. The requirements are hardly onerous or unreasonable. The principal elements of contract review activity are designed to ensure that:

- The scope of the contract is clearly defined.

- The requirements are adequately documented.

- Any variations are identified and resolved.

- The capability to fulfill the contract exists:

    —Technically: the skills exist in-house or can be acquired.

    —Financially: the work can be done for the price.

    —Delivery: the work can be delivered according to requirements and the company has the infrastructure to support the contract requirements.

- Amendments to the contract are properly communicated and effectively handled within the organization.

- Records of contract review activity are maintained.

The Victoria Group

4.4 DESIGN CONTROL

The essential quality aspects of a product—such as safety, performance and dependability—are established during the design and development phase. Thus, deficient design can be a major cause of quality problems. ISO 9001 establishes separate requirements for design review and design verification. The standard includes the following subclauses:

4.4.1 General

4.4.2 Design and Development Planning

4.4.3 Organizational and Technical Interfaces

4.4.4 Design Input

4.4.5 Design Output

4.4.6 Design Review

4.4.7 Design Verification

4.4.8 Design Validation

4.4.9 Design Changes

ISO 9001 Requirements

 The general requirement is for suppliers to establish and maintain documented procedures to control and verify product design to ensure that it meets specified requirements.

4.4.2 DESIGN AND DEVELOPMENT PLANNING

ISO 9001 Requirements

 The supplier shall develop design plans for each design activity. The plans should define each activity and assign the responsibility to qualified personnel who have adequate resources. Design plans should be updated as necessary.

ISO 9000-2 Guidance

 Design activities should be sufficiently specific and detailed to permit effective verification. Planning procedures should take into account:

- Sequential and parallel work activities.

- Design verification activities.

- Evaluating the safety, performance, and dependability incorporated in the product design.

- Product measurement, test, and acceptance criteria.

- Assignment of responsibilities.

QS-9000 Requirements

 Required Skills

As appropriate, design personnel are to have specific, required skills:

- Geometric Dimensioning and Tolerancing (GD&T).

- Quality Function Deployment (QFD).

- Design for Manufacturing (DFM)/Design for Assembly (DFA).

- Value Engineering (VE).

- Design of Experiments (Taguchi and classical).

- Failure Mode and Effects Analysis (DFMEA/PFMEA, etc.).

- Finite Element Analysis (FEA).

- Solid Modeling.

- Simulation Techniques.

- Computer-Aided Design (CAD)/Computer-Aided Engineering (CAE).

- Reliability Engineering Plans.

4.4.3 ORGANIZATIONAL AND TECHNICAL INTERFACES

ISO 9001 Requirements

Design input can come from a variety of sources. The responsibilities and authorities of these sources must be defined, documented, coordinated, and controlled. The basic requirements include the following:

- Identify the interfaces between different groups.

- Document, transmit, and regularly review the necessary information.

ISO 9000-2 Guidance

To function effectively, the information procedure should establish the following:

- What information should be received and transmitted.

- Identification of senders and receivers.

- Purpose of information.

- Identification of the transmittal mechanism.

- Document transmittal records to be maintained.

4.4.4 DESIGN INPUT

ISO 9001 Requirements

Design inputs are usually in the form of product performance specifications or product descriptions with specifications. The requirements include the following:

- Identify all design input requirements pertinent to the product.

- Review the selection for adequacy.

- Resolve incomplete, ambiguous, or conflicting requirements.

Input requirements also include applicable statutory and regulatory requirements. Also, design input "shall take into consideration the results of any contract review activities."

ISO 9000-2 Guidance

A design description document can serve as a definitive reference throughout the design process. It should quantify all requirements as much as possible, with details agreed to between the customer and supplier. The document should identify design aspects, materials, and processes that require development, including prototype testing.

QS-9000 Requirements

4.4.4 Design Input—Supplemental: The supplier shall have appropriate resources for computer-aided product design, engineering, and analysis (CAD/CAE)

4.4.5 DESIGN OUTPUT

ISO 9001 Requirements

Design outputs are the final technical documents used throughout the process, from production through servicing. They can include drawings, specifications, instructions, software, and servicing procedures.

The requirement is to document design output "in terms that can be verified and validated against design-input requirements." The design output must:

- Meet design input requirements.

- Contain or reference acceptance criteria.

- Identify design characteristics crucial to safety.

- Include a review of design output documents before release.

QS-9000 Requirements

4.4.5 Design Output—Supplemental: Evaluation of design output shall include:

- Simplification, optimization, innovation, waste reduction, and environmental conditions.

- Geometric dimensioning and tolerancing.

- Cost/performance/risk analysis.

- Feedback from testing, production, and field.

- Design FMEAs.

4.4.6 DESIGN REVIEW

ISO 9001 Requirements

The standard requires formal documented reviews of design results. Participants at the design reviews shall include representatives from all functions concerned with the design stage and any other specialized personnel, as required. As with most every clause, Subclause 4.4.5 requires the supplier to maintain records of the design reviews.

4.4.7 DESIGN VERIFICATION

ISO 9001 Requirements

The requirement of Subclause 4.4.7 is to establish a design verification plan to ensure that the design output meets design input requirements. In a note to Subclause 4.4.7, the standard lists various design verification activities, such as alternative calculations, comparisons with proven designs, tests and demonstrations, and reviewing the design stage documents before release.

ISO 9000-2 Guidance

In most cases, two or more of these measures are used. Design reviews and/or type-testing may be a regulatory requirement. Design verification should involve personnel independent of those who did the work under review.

ISO 9000-2 includes many questions that the design review can address, among them the following:

- Do design reviews satisfy all specified requirements?
- Are product design and processing capabilities compatible?
- Are safety considerations covered?
- Are the materials and/or facilities appropriate?
- Are components or service elements standardized?
- Are purchasing, production, installation, inspection, and testing plans technically feasible?
- Has software been validated, authorized, and verified?
- Where qualification tests have been performed, did the conditions represent actual use?

QS-9000 Requirements

Design Verification—Supplemental: The supplier shall:

- Conduct a prototype program (unless waived).

- Use the same subcontractors, tooling, and processes whenever possible.

- Test life, reliability, and durability.

- Track completion of performance tests.

(Services may be subcontracted, but technical leadership must be provided by the supplier.)

4.4.8 DESIGN VALIDATION

ISO 9001 Requirements

ISO 9001 has added a separate subclause on design *validation*. Unlike design *verification*, which matches the design output to input requirements, (the producer's point of view), design validation ensures that the product conforms to defined user needs and/or requirements (the customer's point of view).

In a note, the subclause emphasizes that validation may take place not only on the final product but at earlier stages, and multiple validations may be necessary if there are different intended uses of the product. Thus, in addition to the end product, there may be major product components that can be validated from the customer's point of view.

4.4.9 DESIGN CHANGES

ISO 9001 Requirements

Designs may be changed or modified for many reasons. The requirement is for authorized personnel to identify, document, review, and approve all design changes and modifications before their implementation.

ISO 9000-2 Guidance

Design changes in one component should be evaluated for their effect on the overall product. Sometimes, improving one characteristic may have an unforeseen adverse influence on another. The new design output should be communicated to all concerned, and the changes documented.

ISO 9004-1 Guidance

Clause 8.0, Quality in Specification and Design, discusses the specification and design function in detail. The overall design function should "result in a product that provides customer satisfaction at an acceptable price that gives a satisfactory financial return for the organization." The product must be "producible, verifiable, and controllable under the proposed production, installation, commissioning, or operational conditions."

The topics covered in Clause 8.0 include the following:

- Contribution of specification and design to quality.

- Design planning and objectives (defining the project).

- Product testing and measurement.

- Design review.

- Design qualification and validation.

- Final design review and production release.

- Market readiness review.

- Design change control.

- Design requalification.

- Configuration management in design.

These topics expand on subjects covered in ISO 9001. The results of the final design review should be "appropriately documented in specifications and drawings which then form the design baseline." The quality system should provide for a market readiness review to determine "whether the organization has the capability to deliver the new or redesigned product."

Regarding design change control, the quality system should provide a procedure for "controlling the release, change, and use of documents that define the design input and the design baseline (output), and for authorizing the necessary work" to implement changes throughout the product cycle.

ISO 9004-1 stresses the need for "periodic re-evaluation" of the product to ensure that the design is still valid compared to all specified requirements.

QS-9000 Requirements

 Design Changes—Supplemental: Customer must approve all design changes for impact on form, fit, function, performance, and durability.

A Road Map for Design Control

"Design output meets design input," sums up the entire intent of the design requirements. Clause 4.4 can be met following this suggested road map:

1. Plan what is to be done.

2. Document that plan.

(Continued on next page)

(Continued from page 84)

3. Assign someone to review the contract, document the designated person, and ensure that this person feeds data to the task of design input.

4. Create an input specification that includes "applicable statutory and regulatory requirements" and is unambiguous and adequate.

5. Follow-up as progress is made and make sure that the plan is still being followed.

6. Review how the system is working with the involved employees and document the progress. Perform design reviews at previously defined stages in the evolution of the design.

7. At the end of the task, make sure the output matches the input.

8. Conduct tests and keep records. Do these in accordance with the "acceptance criteria" in step 4. above.

9. Make sure that a good method of tracking changes is in place and that everyone understands the rationale for all changes made.

Design Validation

Subclause 4.4.8 contains a requirement for design validation. The standard makes a distinction between verification and validation. The software engineering profession, for example, has long made this distinction. There is a recognition that much in process design verification may well have little direct linkage to the defined requirements. Thus it is possible to have a design project that has been successful in every verification phase but fails to meet the user requirements in any way. This requirement may cause some design organizations to rethink some of their strategies to ensure that their design-proving methodologies take sufficient heed of customer operational environments. It may require some reorganization of test methodologies, even if there is no actual increase in the amount of testing being performed.

Design Changes

Subclause 4.4.9 calls for approving changes and modifications "before their implementation." One important consideration for companies is that they ensure that their systems provide the necessary levels of authorization to allow for implementing essential changes or modifications quickly to meet production or customer needs.

The Victoria Group

4.5 DOCUMENT AND DATA CONTROL

Clause 4.5, which covers procedures to control documents, includes "data" in the title of the clause. A note to Subclause 4.5.1 emphasizes that document and data can be in any type of media, such as hard copy or electronic media.

Does Design Control Apply to Process Design?

Do the requirements of ISO 9001 Clause 4.4, Design Control, apply strictly to product design, or do they also apply to process design? If process design falls within the scope of Clause 4.4, can the requirements be used for service companies or companies that employ continuous processes and typically seek registration to ISO 9002? Three of four experts agreed that Clause 4.4 also applies to process design and that it may be applicable to the service industry and to companies that employ continuous processes, such as those in the chemical industry.

Establishing Design Verification

Joseph A. Chiaramonte, senior staff engineer, Underwriters Laboratories, Inc. , said Subclause 4.4.2 addresses "each . . . activity" with respect to design and development planning. "Suppliers of hardware may use existing processes in the design of new products, or they may design new processes for existing products," he said. "In all cases, the supplier must determine the points at which verification functions must be established throughout design and development activities."

According to Chiaramonte, in certain contractual situations, such as a government contract for military parts, the contract may specify which processes must be baselined and certified. Design verification functions, however, might still be established on the basis of partially manufactured product parameters at various points during critical processes. "Alternatively, process parameters may be used to form the basis for such verification points," he said.

The requirement clearly applies to hardware suppliers, he said, but it also applies to planning design and development activities that are used to produce services, software, or processed materials. "Unless contractual requirements prescribe specific verification points in the design process, the supplier must establish them on the basis of the relevant design input/output criteria, process/product complexity, and other related factors or requirements," Chiaramonte explained.

Clause 4.4 as Guidance in Process Control

The experts said that Clause 4.4 may be used as guidance in process control. They noted that Subclause 4.4.1, General, specifically references the control and verification in the design of product. "Using Clause 4.4 as supplemental guidance is appropriate," the experts explained.

A registrar would not rely on Clause 4.4 to assess process design in a company unless the operation is part of the design and development functions. "The audit body may overlook the product design stipulation in Design Control, General, and make use of Subclauses 4.4.1 through 4.4.6 in the assessment," the experts said. Companies should consider all of Clause 4.4 in the structure and documentation of any design discipline—the experts explained, whether it be product or process.

(Continued on next page)

(Continued from page 86)

ISO 9004-3

William J. Deibler, II, partner, Software Systems Quality Consulting, said that process design falls within the scope of Clause 4.4 to the extent that production engineering provides input to, or controls, product design, and that process is integral to the product offering. He said that Clause 4.2 suggests that "identification and acquisition of . . . processes" are activities that should be addressed by the quality system, but it makes no mention of process design. "This ambiguity about the correct position of process design is addressed in ISO 9004-3, Quality management and quality system elements—Part 3: Guidelines for processed materials, which consistently attempts to replace 'product' with 'product and process,'" he said.

Deibler said that ISO 9004-3 uses the term "design development" to include development of a process design that meets product requirements. "Other paragraphs in the standard emphasize the need for process as well as product reviews during the design development stages," he said.

Process Networking

Serge E. Gaudry, senior consultant, Government Compliance Consultants, Inc., said that Clause 4.4 applies strictly to product design and does not address process design issues. "The key here is the insertion of the Design Review clause in ISO 9000. This implies that documented reviews of the design results be planned and conducted," according to Gaudry. He said that ISO 9000-1 makes it clear that a quality system under ISO 9001 or 9002 must take process networking into consideration during the product design phase.

"Specifically, the organization must highlight the main processes associated with the product design issue," he said. "ISO 9000-1 further states that each process should have an owner in the organization with overall responsibilities and authority to handle interfacing within the organization. One can easily draw the conclusion that this process ownership and assignment of responsibilities applies to every aspect of the quality system."

He said that the ISO 9000 series standards are based on the premise that all work is ultimately accomplished by processes and that every process has inputs that generate results termed "outputs of the process." "The concept of a process as defined in ISO 9000-1 applies equally to 9001 and 9002," he explained. "Under the 1994 revisions these concepts can be utilized by those seeking 9002 registration."

4.5.1 GENERAL

ISO 9001 Requirements

The supplier shall establish and maintain procedures to control all documents and data. These can include external documents such as standards and customer drawings.

Subclause 4.4.8 Design Validation

Clause 4.4.8, Design Validation of ISO 9001, states: "Design validation shall be performed to ensure that product conforms to defined user needs and/or requirements." Is it necessary to make changes to the design, engineering, or the quality system?

QSU's panel of experts agreed that many companies already comply with the requirements of Clause 4.4.8. However, to be in compliance some companies will have to make changes to the way they bring new products to market.

Moderator Harvey S. Berman, manager of Corporate Quality and Reliability, Underwriters Laboratory, said that most companies should have no difficulty meeting the requirement because they already perform design validation.

"We were typically looking at validation as a last step in the verification process," he said. "I don't think [the change] is going to have much of an impact."

He defined validation as ensuring that the product conforms to a defined user need. "If you're not going to do that in your design then you're taking a shot on manufacturing something that no one needs," he said. Berman said that the biggest challenge for most companies will be revising their quality manuals to reflect the added requirement.

Michael P. Enders, principal, American Quality Resources Inc., agreed that many companies already perform design validation in the course of bringing a new product to market, but said they often bypass the validation step when improving on an existing design. This practice may result in a number of disputes between ISO 9000 registrars and clients. "I guess my experience is if a company had the option of comparing a new design to an old design against sending it out for testing," he said, "they'd opt for the first choice because it's far less expensive."

Enders said the disagreements most likely will revolve around whether a design has sufficiently changed to warrant additional testing.

(Continued on next page)

ISO 9000-2 Guidance

Document control applies to all documents and/or computer records pertinent to design, purchasing, production, quality standards, inspection of materials, and internal written procedures. Internal written procedures describe the following:

- How documentation for these functions should be controlled.

- Who is responsible for document control.

- What is to be controlled.

- Where and when is it to be controlled.

(Continued from page 88)

The 1987 version of the standard not only gave companies the option of performing design validation, but gave them several options for performing it. The 1994 version mandated that validation take the form of physical testing. "It seems on the surface that some of the flexibility is removed from the new standard," he said. "It definitely sounds to me like they've broken it out from the theoretical design verification and put it into the empirical, actual testing of product."

Jeffry J. Omelchuck, principal, International Quality Associates, agreed that the impact will vary from company to company. "In general, if the design process has a validation step in it, then there may not be any kind of addition or change required," he said. He said the revision focuses more attention on the need to subject prototypes to testing. The critical question that must be asked is: "Does it actually meet user requirements?" he said. Chuck Rhodes, executive director, Management Resource Group, said problems will arise for companies that have not clearly identified user needs in the past.

"Some companies have commercialization processes well in place. Other organizations have nothing but intuitive processes," he said. "They should not underestimate the complexity of this activity because it affects all departments in an organization."

Rhodes predicted that it will be necessary for many companies to formalize their procedures for validating designs and that the validation will have to be carried out earlier in the design process for many companies that already include it as a step in bringing new products to market. "What it really means is that you cannot really validate a design until you understand the ultimate need that the product is going to satisfy," he said. "This will give reinforcement to the original identification of the need."

ISO 9004-1 Guidance

 ISO 9004-1, Clause 17.0, Quality Documentation and Records, offers the following examples of the types of quality documents that require control:

- Drawings.

- Specifications.

- Inspection procedures and instructions.

- Test procedures.

- Work instructions.

- Operation sheets.

- Quality manual and quality plans.

- Operational procedures.

- Quality system procedures.

It also stresses that "sufficient records be maintained to demonstrate achievement of the required quality and verify effective operation of the quality management system."

QS-9000 Requirements

Reference Documents: Specific types of documents to be included in document control are:

- Engineering drawings.

- Engineering standards.

- Math (CAD) data.

- Test procedures.

- Work instructions.

- Operations sheets.

- Quality manual.

- Operational procedures.

- Quality assurance procedures.

- Material specifications.

Document Identification for Special Characteristics

Special requirements by the customer should be separately identified by the customer's special characteristic symbol. A record is required in 4.16.

4.5.2 DOCUMENT AND DATA APPROVAL AND ISSUE

ISO 9001 Requirements

Documents and data must be reviewed and approved for adequacy before they are issued. To avoid using invalid and/or obsolete documents, companies must establish and make available a master list or similar procedure that identifies the current revision status of the documents. Subclause 4.5.2 also requires that:

- The appropriate documents be available at all relevant locations.

- Invalid or obsolete documents be removed or "otherwise assured against unintended use."

Any obsolete documents retained for legal and/or knowledge-preservation purposes are suitably identified.

QS-9000 Requirements

 **Engineering Specifications:** A procedure is to be established for timely reviews (days, not weeks or months) of all standards and specification changes, with a record of implementation date.

4.5.3 DOCUMENT AND DATA CHANGES

ISO 9001 Requirements

 The basic requirement is to identify changes in documents and/or data and to review and approve the changes. The review and approval process should be performed by the functions/organizations that performed the original review "unless specifically designated otherwise."

Editor's Note: The intention is to ensure that changes made in documents already issued follow the same approval process as conventional industry practice. This would be a system in which a document is issued by a designated originating activity (department) before it passes through one or more approval stages. These stages may be organizationally in the same department or another department. One situation where waiving the approval process might be acceptable is if a supplier had field-installation responsibilities and its qualified personnel were empowered to modify a product or practice on the spot, an approval mechanism might not be available or practical.

ISO 9000-2 Guidance

 Supplier documentation is usually subject to revision. This requirement applies both to internal and external documentation, such as national standards. Organizations should consider the effect that changes in one area may have on other parts of the organization and the actions that should be taken to assess this effect. Other things to keep in mind include:

- Planning the circulation of a change proposal to avoid disruption.

- Timing the change's implementation.

The standard specifically mentions that 'documents of external origin' must be included within the controlled system. This means customer specifications, national and international specifications, regulatory documents, and so on.

4.6.1 PURCHASING—GENERAL

ISO 9001 Requirements

The basic purchasing requirement is to establish and maintain documented procedures to ensure that the purchased product conforms to specified requirements.

Clause 4.5: Document and Data Control

Poor document control frequently lets companies down during certification audits. It is an important discipline. ISO 9001 requires that there be a master list identifying every document that forms part of the controlled system, so that it is clear as to which specific documents make up that system.

The requirement covers all documentation that relates to the quality system itself—not merely manuals, procedures, and work instructions but also external Standards and specifications relating to the product, regulatory compliance documents, customer-supplied documents, and so on. These last documents are the "documents of external origin" referred to in this element. On the other hand, it doesn't include everything. For example, a manager's memo is not typically a "controlled document" within the meaning of the standard.

A master list makes it possible for those using the documents to make sure they have the right ones. In addition, it is useful that copies of all documents be distributed where they are likely to be used. The master list doesn't have to be one, enormous list; it may be a devolved system, where each department keeps its own list. So long as they can all be linked together through an audit trail, it is the company's choice.

Make sure that relevant paperwork and documents are available for people to do their jobs properly. Put a system in place to ensure that when a document changes, all the old copies are removed to prevent accidental use. When changes are made, ensure that either the original author or someone else who has all the relevant information makes them.

Obsolete documents should be "assured against unintended use" by removing them from the system or by other means. It is safe to assume that documents can be readily accessible if they are in some way clearly marked as "not current."

The Victoria Group

ISO 9000-2 Guidance

Planned and adequately controlled purchasing procedures ensure that purchased subcontracted products, including services, conform to specified requirements. Suppliers should establish effective working relationships and feedback systems with all subcontractors.

ISO 9004-1 Guidance

Clause 9.0, Quality in Purchasing, discusses procurement quality in detail. The section states that "purchases become part of the organization's product and directly affect the quality of its product." A procurement quality program should include the following elements:

Document and Data Control Requirements of the ISO 9000 Series Standard

Does ISO 9001, Clause 4.5, Document and Data Control, cover all documents in the company or only documents covering product technical information? What are some of the general requirements of this clause?

The experts agreed that the language used in the clause is drafted broadly. Charles McRobert, president, Quality Practitioners, Inc., stressed that "a document is any means of conveying technical information concerning the product or the quality system from the conceptual phase to installation and servicing."

Controlled Documents

The list of controlled documents includes product drawings and specifications, quality manuals, procedures, process sheet specifications, recipes, formulations, purchase orders, product labels, packaging instructions, product rosters, training records, contracts, and inspection and test criteria.

David Middleton, president, Excel Partnership, Inc., noted that the "scope of the quality system should define the service provided within a company's scope of activities." For example, a customs clearance company should include customer notification under its controlled document plan if informing customers of delays is part of that company's scope of activities. In another example, a manufacturer of general engineered products would not need to control documents covering analysis of wage rates since those documents do not directly impact product quality.

Joseph Tiratto, president, Joseph Tiratto and Associates, Inc. , noted that guidance on this issue can be found in ISO 9000-2, *Generic guidelines for the application of ISO 9001, ISO 9002, and ISO 9003*. Tiratto said ISO 9000-2 states that document control should include "those documents and/or computer records pertinent to inspection of materials and the supplier's internal written procedures which describe the control of documentation for these functions."

Document Changes

Tiratto noted further that Clause 4.5 of ISO 9001 requires controls for the preparation, handling, issuance, and recording of changes to documents. He pointed out that this requirement "applies not only to internal documentation, but also to externally updated documentation."

Middleton pointed out that document control is a common weak link found during third-party audits. He said this problem is usually remedied by identifying "relevant documents" (those that directly impact product or service quality) and by demonstrating document control by date, review status, approval, and master list. For further guidance on the issue, Middleton suggested reading Section 17 of ISO 9004-1.

Clause 4.5.2: What Is an Acceptable Master List?

ISO 9001 Subclause 4.5.2, Document changes/modifications, states: "A master list or equivalent document control procedure identifying the current revision status of documents shall be established and be readily available to preclude the use of invalid and/or obsolete documents." What is an acceptable master list?

The panel of experts agreed that an acceptable master list can be as simple as a handwritten document or as sophisticated as a computerized database. It all depends on the size and dynamics of the company.

Master Lists versus Alternatives

According to James Highlands, president, Management Systems Analysis, the contents of a master list of documents are always determined by the document control procedure used. "In some cases a master list is not required when alternative provisions are established in the document control procedure," he said. In reading the requirements, many companies overlook the option of using an equivalent procedure in place of a master list.

Suppliers use master lists as references or indexes to procedures manuals, he said. Such lists typically include the identifier or number of the document, title, latest revision level, and occasionally the date of issue. It may also include the distribution of the document or the identification of the approving authority for the document. "The key here is [deciding] what alternative controls are applied by the document control procedure," Highlands said.

One example of an alternative control is the procedure for drawings that are controlled by most engineering departments, he said. "A master file of drawings is maintained, which identifies the revisions made or references an engineering change notice that describes the changes."

When the master drawing is revised, it is reviewed and approved by the project manager and is issued to holders of the controlled drawing as identified on a distribution list. Distribution is accompanied by a receipt that must be signed by the recipient and returned with the obsolete drawings. "The number of permutations here is endless," he said. "The true test is whether an individual who is not familiar with the system can read the document control procedure and determine the latest revisions of each document and who holds copies. Whatever data on the master list needed to perform this activity is required."

Types of Lists

Foster Finley, senior consultant, Deloitte & Touche, said the master list serves two purposes: It meets ISO 9000 series requirements for document control and serves as a barometer of the breadth and penetration of a company's quality management sys-

(Continued on next page)

(Continued from page 94)

tems documentation. "It may be in the form of a comprehensive, real-time computer database," he said. "Or it may entail multiple, independent lists."

Finley said an acceptable master list can be any list that identifies documents and current revisions for the quality management system. The creation of such documents, however, is not always feasible.

"The scope, complexity, and sophistication of the company must be established," Finley said. "Along one end of the continuum might be a small company with a very limited product offering. Representative of the other extreme would be a large, vertically integrated manufacturer with an extensive product offering." Preferred vehicles for identifying revisions to first- and second-tier documentation include a table of contents, cross reference, or other matrix, according to Finley.

"The most robust approach to the creation of acceptable master lists is to catalog all quality management systems documentation but to allow segmentation of the lists in any logical manner that suits the company's needs," Finley said. "This alternative gives companies the flexibility to create sublists that are easier to use and maintain."

Another approach is to create and maintain a documentation matrix or master list containing all company documents and their current revision levels. "While this would undoubtedly satisfy the standard, it is not a specific requirement," he said. "Endeavoring to accomplish this in a large, complex organization might undermine the intent of the standard. Such a list might become obsolete so quickly and prove so cumbersome to use that it loses its effectiveness as a source for control."

Where companies rely on computerized databases to catalog all quality management system documentation, Finley said, the focus may shift from preventing inadvertent use of documents to preventing misuse.

Formats for Master Lists

Stephen Hedman, president, Hedman Consulting Services, agreed that a master list must also be easy to understand and effective in preventing obsolete documents from being used when manufacturing a product. Hedman said one system for a small, eight-person metal stamping shop assigns an issue letter to each section of a controlled document blueprint and index page. When the documents are updated, he said, the revision is accompanied with an updated index page. The index page is maintained in a file labeled "current documents." Such a system, he said, probably would not work in an organization with complex processes.

Hedman said that he was most impressed by a network system that controlled all documents electronically. "Anyone could access the document library to check on document status, but only one or two could access to make changes," he explained.

(Continued on next page)

(Continued from page 95)

William E. Cox, president, TQM Consulting, suggests using the table of contents of a procedures manual to keep tabs on documents. "By adding columns for the latest document revision number and date, a user of the manual can readily tell what the most recent revision should be," he said. "A distribution list for the procedures in the manual can be added at the end of the table of contents." Cox said that the table of contents should also be considered as a controlled document and assigned a separate revision number and date. "Knowing what the latest version should be, this tells the user at a glance whether the contents of a manual are up-to-date," he said.

"On the other hand, if each document has a unique, different distribution list that can be incorporated into the procedure itself as a separate section, any procedure revision will then be distributed to the holders of controlled copies of the procedure," he explained. Cox added, however, that this practice should be avoided unless there is a "strong" business reason for each document to have a unique distribution list.

What the Master List Should Include

In general, Cox said, a master list should include the title of each controlled document, holders of each controlled document, and the current revision number and date. He said that controlled documents should include quality manuals, work procedures, detailed work instructions, product specifications, related software files, quality standards, quality plans, inspection instructions, and engineering, assembly, and installation drawings.

Resolve Acceptability before Audit

In any case, Foster Finley said companies should resolve any potential issues over the acceptability of the master list prior to an audit. "The decision of a competent registrar will be guided by the overall effectiveness of a document control procedure in preventing use of nonapplicable documents," he said.

Use of a master list alone does not preclude the inadvertent use of nonapplicable documents, Finley said. "It merely serves as a key element in a document control procedure. Therefore the acceptability of any master list must be judged relative to its ability to serve this purpose."

- Requirements for specification, drawings, and purchase orders.

- Selection of acceptable subcontractors.

- Agreement on quality assurance.

- Agreement on verification methods.

- Provisions for settling disputes.

- Receiving inspection planning and control.

- Quality records related to purchasing.

QS-9000 Requirements

Approved Materials for Ongoing Production: When required, the supplier (your organization) may purchase material from subcontractors that have been approved by the customer.

4.6.2 EVALUATION OF SUBCONTRACTORS

ISO 9001 Requirements

The requirements for the evaluation of subcontractors include the following:

- Evaluate and select subcontractors on the basis of their ability to meet requirements.

- Establish and maintain records of acceptable subcontractors.

- Define the type and extent of control exercised over subcontractors.

ISO 9000-2 Guidance

The supplier may employ several methods for choosing satisfactory subcontractors, including the following:

- Reviewing previous performance in supplying similar products (including services).

- Satisfactory assessment of an appropriate quality system standard by a competent body.

- Assessment of the subcontractor by the supplier to an appropriate quality system standard.

The supplier's quality records should be sufficiently comprehensive to demonstrate the ability of subcontractors to meet requirements. Factors can include the following:

- Product compliance with specified requirements.

- Total cost for the supplier.

- Delivery arrangements.

- Subcontractor's own quality systems

- Performance of subcontractors (should be reviewed at appropriate intervals).

QS-9000 Requirements

 Subcontractor Development: The supplier will perform subcontractor quality system development.

Subcontractor assessment options include audit by:

- Supplier, in accordance with ISO 9001 and QS-9000.
- Customer-approved OEM second party.
- Accreditated third-party registrar.

Note: Responsibility for ensuring quality always remains with the supplier.

Scheduling Subcontractors

- 100 percent on-time delivery performance is required of subcontractors.
- Supplier shall plan and commit accordingly.
- Subcontractor delivery performance and premium freight costs are to be monitored.

4.6.3 PURCHASING DATA

ISO 9001 Requirements

 The requirements for purchasing data include the following:

- Clearly and specifically describe the product ordered in the purchasing document, including where applicable:

  —Type, class, style, grade, or other precise identification.

  —Title or other positive identification and applicable issue of specifications, drawings, process requirements, inspection instructions, and other relevant data, including requirements for approval or qualification of product, procedures, process equipment, and personnel.

  —Title, number, and issue of the quality system standard that applies to the product.

- Review and approve purchasing documents for adequacy of specified requirements.

ISO 9000-2 Guidance

 The purchasing data should define the technical product requirements to the subcontractor to ensure the quality of the purchased product. This can be done by reference to other applicable information such as national or international standards or test methods.

Companies should assign responsibility for reviewing and approving purchasing data to appropriate personnel.

QS-9000 Requirements

Restricted Substances: The subcontractor is to provide materials that satisfy government requirements for:

- Restricted, toxic, and hazardous materials.

- Environmental factors.

- Electrical and electromagnetic requirements.

The supplier is to have a process to assure compliance with these requirements.

4.6.4 VERIFICATION OF PURCHASED PRODUCT

ISO 9001 Requirements

This requirement takes into account the following two situations for verifying that sub-contracted product conforms to specifications:

- Where the supplier verifies purchased product at the subcontractor's premises.

- Where the supplier's customer or representative, by contract, verifies at the sub-contractor's premises and the supplier's premises.

In the first situation, the supplier "shall specify verification arrangements and the method of product release in the purchasing documents." In the second situation, the standard adds two caveats:

- Verification can't be used by the supplier as evidence of effective quality control by the subcontractor.

- Verification by the customer doesn't absolve the supplier of responsibility for providing acceptable product, nor any subsequent rejection by the customer.

ISO 9004-1 Guidance

In Clause 9.0, ISO 9004-1 notes that agreements with the supplier "may also include the exchange of inspection and test data" to further quality improvements. A clear agreement on verification methods can "minimize difficulties in the interpretation of requirements as well as inspection, test, or sampling methods."

Clause 4.6: Purchasing

The relationships between customer, supplier, and subcontractor are fixed in ISO 9000. Hence the section on purchasing is clearly defined as being the purchasing activities of the company applying the standard to its own systems. ISO 9001 calls this organization "The Supplier."

The subcontractor is the vendor, supplier, or person from whom the company obtains materials, services, or personnel impacting the product or service the company sells.

The subcontractor is always the person from whom the supplier is buying product. (Refer to Table 4–1.)

On-site Verification

Where the supplier elects to carry out on-site verification at a subcontractor's premises, the standard requires that the verification arrangements and the method of release be specified in the purchasing documents. This does not require detailed descriptions. It could be a reference to a separate document that was independently supplied, a published product specification that was to be used, and so on.

There remains the option for "customer verification," which can take place at either the subcontractors' or the supplier's site, but as the standard makes crystal clear, acceptance by the customer at either location does not remove the supplier's liability to provide acceptable product, nor does it preclude the customer from subsequently rejecting that same product at a later date.

The Victoria Group

Subclause 4.6.2: Assessing Subcontractors

ISO 9001 Subclause 4.6.2, Evaluation of Subcontractors, states: "The supplier shall evaluate and select subcontractors on the basis of their ability to meet subcontract requirements including quality system and any specific quality assurance requirements." What does "evaluation" mean? How does a supplier set up an effective, documented evaluation system?

Editor's Note: *In this context, the terms evaluation and assessment are synonymous. ISO 8402, in Clause 4.6, uses the term quality evaluation. ISO 8402 states that quality evaluation may be used for qualification, approval, registration, certification or accreditation purposes.*

(Continued on next page)

(Continued from page 100)

Ian Durand, president of Service Process Consulting, Inc., noted that "in the context of ISO 9001, the evaluation has a scope that extends beyond quality requirements to include all requirements the supplier chooses to place on its selected subcontractor" list. He said that quality system and quality assurance requirements noted in the first sentence of Subclause 4.6.2 could include service aspects such as "on-time delivery performance in addition to technical quality history."

In addition, Durand said the supplier might consider other factors such as the financial security, market position, interpersonal skills, and customer satisfaction policies of the subcontractor.

Types of Subcontractors

Ronald Muldoon, senior manager, KPMG Environmental Management Practice, said the assessment of subcontractors depends on the importance of the procured item and the nature of the relationship between the supplier and subcontractor—that is, whether it is long-term or periodic. "For long term relationships the records should concentrate on the performance and the effectiveness of corrective actions to prevent recurrence of nonconformances," he said.

"For periodic relationships, the assessment should concentrate on the capability to produce what is required to meet that particular purchase order." Muldoon said that, "in addition to assessing the subcontractor's ability to meet the quality requirements, [the supplier must assess] the subcontractor's ability to meet the technical, commercial, and schedule requirements of the contract."

Ralph D. Schmidt, director, Thornhill USA, recommends that suppliers make a list of subcontractors and divide it in two—those critical to the quality of the supplier's services and those having minor or no impact on the supplier's quality. For example, he said, vehicles used as rental cars are critical to quality, but the products used to clean them are probably not.

Schmidt said subcontractors on the critical list should be given written quality requirements. "The ideal is a survey for them to fill out to determine their ability to conform," he said. "Critical subcontractors not complying with your needs should be replaced with ones that are capable. If the subcontractor is the only one of its kind, visits, discussions, and audits are in order to assure conformance. You may have to invest money and training in the subcontractor to assure your quality."

He said that a supplier-receiving system is "vital" to tracking subcontractor performance. The system should keep track of receipts, deviations, rejections, complaints, test reports, and certificates of compliance, which in turn become part of the subcontractor performance file. "Subcontractors may furnish suppliers their quality

(Continued on next page)

(Continued from page 101)

manual for review and may be certified or registered under a system," Schmidt said, adding that in such cases supplier audits may be waived if registration is pertinent and the manual covers critical subcontract aspects.

Developing a Supplier Validation and Verification System

To satisfy this clause of the standard economically, Kevin Drayton, senior consultant of Kevin Drayton Associates, suggests companies should have a supplier validation and verification system in place. The system must ensure that the supplier has chosen a company that is capable of producing product to its requirements and that the company has an auditable, documented quality system as part of its overall business strategy. Drayton said the validation and verification system typically includes surveys, on-site audits, capability studies performed by the subcontractor and verified by the supplier, and a review of products and processes via statistical techniques. "It is quite obvious that when a system such as the one described here is implemented, the objective evidence of the subcontractor's ability to meet requirements is quite easy to gather and maintain," Drayton said.

Methods of Assessment

"The wording of the sub-element makes it clear that subcontractors should not usually be selected simply because they are certified to one of the ISO 9000 standards, although in some cases, this may be adequate," commented Roderick Goult, president and CEO of the Victoria Group. Goult said, "it is common for suppliers to 'believe' that ISO 9000 registration is all that is needed from their subcontractors."

Goult said that "all the relevant issues need to be addressed in determining which subcontractors to use including all the subcontract requirements for technical capability, delivery time frames, cost, and possibly post-sales support."

Goult said that if the subcontract items constitute high value purchases, or they have a significant impact on the supplier's own deliverables it is very common for the potential subcontractor to have financial stability also verified. "The extent of control required should take all these issues into account, and may be more than a monitor of performance including regular audits and product verification," Rod noted.

Dennis Arter, owner, Columbia Audit Resources, said other possible assessment methods could include a listing in a register for a particular product line or written responses to a questionnaire. He said the process of evaluating potential suppliers must be documented in an in-house procedure and the results recorded. He noted that this record usually takes the form of an "Approved Subcontractor List," but it may be as simple as a notation or signature in the vendor's file folder.

(Continued on next page)

(Continued from page 102)

Arter said suppliers must have a program for keeping approved subcontractors on and off a list. Arter said subcontractor performance "needs to be evaluated on a periodic basis." He noted that some subcontractors need an annual evaluation, while others might be evaluated every three years.

Required Documentation to Demonstrate Subcontractor Assessments

The panel gave the following examples of documentation to demonstrate that an adequate assessment has been performed of subcontractors:

• Receiving reports, test reports, specifications, and certificates of compliance.

• Copies of complaints, investigations, and implemented changes that have been agreed to.

• The subcontractor's quality manual, recognitions, audits, third-party registration, and scope.

• Periodic scheduled subcontractor reviews, internal and on-site.

• Documentation of completed corrective actions as referenced by the audit reports.

• Statistical evidence of capability and control.

• Periodic assessment of process output and effective implementation of quality plans as supported by the subcontractor ratings program.

Subcontractors Registered to ISO 9000?

Muldoon agreed that when a company is registered to the ISO 9000 series standards, this does not guarantee that its products meet high quality standards. Rather it assures customers that a contract review has been performed as required by Clause 4.3 of ISO 9001 and 9002. "The review ensures that the subcontractor understands the requirements and has the capability to meet the contractual requirements," he explained. "The subcontractor with an ISO 9000 quality system is also more accustomed to doing front-end planning with the bid, providing documentation, and controlling its [own] subcontractors."

In assessing subcontractors holding ISO 9000 registration, a company should focus on those aspects of the purchase order that have not already been covered by the registration audit, according to Muldoon. This strategy is cost-effective for all parties.

(Continued on next page)

(Continued from page 103)

Ralph Schmidt added that although ISO 9000 registration does not guarantee that the product will meet requirements, a list of subcontractors registered under ISO 9001 or ISO 9002 may satisfy a third-party audit.

Guidance for Specific Industries

Joseph Tiratto, president of Joseph Tiratto and Associates, pointed out that guidance documents helpful in this area have also been prepared for specific industry sectors such as chemicals and software. For example, in ANSI/ASQC Q90 ISO 9000: *Guidelines for use by the chemical and process industries,** an assessment of the subcontractor's ability consistently to meet requirements may be based on the following evidence:

- On-site assessment of subcontractor's quality and/or performance data (current and historical).

- Trials or demonstration in the supplier's laboratories or plant. For example, the supplier may have to rely on inspection and testing when a subcontractor's appraisal is not feasible (i.e., spot purchases of bulk material).

- Documented evidence of successful use in similar processes.

- Third-party assessment and registration of the subcontractor's quality systems to an acceptable standard.

Developing a Teaming Relationship

Drayton said assessment and selection of subcontractors is good business practice, in addition to being a requirement of the ISO 9000 series standards. "Materials-driven businesses have long relied on quantity of subcontractors to make up for poor quality," he said. "Today's marketplace is rapidly changing that philosophy. Time and time again it is proven to be more cost-effective to slim down the number of subcontractors and form what are known as teaming relationships with them."

Drayton defined this relationship as one where both supplier and subcontractor understand and respect their interdependence and interreliance. "Rather than the adversarial relationships which characterized the old-style vendor-contractor relationship, businesses who team have quality as a focus and success as a mutual goal," he argued.

* ASQC Quality Press. 800/248-1946

4.7 CONTROL OF CUSTOMER-SUPPLIED PRODUCT

ISO 9001 Requirements

 The requirement for control of customer-supplied product is to establish and maintain documented procedures for verification, storage, and maintenance. Products that are lost, damaged, or unsuitable must be recorded and reported to the customer.

The standard emphasizes that "verification by the supplier does not absolve the customer of the responsibility to provide acceptable product."

ISO 9000-2 Guidance

 Customer-supplied product is any product owned by the customer and furnished to the supplier for use in meeting the requirements of the contract. The supplier accepts full responsibility for the product while in its possession.

Customer-supplied product could be a service, such as the use of a customer's transport for delivery. The supplier should make sure that the service is suitable and that its effectiveness can be documented.

QS-9000 Requirements

 Customer-supplied product, includes customer-owned tooling and returnable packaging.

4.8 PRODUCT IDENTIFICATION AND TRACEABILITY

In some cases, contracts require the organization to trace specific materials or assemblies throughout the process of their development, through delivery and/or installation. Product (and service) traceability refers to the ability to trace the history, application, or location of an item or activity by means of recorded identification.

ISO 9001 Requirements

 Where appropriate, the supplier shall establish and maintain documented procedures for identifying the product during all stages of production, delivery, and installation. To the extent that traceability is a specified requirement, individual product, or batches thereof, must have a unique identification.

Clause 4.7: Control of Customer-Supplied Product

Customer-supplied product, as referenced in Clause 4.7, has a broad scope. It not only refers to items the customer provides to the supplier which are to be incorporated into the product and then returned in final form to the customer, but also to product (which may be a service, see Paragraph 3) "provided for . . . related activities." The customer-supplied item doesn't actually have to go back to the customer for this element to apply.

The software house that is supplying a turn-key package to a client, but is incorporating some hardware, printers, modems, and network systems that are already in place, is in the situation of utilizing customer-supplied product if any of those items or interfaces for them are delivered to the supplier's site for any purpose. The company that receives packaging from a customer, or even wooden pallets used to ship materials, is handling customer-supplied materials, as is the company whose customer leaves trailers in the supplier's parking lot for loading.

The standard clearly states that the company must ensure that the customer-supplied product is up to par. It is the duty of the customer, however, to provide adequate materials, even though the supplier is going to check it out.

Responsibility for Ongoing Maintenance

One area that is often overlooked is the need to identify and define the responsibilities and needs for the support and maintenance of such items. The responsibility for ongoing maintenance activities should be clearly spelled out at the time that customer-supplied product is identified.

The Victoria Group

ISO 9000-2 Guidance

There are many identification methods, including marking, tagging, or documentation, in the case of a service. The identifier should be unique to the source of the operation. Separate identifiers could be required for changes in various aspects of the production process.

Traceability may require identifying specific personnel involved in phases of the operation. This can be accomplished through signatures on serially numbered documents, for example.

QS-9000 Requirements

For QS-9000 Requirements, "Where appropriate" refers to situations where product identity is not inherently obvious.

Clause 4.8: Product Identification and Traceability

The key words in this clause are, "where, and to the extent that, traceability is a specified requirement . . ." The level of traceability required is left to the discretion of the company unless specifically called for under contractual obligations, regulatory requirements, or industry norms. Some companies, such as one that manufactures inexpensive pencils, could decide that there is no traceability requirement at all.

This also allows the company some latitude in determining the type of traceability, encompassing, for example, traceability of materials within the factory and finished product traceability to the customer. Product traceability should never be confused with product identification.

Product identification requirements can also vary widely. They range from identification by serial number and full test and inspection records to virtually no identification at all. Once again, the company must decide, document its decision, and then adhere to it. Companies should realize that if there is an industry norm for identification and traceability, auditors will expect to see that norm being followed. If there is a regulatory requirement, companies are required to follow it.

The Victoria Group

4.9 PROCESS CONTROL

Preventing problems by controlling the production process is preferable to discovering them at the final inspection. Process control activities often include statistical process control methods, procedures for accepting materials into the process, and the proper maintenance of process equipment and essential materials.

ISO 9001 Requirements

The clause requires the supplier to do the following:

- Identify and plan the processing steps needed to produce the product.

- Ensure that the processes are carried out under controlled conditions.

- Provide documented instructions for work that affects quality.

- Monitor and approve necessary processes.

- Observe and stipulate relevant criteria for workmanship, where practical.

- Maintain equipment to ensure continuing process capability.

The last requirement was not included in the 1987 version. The implication is that where process instrumentation is establishing (controlling) quality, it must be calibrated in the same manner as inspection and test equipment.

Special Processes: In a note, the clause says that "Such processes requiring prequalification of their process capability are frequently referred to as special processes." The requirements regarding special processes are:

- Continuously monitor special processes by qualified personnel to ensure that requirements are met.

- Maintain records for qualified processes, equipment, and personnel.

ISO 9000-2 Guidance

The adequacy of production process control should take into account the adequacy of the measurement processes. When effective process control depends upon consistent operation of process equipment and essential materials, the supplier should include within the scope of the quality system the proper maintenance of such process equipment and essential materials.

ISO 9004-1 Guidance

Clauses 10.0 and 11.0 in ISO 9004-1 discuss quality of processes and control of processes, incorporating service businesses as well as manufacturing.

Clause 10.1 discusses planning for controlled production. Controlled conditions include "appropriate controls for materials, approved production, installation and servicing equipment, documented procedures or quality plans, computer software, reference standards/codes, suitable approval of processes, personnel, as well as associated supplies, utilities, and environments."

Clause 10.0 recommends that companies do the following:

- Conduct process capability studies to determine the potential effectiveness of a process.

- Develop work instructions that "describe the criteria for determining satisfactory work completion and conformity to specification and standards of good workmanship."

- Verify the quality status of a product, process, software, material, or environment.

- Verify the capability of production processes to produce in accordance with specifications.

- Control and verify auxiliary materials and utilities, "such as water, compressed air, electric power, and chemicals used for processing," where important to quality characteristics.

Clause 11.0 offers guidance in the areas of:

- Material control and traceability.

- Equipment control and maintenance.

- Special processes.

- Documentation of work instructions, specifications and drawings.

- Process change control.

- Control of the verification status of material and assemblies (see ISO 9001, Clause 4.12).

- Control of nonconforming materials.

ISO 9000-2 Guidance

 ISO 9000-2 emphasizes that all products are produced by processes. Special processes are those whose results cannot be fully verified by subsequent inspection and testing of the product and where processing deficiencies may become apparent only after the product is in use. Special processes are particularly common in producing processed materials. Critical product quality characteristics in this area include the following examples:

- Metal parts (strength, ductility, fatigue life, corrosion-resistance following welding, soldering, etc.).

- Polymerized plastic (dyeability, shrinkage, tensile properties).

- Bakery products (taste, texture, appearance).

- Correctness of financial or legal documents/software.

Special processes may require:

- Comprehensive measurement assurance and equipment calibration.

- Statistical process control and special training.

Clause 11.4, Process-control management of ISO 9004-1, notes that process variables should be monitored, controlled, and verified at appropriate frequencies to assure:

- Accuracy and variability of equipment used to make or measure the product.

- Skill, capability, and knowledge of workers.

- Accuracy of measurement results and data used to control the process.

- Special environments, time, temperature, or other factors that affect quality.

- Certification records that are maintained for personnel, processes, and equipment.

For more information regarding special processes in the chemical industry, see Chapter 20.

QS-9000 Requirements

Government Safety and Environmental Regulations

Supplier shall have processes to ensure that applicable regulations, including those on hazardous materials, are met. Where appropriate, certificates or letters of compliance will be required.

Designation of Special Characteristics

Documentation is required to show that the supplier is meeting specific customer requirements. Special characteristics may affect safety, regulation compliance, fit, appearance, or effect on later manufacturing operations.

Preventive Maintenance

Supplier shall develop a planned preventive maintenance system requiring:

- Procedures
- Predictive maintenance
- Schedule
- Replacement parts

4.9.1 PROCESS MONITORING AND OPERATION INSTRUCTIONS

Supplier shall develop and maintain explicit instructions called for in the *Advanced Product Quality Planning and Control Plan* Reference Manual. Instructions are to include, as appropriate:

- Operation name and number keyed to process flowchart.
- Part name and part number.
- Current engineering level/date.
- Required tools, gauges, and other equipment.
- Material identification and disposition instructions.
- Customer/supplier designated special characteristics.
- SPC requirements.
- Relevant engineering and manufacturing standards.
- Inspection and test instructions (see 4.10.4).
- Corrective actions instructions.
- Revision date and approvals.

- Visual aids.

- Tool change intervals and set-up instructions.

4.9.2 PRELIMINARY PROCESS CAPABILITY REQUIREMENTS

Studies are required for new processes, to meet customer requirements. If no requirements are specified, then a Ppk target of ≥ 1.67 should be achieved. (See the *Production Part Approval Process* Reference Manual.)

Limitation of attributes data prevent their use in statistical analysis.

4.9.3 ONGOING PROCESS PERFORMANCE REQUIREMENTS

- Cpk ≥ 1.33 target for stable processes.

- Ppk ≥ 1.67 target for unstable processes that meet specification and have a predictable pattern.

- Significant changes should be noted on control charts.

- Customer can permit revision of Control Plan when high degree of process capability is indicated.

- A reaction/corrective action plan is required when the process is not stable or capable.

- Continuous improvement is required, particularly on special characteristics.

4.9.4 MODIFIED PRELIMINARY OR ONGOING CAPABILITY REQUIREMENTS

The Control Plan is to be annotated if customer requires higher or lower capability requirements.

4.9.5 VERIFICATION OF JOB SETUPS

- Job set-ups are required to show that all produced parts meet requirements.

  —Documentation for set-up personnel.

  —Last-off part comparison recommended.

- Some customers may require statistical verification.

4.9.6 PROCESS CHANGES

- Changes to the process generally require prior customer approval. Refer to *Production Part Approval Process* Reference Manual, and to unique customer requirements.
- A record of "change effective" dates is to be maintained.

4.9.7 APPEARANCE ITEMS

Supplier shall provide:

- Adequate/suitable lighting in inspection areas
- Master standards for appearance items, as needed
- Adequate care of physical standards and equipment
- Qualified personnel.

Clause 4.9: Process Control

Suitable Maintenance of Equipment

A requirement for "suitable maintenance of equipment to ensure continuing process capability" is included in the standard. Many companies have formal, planned maintenance programs and activities, but these are rarely included in the management system. ISO 9001 requires them to be incorporated. For those who have no such program, creating one need not be complex or onerous. The planned maintenance will need to be laid out, the activities defined to a level appropriate to the skills and training of the maintenance staff, and records kept of the work performed.

It should be stated where the appropriate maintenance is "run it until it breaks and then fix it." It is after all the company's system-not the auditor's. Examples of "run until it breaks" equipment could include process control computers, some forms of automatic test equipment, hand tools, and similar items. Where "routine maintenance" is basically operator-driven, such as lubricating and cleaning metal-working equipment, then this activity can now be referenced in work instructions.

The term *suitable* is open to interpretation. The supplier has the absolute right to determine suitability, but a process-knowledgeable auditor equally has the right to question the program that has been defined. The key to the success of this requirement lies in auditors recognizing that they have to prove "unsuitability" by providing evidence of

(Contineud on next page)

(Continued from page 112)

adverse effects on deliverables before a noncompliance can be written up, and in the company making sure that the "continuing process capability" requirement is adequately covered.

Work Instructions

Work instructions, which are referred to in Clause 4.9: Process Control, can take the form of anything from a representative sample to a detailed, written document. A work instruction can be a videotape. It could be a model of the work on display by the operator-a common technique in high-volume electronic manual assembly that can work very well.

It is important to assess the training and skills of employees when evaluating where to make use of work instructions. In determining their use, the standard cautions companies to make the decision based on consideration of "where the absence of such instructions would adversely affect quality." In assessing whether an employee knows how to perform his or her job correctly, there are three possible ways this can be demonstrated:

- A work instruction exists that details these responsibilities.

- Records exist proving the individual was hired with a particular skill.

- The employee has received on-the-job training; records exist to document the training and prove it has been performed.

The scope of Clause 4.9 also covers servicing activity, so when such work lies within the registered scope of a company, all the requirements of this clause need to be applied to the servicing activity.

The Victoria Group

Special Processes and Chemical Producers

Does the reference in note 16 of ISO 9000 to special processes requiring prequalification encompass all production processes used by chemical producers? Some auditors have affirmed that it does, basing their decision on the interpretation that all processes are "one, large, special process."

The experts agreed that the reference to special processes in Clause 4.9 could cover all the production processes used by a chemical producer. Robert W. Belfit, president,

(Continued on next page)

(Continued from page 113)

Omni Tech International, Inc., however, said he interprets this clause differently. Belfit suggested that when proceeding through the registration process, a chemical producer should interview potential registrars to ascertain the possible interpretation. (See Chapter 20 for Belfit's article on the chemical industry.)

Belfit cited the example of a product that is "manufactured under as much control and testing as possible, but requires the shipment be delayed until the customer has utilized the product in his process, and therefore releases the product for shipment."

Belfit said these special processes are generally not run on a regular basis, and "therefore the processing parameters and testing parameters have not or cannot be defined precisely to ensure that the product will perform in the customer's application." He concluded that "special processes in the chemical industry imply that it is not possible through process control, in-process testing, or final testing to establish whether the product will perform in the customer's application."

On the other hand, Belfit pointed out, "the production of benzene, styrene, ethyl alcohol, polystyrene, or polyvinyladine chloride are processes that are run on a continuous basis. These are, in reality, commodity products." He said the performance of these products in the customer's application is predictable based on the process controls, the end-process analysis, and/or the product analysis of the final product. "These products are covered by Clause 4.9 in the general statement under process control," Belfit said.

Prevention versus "Find-and-Fix"

Ian Durand, president, Service Process Consulting, Inc., noted that examining the central theme of the ISO 9000 series is important in interpreting the reference to special processes. He said the overall emphasis of the ISO 9000 series is on "preventing quality problems before they occur, rather than relying on 'find-and-fix' approaches." For this reason, Durand said, "it is not unusual for registrars to look for, and prefer to find, attention given to controlling all processes that affect the quality of the "total market offering," i.e., both the tangible goods and accompanying services."

Durand noted that "in the real world there are always trade-offs between process control and inspection and testing." He said Clause 4.9, Process Control, and 4.10, Inspection and Testing, should be considered complementary. In establishing the balance between the two approaches, Durand said such factors as the feasibility of subsequent inspection, the relative effectiveness and costs, and the specific processes and products being considered should be evaluated.

To illustrate, he noted that in some chemical industries, skilled operators assess color, granularity, texture, and handling characteristics to complement process control. Durand concluded that assessing relevant factors and establishing a balance between

(Continued on next page)

(Continued from page 114)

process control and inspection requires working knowledge of the industry and the specific processes and products under consideration.

For example, Durand said that keeping levels of airborne contaminants below specified thresholds during production of solid-state devices is essential. "Complete reliance on inspection and testing is generally not a viable alternative to cleanliness and sanitation during food preparation either," Durand added. He said that both of these examples are types of chemical processes.

Chemical Processes as Special Processes

Terry Heaps, audit program manager, Vincotte USA, Inc., also agreed that a case could be made for including special processes as part of all production processes. Heaps noted that "confusion may exist, if any does exist, in the manner with which auditors approach special processes, since there is little difference between the requirements for special processes and process control in the ISO 9001 and ISO 9002 standards."

Heaps said ECIC guidelines state that chemical processes may be considered special processes for a variety of reasons, including the following:

• A characteristic can be measured only during the process and not in the finished product.

• A characteristic of the product changes (matures) after the product has been delivered.

• The complete characteristics of a product are not known.

• There is no satisfactory method of measuring a product characteristic.

"There may be a greater emphasis on the results of the in-process inspection and testing and calibration of the equipment used to make or test the product than may be required for a conventional process," Heaps said. Heaps cited other ECIC guidelines concerning customer requirements. For example:

• A specification is agreed for certain characteristics inspected or tested during the process.

• Before accepting the contract, the product is evaluated after use in this product or process.

• The process and/or the source of raw materials is not changed without the customer's agreement.

• Specified statistical process control methods are used.

(Continued on next page)

(Continued from page 115)

Special skills, capabilities, and training personnel may be needed to meet any additional quality requirements.

Joseph Tiratto, president, Joseph Tiratto and Associates, Inc., said that "processes of the chemical process industry are generally considered special processes." He cited ECIC guidelines as a reference. In addition, Tiratto cites guidelines in ANSI/ASQC Q90/ISO 9000: *Guidelines for Use by the Chemical and Process Industries.** These guidelines include the following:

- Equipment used to make or measure product.

- Operator skill, capability and knowledge.

- Environmental factors affecting quality.

- Records of qualifications.

*ISO 9000 Guidelines for the Chemical and Process Industries, second edition, is available from ASQC Quality Press, 611 East Wisconsin Avenue, Milwaukee, Wisconsin 53201, 1-800-248-1946.

Clause 4.9: Process Control—Does Suitable Maintenance of Equipment Include Calibration?

Clause 4.9, Process Control of ISO 9001, states: "The supplier shall identify and plan the production, installation, and servicing processes which directly affect quality and shall ensure that these processes are carried out under controlled conditions." A list of controlled conditions includes "suitable maintenance of equipment to ensure continuing process capability." Does this refer to calibration, and if so what type of equipment is affected?

QSU's panel of experts was split on whether calibration is included by the phrase: "suitable maintenance of equipment." Moderator Robert Kennedy, president, the Kennedy company, said that subparagraphs D and G of Clause 4.9 both have to do with making certain that nonhuman assets are performing properly. Calibration, he said, is included in so far as it pertains to the truthfulness of measuring or test equipment in use.

"Calibration is done to perform a lie-detector test," he said. Process control is much more encompassing than that. "It really has to do with the equipment functioning prop

(Contineud on next page)

(Continued from page 116)

erly in all respects, the availability of equipment as specified by whatever their production plan calls for, and the maintenance of that equipment."

Kennedy said that, although calibration is addressed in much greater detail in Clause 4.11.2 (B), a company's failure to calibrate equipment properly or to use calibrated equipment may also violate subparagraphs D and G of clause 4.9.

Larry Bissell, vice president of Management Standards International Ltd., agreed that calibration is covered to some extent by Clause 4.9. Process capability, he said, is the overall variation with the process of the measuring device compared with the acceptable range. Bissell said that the primary purpose of Clause 4.9 is to address the need for preventive maintenance on equipment used to produce the product. "An additional impact is on the calibration requirements," he said. "In order to ensure that the equipment continues to be capable . . . that same equipment must be maintained in a calibrated state."

He said that Clause 4.11.2 requires companies to identify all inspection, measuring, and test equipment, including measurement devices, that can affect product quality. The pertinent language in Clause 4.9 is merely for clarification, he said.

"The primary purpose of [the clause] is to address the need for preventive maintenance on all equipment, but as a subset of that primary need is the recognition and the need to include calibration of measurement devices on the equipment."

Most companies, he said, will not have to do anything different under the 1994 version of Clause 4.9. "If in fact organizations took the shortcut route or a minimum approach, then they may be impacted," he said.

Graham Cartwright, managing consultant, MRA International, said that calibration is not covered. "It does refer to process equipment," he said. "Definite evidence of the provision of control of maintenance of process plant and equipment is now required." Cartwright said the evidence would take the form of maintenance procedures, schedules, and records relating to those items that affect process capability. "Assessors will concentrate on the outputs of these maintenance systems rather than the details of the system itself," he said.

The experts noted that equipment maintenance refers to all production equipment and that testing equipment, which is covered by Clause 4.11.2, deals specifically with the calibration of test and inspection equipment.

Kennedy added that it is also important to remember that a given piece of equipment may not be used solely for measurement and test or for production. It is now common for the same piece of equipment to serve both of these functions, according to Kennedy.

4.10 INSPECTION AND TESTING

This clause looks at the following three areas of inspection and testing:

- Receiving

- In-process

- Final inspection.

4.10.1 GENERAL

ISO 9001 Requirements

Subclause 4.10.1, General, states that the "supplier shall establish and maintain documented procedures for inspection and testing activities" to ensure that product requirements are met. This includes documenting the inspection and testing procedures in the quality plan or documented procedures.

QS-9000 Requirements

Acceptance Criteria:

- Zero defects shall be the acceptance criterion for attribute sampling plans; other situations require customer approval.

- Accredited laboratories are to be used if required by customer.

4.10.2 RECEIVING INSPECTION AND TESTING

ISO 9001 Requirements

Receiving inspection allows suppliers to verify that subcontractors are fulfilling their contractual obligations. The supplier is required to do the following:

- Ensure that incoming products are not used or processed until they have been inspected or otherwise verified.

- Verify in accordance with the quality plan and the documented procedures.

ISO 9000-2 Guidance

This Subclause does not imply that incoming items must be inspected and tested if the supplier can use other defined procedures that would fulfill this obligation. These defined procedures should include:

- Provisions for verifying that incoming items, materials, or services are accompanied by supporting documentation.

- Provision for appropriate action in the event of nonconformities.

ISO 9004-1 Guidance

Clause 9.7, Receiving Inspection Planning and Control, notes that the "extent to which receiving inspection will be performed should be carefully planned. . . . The level of inspection should be selected so as to balance the costs of inspection against the consequences of inadequate inspection."

Clause 9.8, Receiving Quality Records, stresses that appropriate records should be kept to "ensure the availability of historical data to assess subcontractor performance and quality trends." Companies should also consider maintaining "records of lot identification for purposes of traceability."

QS-9000 Requirements

Incoming Product Quality: one or more of the following methods shall be used:

- Statistical data.

- Receiving inspection and/or testing.

- Assessment of subcontractor by second or third party.

- Part evaluation.

- Warranty or certifications (if in conjunction with one of the above).

4.10.2.3 RELEASE FOR URGENT PRODUCTION PURPOSES

ISO 9001 Requirements

The supplier is required to positively identify and record incoming product to permit recall and replacement, if necessary.

ISO 9000-2 Guidance

The release of incoming product subject to recall should generally be discouraged as a matter of good quality management practice. There are two exceptions:

- An objective evaluation of quality status and resolution of any nonconformities can still be implemented.

- Correction of nonconformities cannot compromise the quality of adjacent, attached, or incorporated items.

The supplier's procedures should accomplish the following:

- Define responsibilities and authority of people who may allow incoming product to be used without prior demonstration of conformance to specified requirements.

- Explain how such product will be positively identified and controlled in the event that subsequent inspection finds nonconformities.

4.10.3 IN-PROCESS INSPECTION AND TESTING

ISO 9001 Requirements

The supplier is required to do the following:

- Inspect and test product as required by the quality plan or the documented procedures.

- Hold the product until the required inspection and tests have been completed.

The exception is when the product is released under positive recall procedures. The release under positive recall procedures, however, would not preclude the inspection required above.

ISO 9000-2 Guidance

In-process inspection and testing applies to all forms of products, including services. It allows for early recognition of nonconformities.

Statistical control techniques are commonly used to identify product and process trends and prevent nonconformities. Inspection and test results should be objective-including those carried out by production personnel.

ISO 9004-1 Guidance

Clause 12.2, In-Process Verification, lists the following types of verification checks:

- Set-up and first-piece inspection.
- Inspection or test by a machine operator.
- Automatic inspection or test.
- Fixed inspection stations at intervals through the process.
- Patrol inspection by inspectors monitoring specified operations.

4.10.4 FINAL INSPECTION AND TESTING

ISO 9001 Requirements

Regarding final inspection and testing, the supplier is required to carry out all specified final inspection and tests, including those specified either on receipt of product or in-process. No product shall leave the company until every activity specified in the quality plan or documented procedure has been satisfactorily completed.

ISO 9000-2 Guidance

 Final inspection involves the examination, inspection, measurement, or testing upon which the final release of a product is based. Release specifications should include all designated release characteristics.

ISO 9004-1 Guidance

 Clause 12.3, Finished Product Verification, lists the following two types of final production verification:

- Acceptance inspections or tests.

- Product quality auditing.

Acceptance inspections are used to ensure that the "finished product conforms to the specified requirements." Product quality auditing that is performed on representative sample units may be either continuous or periodic.

QS-9000 Requirements

 Layout Inspection and Functional Testing: Requirements are established by the customer.

4.10.5 INSPECTION AND TEST RECORDS

ISO 9001 Requirements

 The supplier is required to establish and maintain records that indicate whether the product has passed inspections and test procedures. When the product fails to pass an inspection and/or test, the procedures in Clause 4.13 for nonconforming product apply. Records are to identify the inspection authority responsible for releasing product. It includes a reference to Clause 4.16.

ISO 9000-2 Guidance

 Inspection and test records facilitate assessment according to specifications and are useful for regulatory requirements and possible product liability problems.

ISO 9004-1 Guidance

 Clause 17.0, Quality Records, discusses in detail procedures for establishing and maintaining quality records, including inspection and test records.

Clause 17.3, Quality Records Control, notes that all documentation should be maintained in "facilities that provide a suitable environment to minimize deterioration or damage and to prevent loss."

Clause 4.10: Inspection and Testing

ISO 9001 quite rightly places a good deal of emphasis on the importance of inspection and test as a means of demonstrating compliance with specifications.

Contrary to a commonly held view, the standard does not emphasize inspection and test as a means of controlling quality but rather as a method of assessing compliance. In terms of evaluating the effectiveness of the overall system, this apparently philosophical point is of considerable importance.

Paragraph 4.10.1 includes a generic requirement to "establish and maintain documented procedures for inspection and testing activities . . . to verify that specified requirements . . . are met." These requirements may be in the generic system or may be contained in specific documentation such as a quality plan; the process of assurance may place the responsibility on a subcontractor. In these situations the responsibility of the supplier is one of assuring adequate controls exist at the subcontractor's premises.

Controlling quality and meeting specified requirements are quite different concepts requiring different methodologies. The standard also recognizes that the approach to meeting specified requirements will vary significantly between receiving, in-process, and final inspection, and test activities.

The entire process of inspection and test is fundamental to the demonstration of compliance by the provision of objective evidence of compliance. No amount of process control can demonstrate product compliance per se; but the Standard leaves open to the user the approach to be taken and where the responsibility and authority should lie. The records of inspection and tests are required to identify the authority responsible for the release of the deliverable.

The common thread remains the same-decide what to do, write it down, do it in the manner defined, and keep records to prove that it has been done.

The Victoria Group

Must Product Shipment Await Testing?

Based on statistical control of process and historical product test results or on per-formance, is it acceptable to release a product before all testing has been com-pleted? For example, test results normally take up to one month, and there is insufficient storage space for more than a week's production of bulk product. If so, must the customer be notified, and under what ISO 9001 clause should notification take place?

The panel of experts agreed that there are times when product may be shipped prior to completion of testing, but they disagreed as to the circumstances under which that shipment may take place.

Elizabeth A. Potts, president, ABS Quality Evaluations, Inc., said there is little room for interpretation. "The standard means exactly what it says," she explained. "Specified inspections and tests need to be completed and documented prior to shipment. At the time of contract, certain inspections and tests are explicitly or implicitly agreed to by the parties to the contract."

The supplier would be deviating from the terms of the agreement by shipping before completion or documentation of all tests, according to Potts. "The supplier must get consent of the other party to the agreement, namely the customer, to do so," she said.

Potts said issues such as limited storage or lengthy reliability testing would have to be addressed by the customer and supplier well in advance of shipment. "These issues should be addressed in the supplier's quality system to meet the requirements of Clause 4.3, Contract Review," Potts said.

Shipping with Notification

Bud Weightman, president, Qualified Specialists, Inc., acknowledged that Subclause 4.10.4 prohibits a product from being dispatched until the required inspection and test-ing are completed. However, he said, the supplier may elect to ship before that with proper notification.

"If the supplier elects to do so he has the obligation to notify the customer under the requirements of ISO 9001, Clause 4.3," he said. The clause states: "The tender, con-tract, or order shall be reviewed by the supplier to ensure that . . . the supplier has the capability to meet contract or accepted order requirements."

Weightman said "capability" refers to the supplier's ability to deliver the product to the customer with all of the test results at the time of the shipment. The customer should provide documented evidence that it will accept the product without complete test results at the time of shipment, he added.

(Continued on next page)

(Continued from page 123)

"The supplier's decision to ship product with incomplete test results should be based upon historical product test results, product performance data, and documented evidence of statistical process control, which could be submitted to the customer as an added assurance that the product test results will comply with stated requirements," Weightman explained.

Contract Requirements for Extended Testing

Dan Epstein, senior advisor, Quality Management Consulting Services, Inc., said there are circumstances that would allow shipping the product prior to completing the testing, but the example of limited space is not one. "Contractual obligations may require extended testing in the form of endurance, reliability, or life testing," he said. "In addition, satisfactory completion of these tests may determine product acceptability."

Moreover, percent defect allowable (PDA)-the customer-defined acceptable quality level-and Six Sigma requirements-a statistical quality level equating to approximately three defects per million-may be contractually imposed, Epstein said. These allow a supplier to compute an acceptance limit only after a period of time, possibly after many shipments. The customer, he said, may have specified a delivery schedule that is in conflict with the supplier's ability to determine acceptability. Potts notes that both partners should understand that this is an issue at the time of contract and should address it under contract review procedures.

If the original contract does not provide any of these conditions (i.e., purchase order, performance specification, etc.) and the customer is expecting products and services from an ISO 9000 registered facility, Epstein said, it is the obligation of the supplier to notify the customer of the system noncompliance under ISO 9001 Clauses 4.2 and 4.3. "Ideally, every attempt should be made to avoid these sorts of problems during the contract review process of Clause 4.3," Epstein said.

Shipping and Traceability

Stephen S. Keneally, president, Scott Technical Services, Inc., said the clause appears to require that all testing be completed prior to shipping. Nevertheless, many industries, such as aerospace, defense, and integrated circuits, routinely ship and assemble while samples from the production lot continue to be tested, he said.

"Depending on the product and industry, elaborate systems of maintaining traceability are used in the unlikely event if 'life testing' or long-term reliability or environmental tests result in the product being rejected and the lot being recalled," Keneally explained.

In regulated industries such as defense electronics, medical devices and pharmaceuticals, the traceability and testing requirements are part of the product approval cycle,

(Continued on next page)

(Continued from page 124)

according to Keneally. He said any changes to the manufacturing, inspection, and testing process would require re-approval by the appropriate regulatory agency or the customer.

In commercial industries, he said, the revised quality plan should reflect reduced inspection, skip-lot inspection, or periodic versus 100 percent lot-by-lot testing. It should also reflect an analysis of whether statistical process control data and historical product test results justify changes.

"The manufacturer develops the quality plan and test procedure and has the option to change them when warranted," Keneally said. "It is common sense and good business practice to reduce nonvalue-added activity when processes are under control and statistical or other data support reduction of inspection or testing activities. Obviously, sufficient history should be available before radical changes are made."

Keneally said customers need only be notified if their contract or specification indicates what tests are to be performed or if catalogs and other sales literature reflect specific tests, with the implication being that suppliers conduct such tests in every case.

According to Keneally, Clause 4.13, Control of Nonconforming Product, establishes the supplier's responsibility to notify customers when product is found to be nonconforming. "It is the supplier's responsibility to notify customers should a problem with the product come to their attention through other uses or applications," he said. "The supplier is responsible for notifying customers of the possibility of a problem so that inadvertent use or installation is prevented."

4.11 CONTROL OF INSPECTION, MEASURING, AND TEST EQUIPMENT

4.11.1 GENERAL

ISO 9001 Requirements

The supplier is required to do the following:

- Establish and maintain documented procedures to control, calibrate, and maintain the inspection, measuring and test equipment to demonstrate conformance of the product to requirements.

- Use equipment in a manner that ensures that measurement uncertainty is known and is consistent with the required measurement capability.

- Check and recheck the capability of any test software or test hardware used as forms of inspection.

• Make technical data pertaining to measurement devices available when required by the customer.

4.11.2 CONTROL PROCEDURE

ISO 9001 Requirements

Subclause 4.11.2 spells out in detail the requirements for testing accuracy, calibration of equipment, handling of equipment, and documentation of the checking procedures. These include the following:

• Identify necessary measurements, the accuracy required, and the appropriate inspection, measuring, and test equipment.

• Identify, calibrate, and adjust all equipment.

• Establish, document, and maintain calibration procedures.

• Ensure that equipment is capable of required accuracy and precision.

• Identify equipment to indicate calibration status.

• Maintain calibration records.

• Assess and document the validity of previous inspection and test results when equipment is out of calibration.

• Ensure suitable environmental conditions for calibration, inspection, measurement, and testing.

• Ensure accuracy and fitness for use when handling, preserving, and storing equipment.

• Safeguard inspection, measuring, and test facilities.

ISO 9000-2 Guidance

Clause 4.11 addresses the suitability of the equipment used in monitoring quality. ISO 10012-1, *Quality assurance requirements for measuring equipment-Part 1: Management of measuring equipment,* offers guidance for the management of measuring equipment. However, the guidance in ISO 10012-1 does not add to or otherwise change the requirements in ISO 9001, except where conformance to ISO 10012-1 is required.

Measurements may include less tangible instruments, such as polling, questionnaires or subjective preferences.

The requirements of this clause also should be applied to measurements subsequent to producing and inspecting a product (e.g., handling, storage, packaging, delivery, or servicing).

ISO 9004-1 Guidance

In Clause 13.2, ISO 9004-1 notes that the control of measuring and test equipment and test methods should include the following factors, as appropriate:

- Suitable specification and acquisition.

- Initial calibration prior to first use in order to validate required bias and precision.

- Testing of software and procedures controlling automatic test equipment.

- Periodic recall for adjustment, repair, and recalibration to maintain required accuracy in use.

- Documentary evidence that covers instrument identification, calibration status, and all handling procedures.

- Traceability to accurate and stable reference standards.

QS-9000 Requirements

Inspection, Measuring, and Test Equipment Records: Calibration/verification records shall include:

- Revisions following engineering changes.

- Gauge conditions and readings received.

- Customer notification of suspect material.

To include gauges owned by employees.

Measurement System Analysis: Statistical analysis of variability of equipment, e.g., gauge repeatability and reproducibility, is required. (See the *Measurement System Analysis* Reference Manual.)

Clause 4.11: Control of Inspection, Measuring, and Test Equipment

Clause 4.11 discusses control of inspection and test equipment and calibration. The standard requires that the supplier establish and maintain documented procedures to "control, calibrate and maintain inspection, measuring and test equipment (including test software) used by the supplier to demonstrate the conformance of product to the specified requirements. "

(Continued on next page)

(Continued from page 127)

It is easy to misread the requirement laid down in this section. It does not require calibration of all equipment, only equipment used to "demonstrate the conformance of product to the specified requirements." Equipment that is used for in-process measurement need not be calibrated, provided that the measurements made are not the last opportunity to record some item that is a deliverable specification. The same measurement, repeated at final release test or inspection, must be made on calibrated equipment. Remember the words "demonstrate conformance." If any measurement taken in-process is part of demonstrating conformance, or in other words, measuring a specified deliverable, the equipment used must be controlled.

In one example, an audit team demanded that each of 35,000 gauges and meters at a certain refinery be calibrated in a manner traceable to national standards. The company correctly calibrated only those gauges that influenced the quality of the product and were used to demonstrate conformance to specified requirements.

Demonstrating conformance is not confined to a single site in the company. Test and inspection equipment must be under traceable calibration control at those locations or places in the process where conformance can be demonstrated.

Portable Transfer Standard

External calibration to recognized national standards by the National Institute for Standards and Technology (NIST) or a recognized, accredited laboratory or test house is an expensive business. Normal practice, therefore, is to have a certain amount of the critical equipment calibrated externally, and then to use this equipment to calibrate the rest. This is known as using a "portable transfer standard."

It is an economical way of establishing calibration of all test and inspection equipment without breaking the bank. When adopting this approach, it pays to remember that the equipment providing the "portable transfer standard" should whenever possible be an order greater in its measurement capability than the equipment being checked. This assures the accuracy of the secondary equipment measurement capability.

In summary, the requirement is to identify equipment that assures compliance; uniquely identify each piece of equipment; define the accuracy and precision required of each piece of equipment; define the calibration method; define the reference standard; keep records of calibration performance; be able to identify the calibration status of any measuring equipment; and know what to do about previously tested material if an item is found out of calibration.

ISO 10012-1, *Quality assurance requirements for measuring equipment-Part 1: Management of measuring equipment* is now referenced as a guidance document for structuring calibration systems, but this will be excessive for most companies unless they operate a full calibration laboratory system.

The Victoria Group

How Far Do You Take Calibration Programs?

Do the requirements of Clause 4.11 imply that all inspection, measuring, and test equipment must be put on a calibration program, or does it merely require the calibration of certain equipment?

In general, the panel of experts agreed that the standard does not require all inspection, measuring, and test equipment to be put on a calibration program. However, one panelist, Robert W. Belfit, Jr., president, Omni Tech International Ltd., disagreed, maintaining that no such equipment should escape periodic scrutiny. Nevertheless, he said, there are varying levels of calibration, and not all equipment need be subjected to the same degree of scrutiny. "Why would one want to produce data and not know the accuracy and precision?" asked Belfit. "The question should be, how do we keep the measurement process under process control?"

Level of Calibration Required

Belfit said the level of calibration required varies with each piece of equipment. He said it is important to consider the following factors:

- What the measurement will be used for.

- The required tolerance versus equipment capability.

- The ruggedness of the equipment.

- Working conditions.

- Frequency of use.

- Possible malfunctions.

- Whether the measurements will be supported by other data.

- Whether the measurements will be used to support a specification or claim.

Bud Weightman, president, Qualified Specialists, Inc., said it is not necessary to put all such equipment on a calibration program provided "objective evidence exists" to substantiate that the supplier made a "conscientious" decision. "It is the supplier's responsibility to identify those characteristics of the design and processes which could have a direct effect on quality," he said. "Characteristics could include specific product dimensions and process-related elements that require an inspection, test, or measurement to verify conformance to the stated requirements of the design output."

Calibration to Demonstrate Conformity

Roderick Goult, president and CEO, The Victoria Group, said the standard is confusing with respect to calibration. "Handle with care. Formal, traceable calibration is very expensive," he said. "Only do what you absolutely have to. The rest can be validated [as against being calibrated] by checking it against formally calibrated equipment; in

(Continued on next page)

(Continued from page 129)

other words, by using a portable transfer of the standard." Goult said there are two apparently confusing statements contained in ISO 9001, Clause 4.11. The first requires the calibration of equipment used to demonstrate product conformance to specified requirements, and the second, in Clause 4.11.2 b), deals with the calibration of inspection, measuring, and test equipment that "can" affect product quality.

In the first statement, he said, only measuring equipment used for conformance testing need be formally calibrated. However, the last opportunity to assure compliance is commonly interpreted to be very early in the process, possibly even at receiving inspection, he noted. The second statement implies something completely different, Goult said. "The escape from massive and pointless expense comes from the use of the word 'can,'" he said. "Who is to define what can or cannot affect product quality? That is for the supplier to judge, and unless the decision is clearly flawed, the auditor is not in a position to argue."

Peter M. Malkovich, president, East Concord Associates, agreed that use of the word "can" is significant. "In other words, if the equipment does not control and/or verify quality, it does not have to be included in the calibration program," he said. "Hence, there is equipment that may not affect quality and equipment that may be used as a work aid or indicator only. Such equipment can be excluded from the calibration program and should be identified as 'Not Calibrated' with the reason why it is not included." For example, he said, it is not necessary to calibrate gauges and equipment used to monitor the condition of the plant or to indicate that something is operating. They still must be checked periodically to establish that they do work.

Malkovich said firms should calibrate all equipment that affects quality and other equipment used for important measurements. Nevertheless, he said, firms must be prepared to demonstrate why excluded equipment does not affect quality.

Labeling Uncalibrated Items

Weightman suggests that companies back up their decision not to calibrate certain pieces of equipment through documentation. Possible examples include affixing "calibration not required" stickers to them; stipulating which types of equipment will be used in design output documents; including an evaluation of the inspection, measuring, and test equipment in-process capability studies; and having procedures that document which types of equipment will be calibrated and which will not.

He also suggests that companies evaluate "all potential sources" where inspection, measuring, and test equipment is in use to determine their potential effects on quality. Such sources include equipment owned by the firm, equipment owned by employees, equipment on loan, equipment provided by the customer, and equipment used in vendor processes or operations.

Preventive Maintenance Requirements: Yes or No?

ISO 9001, Clause 4.11, Inspection, Measuring and Test Equipment, has been interpreted to include requirements for "preventive maintenance" of equipment. However, the only clear requirement is for calibration. Does a requirement exist for preventive maintenance?

All four experts agree that ISO 9001 does not specifically require preventive maintenance of equipment. However, experts' commentary ranged from implied requirements to a requirement under special circumstances.

A Built-In Requirement?

"I believe that ISO 9001 has a 'built-in' requirement for preventive maintenance, although preventive maintenance is not specifically spelled out," commented Bud Weightman, president, Qualified Specialists, Inc.

He said ISO 9001, Clause 4.11.2 b), states that the supplier shall "identify all inspection, measuring and test equipment that can affect product quality, and calibrate and adjust them at prescribed intervals, or prior to use, against certified equipment . . ."

Joseph DeCarlo, operations director, East Coast, Excell Partnership, said that ISO 9001 Clause 4.11 only outlines requirements for calibration of equipment and not for preventive maintenance of equipment. "Therefore, while preventive maintenance is certainly part of any sound quality system, it is not an ISO 9001 requirement," DeCarlo said. "If, however, a manufacturer specifies preventive maintenance in their own procedures, i.e., quality system, then it does become a requirement that an ISO auditor will look for."

Joseph Tiratto, president of Joseph Tiratto and Associates, Inc., agreed that ISO 9001, Clause 4.11 does not specifically require preventive maintenance. "However," Tiratto said, "the emphasis of ISO 9001 is to be proactive rather than reactive."

Calibration Interval

Weightman argues that the process of assigning the calibration interval is critical. He pointed out that a supplier should compare its inspection, measuring, and test equipment to the measurements to be made and designate an appropriate calibration interval (assuming all other Clause 4.11 requirements have been met).

If adequate history of the equipment is available and all other Clause 4.11 requirements have been met, Weightman said, "this [calibration] interval should be based on the inspection, measuring, and test equipment's stability, known degree of drift, degree of usage, and environment," where the inspection, measuring, and test equipment is being used.

(Continued on next page)

(Continued from page131)

More Than Just Calibration

Dr. James Lamprecht, author and consultant, certified ISO 9000 auditor, emphasized that Clause 4.11 addresses more than just calibration. "The precision and accuracy of some measuring and test equipment is also required," but he asked that "in many cases it would be prohibitively expensive to conduct capability analysis on all instruments." Suppliers also may not know how a capability study or accuracy study could be conducted on a particular set of gauges or instruments.

"Common sense and experience is relied upon in most cases to determine if a [process] instrument as opposed to a lab instrument is misbehaving,'" Lamprecht said. "For new instruments you can always rely on the vendor's documentation. In many cases, if you subcontract your equipment calibration/maintenance program, your supplier will-should-provide you with the necessary data."

Lamprecht said determining how many instruments should be placed under "control" is hard. According to Lamprecht, nearly all laboratory instruments have to be under control, but process monitoring instruments are more difficult to manage.

He said the ISO 9001 standard requires a supplier to demonstrate the conformance of product to the specified requirements." Taken to the extreme, this requirement might include the office thermostats. However, temperature sensors, pressure valves, flow meters, and pH meters used to verify specifications are important instruments that must measure to required accuracy, Lamprecht said.

"Remember that while an instrument does not specifically fall under the scope of the ISO 9000 series, it might well have to be controlled as per OSHA [Occupational Safety and Health Administration] or EPA [Environmental Protection Agency] requirements," Lamprecht concluded.

4.12 INSPECTION AND TEST STATUS

ISO 9001 Requirements

The supplier is required to do the following:

- Identify the inspection and test status of the product throughout production and installation to ensure that only acceptable product has been used.

- Identify the inspection authority responsible for the release of the conforming product.

The test status must be indicated by suitable means.

ISO 9000-2 Guidance

The status should indicate whether a product has

- Not been inspected.

- Been inspected and accepted.

- Been inspected and is on hold awaiting decision.

- Been inspected and rejected.

The most certain method of ensuring status and accurate disposition is physically to separate these product categories. In an automated environment, however, other methods can be used, such as a computer database.

ISO 9004-1 Guidance

Clause 11.7 addresses control of verification status. Identification of verification status should be by suitable means, "such as stamps, tags, notations, or inspection records that accompany the product, or by computer entries or physical location." Identification should be capable of indicating the following:

- Verified versus unverified material.

- Acceptance at the point of verification.

- Traceability to the unit responsible for operation.

QS-9000 Requirements

Note: The control plan may be the quality plan.

- **Product Location:** The physical location of the product in the production process is not sufficient to indicate inspection status, unless it is clearly obvious, such as with an automatic process.

- **Supplemental Verification:** The customer may set additional requirements.

4.13 CONTROL OF NONCONFORMING PRODUCT

4.13.1 GENERAL

ISO 9001 Requirements

The supplier is required to establish and maintain a procedure to prevent the inadvertent use or installation of a nonconforming product. The nonconforming product should be segregated where practical.

Clause 4.12: Inspection and Test Status

The main requirement in the context of this paragraph is to clarify the breadth of application. The identification of inspection and test status means more than the simple question, "Has it passed this test or that inspection?" It would be lunacy to get into a position where no one knew what had been tested and inspected and what had not!

This is only part of the story-the full requirement is that it should be possible to identify any element at any stage of its progress through the process, within the framework of the way that the system is established. In other words, if the process is batch-oriented, then it is possible that a single item may get separated from a batch, and if it is pre-inspection it may not be possible to tell which batch it is from. This item should be segregated until its status has been determined. Once the batch has been inspected, then it must be possible to tell that the batch has been inspected. This requirement continues throughout the process, from receiving to dispatch.

The Victoria Group

ISO 9000-2 Guidance

A nonconforming product—either an intermediate or final product or service—is one that fails to meet specifications. This applies to a nonconforming product that occurs in the supplier's own production as well as nonconforming products received by the supplier.

The procedures for controlling a nonconforming product should include the following:

- Determining which product units are involved in the nonconformity.

- Identifying the nonconforming product units.

- Documenting the nonconformity.

- Evaluating the nonconformity.

- Considering alternatives for disposing of the nonconforming product units.

- Physically controlling the movement, storage, and processing of the nonconforming product units.

- Notifying all functions that may be affected by the nonconformity.

QS-9000 Requirements

Suspect Product:

- This clause applies to suspect product as well as nonconforming product.

- The "control plan" may replace the quality plan.

4.13.2 REVIEW AND DISPOSITION OF NONCONFORMING PRODUCT
ISO 9001 Requirements

The supplier is required to do the following:

- Define the responsibility for review and authority for the disposition of nonconforming product.

- Document the disposition of the product.

Nonconforming products may be the following:

- Reworked.

- Accepted without repair by concession of the customer.

- Regraded for alternative application.

- Rejected or scrapped.

ISO 9000-2 Guidance

Suppliers should consider the procedures in Clause 4.13 in relationship to the risk of failure to meet customer requirements. Actions (a) through (d) in Subclause 4.13.2 all carry degrees of risk. In the long term, action (d) may carry the lowest risk.

ISO 9004-1 Guidance

ISO 9004-1, Clause 14.0, Control of nonconforming product, includes the following guidance:

- Procedures to deal with nonconforming product "should be taken as soon as indications occur that materials, components, or completed product do not or may not meet the specified requirements."

- The persons who review nonconforming items "should be competent to evaluate the effects of the decision on interchangeability, further processing, performance, reliability, safety and aesthetics."

- "A decision to accept product should be documented, together with the reason for doing so, in authorized waivers, with appropriate precautions."

4.13.3 CONTROL OF REWORKED PRODUCT
QS-9000 Requirements

- Rework instructions shall be available to, and in use by, appropriate personnel.

- A plan for reducing the amount of nonconforming product is to be in operation and progress tracked.

- Product supplied for service applications is to have no visible evidence of rework without approval.

Rework produces an item which is in every way indistinguishable from a "first-time through" acceptable one.

Repair makes the item meet requirements but is different in some way, e.g., welded, from the original design.

4.13.4 ENGINEERING APPROVED PRODUCT AUTHORIZATION

Changes in product or progress require customer approval and apply also to subcontractor purchases. Specific product identification and records are to be kept, including the time interval during which change is authorized.

Clause 4.13: Control of Nonconforming Product

ISO 9001, Clause 1.1, states "the requirements specified are aimed primarily at achieving customer satisfaction by **preventing** nonconformity at all stages, from design through servicing" (emphasis added). Fortunately, the authors of this standard were realists and, despite the stated intent of Clause 1.1, it *is* recognized that even in the best of all possible worlds things still go wrong. That being the case, there is a need to design the system so as to prevent the "unintended" use or installation of **any** nonconforming product or service. There must be a clear, unequivocal method of making sure that nonconforming product is properly identified and isolated until such time as the procedures that have been created to manage the review and disposition of this unacceptable product or service have been put into effect.

The procedure to be followed must be defined. Questions to guide the process include the following:

- Who has the authority to sentence nonconforming product?

- How is the review to be carried out?

- What are the options for disposition?

- Are processes and authorities the same across the entire company, or are there different authorities and responsibilities in various areas of the operation, from design to after-sales service?

All of the above must be defined and documented.

The Victoria Group

4.14 CORRECTIVE AND PREVENTIVE ACTION

Clause 4.14 places *more* emphasis on prevention. Thus, the required corrective and preventive actions are identified in separate subclauses, as described below. Corrective action is directed toward eliminating the causes of *actual* nonconformities. Preventive action is directed toward eliminating the causes of *potential* nonconformities.

4.14.1 GENERAL

ISO 9001 Requirements

The basic requirement is to establish and maintain documented procedures for implementing corrective and preventive action. Actions taken should be appropriate to "the magnitude of problems and commensurate to the risks encountered."

The supplier must implement and record any changes in documented procedures that result from corrective and/or preventive actions.

QS-9000 Requirements

Problem-Solving Methods

For product which has been identified as nonconforming by the customers, the supplier is to follow the problem-solving methods prescribed by the customer.

4.14.2 CORRECTIVE ACTION

ISO 9001 Requirements

The supplier's procedures for corrective action are required to include the following elements:

- Effectively handle customer complaints and nonconformity reports.
- Investigate and analyze the problem and record the results.
- Determine the effective corrective action.
- Ensure that corrective actions are taken effectively.

ISO 9000-2 Guidance

This clause explains what an organization must do when things go wrong. Analysis of nonconformities can be performed by using inspection and test records, process monitoring, audit observation, and all other available feedback methods. Corrective action procedures should include the following:

- Establishing responsibility for taking corrective action.

- Defining how the action will be carried out.

- Verifying the effectiveness of the corrective action.

Procedures should also take into account nonconformities discovered in a product that has already been shipped and designated as satisfactory.

ISO 9004-1 Guidance

 According to ISO 9004-1, Clause 15.2, Assignment of responsibility for corrective action, the "coordination, recording, and monitoring of corrective action related to all aspects of the quality system should be assigned within the organization. The analysis and implementation may involve a variety of functions, such as design, purchasing, engineering, processing, and quality control."

A problem affecting quality "should be evaluated in terms of its potential impact on such aspects as processing costs, quality-related costs, performance, reliability, safety, and customer satisfaction."

QS-9000 Requirements

Returned Product Test/Analysis

 Parts that are returned from the customer are to be analyzed, with records kept and appropriate customer action and process change taken.

4.14.3 PREVENTIVE ACTION

ISO 9001 Requirements

 Subclause 4.14.3 lists the following key steps for preventive action:

- Use all available information, such as work processes, audit results, quality records and customer complaints to detect, analyze, and eliminate potential causes of nonconformities.

- Determine a method for preventive action.

- Initiate preventive action and ensure that it is effective.

- Submit any relevant information on actions taken for management review.

4.15 HANDLING, STORAGE, PACKAGING, PRESERVATION, AND DELIVERY

ISO 9001 Requirements

 The requirements in this clause include the following:

- Establish and maintain documented procedures for handling, storage, packaging, preservation, and delivery.

Clause 4.14: Corrective and Preventive Action

The clause pertaining to nonconforming products is followed, logically, by a clause on corrective and preventative action. Often the weakest part of quality systems, corrective action loops are frequently designed only to address the immediate problem while failing to act to avoid its recurrence. Another common problem is that they often deal only with matters of processes, products, or services while overlooking the system. ISO 9001, Clause 4.14 addresses all three. The standard requires a rigorous examination of all the quality data and records to detect and remove all potential as well as actual causes of nonconformance. This is proactive quality, not reactive.

The division of Clause 4.14 into corrective and preventive actions reinforces the primary intent of the standard, which is preventing nonconformity at all stages.

Product, Process, and System Investigation

Investigation of nonconformance has to operate on three levels: "investigation of the causes of nonconformities related to product, process, and quality system . . ." (4.14.2 (b)). Any one of the three can be the cause of a nonconformance and therefore require corrective action. Companies with systems that only apply corrective actions to product issues must extend that system to cover the other two sides of the conformance triangle.

Preventive Action

Subclause 4.14.3, Preventive action, creates extended requirements for preventive actions, and all such actions must be submitted to management for review. The Standard is seeking to ensure that management becomes fully engaged in operating the system.

Most companies will need to create procedures specifically designed to address the requirements of Subclause 4.14.3. Companies should carefully consider the matter of the "comprehensive analysis of all available data." It is very easy to end up with a procedure that, while being very comprehensive in its coverage, requires too much time and effort to fulfill. The preamble statement in 4.14.1 must be the guide: that actions "shall be to a degree appropriate to the magnitude of problems and commensurate with the risks encountered."

The Victoria Group

- Provide a method to prevent damage or deterioration.

- Provide secure storage and stipulate appropriate receipt and dispatch methods.

- Control packaging, packing and marking processes.

- Provide appropriate methods for preserving and segregating products when they are under the supplier's control.

- Protect product quality after final inspection and test, including delivery to destination.

ISO 9000-2 Guidance

The requirement applies to incoming materials, materials in process, and finished product. The procedures should provide proper planning, control, and documentation.

The handling methods should include provision for the transportation unit such as pallets, containers, and conveyors, to prevent damage. Another factor to consider is the maintenance of the handling equipment.

Suitable storage procedures should take into account the following:

- Physical security.
- Environmental control (temperature and humidity).
- Periodic checking to detect deterioration.
- Legible, durable marking and labeling methods.
- Expiration dates and stock rotation methods.

The packaging procedures should:

- Provide appropriate protection against damage, deterioration, or contamination as long as the material remains the responsibility of the supplier.
- Provide a clear description of the contents or ingredients, according to regulations or to the contract.
- Provide for checking packaging effectiveness.

For some products, delivery time is a critical factor. Procedures should take into account various types of delivery and variations in potential environmental conditions.

ISO 9004-1 Guidance

Clause 16, Post-production activities, emphasizes the need for a documented system for incoming materials, materials in process, and finished goods. Section 16 also refers to the need for proper and complete installation instructions.

QS-9000 Requirements

- Inventory—An inventory management system shall be used to optimize inventory turns and level and ensure stock rotation.
- Customer Packaging Standards—packaging requirements are established by customer.

- Labeling—labeling requirements are established by customer.

- Supplier Delivery Performance Monitoring—systems are to support 100 percent on time shipments. When not met, corrective action to improve delivery performance is necessary, with notification of customer.

  —Lead times are to be established and met.

  —Performance to delivery requirements are to be tracked.

  —Customer is to specify transportation mode, routing, and containers.

- Production Scheduling-current orders are to drive production scheduling. Small lots (synchronous one-piece flow) are encouraged.

- Shipment Notification-on-line transmittal of advanced shipment notification is required, with a backup method available.

Clause 4.15: Handling, Storage, Packaging, Preservation, and Delivery

Clause 4.15 is fairly straightforward. All the clause asks is that these operations be managed in a way that keeps the product secure from deterioration, loss, or damage from the start of the process until the responsibility of the supplier passes to someone else. The usual problem here is that companies tend to forget that these activities take place throughout the entire process, not merely at the end.

There is also a requirement to ensure that items kept in storage for any period are regularly reviewed to ensure that deterioration is not occurring-control of shelf-life by the use of FIFO (First In, First Out) techniques being one example of how this is achieved. In other cases, it may mean that special provisions have to be made for long-term storage, such as the use of desiccants or other means of product protection. In the case of software, the nature of the storage medium is volatile, and this fact must be taken into consideration when storage facilities are arranged.

Designated Storage Areas

Clause 4.15 requires that "The supplier shall use designated storage areas or stockrooms . . ." The intent is that the product be safe from accidental damage.

The Victoria Group

Subclause 4.15.3: Do Secure Storage Areas of Stockrooms Mean a Locked Area?

Editor's Note: ISO 9001:1987 stated in Subclause 4.15.3 that the supplier "shall provide secure storage areas or stockrooms . . ." The wording of ISO 9001:1994 was changed to "The supplier shall use designated storage areas or stockrooms . . ." Nevertheless, the question regarding the need for a secure storage area has been a subject of discussion and various interpretation. As background concerning the need for this change, we are reprinting verbatim the discussion from the first edition of The ISO 9000 Handbook, *which references the 1987 standard, concerning this issue.*

ISO 9001 Subclause 4.15.3, Storage, requires suppliers to provide secure storage areas or stockrooms to prevent damage or deterioration of product pending use or delivery. Does secure mean a locked area, or does it merely refer to proper training and discipline for appropriate personnel?

All four experts agreed that use of the word "secure" does not necessarily connote a locked room. Ian Durand, president, Service Process Consulting, maintains this interpretation is also in keeping with proposed revisions to ISO 9001 and 9002 external quality assurance standards. Those revisions may ultimately reduce the range of interpretations.

Charles McRobert, president, Quality Practitioners, Inc., said there is no shortage of expert opinions on the topic. "There are as many interpretations of secure storage as auditors," he said. "Some look for walls, gates, padlocks, and guardian staff to dispatch product."

Unfortunately, traditional interpretations lead to a clash of philosophies in auditing a facility where just-in-time is employed. "Efficiency in today's manufacturing world requires a more sophisticated approach to the locked stockroom of yesterday," McRobert said.

Good Business Practices

"As an auditor, I accept both the traditional approach or the open concept. My approach to either case is to determine if the employees are following the concepts of material management as determined by their company and are exercising the self-discipline required to make any system successful," he said.

Durand supported this view and emphasized that the ultimate arbiter for interpreting the standards must be good business practice. If a proposed interpretation or response does not support prosperity of the company in the long term, it usually means the standard's intent is being misconstrued or viable alternatives have been overlooked. Durand said this basis for interpretation is clearly stated in Section 5 of ISO 9000: "In

(Continued on next page)

(Continued from page 142)

both [contractual and noncontractual] situations, the supplier's organization wants to install and maintain a quality system that will strengthen its own competitiveness and achieve the needed product quality in a cost-effective way."

Joseph Tiratto, consultant, Joseph Tiratto and Associates, Inc., said the storage clause does not specifically require storage areas or stockrooms to be locked. However, regulatory requirements may specify that hazardous material be kept in locked areas.

Durand noted that the clause applies throughout the product realization process, starting with receipt of incoming material, parts, or subassemblies and continuing until the product is ready for delivery. Response to the requirements of 4.15.3 must be integrated with those of related clauses, such as inspection, test status, and control of nonconforming product.

Tiratto said the storage clause merely requires storage areas or stockrooms to be so designated and be equipped and operated to prevent product damage or deterioration prior to use or delivery. "Control procedures shall stipulate the methods and authority for receipt and dispatch to and from these areas," Tiratto said. "Procedures are also to include the control and shipment of a product that has a shelf-life, expiration date, and any environmental condition. Storage employees are to be included in training procedures established in accordance with ISO 9001, Clause 4.18."

Robert Bowen, president, r. bowen international, inc., agreed that a locked storage area is not required. He interpreted the essential features of secure storage to mean defined, controlled material flows and accurate inventories and work practices aimed at preventing damage or deterioration of inventory.

Advice to Companies

The intent of the clause is to ensure that appropriate items are available when needed, according to Bowen. In developing practices to prevent damage or deterioration, he said, many companies could take advantage of existing cycle count programs, which randomly sample inventory accuracy by requiring a clerk to count selected part numbers. This is one of many acceptable ways to fulfill the requirement to "assess the condition of product in stock at appropriate intervals."

"The duties of this clerk can easily be expanded to accommodate ISO 9000 requirements," Bowen said. "As the clerk counts inventory, he can physically examine stock for damage or deterioration. Upon completion, a simple notation on the cycle count record indicates the status of damage or deterioration. These cycle count records become the basis for corrective action of damaged goods and act as the quality record to demonstrate achievement of Subclause 4.15.3."

4.16 CONTROL OF QUALITY RECORDS

ISO 9001 Requirements

The supplier is required to do the following:

- Establish and maintain documented procedures for handling, maintaining, and disposing of quality records (including pertinent subcontractor quality records).

- Store records effectively and prevent loss or damage.

- Establish and record retention times of quality records.

- Make quality records available for evaluation by the customer or its representative.

All quality records shall be legible and identifiable to the product involved. In a note, the standard emphasizes that records can be hard copy, electronic, or any other media.

ISO 9000-2 Guidance

The purpose of quality records is to demonstrate required quality and the effectiveness of the quality system. Quality records are referred to throughout ISO 9001. Effective quality records contain direct and indirect evidence that demonstrates whether the product or service meets requirements.

The records should be readily accessible. They may be stored in any suitable form, either as hard copy or on electronic media.

Sometimes customers may be required to store and maintain selected quality records that attest to the quality of products (including services) for a specified period of the operating lifetime. The supplier should provide such documents to the customer.

International standards do not specify a minimum time period for retaining quality records. Suppliers should consider the following:

- Requirements of regulatory authorities.

- Product liability and other legal issues related to record keeping.

- Expected lifetime of the product.

- Requirements of the contract.

Aside from these considerations, retaining records five to seven years is common practice.

ISO 9004-1 Guidance

Clause 17.2, Quality Records, in ISO 9004-1 gives the following examples of quality records that require control:

- Inspection reports.

- Test data.

- Qualification reports.

- Validation reports.

- Survey and audit reports

- Material review reports.

- Calibration data.

- Quality-related cost reports.

QS-9000 Requirements

 Records retention:

- Production parts approval, tooling records, and purchase orders must be retained one calendar year after time requirements for production and service. These must include production parts approvals, tooling records, and purchase orders.

- Quality performance records are to be retained one year beyond the year created.

- Internal quality systems audits and management reviews are to be retained for three years.

- The above record retention requirements do not supersede customer or government regulations.

- The specified periods are "minimums."

Superseded parts:

- The "new part" file is to contain documents from superseded parts if they are required for a new part qualification.

Clause 4.16: Control of Quality Records

The ISO 9001 standard continually refers to the need for records to demonstrate completed actions. Clause 4.16 does not call for the creation of additional records, but points out that records must be identified, sorted, stored, and maintained in a manner that makes them easily accessible.

(Continued on next page)

(Continued from page 145)

The difference between the quality records filing system and any other in a business is that a great deal of the information contained in these records should be decentralized to be instantly available when required. This means that the maintenance of the defined quality records will need to be monitored by the process of internal audit to ensure that everyone is doing what they should. It is easy to become paranoid about fulfilling the requirements of this element, which are quite basic.

Contrary to popular belief, the clause does not require fireproof safes, bank-vault storage, microfiche, or other similar methods of storage. The level of protection required for records depends upon the nature of the business, as well as any contractual or statutory requirements. The records connected with the construction of a nuclear power station need to be kept somewhat longer, and under more secure conditions, than the records for making a compact disc.

Some registrars have definite ideas about record-retention times. It is a good idea to establish whether the company's record-retention policy agrees with the registrar's demands during an initial meeting. The record-retention policy should be clarified and agreed upon before the audit team arrives on-site.

Procedures for Access

The methodology in use must provide for appropriate access to records as required. Specific references may be required for access arrangements, particularly when records have to be maintained for extended periods of time or in compliance with specific contractual requirements.

The Victoria Group

Clause 4.16: Electronic Control of Documents

How can electronic documents be controlled to meet the requirements of ISO 9000 series standards?

All four experts agreed that generating and tracking documents electronically is allowed by the ISO 9000 series standard, according to the note in Clause 4.16 that states "Records may be in the form of any type of media, such as hard copy or electronic media."

Advantages of Electronic Media

Robert D. Bowen, president, r. bowen international, inc., said that ISO 9000 series document control principles apply equally to all media; he also pointed out four distinct advantages of electronic media:

- Accuracy: Immediate on-line review of proposed changes by all knowledgeable persons and a transaction history file showing the date and nature of changes.

- Authenticity: Secure sign-on functions to ensure controlled access to read and write functions.

- Completeness: On-line edit-checks that ensure all required information is complete before a document is released.

- Currency: Instantaneous removal of all obsolete documents. Uniform start-up of all concerned persons when initiating procedures or changes.

Other strengths noted by Bowen include immediate update of suppliers' documents through electronic data interchange. In addition, "many organizations find it easier to establish a planned review system of quality-related documents if those documents are on-line." He noted that setting up a database reminder to review documents at specified, agreed-upon intervals is easy to accomplish.

Charles McRobert, president, Quality Practitioners, Inc., said that he had worked with "nearly paperless" companies whose document-control systems were excellent. He agreed with Bowen's contention that electronic control of documents makes review and approval of documents and highlighting of changes easier.

McRobert noted that "some auditors with misguided zeal have requested hard copies of all controlled documents with approval signatures." He suggested that any company whose auditor suggests this approach "immediately seek relief" from this requirement.

Document Control

Ian Durand, president, Service Process Consulting, Inc., pointed out that controlled documents can include mechanical assembly drawings, circuit schematics, process

(Continued on next page)

(Continued from page 147)

flowcharts, physical reference samples, pictures of reference samples, or videotapes illustrating proper work methods.

He agreed that using electronic media for document control has many distinct advantages. He said the complexity of such systems should not be a roadblock, pointing to the demanding access and control requirements of electronic fund transfer financial-control systems and security and administration systems.

Durand noted that a number of software programs are currently available to handle documentation and quality records of an ISO 9000-based quality system.

Joseph Tiratto, president, Joseph Tiratto and Associates, Inc., agreed that documents can be generated electronically in accordance with ISO 9000-2, *Quality management and quality assurance standards-Part 2: Generic guidelines for the application of ISO 9001, ISO 9002 and ISO 9003.*

He noted, however, that in accordance with ISO 9001, Clause 4.16, Control of quality records, document control should include written procedures that describe the following:

- How documentation should be controlled.

- Who is responsible for the control.

- What is to be controlled.

- Where and when the control is to take place.

Tiratto said the procedures should also include back-up provisions for electronically stored documents that are readily retrievable. He said back-up provision can also be electronically processed.

4.17 INTERNAL QUALITY AUDITS

ISO 9001 Requirements

The supplier is required to do the following:

- Establish and maintain documented procedures for internal quality audits of the quality system.

- Schedule audits according to the status and importance of the activity.

- Carry out audits according to documented procedures.

- Record audit results and communicate them to the appropriate personnel.

- Perform timely corrective action.

- Record the effectiveness of the corrective action in follow-up audit activities.

Internal quality audits are to be carried out by personnel independent of those directly responsible for the activity being audited. In a note, the clause reminds companies that the results of internal quality audits form an integral part of management review activities. The clause also refers companies to the guidance included in ISO 10011.

ISO 9000-2 Guidance

The purpose of an audit is to make sure the system is working according to plan, to meet regulatory requirements, or to provide opportunities for improvement. Auditors should be selected and assigned according to the criteria contained in ISO 9001, Subclause 4.1.2.2.

Internal audits may also be initiated for other reasons, including the following:

- Initial evaluation of a system for contract reasons.

- When nonconformities jeopardize the safety, performance, or dependability of the products.

- Verification of corrective actions.

- Evaluation of a system against a quality system standard.

ISO 9004-1 Guidance

Clause 5.4, Auditing the quality system, in ISO 9004-1 suggests that companies formulate an appropriate audit plan that covers:

- Planning and scheduling the specific activities and areas to audit.

- Assigning qualified personnel to conduct audits.

- Documenting procedures for carrying out the audit, reporting the results and agreeing on corrective action.

Companies should submit documented audit findings, conclusions, and recommendations to management on reporting and follow-up matters. The items that should be covered include:

- All examples of nonconformities or deficiencies.

- Appropriate and timely corrective action.

QS-9000 Requirements

Inclusion of Working Environment: Internal audits shall include working environment as an audit element.

Clause 4.17: Internal Quality Audits

Internal quality audits are the mainstay of system conformance. The quality system audit is a powerful tool for continuous improvement. The standard requires a planned, systematic, ongoing process of audits to ensure that the documented system is effectively implemented and that corrective actions are taken in a timely manner. The audit process ensures that the system is working as planned and that corrective action is taken when it is not.

It is normal to expect that every area will be audited at least once a year, with areas which receive bad audit reports receiving more frequent scrutiny. The full audit plan should be properly documented-as should the audit reports-complete with details of the effective implementation of any corrective and preventive actions. Follow-up activities are required to be performed, which "shall verify and record the implementation and effectiveness of the correction action taken." In a note, the standard adds a reminder that "the results of internal audits form an integral part of management review activities." This reinforces the requirement of Clause 4.14, Corrective and preventive action, to refer system changes resulting from corrective and preventive actions to management review.

The Victoria Group

Formal Training Requirements for Auditors?

ISO 9001, Clause 4.17, Internal quality audits, has been interpreted to mean formal audits by staff trained to ISO 10011 by outside trainers. Does an actual requirement for formal training exist?

No specific requirement for training exists, according to both Ian Durand, president, Service Process Consulting, Inc., and Robert Bowen, president, r. bowen international, inc. Durand points out that the only explicit requirement found in ISO 9001, 9002, or 9003 relating to auditors is found in Subclause 4.1.2.2, which states that the supplier shall ". . . assign trained personnel for management, performance of work, and verification activities including internal quality audits."

Bowen cites Clause 4.18 in ISO 9001 and Clause 4.17 in ISO 9002. He said the qualification requirement indicates that the appropriate education, training, and/or experience is needed. He said the "specific definition of these categories must be determined by each organization."

(Continued on next page)

(Continued from page 150)

Joseph Tiratto, president, Joseph Tiratto and Associates, Inc., agrees that ISO 9001 does not specifically require that internal audit staff be trained by outside trainers. He points to Clause 4.18 and Subclause 4.1.2.2 of ISO 9001 to support his argument. Tiratto noted that several organizations have developed internal auditor training courses. *He said the Institute of Quality Assurance in the United Kingdom is in the process of establishing requirements for certification of internal auditor training courses.*

Basic Requirements

Bowen listed the basic requirements of most internal auditors. The list includes the following:

- Demonstrated understanding of the general structure of quality systems.

- Demonstrated understanding of a company's proprietary quality system.

- Professional understanding of auditing techniques such as audit planning and audit checklists.

- Excellent communication skills.

He said that most organizations are unlikely to have sufficient in-house understanding of the ISO 9000 series standard, so formal outside training is usually required.

Additional Requirements

Durand emphasized that additional audit staff requirements could be a good decision for some companies. "A company might decide that auditing practices described in ISO 10011 might make business sense," he said. By the same token, Durand said that sending a few "well-chosen people to outside audit training makes good sense to assure that those performing internal audits have developed adequate auditing skills."

Durand concluded that the "cardinal rule in designing a quality system based on the ISO 9000 series is to act in the best long-term interests of the business." He said going beyond the minimum requirements of the standard may be the best business decision.

4.18 TRAINING

ISO 9001 Requirements

The supplier is required to do the following:

- Establish, maintain, and document procedures to identify training needs.

- Provide appropriate training for all personnel performing activities affecting quality.

- Maintain records of training.

ISO 9000-2 Guidance

Training is essential to achieving quality. Training should encompass the use of and underlying rationale for the quality management approach of the supplier. The training process should include the following:

- Evaluate the education and experience of personnel.

- Identify individual training needs.

- Provide appropriate training, either in-house or by external bodies.

- Record training progress and update to identify training needs.

ISO 9004-1 Guidance

Clause 18.0, Personnel of ISO 9004-1, discusses the training, qualification, and motivation of personnel. Companies should consider training "all levels of personnel within the organization performing activities affecting quality." This includes "newly recruited personnel and personnel transferred to new assignments."

The various levels of personnel within a company require specialized training. Executives and management require training in the "understanding of the quality system together with the tools and techniques needed" to operate the system.

Training for technical personnel "should not be restricted to personnel with primary quality assignments, but should include assignments such as marketing, procurement, and process and product engineering." Process supervisors and operating personnel should receive thorough training, including instruction in the following:

- Proper operation of instruments, tools, and machinery.

- Reading and understanding documentation.

- Relationship of their duties to quality and safety in the workplace.

- Certification or formal qualification in specialized skills (such as welding).

- Basic statistical techniques.

Clause 18.3, Motivation, looks at efforts to motivate all personnel in the organization. An effective motivation program focuses on elements such as the following:

- Communicating to all employees an understanding of their tasks and the advantages of proper job performance.

- A continuous quality-awareness program.

- A mechanism for publicizing quality achievements and recognizing satisfactory performance.

QS-9000 Requirements

Training as a strategic issue:

- Since training affects everyone, it should be considered as a strategic issue.

- Training effectiveness is to be evaluated.

Clause 4.18: Training

The training requirements of ISO 9001/9002 are very general and take the global view of quality—that is, every person in any company performs "activities affecting quality." The choice of training required is made on the basis of appropriate education, training, or experience. The company is left to decide what is appropriate, with the exception of certain regulated areas where there are statutory requirements for training.

The company is then asked to record its training decisions, follow them up, and make sure that the training continues to be appropriate throughout the individual's career. Many companies do an extremely good job of providing extensive employee development training that is carefully and completely documented, but then forget about the need to support a person's current activities with appropriate training. The standard is specifically interested in the training related to those activities currently being performed by the employee. It is most important to capture and record the details of "on-the-job" training. The basic purpose is to ensure that the next person to be trained for that same task receives the same training, thereby ensuring consistency in performing the task.

The Victoria Group

Documenting Personnel Qualifications

Most quality systems (especially to ISO 9000) are adopted after a company has been in business for quite some time and where employees have been working at a job, possibly for years. How should personnel qualifications be documented as per Clause 4.18?

The panel of experts agreed that companies have a certain amount of discretion in documenting personnel qualifications. Ira Epstein, vice president of Government Services, STAT-A-MATRIX, mentioned the following two aspects of documentation that must be considered:

- A requirement for written procedures that identifies training needs.

- A requirement for records of training that has been accomplished.

Identifying Training Needs

Training needs may be dictated by external sources such as government or professional societies, Epstein said. Examples include training requirements for workers in the nuclear reactor industry or workers in the medical industry. Other training requirements may be established by the customer in a contractual situation.

"Contracts should be reviewed carefully to determine if training requirements are specified," he said. An example might be the requirement for training in nondestructive inspection cited in many Department of Defense contracts. "Many of the training requirements from these two sources often require certification and recertification of personnel," he said. The most common source of training needs comes from internal management, which typically is most familiar with the job and with any necessary training, he noted.

Raymond P. Cooney, Ph.D., of Cooney and Mori Associates, said the supplier determines the appropriate mix of education, training, and experience needed for personnel to perform certain tasks and to ensure that personnel are properly qualified. He said that companies frequently upgrade or "tighten-up" training and qualification procedures when implementing ISO 9001 or 9002. "If employees' experience on the job qualified them without needing to go through the new system, a note to that effect in the employees' personnel files or other appropriate place would be adequate documentation."

Experience as a Substitute for Education and Training

Epstein noted that experience may be substituted for education and training at the discretion of management, provided this does not conflict with education and training requirements imposed by customers and other external sources. In cases where experience is accepted as a substitute, training records must be appropriately noted, possibly by a brief narrative of the person's experience.

(Continued on next page)

(Continued from page 154)

"The subject of training must be considered in context with the rest of the quality system and not in isolation," Epstein said. "If an audit of the quality system is being accomplished, training should be one of the last elements of the system to be audited. It would be difficult if not impossible to make a valid judgment of the adequacy of training without making a judgment of the adequacy of the rest of the quality system."

Epstein added that training procedures, training, and related documentation may appear satisfactory but will raise an auditor's suspicions if there are high levels of nonconformities in other system problems.

Training and Testing Criteria

Dean Stamatis, Ph.D., president, Contemporary Consultants Co., said the intent of the clause is to establish requirements for training, certifying, and recertifying employees involved in performing critical and specialized functions at a given organization. "It covers all employees performing routine, critical, and specialized functions related to deliverable items. This includes both the management and nonmanagement personnel. In addition to this general impact the clause may also include requirements specified by the customer's contract."

"As part of the audit and the Clause 4.18 requirements it is also essential to look at the responsibility for developing training and testing criteria," Stamatis said. "Here, the auditor looks for consistency in operating the training sessions for normal/generic, special, certification, and recertification sessions."

Stamatis said the criteria for such an evaluation probably is contained within the quality manual under the heading for the quality management and administration department. He said auditors may also be interested in addressing customer-imposed requirements while auditing this section.

Records of Training

"The format for both procedures and records are not specified in ISO 9001," Epstein said. "Format is determined by the supplier to satisfy his needs, his customers' needs, and the needs of society. The extent of documentation and the format should be as simple as possible and yet satisfy the above needs."

Dean Stamatis lists the following information as typical with respect to appropriate records:

• Employee's name and identification number.

• Employee's department.

• Date and duration of training.

(Continued on next page)

(Continued from page 155)

- A check mark for certification or recertification.

- Location of training.

- Type of training.

- Name of the course.

- Certification number.

- Function of the certification.

- Expiration date.

- Name of the instructor.

Auditors may also be interested in viewing records and verification documents of training for employees who are unable to attend or be present at a scheduled session. "Although the concerns of Clause 4.18 may be very frightening to some companies, in reality they are nothing more than a substantiation of the quality system that the company itself has defined in its own quality manual," Stamatis said.

According to Raymond Cooney, "the system for keeping qualification records need not—should not—be rigid or onerous. The simplest, easiest, most flexible system that works is the best. Qualification records can take many forms."

Cooney said that qualification records must have credibility and utility. "This implies two things," he said. "Responsible people must sign off on the records. Oftentimes this means both the employee and appropriate management. Secondly, records need to be readily accessible to those who need to use them."

Good business practice, he said, dictates that a supplier be able to answer the question: How do we know that a person doing or being told to do a job is qualified to do it?

According to Stamatis, the following documentation or procedures are required before verification: a quality assurance certification or recertification notice, a quality certification record, a quality certification card, and a list of attendees for training. He said those documents usually can be found in the trainee's department, the training department, or the quality department if the training is quality-related.

Job Descriptions

Kirk Eggebrecht, executive manager, Geo. S. Olive & Co. LLC., said it is important to consider any single element of ISO 9001 in the context of the document as a whole. "The supplier's specific procedures for addressing training and work assignments will be a function of the approach used to address other ISO 9001 requirements, such as Design control (4.4), Process control (4.9), Internal quality audits (4.17), and many others," he said.

(Continued on next page)

(Continued from page 156)

"In addressing these requirements, many suppliers conclude that job descriptions are needed for each unique position in the organization." Typically, he said, these job descriptions prescribe the specific skills, training, and experience required for that position. "The make-up of these job descriptions then determines the make-up of the information needed in each employee's personnel and training file," he said.

"Naturally, the key to any quality system is that there is a match between the job requirements as documented and the employee's qualifications as documented," Eggebrecht said. "If there is a mismatch, the supplier should have procedures implemented to provide the appropriate training and supervision until the prescribed qualifications are met."

In the case of employees who have been performing a job for years without any formal training, he said, registrars allow suppliers to "grandfather in" experienced workers, exempting them from prescribed training or educational requirements. However, to qualify for this treatment, the employee must have demonstrated a proven capability.

New Employees

New employees should be trained by a designated instructor, and that training should be documented. There should be a sign-off procedure authorized by a designated person or position, who may or may not be the instructor, Eggebrecht said. When hiring skilled employees, companies must have specific procedures that address the employee's placement in the organization along with a transition period for them to become familiar with the position.

"The same amount of care, control, and effort should be taken in recruiting and transitioning new employees as is prescribed in training and transferring existing employees," he said.

4.19 SERVICING

ISO 9001 Requirements

With the 1994 revisions to the ISO 9000 series standard, the clause on servicing is also included in ISO 9002. The basic requirements call on the supplier to do the following:

- Establish and maintain documented procedures for servicing (when required by contract).

- Verify and report that servicing meets specified requirements.

ISO 9000-2 Guidance

In planning procedures for servicing, suppliers should:

- Clarify servicing responsibilities.

- Plan service activities (supplier or externally provided).

- Validate design and function of necessary servicing tools and equipment.

- Control measuring and test equipment.

- Provide suitable documentation and instructions.

- Provide back-up technical advice, support, and spares or parts supply.

- Provide competent, trained service personnel.

- Gather useful feedback for product or servicing design.

ISO 9004-1 Guidance

Clause 16.2, After-sales servicing, notes that instructions for products should be comprehensive and supplied in a timely manner. Instructions should cover assembly and installation, commissioning, operation, spares or parts lists, and servicing of products. In the area of logistical back-up, responsibility should be clearly assigned and agreed among suppliers, distributors, and users.

QS-9000 Requirements

Feedback of Information from Service: Information from service shall be communicated to manufacturing, engineering, and design.

4.20 STATISTICAL TECHNIQUES

ISO 9001 Requirements

The standard includes two subclauses: 4.20.1, Identification of need, and 4.20.2, Procedures. The requirements include the following:

- Identify the need for statistical techniques to establish, control, and verify process capability and product characteristics.

- Establish and maintain documented procedures for statistical techniques.

ISO 9000-2 Guidance

Statistical techniques are useful in every aspect of an organization's operation. Useful statistical methods include:

- Graphical methods to help diagnose problems.

Clause 4.19: Servicing

Quality applies to after-sales servicing just as much as to the prime supply of any goods or services. So the standard requires that procedures be established for ensuring that the specified requirements of the servicing operation are achieved. These may be specified by direct contract, implied contract, or statutory provisions. The complexity of the system created to control and monitor the service function will depend upon how significant a portion of a business it may be, and can range from a simple document to cover the occasional after-sales return to a complete, independent quality system to cover the postcontract service and maintenance of a product through its entire life cycle.

The requirement is to have documented procedures for handling the activity, determining the requirements to be met by products before being returned, and maintaining records of the activity. These records are also referenced in Clause 4.14 as a potential source of data for identifying possible preventive activities.

The Victoria Group

- Statistical control charts to monitor and control production and measurement processes.

- Experiments to identify and quantify variables that influence process and product performance.

- Regression analysis to provide quantitative models for a process.

- Analysis of variance methods.

The documentation resulting from these methods can be used to demonstrate conformance to quality requirements.

ISO 9004-1 Guidance

In Clause 20.0, Use of statistical methods, ISO 9004-1 suggests that the application of statistical methods may include:

- Market analysis.
- Product design.
- Reliability specification, longevity/durability prediction.
- Process control/process capability studies.
- Determination of quality levels/inspection plans.

- Data analysis/performance assessment/defect analysis.
- Process improvement.
- Safety evaluation and risk analysis.

QS-9000 Requirements

Selection of Statistical Tools: Statistical tools to be used are to be identified during quality planning, and shall be included in the control plan.

Knowledge of Basic Statistical Concepts:

- As appropriate, concepts such as variability, control (stability), capability, and overadjustment should be understood.
- Consult the *Fundamental Statistical Process Control* Reference Manual.

Clause 4.20: Statistical Techniques

The thoughtful use of statistical processes can result in significant benefits. What is most often overlooked is that Clause 4.20 not only refers to the commonly recognized techniques of statistical process control (SPC) and inspection sampling, but also to all other analysis techniques that are statistically based-from the simple Pareto chart through to the most sophisticated design tools like finite element analysis and failure mode and effects analysis (FMEA).

The standard requires that the method used be appropriate and valid. It is the latter requirement that is most often not met, particularly in the use of statistical sampling techniques. The auditor should establish that any sampling being performed is carried out against known risks.

There is no requirement for the use of statistical methods, but when they are used, it is essential that staff involved in their use understand what they are doing and why. This will often require training.

Remember that as statistical techniques will be used throughout the organization, it will be necessary to ensure that there is some formal identification process that has a global effect. In many cases, it may be sufficient to attach to certain jobs the responsibility for identifying a need for statistical techniques and then defining the techniques to be used.

The Victoria Group

4.20.1 Identifying the Need for Statistical Techniques

Clause 4.20.1, Identification of need (statistical techniques), of the ISO 9001 external quality assurance standard requires companies to identify the need for statistical techniques for establishing, controlling, and verifying process capability and product characteristics. What does this entail?

The panel of experts agreed that each company must determine what type of statistical techniques, if any, are appropriate based on customer requirements, industry practice, and overall cost in relation to the product.

Ira Epstein, Vice President for Government Services with Stat-A-Matrix, an ISO 9000 training and consulting firm, said most companies rely on some form of statistical techniques.

"Since it is practically impossible for most organizations to produce a product or service of any complexity without using statistical techniques, this paragraph is generally applicable in most suppliers' operations," he said. Epstein said company management must review all activities within the organization and then determine if the appropriate statistical techniques are being used.

"In this approach, the supplier has identified the fact that he is using statistical techniques and also has identified the need for such," according to Epstein. "Once these statistical techniques are identified through this search of actual operations, then the supplier is responsible for documenting these procedures. In many cases, suppliers are surprised to find that they actually employ more statistical techniques than they initially perceive."

Tim Purnell, president, Quality Certification Bureau Inc., an ISO 9000 registrar, said the use of statistical techniques varies by industry.

"Obviously, the application of statistical techniques is going to be a little bit different in the production of different industries," he said. "What the auditors will do is to assess that from the determination of the supplier and make sure that (the statistical techniques used) are applied properly."

Purnell said companies should look for areas in which statistical techniques will assist them to ensure that their products or services meet customer requirements at a reasonable cost with relation to the product. "Obviously, there's our obligation as auditors to make sure they are meeting industry standards."

Robert Kozak, certification manager, Entela Inc., Quality System Registration Division (QSRD), an ISO 9000 registrar, agreed that the use of statistical techniques should be determined by customer requirements.

(Continued on next page)

(Continued from page 161)

He suggested that companies consult the ISO 9004-1 guidance standard to gain a better understanding of the requirements with respect to statistical techniques. Typically, he said, companies employ these techniques on product designs to predict reliability, longevity and durability, as well as process capability. "Another [use] may be for a determination of quality levels and sampling plans," he added. "And a very important one is safety evaluation and risk analysis when looking at regulated products."

Robert Sechrist, an auditor with TUV Essen, an ISO 9000 registrar, said auditors look for objective evidence that companies have conducted a review of their statistical techniques.

"We usually look for what they have in place," he said. "If they don't have any statistical techniques in place, we discuss with them how they determined that they don't need them." Sechrist said auditors routinely ask for company procedures regarding statistical techniques.

"We would look at whatever technique it is and make sure that it is properly used," he said. "We have run into companies that misused the control charts, for example. They did not calculate the parameters properly. It failed this clause of the standard."

To some extent, he said, the identification of needs also reflects on a company's training program. "The standard says the supplier shall identify the need," he explained. "It's important for the company to know what they're looking for."

Note: Additional QS-9000 requirements are contained in three supplemental clauses:

- Production part approval process
- Continuous improvement
- Manufacturing capabilities

These requirements, plus company specific requirements, appear in Chapter 15, beginning on page 578.

*The paraphrased QS-9000 requirements in Chapter 4 are taken from *The Memory Jogger*™ *9000* published by GOAL/QPC, 13 Branch Street, Methuen, MA 01844, (800) 685-3900, Robert W. Peach and Diane Ritter authors.

THE REGISTRATION AND AUDIT PROCESS

Chapter 5   Registration and Selecting a Registrar

Chapter 6   The Audit Process

Registration and Selecting a Registrar

The two articles in this chapter will help you decide which registrar to choose and will provide a basic understanding of the ISO 9000 registration process.

The first article, written by Bud Weightman, will help you assess the positive and negative aspects of each registrar you are considering. Mr. Weightman's article will help you choose a registrar that knows your industry and is competitively priced. The guidelines and questions address the following issues:

Making the registration decision.

Marketplace considerations.

Registrar fees.

The importance of a registrar's financial stability.

Conducting background checks on registrars.

Auditor qualifications.

Audit team selection.

Getting a list of the registrar's clients.

Halting the registration process.

The second article, written by Elizabeth Potts, president of ABS Quality Evaluations, Inc., describes the registration process and includes the following topics:

Completing your application.

Documentation review.

The preassessment.

The actual assessment.

Achieving registration.

Maintaining registration through surveillance.

The costs and time involved.

HOW TO SELECT A REGISTRAR®

by R. T. "Bud" Weightman

INTRODUCTION

ISO 9000 quality system registration has skyrocketed since the early 1990s. More than 127,000 registrations have been issued internationally with at least 10,000 registrations recorded in North America as of the summer, 1996. (See Chapter 19 for more statistics on the growth of registrations.)

As the number of companies seeking ISO 9000 registration has grown, so have the ranks of third-party ISO 9000 registration services. What does this choice of services mean for a company in the market for a registrar?

The best advice—be cautious! Choosing a registrar should be viewed as a long-term commitment, because registrars usually require clients to sign a three-year contract. Asking the right questions will ensure that your registrar is appropriate for your business. ISO 9000 registrars appreciate better than most that an ounce of preventative action is worth a pound of cure. This chapter will help your company make an informed decision.

Registrars Are Part of the System

The International Organization for Standardization (ISO) developed ISO 9000 in large measure to meet a demand by business seeking a voluntary, internationally accepted quality management standard. Your accredited registrar is part of this international system of consensus standards and guidelines that ensure companies fully comply with the requirements of the standard. Choosing a third-party registrar from the more than 70 U.S.-based companies requires careful consideration of your company's goals.

The first consideration is accreditation status. Registrars, like the companies they audit, face the scrutiny of an audit by an accreditation body. In the United States, the accreditation body is the Registrar Accreditation Board (RAB) (see Chapter 17 for more information on accreditation issues). In Europe, registrars may seek accreditation from a handful of bodies, including the Dutch Council for Accreditation (RVA) in the Netherlands and the United Kingdom Accreditation Service (UKAS) in the Great Britain. The RAB and other accreditation bodies evaluate registrars principally against the requirements of a European standard called European Norm (EN) 45012, "General Criteria for Certification Bodies Operating Quality System Certification" (see Figure 5–1 for an explanation of these criteria).

Figure 5–1

EN 45012 Requirements

Accredited registrars must meet the requirements of The European Norm (EN) 45012, "General Criteria for a Certification Body Operating Quality System Certification," which describes how an entity offering quality system registration services must operate. For a detailed description of these requirements, see Chapter 17.

The requirements describe all aspects of activity from organizational structure to the withdrawal of registration certificates. Here are some highlights of these requirements. Registrars must:

• Operate and administer the service in a nondiscriminatory manner.

• Be impartial.

• Describe the responsibility and reporting structure of the organization and in particular the relationship between the assessment and certification functions.

• Provide a documented statement of its certification systems, including its rules and procedures for granting certification.

• Employ competent personnel and keep training records.

• Maintain a system for the control of all documentation related to the certification system.

(Continued on next page)

(Continued from page 167)

- Maintain and store records appropriately.

- Maintain documented procedures and quality systems.

- Maintain a quality manual and quality policy statement.

- Ensure confidentiality.

- Maintain a list of registered suppliers.

- Have an appeals procedure.

- Conduct internal audits.

- Control the use of its certificates.

- Maintain records of complaints from suppliers and maintain procedures for withdrawal of certificates.

Accredited versus Nonaccredited Registrars

A few companies offer ISO 9000 registration services, but have not been formally accredited by any international body such as the RAB, RVA or UKAS. While these nonaccredited registrars may offer technically adequate services, your customers are the real arbiters of the value of these certificates. Here is a list of questions to ask these registrars:

What are the benefits of selecting a nonaccredited registrar? How many companies have found value in using a nonaccredited registrar and how many companies have switched from a nonaccredited registrar to gain the marketplace benefits of an accredited registrar?

Another factor to consider is a notified body status (see Chapter 16 for additional details on the European Union). Some registrars have been selected or "notified" by the appropriate European Union member state to ensure that certain products meet specific performance requirements described in regulations called directives. Not all products fall into this strictly controlled category. However, if the product your company manufactures requires a "notified body" to perform the ISO 9000 registration audit, then these questions are appropriate for your registrar.

Finally, not all registrars offer or plan to offer registration to new or developing standards and codes of practice. Some of these standards and/codes of practice include:

- ISO 14000 for environmental management.

- QS-9000 requirements for the automotive industry.

- The Food and Drug Administration's Code of Good Practice requirements for medical devices.

- ISO 13485 and ISO 13488 for medical devices.
- Occupation health and safety management systems.

SELECTING AN ACCREDITED REGISTRAR

The importance of selecting a registrar cannot be overemphasized. The agreement is a long-term commitment, and changing registrars can be costly. You should give the project of choosing your registrar the attention it deserves. Here are some suggestions that will give you a good start toward selecting an appropriate registrar:

- Include adequate time for registrar choice in your overall plan for ISO 9000 registration.
- Obtain a listing of all potential registrars (see Figure 5–2 for sources).
- Identify key questions to ask potential registrars, using this chapter as a starting point.
- Identify all possible cost factors and address these with a potential registrar.
- Finally, create your own weighted scoring mechanism for potential registrars based on the needs of your company, marketplace, and location(s). This article includes some suggestions on how to create this evaluation tool (see Figure 5–3).

The following questions are grouped into interest areas beginning with the registrar accreditation. These questions are not intended to address every company's marketplace needs. However, getting answers to these questions will help ensure that you make the best registrar choice possible.

Figure 5–2

Where to Get a List of Registrars

Registrar Accreditation Board (RAB)—lists only Registrars accredited by the RAB. Contact the American Society of Quality Control (ASQC) at 1-800-248-1946 and ask for the RAB list of accredited registrars.

The U.S. National Institute for Standards and Technology (NIST)—lists all known Registrars, both domestic and international, which are operating in the United States. The listing name is the "North American Quality System Registration Organizations" (NAQSRO) list. Contact the NIST at 301-975-4039.

Irwin Professional Publishing—the publishers of this book will provide a basic list of ISO 9000 registrars free of charge. Irwin also offers a complete description of all of North American registrars for a nominal charge. 703-591-9008.

Figure 5–3

How to Use Your Information

Once you have collected key information for your registrar choice, you'll need to put it in a form that will be useful. Here is one way to organize registrar responses.

1. List your key questions and answers in a matrix with each registrar's response in an adjoining column. Try using a word processing table, spreadsheet, or database format.

2. Create your own weighted scoring mechanism for the matrix based on your company's needs, its locations, marketplace, and other factors. Your best set of registrar choices should rise to the top of this list.

Accreditation Status

A registrar's accreditation(s) to perform ISO 9000 registrations may be vital to the value of your certificate. These basic questions will provide information you and your customers should know about your chosen registrar:

- What countries and/or entities have accredited the registrar? Is the registrar accredited by a single or multiple countries and/or entities? If the registrar's accreditation is from a single country or entity, what mutual recognition agreements are in place to recognize the registrar's registration marks?

- What additional costs are incurred when multiple registration marks are required?

- Does the European Union recognize the registrar's accreditation? Some customers may require a European-based accreditation.

- What other industry audits are performed and under what circumstances? If the registrar is contracted by an industry group to perform registration activities to ISO 9000 and other industry standards, is the industry group for whom the registrar works accredited?

Marketplace Questions

Depending upon the specific industry or marketplace for your product, the registrar choice may be narrowed down significantly by examining how your industry and specific customers value and use the ISO 9000 registration process. Some overall marketplace considerations include:

- Does your customer or industry require or prefer a registrar that specializes in a specific industry?

- Does an EU directive require registration by a notified body?

- Does the marketplace prefer a specific registrar for a geographic area?

Other marketplace questions are more difficult to answer. However, the value of registration is tied to its value in the marketplace, even if many internal benefits exist for companies seeking ISO 9000 registration.

Competitor Questions

Some companies choose the same registrar and ISO 9000 standard as their competition. This may or may not be appropriate for your business.

- Is registration to your competitor's scope of operations and ISO 9000 series number appropriate? For example, it would be a mistake to choose registration to ISO 9002 if product design is clearly part of your company's operations. If your competitor did choose ISO 9002 when ISO 9001 was clearly required, perhaps choosing ISO 9001 and another registrar will provide a better market advantage than your competitor.

Some customers (and even your competition) may be uncomfortable with registrars who offer separate consulting services despite the existence of international guidelines intended to prevent conflict of interest.

ISO/International Electrotechnical Commission (ISO /IEC) Guide 48 states in part, "an organization that, directly or through the agency of subcontractors, advises a company how to set up its quality system, or writes its quality documentation should not provide assessment services to that company, unless strict separation is achieved to ensure that there is no conflict of interest."

The auto industry's requirements for its suppliers, QS-9000 (see Chapter 15 for more on this requirement), state that "organizations that have provided quality system consulting services to a particular client are not acceptable as registrars for that client, nor may they supply auditors. This restriction includes subsidiaries or affiliates of the same parent company."

If a registrar is part of a larger company that offers both consulting and registration services, here are some key questions to ask:

- How are registration and consulting activities separated? Does a clear division of responsibilities exist between the services?

- Are any of the registrar's personnel involved in consultation, including governing board member(s), registration board member(s), principal(s), owner(s), or stockholder(s)

- Do contracted audit personnel also work as independent consultants? If so, how does the registrar deal with conflict of interest questions?

- Does the registrar occasionally use its consulting personnel to supplement its auditing staff?

- What specific guidelines ensure that the registrar and consulting activities are truly independent? Any policy should specifically state:

—That the registrar has no involvement in consultation activities.

—That individuals involved in the registration activity, including those acting in a managerial capacity, have not been involved in any consultancy activities with a specific supplier, or any company related to that supplier, within the last two years.

—That the company does not jointly market consultancy and accredited services and that marketing material, written or oral, does not give the impression that the two activities are linked. The policy should also state that the registrar will in no way imply that passing a registration audit will be easier or less expensive if its company consultancy services are used.

—That the registrar's assessors are not permitted to advise or give consultancy as part of an assessment.

Background Questions

Basic background information about a registrar plays an important part in registrar choice. Not all registrars may be qualified to work in your business area. While adding an industry scope is part of a registrar's business activity, don't assume that the registrar will be able to audit your company. A registrar's financial security or customer satisfaction record should also not be taken for granted.

Sometimes obvious questions are the most important. Here's a good starting list:

• How long has the registrar been in business?

• How many registrations has the registrar performed? Is a list available along with contacts and telephone numbers?

• Does the registrar provide potential clients a complete description of its registration system and provide policies on certificate suspension, withdrawal, and cancellation?

• How are the registrar's clients notified of rule changes, and are comments permitted? How long do clients have to implement changes?

• Does the registrar require notification of any applicable customer complaints?

• Will the registrar grant quality system registration to ISO 9001, 9002, or 9003? (Some registrars will not grant registration to ISO 9003.)

The standard industrial classification code (SIC) is used in the United States to identify business type. Registrars are accredited to work in specific business areas as defined by SIC codes. It is important to know that a registrar is accredited to provide registration services for your business area. Appropriate questions include:

• Is your company's business SIC included in the registrar's accredited scope of operations? What internal expertise does the registrar have to audit your industry sector? Will the registrar help its clients identify its proper SIC?

- What limitations exist to list your company under multiple SIC codes, and how will these listings be published?

Financial Security

A registrar's financial health is important to the long-term validity of your registration certificate. Here are some key questions:

- Does the registrar provide financial statements when requested?

- Does the registrar have contingency plans in the event of business failure? What additional costs might be expected in the event of business failure or bankruptcy?

Operational Concerns

- Does the registrar have a memorandum of understanding with other registrars? A few registrars have mutual recognition arrangements which may expand the reach of your registration.

- In which state or country is the registrar incorporated. This may affect future legal proceedings and claims.

- Does the registrar subcontract any of its registration activities to other organizations? If so, does this service follow the registrar's policies and procedures? You have to decide if this arrangement meets your needs.

- Does the registrar allow the use of its symbol or logo? What are the restrictions/requirements governing its use?

Current Customer Questions

- Was your registrar experience positive or negative?

- What would you change about the registration process?

- Do you expect to renew your contract with the registrar?

Internal Operations

Does the registrar have a confidentiality agreement with:

- Employees?

- Contract assessors?

- Subcontracted organizations/personnel?

- Members of its governing board?

- Members of its registration committee?

Registration Costs

Determining how much registration will cost your company depends on a number of factors. Some of the most basic cost considerations that will affect the cost of registration include:

- Size of an organization.
- Number of facilities.
- Number of employees.
- Type of business.
- ISO 9000 series number selected (i.e., 9001, 9002, 9003).
- Scope of registration, (i.e., one product, a product line, an entire product family).
- Location of a facility/facilities.
- Application fees.
- Preparation and initial visit.
- Review of quality system manual.
- Review of revisions to the quality system manual.
- Initial visit and number of auditors sent.
- Preassessment charge.
- Assessment charges.
- Number of auditors sent by the registrar specifically for the assessment.
- Registration charges.
- Report writing
- Surveillance fees.
- Listing fees.

Other Important Costs

Some cost questions are easy to determine, and you should have little trouble getting a definitive answer. However, getting cost estimates for other registration scenarios requires careful questioning:

- Are discounts available for registering multiple sites? How will these multiple site audits be performed? Will all facilities receive a stand-alone certificate? Will one certificate be issued for all locations? What are the dangers, if any, of using one certificate to cover multiple locations?

- What is the policy for revolving site assessments? A revolving site assessment could mean significant savings if your company is planning to register multiple sites. Revolving site assessments allow a registrar to register multiple sites after complete audits of only a few sites in the group slated for registration. The remaining sites are audited during periodic surveillances.

- What is the cost of modifying the scope of a registration?

- What is the cost of reassessment after the expiration of the original registration? Will the cost be the same as the initial assessment? Will it take as long as the initial assessment?

- Is surveillance included in the registration fee? How often will surveillance be performed over the life of the registration? How many quality system elements are covered during each surveillance? How long will each surveillance (or periodic inspection) last?

- How are clients charged for corrective action implementation follow-up visits?

- What are the cancellation charges if your company decides not to move ahead with registration?

- How will your company be billed? Is it by the man-day or by the hour? Is overtime included? Is travel time for auditors included in the cost estimates? What does the registrar consider to be reasonable travel and lodging expenses?

- Are the auditors all based in the United States or will international travel be required?

EVALUATING PREASSESSMENT

A preassessment can be one of the most valuable exercises prior to an actual registration audit. Make sure your company gets complete value out of the process. A useful preassessment should not be a simple "walk-through" that identifies only superficial quality system problems. Whether the preassessment is an internal audit function, performed by an external consultant, or performed by the same registrar handling the actual registration audit, you should evaluate the preassessment choice based on these factors:

- The number of ISO 9000 preassessments performed by the consultant or registrar and the business areas in which these audits were performed.

- The adequacy of training and the consultant's or registrar's auditing experience, including proper auditor certifications.

Companies sometimes request a registrar preassessment "to get to know the registrar" or "understand the registrar's point of view". Whatever the reason for selecting your preassessment method, the following should be considered:

- How does your customer view your choice of a preassessment body? Does your customer consider such activity a conflict of interest?

- Do you consider registrar recommendations important? If so, this would be a conflict of interest for a registrar, as recommendations could be perceived as consulting.

- Will problems areas found during preassessment get special attention during the actual registrar ISO 9000 audit?

- How does the registrar interpret the standard? Are these interpretations based on internal procedures or accreditation body recommendations? Do these interpretations serve the interests of your business?

- Do you expect to gain internal auditing experience and guidance on interpreting the standard during your preassessment? Will your registrar include role playing during a "mock" assessment?

- How do you expect the preassessment findings to be categorized? From most significant to least important? Will the registrar provide this list?

ASSESSMENT AND REGISTRATION

The number of days that your registrar spends auditing your facility can greatly affect the final cost. Registrars generally follow published guidelines which define the approximate amount of time the registrar should expect to spend on-site, including follow-up activities (see Table 5–1). Here are some other important questions concerning the actual registration process:

Table 5–1

Number of Employees	Initial Assessment (Man days)	Subsequent Annual Visits (Man days)	Reassessment Visits (Man days)
1–4	2	1	1.5
5–9	2.5	1	1.5
10–19	3	1	2
20–29	4	1.5	3
30–59	6	2	4
60–100	7	2	4
100–250	8	2.5	5
250–500	10	3	6
500–1000	12	4	8
1000–2000	15	5	10
2000–4000	18	6	12
4000–8000	21	7	14

- What is the registrar's pass/fail rate?

- What are the most common problems and reasons for failing an assessment?

- How far in advance must assessments be scheduled? When is the next date for scheduling an assessment? How long will the assessment take?

- What are the fees, if any, for canceling or postponing a scheduled assessment?

- How long will the certificates be effective (generally one to three years)?

- Will a controlled quality system manual be required for submission? How long will it take to review this document?

- How are clients notified of quality system omissions or deficiencies, and how much time is allowed for making modifications?

- Must quality system manual amendments be corrected and implemented prior to the final assessment?

- Once accepted, will you be required to submit a quality system manual for review and approval prior to making and implementing any revisions?

- Does a quality system have to be 100 percent implemented to receive registration?

- Will you be notified of any deficiencies in a quality system before the assessment team leaves the site? If so, will the notification be verbal or written?

- How much time is given to correct identified deficiencies?

- Will a reassessment or partial assessment be performed to verify corrective action implementation of deficiencies identified during the initial assessment?

- Will changes/revisions in a quality system manual necessitate a reassessment?

- Will a reassessment be required if a modification to the registration scope is requested?

- What is the frequency of the periodic surveillances? How many quality system elements are covered during each surveillance? Will a surveillance schedule be provided?

AUDITOR QUALIFICATION

Make sure you understand the registrar's auditor qualification and certification program. A recognized auditor qualification/certification program helps produce consistent audit performance for the suppliers. These schemes include:

- International Register of Certified Auditors (IRCA) (e.g., Provisional Assessor, Registered Assessor, or Registered Lead Assessor).

- American Society for Quality Control (ASQC)—Registrar Accreditation Board (RAB), accreditation scheme (e.g., quality system provisional auditor, auditor, or lead auditor).

The IRCA administered scheme is backed by the Governing Board of the UK National Certification Scheme for Assessors of Quality Systems. This system has worldwide recognition. The RAB scheme has been recently introduced and should parallel the IRCA scheme.

The International Auditor and Training Certification Association (IATCA) is developing an auditor certification program with the guidance of 12 ISO 9000 auditor certification and course accreditation bodies.

Another benchmark of auditor training is participation in a 36- to 40-hour ISO 9000 lead auditor training course. The course may or may not be registered to a national scheme (such as IQA or the RAB-administered registered course).

Here are some points to consider when evaluating a registrar's auditor certification requirements:

- Does the registrar require its auditors to be certified to a national or international scheme? If not, does the registrar's internal training program/certification follow a specific scheme that may or may not be affiliated with a national scheme?

- Is a registrar's quality system auditor acceptance based upon the ASQC Certified Quality Auditor (CQA) program? If so, the CQA certification should be followed by an RAB-approved 16-hour ISO 9000 series course followed by an approved test. (See Chapter 17 for more information on auditor certification issues.)

Audit Team

Questions to ask a registrar about its audit team include:

- What experience, training, and education are its auditors required to have, and are these verified? Does the training include ISO 9000, company procedures, and policies? Are the registrar's standards and auditor criteria recognized by the European Union?

- What responsibilities are given each member of an audit team, from the lead auditor to an auditor in training? Will at least one member of the team be familiar with your specific business?

- What rights do clients have concerning auditors? Are qualifications available for review? May clients request a change in the makeup of the audit team?

- Will the same audit team perform the initial audit as well as all subsequent surveillance? You should consider how this may affect customer perceptions of your registration. How thorough will these auditors be during future audits?

OTHER CONSIDERATIONS

Supplier List

If the registrar offers a list of suppliers or manufacturers it has certified, make sure the following questions are answered:

- How current is the list, and what information is included with it (scope should be included)? Does the registrar charge its clients for the list, and how should clients expect to receive the list?

- What SIC codes will be listed, and what organizations (in addition to the registrar) will publish the list?

Halting the Process

Although not common, the registration process may be stopped by either the client or the registrar for a variety of reasons. Before you begin the process, you should discuss this possibility with the registrar so that policies will be understood. Furthermore, sometime a registration certificate is withdrawn. Here are some key questions to ask:

- What is the registrar's policy regarding the suspension, withdrawal, or cancellation of the quality system registration? How are clients notified? Are these changes in registration status published? Will any pre-paid fees be returned?

- Will the registrar withdraw or cancel the quality system registration if a product, process, or service is not supplied for an extended period of time? Ask the registrar to define the rules.

Small Business Considerations

Small businesses have been concerned for years about the high cost of ISO 9000 registration. Ask your registrar if any special programs exist for small business. Generally, registrars offer a flat rate based on the number of employees or some other sliding scale program. However, these reduced costs do not mean that a small business audit is any less rigorous than a larger company audit or will require fewer auditors than needed to meet international guidelines. Here are the basic questions to ask:

- What are the restrictions of a fixed-fee registration, and is a fee schedule available listing company size and approximate costs? How many assessment days will be required, and what are the costs for follow-up assessments?

- How can other costs such as travel related costs for the auditor be reduced?

STEPS IN THE REGISTRATION PROCESS

by Elizabeth Potts

INTRODUCTION

This article describes what to expect during the registration process. Regardless of the registrar selected, registration to ISO 9000 generally consists of the following six basic steps:

1. Application

2. Document review

3. Preassessment

4. Assessment

5. Registration

6. Surveillance

This article also considers the time and costs of registration.

INTEGRATED MANAGEMENT SYSTEMS

When the first edition of the *ISO 9000 Handbook* was published, making a decision to become ISO 9000 registered was in some ways less complicated. Fewer than 200 ISO 9000 certificates had been issued in the United States. Companies seeking registration for the most part were doing so for very practical reasons. They could afford to keep their efforts very focused on achieving ISO 9000 compliance.

In 1996, this focus on a single standard may be a luxury for most companies operating in the global marketplace. The Big Three (Chrysler, Ford, and General Motors) have developed their own requirements based on ISO 9000 called QS-9000 (see Chapter 15). QS-9000 includes all of ISO 9001 (clauses 4.1 through 4.20 verbatim). While a company implementing QS-9000 is still required to register to ISO 9000, additional sector specific and individual automaker requirements must be met to comply with the requirements.

The ISO 14000 series of standards for environmental management, a close cousin of ISO 9000, is scheduled for publication in the third quarter of 1996. In addition to these standards, other complementary management standards are being developed, or considered for development including a standard for occupational health and safety and one for medical devices (see Chapters 12 and 14 for further information on these topics). These developing standards will add another level of complexity to registration decisions.

Leaving aside these developing standards, many companies will soon face the possibility of registering to ISO 14000 in addition to ISO 9000, QS-9000, or perhaps all three documents. Companies must now tackle the complex task of integrating these requirements into a single management system. (See Chapter 12 for details on integrating and implementing ISO 9000, QS-9000, and ISO 14000.)

ISO 9000 REGISTRATION

Most registrars require a completed application, sometimes called a contract, to begin the registration process. The application should:

- Define the rights and obligations of both the registrar and the client.

- State the registrar's access rights to facilities and necessary information.

- Address liability issues. *Note:* Companies considering integrating ISO 14000 registration into their system should pay close attention to the issues of confidentiality and liability. (See Chapter 12 for a complete discussion of the legal issues surrounding ISO 14000 registration.)

- Define confidentiality policy and advise the client of the right to appeal a decision and/or file a complaint.

- Offer instructions for the use of the registration certificate and associated marks.

- Define conditions for terminating the application.

Selecting a Registrar

First, make sure that the registrar's accredited scope of operations covers the business to be registered. Different methodologies are used to identify a registrar's accredited scope. In the United States, Standard Industrial Classification (SIC) Codes are used. In Europe, a similar system of codes, Nomenclature Generale des Activites Economiques Dan les Communautes Europeenes (NACE), is used to define a registrar's operational scope.

Scope definition is critical to the success of an organization's registration effort. The registrar should be willing to work with the organization to define the extent of registration and how it will be achieved.Without a clear understanding of the project's scope everyone from the company seeking registration to the marketplace will be affected. Companies that must also seek QS-9000 approval will discover that careful scope definition enables the registrar to more efficiently audit for both ISO 9000 and QS-9000 compliance using auto industry guidelines. *Note:* Registrars must receive separate accreditations to offer QS-9000 and ISO 14000 audits. A company that expects to develop an integrated standards management program and use the same registrar should carefully consider registrar choice.

Registrars, like any other business entity, have different internal policies. For example, some registrar policies make it difficult to use two different registrars at the same facility. A restrictive policy may cause delays and add cost for a large company with centralized functions, such as purchasing or design, that elect to use more than one registrar.

Document Review

Once the application is completed listing basic information on the company's size, scope of operations, and desired time frame for registration, the registrar typically asks the company to submit documentation of its quality system.

Most registrars are interested in first reviewing the quality manual that describes the existing quality system. Every company procedure is not required to be in the manual only referenced. The manual is compared to the appropriate ISO 9000 series standard (or QS-9000 requirements if applicable) to determine compliance. Some registrars prefer to perform an on-site document review. However, an off-site quality manual review (preferred by most registrars) saves travel costs and the internal costs of hosting the registrar.

The costs for this manual review should be discussed up front along with circumstances that would require a follow-up review. For example, if the company adds an additional site under the same quality system, some registrars may require a second review. Quality manuals that have been extensively revised often require new reviews.

Although the quality manual is not expected to fully define the details of an organization's quality system, an effective manual will provide enough information to allow the registrar to determine if a quality system exists. Companies are required to describe the structure of the quality system and refer to supporting procedures. Questions raised during the document review frequently refine the scope of the registration and ensure that the company sets realistic certification goals.

Preassessment

Most registrars either recommend or require a facility preassessment. To some registrars preassessment means a complete assessment that determines the current status of a company's operations. To others it is an aid in audit planning (number of auditors required, auditor-days required) and used to determine preparedness for a full assessment. For some registrars preassessment means a client paid sales visit and facility tour.

Remember, not every company will need a preassessment, but every company should carefully consider the benefits. Years of data indicate that a preassessment is the best way to assure a successful initial audit. A preassessment can identify major system deficiencies (or inadequate documentation) before a full assessment. Preassessment will increase a company's chances of passing a full assessment on the first attempt. On the other hand, the preassessment may point out that a company has overprepared. For example, the preassessment may allow a company to discover that documentation it has prepared is not required for registration.

Finally, overall costs are often reduced as a result of preassessment. The registrar may determine that the final audit will require a smaller audit team or fewer auditor days. However, the company should make this preassessment decision based on its own agenda and business goals. A company must consider that if a registrar requires a complete quality system preassessment, another full registration assessment will be required, adding internal and external expense.

One preassessment practice, however, is universal among registrars: Providing consulting services to the client company during a preassessment and final audit phase is strictly forbidden. Evaluating the adequacy of the supplier's quality system and documentation to meet the requirements of the ISO 9001, ISO 9002, or ISO 9003 standard (or QS-9000 if appropriate) is allowed. However, the registrar cannot provide substantive advice and guidance to the company on system implementation. Use an internal or external consultant to provide this guidance as needed.

It is interesting to note that the Big Three automakers and others who support QS-9000 initiative have clearly defined what is meant by consulting activities. Under these rules, delivery of consulting activities by a registrar or any related organization precludes that registrar from providing registration services. A company may not be able to use its registrar of choice if this boundary is crossed.

Note to companies considering ISO 14000 registration: The preassessment (document review/ initial assessment) phase of an ISO 14000 audit is more in-depth than the document review associated with an ISO 9000 assessment. Review of the organization's environmental management system, including its stated environmental impacts, is conducted and the results are used to plan the assessment phase in detail. For additional details, see Irwin's *Implementing ISO 14000,* by Tom Tibor with Ira Feldman (1996).

Full Assessment

A full assessment is conducted after the registrar determines that the company's documented quality system conforms to the requirements of the selected management system standard. Typically, two or three auditors spend from two to five days at a facility. The duration of the audit depends on the size and complexity of the company's operations. Challenging or questioning an overly short or long audit is perfectly appropriate.

The Big Three automakers have tried to standardize this process by clearly defining the duration of QS-9000 audits via the International Auto Sector Group (IASG) sanctioned interpretations. These audit times are typically 30 percent longer than a comparable ISO 9000 audit.

Before beginning its audit, registrars conduct an introductory meeting with company management and request that auditor escorts be assigned. At the end of the audit cycle, a closing meeting is held to communicate to management any system deficiencies discovered. Registrars often leave a final report containing the audit team's recommendations. Some of these recommendations are binding, but others require further internal review and disposition by the registrar. The client should completely understand the implications of the audit team's recommendations.

During the audit, most registrars review any findings daily with the client. The client may wish to respond to a stated deficiency and should not feel constrained about expressing an opinion regarding the validity of the findings. However, in most cases, all detected deficiencies, even if rectified during the audit, will be reported. Auditors will also interview all levels of company personnel to ensure the quality system as documented in the quality manual and supporting procedures has been fully implemented.

Note on QS-9000 audits: In the case of QS-9000 audits, the registrar will conduct the assessment in accordance with the Appendix B Code of Practice.

Registration

Three outcomes of an audit are possible: approval, conditional/provisional approval, or disapproval.

Approval

A company can expect to become registered if it has implemented all the elements of ISO 9001, ISO 9002, or ISO 9003 (or QS-9000 or ISO 14000 if appropriate) and only minor deficiencies are detected during the assessment. It should be noted that all deficiencies must be closed out prior to issuing a QS-9000 certificate.

Conditional or provisional approval

A company will probably be either conditionally or provisionally approved if:

- It has addressed all the elements of the standard and has documented systems, but perhaps not fully implemented them.

- A number of deficiencies detected in a particular area show a negative, systemic trend.

Conditional approval requires the company to respond to any deficiencies noted during a specific time frame defined by the registrar. The registrar may elect to perform an on-site reevaluation or accept the corrective action in writing and review the implementation during subsequent surveillance visits.

Disapproval

The final possibility is disapproval, which usually occurs when a company's system is either well documented but has not been implemented, or when entire elements of the standard, such as design control, internal auditing, corrective action or process control, have not been addressed. A comprehensive document review or an in-depth preassessment should identify either problem before the final audit. A disapproval recommendation requires another comprehensive reevaluation by the registrar before it can issue a registration certificate.

Once a company is registered, it receives a certificate and is listed in a register or directory published by the registrar or another organization. (Irwin Professional Publishing publishes a comprehensive directory of these ISO 9000 registered companies.) The company should also receive guidance for use of the ISO 9000 certificate and associated quality marks. The client should also understand the registrar's policy for publishing registrations, including actions taken when registration is suspended or withdrawn.

Surveillance

The duration and/or validity of its registration is important for a company to understand. Some registrars offer registrations that are valid indefinitely, pending continuing, successful surveillance visits. Others offer registrations valid for a specific time, such as three or four years.

Most registrars conduct surveillance semiannually or annually. The client should clearly understand the registrar's surveillance policy. Some registrars conduct a complete reassessment at the end of the registration period, while others conduct a less thorough assessment that is more than a surveillance audit but less than a complete re-audit. A six-month surveillance schedule will seldom entail a full reevaluation at the conclusion of the registration. An annual surveillance audit policy usually requires complete reassessment when the certificate expires. Companies seeking to meet the QS-9000 requirements should note that the automakers have mandated a six-month surveillance policy.

The surveillance visits are designed to ensure that a demonstrated quality system remains in place. The internal quality audit (required by ISO 9001, Clause 4.17) and its review by management (required by ISO 9001, Subclause 4.1.3) are mechanisms that drive this process. (Refer to Chapter 6 for further information on the audit process.)

However, rigorous documentation and deployment of an existing system should not stifle continual improvement. Changes toward continuous improvement are encouraged, but any change in the system should be specifically documented and may require that the registrar be informed of these changes. Some registrars require notification of major changes to the quality system, while others require that the client apprise them of all changes. The client should clearly understand the registrar's policy and the possible profound impacts on registration maintenance.

Again, companies seeking to meet QS-9000 requirements should note the registrar is required to identify opportunities for improvement noted during the assessment. The automakers strongly encourage a company's efforts at achieving continuous improvement. The ISO 14000 standard specifically requires a continuous improvement system to be in place.

TIME AND COSTS OF REGISTRATION

Achieving registration is not guaranteed no matter how diligently a company pursues the goal. Management commitment and dedicated implementation efforts are required.

The time required to implement an ISO 9000 quality system depends on the company's current status, its commitment to the implementation of the system and its resources. A realistic estimate (if a company is starting with no system or a poorly documented system) is 18 to 24 months. The time required for actual registration depends upon the number of deficiencies detected during the preassessment, document review and/or initial assessment.

It is important to determine registrar resources before selecting a registrar to ensure that the organization can meet the client's goals and deadlines. Registrar schedule lead times can vary dramatically. Current registrar lead times range from one or two months to over a year.

The growing pressure for QS-9000 registration will likely make choosing an appropriate registrar difficult for some companies. Chrysler has given all production and part suppliers the requirement to meet QS-9000 by July 31, 1997, and General Motors suppliers must meet the requirements by December 31, 1997. Ford, as of the printing of this *Handbook,* has not set a deadline for meeting QS-9000 requirements.

Costs

Many costs are associated with registration. Actually developing and implementing the management system is the first cost. (A survey of 7,000 ISO 9000 registered companies conducted by Irwin Professional Publishing and Dun and Bradstreet, published in March 1996, indicated that companies spent $187,000 on average to complete registration.) A company may elect to use internal resources to implement the system, or rely on the services of an outside consultant, or combine both approaches. (Refer to the *ISO 9000 Survey: 1996* published by Irwin Professional Publishing in Chapter 19 for more specific registration costs.)

When selecting a registrar, companies should assess actual costs of the registration process, including cost estimates for the following:

- Application and document review.
- Preassessment visit.
- Actual assessment.
- Costs associated with issuing the registration and writing the report.
- Surveillance visits.
- Reevaluation visits (if required).

A company should also consider that some registrars require application fees, listing fees, and registration fees in addition to those costs normally associated with registration. Reimbursed expenses can also vary from registrar to registrar. Some registrars are reimbursed at cost, some charge a fixed fee, and others operate on a per diem basis. These costs must be taken into account in order to calculate the full cost of the registration.

The company should also understand registrar surveillance visit costs and the number of days expected for these visits. Other important, baseline information includes how long the registration is valid and any costs associated with full or partial reassessment.

Finally, the client should feel comfortable to openly discuss any issue, such as scheduling and qualifications of audit team members. Although it cannot act as a consultant, the registrar should be willing to guide the client through the registration process. Remember that companies seeking ISO 9000 registration are clients of the registrar and should feel comfortable complaining. However, clients must realize that the registrar operates in accordance with the requirements of its accreditors. Both the registrar and the client share the responsibility of seeking to continuously improve processes and operations.

How to Publicize Your ISO 9000 Registration

Companies achieving ISO 9000 registration are understandably happy about their achievement. They want their customers and potential customers to know about their accomplishment. In fact, according to a survey of registered companies conducted by Irwin Professional Publishing and Dun and Bradstreet published in March 1996, nearly all the companies responding to the survey expect to publicize their ISO 9000 registered status (See chapter 19 for more information on the survey).

The enthusiasm to promote has led to confusion and in some cases to deliberate misrepresentation of the meaning of an ISO 9000 certificate. ISO in Geneva has tackled this problem with an official publication called *Publicizing Your ISO 9000 Registration*. Along with numerous examples of improper and confusing uses of the ISO 9000 registered mark, the pamphlet explains how a company can avoid these improper uses of their certificate.

Pamphlets are available through ISO in Geneva or in the United States, through the American National Standards Institute (ANSI)–see resource section of this book for contact information.

Maintaining Client Confidentiality

James P. O'Neil

The very nature of ISO 9000 registration allows registrar personnel access to the most intimate workings of a company's business processes, even if those processes are proprietary. Many clients are naturally apprehensive about this part of the registration process and need assurance that proprietary information will not be revealed to competitors.

(Continued on next page)

(Continued from page 187)

National accreditation requirements for registrars address this concern by mandating a strict security review for those associated with a registrar.

The security requirements begin at the top of the registrar's organization. In addition to staff members directly involved in day-to-day activities, all directors and certification boards associated with a registrar's business must complete confidentiality agreements. Auditors, those most closely associated with the client company, must pass strict security clearances before joining a registrar's organization. Subcontracted auditors are required to sign a confidentiality agreement in addition to passing security screening.

Finally, any visiting representative of a national accreditation organization seeking access to the registrar's client files must endorse a copy of the registrar's confidentiality agreement.

The United Kingdom goes one step further to protect client files. National accreditation in that country requires that client files be protected by an intruder alarm system monitored 24 hours a day.

Confidentiality and Certification

The responsibilities of an accredited registrar and the requirements for the preservation of client confidentiality are fully described in ISO guidance documents. Activities related to performing an ISO 9000 audit are described in guidance document ISO 10011.

In addition, standards maintained and set forth by the International Register of Certified Auditors (IRCA) in the United Kingdom and the American National Accreditation Program for Registrars of Quality Systems, which is operated jointly by Registration Accreditation Board (RAB) and American National Standards Institute (ANSI) in the United States, assure high competence for assessors. Experienced assessors minimize the subjective nature of auditing and objectively record a noncompliance without direct reference to any proprietary process.

An experienced auditor develops high standards and effective methodologies for dealing with proprietary issues. Professional conduct requirements of the IRCA and ANSI/RAB-registered assessor schemes also address these issues and assure that quality auditing techniques are employed.

Special Cases

Not all proprietary issues are the same, and some special rules govern each category of business activity. These include:

(Continued on next page)

(Continued from page 188)

- **Industrial security.** Companies that provide products or services (e.g., soft drink manufacturers seeking to protect a specific formulation).

- **Personal security.** Companies that provide personal services to the community (e.g., doctors or lawyers with access to sensitive information concerning individual citizens).

- **Civil security.** Companies providing business services that protect property (e.g., burglar alarm installers with access to sensitive coding or entry information).

- **National security.** Companies providing services that protect the nation (e.g., manufacturers of strategic or conventional weapons whose information may be valuable to foreign interests).

Organizations involved in the industrial and personal security sectors are easily protected by client agreements. Auditors working in the civil and national security sectors must be full-time employees of the registrar who have been subjected to official background investigations.

In the United Kingdom, these employees must sign the Official Secrets Act and be "positively vetted" by the Ministry of Defense, or undergo a security screening in accordance with the requirements of BS 7499: Part 1: 1991: Appendix B.

A confidential, secret, or top secret security clearance by the Department of Defense is required in the United States to audit in these sensitive areas. In addition, each client file is kept in a locked, fireproof safe under the direct control of the registrar's chief executive officer.

Confidentiality is taken very seriously by accredited registrars. The current system virtually assures a company that its proprietary processes will be protected during an ISO 9000 audit.

The Audit Process

6

This chapter contains two articles related to the audit process. The first one, by David Middleton, describes the internal quality audit process, including the following:

- Defines an audit and the auditor's role.

- Discusses the phases of an internal audit, including the following:

    —Planning an audit.

    —Executing it.

    —Reporting your findings.

    —Applying the corrective action process.

The second article, by Roger Pratt, discusses communication techniques helpful in both internal and external audits. These techniques will help make the actual audit proceed more smoothly. His article discusses the following issues:

- Putting the auditee at ease.

- Helpful techniques to try.

- What to do during unusual situations or conflicts.

- Ethics involved with auditing.

INTERNAL QUALITY AUDITS

by David Middleton

WHAT IS AN AUDIT?

This section is designed to give a brief overview of the key steps in developing, conducting, and reporting an internal quality audit.

An audit is the process of comparing actions or results against defined criteria. Internal audits are an integral part of any management system, whether it is focused on quality, safety, the environment, or any other business element. A management system audit compares the implementation and effectiveness of the system against a standard as well as against its own internal criteria, as defined in policies, procedures, and other documentation.

A quality audit is defined in *ISO 8402, Quality management and quality assurance—Vocabulary*:

Quality Audit

A systematic and independent examination to determine whether quality activities and related results comply with planned arrangements, and whether these arrangements are implemented effectively and are suitable to achieve objectives.

Notes

1. The quality audit typically applies to, but is not limited to, a quality system or elements thereof, to processes, to products, or to services. Such audits are often called "quality system audit," "process quality audit," "product quality audit," or "service quality audit."

2. Quality audits are carried out by staff not having direct responsibility in the areas being audited but, preferably, working in cooperation with the relevant personnel.

3. One purpose of a quality audit is to evaluate the need for improvement or corrective action. An audit should not be confused with surveillance or inspection activities performed for the purposes of process control or product acceptance.

4. Quality audits can be conducted for internal or external purposes.

ISO 9001, Clause 4.17, also offers a definition of internal quality audits:

The supplier shall establish and maintain documented procedures for planning and implementing internal quality audits to verify whether quality activities and related results comply with planned arrangements and to determine the effectiveness of the quality system.

Internal quality audits shall be scheduled on the basis of the status and importance of the activity to be audited and shall be carried out by personnel independent of those having direct responsibility for the activity being audited.

The results of the audits shall be recorded [see Clause 4.16 in Chapter 4] and brought to the attention of the personnel having responsibility in the area audited. The management personnel responsible for the area shall take timely corrective action on deficiencies found during the audit. Follow-up audit activities shall verify and record the implementation and effectiveness of the corrective action taken. (See 4.16.)

Notes:

1. The results of internal quality audits form an integral part of the input to management review activities. [See Subclause 4.1.3 in Chapter 4.]

2. Guidance on quality system audits is given in ISO 10011.

Simply stated, an internal audit evaluates a company's quality management capability to determine the following:

- Does a system exist?

- Is it implemented?

- Is it effective?

Internal quality audits should be scheduled regularly, but some flexibility may be used in determining frequency, according to guidance document ISO 10011-1. (See sidebar for an explanation of ISO 10011, Parts 1, 2, and 3.) A company should consider any changes to its quality system (including changes in management, policy, or technology) and any corrective actions taken for previous audits. An internal audit system can also be used to support ISO 9000 registration, to provide a basis for improving an existing quality system, or to ensure that regulatory requirements are met.

THE ROLE OF THE AUDITOR

The role of the auditor is to examine whether or not a company or department is meeting the requirements of a declared quality assurance standard and, by collecting objective evidence, verify that the system is implemented and effective. Determining the system's effectiveness is difficult, but it is key to complying with ISO 9001. By interviewing personnel and witnessing activities, you can identify whether there is a system and whether it is being followed, but without a frame of reference as to what you expect from an effective system you cannot evaluate effectiveness.

The auditor's role is not merely to report facts. An auditor obtains information from a variety of different people and interprets the data to make an informed judgment about the effectiveness of the quality system.

Many people think that an audit is a primary policing function that ensures compliance with a set of defined criteria or rules. This perception is often shared by auditors, particularly if they are part of a newly developing audit program within a company. If an audit program has ill-defined objectives or the auditees misunderstand the purpose of the audit, then the "policing" aspect of the audit dominates over evaluating the effectiveness of the quality system.

An example of this misunderstanding is the "gotcha" reaction by an auditor who finds a noncompliance. Unfortunately, some auditors do a little dance of glee and wave a piece of paper, announcing, "It's wrong, it's wrong!" Although this is perhaps the worst-case scenario, an auditor should never express satisfaction in finding a noncompliance.

Involving managers and section heads in the internal audit process means convincing them of its value—convincing them that it won't become a fault-finding exercise. People won't believe that the audit system is designed to improve the process unless they see it; as such, your company's first audits may also be its most critical. Audit preparation should begin with educating everyone in the organization on the following topics:

- What the audit process will involve.

- The audit's expected benefits.

- How it can be used to measure the effectiveness of the links between internal customers and suppliers within an organization.

ISO 10011: Parts 1, 2, and 3

ISO 10011 offers guidance for establishing a quality audit system, and while the requirements are not mandatory, they are an excellent resource for establishing consistent audit practice worldwide. *ISO 10011, Guidelines for auditing quality systems* consists of the following three parts, which are explained in more detail below:

- Part 1: *Auditing*

- Part 2: *Qualification criteria for quality systems auditors*

- Part 3: *Management of audit programs.*

The three parts of ISO 10011 complement the requirements of ISO 9001 by providing consistent guidance for one critical area of implementing the standards, namely auditing.

As earlier chapters have emphasized, the key role of audits is further strengthened in the Phase 1 revisions. Requirements for internal audits are now present in all of ISO 9001, ISO 9002, and ISO 9003.

ISO 10011-1, Auditing

Part 1 provides guidelines for actually conducting an audit of an organization's quality system, including establishing basic audit principles, criteria, and practices. It also provides guidelines for establishing, planning, carrying out, and documenting quality

(Continued on next page)

(Continued from page 194)

systems audits. In addition, it includes guidelines for verifying whether a quality system exists, whether it has been implemented, and whether it is able to achieve the defined quality objectives.

ISO 10011-1 is general enough that users can tailor the guidelines to suit their needs. The following definitions are from ISO 10011-1.

Auditor (quality). A person who has the qualifications to perform quality audits.

Notes:

1. To perform a quality audit, the auditor must be authorized for that particular audit.

2. An auditor designated to manage a quality audit is called a lead auditor. (Clause 3.3)

Client. A person or organization requesting the audit.

Notes:

The client may be:

a. The auditee wishing to have its own quality system audited against some quality system standard;

b. A customer wishing to audit the quality system of a supplier using his own auditors or a third party;

c. An independent agency authorized to determine whether the quality system provides adequate control of the products or services being provided (such as food, drug, nuclear or other regulatory bodies);

d. An independent agency assigned to carry out an audit in order to list the audited organization's quality system in a register. (Clause 3.4)

Auditee. An organization to be audited (Clause 3.5).

During internal quality audits, the client and the auditee are the same person; the term "organization" can mean a company, division, branch, department, group, section, or function.

ISO 10011-2, Qualification Criteria for Quality Systems Auditors

To ensure that quality systems audits are carried out effectively and uniformly as defined in Part 1, minimum criteria are required to qualify auditors. Part 2 describes these minimum criteria. They include education and training, experience, personal attributes, management capabilities, and competency. ISO 10011-2 also provides a method to judge a potential auditor's compliance with the criteria.

(Continued on next page)

(Continued from page 195)

ISO 10011-3, Management of Audit Programs

Companies that conduct ongoing quality systems audits should establish a way to manage the process. Part 3 describes the activities that should be addressed by such an organization and offers basic guidelines for managing them. This guidance document addresses the organization itself, applicable standards, staff qualifications, the suitability of team members, the monitoring and maintenance of auditor performance, operational factors, joint audits, audit program improvement, and a code of ethics.

In addition to the definitions given in ISO 8402 and ISO 10011-1, the following definition applies:

Audit program management. Organization, or function within an organization, given the responsibility to plan and carry out a programmed series of quality systems audits (Clause 3).

PHASES OF THE AUDIT: PERC

All types of Audits have four basic phases: **P**lanning, **E**xecution, **R**eporting, and **C**orrective action.

The auditee's perception of the audit is significantly affected by poor planning, inadequate execution, confused reporting or corrective action records. As an internal auditor, you should emphasize each of these four areas to ensure a professional approach.

Planning

Sufficient planning is especially important. Spending time in this particular phase will reap benefits when trying to ensure a smoothly run audit process. Planning an audit involves six steps:

1. Select a skilled and capable audit team.

2. Confirm the audit's objective and scope together with the specific quality assurance requirements with the auditee.

3. Identify information sources on which to base the audit, including the quality system standard itself, the quality manual, procedures, etc.

4. Develop an audit plan.

5. Confirm the plan with the auditee.

6. Develop checklists.

Selecting the Team

According to ISO 10011-1, whether an audit is carried out by a team or an individual, a lead auditor should be placed in charge. The audit team is best assembled from a diagonal cross-section of the organization.

People from all levels in the company who perform a variety of tasks from different departments should be included. This team will then audit a function of the company where they are not directly responsible; e.g., marketing should audit manufacturing, sales should audit design, etc. Also, they should have some training in auditing techniques and ethics and remain free of bias. (See Chapter 17 for more information on auditor qualifications.)

Objective and Scope

Defining the objective and scope of the audit involves answering the following four questions:

Where am I auditing from?

What am I auditing to?

At what point in the process am I starting?

At what point in the process am I finishing?

One of the ways to identify the scope of the audit is to obtain the relevant documentation, review it, identify the beginning and ending points of the audit, and then use that initial review to prepare a checklist.

However, another strategy is to develop a more proactive approach to preparation; that is, get the auditees involved. It is often more effective to sit down with a department manager, supervisor, or section head to identify their key processes and then use that as the foundation of your checklist. A checklist should never be a "secret weapon" used to find fault. Rather, preparing and developing the checklist should be part of an open audit process.

Information Sources

Identify the sources of relevant information that should be used in selecting the audit sample to ensure a balanced view of the company's operation. An audit sample is a sample of the documents or processes you may wish to audit; e.g., in purchasing, how many purchase orders will you look at—5, 10, 15? In manufacturing, which process will you audit? It is the objective evidence evaluated during the audit itself.

Sources of information from which to develop the audit plan/checklists include:

- Quality manual/procedures.
- Management priorities.
- Quality problems.

- Previous audits/outstanding corrective action.

- Product information.

- Experience of the auditors.

Develop an Audit Plan

Your audit program should identify the duration of the audit, the areas of the organization that will be subject to assessment, and the people who should be available to answer the auditor's questions. Some of the key issues in developing the program are:

- **Is it well planned?** Have you thought through the process? Have you identified a beginning and an end? Can you take a sample from the system that will enable you to follow the process in a logical way?

- **Have you set achievable objectives?** Can you verify that something is actually happening? Can you find evidence of an effective system, or are you wasting time following dead ends and trying to establish a pattern? Can you verify that the system or process you are examining is working? If so, how?

For example, say you are in the receiving area and are told that product is tested to a certain specification. How can you verify that the specification is up to date? You can see a date on the specification, but it might not be current. A methodology is necessary to identify a document and to determine whether or not it is up to date. It is far better to identify those difficulties at the audit preparation stage than during the audit itself.

- **How long is the audit?** Is the audit to last one day, two days, one week? The length depends on the objective and scope of the audit.

The audit itself has tight time constraints. It is important to keep the audit plan on schedule and yet maintain the flexibility to follow up on any leads that develop. Therefore, you must have a planned approach and a firm idea of how to conduct the audit. During the audit it is too late to make decisions about what and where the sample is to be taken. However, a plan that is changed for valid reasons is far better than no plan at all.

Confirm the Program with the Auditee

Be sure to confirm the audit program, including the dates, time, and schedule with the auditee.

Develop a Checklist

Details of the sample and where it will be taken are on the audit checklist. The plan is the audit strategy that identifies what areas will be examined and when, whereas the checklist provides the tactical component by identifying how the company complies with ISO 9001 or 9002.

Thought must be given to structuring the checklist to achieve the stated objective of process improvement. Checklist questions need to be open-ended ones that will enable the auditee to explain the process and show how that process is documented. Whether the documentation is in a procedure or a flowchart does not matter, as long as it is reflected in the actual process. (See the next section of this chapter for more information on phrasing questions.)

Execution

The actual execution of the audit consists of a number of distinct events:

- The opening meeting.
- The audit itself: collecting and verifying information.
- Recording the discrepancies.

The Opening Meeting

While an opening meeting for an internal audit is far less formal than an opening meeting held for a supplier audit, it still requires preparation. Follow an agenda to ensure that all necessary points are covered in as short a time as possible. Items you should consider putting on the agenda include:

- In large organizations it may be necessary to introduce the auditor(s), but departments in medium-sized and small organizations might already know the team.
- Explain the audit's purpose and scope as well as the range of activities to be reviewed.
- Confirm that the details of the program are acceptable to the auditee and that the necessary employees are available at the scheduled times.
- Confirm the status of the procedures and any relevant documents prior to the actual physical audit; clarify any ambiguities.
- Explain the manner of identifying and recording nonconformances.

It is also good practice to list those present at the opening meeting.

Collecting Information

The purpose of the audit is to collect objective evidence regarding the effectiveness of the company's quality system. It is a dynamic and practical tour through the company's quality management system along a path prescribed by the auditor's program and checklists.

The team leader's responsibilities during the audit include the following:

- Introduce yourself and the team to the section/department manager.
- Develop a rapport with the auditee.

- Explain what you want to see.

- Focus on the process/understand the objectives.

- Investigate as much as necessary.

- Get the auditees involved.

- Satisfy your sample.

- If you don't find any problems, don't panic—some processes might be correct.

You can gain much information by interviewing the staff, observing activities, or documenting evidence found in a company's records. Staff interviews should not be limited to department heads and senior managers—everyone in the company has a part to play within the quality system.

When an employee tells you about another employee, however, that information is hearsay evidence and is unacceptable. During one warehouse audit, the manager, who was under increasing pressure from the auditor, finally burst out in exasperation, "Well if you think this place is bad, just wait until you get to Sales!" You cannot use such claims as a basis for a discrepancy. You can, however, use the information gained to check if a discrepancy indeed exists. But without the "hard" evidence of your eyes, documented proof, or a statement from the person responsible for a particular activity, you must give the auditee the benefit of the doubt.

In gathering information, an auditor should ask open questions that cannot be answered by a simple yes or no. For example:

"Does this company have a procedure to define the quality audit activity?"

can be rephrased as,

"Can you explain how your audit procedure works?"

Yes-and-No questions do not allow individuals to elaborate on their work and do not give the auditor confidence that the employees understand their operations. Open questions allow you to ascertain what is not recorded in the procedures and to determine the level of understanding of the people who are responsible for undertaking various functions.

To verify facts, it is permissible and indeed desirable to ask several people the same question to ensure a consistent response. Do not underrate the use of silence. When you think you have received an incomplete answer, you can encourage the auditee to provide more information by using body language or simply remain silent. In general terms, people are anxious to avoid a silent pause and will tend to provide more information to fill that gap.Remember, however, that it is not necessary to always find something wrong; there are areas that might actually be under control.

Verifying Your Observations

Auditors have to examine samples of documents, equipment, products, and so on to verify their observations. These samples are part of the audit sample, and the auditor determines the size. However, it is not prudent to select only one sample of a system, nor is it possible to select tens or even hundreds of samples. If one sample is incorrect, it would be wise to take another sample to determine whether it is an isolated occurrence or a larger problem.

When following an audit trail or selecting samples for examination, politely insist on selecting the sample rather than asking the auditee to do so at random. The samples taken by auditees are rarely, if ever, random, and likely will be the information that the auditee wishes you to see rather than that you might wish to select.

An empty file folder may mean that the record is currently in use, or it could mean the auditee does not want you to examine it. Remember that it is your audit. If a piece of information is missing, you have the right to ask for it, but try to be polite and to remain objective. Also, avoid unduly delaying the audit process by asking for information; you can always return to that point when you have more time.

Nonconformities

Many different words are used when referring to nonconformities within a company's system—discrepancy, deficiency, and finding, for example. All mean the same thing—in effect, they are the "non-fulfillment of a specified requirement" (ISO 8402, Clause 2.10). The International Standard ISO 8402 standardizes the term nonconformity, but it will take some time for this to be adopted on a widespread basis.

Recording Nonconformities

When you identify a nonconformity and can trace and reexamine it to reveal the scale of the problem, stop and record those facts. You do not have to list every single occurrence of a problem, but state that what you witnessed is repeated in other records or in other areas. The audit is very much a "show me" exercise that looks for factual evidence. In this respect, a nonconformity report is a concise record of the facts relating to the nonconformance.

Ideally, a nonconformity should be recorded and signed by the auditee as agreed at the point of identification. You should explain to the auditee that such an acknowledgment does not necessarily mean that you will issue a corrective action request but that you must evaluate the finding in the context of the entire audit.

How you record those facts is up to you. Remember that phrasing the nonconformity requires some care on the part of the auditor. To stop the audit to write the details in full, well-structured English can destroy the pace and timing of the audit. The actual written nonconformity may not be presented until later in the day.

Some auditors have used recording devices during an audit. In one instance, an auditor decided the best way to do this was to cover his head with his coat, hide in the corner, and talk into the machine. This activity caused some consternation on the shop floor (not to mention various ribald comments) and isn't recommended.

When recording a nonconformity, the statement should be in a format that can be understood by both the auditor and the auditee, including those members of the department who may not have been present when the nonconformity was identified. Adequate references should be included to allow the department to reexamine the observations after the auditor has left the department.

The recorded nonconformity should include the following information:

- Where the nonconformity was found.
- An exact observation of the facts surrounding the discrepancy.
- The reason why the facts constitute a nonconformity.
- Sufficient references to allow traceability.

When writing the nonconformity, remember the following suggestions:

- Use local terminology, i.e., the language of the department.
- Make the information easily retrievable for future reference.
- Make it helpful to the auditees.
- Make it concise, yet complete.

Reporting

Just as a company has a customer for its product or service, an auditor has a customer for the results of the audit process. With an internal audit, the customer is likely to be the company's own quality assurance manager and the department head of the area under examination. As such, the format of the final report and tone of the closing meeting must be structured to meet the requirements of the department or function.

Before the closing meeting, the audit team should meet to evaluate the information they found during the audit and to ensure validity of nonconformities. They should then record their findings in a nonconformity report (see Figure 6–1). At this point, only the "Nonconformance" section should be filled out.

The Closing Meeting

Whether an audit has been an internal or external assessment, the auditor/audit team should meet with the department and/or company management to confirm the results and to identify the subsequent actions required. For an internal audit, discuss the corrective action with the appropriate department manager, who is more involved in the process.

Figure 6–1

Nonconformity Report

NONCONFORMITY REPORT		
Department/Area Audited:		
Department Representiative:	**Auditor:**	**Nonconformance Report No.:**
NonConformance:		
Cause Identification/Proposed Corrective Action:		
Agreed Time Implementation:	**Responsibility for Action:**	**Dept. Rep's Signature:**
Auditor's Signature:		**Date:**
Entered in CAR Log by Quality Manager: **Signature:**		**Date:**
Corrective Action Completed Satisfactorily: ❐ **YES** ❐ **NO** **Comments:**		
	Signature:	
©Excel Partnership, Inc. 1993.		Form Number

The closing meeting should not last longer than 30 to 45 minutes, and the following items should be covered:

- Thank the auditees for their hospitality and assistance.

- Remember to record the names of the attendees of this meeting.

- Confirm the scope of the audit.

- Identify the audit standard and revision status of the company's documented quality system.

- Discuss any corrective action requests (CARs) in the nonconformity report.

- Explain that the audit has been a sampling exercise of the company's quality system, and therefore the fact that noncompliances have not been identified in a particular area does not mean that none exist.

- Ask whether any points need to be clarified.

- Confirm future actions.

The last item is an important—if not the most important—element of the audit process. Both internal and external audits are a complete waste of time unless action is taken to correct failures that have been identified within the quality system (see Corrective Action Chapter 4). The auditee may provide objective evidence that nullifies your claim of nonconformity. The auditor/audit team should evaluate the claim and record it if it is valid. In addition, the auditee might refuse to acknowledge one of your CARs, in which case the auditor should record the refusal. The auditor should leave the corrective action request forms with the department manager at the end of the meeting.

The Formal Audit Report

The formal report should do the following:

- Assure a customer or independent third party that the company's quality management system is periodically checked for effectiveness.

- Evaluate the adequacy of the company's quality system as compared to its past performance.

- Identify the areas of the company's quality management system that need improvement.

- Assign responsibilities and apply timetables to monitor the progress of the corrective action.

No auditee expects you to write a dissertation, but your report should provide enough detail to validate your conclusions. A summary is a useful way to assemble these facts. A summary statement should contain the following applicable information:

- The department audited.

- The audit's scope/objective.

- The duration and extent of the audit and the dates it was conducted.

- The standard against which the auditee was audited.

- The total number of discrepancies and where they were found.

- Areas/functions where there were no nonconformities.

- The effectiveness of the system.

- Recommendations for corrective action.

- Report distribution list.

Auditing, by its very nature, looks for areas of noncompliance, and it is important for the auditor to try to provide a balanced report that identifies the positive as well as the negative aspects of a company's systems. Therefore, identify and record acceptable elements of the system. The report should also identify—either by reference to the plan or checklists or within the narrative of the report itself—the areas that were visited and the samples that were taken. It should be possible for someone to retrace the auditor's steps and examine the same evidence by referring to the audit report.

Corrective Action

An audit uncovers areas where the system is not functioning in accordance with management's objectives or in regard to the quality standard itself. As such, it can identify the illness, but does not provide the cure. Auditing for the sake of producing a report will serve little purpose. It must be followed by effective corrective action. It is the responsibility of the audited group, not the auditor to determine appropriate corrective action for each nonconformity.

In the nonconformity report mentioned earlier (Figure 5–1), there is a space to record the proposed corrective action. A corrective action/follow-up procedure should include the following elements:

- Auditee and auditor identify and agree to the details of the nonconformity.

- Agree to the corrective action.

- Agree to timetables and dates to accomplish the following:

  —Resolve the problem

  —Implement the solution

  —Evaluate the effectiveness

  —Re-audit to confirm completion (i.e., close out) of the program

- In the case of serious nonconformities generally associated with observed failures of system or products, it will be necessary to re-audit an activity to verify that the corrective action has been implemented and is effective.

A third-party registrar will expect to see evidence that the above points have been addressed.

The corrective action program requires some paperwork—not to create undue bureaucracy, but to provide traceability demonstrating the outcome of an audit, identify those responsible for its resolution, monitor progress, and initiate the "closeout" of the non-conformity.

Another important feature of the corrective action system is a management review (refer to ISO 9001, Subclause 4.1.3) of actions taken. This summary of corrective actions can form the basis on which to judge the entire quality system. It is important to try to quantify the benefits derived from the corrective action program. This can be in the form of increased customer confidence, fewer complaints, or operational cost savings.

Some departmental managers within an organization may not respond positively to the need to implement corrective action. A system for escalation to more senior management should be built into the audit procedures when managers fail to act upon audit findings.

Corrective action and follow-up includes the following tasks:

- Identify the discrepancy.
- Raise and issue a corrective action request.
- Develop timetables with the auditee.
- Evaluate the corrective actions taken.
- Maintain accurate records to verify the corrective action has been completed.
- "Close out" completed corrective actions requests in the records.
- Escalate the issue to senior management if the auditees are not fulfilling their duties.

The corrective action program and its implementation is a highly visible part of the total quality system, and management must ensure that their actions demonstrate their commitment to the company's quality improvement objectives.

These audit phases are similar regardless of the type of management system audit being conducted. In fact, many organizations choose to integrate their internal audit systems for quality, safety, and/or environmental management. Chapter 12 provides additional details on the degree of integration possible between different management systems.

Throughout the audit process, keep in mind the ultimate goal of a management system audit: to determine whether the system is implemented effectively and is suitable for achieving your goals. The combined audit and corrective action process, if done in an effective and positive way, can be a powerful tool to evaluate the systems and drive improvement.

Auditing the Internal Audit System

By Bud Weightman

Must a company's internal audit system be audited itself under ISO 9000 series standards?

ISO 9001 refers to the requirements for a supplier's quality system. It holds that a supplier shall carry out a compliance system of planned and documented internal quality audits to verify whether quality activities comply with planned arrangements and to determine the effectiveness of the quality system.

There are 20 basic elements in the standard, all of which are applicable to the supplier. Any of these elements is supported by all of the others; you can't take away one of them and say it doesn't apply. Therefore you must also do internal audits of the audit system.

If the quality department, for example, is charged with internal audit scheduling and performance, get a qualified individual from another department, possibly manufacturing, to audit the quality department. In such a case manufacturing would verify, by using an audit procedure or checklist, that the quality department has implemented the applicable controls established by the quality manual. This is most likely supported by a lower-level procedure and further requirements.

Additionally, manufacturing would verify the audit system requirements and that those personnel responsible for performing the company's internal audit system were qualified and trained to do so and were actually independent from the activities they have audited.

Without such a system in effect, the auditors (in this case the quality department) would get a free ride. Every individual and every department must be counterchecked. That includes the department or individuals charged with performing the audits on everyone else.

INTERVIEW NOT INQUISITION: SUCCESSFUL COMMUNICATION TECHNIQUES

by Roger C. Pratt

This article highlights some interviewing skills that auditors—both internal and external—can use to help smooth communications with the auditee. While it is written from an auditor's point of view, auditees can learn from it as well.

INTRODUCTION

Although part of any auditor's time is spent gathering information from documents, a more significant portion is spent gathering information from people. This investigative field work is the core of the audit. It is how auditors make observations, collect data and interview employees. Therefore, much of this article addresses the auditing nontangibles, so to speak, including interviewing techniques, body language, and ethical issues.

This article restates the golden rule, with a little poetic license: *audit others as you would like to be audited.*

PUTTING THE AUDITEE AT EASE

The announcement that an organization is going to be audited or that a particular function is going to be surveyed automatically creates a fear of the unknown in those being audited. Therefore, auditors must recognize that their early activities will be suspect. Even though the auditees are aware of the ground rules and the scope of the audit, they may be concerned that they will be singled out and that areas under their responsibility will be found lacking.

Change of any sort can be threatening, causing a defensive or hostile attitude. The auditee will probably spend more time justifying the status quo than listening to suggested improvements. The auditor must be sensitive to the anxiety of all auditees and consider the audited individual's personal philosophies, motivational characteristics, and individual objectives. An auditor should be able to counter any defensiveness by accepting different methods of compliance to a specific requirement; auditors must avoid the "my way or the highway" mentality.

To alleviate some of this stress, the audit team should arrive at the designated locations on time, mentally and physically prepared to audit. The audit team should appear cohesive and have a leader who sets the tone and pace of the audit. The team should be enthusiastic, unbiased, and confident in their ability to assist the organization.

INTERVIEW/COMMUNICATION TECHNIQUES

How auditors ask questions will affect the amount and quality of the information they receive. Preparation is the first and most important step: If the auditor knows in advance what should be discussed, the auditor is more likely to ask appropriate questions.

It is up to the auditor to establish an initial atmosphere of trust and open communication. The goal is to obtain as much valid information as possible in the shortest time possible. Some potential conditions that affect the initial interactions include the following:

- The auditee's perceptions of the audit process: "This is a waste of time; I am being evaluated only to fulfill requirements."

- Auditee's initial feelings of fear and skepticism: "Will this clear up the problems; will I come out of this looking OK?" (The auditee might have certain predispositions based on past experiences.)

The auditor must recognize that these factors affect the audit process. Auditors should emphasize that they are there to audit the system or program, not the person. Auditors should explain to the auditee's manager that the audit is not an inquisition, that it will identify potential problems and assist in correcting them.

Consequently, auditors should take notes throughout the interview process; memory is unreliable, at best. Note-taking may create brief silences, which can induce stress, but most auditees will be comfortable if they understand that notes are being taken to ensure an accurate record.

The importance of listening cannot be overstressed. It is difficult to gather information while talking; auditors should not formulate new questions when the individual is responding to the previous ones, and they should listen for more than the "bottom line." It is important to let the auditee respond with as much detail as possible to get the needed information. The best way to accomplish this is to first ask open-ended questions and then move to close-ended questions to get clarification of details.

Questions that can be answered yes or no should be kept to a minimum. An example of a yes-or-no or close-ended question is, "Do you perform reviews according to your project management procedure Number 51?" The reply would undoubtedly be "Yes." One way to rephrase the question in an open-ended format is: "I've read your Procedure 51, which indicates that a particular type of review process is performed. Explain to me in detail how you implement that procedure."

It is important for auditors to communicate at the same responsibility level and knowledge of the person they are interviewing. The discussions may range from quality philosophy with top management to specific manufacturing techniques with the worker on the production line.

Several clarification techniques can be used to make sure that the information received from interviewees is clear and complete:

- Probing: Using follow-up questions to further explore an auditee's response.

- Paraphrasing: Repeating and rewording important points.

- Summarizing: Recapping and repeating a set of major points to make sure all the important information has been noted.

Using these techniques the auditor can demonstrate that he or she is a good listener and a professional. In addition, the techniques give the auditee a chance to fill in any missing or misunderstood information.

GENERAL CONSIDERATIONS

It is appropriate in a business situation to shake hands with all individuals when they are introduced, both males and females. Auditors should also try to use the person's name occasionally in conversation. This recognizes the auditee as an individual and facilitates the free exchange of information.

Auditors should use appropriate body language to show that they are listening and receiving the message that the individual is sending. This means good eye contact, head nods, and such, as needed. Eye contact should be maintained about 25 percent of the time, as a rule. Too much eye contact, however, makes the auditee feel uncomfortable, and too little makes the interview impersonal. The auditor should smile when appropriate, for it is possible for the auditor to concentrate to such an extent that he or she appears unfriendly.

The audit team's dress sets the tone for the level of professionalism perceived by the audited organization. A suit projects a power image, while blue jeans set a casual tone. Dress should be appropriate to the organization that is being audited. It would be inappropriate, for example, to conduct an audit of construction activities in a three-piece suit or a silk dress.

DEALING WITH UNUSUAL SITUATIONS OR CONFLICTS

The goal during the interview portion of the audit is to gain information in the best atmosphere possible. There may be times, however, when conflicts or difficult situations arise. This includes situations when the auditee repeatedly fails to answer a question or answers inconsistently, tries to dominate the situation, or rambles on in an irrelevant monologue. In these cases, it is best to directly confront the individual and redirect the conversation. The auditor should be persistent and not allow intentional or unintentional avoidance of a topic; it is important to demonstrate control in these types of situations.

It is possible that personnel who are anticipating the audit may develop data, statistics, or other information and offer it to the auditor as evidence of a previous corrective action or as evidence that no problems exist. Some individuals are skilled at using statistical data deceptively. Such data should be used and accepted only if the auditor is convinced that it is valid and appropriate. Usually, obtaining validations from other sources is necessary before such information may be accepted and used in the audit.

Occasionally, auditees will resort to tactics such as showing new products under development, giving tours of the plant, or taking long lunch hours to divert the auditors from their planned activities. Auditors should resist such obvious diversionary tactics.

If a facility is in trouble, and particularly if there are management problems, there may be sincerely ethical people who will indicate an interest in talking to the auditor. This does not mean that these volunteers are necessarily correct in their analysis of the situation. Care must be exercised so as not to be distracted by the side issues. On the other hand, this information should not be ignored and, with proper validation, may provide valuable feedback on system effectiveness.

ETHICS

When dealing with the audited organization, there are some key principles to keep in mind that take the above factors into account.

- *Maintain the self-esteem of the individual.* Refrain from making cutting or sarcastic remarks. This is important in building rapport between the auditor and the auditee, maintaining an atmosphere of openness and trust, and encouraging the flow of information.

  Showing empathy and understanding also helps build rapport, easing tension so that the auditor can gain information that would not have been gotten otherwise. To show empathy and understanding, the auditor should listen for both facts and emotions in the interview with the auditee. Using the technique of "reflecting," the auditor states the emotional content of what the auditee has said to show that the auditor recognizes and understands the auditee. An example might be, "You seem dissatisfied with the manner in which that procedure was implemented."

  Also, the auditor should try not to show shock, dismay, or surprise if the auditee reveals potentially damaging information; simply indicate that the facts were understood and communicate an appreciation of openness and honesty.

- *Check findings and observations against the "so what?" reaction.* This means measuring the perceived problem against potential consequences or risks if it is not corrected. If the consequences are small or nonexistent but are symptoms of a larger system problem, the auditor should investigate. This information can then be used as facts supporting that more general finding.

  The auditor should maintain a conscious objectivity toward the subject being evaluated. Previous practices or personal beliefs can prevent a full understanding of existing conditions. The auditor will be confronted many times with conjecture, suggestions, and leading or distracting opinions expressed by those being contacted. It is essential that the auditor keep his or her personal opinions private and concentrate on elements of observed fact.

- *Concentrate on the relevant facts.* The situation should be evaluated in sufficient depth so that the root cause can eventually be determined. It is not, however, the responsibility of the auditing organization to determine the specific source of the problem or to place blame.

Surprises should not be a part of an auditor's evaluation. An ethical audit is not the place for cloak-and-dagger tactics, witch hunting, or identifying situations at a critical and embarrassing time (a "gotcha"). These practices violate auditor ethics.

All reported observations and recommendations—including the discussion and supporting data for such recommendations—should be stated in the impersonal tense. Avoid using names; substitute instead a definition of the functions that were evaluated and/or the level of the persons in that function.

The auditor should comply to the greatest extent practicable with the customs of the audited facility. This includes compliance with working hours, mode of dress, observance of lunch hours, and other facility requirements or customary procedures.

Give the benefit of the doubt to the audited organization. When there is significant doubt in the mind of the auditor as to the verifiable facts or the correctness of the auditor's recommendation, the item should be carefully evaluated with other members of the team and the team leader. If, in further evaluation, the item continues to be in doubt, it should be dropped.

CONCLUSION

Auditing is not a simple task, for an auditor must gather factual information while at times using "intangible" techniques to effectively deal with an organization's employees. It is important that the auditor learn how to interview the auditee, how to handle unusual situations, and conduct an ethical audit. The auditor must also be professional, preserve the auditee's self-esteem, and inform the auditee of all information gathered during the audit (i.e., create no surprises). Remember the "amended" golden rule: *Audit others as you would like to be audited.*

IMPLEMENTING ISO 9000

Chapter 7   A Basic Guide to Implementing ISO 9000

Chapter 8   Quality System Documentation

Chapter 9   A Quality System Checklist

Chapter 10   Using ISO 9000 in Service Organizations

A Basic Guide to Implementing ISO 9000

7

by Ian Durand, April Cormaci, and Roderick Goult

The requirements of the quality assurance standards ISO 9001 and ISO 9002 reflect a responsible, sensible, practical way to run a company or an organization. They "are aimed primarily at achieving customer satisfaction by preventing nonconformity at all stages" of product realization covered by the respective standard (from Clause 1 of the standards).[1]

As previous chapters have discussed, ISO 9001 covers all stages of product realization from review of customer contracts or orders through design, purchasing, production, delivery, installation, and servicing (where servicing is understood as product support at any point). ISO 9002 is identical to ISO 9001 except that it does not include requirements for design control.

The third quality assurance standard, ISO 9003, contains a subset of the requirements in ISO 9002 that focuses on final inspection and test of delivered product, and even the requirements in these areas are abbreviated. In practice, its limited coverage makes ISO 9003 a feasible choice for few companies and provides significantly reduced confidence to customers and far fewer internal benefits.

Managing quality to achieve the objective stated in Clause 1 of ISO 9001 and ISO 9002 requires a systematic, structured approach which should couple a complete, accurate understanding of customer requirements with the organization, resources, and procedures to translate requirements into satisfactory creation and delivery of goods and services

---

[1] "Product" refers to hardware, software, processed materials, services, or a combination of these, and "nonconformity" to nonfulfillment of specified requirements.

with a predictable level of quality. This means a quality system that is documented, consistently followed, monitored, and improved as necessary.

For many companies, putting a quality system into effect that satisfies the requirements of ISO 9001 or ISO 9002 in a manner that yields significant internal benefits will take at least 12 to 18 months. The cost and time involved qualifies the effort as a major project, comparable to expanding physical facilities, converting to a new management information system, or designing and introducing a new product line. In fact, ISO 9000 implementation requires the same essential activities as any other major project:

- Management commitment.

- Accurate assessment of current status.

- Careful planning.

- Effective design, development, and operation.

- Progress monitoring and appropriate adjustments.

- Good project management.

This chapter begins with a brief discussion of the benefits companies seek and derive from implementation. The rest of the chapter describes a proven ISO 9000 implementation process that integrates the substantial body of experience and knowledge accumulated by diverse companies over the years. The description is divided into several sections:

- Access Ramps: Guidelines for Success, which summarizes major elements that can support and foster successful implementation.

- Potholes and Roadblocks: Common Quality System Shortcomings, which summarizes those quality system elements where companies typically fall short of satisfactory implementation.

- The Journey: A Proven Implementation Process, which provides an overview of a proven, eight-phase implementation process. Each of eight subsections discusses one of the phases, including objectives and selected major activities. Other chapters in this book address selected implementation activities in more detail, and those chapters are referenced in this chapter where appropriate.

- Formal Registration Assessment and Surveillance, which briefly discusses preparing for the formal registration assessment and surveillance.

WHY DO COMPANIES TAKE THE ISO 9000 JOURNEY?

Companies typically report three primary motivations for implementing ISO 9000 quality management systems: customer pressure, competitive advantage, and improved internal operations. For most companies, it is likely that external forces—customer pressure and competitive advantage—are the initial catalysts for action. As companies implement ISO 9000 quality systems, they often experience increased effectiveness and efficiency of internal operations.

Improved business results come from both internal savings from more efficient systems and increased market opportunity resulting from "registered firm" status, better performance, and higher-quality products. (See Chapter 19 for results of surveys and discussions of the use of the ISO 9000 standards around the world.)

Note that pursuing quality system registration solely to receive a certificate that can be shown to customers or to obtain a market advantage is a self-defeating strategy. It may yield short-term benefits, but has a long-term liability attached. Repeat business requires ongoing customer satisfaction and loyalty. If customers are not satisfied, they will take their business elsewhere regardless of a company's registration status.

This chapter assumes that the reader has the broader perspective of industry leaders: commitment to implementing a quality system that will improve business in tangible terms such as higher product quality, better customer service, faster response times, and lower costs.

The Customer Requires It!

In general, customer pressure to demonstrate compliance to the quality system requirements in ISO 9001, ISO 9002, and, to a lesser extent, ISO 9003 has emerged as a major motivator for many companies selling to industrial and commercial customers. Demonstrated compliance to one of these quality assurance standards is an expected part of doing business in Europe for many industries. Demonstrated compliance to ISO 9001 and ISO 9002 is fast becoming an expectation of commercial and industrial customers in the United States, as well as major procurement and regulatory agencies.

Currently, there are only two accepted approaches to demonstrating compliance: (1) direct, second-party audits by the customer and (2) formal assessment and registration of a company's quality system by an independent, third-party registrar. Leading U.S. manufacturers of electronic components, chemicals and allied products, and industrial machinery, as well as service companies that support industrial and commercial customers, are pursuing registration as a way to develop and maintain widespread customer confidence in their goods and services.

Increasingly, major industrial customers are moving from second-party customer audits to requiring their subcontractors to demonstrate compliance to the quality assurance standards via third-party registration, as tangible evidence of commitment to quality products.

What Is the Competition Doing?

Having a registered quality system impacts a company's competitive position. If major customers require registration, a company will lose business if it fails to achieve registration. If major customers are not requiring registration (even if, in fact, they are unlikely ever to require it), a company may gain business by exhibiting a proactive approach to maintaining and improving customer confidence and achieve market differentiation from competitors.

Remember, though, that competitive advantage is usually temporary. In the current marketplace, competitors are often close behind. In addition, while registration may favorably impact an initial decision to purchase a company's products, continued use depends on actual experience with the company and the products.

We Want to Do Better . . .

Registered companies routinely report these improvements in internal company operations:

- Better documentation that results in:

  —increased knowledge of job responsibilities and activities

  —perpetuation of best operational practices

  —consistency in operations.

- Increased productivity as the result of:

  —lower variable costs due to reduced error, scrap, and rework

  —decreased cycle time and increased process capability; process simplification and improvement frees up surplus capacity that was previously dedicated to handling waste and doing rework

  —lower life cycle costs for raw materials and off-the-shelf goods and services—a product of more effective relationships between companies and subcontractors that result from fulfilling requirements related to purchasing.

- Positive cultural change because:

  —employees know exactly what their jobs entail

  —employees have the competencies and documentation they need to perform their jobs

  —employees are more aware of the impact of their jobs on the quality of products.

These and other improvements can occur even if a company does not pursue registration. Achieving registration does not yield results; effective implementation and maintenance of the quality system does.

How do these improvements happen? Implementing a system that satisfies ISO 9001 or ISO 9002 involves *all the jobs* in the entire company or facility that affect the fulfillment of customer requirements. And it requires companies to address *day-to-day issues* that impact the quality of products.

Consider the potential impact of just the subset of quality system elements required by ISO 9001 and ISO 9002 summarized in Table 7–1:

Table 7–1

Quality System Elements That Support Continuous Improvement

- Formal management involvement (Subclause 4.1)
- Quality planning (Subclause 4.2)
- Preparing documentation (all subclauses)
- Process control (Subclause 4.9)
- Corrective and preventive action (Subclause 4.14)
- Internal quality audits (Subclause 4.17)
- Training (Subclause 4.18)

Formal Management Involvement.   The standards require that senior management define a quality policy, set quality objectives, and hold formal, regular reviews of quality system operation and results.

These activities give a company several advantages:

- Fulfilling the management responsibilities is a visible confirmation of a management direction that is the basis for all company operations related to the quality system.

- The personal involvement of senior managers in establishing a quality policy and objectives, evaluating the quality system, making resources available, and setting priorities that support maintenance and improvement of an effective quality system is a significant demonstration of commitment.

- From a business perspective, the more a company knows about its operations and the more current its information, the more effective operations, tactical planning, and strategic planning can be.

Most of the performance indicators that management typically reviews are "lagging indicators"—that is, they depend on historical data. And, therefore, the response to the indicators must be reactive. While an ISO 9000 quality system provides these same indicators, it also gives management two additional tools: currency of information through regular, disciplined reviews, and the opportunity to build some "leading indicators" into the review process. This enables management to take action to prevent problems. For example, a regular review might look at trends in warranty costs or volumes and content of telephone inquiries to a product hotline, which could indicate the need for preventive action to nip a prospective problem in the bud.

Quality Planning.   The standards require a formal process to identify and communicate the way specific practices, resources, and sequences of activities combine to meet customer requirements for a particular product, project, or contract. A company must have evidence that quality planning has taken place; often this is done by developing quality plans. Planning for quality reduces confusion about goals, responsibilities, and acceptable measurement approaches, which results in more consistent products as well as less design, development, and implementation time.

Documentation.   The discipline of analyzing the way a business works that is necessary to create quality system documentation can provide *outcomes:* greater understanding of actual operations, guidance for making changes to operations, increased consciousness of the impact of every employee's actions, and increased interdepartmental and intercompany communications. More effective documentation clarifies what is expected of employees, increases the probability of consistent operation, and is a way to capture the best practices of a company.

Process Control.   All processes involved in meeting customer requirements must be identified, planned, and controlled. For many companies, responding to the requirements for this element stimulates a rigorous review of process design, management, and control beyond that done in the normal course of business. Process investigation increases communication and awareness of everyone's contribution to quality. Furthermore, the increased knowledge about the way the company *actually operates* provides a basis for making effective decisions as well as a baseline for evaluating improvements.

Corrective and Preventive Action.   The requirements related to this element go well beyond deciding what to do with "nonconforming product." They require a thorough investigation and analysis of the current and potential problems to determine root causes, actions to be taken, and confirmation of solution effectiveness. As a result, processes become more efficient and the quality system becomes more effective.

Internal Quality Audits.   Internal auditors act as surrogates for senior managers and customers to assess how well the quality system satisfies the relevant quality system standard; to what extent daily practice consistently follows the documented system; and how well quality objectives are achieved.

Regular quality system audits are a company's single most powerful tool for both effective quality system implementation and maintenance and for driving continuous improvement. Auditing every aspect of a company's quality system detects any changes required in practices as well as documentation and can help avoid potential problems.

Training.   One of the serious shortcomings in many companies is the lack of attention to those things that employees need to do their jobs properly. Effective documentation addresses one aspect of employee needs. The other is sufficient training, qualification, and experience (collectively called "competencies" in this chapter) for every job that impacts quality. The standards require that competencies be defined for all relevant jobs; that the competencies of all employees in those jobs be evaluated; and, where needs are identified, that plans to meet the needs are created and fulfilled.

Employees with the combination of effective documentation and appropriate competencies are highly likely to do their jobs properly almost all of the time.

ISO 9000 AS A MILESTONE AND A FOUNDATION

ISO 9001 and ISO 9002 address the primary aspects of a company's operations that support the delivery of consistent products that meet customer needs. They also facilitate continuous improvement through such requirements as management review, data collection and analysis, and corrective and preventive action. In addition, the standards require elements that are part of more complex and comprehensive systems; for example, quality planning and managing process control.

Many companies pursue ISO 9000 implementation as part of a larger corporate strategy. In particular, a company might use a quality system based on ISO 9001 or ISO 9002 as the foundation for a more comprehensive quality management system, such as those outlined by many national quality awards and other quality management approaches. Often, quality system registration is a planned milestone on a longer journey. (Chapter 13 compares an ISO 9000 quality system, the U. S. Malcolm Baldrige National Quality Award criteria, and other quality management approaches.)

ACCESS RAMPS: GUIDELINES FOR SUCCESSFUL ISO 9000 IMPLEMENTATION

Many different implementation approaches have been advocated. Some work well, others create an attractive facade, but fail to deliver the potential benefits. All successful approaches consider the guidelines summarized in Table 7–2 and discussed below. It is important to note that every approach will require some adaptation to accommodate the particular characteristics of a company.

Table 7–2

Guidelines for Successful Implementation

- Visible, continuous senior management commitment is essential.
- Only do what makes good, long-term business sense.
- Plan early and plan well.
- Use what you have to the extent possible.
- Use appropriate external resources.
- Document everything you do.
- Be prepared to adjust plans as you proceed.
- Involve employees.

Senior Management Commitment is the Single Most important Element in Implementation and Maintenance of an ISO 9000 Quality System

Lack of visible, continuous commitment from senior management makes successful implementation highly unlikely; efforts will certainly be painful and probably fail.

Only do What Makes Good, Long-term Business Sense

ISO 9001 and ISO 9002 are intended to help you run your operations effectively and efficiently. They are not intended to create paperwork and bureaucracy. This means that every procedure and record you have should serve a useful and valid business purpose, and employees should understand that business purpose.

In addition, the standards are intentionally brief, allowing for the flexibility in application necessary for diverse industries and organizations. You can and should interpret the standards in ways that make sense for your company, and choose a registrar who understands your business, its operations, and needs.

There are, however, advantages to making clear connections between the aspects of your quality system and the elements required by ISO 9001 or ISO 9002. If you are seeking registration, this will make it easier for the registrar's auditors to assess your system. From an internal perspective, clear connections will help you verify that you have covered all the appropriate elements.

Interpreting the requirements in the standards may be somewhat more difficult for service and nontraditional organizations than for hardware manufacturers. But thousands of service companies have done it successfully. (See Chapter 10 for a discussion of using the standards in service organizations.)

Plan Early and Plan Well

This is a lesson we all learn in our personal lives. It is no less true in business. In the implementation process described in this chapter, planning and organizing the project is a substantial effort itself. Use of resources, the expediency of the process, and the level of employee satisfaction are influenced heavily by how well the project is planned.

Use What You Already Have to the Greatest Extent Possible

Most companies have some form of a quality system in place. In larger companies with more complex products, quality system elements such as process control, corrective and preventive action, and inspection and testing may be more heavily documented and approached more systematically than other elements. The standards do not intend that you scrap anything you have that works.

The sensible approach is to determine what you need, evaluate what you have, and use what works to the maximum possible extent. For example, if your senior management holds formal, regular senior management reviews of business results already, add quality system results to the regular agenda. If you have a document control system in place, just ensure that it meets the requirements of the standards; if not, it may only require adjustment, not replacement. In addition to reducing the amount of work you may have to do to satisfy the requirements, using existing systems and practices makes it easier to implement a new quality system.

Use Appropriate External Resources

Companies now implementing ISO 9000 quality systems have the advantages of the accumulated experience, knowledge, and wisdom of companies that have gone through it. You are not, in most cases, breaking new ground. Here are just some of the resources you can tap:

- Courses, books, videos, software tools.
- Credible consultants who specialize in helping companies implement ISO 9000 systems.
- Trade publications, such as *Quality Systems Update* and the *ISO 9000 News.*
- Registered companies.
- The Internet (search on "ISO 9000").
- Your chosen registrar, if you are pursuing registration.

Document Everything You Do

In the implementation process, you will be constantly planning or doing. Keeping records and updating them as necessary provides:

- Continued visibility of the implementation project.
- Guidance for the effort.
- A baseline against which to check progress.
- A record of what happened for future use and reference.

Be Prepared to Adjust Plans as You Proceed

Good planning at the beginning of the project (which, as you will see, includes a realistic assessment of current status) will significantly reduce the adjustments downstream, but it will not eliminate them. At the company level, business forces may change. As you proceed through implementation, you may identify major areas for improvement.

On the other hand, you may find that many of your operations fit into an ISO 9000 quality system better than you expected. ISO 9000 implementation is like a motor journey across the country—your carefully planned route will encounter construction and roadblocks as well as that new cut-through road that is not on the map but takes hours off the trip! Make progress assessment a regular event, and make changes as necessary.

Involve Employees

Involving employees as appropriate throughout the project affords you several benefits:

- It allows you to take advantage of employee knowledge and experience. Involvement might be in the form of acting as internal auditors, training other employees, or helping with implementation activities. Only the employees who work in the system know what really happens on a daily basis, and accurate information is critical to successful implementation. By the way, involving employees in analyzing, redefining, and documenting their jobs, and giving them information about their roles in the larger business, gives them ownership for part of the quality management system and supports employee empowerment.

- Your employees are ultimately responsible for making the system work on a day-to-day basis. Without their support and cooperation, the best designed quality system is just so much paper.

- Involving employees is one way to communicate information, for example, about the company effort, the requirements of the standards, and new procedures.

- Using teams of employees to accomplish some implementation tasks can introduce new ways of operating that can lead to positive and effective results.

POTHOLES AND ROADBLOCKS: COMMON QUALITY SYSTEM SHORTCOMINGS

Barriers to successful quality system implementation (summarized in Table 7–3) fall into two major categories:

- Problems in the *overall approach to the implementation project*. You will recognize these examples of problems in this category; they relate directly to the guidelines for success discussed in the previous section:

    —lack of management commitment

    —not involving employees

    —inaccurate or incomplete assessment of actual operations, compounded by an unwillingness to acknowledge problems

    —inadequate training.

- *Inadequate coverage or implementation of ISO 9000 and 9002 Quality Systems elements*. The rest of this section discusses the requirements of those elements that seem to cause difficulties for many companies. These are:

    —reviews of the quality system: Management review (Subclause 4.1.3) and Internal quality audits (Subclause 4.17)

—Contract review (Subclause 4.3)

—Document and data control (Subclause 4.5)

—Control of inspection, measuring, and test equipment (Subclause 4.11)

—Corrective action (Subclause 4.14.1)

—Training (Subclause 4.18)

(Chapter 4 discusses the requirements of ISO 9001 and ISO 9002.)

Table 7–3

Common Quality System Shortcomings

- Overall approach to the implementation project:
  - —Lack of management commitment
  - —Not involving employees
  - —Inaccurate or incomplete assessment of actual operations
  - —Unwillingness to acknowledge problems
  - —Inadequate training
- Management reviews (Subclause 4.1.3):
  - —Not taking action based on quality system review
  - —Not following up on actions taken
- Internal quality audits (Subclause 4.17):
  - —Inadequate training of auditors
  - —Auditors not independent of area being audited
  - —Not taking action based on findings
  - —Not following up on actions taken
- Contract review (Subclause 4.3):
  - —Incomplete customer requirements
  - —Inadequate evidence of review
  - —No mechanism to resolve ambiguities
  - —No mechanism to ensure company can meet requirements
- Document and data control (Subclause 4.5):
  - —Obsolete versions available for use
  - —Documents and data not available where needed
  - —Documents and data not accessible
  - —Inadequate review and approval of changes
- Control of inspection, measuring and test equipment (Subclause 4.11):
  - —Inadequate identification of relevant equipment
  - —Ineffective calibration control system
- Corrective action (Subclause 4.14.1):
  - —Inappropriate responses to nonconformances
  - —Not addressing past occurrences of non-conformance
  - —Incomplete corrective action processes
  - —Inadequate records
  - —Inadequate identification of root causes of problems
- Training (Subclause 4.18):
  - —No systematic method to identify and record competencies needed
  - —Ignoring on-the-job training
  - —No systematic method to assess employee competencies
  - —Incomplete records
  - —Inaccessible records

Reviews of the Quality System

Management reviews (Subclause 4.1.3) and internal quality audits (Subclause 4.17) are critical elements in maintaining the quality system: they are the vehicles for confirming effectiveness and suitability of the quality system and driving improvements.

The major problems companies encounter in fulfilling the requirements for periodic, formal *management reviews* are taking action based on the review and following up on the results of those actions. Management reviews are opportunities to reallocate resources, adjust the direction of the quality system based on business changes, identify improvements, and confirm the effectiveness of actions taken (for example, corrective or preventive action). Management reviews are also one of the major ways through which management can demonstrate commitment and involvement in the ongoing quality system.

Internal quality audits must be conducted on a regular basis by people who are independent of those having direct responsibility for the areas being audited. Typical inadequacies related to internal quality audits include:

- *Inadequate training of internal auditors.* Auditors must be trained in both content and process. They must understand the intent and the requirements of the quality system element(s) they are auditing and know the documentation to examine and the questions to ask.

  They must also have fairly well-developed skills in performing the audit; for example, how to ask relevant questions and present findings to those who are responsible for the area being audited. Well-trained auditors who have other job assignments can also serve as local experts in their own departments.

- *Auditors who are not independent of the area being audited.* This is an essential ingredient for a valid audit. Auditors may have difficulty being effective if they are part of the area being audited. They may face resistance from supervisors and colleagues. It may be difficult for an auditor to be objective when he or she knows the operations so well. And an auditor may feel it is a risk to highlight internal problems for senior management.

- As with management reviews, internal quality audits are not useful if companies *do not take actions* based on the findings *and follow up on the results to ensure problems are resolved.*

Contract Review

Subclause 4.3 requires a formal review of customer requirements to ensure the requirements are understood, that the company can meet them, and that a mechanism exists to respond to changes in requirements. This applies to a statement of customer requirements in any form: purchase orders, verbal instructions, contracts, or requests for proposal (called "tenders" in ISO 9000).

Typical inadequacies include:

- *Incomplete customer requirements.* Companies often make too little effort to truly understand their customers' requirements, and they end up with incomplete or ambiguous ideas as to what is required. This situation is further damaged when companies make assumptions regarding unclear requirements rather than going back to a customer for clarification. With inadequate input, the likelihood of a satisfactory outcome becomes a matter of guesswork and luck.

- *Inadequate or no evidence that a review of initial requirements or changes has taken place.* All too often, it is assumed that what is on the internal order form is correct, even when it does not appear to make sense. A thorough review process must result in improved data collection from customers and improved linkages internally between those who take in customer requirements and those tasked with filling those requirements. Often, what comes in and what goes out are not the same.

- *No mechanism exists to resolve ambiguous requirements or negotiate changes in requirements.* Without a formal system through which data that does not appear to make sense can be identified, queried, and resolved, changes are made by guesswork. The result is rarely good.

- *No mechanism exists to make sure that the company can meet all the requirements.* The review process needs to be holistic in scope—all aspects of the ability to fulfill customer requirements need to be addressed. The requirements must be clear and unambiguous, and personnel performing contract review must either have the technical skills necessary to evaluate the company's competency to meet the requirements or have (and use) access to people with those skills. It should also be remembered that fulfillment of requirements often goes beyond initial deliverables and extends into product support on a medium- or long-term basis. The infrastructure of the company must be adequate, or capable of being built up, to meet the needs.

Document and Data Control

Subclause 4.5 requires documented procedures for controlling all documentation and instructive data that relates to the quality system and product quality. The intent of the requirement is not only to ensure that employees are working with correct information, but to prevent unauthorized changes to that information and obsolete information from being used.

Typical inadequacies include:

- *Obsolete versions* of procedures, work instructions, specifications, engineering drawings, etc., left in areas where they are *available for use.*

- Copies of *documents and data not available* where they are needed.

- Copies of *documents and data not accessible* when they are needed. For example, local documents locked in a supervisor's office outside of "normal working hours" when there is a late shift staff.

- *Changes* made to documentation or data that *have not been reviewed and approved by the appropriate authority.* For example, handwritten changes made on controlled documents and used to guide daily work.

Technical and engineering staffs often have difficulty with document and data control that stems from their operational needs. Staff members, quite legitimately, need to annotate copies of drawings and specifications as they develop technical changes, ideas for production aids, work instructions, test plans, and so on. This practice is totally acceptable provided the document is clearly marked with appropriate wording, such as "uncontrolled copy—not to be used for production purposes," and the meaning of the marking is clear to all employees with access to the documents.

In production environments, it may be necessary at times to transfer an urgently needed change to a controlled document. This is acceptable if:

- There is a documented, formal procedure for this circumstance, including identification of authorized approval personnel.

- The controlled document is signed and dated by the authorized person, typically supported by a change note, waiver notice, or other formal temporary change approval.

- The effectiveness of the temporary change is limited by time, quantity, or contract, which is indicated on the controlled document or supporting documentation.

- Temporary changes are reflected across all relevant operations.

In design and development areas, one of the most commonly encountered difficulties is the use of outdated copies of national standards and other externally controlled specifications. Design and development staff often store these documents for future reference, which then turn up later, well out of date!

Control of Inspection, Measuring and Test Equipment

Subclause 4.11 is the second longest subclause in ISO 9001 and ISO 9002, emphasizing the importance of ensuring that equipment used to make *quantitative* measurements that are the *basis for product acceptability* provides valid data.

Typical difficulties companies have fall into two categories:

- *Not identifying the appropriate relevant equipment.* Typical problem areas include:

  —Equipment used to make measurements at an early stage in product realization is not identified as relevant even though the measurements may be critical to the achievement of the final product requirements. The problem arises when the measurements, either are not or cannot be checked at a later stage.

  —Different departments purchase their own test equipment and do not report the purchase to the individual responsible for calibration control. One way to prevent this is to require the purchase requisition for test or measuring equipment to be signed by the calibration control manager. Another approach is to require that all incoming measuring equipment be routed through the calibration area for receiving inspection.

—Equipment is often overlooked; for example, items used by field service personnel, or employee-owned equipment.

—In some environments, highly sophisticated methods of process control, such as statistical process control (SPC), are used almost to the complete exclusion of any testing. Here, a company must consider very carefully which of the control measures are, in fact, critical to achieving the final requirements for the product and calibrate equipment accordingly.

—Design and development departments make measurements that affect final products. Equipment used for design verification and validation must be under calibration control.

—Equipment or devices that are used for "indication only" or only as a means of monitoring a process need not be calibrated to any form of traceable standard. So, for example, a temperature gauge that is only used to indicate when to adjust settings for a chemical process need not be under calibration control as long as the process performance is monitored and controlled by equipment that is under calibration control.

• *An ineffective calibration control system.* Many companies end up with a great deal of equipment that must be under calibration control, and it is important to ensure that all calibrated equipment can be efficiently tracked with a sound recall system when equipment becomes due for calibration.

Some common indications of an inefficient system are outdated calibration stickers and labels, equipment that has missed a calibration interval completely, or calibration certificates that carry no valid information that would allow traceability to the appropriate national or international standard.

To protect against these problems, the quality system must ensure that:

—A sound foundation exists for the calibration control system. Several useful software programs exist that will handle the date-dependent recall requirements as well as the instrument detail, the calibration certificate records, and, if required, the associated procedures contained within interlinked text files within the database. Some of these programs can also store pre- and post-calibration measurements to assist in defining calibration intervals.

—The individual with assigned responsibility for managing the calibration system has the freedom and authority necessary to make calibration control work. This may, from time to time, involve withdrawing equipment from use when it is due for calibration review. Too often, the responsibility for managing the calibration control system is delegated, but not the authority to make it work.

—Equipment users appreciate the criticality of calibration control. Whenever inspections, measurements, or tests are made with relevant equipment, validation of equipment calibration status should be the first item on the report. The second should be an annotation that work was stopped and the appropriate person contacted if equipment was found out of calibration.

Corrective Action

Subclause 4.14.1, Corrective action, requires a systematic approach to problem identification, analysis, and solution, with the objective of ensuring that a company formally addresses all nonconformances that pose a *significant threat* to the quality of products or the safety of employees or product users.

Frequent inadequacies include:

- *Responding to every nonconformance* with a corrective action process. Not every problem needs formal assessment and resolution. In fact, the first step in any effective corrective action process should be to determine (using a defined set of criteria) the severity of a problem and the need for further action. For example, defects or failures within the defined range of acceptability would not trigger corrective action. In products related to personal safety or the environment, the criteria might require that *all* nonconformances be subject to formal corrective action. Employees need clear direction on when to initiate corrective action.

- *Not identifying and addressing past occurrences* of a problem. Underlying the corrective action requirements is the necessity to address problems in four time periods: the past, the present, the near term, and the future. Suppose a critical gauge is discovered to be out of calibration. An immediate (*present*) remedial action could be to take the gauge out of use. There may be *near-term* actions that can be taken until the gauge can be recalibrated or replaced.

  In most cases, however, present and near-term actions are not sufficient. What about the *past*—those products that have been produced or delivered between the time the gauge lost acceptable accuracy and the time the problem was detected? At the same time a company is addressing the first three time periods, it should be investigating the root cause of the problem and instituting actions to prevent *future* occurrences of the same problem.

  If the quality system is effective, substantial information exists from which a company can infer past occurrences. Quality records of inspection and test status should have identified nonconforming product; identification and traceability systems can help identify which products were likely to have been tested by the offending gauge. Even though the "trail" exists, however, companies often do not take action with regard to past occurrences until a much bigger problem results.

- *Incomplete corrective action process.* An effective correction process is a closed loop. Once root-cause analysis has been done, it is vital that the four steps in the well-known "plan-do-check-act" cycle be completed: plan the action to be taken after identifying and evaluating possible solutions, take the action, confirm that the action was effective, and continue to monitor results to ensure that the problem has been solved. Figure 7–1 shows a basic corrective action process.

Figure 7–1

Sample Corrective Action Procedure

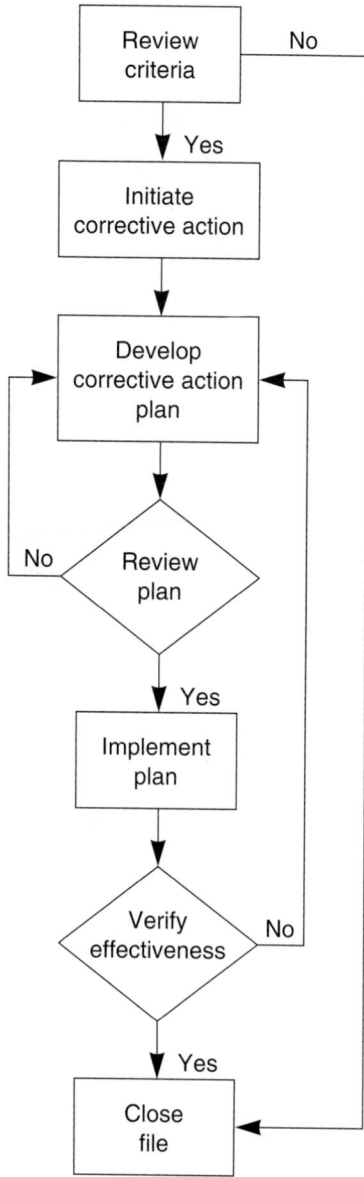

- *Inadequate records.* Auditors (internal and external) will look for records indicating that the corrective action process was completed, as well as examining records to determine if the problems are recurring.

- *Not determining the real root causes of problems.* Problems can be caused by operations (for example, processes, incoming materials, tools, systems, or working conditions), documentation (adequacy or currency), or personnel situations (for example, training or resource allocation).

Training

Subclause 4.18 requires that a company identify the competencies (skills, knowledge, experience) for *each job that affects the quality of products*, assess competencies of employees performing the jobs, develop plans to close the gaps, and keep records. The requirements do not include training or experience related to career or personal development unless it has some impact on the work an employee currently performs to support the quality system.

Typical inadequacies span the requirements, including:

- *No systematic method to identify and record the competencies needed to perform jobs.* The ubiquitous "job descriptions" are routinely lacking in the identification of competencies.

- *Not including on-the-job training*, which often contributes substantially to the development of employee competencies. Many companies only consider training that takes place off company premises or in a classroom as "real" training. Yet, in most companies, the amount of this off-the-job training is minimal in comparison with the amount of on-the-job training that is carried out.

- *No systematic method to assess individual employee competencies,* especially if on-the-job training is the educational vehicle.

- *Incomplete records.* Many training records do not include:

  —relevant training received by all employees at all levels within the company—no one is exempt, including executive management

  —on-the-job training

  —sufficient detail, such as which skills were imparted at what time and by what vehicle

  —the person(s) with authority to sign off that training has been adequately completed

  The records need not be complicated; a training syllabus for a particular job may simply be a list of skills and experience, and an employee record may simply be an identical list with spaces to indicate skill or experience attained, the date, and an authorized signature.

- *Inaccessible records.* Training records should be readily accessible to those who need them on a daily basis—this usually means line supervisors. If records are not readily accessible to those making assignments, work may be assigned to employees who are not appropriately trained and qualified.

    The records may be maintained in any form and accessible by any means that works for the company. If training records are maintained by a personnel department, it is important to keep them separate from employees' personal records (such as salary administration, performance reviews, medical records, and attendance records). Personal records are confidential information, not part of quality system records and not to be seen by any internal or external auditor.

THE JOURNEY: A PROVEN IMPLEMENTATION PROCESS

The eight-phase implementation process summarized in the rest of this chapter incorporates the experiences of dozens of companies of all sizes and providing all types of products. Used effectively, it will result in a quality system that satisfies the requirements of ISO 9001 or ISO 9002 and provides significant internal operational benefits. In addition, at various stages, you can make decisions that can result in an even more comprehensive system and increased operational benefits.

The process was designed with medium-sized companies in mind. If your company is very small or very large, you may need to make adjustments. For example, a very small company may have to pay more attention to leveraging resources. A very large company may have more organizational issues. All companies are unique, however, and will probably have to tailor the activities to some extent to fit the situation.

Overview

Figure 7–2 shows the eight distinct but interrelated phases involved in productive implementation of an ISO 9000 quality system. You will notice that the phases closely parallel developing and introducing a major new product line:

- Seek management commitment for expenditure of resources and for support.

- Form a project management team, assess the current situation to facilitate the design of the product line, and develop a project plan.

- Design the product line.

- Build the product line.

- Market the new product.

- Assess the success of the product, making adjustments as needed.

The ISO 9000 implementation process assumes you will establish a cross-functional *project team* to plan and guide the entire effort, with demonstrated support from a *steering*

Figure 7–2

A Proven Implementation Process

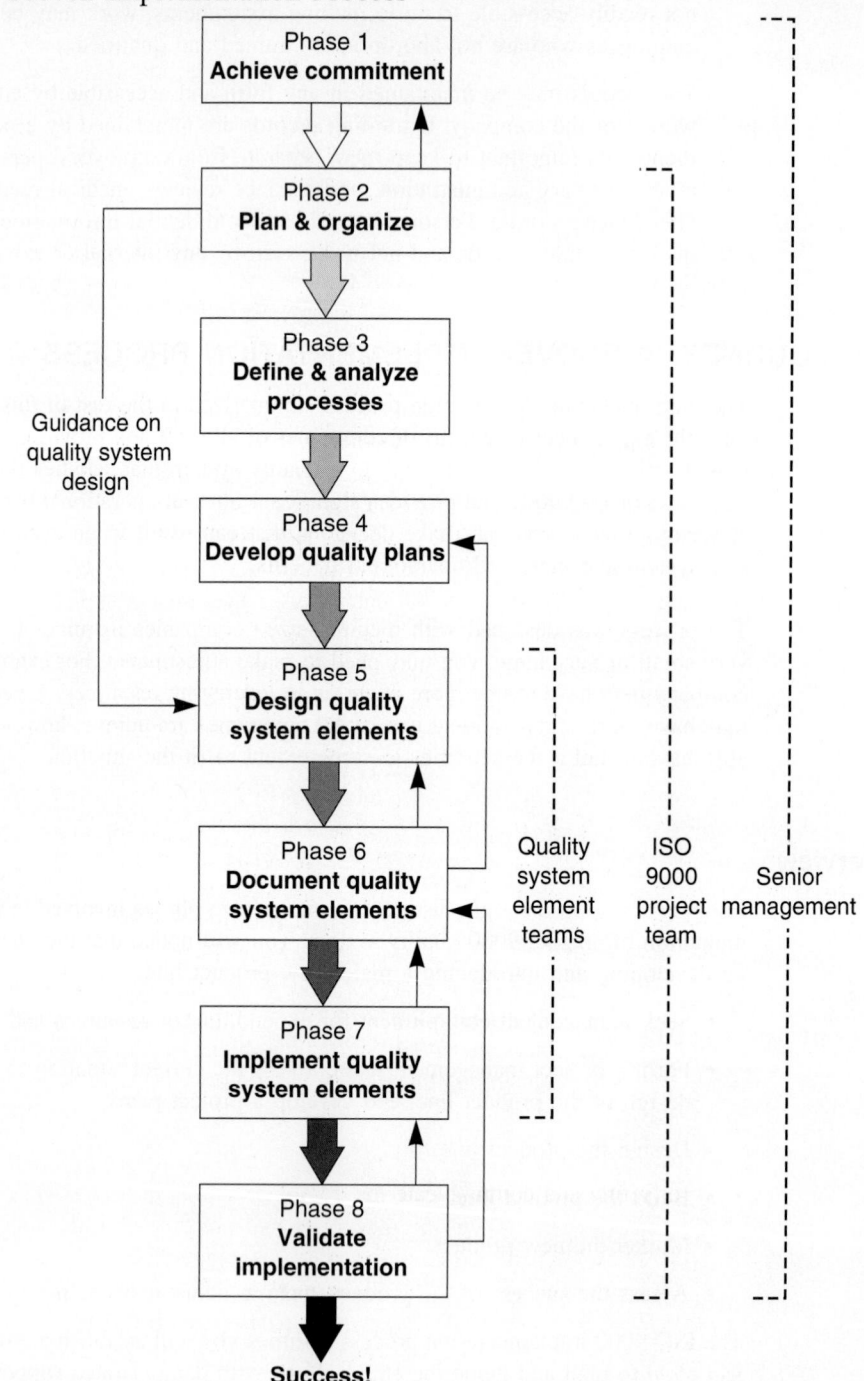

committee composed of senior management. Smaller *element teams* composed of knowledgeable employees will actually design the quality system. (See the dashed lines in the figure.) If you are a small company, project team and steering committee members may be the same people.

Phase 1 is perhaps the most critical: Pursuing implementation without support and demonstrated commitment throughout the effort from all levels of management is not likely to succeed.

In Phase 2, the project team comes together, is educated on the requirements of the chosen standard, assesses the status of the current quality system against the standard, and develops a detailed plan of the way the project will be organized, staffed, and managed. Once the initial work is underway, the team moves into a project management role until implementation is complete.

Phases 3 through 7 build on each other. A thorough understanding of major business processes (Phase 3) is necessary for effective quality planning, often captured in "quality plans" (Phase 4). The results of quality planning, in turn, are essential for the design or refinement of most quality system elements (Phase 5), along with other guidance provided to the element teams by the project team. Unless the elements of the quality system have been designed and validated before they are documented (Phase 6), there is a likelihood of repeated rewriting, preparing too much documentation, and creating non-value-adding bureaucracy. And, of course, implementation (Phase 7) is impossible without appropriately documented procedures. In practice, final changes to documentation and closing any remaining gaps in implementation overlap with validation (Phase 8).

Figure 7–3 shows a typical timeline for an ISO 9000 implementation project, not taking into account the highly variable time to achieve management commitment. The figure emphasizes that, in practice, *activities are done in parallel to the extent possible.* One advantage of parallel activity is that it *minimizes time to completion.* For example, Phases 3 and 4 are in progress while the project team is completing activities that will support Phase 5 (see the arrow connecting Phase 2 and Phase 5 in Figure 7–2).

More notably, parallel activity allows your company to *let the quality system work for you* during the project. Note that the timeline indicates that some design, documentation, and implementation activity begins early in the project and parallels Phases 2 through 4. The quality system elements addressed early are:

- *Management responsibility* (Subclause 4.1). Senior management should develop or refine and communicate the company quality policy and quality objectives by the beginning of Phase 2; the policy and objectives will be inputs to many downstream activities, particularly quality planning (Phase 4). In terms of facilitating the implementation project itself, procedures for formal management review of project progress and quality system results are critical for keeping the project on track and moving forward—they provide opportunities to redirect effort, reallocate resources, and take corrective and preventive action. *A comprehensive,*

Figure 7–3

Typical Timeline for Implementation (circled numbers = Phases)

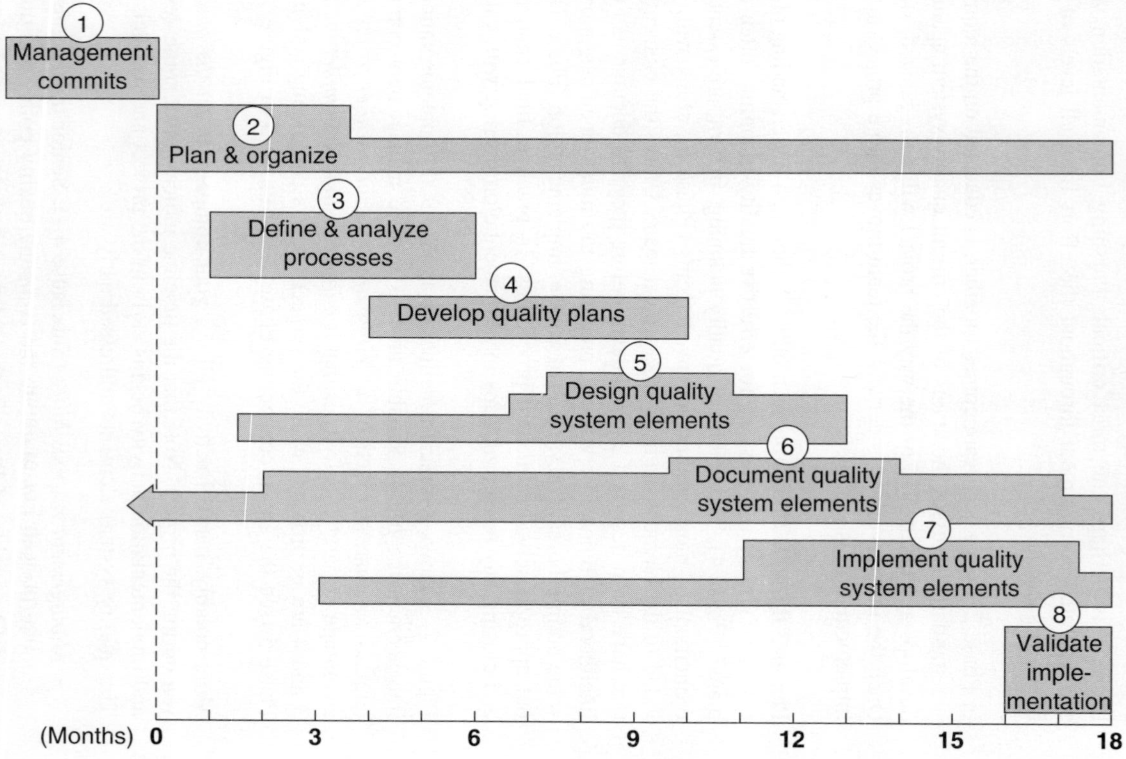

(Months)

formal management review should be held at least once every few weeks starting at the beginning of Phase 2 and continuing throughout Phase 8.

- *Common quality system elements:* Designing, documenting, and implementing elements that span the quality system very early can benefit you two ways: (1) these elements give you tools that you can use throughout; and (2) your company gets practice using the procedures as well as gathering data that can help validate effectiveness and guide later actions. In addition, of course, you are fulfilling requirements of the standards. Common quality system elements include:

  —Internal quality audits (Subclause 4.17) can be used to evaluate the completeness and effectiveness of quality system elements as they are implemented.

  —Document and data control (Subclause 4.5) helps you manage the documentation you create and refine throughout the project.

  —Corrective action (Subclause 4.14) procedures can be the vehicle for refining design, documentation, and implementation of quality system elements, as well as for addressing nonconformances identified during validation assessment.

  —Control of quality records (Subclause 4.16) can help establish a history of quality system operation that can be the basis for evaluating progress, confirm the effect of corrective actions, and provide evidence and information for internal quality audits, validation assessment, and formal registration assessments.

Additional information about implementation appears in Chapter 4, "Implementing ISO 9000—Integrating QS 9000," and Chapter 10, "Using ISO 9000 in Service Organizations."

Pervasive Issues throughout the Implementation Project

Each of the next eight subsections describes one phase of the implementation process, including objectives and selected major activities. While the discussions give a comprehensive summary, *they do not plumb the depths* of the process. For example, these major activities and considerations are not specifically addressed here, but you must address them during the implementation project:

- *Training and education.* Implementing a new or substantially enhanced companywide system is likely to require that employees have new and different knowledge and skills:

  —Some knowledge and skills are specifically related to the ISO 9000 implementation project. For example, senior managers will need to understand the requirements of the relevant standard, their role in the quality system, the extent of the potential company effort required, and the place of quality system registration.

  Other employees will need training in assessing current documentation and conducting internal audits of the quality system. If you use teams, team members may need training in working in teams and holding effective meetings.

The project team may need training and support in project management tasks or conflict management.

—Additional knowledge and skills are related to new procedures employees will be following in their daily jobs.

The important point here is that, throughout the implementation project, the project team must identify the knowledge and skill needs of all employees and ensure that those needs are fulfilled. Education and training is most effective if it is provided at the time employees can use it.

- *Skill-documentation balance.* Companies typically rely on two mechanisms to perpetuate operations: knowledge, experience, and skill of employees; and documentation (policies, procedures, instructive data). The standards intend that you determine and realize the appropriate balance between these mechanisms for all jobs that affect the quality of your products. At successive phases of the implementation project, you should perform activities that are designed to achieve that balance. Figure 7–4 illustrates the link between skill and documentation. The activities are only briefly discussed in this chapter.

- *Communication strategy.* A responsibility of the project team is to develop and implement a companywide communication program about the implementation project. As with education and training, information should be delivered when it is useful for employees. Publicizing the implementation project and then providing no information on progress for several months is not going to garner support or enthusiasm among employees. On the other hand, a well-conceived flow of pertinent information can keep momentum going.

Figure 7–4

Skill-Documentation Balance

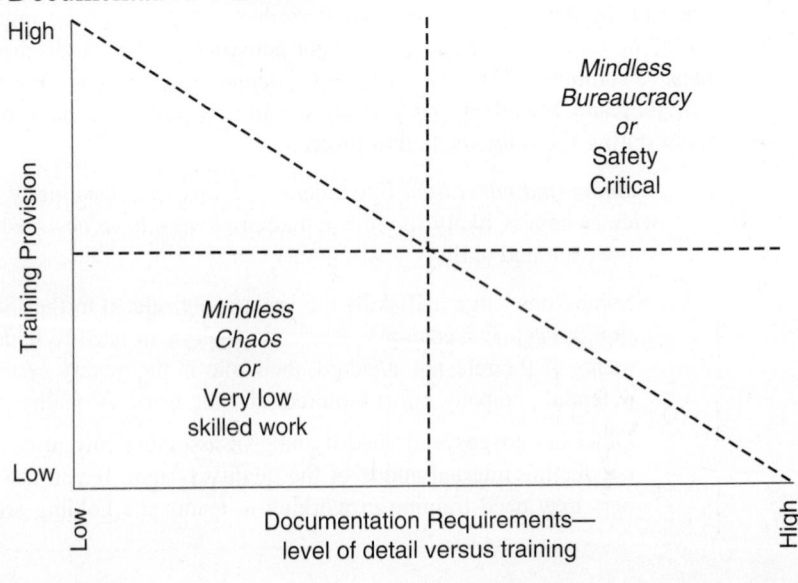

Note that involving employees in the implementation project is a form of communication and can benefit the project greatly, as discussed in an earlier section.

- *Reevaluation of implementation strategy.* Implementing an ISO 9000 quality system is often similar to taking a trip in uncharted territory—you can never quite predict what you might find. Sometimes companies find substantial opportunities for change and improvement. Some of the phases, in fact, include an activity where you consider whether to continue working with what you currently do and just fill in the gaps or adjust the project schedule to make major changes.

THE EIGHT PHASES

Phase 1: Achieve Commitment

Phase 1 creates the environment and the basic guidance statement for the entire implementation project. The objectives of Phase 1 are:

- To gain adequate confidence that managers at all levels, but especially senior managers, consider implementing an ISO 9000 quality system as sufficiently important to warrant investing the necessary resources over the period of the project.

- To develop a "Statement of Purpose"—a written description of the objectives, expected benefits, scope of the quality system, and priority of the project.

Table 7–4 lists the major activities in Phase 1, and the rest of this section briefly discusses selected activities.

The Critical Role of Managers. All levels of management—company executives, other senior managers, and line or operational managers (often called middle managers) who are accountable for the various departments or functions in your company—must

Table 7–4

Phase 1: Achieve Management Commitment

Senior management commitment is the single most important aspect of ISO 9000 implementation. Without demonstrated commitment, further efforts will be painful and probably fail.	
Activity:	*Who Does It:*
• Identify and clarify priorities	Senior management
• Recognize the ISO champion	Senior management
• Define the effort	Senior management and ISO 9000 champion
• Educate management	Senior management and ISO 9000 champion

support and be involved in quality system implementation. Also, strongly consider including the leaders of major employee unions to the extent appropriate for your company.

The time and money involved demand that the decision to implement an ISO 9000 quality system (and, if appropriate, pursue registration) be part of your company's *long-term business strategy*. This translates into a need for demonstrated commitment from *senior managers*. Even when implementation is a corporate mandate, senior managers may not be aware of the effort and cost needed.

The requirements of ISO 9001 and ISO 9002 impact almost everyone in a typical company, so internal preparation will not succeed without sincere effort from the *managers of each department* in your company. Your company must ensure that middle managers:

- Understand the reasons for implementing an ISO 9000 system and the expected positive impact on their operations.
- Understand the priority senior management has given implementation and its place among their other priorities.
- Have guidance, support, and recognition for their contributing efforts.

You may not gain early commitment from all middle managers, but you must have, at least, a willingness to participate in order to continue with the project.

Specific actions *senior managers* can take to demonstrate commitment are summarized in Table 7–5. As a result of demonstrating these behaviors, senior management involvement will become a permanent part of the quality system. In addition, these actions are likely to influence other management levels to support the ISO 9000 project. Specific actions *all management* can take to demonstrate commitment are summarized in Table 7–6.

Table 7–5
Demonstration of Senior Management Commitment

- Establish and maintain quality system implementation as one of a relatively few top priorities.
- Communicate the vision for the future of the company and describe the way quality system implementation contributes to that vision.
- Form an executive steering committee.
- Fulfill the leadership responsibilities required by the standards:
  —establish and communicate a quality policy
  —establish and communicate quality objectives
  —identify responsibilities and authority for quality across all jobs
  —allocate resources to support the quality system
  —appoint a management representative
  —conduct formal, regular reviews of quality system operation and results.

Identify and Clarify Priorities.   It is unlikely that any company will receive positive dividends from the effort invested in implementing an ISO 9000 quality system without that project being a personal priority for its senior managers. It is helpful if senior management can articulate its hierarchy of priorities. This is especially effective in the form of a written statement that can be distributed. It is all right if your company has more important priorities than implementation (and, possibly, registration), as long as everyone understands the situation, and the schedules and resource allocations reflect the true priorities.

Recognize the ISO 9000 Champion.   Every successfully registered company can point to a champion for ISO 9000 implementation who started the project and kept it going. The champion might be self-appointed, designated by senior management, or a member of senior management. Often, the champion leads the ISO 9000 project team and serves as the management representative required by Subclause 4.1.2.3 of the standards.

An effective champion must have the support of senior management and be recognized as the focal point for the ISO 9000 project. Familiarity with the company's business and good communication and project management skills are also important. *The most critical element, however, is that the ISO champion have adequate time—as much as 40-50 percent of his or her time in small companies, full time in large companies during the implementation project.*

Define the Project.   Just as you would do for any major company project, you must define a clear purpose behind implementing an ISO 9000 quality system. Creating a statement of purpose has several benefits:

Table 7–6

Demonstration of Commitment from All Managers

- Make implementation a priority.
- Assign resources (for example, personnel and training) to develop, implement, and maintain a quality system, and adjust them as necessary based on periodic reviews of progress, while, at the same time, ensuring that the business continues to run.
- Identify and remove barriers such as interdepartmental communication.
- Actively promote examination and improvement of the quality system by forming problem-solving teams (including at the senior management level), perceptive listening to suggestions, and promptly implementing solutions.
- Periodically review the status and results of quality system implementation and operation and publicly recognize progress.
- Reject activities that do not support implementation and maintenance of the quality system.

- During development of the statement, issues related to implementation surface, and you have an opportunity to clarify and make decisions on them.

- A completed statement provides focus and guidance for the entire project.

- During the project you may have to respond, for example, to changes in the external business environment, internal priority shifts, and opportunities for process improvement. The statement can be invaluable when the inevitable competition for resources develops. But, it may also be necessary to adjust the information in the statement as implementation progresses.

- A statement signed by senior management, circulated to all managers, and updated as necessary is a visible and continuing demonstration of management commitment to the project.

Figure 7–5 is a sample statement of purpose. (TecDoc Services, Inc., is pursuing registration, and has chosen to call its statement of purpose a "prospectus for registration.") Four types of information are included, and discussed below:

- Objective: target date, criteria for successful effort or benefits expected, and standard to be used.

- Intended scope of quality system.

- Position of implementation among company priorities.

- Priorities of implementation needs.

Objective: What Do You Want to Accomplish by When? ISO 9000 implementation is not a quick fix. It is a long-term company commitment. For some companies, it may require a totally new perspective on operations, with new roles for managers and other employees.

Expected Benefit(s). The information included here is based on your motivation for implementation. If registration is your goal (at least for this particular effort or part of the effort), and you are pursuing registration to meet customer requirements, you should also indicate which customers are requiring registration and by when. Will registration be important in other countries? Which countries?

If you are primarily pursuing positive business results, you need to indicate the criteria by which you will know you have achieved these results, for example, reduction of time to respond to customer orders by 50 percent, reduction in number of defects found to one in a million, or decreased operating expenses to 1 percent of profit. What benefits can senior management expect to receive as a result of the investment in time and money? You may, in fact, be pursuing more than one result, as TecDoc, Inc., is doing.

Your motivations can have an impact on target dates. For example, if the primary motivation is to improve internal operations to meet anticipated future needs, your time frame may be more flexible than a company pursuing registration by a customer-determined date.

Figure 7–5

Sample Statement of Purpose

TecDoc Services, Inc.
PROSPECTUS FOR REGISTRATION

Objective

By mid-19XX, TecDoc Services, Inc., will implement a quality management system that satisfies the requirements of ISO 9001 and achieve registration.

By achieving this objective, we expect to:

- Meet compliance requirements expected within the next five years from U.S. government agencies, and software development companies, computer manufacturers, and computer distributors in the United States and the United Kingdom.

- Increase the efficiency of our processes so we can reduce cycle time from design to market by 20%.

- Increase the effectiveness of our processes so we can meet customer needs better and reduce rework.

- Establish an operational style that involves all employees and supports continuous process improvement.

- Give senior management more and better information as a basis for making strategic and operational decisions.

Scope of Registration

All five U.S. facilities will be registered as a unit since we use the same quality system in all of them. Registration will cover the design, development, production, and servicing of reference documentation for application software. It will also include customer invoicing.

Company Priorities

Achieving ISO 9001 quality system registration is the third priority for TecDoc. First priority is completing the development and introducing our new on-line documentation service enhancement (scheduled for mid-19XX). Second priority is installing our new project management system (scheduled for December 19XX).

Registration Priorities

Is is *essential* that our registrar have credibility in the United States and the United Kingdom.

It is *highly desirable* that all of our U.S. facilities be registered as a unit. Although, since each facility has an independent quality system, each could be registered separately and in sequence.

It is *desirable* that the TecDoc quality system include customer invoicing. This is a specific request of two of our U.S. customers, although they will not stop doing business with us if we do not include it when we are first registered.

Signed by:

John Smith

John Smith, President and COO

December 2, 19XX   on behalf of the TecDoc ISO 9000 Steering Committee

Target Date. Before setting a target date for achieving your objective(s), it is important to consider several factors:

- Motivation for implementation.

- Other company priorities, identified and articulated by senior management.

- Company operating styles. (Is teamwork a normal way of approaching tasks? Does senior management typically get involved?) Some company characteristics (such as widespread employee involvement in all operations) often provide more support for implementation than others (such as lack of formal communication channels).

- Size of company or facility, physical plant, and number of employees.

- History of management involvement in companywide initiatives.

Remember also that most companies require 12 to 18 months to implement an ISO 9000 quality system. And if registration is your goal, consider that most registrars are reluctant to conduct a formal registration assessment until the quality system has been reasonably stable and in operation for about six months.

Standard. There are really only two meaningful choices for the quality assurance standard upon which to base your quality system: ISO 9001 or ISO 9002. Remember that the standards are identical, except that ISO 9001 includes control of the design function and ISO 9002 does not.

The definition of design can be very broad. Universities can design training. Banks and insurance companies can design service offerings. Software companies can design computer programs. Hardware companies can design motors. They can, but they do not necessarily.

In choosing a standard as the basis for the quality system, companies should consider two questions:

- How much control over the design of the product is resident in the company offering the product? (See Figure 7–6.)

- In the realization process from concept to dispersion of the product, to what extent does the design of the product determine the delivered quality and performance of the product?

In general terms, if a company has complete control over design, and design is a major factor in ensuring quality, then ISO 9001 is the only choice. If a product has been an integral, unchanging part of a company for many years, and the principal factor determining quality is, for example, the manufacturing, delivery, and/or installation processes, then ISO 9002 is the appropriate choice.

Sometimes, the decision is influenced by the amount of design. A company that performs major design efforts involving months of work strongly indicates the use of ISO 9001, especially if the design work is undertaken in response to customer requirements.

Figure 7–6

The Concept to Disposal Continuum

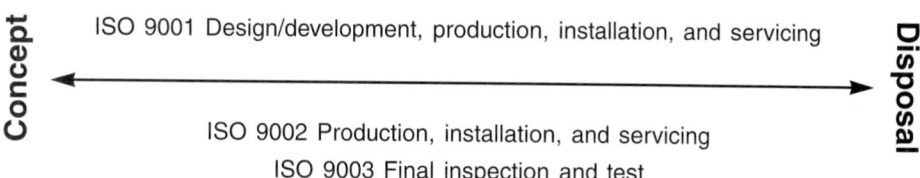

Minor design changes to standard products that involve only a few hours of design time generally do not require ISO 9001.

However, companies seeking registration by registrars accredited in the United Kingdom have an additional consideration. In 1992, the National Accreditation Council for Certification Bodies (NACCB), now known as the United Kingdom Accreditation Service (UKAS), issued a clear policy to the effect that certificates of compliance (called certificates of registration in the United States) to ISO 9002 should not be issued to companies whose customer orders require them to undertake specific design efforts. At this time, UKAS is the only accreditation body with a specific policy about ISO 9001 registration.

If you need to include design control, there are some options on timing:

- You can start with ISO 9001 and include control of the design process from the beginning.

- You can be initially registered to ISO 9002 and expand the scope to cover design activities later, with the concurrence of the registrar and within a defined time frame.

Intended Scope of the Quality System. Scope has three aspects: products, facilities, and quality system elements.

Products. The ISO 9000 standards define product as hardware, software, processed materials, services (where service could be knowledge), or any combination of these. For a hardware manufacturer or company that prepares processed materials, primary products are easier to define.

Some examples of products for service organizations include: for a distributor, packaging and shipping parts to customers; for a legal firm, applied knowledge of corporate law. For many companies, the product or product line is a combination of tangible goods and services. For example, the market offering of a software development company might be a combination of software programs, reference manuals, and technical support available by telephone.

Your quality system can cover one or more specific products or product families, not necessarily all the products your company produces. In deciding on the products to be covered by your quality system, the first consideration must be motivation for implementing an ISO 9000 quality system. For example, if a customer requirement for registration comes from a particular product sector, a company might decide to limit its ISO 9000 implementation to the quality system that supports that product line.

A company that makes multiple products for very different markets might choose to pursue registration for the quality system that supports a product for which customer pressure is high first, and then later expand to include other products over a period of years. For example, consider a plastics molding company that produces high-precision items for the telecommunications industry, small plastic toys, and a component part used in the manufacture of commercial laptop computers.

The high-precision telecommunications products will probably face customer and competitive pressure for registration because of stringent quality needs; the computer industry is facing an increasingly competitive marketplace and becoming more selective about subcontractors, leading to pressure for registration. The plastic toy business has only one market pressure—price. Also, the quality standards are not the same as on the telecommunications items or computer components.

This company may decide to pursue registration that covers only the quality system for telecommunications items. If the company uses the same system for computer components, it might choose to include these products. If not, the company would have to weigh the relative urgencies for each, and perhaps do them in tandem or sequentially.

Facilities. Determining product(s) coverage will dictate which facility(ies) to include in your quality system. Increasingly, companies are including more than one facility in registrations where either (1) the quality system activities span facilities, or (2) all facilities use the same but independent quality systems.

Quality System Elements. What activities are included in your quality system? For example, control of customer-supplied product (covered by Subclause 4.7) may not apply to many manufacturing companies, but it is often a critical factor for service organizations. Servicing activities (covered by Subclause 4.19) such as technical support or maintenance of products may be part of your market offering; for some companies it may be the market offering!

You may decide to include activities not explicitly required by the standards, but seen by your customers as part of your product, for example, invoicing. These activities must be treated in the same way as the quality system elements covered by the standards. Also remember that any quality system element fulfilled by subcontractors but under your control (for example, calibration of measuring equipment) must be included. Agreeing on the scope of your quality system before you begin implementation will help define tasks later.

Position of Implementation among Company Priorities. The position of ISO 9000 quality system implementation among all the major efforts within a company should be articulated, documented, and communicated by senior management. ISO 9000 implementation need not be first priority, but employees must know where it stands. To be believable to the employees, especially those managers who must commit resources, the rhetoric of senior managers must be closely aligned with their behavior.

Implementation Priorities. If you are pursuing registration, this section should identify the priorities for each need you listed—is it essential, highly desirable, or only preferable for a need to be met? For example, is credibility of your registration in a particular country essential, or only preferable? In some cases, you may have to supplement the information with reasonably detailed plans and estimated costs.

If your motivation for ISO 9000 implementation is to improve the operations across your company and products, this would be the place to indicate any circumstances that impact your progress. For example, a company might state that it is desirable to implement an ISO 9000 quality system in all its facilities, but essential that it be done in one or two specified facilities.

Educate Management. Obtaining management commitment and support is not always easy. When customers demand registration, most senior executives recognize the sensible course of action. When implementing an ISO 9000 quality system is not driven by outside forces, however, the effort has to be sold to them in the same way as any other venture that requires an investment of time and money. The key points to emphasize are those related to what is most important to management. Here are some examples:

- An ISO 9000 system will:

  —improve the ability to collect quality-related data to use as a basis for improving both the quality of products and reducing expenses to produce them.

  —enhance the ability to develop stable processes and eliminate costly mistakes.

  —improve overall business efficiency by eliminating waste and redundancy.

  —drive continuous improvement through the use of a formal corrective action process.

- Registration enables a company to demonstrate its commitment to quality to current and potential customers.

- Registration may provide opportunities to enter markets only open to registered companies.

The ISO champion might present this and other information from a variety of sources using several channels. Some sources are:

- Comprehensive surveys of registered companies that provide information on costs, benefits, and implementation experience.

- Registered companies, especially those that have gone through one or more surveillance audits.

- Case histories published in trade journals and books.

Your champion can use different channels to provide information:

- Circulate drafts of the statement of purpose as a basis for engaging managers in discussions of costs and benefits.

- Arrange for or lead presentations on the significance of the standards and on registration in your business or industry, perhaps at regularly scheduled management meetings. An executive sponsor for implementation might be very effective at positioning the presentations. Often an external consultant has more credibility at this stage than internal resources.

- Schedule a full-day workshop for managers and other key employees that includes an opportunity for participants to complete a preliminary assessment of current operations against the requirements of ISO 9001 or ISO 9002. This can be a powerful experience in terms of understanding the requirements and the potential benefit of an ISO 9000 quality system, as well as the potential effort implementation might require.

- Arrange for attendance at public seminars or distribution of books and videos.

Phase 2: Plan and Organize

Just as the quality of a delivered product reflects the effort put into designing it, the effectiveness of your quality system will reflect the level and kind of effort you put into planning and organizing its implementation. The objective of Phase 2 is:

- To establish a structure, guidance, and processes that will guide the implementation of an effective ISO 9000 quality system.

Table 7–7 lists the major activities in Phase 2, and the rest of this section briefly discusses selected activities.

Assess Current Status.   Successfully planning and managing the implementation of an ISO 9000 quality system requires a continuing flow of information in two major areas: where you want to be (defined in the statement of purpose) and where you are today. An initial assessment of current status should not only tell you what needs to be done to comply fully with your chosen standard, but how best to organize.

The most helpful initial assessment has two parts:

- A gap analysis—an assessment of compliance of your current system relative to the requirements of your chosen standard. A good gap analysis will indicate what quality system elements are already in place in your system, what elements need to be enhanced or established, where work is needed, and the extent and kind of work.

Table 7–7

Phase 2: Plan and Organize

(Months 0–18)	
Activity:	*Who Does It:*
Set up the project	
• Assess current status	Project leader and steering committee
• Establish project structure	Project leader and steering committee
• Educate project team members	Project leader
• Develop project plan	Project team
Manage and support the project	
• Develop guidance on quality system design	Project team
—Document control	
—Tradeoff guidelines on employee qualification and documentation	
—Document preparation	
• Begin to design, document, and implement common quality system elements	Project team
• Begin registrar selection	Project team and steering committee

- An objective assessment of the environment in which your quality system must operate will indicate *the way to do the work*, and how to design the system.

To be most effective, your quality system must be designed to be compatible with the existing culture, management style, and organizational climate in your company.

The operational environment must also be a factor in the way you organize the ISO 9000 implementation project, including the way you conduct the initial assessment. For example, in a company where employees and senior managers rarely interact, where the manager of the quality department is considered responsible for quality, and where resistance to change is high, it may be quite difficult to develop the kind of enthusiasm and commitment necessary for employees to contribute meaningfully to the initial assessment. In terms of implementing the quality system, employees who typically wait for direction and do not take initiative will not respond well to participative management approaches or working in self-directed teams; in this case, you will probably need to rely on managers to direct and coordinate most of the work.

In companies where employee involvement in daily decision making is widespread and opportunities for change are sought and embraced, you can expect more support, enthusiasm, and valuable contribution from employees.

If department managers are accustomed to autonomy and flexibility, they will not respond well to strong, central direction from a steering committee of senior managers.

The key consideration is that characterizing the existing environment should precede any attempt at performing a gap analysis.

An organizational assessment can be the responsibility of a group of managers or an individual. Frequent choices for individuals are the manager responsible for the quality function in a company (if there is one) or the leader of the ISO 9000 project.

You have options for the people to conduct the assessment; for example:

- Properly qualified specialists hired from outside of the company.

- A small group of trained internal auditors.

- An adequate sample of employees representing all of the departments covered by the quality system.

The most effective approach is a probably a hybrid. A properly qualified ISO 9000 consultant has the advantage of intimate knowledge of the requirements of the standards and experience with other companies. He or she can give you a comprehensive assessment from the outside looking in and from the perspective of an external auditor. Good consultants usually also assess operational characteristics and advise companies accordingly.

Employees who do the bulk of the work in a company and have some understanding of workflow and practices beyond their immediate jobs can give you an assessment from the inside looking in, that is, they can tell you what *actually* happens on a day-to-day, minute-to-minute basis. An integration of the results of a comprehensive external audit and a well-planned, well-conducted internal assessment provides an excellent basis for planning.

Operational Environment.   Three aspects of operational environment are especially important considerations in planning and conducting a compliance assessment and designing your quality system:

- *Quality culture.* Who is held accountable for quality, and how are decisions about quality made?

- *Environment for change.* What stimulates change? How does the company respond to the need for change?

- *People-procedure orientation.* Companies typically rely on two mechanisms to perpetuate the operational environment: knowledge, experience, and skills of the employees; and documentation (policies and procedures). The people-procedure orientation that exists in your company will have a significant effect on the amount of documentation required to satisfy your chosen standard. If this characteristic of your operational environment is ignored, you face the danger of documenting too much, resulting in unnecessary work and rework, and possible lengthy delays during implementation project.

The prudent project manager will perform a preliminary assessment of at least the quality culture and organizational environment for change before attempting a full gap analysis.

Gap Analysis.   The standards are based on the premise that the quality system should be formal, planned, systematic, and documented appropriately. As a result, three dimensions should be considered in an initial compliance assessment for each quality system element:

- A definition of expected skills, knowledge, and competence for people performing each activity that affects the quality system (referred to as "training" in the standards).

- Documented procedures for each activity that complement expected skills and knowledge.

- Consistent adherence to documented procedures.

Two other important factors for compliance assessment are:

- Recognition of where variation exists among functions or departments in defining skills and knowledge, in the extent and effectiveness of documented procedures, and in consistency of adherence.

- Objectivity of the assessment.

In summary, *a useful compliance assessment involves an objective analysis of the degree of compliance along three dimensions, across all affected departments, and for each of the requirements in the chosen standard.* Without the resulting level of detail and accuracy, it is virtually impossible to develop a reasonable estimate of the amount of effort and time that will be needed for implementation.

Establish the Project Structure.   You need not have a particular operating style to implement a quality system that satisfies the requirements of ISO 9001 or ISO 9002, although companies that normally have widespread employee involvement have advantages. It is important that you recognize your company's operating style and management style and plan with them in mind.

The three-level team approach to organizing shown in Figure 7–7 is common in large and medium-sized companies:

- The most senior manager and his or her direct reports form an ISO 9000 steering committee which will monitor and support the project. During the implementation project, the steering committee should hold comprehensive reviews of progress and quality system results at least once every few weeks.

Figure 7–7

A Common Project Structure

Once the system is in place and validated, members will hold reviews according to the review procedures they developed to meet the requirements for management responsibility in the standards (Subclause 4.1).

- The steering committee creates a cross-departmental or cross-functional ISO 9000 project team to set up and manage the project. Project team members are often middle managers from all areas of the company whose activities impact the quality of your products. In addition to project management, project team members are in a good position to design, document, and implement quality system elements that span the quality system. These elements should be put in place early so they can be used during implementation.

  The *project team leader* should be a member of the steering committee and is often the ISO 9000 champion. Members typically spend one-half day per week for the duration of the project, with the exception of the project leader, who may spend up to full time.

- The design, documentation, and implementation of ISO 9001 or ISO 9002 quality system elements that relate to specific line functions or support line functions is often done by element teams of nonmanagement employees knowledgeable in company practices related to the particular element. The leader of each element team should be a member of the project team. Element teams are not formed until they are needed. Members typically spend one-half day per week while they are working.

The cross-linking of teams—project team leader serving on the steering committee, project team members leading element teams—supports coordination and integration throughout the project.

Table 7–8 is a summary of the project responsibilities throughout implementation.

Develop a Project Plan.   Good business sense dictates that you prepare a formal, documented plan to manage a project of the magnitude of ISO 9000 implementation. A comprehensive project plan covers at least these topics:

- Project definition (based on statement of purpose).

- Project structure.

- Responsibilities of employees involved.

- The results of the operational environment and gap analysis. You cannot develop a realistic schedule and a reliable estimate of resource needs without knowing the amount of work you need to do, nor can you develop an approach to doing it.

- Schedules, including management reviews and milestones.

- Anticipated resource needs.

- Constraints and other issues that may affect internal preparation.

Develop Quality System Guidance.   The project team should begin to develop and document guidance on issues that impact the design and documentation of quality system elements. The results of these activities will be needed at various points during the implementation project. Specifically, this is the time to begin the activities described in the next few paragraphs.

Develop a document and data control system which will ultimately fulfill the requirements of Subclause 4.5 and assures that employees and managers have confidence that they have the most recent, accurate versions of all documents and data they need to do their jobs. This system should be in place in some form before any quality system elements are documented.

Determine guidelines for tradeoffs among qualification, training, and documentation for all activities that have a direct impact on meeting quality requirements. For jobs where

Table 7–8

Project Responsibilities

(Numbers in parentheses refer to subclauses in ISO 9001 and ISO 9002.)	
Steering Committee	
Accept ownership for the quality system	
Clarify company priorities	In Phase 1
Approve project definition (statement of purpose)	
Set up project (with project team)	In Phase 2
Design, document, and implement procedures to satisfy Subclause 4.1	Begin in Phase 1; end by beginning of Phase 4
Assign resources	
Conduct management reviews	Throughout all phases
Remove barriers and reject nonsupportive activities	
Project Team	
Set Up Project:	
Assess current status (with steering committee)	
Establish project structure (with steering committee)	
Educate project team members	In Phase 2
Identify responsibility for quality system elements	
Develop project plan	
Select registrar (with steering committee)	Begin in Phase 2; finish 1/3 of the way through the project
Form element teams and educate members	In Phase 5
Manage and Support Project:	
Plan and manage all activities in implementation project	
Identify resource needs and arrange for resources	
Oversee project progress and remove barriers to progress	
Update project plan as necessary	Throughout all phases
Make periodic progress reports to steering committee	
Ensure appropriate education for all involved employees	
Develop and implement a communication program	
Develop guidance on quality system design:	
—Document control system (4.5)	Begin in Phase 2; end by beginning of Phase 4
—Document preparation guidelines	
—Policy on tradeoffs between employee qualification and documentation	Begin in Phase 2; end by beginning of Phase 5
Design, document, and implement common quality system elements:	
Documentation structure and format (4.2) Internal quality audits (4.17)	
Corrective and preventive action (4.14) Training (4.18)	Begin in Phase 2; end by beginning of Phase 6
Control of quality records (4.16)	

(Continued on next page)

Table 7–8 *(concluded)*

Set priorities for designing quality system elements	Begin in Phase 2; end early in Phase 5
Ensure consistency and compatibility of design and implementation of quality system elements	Phases 2 through 7
Plan, coordinate, and participate in process analysis	In Phase 3
Plan, coordinate, and participate in quality planning	In Phase 4
Prepare the quality manual	In Phase 6 or Phase 7

Element Teams

Design, document and implement relevant quality system elements:

Contract review (4.3)
Design control (4.4)
Purchasing (4.6)
Control of customer-supplied product (4.7)
Product identification and traceability (4.8)

Process control (4.9)
Inspection and testing (4.10)
Control of inspection, measuring and test equipment (4.11)
Inspection and test status (4.12)
Control of nonconforming product (4.13)

Handling, storage, packaging, preservation and delivery (4.15)
Servicing (4.19)
Statistical techniques (4.20)

} Throughout Phases 5, 6, and 7

Adjust approach to ensure harmony with other elements — Throughout Phases 5, 6, and 7

Report progress and issues to project team — Throughout Phases 5, 6, and 7

your company uses temporary help, or entry-level jobs, a guideline might be to rely on detailed, written instructions for employees. In other jobs, employees might need to be highly experienced, and procedures may not capture the essence of how to detect deviation from acceptable quality. For example, processed material might need to have a certain texture and "feel" to satisfactorily pass through to downstream processing stages. In yet other jobs, acceptable performance may require employees to have completed training or received a certificate of qualification.

The people-procedure orientation addressed earlier as part of assessing operational environment will provide a starting point. The tradeoff guidelines need to be in place before any quality system elements are designed, documented, or implemented. As the project moves through other phases, these guidelines will be refined and expanded into defined skill and knowledge requirements for every job.

Develop document preparation guidelines. In addition to the decisions related to training and documentation tradeoffs, the project team should develop an overall documentation structure; acceptable formats and media, perhaps building on existing documentation; writing guidelines; and the review and approval mechanisms (as an adjunct to document and data control procedures). These activities should be completed before any quality system elements are documented.

All of your documentation taken together must form a coherent whole. Many companies choose a structure such as the one depicted in Figure 7–8:

- A high-level *quality manual* that outlines the quality system, including the structure and location of related documents such as quality plans or procedures.

Figure 7–8

A Common Documentation Structure

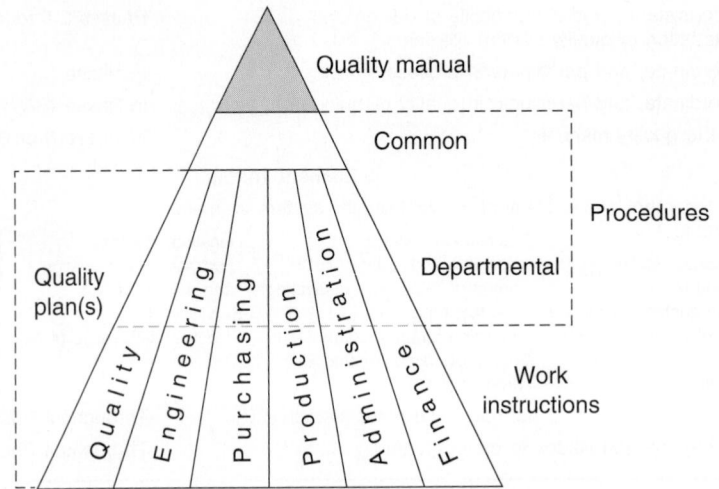

- *Procedures* that cover common quality system elements that span all operations in the company (such as corrective and preventive action), and quality system elements that are specific to departments or functions (such as purchasing or design control). Note that you will also have procedures for keeping *quality records*.

- *Work instructions* for instances where specific, detailed steps must be followed to assure the quality of the output.

- *Quality plans* or other vehicles for capturing the results of quality planning.

(See Chapter 8 for a detailed discussion of documentation.)

Set priorities for working on quality system elements. Setting and refining priorities is an ongoing responsibility for the project team. In the first few weeks of the implementation project, the steering committee (as an element team) should begin work on designing, documenting, and implementing procedures to address *management responsibility* (Subclause 4.1).

In the first few months, the project team (acting as various element teams) should begin work on common quality system elements that can be helpful throughout implementation; that is, *Corrective and preventive action* (Subclause 4.14), *Control of quality records* (Subclause 4.16), *Internal quality audits* (Subclause 4.17), and *Training* (Subclause 4.18). The project team can set preliminary priorities for working on other elements based on the results of the gap analysis and will refine them based on further in-depth analyses of the elements by element teams (discussed under Phase 5).

Ensuring consistency and compatibility in the design, documentation, and implementation of all quality system elements.

Ensuring appropriate education for employees working on the project and the larger employee body as the project proceeds.

Select a Quality System Registrar (If Registration Is a Goal).   Given that it may take several weeks to a few months to select a registrar systematically, start during the first few weeks of your implementation project so that you have made your selection no later than a third of the way through. Starting early gives you an opportunity to select a registrar who meets your needs and set up a relationship. It assures that your registrar will have time available when you are ready. And establishing dates for a pre-assessment and the formal assessment dates well in advance can help maintain momentum in your company toward registration.

Because of the potential for conflict of interest, registrars are prohibited from consulting. However, they do help established clients by confirming interpretation of the requirements in the standards. And, if asked, they usually will share their experiences of workable approaches seen in other companies. So, the earlier you establish a relationship, the longer your registrar can serve as a resource.

The basic steps in selecting a quality system registrar are listed in Table 7–9. (See Chapter 5, "Registration and Selecting an ISO 9000 Registrar"; Chapter 17, "Registrar Accreditation"; and "Appendix D—List of Registrars.")

Table 7–9

Select a Quality System Registrar

1. Review statement of purpose
Ensure accuracy and completeness
2. Identify candidates
Determine accreditations
Determine stability
Determine industry competence
Identify other circumstances that affect choice
3. Gather information
Assign responsibilities to project team members
Request information packets
Interview registrars
Contact registered companies
4. Choose registrar
Select final candidates
Conduct formal discussions
Come to consensus in project team
Obtain approval from steering committee
Negotiate contract with registrar

Phase 3: Define and Analyze Processes

Every company has ways of doing work called processes and has systems to help guide work through processes. For work to flow smoothly and efficiently, systems must support processes. If you have ever been told "I'm sorry, but our systems don't allow us to do that," or worked in situations where formal systems, often computer based, seem to be an obstacle to getting the job done, you will understand the impact when systems do not support processes.

To be effective, your quality system must be designed to support your companywide business processes. In fact, *you cannot do quality planning* for your quality system (required by Subclause 4.2.3 of the standards and Phase 4 of the ISO 9000 implementation project) *without understanding your major business processes.*

Phase 3, then, has one primary objective:

- To understand the processes you use to create and deliver products, in preparation for quality planning.

Table 7–10 lists the major activities in Phase 3, and the rest of this section briefly discusses selected activities.

Define Business Processes.   All companies have two major processes that involve most departments or functions and have an interface with external customers, although the processes may be informal and followed in a highly variable manner from one employee to another:

- Stimulating customer inquiries (for example, by advertising, direct mailings, sales calls) and responding to customer inquiries (for example, responding to requests for a proposal, and preparing quotations and estimates). The results of this process are customer orders in one form or another.

- Designing, producing, delivering, installing, and servicing products (or whatever subset of these activities is appropriate for the company) to satisfy customer orders.

Table 7–10

Phase 3: Define and Analyze Processes

(Months 1–6)	
Activity:	*Who Does It:*
• Define processes	Project team and knowledgeable employees
• Characterize process interfaces	Project team and knowledgeable employees
• Measure overall process performance	Project team and knowledgeable employees
• Modify processes	Project team and knowledgeable employees

Some companies have a third major process:

- Developing and introducing new products.

A quality system that satisfies the requirements of ISO 9001 or ISO 9002 will cover part of the first process, since both standards include requirements related to responding to customer requests for quotations and proposals as part of contract review (Subclause 4.3), and most of the second process. If you are pursuing registration to ISO 9001, you will address all of the second process and the third process (since ISO 9001 includes design control, required by Subclause 4.4).

Create Process Charts.   In Phase 3 of the implementation project, you essentially describe your major business processes. You might start with "designing, producing, delivering, installing, and servicing products" (or whatever subset of these activities is appropriate for your company). One advantage of beginning with this process is that it is easy to see a direct relationship between it and the quality of your products.

Flowcharts and block diagrams are excellent, proven vehicles for describing or creating "pictures" of processes. Effective "process charts":

- State the *purpose* (or value) of the process.

- Show what *actually happens.*

- Identify and include all the *major activities* (or departments) *involved* in the process.

- Identify and indicate (for example, with arrows) all the *interfaces:*

  —between departments

  —with external suppliers

  —with external customers

  —with systems (for example, computerized inventory systems).

- *Capture the flow of both the product and information about the product* (for example, customer requirements, technical specifications, purchase orders, and routing sheets).

Companies usually create levels of process charts, with each succeeding level covering a narrower scope but in more detail:

- The *most general* charts describe *companywide processes* at a high level.

  Figure 7–9 is a flowchart of a companywide process for the design, production, and delivery of children's clothing. Customers (shown on the right side) have requirements related to that clothing. Purchasers may have requirements related to characteristics of the material to be used and available sizes and colors. Distributors and retailers have additional requirements related to number of pieces

Figure 7-9
Sample Major Business Process

Specialty Clothing, Inc.

Purpose of process: to design, produce, and deliver children's clothing.

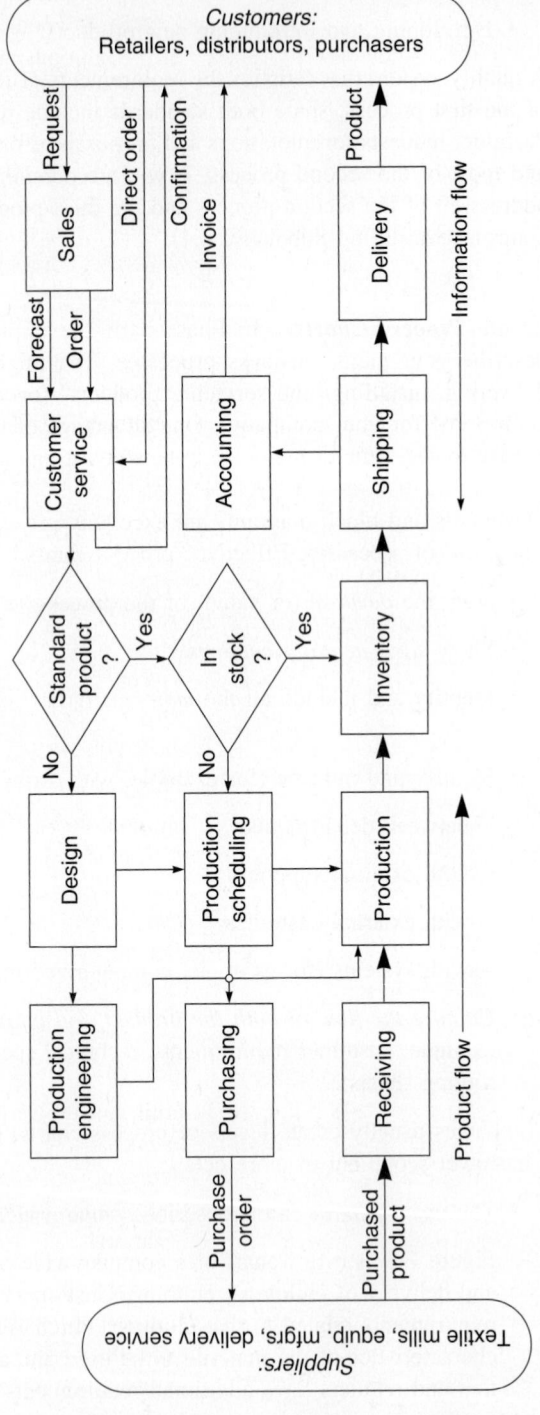

provided, distribution of sizes and colors, packaging, delivery intervals, delivery locations, and provisions for returning goods. For Specialty Clothing to meet customer needs, the company will need goods and services from external suppliers (some of whom are shown on the left side of the figure).

The process Specialty Clothing involves all departments, performing activities such as order taking, designing clothing, purchasing goods and services, producing patterns, controlling inventory, scheduling production, cutting cloth, sewing clothing, inspecting finished articles, sending invoices to stores, and managing returns. Each one of these departments requires something from one or more other departments. For example, production scheduling needs information from sales and inventory; cutters need material from suppliers or from inventory as well as accurate, complete patterns; employees who sew clothing need appropriately cut material, accessories, and instructions. The extent to which the finished article of clothing meets customer requirements is directly related to the extent to which internal departments meet each other's requirements.

- *Second-level* charts show the flow of product and information *within each of the departments* or activities (boxes on Figure 7–9) identified on the companywide chart. You may have existing department-level charts that can serve as the basis for these second-level charts.

- *Third-level* charts could show even more detailed work activities if necessary. This level for Specialty Clothing's customer service department may describe the steps involved in accessing an inventory data base, calculating prices, or entering an order into an ordering database.

The best way to create accurate and complete process charts is to call in the experts— the people who work in the process. They are the only ones who know what really occurs on a daily basis. If members of an ISO 9000 project team represent all major activities, they might be the appropriate people to create high-level, companywide process charts. If not, arrange for representatives of all relevant departments to be involved. In either case, charts should be reviewed by several people who work in the process. To create the more detailed second- and third- level charts, managers could work with the employees in various departments.

Be prepared to revise the charts until agreement is reached. *You cannot continue with process analysis until you have an accurate picture of your operations.* If you cannot reach consensus, indicate the problems on the charts and investigate. For example, send a representative to watch what actually happens. In slow processes, it may be useful for someone to walk a job all the way through the process.

Establish Change Control.   During process charting, employees often bring up problems they experience on a daily basis. In addition, the charts themselves may show, for example, departments that do not connect but should connect, reports that are generated but do not go anywhere, or raw materials that are not used.

Record ambiguities and problems, but *do not make changes to processes while you are creating charts.* It is wise, however, to channel the natural desire of employees to fix problems immediately into developing a *formal mechanism for review and approval of proposed changes and for updating of process charts* early in process analysis. In other words, process charts should be under document control!

Characterize Process Interfaces.   Characterizing interfaces involves three activities. For each block on a process chart, at any chart level:

- *Identify what goes into (inputs) and comes out of (outputs) the process block that is related to the quality of products.* Inputs and outputs may be tangible, such as a piece part, a software program, or a chemical compound. They may be information, such as entries into a database or a financial report.

- *Identify the customers for outputs and the suppliers for inputs.*

- *Validate inputs and outputs directly with customers and suppliers.* For internal relationships, representatives of supplier department representatives can meet with representatives and other employees from internal customer departments to *verify that the outputs from their departments match up with the inputs their customers receive.* Discrepancies should also be resolved during these meetings. Any discrepancies that remain may require further investigation. Record for future attention those that cannot be resolved at all. Some of these will provide opportunities for future improvements.

Department representatives who collaborated in creating process charts should be knowledgeable enough to develop preliminary lists of inputs and outputs.

Table 7–11 shows sample inputs and outputs for the customer service department of Specialty Clothing, Inc.

Measure Overall Process Performance.   An accurate picture of the activities, inputs, and outputs of your business processes is only half of the information you need to understand your processes. You also need to know how well they operate.

In Phase 4, as an input to quality planning, you will talk to external and internal customers about their detailed requirements for the outputs at each of the process interfaces. For now, you are interested in some *general measures of overall efficiency and effectiveness* of companywide processes and department-level processes.

Measurements of process performance are a cornerstone of an effective quality system. Not only do they tell you how well your processes are operating currently, they also give you critical information for improvements:

Table 7–11

Sample Inputs/Outputs Matrix

For customer service dept.

Suppliers	Inputs	Outputs	Customers
Distributors/ retailers/ sales dept.	Orders (form C-95)	Interface with company	Distr./retailers
		Price and del. date info.	Distr./retailers
		Order confirmation	Distr./retailers
Sales	Forecasts	Order status	Distr./retailers
Sales	Credit information	Requests for special orders	Design dept.
Design and sales	Special pricing	Entries into central data base	All depts.
Design and sales	Special delivery	Requests for credit checks	Sales dept.
Inventory	Inventory info.	Packing slips	Shipping dept.
Sales	Product line info.		
Central data base system admin.	Computer support		

- Knowledge of current process performance gives you a *baseline* against which to measure progress and the effects of changes you introduce during corrective action, preventive action, or process management and improvement.

- Without measurements of performance, you have few *tangible clues to the sources of problems* or their priority. Measurements are a primary tool for identifying which problems are the most important to address (perhaps because they relate to customer requirements or have a wide impact) and the magnitude of the problems.

In creating overall process measures (or checking those you have in place) for Phase 3, consider that customers (external and internal) almost always ask these four questions:

1. Did I receive what I asked for and expected?

2. Was it complete and accurate?

3. Did I receive it when I asked to receive it and when I expected it?

4. Am I satisfied with the way I was treated during interactions with the company?

And managers typically ask:

5. Are we meeting our commitments? (Commitments might be on-time delivery, customer waiting time, meeting production schedules, or shipments to forecasts.)

6. What is the error rate? (Error rate might mean yield, scrap, rework, or returns.)

7. Is the productivity of the organization at expected levels?

Table 7–12 summarizes the attributes of a good measurement system. Table 7–13 lists the considerations that you should address for every measure you put in place, and Table 7–14 shows a sample measure Specialty Clothing might put into place in Phase 3 of implementation.

Modify Processes. Having created an accurate picture of your work processes and how well they perform, you are now at a decision point. You can:

- *Move on to Phase 4,* Develop Quality Plan(s).

- *Make obvious, localized changes to your processes* before or as you proceed with the implementation project. When there appear to be small, obvious, localized changes that would improve workflow and address customer needs, make them using your change control system. Then use your measurement system to see if the changes have an effect, and ask the people who work in the process and customers to critique the changes. Continue updating process charts, input and output lists, and characteristics, and making small changes to processes until you experience an acceptable degree of smoothness and responsiveness in your processes.

- *Address complex situations whose impact is widespread.* During process analysis, you may have found, for example, that the processes for handling customer

Table 7–12

Attributes of a Good Measurement System

- Measures process performance, not employee performance.
- Measures process effectiveness—what affects internal and external customers, usually related to timeliness, accuracy, completeness, and service.
- Measures process efficiency—how well your internal operations work; for example, measures of resources required per unit output and internal error rates or levels of rework.
- Is related to company's quality objectives.
- Uses only the measures needed.
- Measures what affects downstream operations.
- Is easy to understand for employees who collect and use measures.
- Measures what you think it measures.
- Uses appropriate methods of data collection and analysis.
- Includes qualitative and quantitative measures.
- Is dynamic, that is, changes as appropriate.

orders is workable but cumbersome because the computer system used to support these activities requires multiple, redundant entries. Changing the computer system to streamline the processes may be a significant undertaking and not be the highest priority during quality system implementation. Pursuing this option will lengthen the time it takes your company to complete the ISO 9000 implementation project.

Note that *before you make major changes to processes, you should have had your measures in place for long enough to have confidence that you know the normal capability of your processes.* If your measures were in place before you began the implementation project, you probably have this confidence. If you put new measures in place or modified existing ones significantly, be sure you have enough data collected to understand adequately your normal process performance.

Table 7–13

Considerations in Establishing Measures

For each measure you have in place, ensure that you have determined and documented:

Purpose:

— Why the measurement is being made

— How measurement results are used

Collection method:

— What to measure

— How often to measure

— Who measures

— How measurement is done

— Variable factors to be recorded

Analysis method:

— Who analyzes

— When analysis is done

— How analysis is done

— How output is presented

Review method:

— Who receives analysis

— When do they receive it

— What are they expected to do with analysis results

Feedback method:

— Who supplies feedback

— How usefulness and use of the measure are determined

— How satisfaction with *analyses* and *presentation* are determined

Table 7–14

Sample Measure

Customer:	Retail stores
Product:	Clothing
Req't:	Actual delivery date matches promised delivery date
Collection	
Measure:	Deviations from promised delivery date
Frequency:	Every order
By:	Designated customer service rep.
How:	Compare promised dates given and actual delivery dates; record deviations on Form CS-101 manually
Other Factors to Be Recorded:	Equipment breakdown, severe weather problems, database crashes
Analysis	
By:	Customer service supervisor
How:	Control chart of number of deviations, plotted monthly
	Control chart of absolute amount of deviations, plotted monthly
	Pareto chart of causes of deviations when more than 1 percent of deliveries do not match promised dates
Review	
Distribute to:	Sales vice president
When:	10th of each month
Action:	Review and discuss trends with customer service supervisor and shipping supervisor monthly

- *Move assertively and directly into process management and improvement.*[2] Process management has two major aspects: (1) *planning and managing daily operations* so that you maintain current process performance, and (2) *identifying opportunities* for improving processes to better meet customer requirements and improve process efficiency, and prioritizing and *pursuing identified opportunities* to drive processes to a new, higher level of performance.

ISO 9001 and ISO 9002 implicitly support process management, although they do not use the term nor do they explicitly cover all the elements necessary. However, control-

[2] An excellent book on process management is *Process Quality Management and Improvement Guidelines*, Issue 1.1, published by AT&T. These guidelines are part of an impressive AT&T Quality Library, a collection of books and resources to help build quality into every aspect of doing business. Each volume addresses a particular need or area of interest from leading quality initiatives through application of quality management tools and techniques. For a brochure describing the library: from the United States, call 1-800-432-6600 or fax your request to 1-800-566-9568; from Canada, call 1-800-255-1242; from other countries, call 317-322-6416 or fax your request to 317-322-6699.

ling processes (required by Subclause 4.9) is the primary activity in maintaining process performance on a daily basis and provides data for preventive and corrective action. Preventive and corrective action (required by Subclause 4.14) require data collection, data analysis, and taking action that should lead to process improvement.

Pursuing this option will require more widespread process definition and analysis and greater involvement of people. It will lengthen the time it takes your company to complete the implementation, but it can pay substantial dividends on the invested effort.

The choice of options is a strategic business decision that should be made by your ISO 9000 steering committee.

Phase 4: Develop Quality Plan(s)

ISO 9001 and ISO 9002 require that your company does *quality planning*; that is, that your company identifies the way specific practices, resources, and sequences of activities combine to meet quality requirements for products, projects, or contracts. The standards do not specify a vehicle for capturing the results of this activity, but you *must have evidence* of having done it. Many companies capture their results in *quality plans*. This section will use the term quality plan with the understanding that it means whatever mechanism you use to capture the results of quality planning.

The objective of Phase 4 of implementation, then, is:

- To create and document an integrated overview across all departments and functions of the way specific practices, resources, and sequences of activities combine to meet customer requirements.

Table 7–15 lists the activities in performing quality planning, and the rest of this section briefly discusses selected activities. You might also want to refer to ISO 10005, Quality management and quality system elements—Guidelines for quality plans.

Decide How Much to Do.   If you have a system of measures and controls in place that functions quite well in ensuring consistent product quality, and you have adequate records to prove it, you may only need to document what you already do to complete Phase 4. If not, the amount of quality planning you need to do and the number of quality plans you need depend on the nature of your products and your customers. For example:

- If you have a single product line, even if you have many similar products, and none of your customers impose unusual requirements, the way you meet the quality requirements for each product will probably be very similar. Here, you probably only need one, generic quality plan, possibly with small variations to account for minor differences among products.

   Note, however, that if different facilities that create and deliver the same product have quite different approaches to ensuring quality, you would need a different

Table 7–15

Phase 4: Develop Quality Plans

(Months 4–10)	
Activity:	*Who Does It:*
• Determine effort required	Project team
• Determine quality requirements	Project team and knowledgeable employees
• Translate requirements into factors to be controlled	Project team and knowledgeable employees
• Choose control limits	Project team and knowledgeable employees
• Establish measures and control methods	Project team and knowledgeable employees
• Document the quality plan(s)	Project team and knowledgeable employees
• Modify processes	Project team and knowledgeable employees

quality plan for each facility. For example, one facility may rely heavily on inspection by a separate department; another may combine process control with operator inspection.

- If you custom design products, or your customers have special requirements, you may need a separate quality plan for each project or contract.

- You may have a hybrid situation where you develop a generic quality plan to which you make additions, changes, or deletions to meet the needs of particular orders or contracts as they come in.

Determine Quality Requirements.   In Phase 3, you identified overall process measures to help you determine the capability of your processes based on your experience. In Phase 4, you supplement them (and possibly modify them) with more detailed measures based on specific quality requirements from your customers.

Quality requirements are attributes of a process output that internal and external customers expect and need in order to use the output, stated in quantitative terms to the extent possible. Note that quality requirements can also come from your own company, which might have internal standards for process and product performance based on strategic business decisions.

Determining quality requirements involves this set of activities designed to provide suppliers with an absolutely clear understanding of customer requirements: identifying, clarifying, and prioritizing requirements; coming to consensus about realistic performance; and documenting the results.

For each customer-supplier interface on your process charts, the *designated representative(s)* of the supplier uses whatever vehicles are appropriate (for example, face-to-face interviews, focus groups, questionnaires, or trials) to gather quality requirements from customers.

It is most useful if you start determining quality requirements at interfaces with external customers and work back through your processes. This way, internal customer organizations have an opportunity to react to requirements that they might not have fully appreciated before.

Those departments that interface with external suppliers should take the initiative as customers and give the suppliers documented requirements, feedback on current performance, and specific problems.

Customer requirements are often stated as:

- *Qualitative needs and expectations*, such as "I need legible order confirmation."

- *Perceived current performance*, such as "Your deliveries are late half of the time."

- *The biggest problems*, such as "I can't ever get through on your 800 number."

- *Acceptable performance in the short term*, such as "I need the data by 10:30 A.M. every Monday."

How do you know what "legible" means or what "efficient" deposit handling is for a commercial bank? A 98 percent on-time delivery may seem very good until it becomes clear that the 2 percent late delivery stops the flow of work in an internal customer's department.

Statements such as the examples above must be clarified so you can develop measures of how well you meet them. Some approaches to identifying and clarifying requirements are summarized in Table 7–16.

Make sure you include any regulatory requirements, such as those related to environmental performance and occupational health and safety. For example, unintended byproducts from your processes may have to be controlled. Regulatory requirements are usually quantitative and specific.

You may be overwhelmed by the number and extent of requirements you receive. Trying to address *all* customer requirements can, therefore, be frustrating, not particularly successful, and not necessary! So, during identification and clarification, determine customer priorities. Then, s*tart with the higher priority requirements for quality planning.* You can deal with the others later as part of process improvements.

Table 7–17 shows examples of clarified and prioritized company-level quality requirements for Specialty Clothing. You can use any format or process for capturing requirements that makes sense for you.

Table 7–16

Identify and Clarify Quality Requirements

- *Rephrase qualitative requirements.* For example, determine that "legible" means "either typewritten or a hard-copy printout from a computer."

- *Validate quantitative requirements.* Customers might say they want a one-second response time when they strike a key at a computer terminal. A simple trial might show that, in reality, customers become annoyed if they have to wait longer than one-tenth of a second.

- *Investigate problems and frustrations.* Determine whether customer complaints about cost mean the cost is too high, invoices are inaccurate, or the charges are reasonable but there are frequently large differences between estimated and final costs.

- *Negotiate interim quality levels* if your processes are not currently capable of meeting customer requirements. (By the end of Phase 3, you will have some information on the capability of your processes.)

- *Negotiate requirements that conflict with business direction.* Perhaps a customer wants paper copies of orders while your department is moving to entirely electronic transmission of orders.

- *Identify requirements based on process expectations.* A customer who wants paper copies of orders might not realize that you can generate and send an electronic order, which would, in fact, work well for the customer.

Translate Requirements into Factors to Be Controlled.   During this activity, you translate each documented quality requirement into one or more product or process factors that must be *controlled to meet customer requirements.*

Identify Factors. Note that some requirements can be controlled directly. For example, satisfying tight requirements for shaft endplay in geared servo motors might be controlled directly by limiting tolerances of individual components and operator adjustments during assembly. Other customer requirements must be controlled indirectly; for example, the amount of time airline customers spend waiting to be served. Meeting frequent flier expectations of a 10-minute maximum airport check-in time may translate into controlling the number of service agents on duty and an adequate level of computer response time. Meeting requirements for moisture content of powdered chemicals may mean controlling environmental conditions.

From a review of companywide process charts and process measures, you can identify factors that have an obvious impact on satisfying customer requirements. For example, the external customers of Specialty Clothing require shipments that:

- Are received on the promised date.

- Contain the correct number of items.

- Contain no defects in materials, workmanship, or packaging.

- Exactly match the order as acknowledged.

Table 7–17

Examples of Clarified and Prioritized Company-Level Quality Requirements

Customers: Distributors and retail stores

Output	The Customer Said:	Quality Requirement (the Customer Meant)	Priority
Orders received	"Correct"	Colors, styles, sizes of clothing received match order as acknowledged	1
		No defects in materials, workmanship, or packaging (specific defects listed)	1
		Delivery date matches estimate	2
		Number of items delivered matches number of items ordered	1
Interface with company	"Responsive service"	Credit checks completed in three business days	8
		Hours of operation from 10 A.M.–9 P.M. Mon.–Sat.	5
		Avenue for expediting orders*	7
	"Quick problem resolution"	Problems with orders resolved in one business day	6
	"Courteous and knowledgeable"	Employees can give info. on product lines, prices and deliv. times for standard products upon request	3
		Employees ask correct questions for special orders	4
Price	"Correct"	Final price within 2 percent of estimate	3
Invoices	"Timely"	Received by customer within three business days of order delivery	10
	"Accurate"	Price and quantity info. match order	9

* Requests for an avenue for expediting indicates that customers are not confident in the ability of the company to meet delivery dates, which, in turn, indicates inefficient and ineffective processes.

For standard products, the factors that influence the ability of the company to meet these requirements might include:

1. Accurate information on physical inventory.

2. Realistic commitment dates for deliveries by customer service.

3. Timely receipt of legible copies of orders by the inventory department.

4. Freedom from defects of materials, workmanship, and packaging.

5. Accurate stocking of newly produced product by the inventory department.

6. Timely and accurate picking of orders by the inventory department.

7. Timely and accurate packing, labeling, and shipping by the shipping department.

8. Appropriate choice of shipper for delivery by shipping.

9. Performance to agreed shipping intervals by the shipper.

Note that the technical quality of the products (that is, freedom from defects) is a small part of the total product quality from a customer perspective. Many factors previously viewed as internal operational considerations may have a direct bearing on responsiveness to customer requirements, and therefore fall under the purview of the quality system. Virtually every internal department and external supplier directly or indirectly influences how well the end customer requirements related to receipt of shipments are met.

You can determine more subtle factors that influence performance relative to customer requirements from people who work in each department. For example, orders from retail stores might follow seasonal patterns, and the resulting peaks in orders have the potential for processing delays in the inventory and shipping departments.

Purchases from suppliers in the Far East might be subject to shipping delays. The availability of designs and patterns for new styles may depend on annual events such as fashion shows and buyers' conventions. Some types and colors of cloth may be more susceptible to defects than others.

All the factors should be cataloged as they are identified and assigned to the department with primary responsibility.

Prioritize Factors. You will identify two categories of factors:

- Factors that can and should be measured and actively *controlled*; for example, timely and accurate picking of orders by the inventory department. *These factors are the primary focus of quality planning.*

- Factors that only need to be *monitored,* with active intervention only if performance becomes a problem. Examples might include factors outside of the direct control of the company (such as the performance of external shippers), factors strongly dependent on computerized systems (such as up-to-date inventory information), and factors that have not proven to be troublesome in the past (such as timely receipt of legible orders by the inventory department).

  These factors should be recorded as part of the quality plan, with the frequency and depth of monitoring dependent on the likelihood of a problem developing and the potential impact on customer satisfaction.

If you have many factors to be controlled, assign them priorities, based, for example, on the degree of impact on customer requirements, relative priority of the impacted customer requirement, cost in time and money if problems go undetected until later in the process, and importance of the factor in limiting the flow of work and products through processes.

Select Factors. Based on past experience, you might feel that all of the identified factors, taken together, are more than are needed for efficient control of quality relative to requirements. A good way to assess whether the high-priority factors identified so far are appropriate is to map all the factors onto the company-level process charts.

To assess the *adequacy of the proposed factors* for control, identify the 10 to 15 most significant quality problems in the last year and ask:

- Would the controlled factors identified so far have detected the problems with high probability?

- Would the controlled factors identified so far have detected the problems early enough in the process to contain the problem and avoid significant losses of money or time to meet customer requirements?

- Are all internal or external customer requirements adequately covered by these factors?

If the answer to any of these questions is no, determine the minimum number of additional control factors needed to contain the problems.

To *evaluate the possibility that an excessive number of factors have been proposed*, ask:

- Is there any duplication among the control factors as proposed?

- If each proposed factor was individually changed from "controlled" to "monitored," what would have been the worst that could have happened over the last year?

Use the answers to these questions to reduce the number of controlled factors to the minimum that makes good business sense. That is, establish a prudent balance between risk of problems and costs of control.

Choose Control Limits.   With a comprehensive set of customer requirements and a list of the factors to be controlled or monitored in hand, you can now define the desired performance for each factor. Defining desired performance is usually done by defining *control limits* (or acceptable ranges) and acceptable *probabilities of staying within those limits*. Any tendency for performance to go outside of these limits should trigger interventions to return performance to acceptable levels.

Overall guidance for setting control limits should be found in the quality policy and the quality objectives required by Subclause 4.1.1 of the standards and discussed in Phase 1. The specific control limits chosen will be influenced by, for example, company reputation desired, priority of customer requirements, cost-benefit tradeoff to fully meet customer requirements, and operational feasibility.

Most companies will be able to set reasonable trial control limits based on *past experience*. The collective wisdom and experience of people working in the process, if appropriately used, can yield fully acceptable control limits on individual factors.

You can choose among several approaches to *assessing the adequacy* of the control limits proposed; for example, computer simulation, successful past experience, small-scale pilot studies, tracking performance with control limits in place, or a combination. Whatever approach the project team chooses should include tracking performance of the full-scale quality system and making adjustments as necessary.

Once the project team receives acceptance from all affected departments, the steering committee should approve the final proposed control limits.

Establish Measures and Control Methods.   The discussion of Phase 3 commented on the value and attributes of good measures and considerations in establishing measures. Those comments apply to Phase 4 as well, where you are instituting much more detailed measures.

The *types* of measurements to be made can almost always be readily determined from the nature of the factor to be controlled or monitored:

- Measurements of *accuracy* might be as simple as counting the number of items; they might require measuring physical parameters such as dimensions, weight, or volume; or they might require assessing the performance of specialized medical diagnostic equipment.

- Measurements of *timeliness* might be on a scale of seconds, hours, or days. Sometimes an indirect measurement of timeliness is the size of the backlog in a queue.

- Measurements of *completeness* usually involve comparisons with a standard reference. For example, documents in an application for life insurance can be compared to a standard checklist.

- Measurements of *service* are almost always more subjective, although the factors may have been translated into one of the other three types of measurements which are, typically, more objective. For example, customers may consider good service as receiving answers to their questions without being referred to someone else or being called back. In this case, a measure of the percentage of calls that require referrals or callbacks would be a good proxy for level of service.

Whether the purpose is active control of the factor or only monitoring compliance the following considerations should dictate the *frequency* of measurement and analysis:

- Variability of the factor.

- Well-established cyclical patterns.

- The speed with which control actions can be taken.

- Difficulty or expense of data gathering for the measure.

- The availability of early warning signals that might be helpful.

The secret to *useful analysis* of measurements is to respond to the needs of users—the people who need to make decisions based on the data. Use visual presentation of results if possible. Keep it simple. Make it fast to perform. *The goal should be to keep the measurements and analysis as simple and as straightforward as possible while still achieving the desired overall performance.*

In addition, you should consider the employee skill and documentation tradeoff policies developed in Phase 2 as an input to establishing the way measurements are made and analyzed. Suppose one guideline is to minimize paperwork and rely on the skill and experience of the employees to the maximum extent possible.

If the control mechanism for waiting time in a check-in line is to adjust the number of active service agents, a short training session (possibly coupled with a brief policy statement on intended performance and general guidelines on typical trigger points) may be all that is needed. If collecting and analyzing data for a simple history chart of processing times is to be done on a daily or weekly basis, it might make sense to rely on training rather than a work instruction; training could be as simple as showing someone a copy of a recent analysis, telling them what to, and verifying that they can do it correctly. If the analysis is done infrequently, it may be necessary to have some elementary training documentation that could be used as a memory refresher if needed.

For each factor, record the decision on the mix of skill, training, and documentation and the reasoning behind the decision, for use in Phases 5, 6, and 7. At this point, you may have identified existing training or procedures that will support the quality plan. Or you may have decided on the need for additional training or procedures. You may be able to assign a unique number to needed procedures even if they are not yet prepared. Reference to these procedures and training should be included in your quality plan as soon as they are identified. This reference will prove invaluable during later phases of preparation.

Document Quality Plan(s). Documenting the quality plan will be an ongoing activity throughout most of Phase 4 and beyond, as the quality plan matures. This means that the document control system must be in place by early in Phase 4.

As you are developing the list of controlled factors, corresponding control limits or acceptable ranges, and control methods, it will be important to record them as part of your evolving quality plan. The exact format is not as important as finding a way to relate all of the aspects of the quality plan in a manner that is comfortable for the people in your company. Figure 7–10 shows one possible format. Note that the sample in the figure:

- Has redrawn the company-level process chart in a vertical format.

- Reiterates the factor, whether it is controlled or monitored, its priority, and the related customer requirement.

Figure 7–10

Part of a Sample Quality Plan

Department	Factor	Priority (L, M, H)	(C)ontrol or (M)onitor	Related Customer Requirement
Sales	Credit check completion time	L	M	Credit checks completed in three bus. days
Customer service	Est. delivery dates	H	M	Delivery date = order
	Product knowledge	M	M	Get product info. on request
	Dispatch of orders	H	C	Delivery date = order
	Problem resolution time	M	M	Problem resolved in one bus. day
Production sched.				
Purchasing	Purchase order accuracy	H	M	Delivery date = est,; colors and styles = order
	Ongoing supplier qualification	M	M	Delivery date = est.; colors and styles = order
Receiving	Received goods accuracy	M	C	Colors and styles = order; delivery date = est.
Production	Defects in finished goods			
	—material	H	C	0 defects in material
	—workmanship	H	C	0 defects in workmanship
	—packaging	H	C	0 defects in packaging

To Shipping

To inventory

To inventory

Acceptable Range	Control/Monitor Method	Reference
>80 percent within three bus. days	Monitor weekly perf. on history chart	Procedure SA-1
<5 percent orders challenged by inventory dept.	Train cust. service agents; use Wizard sys. for inventory control; monitor weekly perf. on history chart	Training course HR-1; proc. MIS-1; proc. CS-1
>90 percent of requests answered without callback	Train cust. service agents; monitor weekly perf. on hisoty chart	Training course HR-2; Proc. CS-2
>90 percent within one bus. day; 100 percent within two bus. days	Measure >1 day old every day and adjust staff to keep below 10 percent	Proc. CS-3
>75 percent within one bus. day	Monitor weekly perf. on history chart	Proc. CS-4
<0.5 percent of purchase orders changed because of errors	Monitor weekly perf. on history chart	Training course HR-3; proc. PU-1
>90 percent supplier perf. on delivery times;	Monitor weekly perf. on history chart	Training course HR-4; proc. PU-2; PU-3
<1 percent errors on colors and styles	Monitor weekly perf. on history chart	
<1 percent errors on colors and styles	100 percent incoming inspection against purchase order	Training course HR-5; proc. SH-1
<0.05 percent of items found with defects after leaving production dept.	Combination of operator training, process control, and operator inspection	Training courses HR-6 thru HR-10; proc. PR-1 thru PR-20

- Indicates responsibility for measuring, monitoring, and controlling activities by positioning the factor next to the responsible department or function in the process chart.

- Lists the training or procedures that describe how to accomplish control or monitoring.

Modify Processes. Quality planning stimulates thinking about your business processes and how well they operate in ways that you may not have done before. An inevitable consequence is additional insight into potential improvements that may range from minor adjustments to major reengineering.

These insights should certainly be recorded for review and consideration. Any further action (large-scale improvement efforts, moving into process management and improvement) you take at this point and its implications should be carefully reviewed by the steering committee.

Phase 5: Design Quality System Elements

Phase 5 has two primary objectives:

- To develop action plans for the design, documentation, and implementation of each quality system element.

- To design/refine and validate procedures that support each element.

Table 7–18 lists the major activities in Phase 5, and the rest of this section briefly discusses selected activities.

Establish Element Teams. Element teams will design, document, and implement your new or refined quality system. All element team members should have some specialized

Table 7–18

Phase 5: Design Quality System Elements

(Months 1–13)	
Activity:	*Who Does It:*
• Establish element teams	Project team
• Perform in-depth gap analyses	Element teams
• Refine priorities for action	Project team
• Develop action plan for each element	Element teams
• Design documentation	Element teams and line departments
• Validate overall design	Element teams

knowledge, experience, or position that makes them appropriate people to work on specific quality system elements:

- The steering committee should act as the element team for *Management responsibility* (Subclause 4.1). The design, documentation, and implementation of this element should have been completed by the beginning of Phase 4 so that the quality policy, quality objectives, and relevant procedures can be used throughout the rest of the implementation project.

- The project team is in the best position to act as the element team for the elements that affect the entire quality system: *Quality system* (4.2), *Document and data control* (4.5), *Corrective and preventive action* (4.14), *Internal quality audits* (4.17), *Control of quality records* (4.16), and *Training* (4.18). The design of these elements should be completed before Phase 5 so that the related procedures can be used during design, documentation, and implementation of the other elements.

- Representatives of line departments who are knowledgeable about relevant work are in the best position to participate on teams for elements that are related to the activities of line departments. In many companies without formal quality systems, the knowledge and motivation of astute lead people and first-line supervisors is the primary reason that requirements for quality are met. These employees have a good perspective on what procedures work well, and the kinds of documentation that will be effective for other employees.

Perform In-Depth Gap Analysis.   The gap analysis conducted in Phase 2 as part of the initial company assessment provided a good overview of the status of your quality system relative to the requirements of your chosen standard. For planning purposes, that analysis was effective. In Phase 5, each element team identifies and reviews all the current practices and documentation related to its quality system element in detail. The results are summarized in a report that indicates the status of all relevant practices and documentation relative to both the relevant ISO 9000 standard and relevant quality plans. Figure 7–11 is a simplified diagram of this activity.

Refine Priorities.   With the summary reports in hand, the project team refines its priorities and relative time frames for each quality system element, using these primary criteria:

- *Applicability across the quality system.* If the project team and senior management have not completed the design, documentation, and implementation of their respective quality system elements, those elements must have top priority now.

- *Amount of work involved.* This criterion has two dimensions. One is sheer amount of *effort.* For example, you may need to make major refinements to most of the practices and documentation for an element that exists in some form in your quality system but does not fulfill the ISO 9000 requirements. Satisfying the requirements for *Control of inspection, measuring and test equipment* (Subclause 4.11) falls into this category in many companies.

A second dimension is related to *time*. While the actual effort required may be minimal, calendar time to verify that procedures work and gain confidence in your process control may take some time.

Satisfying the requirements for Design control (Subclause 4.4 in ISO 9001) falls into this category.

- *Resources* available. This criterion also has two dimensions. One is related to *number of employees available. The* second is related to *expertise*—when can you ensure that the appropriate employees for a specific element will be available to work on the element.

Figure 7–11

Performing an In-Depth Gap Analysis

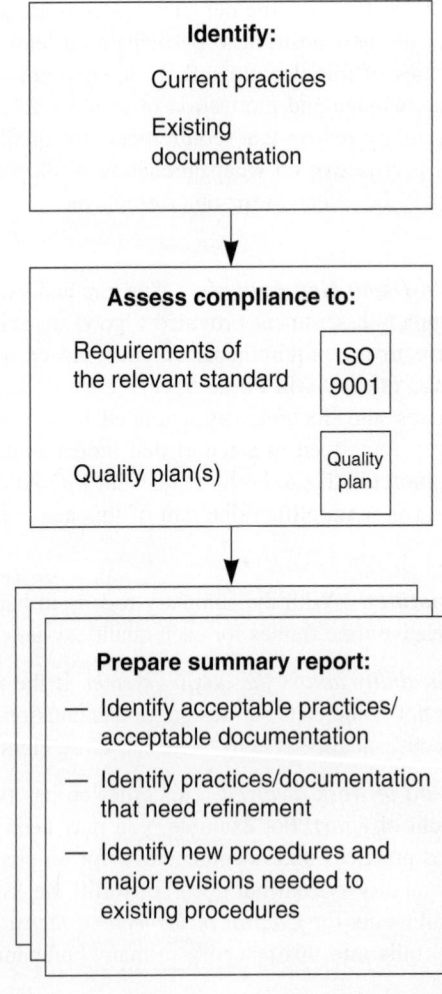

Plan Actions.   Element teams use the results of the in-depth gap analyses and the priorities set by the project team as the foundation for action plans that detail the approach to designing (or refining), documenting, and implementing each quality system element. Action plans should be documented, put under document control, and approved by the project team. The discipline of a written plan forces clarity of thinking, provides an excellent vehicle to assure consensus among element team members, and gives element teams a roadmap to follow, even if membership of the teams changes.

All action plans need not be at the same level of complexity or detail. You may not need a plan at all; it depends on how much work needs to be done. (Table 7–19 summarizes options for action plans.) For example:

- If an element team finds that practices are adequate and documentation current, it need only make sure the procedures are under document control. Effective procedures should not be revised only to match new format guidelines at this time. When a substantive change occurs in the procedure, the format can be revised if necessary using the document control system.

- If a team finds large numbers of drawings or product specifications that are not under document control and may be outdated but are not used frequently, the team need not necessarily plan to rework all of them now. The document control system should provide guidelines for handling this situation; perhaps a guideline will be that each specification will be reviewed and approved or revised, and assigned a document number when it next needs to be used.

- When a substantial gap exists that requires a full, formal action plan, work will probably not proceed in a straightforward, sequential manner. It will more likely be an iterative process. In developing an action plan in this case, the schedule should recognize that extra time will be needed.

A typical, full action plan would include these seven main topics:

- Background:

    —application of element in company (all activities, all departments)

    —analysis of work to be done (detailed description)

Table 7–19

When Do You Need a Quality System Element Action Plan?

If the gap analysis indicates:	Then:
Practices adequate, documentation current, only minor refinements needed	Formal plan is probably not necessary
Practices adequate, documentation inadequate	Abbreviated plan probably sufficient
Practices inadequate/nonexistent, documentation inadequate/nonexistent	Full, formal action plan needed

—potential obstacles, constraints, and concerns.

- Ongoing coordination needs with other element teams (for example, other elements that may impact or be impacted by the relevant element and common issues among elements, such as training).

- Plan for element design and validation. (See Table 7–20.)

- Plan for element documentation. (See Table 7–21. The subsection on Phase 6 discusses selected activities in this plan.)

- Plan for element implementation. (See Table 7–22. The subsection on Phase 7 discusses selected activities in this plan.)

- Refined time and effort estimates.

Design Documentation

Propose detailed skill levels. Designing or refining new practices and procedures must start with proposed skills and knowledge needed for each activity in the quality system. Without this starting point, the element teams are not likely to meet employee needs.

Sources of skill and knowledge information are:

- *Employees who do the activities* are in a good position to tell element teams what it takes to do their jobs and what would help them do their jobs better.

- *Line managers* within the departments or functions that are part of the quality system.

- The training and procedures identified in *quality plans*.

- The results of the *people-procedure analysis* done by the project team during initial assessment in Phase 2.

Table 7–20

Typical Plan for Element Design

- Provide information, education, and training to involved employees
- Design procedures:
  - —Propose detailed skill levels and knowledge
  - —Identify options for fulfilling requirements
  - —Analyze options and make a selection
  - —Outline existing, modified, and new practices and procedures
  - —Conduct trials of modified practices and procedures, analyze results and revise
- Validate designs and submit to project team for review and revise as necessary
- Revise practices and procedures as necessary to address results of documentation, implementation, and validation assessment (Phases 6, 7, and 8)

Line managers are ultimately responsible on an ongoing basis for establishing skill and knowledge needs. Therefore, it makes sense for element team members to work with line managers on developing the specific skill and knowledge requirements for each job and capturing the agreement in writing. Anticipate that the details may change as a result of outlining and validating new or refined procedures. This is normal.

Documented definitions of skill levels and qualifications are a requirement of the standards and must be put under document control. Even at this proposal stage, it will be useful to know what version or revision of skill level is being discussed or used.

Design Procedures. Once the proposed skill and knowledge requirements have been agreed upon, an element team can design new or refined procedures to meet the requirements of the standard and the intent of the quality plans.

A good approach to developing effective procedures is:

- *Review inputs*: strengths and shortcomings shown in the in-depth gap analysis, proposed skill levels, quality plans.

- *Identify options.* Brainstorming with selected employees who do related work is a good way to identify options for fulfilling the requirements, both in terms of content of the procedures and presentation of the material. This approach might be supplemented with informal interviews or focus groups. Capture what you learn on sketches, charts, diagrams, bullet lists, and pictures, anything that is easily generated and communicated for later use.

- *Analyze options and make a selection.* If possible, the same people who proposed the options during brainstorming should be involved in analyzing and evaluating the options. The participants might take the most viable options back to their

Table 7–21

Typical Plan for Element Documentation

- Provide information, education, and training to involved employees
- Review guidance on quality system elements from project team
- Carefully review relevant results of Phase 5
- Develop/refine documentation
- Conduct review by sample users, analyze results, and revise as necessary
- Conduct trials under actual use conditions, analyze results, and revise as necessary
- Conduct adequacy audits and revise as necessary
- Obtain approval
- Revise procedures as necessary to address results of implementation and validation assessment (Phases 7 and 8)

Table 7–22

Typical Plan for Element Implementation

- Identify employee competencies needed
- Identify employee competencies possessed
- Arrange for training, education, experience, or qualification
- Identify implementation strategy(ies) and put into effect
- Perform internal compliance audit and take corrective action
- Track performance
- Revise design, documentation, or implementation as necessary

workgroups to gauge perceived efficiency and feasibility and employee acceptance. During evaluation, consider such things as:

—compatibility with existing company systems

—compatibility with company operating style

—compatibility with skill levels

—available resources

—available time

The goal is to create procedures that work, are efficient, and most people can live with. In the ideal case, the activities required to satisfy the requirements for quality and the relevant ISO 9000 standard should seem so natural and straightforward that the quality system almost seems invisible to most employees.

- *Outline procedures.* Once an option has been selected, element team members should outline the procedures in just enough detail so they can explain them to employees who have *not* been involved in creating and evaluating options. Each outline should include a description of the proposed format and media and include rough sketches of any figures or flowcharts. If tables of settings or operating parameters are to be included, it may be useful to include those as well, although they do not have to be in final form.

- *Trial procedures.* The goal of conducting a trial of outlined procedures is to confirm that the proposed procedures are basically effective, will work, and will be acceptable to the users *before investing significant time and effort in writing full-blown procedures.* Trials should identify most needed modifications. A small group of employees might be asked to try following the proposed procedures for a week or two to identify better which aspects they like and which are not comfortable or helpful.

- *Revise options.* Using the results of the trial, revisions of either the approach or the information in the outlined procedures should be incorporated to achieve full acceptance. If the revisions are extensive, another trial may be prudent. If revisions to the proposed skill levels seem appropriate, these will need to be negotiated with the departmental managers before changing the proposed procedures.

Work instructions to support procedures are the responsibility of the relevant line department. Element teams should come to agreement with line managers on what work instructions, if any, are needed and will add value. And element teams should provide guidance to departments in developing work instructions. In Phase 6, element teams will be responsible for reviewing and approving any work instructions that are developed, since all work instructions must be compatible with related procedures and be referenced in those procedures.

Validate Element Design.   Implicit in the objectives for Phase 5 is the need to validate that the proposed combination of skill levels and procedures:

- Complies with the requirements of the relevant standard.

- Satisfies the intent of the relevant quality plan.

- Is efficient from a business perspective.

- Is compatible with proposals from other element teams.

- Is acceptable to most of the employees who have to work with the procedures.

The element team will need to decide how to validate the overall design of each element to minimize the preparation time and effort within the constraint of limited resources. The validation approach may well vary from one element to another based on, for example, the size of the company and the amount of change required, as shown by the in-depth gap analysis.

Small companies might have good informal communications and the flexibility to respond quickly to needed improvements uncovered during the full documentation and implementation phases. They may decide that little formal validation is needed beyond the trials of the outlined procedures, even if there is substantial documentation of procedures and work instructions.

Larger companies with extensive changes to their quality system should probably plan more formal and extensive validation studies, especially if the quality system will be used at several facilities. This might take the form of a phased implementation of the quality system in selected locations or departments with systematic evaluations of acceptance and effectiveness in fulfilling the performance specified in the quality plan(s). While this would not avoid possible revisions to the documentation, it would save on training and implementation in other departments or sites.

The results of validation should be reflected in procedure outlines and proposed skill and knowledge levels that are the basis for Phase 6.

Phase 6: Document Quality System Elements

In Phase 5, each element team developed a plan for documenting its element (that was shown in Table 7–21). Now that plan is put into effect with this objective:

- To ensure that documentation for each quality system element is appropriately developed, reviewed, and approved.

Based on the priorities established by the project team in Phase 5 and the actual effort necessary to design quality system elements, all elements will not be documented at the same time.

Documentation responsibilities are split:

- Developing *procedures* is the responsibility of the element teams. However, depending on the contents of the plan for documentation created in Phase 5, the actual document development or refinement may be done by other knowledgeable employees with support and review by element teams.

- Developing *work instructions and job aids*, if needed, is the responsibility of line departments in collaboration with the element teams.

- Developing the *quality manual* is the responsibility of the project team and is done after all other documentation is completed.

Table 7–23 lists the major activities in Phase 6, and the rest of this section briefly discusses selected activities. The discussion here focuses on the activities; see Chapter 8 for a discussion that focuses on the documentation itself.

Table 7-23

Phase 6: Document Quality System Elements

(Months 0–18)	
Activity:	*Who Does It:*
• Review quality system guidance	Element teams and other knowledgeable employees
• Develop or refine documentation	Element teams and other knowledgeable employees
• Trial documentation	Element teams and other knowledgeable employees
• Perform adequacy audits	Project team, document users, line managers, internal auditors
• Obtain approval	Element teams
• Develop quality manual	Project team
• Revise as necessary to address results of implementation and validation assessment	Element teams and other knowledgeable employees

Develop or Refine Procedures.   At this point, element team members or other designated employees prepare full drafts of all new or revised procedures. Element teams will need to refer to this information to develop drafts:

- Proposed skill and knowledge levels from Phase 5.

- Validated outlines of procedures from Phase 5.

- From the project team: documentation structure, acceptable formats and media, writing guidelines, document review and approval procedures (such as adequacy audits, management reviews, and project team reviews).

- Document control procedures.

At the same time, designated department members prepare any additional work instructions or job aids required. All work instructions and documented job aids, whether new or already existing, should be properly referenced in the related procedures.

Trial Documentation.   Two levels of trial are useful in obtaining preliminary assurance that documentation is comprehensible, usable, and effective:

- *Review by sample users.* Arrange for review of draft procedures and work instructions by a sample of users that includes both employees who were involved in design validation and some who were not.

  Employees with a range of time on the job might provide additional insights. For example, relatively new employees can identify critical missing information. Experienced employees may have suggestions on ways to deal with contingencies.

- *Use under actual operating conditions.* After incorporating changes in the drafts from the review by sample users, run a trial where actual users follow the completed procedures under working conditions.

The *timing* of trials should make sense for the quality system element. For example, the procedures for elements that affect several departments should to be trialed in each of the departments, although not necessarily at the same time. For some procedures, it might be advantageous to trial the procedure and related work instructions at the same time. Some interdependent procedures may have to be developed and trialed together or sequentially. Some entirely new procedures for major activities might require lengthy, individual trials.

Use any means that makes sense to *gather data* from these trials. Element team members and line departments might observe procedure use or obtain information through debriefings, user critiques, and in-place measures. It is particularly helpful to look for how well procedures and work instructions respond to exceptions and problems. This is frequently when weaknesses are most obvious.

After each trial, revise the documentation accordingly. Continue to run trials, analyze results, and revise documentation until you are confident that it is effective. Of course, each revision, starting with the first draft, should be under document control.

Conducting trials not only results in improvements to the documentation, it starts putting the quality system in place by familiarizing employees with new and refined ways of doing their jobs. Understanding and acceptance of the final procedures and work instructions by the bulk of employees will make or break the system.

Perform Adequacy Audits.   Once the trials are completed and revisions made, procedures should proceed through three levels of audit and approval (summarized in Table 7–24):

- The project team should audit draft procedures for *consistency* and *compatibility* with other documentation, adherence to documentation guidelines, and *adherence* to the intent of the quality plans. Element teams should perform a comparable review of all work instructions and other detailed guidance referenced in the procedures they develop.

  In most companies, the element teams will not all be preparing procedures at the same time, so it will be difficult for the project team to compare drafts. The project team will, therefore, have to anticipate possible compatibility issues all through the design and documentation phases.

- Affected departments should audit draft procedures and work instructions for *usefulness* and *understandability,* using document review procedures developed by the project team and approved by the steering committee.

- Internal auditors should audit procedures and work instructions for *adequacy relative to the requirements* of the particular ISO 9000 standard. If inadequacies are found, this would be a good time to implement formal corrective action to gain experience with corrective action procedures.

Adequacy audits should be performed in the sequence listed above, After each one, the documentation should be revised and submitted to the next audit.

Approve Documentation.   After all adequacy audits have been conducted and resulting revisions made, a final draft should be submitted to the designated authority for approval. The authority may be the project team or the steering committee.

Table 7–24
Types of Adequacy Audit

For:	By:
Consistency and compatibility with other elements, adherence to quality plan, adherence to documentation guidelines	Project team and element teams
Usefulness and understandability	Users
Usefulness and understandability	Management in affected departments
Fulfillment of ISO 9000 requirements	Internal auditors

Develop the Quality Manual.   Subclause 4.2 of ISO 9001 and ISO 9002 requires a quality manual that outlines the structure of quality system documentation. Given this description, the project team cannot complete the quality manual until all quality system procedures and supporting documentation have been developed; that is, at the end of Phase 6 at the earliest.

The primary guideline for developing documentation applies to quality manuals as well: *Have specific users and specific needs in mind for the quality manual before developing it.*

Keep in mind that, from an internal or external auditor perspective, it is quite helpful to have the organization of the quality manual parallel the structure of the relevant standard or to include a cross-reference matrix. It is also important to identify early in the manual any areas where you have "tailored" the quality system to your company. So, for example, the quality manual should declare anything novel or unusual about the quality system, as well as any exclusions of quality system elements or requirements.

Following are examples of considerations you might have in preparing quality manuals.

In small companies with a limited number of procedures and no need to give copies of the manual to customers, there may be no need for a quality manual beyond a collection of procedures with an index. (Note that if the quality manual is to be a collection of procedures with an index, the project team might delay completion until the end of Phase 7, Implement Quality System Elements.)

In large companies with a substantial number of procedures, the quality manual may be an overview of the quality system with lists of related procedures. Some companies prepare a quality manual of this type and collect all procedures in a separate procedures manual.

If the quality manual is primarily for customers, a condensed summary of your quality system would be appropriate.

If you are going to use the quality manual to orient new employees, you might include the company quality policy, company-level process charts, and a condensed overview of the quality system with lists of relevant procedures.

Phase 7: Implement Quality System Elements

Phase 7 has three objectives:

- To fully deploy all quality system elements as designed and documented throughout the relevant parts of the company or facility.

- To ensure consistent adherence to quality system policy, procedures, and work instructions.

- To demonstrate the effectiveness of the entire quality system.

Making your quality system fully operational involves everyone. *Line managers* have a key role since they are responsible for ensuring that employees have the skills and knowledge they need to use new or refined procedures and integrating the procedures into their daily operations. *Employees* must use the procedures. *Element teams* must ensure that the procedures are usable and employees are trained on using them. The *project team* has overall responsibility for quality system implementation. *Most important, senior management must be visible and active throughout implementation, as for any change that involves the entire company or facility.*

Based on the priorities established by the project team in Phase 5 and the actual effort necessary to design and document quality system elements, all elements will not be implemented at the same time.

Table 7–25 lists the major activities in Phase 7, and the rest of this section briefly discusses selected activities.

Ensure Competencies.   For each quality system element, skill and knowledge requirements (competencies) for each job were defined by line management as part of element design in Phase 5. The level of detail in the procedures outlined in Phase 5 and drafted in Phase 6 was explicitly balanced with these agreed-upon competencies that employees would need in order to use the procedures. Before putting procedures into effect, then, element teams must ensure that employees do, indeed, have the competencies.

Three steps are required to ensure competency:

- *Assess employee competencies.* To satisfy the requirements of ISO 9001 or ISO 9002, line managers must *formally* assess the competencies possessed by each employee relative to the defined competencies required.

Table 7–25

Phase 7: Implement Quality System Elements

(Months 3–18)	
Activity:	*Who Does It:*
• Refine implementation strategy	Project team and element teams
• Ensure employees have necessary competencies	Line managers
• Put procedures into effect	Line managers and element teams
• Perform compliance audit	Internal auditors
• Track performance	Project team
• Revise as necessary	Project team, element teams, and line managers

- *Maintain records.* Line managers must keep records for each employee of the activities he or she is competent to perform. These records should be kept readily accessible so work assignments are not made to employees who do not satisfy competency requirements without direct oversight by someone who does.

- *Address competency gaps.* If an employee does not possess all of the required capabilities, line managers must establish an action plan to fill in the gaps and provide for satisfactory oversight of the employee while doing the work until all capabilities are satisfactorily demonstrated and recorded.

Put Procedures into Effect.   Putting procedures into effect has two aspects:

- *Training.* Employees (including managers) must have training in using new or revised procedures, including keeping any records required. The implementation plans developed by the element teams in Phase 5 should have included plans for training, so now it is a matter of following the plans, with any refinements necessary. (Note that training is one aspect of implementation that element teams and the project team should integrate as appropriate across all quality system elements, so that employees learn all they need to know at once rather than in an uncoordinated manner over time.)

  It makes sense for training to be given by line managers or other designated employees with support from either the element team or the project team. If the procedures are truly user friendly, formal training needs should be minimal. Note that once the quality system is fully operational, subsequent training becomes an ongoing line management responsibility.

- *Using the procedures.* In parallel with evaluating competencies and correcting any gaps, line managers should be changing work practices to follow all relevant procedures and work instructions. The approach to implementing procedures in a department or function can be different across a company or facility.

  For example, for those aspects of the quality system that are critical to high-priority customer requirements, a period of *dual operation* (new or refined procedures used alongside existing procedures for a defined interval) might be useful or necessary. Where there is a high penalty for failure, dual operation might be a way to build confidence in the new approach. Foster departmental ownership in the quality system by allowing flexible approaches to implementation as long as the results are consistent with management objectives and the intent of the system as designed and documented.

Perform Compliance Audit.   Within two weeks after employees have been trained on procedures, internal quality auditors should perform a *compliance audit* to determine if practice conforms to relevant procedures and work instructions. It is quite likely that nonconformances will be found that trigger corrective action. Any nonconformance should be resolved using formal corrective action procedures. The results of these audits will not only help detect any start-up problems, but continue to give the auditors practice and build the required track record for internal auditing.

Track Performance.   Use of procedures under actual operating conditions *over time* is the only true way to fully validate quality system design, documentation, and implementation. The amount of time required to evaluate effectiveness of all elements in the system might take months. (Keep in mind that registrar auditors will require evidence of successful operation over time.) For this reason, gathering information to track overall performance needs to be planned well in advance.

Beyond internal audits of the quality system, two primary methods for *tracking performance* are:

- Your *measurement system,* put in place in Phases 3 and 4, should tell you:
    - —if quality objectives are being achieved
    - —if quality objectives are not being achieved, how well the quality system is working
    - —if the quality system is equally effective in all departments or functions
    - —if the quality system is being followed consistently over time, including all shifts of employees
    - —where problems are being experienced.
- *Feedback from employees,* obtained through any effective mechanism, can:
    - —provide subjective data that supports objective data
    - —complement objective data, that is, give you additional information not available from measurements
    - —conflict with objective data—in this case, for instance, your measurements may show improvement, but employees report feeling stress when they use certain procedures over a length of time or at certain periods, so, you may have some long-term considerations that have not been identified previously.

Negative or conflicting *results*, either operationally or from employee feedback, indicate that something is wrong with the design, documentation, or implementation of the relevant quality system element. Element teams and the project team can have various responses to these results depending on the suspected cause:

- If the *design is not fully effective*, identify the changes needed, make the changes, and try the new design with users.
- If *changes impact documentation*, make changes through your document control procedures.
- If *implementation is not effective*, identify what actions are needed (for example, more training, changes to documentation), develop plans to meet the needs, and follow through with the plans.

When the project team is confident that all quality system elements are in place and are operating effectively (illustrated in Figure 7–12), it is time to conduct a formal validation of the quality system—Phase 8.

Phase 8: Validate Implementation

Phase 8 has one objective:

- To provide high confidence that your quality system adequately covers the defined scope of your company's quality system, and, if pursuing registration, that the system will be found in compliance during the formal registration assessment.

Validation involves a formal assessment of your quality system by *external experts*. For companies pursuing registration, this validation phase is usually called a preassessment. While a preassessment is not a requirement for successful registration, there are distinct advantages.

A successful preassessment gives you confidence that the formal registration assessment is unlikely to uncover any major shortcomings in your quality system. And if minor problems are found that require changes in documentation or implementation, these can be accommodated in an orderly way.

As a side benefit, a preassessment gives your management and other employees practice with the external audit process, which can reduce stress and anxiety. Not only is preassessment relatively informal, without a sense of pass or fail, it also significantly improves the odds that the formal audit will be successful. Today, preassessment is widely seen as an insurance policy and is commonly practiced.

Figure 7–12

A Fully Implemented Quality System

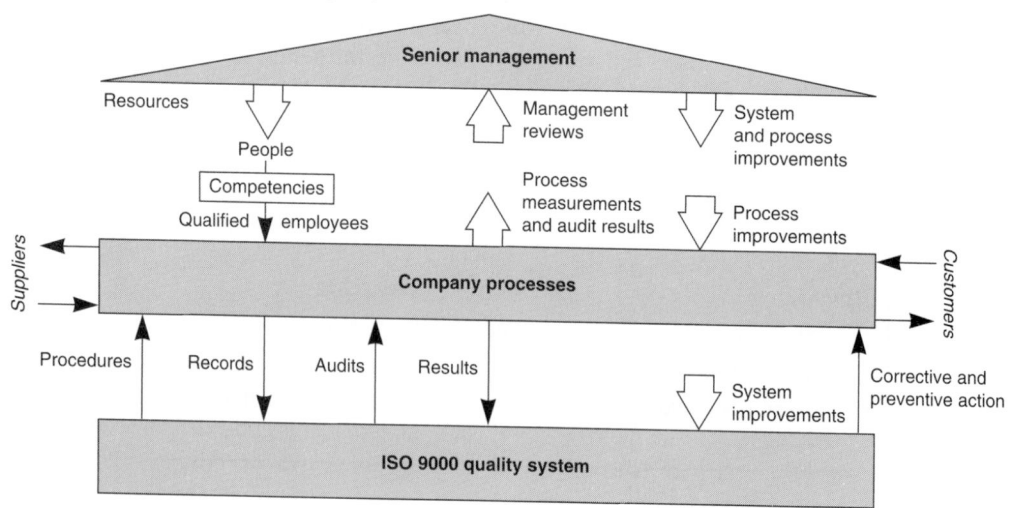

Table 7–26 lists the major activities in Phase 8, and the rest of this section briefly discusses them.

Arrange for Systemwide Assessment.   A validation assessment should be performed by qualified, external ISO 9000 experts against the requirements of your chosen ISO 9000 standard and the defined scope of your quality system. It is not a good idea to rely completely on internal resources for the assessment. Normal project management, management reviews, and internal audits are important, but they are considerably less rigorous than an external assessment.

If your company is pursuing registration, you can take either of two approaches to "preassessment." One is to have your chosen registrar perform the task, and the other is to use external consultants. Both approaches have their advocates and their pros and cons. The advantage of using your registrar is that it gives the registrar an opportunity to get to know your company, your management representative, and your quality system. The preassessment can be a good time to build rapport with the registrar's audit team, and to get past the all-too-frequent feeling that the auditors are hostile forces from whom you must shield your company.

A potential disadvantage is that the registrar's auditors may have preconceptions about your quality system based on the preassessment that may not be valid by the time the formal registration assessment takes place. In addition, a successful preassessment may not be valid by the time the formal registration assessment takes place. In addition, companies with successful preassessments may assume that they will automatically be successful with the formal registration assessment, which is not necessarily the case.

Using an external consultant also has advantages. A registrar cannot provide advice or guidance on how to fix nonconformances that are uncovered during the preassessment; whereas such advice is the primary reason for using consultants to conduct the preassessment. In addition, the preassessment is likely to be extremely thorough because the consultant's success depends on ensuring your success, that is, making sure that everything is in place and working well before the registrar comes in for the formal assessment.

Table 7–26

Phase 8: Validate Implementation

(Months 17–18)	
Activity:	***Who Does It:***
• Arrange for validation assessment	Project team
• Conduct assessment	Qualified external expert or registrar
• Address nonconformances	Steering committee, line management, project team, and element teams

A potential disadvantage is that a consultant may not interpret or approach the requirements of the standards in the same way as your registrar (although you should have addressed this, in part, during registrar selection). In addition, the registrar's audit team will have little or no familiarity with your company or your quality system before the formal assessment, a situation that may have implications for the assessment.

Most validation assessments take place after the total quality system has been implemented. However, some companies pursuing registration ask their registrars to conduct a preassessment during Phase 7 as an objective check that the project is on target.

Conduct Validation Assessment.   Three general questions will be in the mind of an auditor during an assessment (summarized in Table 7–27):

1. Is the quality system, as documented, adequate to satisfy the requirements of the chosen standard?

   Auditors will typically address this question by carefully reviewing documentation, usually your quality manual, evidence of quality planning, and selected procedures.

2. Is there evidence that all employees consistently follow the documented procedures?

   To address this question, auditors will observe your workplace and interview employees.

3. Is the quality system as designed, documented, and implemented effective in meeting customer requirements by preventing nonconformances?

   While gathering information and insight into the degree of compliance to documented procedures, the auditors will look for signs that the quality system is not fully effective. Customer complaints; significant amounts of nonconforming material, whether in receiving, in-process, or finished product; large amounts of work in process; rework; special handling; expediting; evidence of interdepartmental squabbling; and attitudes of "fix-the-blame" or scapegoating could all be clues to an ineffective quality system.

Table 7–27

Auditor Concerns

1. Is the quality system, as documented, adequate to satisfy the requirements of the chosen standard?
2. Is there evidence that all employees consistently follow the documented procedures?
3. Is the quality system as designed, documented, and implemented effective in meeting customer requirements by preventing nonconformances?

Table 7–28 lists more specific questions and requests that employees can expect from auditors during validation assessment.

Address Nonconformances.   At this point, your company's management representative should be assuming most of the responsibility for ensuring that the quality system is "established, implemented, and maintained" in accordance with your chosen standard. Of course, the project team should continue to be a major resource. Element teams will also have a continuing role if any nonconformances found relate to the design or documentation of a quality system element.

Table 7–28

Typical Auditor Questions and Requests

1. Show me how you plan for quality.
2. a) What are the critical quality-related characteristics of your products?
   b) What are the critical parameters for your processes that affect quality?
   c) Show me the way you identified the statistical techniques required to establish, control, and verify these product characteristics and process capabilities.
   d) Show me the results of applying the statistical techniques.
3. a) Show me your definitions of competencies required for each type of activity that affects the quality of your products.
   b) Show me your records of qualifications for the employees who perform these activities.
4. a) Show me records of all of the corrective actions that have been initiated during the last six months.
   b) Show me the results of your preventive actions during the last six months.
5. Show me the reports from internal quality system audits that have been conducted during the last six months.
6. a) What are your quality objectives?
   b) How do you measure performance relative to these objectives?
   c) Show me the trends of these measurements over the last six months.
7. Show me the records of your last four* management reviews of the quality system.
8. Show me your process for identifying all controlled documents and data.
9. Show me your list of inspection, measuring, and test instruments scheduled for calibration this month.
10. Show me how you identify approved subcontractors.

*Editor's note: some experts suggest only two in this scenario.

Common Nonconformances

A good validation assessment should identify any major nonconformances in your quality system design, documentation, or implementation. The most likely nonconformances fall into these categories:

- *Insufficient evidence of quality system effectiveness.* This is usually the result of not enough measurement data or records to confirm that design, documentation, and implementation are effective.

- *Inadequate implementation.* For example, an employee may not be able to tell an auditor the company's quality policy or immediately produce the procedures or work instructions that describe his or her job.

- *Insufficient coverage* of a quality system element in procedures. An earlier section discussed several quality system elements whose insufficient coverage can be roadblocks to successful implementation.

Addressing Nonconformances   If a nonconformance comes as a surprise to anyone in a leadership position, it indicates a problem in management understanding or communication. And it is objective evidence that the quality system does not yet meet the requirements of the chosen standard. The standards do not require a complete absence of problems, but they do require knowledge of what problems exist and documented corrective action plans.

Note that sometimes the steering committee, management representative, and project team make a conscious decision to conduct a validation assessment before implementation is complete as a way to confirm progress. In this case, they should know about nonconformances in advance and fully expect them to be identified during the assessment.

The response to nonconformances will, therefore, depend on the type of nonconformance and management's awareness of its existence.

For *insufficient evidence of quality system effectiveness* or lack of adequate records to prove compliance:

- Your response needs to assure that the appropriate evidence is available at a defined time (for registration, at the time of formal assessment). If management is aware of the need, resolution should be part of the overall project plan. If management is not aware, a corrective action plan must be prepared, approved, and implemented.

For *inadequate or inconsistent implementation*:

- Conscientious planning and management by the project team, thorough compliance audits by internal auditors, and adequate oversight by the steering committee during management reviews should provide good insight into the degree and consistency of implementation.

If the reported nonconformances are known in advance, the project plan should already have provisions to assure complete implementation by a defined time (for registration, by the time of formal assessment).

If senior management is not aware of nonconformances in implementation, it signals a major flaw in management of the ISO 9000 project. It also suggests that line managers involved either do not fully support the project, do not adequately appreciate the priority of the project, or do not adequately understand the quality system.

Dealing with this situation is a steering committee responsibility and should be addressed the same as any other problem in management communications and accountability. The steering committee might delegate specific actions to the project team or the management representative as part of an overall solution.

You should delay formal registration assessment until the management problem has been resolved and the resolution verified.

For *insufficient coverage* of quality system element:

- It is generally not good business practice to enter into validation assessment with known shortcomings in the design or documentation of a quality system element because it reduces the value of the assessment and costs money.

  If the assessment uncovers unrecognized problems with element design or documentation, it generally means that there has been a flaw in planning or management of the project. Possible causes include:

  —inadequate or misinformed education and training of project team and element team members.

  —lack of adequate time and resources either in element teams or line departments.

  —shortcomings in internal audit training or procedures.

  The actions you take to resolve the problem will depend on the underlying causes. The project team, under the leadership of the management representative, has the primary responsibility for assuring satisfactory resolution of the nonconformances, although the actual work will probably be done by the appropriate element teams.

  You should delay formal registration assessment until the element designs and documentation have been corrected, implementation has been completed, and sufficient historical evidence of effectiveness has been collected.

FORMAL REGISTRATION ASSESSMENT AND SURVEILLANCE

Overview of the Registration Process

The formal registration process, depicted in Figure 7–13, begins after implementation is complete. If no preassessment is conducted, the lead auditor assigned to your company may make an introductory visit to the facility(ies) to be audited for the purpose of meeting employees, seeing the layout, and understanding the company's organization and

Figure 7–13

Formal Registration Process

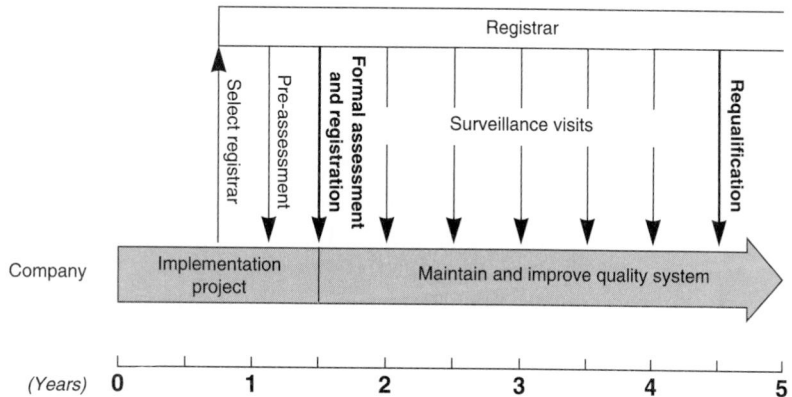

workflow. This information can also be helpful in setting up an appropriate audit team and an effective audit.

During the *initial formal audit*, the registrar's audit team conducts a formal audit of the documentation and deployment of your quality system. First, the audit team reviews quality system documentation, often off-site and before visiting the facility. Once the auditors are satisfied that the documentation meets the requirements of the relevant standard, they conduct an on-site visit to the facility(ies).

The length of the visit may range from one to several days depending on several factors, such as the size of the facility, range of product families, level of automation, number of employees, and complexity of the quality system. During the on-site visit, auditors observe the workplace and interview employees, as well as requesting samples of quality records and other information. During the audit, the team verbally reports its findings on a daily basis. Shortly after the end of the assessment, the team provides a formal, written report. If the audit team finds compliance to the requirements of the relevant standard, it recommends *registration* to the registrar, and, almost always, a "certificate of registration" is issued.

After successful registration, the registrar conducts periodic *surveillance audits,* typically every six months. The registration certificate has a limited life, usually three years. At the end of that period, your quality system must be *requalified*. Surveillance audits and requalification are intended to assure that your registered quality system continues to comply with the specified scope and requirements of the *latest revision of the chosen standard*. Registering an ISO 9000 quality system is a long-term commitment, including ongoing maintenance. Remember also that registrar audits happen infrequently. It is the constant vigilance of employees that ensures the ongoing success of the quality system. Most important of all is management's constant, unwavering commitment.

Presumably, your company will have selected and contracted with a registrar by one-third of the way through implementation. During negotiations, many details of formal assessment, surveillance, and requalification will be addressed and resolved. Chapters 5 and 6 discuss registration and the audit process in detail.

Responsibilities of Company Management

One of the most important aspects of formal registration assessment is that senior management fully understand what is going to happen during the audit and, equally important, what is not going to happen. Unlike many auditors, such as those from the U.S. Environmental Protection Agency, the Food and Drug Administration, or the Occupational Safety and Health Administration, the registration audit team cannot threaten the survival of the company. No lawsuits or suspension of shipments or operations can follow their visit. If you are pursuing registration in response to customer pressure or for competitive advantage, though, it is in your best interests to fully support the audit.

ISO 10011, Part 1, Guidelines for auditing quality systems—Auditing, identifies six responsibilities of company management during an audit, which are discussed in the next several paragraphs.

"Inform relevant employees about the objectives and scope of the audit." Since most everyone in the company or facility is involved in some way in the operation of the quality management system, any of them could be subject to an auditor's activity. Time invested in teaching employees how to respond to auditors is time well spent. Explaining the process in advance can reduce the nervousness many people will experience and help make things move more smoothly.

The management representative should meet with senior management to explain the scope of corporate activities that will be explored by auditors. Members of the project team should explain the auditing process to line managers, who can pass the information on to their staffs.

Every employee should have a clear grasp of these fundamental issues:

- The quality system, not the staff, is under scrutiny.

- The employee should have all the necessary documentation readily available, including reference materials.

- When asked questions, the employee should respond honestly and concisely to the questions asked, not to the questions he or she thinks the auditor asked.

- The employee need not volunteer information beyond that requested.

  If a question is outside the employee's areas of responsibility, the employee should say "I don't know" or "That isn't something I am involved with." Most people hate to admit ignorance and often try to answer questions about things that are not their responsibility. Such answers are frequently wrong, and this can confuse the auditor and waste time.

- It is perfectly acceptable for the employee to say "I don't understand your question" when that is an honest response, or ask for clarification of a question. Some auditors have difficulty phrasing questions in everyday language, or they may use terms with which they are familiar rather than terms used in the company. Sometimes auditors use formal language directly from the relevant standard, and some employees may not understand it. The term "nonconforming material" is often not used in companies, especially service organizations.

- Auditors will take notes, ask for their written observations to be signed by employees, and sometimes ask for copies of the objective evidence they have seen. The employee should understand that taking notes does not necessarily mean there is a problem; auditors note conformances as well as nonconformances, and objective evidence works both ways.

"Appoint responsible members of the staff to accompany members of the audit team." Your company should arrange for guides for the audit team who know the quality system and company operations. Staff members who are trained and active internal auditors are excellent choices. Not only can you gain good intelligence from these guides as the audit progresses, but they will be able to ensure that any misunderstanding or miscommunication between an auditor and an interviewee is quickly corrected.

Make sure that guides are aware that their role does not include answering for others, interrupting, or trying to influence the auditor unless something is going wrong. A guide who is helpful and cooperative can make the auditor feel relaxed and comfortable in a strange environment.

"Provide all resources needed for the audit team in order to ensure an effective and efficient audit process." Auditors will need to exchange information, trace requests, and discuss findings among themselves. Provide a convenient and comfortable room for these discussions, including access to facilities such as telephones, a fax machine, and a copier. Arrange for refreshments and offer to provide lunch, which saves time for everyone.

"Provide access to the facilities and evidential material as requested by the auditors." The audit team will need access to all areas of the company that influence the activities outlined in your company's registration statement of scope. Do not refuse access to areas by claiming commercial confidentiality. Everything auditors see during the audit is treated as confidential, and protecting this information is part of the code of conduct of the Certified Auditor (United States) or the Certified Assessor (United Kingdom).

If you want certain areas to be off-limits, you must declare them and reach an agreement with the registrar before the auditors arrive. During the audit, you will also be asked to produce records. Declining to do so will damage the audit process, particularly in the area of training.

"Cooperate with the auditors to permit the audit objectives to be achieved." The registrar wants your company to succeed. Audit teams usually arrive with a positive attitude, seeking reasons for success, not reasons for failure. In addition, the registrar is a sole-source, long-term subcontractor to your company; it is important that the relationship gets off to a good start. Make sure that everyone in the facility or company knows when the audit team will be on-site, including security staff!

The company undergoing an audit is free, at all times, to challenge findings and should not hesitate to do so if there is any doubt that a report of nonconformance is properly supported by objective evidence. Occasions can arise where an honest difference in judgment exists between the company experts and the auditors about the adequacy of the quality system as designed, documented, implemented, and operated.

If a full discussion with the lead auditor does not produce a satisfactory agreement, a prudent course of action is to seek the advice of a knowledgeable third party, usually an external consultant. If the consultant supports the judgment of company experts, consider an appeal to the registrar directly, or its accrediting body.

"Determine and initiate corrective actions based on the audit report." If the final audit report identifies corrective actions that must be taken, your company should respond promptly and effectively. The audit team leader will expect to see that you have taken corrective action, the action was effective, and you have taken action to prevent recurrence.

CONCLUSION

Implementing a quality system based on ISO 9001 or ISO 9002 will be one of the major challenges in a company's history: it requires long-term, visible senior management commitment; it involves the entire company; it affects the work of all employees; it requires extensive and sometimes painful self-examination; and it can result in substantial changes in operations that may have been static for many years. It can also mean better satisfied customers, an improved competitive position, more efficient internal operations; and a better trained, more informed, supportive workforce—the keys to business success!

Quality System Documentation

8

by Roderick Goult

This chapter discusses methods for documenting a management system. It covers the following topics:

- The importance of documentation.
- Documentation development.
- System documentation and ISO 9000.
- The structure of documentation.
- Implementation of the system.
- Review of the system.
- Typical structure and content of a quality manual.
- Creating procedures:
  - —Procedure planning and development.
  - —Procedure structure and format.
  - —Procedure administration and control.

Appendices to the chapter include a sample quality manual and examples of procedures. The second article in this chapter, by documentation expert Eugenia K. Brumm, Ph.D., CRM, shows how properly structured, implemented, and maintained records can avoid costly legal proceeding.

INTRODUCTION

A formal, documented, management system sets out the processes which control activities within the company that affect the quality of its deliverables. The objective is to make the operational process substantially independent of individuals, so that any appropriately experienced and trained person could make the system work. In short, a well-documented system should allow an established process to continue regardless of a key employee taking the day off or leaving the company suddenly.

A well-documented, effective management system communicates a number of important issues to the entire workforce. These include:

- Management objectives.
- Operational and quality policies.
- Individual employees' responsibilities and authorities.
- Operational procedures.

A management system addresses many of the issues concerned with managing the 'white space' between functional departments. It helps ensure that vertically stratified organizational structures (the typical structure of most companies) do not get in the way of horizontally organized processes that fulfill customer requirements.

The existence of documentation formalizes the operational system of the company. The result is consistency of actions and a common understanding of the processes involved. The system clearly defines authorities and responsibilities. It creates verifiable activities and objective evidence that allow for an audit process to be instituted. It allows management to communicate clearly to all staff when a change in quality policy is required.

Documentation also helps with the induction and training of new personnel. It ensures that new staff will always be trained the same way and promotes consistent performance when personnel changes occur.

WRITING DOCUMENTATION

Any quality management system must belong to its creators. It must therefore be written by the company personnel who are going to use it and not by any other individual or organization. A competent management system will create many real benefits for a company, including lower rework costs, lower-inventory costs, shorter turnaround times, higher levels of customer satisfaction, and ultimately a better bottom line.

The process of obtaining ISO 9000 registration is secondary to the benefits of having an effective management system. The key issues to remember are: (1) that the principal customers for the management system are the company staff, and (2) that every document in the system must add value to the company. If a document or activity fails to add value, then it should be discarded. Nothing should ever be done exclusively to satisfy an auditor.

Documentation need not be excessive. A system should be well planned, simple, clear, concise, and well controlled. It should not be a vast, paper-generating process. Quality system documentation and quality manuals do not have to follow any standard nor conform to any specified format. Every department can take its own approach to documentation; however, following a common pattern and style helps the end user find needed information

Quality Manual Layout

Quality manuals do not need to be written in formal language nor follow the element numbering and names of the chosen ISO model. Quality system documentation should reflect the way a company operates and keep the end-user in mind. The bottom line— keep it simple.

DEFINING THE SCOPE OF THE SYSTEM

The first requirement for system documentation is to determine the scope of products or services the system must address. If certain products or services are not to be included within the registration process, then the company has the option of not documenting them.

The required scope of the system can be determined by providing answers to a few simple questions:

1. What products does the company wish to have covered by the registered management system?

   Create a list of those items, looking at them from the customer's perspective rather than the company's. *Remember the ISO 9001 definition of 'product'—see ISO 9001 Paragraph 3.1, and the attached notes.*

2. What are the processes, from initial customer inquiry to final hand-off to that customer, which impact on the quality of the deliverable for each of the products identified in the answer to question 1?

The first question provides the bones of the scope statement required for registration, and the answer to the second question provides the boundaries of the management system. Everything within these boundaries requires formalized documentation so that the activities are performed in a consistent manner and ensure that the deliverable will satisfy the customers' needs and expectations.

System Documentation and ISO 9000

Remember that the management system exists to benefit the company; ISO 9001 or ISO 9002 provide, as their names imply, a 'model' for such a system. They are not unalterable dictates. Not every paragraph in the standard will necessarily apply to every company—whole sections of the standard can be written out as 'not applicable' if the activities described genuinely do not fit a particular business system. However, it is always necessary to provide a registrar with convincing reasons for these omissions. A section of the standard cannot be ignored because the company does not want to be bothered by its particular requirements.

A list of typical system documentation for ISO 9001 is shown in Box 1, but note that it is a typical list, not mandatory. Individual systems may contain many more or many less items.

Box 1

Typical Quality System Documentation in ISO 9001

This list identifies those subjects which ISO 9001 requires be formally documented in the procedures which form the management system based upon this Standard. Linkages which are often missed; linkages between these elements are identified within the text. The list is not to be considered exhaustive; however, recommendations are included occasionally that exceed the basic requirements of the standard.

4.1 Management Responsibility

- The quality policy, including the executive management commitment to quality.
- The objectives for quality.
- Records of any product quality problems.
- Responsibility, authority, and interrelationship of all personnel within the management system scope.
- Management representative responsibilities and authorities.
- Management reviews, frequency of meetings, agenda, minutes.

4.2 Quality System

- System procedures that document the quality system and cover all aspects of the scope and all relevant parts of the Standard.
- Outline structure of documentation.
- Quality planning—either discrete quality plans or within the system documentation.
- Identification of quality records.

4.3 Contract Review

- Procedure for review of every contract.

(Continued on next page)

(Continued from page 306)

- Procedure for handling contract amendments and circulating necessary data to the appropriate departments—define 'appropriate.'

- Records of the contract review activity.

4.4 Design Control

- Process for assigning design tasks, responsibilities, and authorities.

- Definition of appropriate organizational and technical interfaces within and outside the design department.

- Defined design process, which includes methods for the formal handling of:

  —Design input and output (link back to contract issues).

  —Design reviews at predetermined stages in the process.

  —Design evolution methodologies.

  —Verification of design output.

  —Validation of design output.

  —Control of design change.

  —Records of all significant design activities, including:

    - planning

    - formal inputs

    - formal outputs

    - regulatory/statutory compliance

    - reviews

    - verification

    - validation

    - changes

4.5 Document and Data Control

- Clearly defined structure and road map.

- Defined processes, responsibilities, and authorities for:

  —The development, approval, and issue of documents.

(Continued on next page)

(Continued from page 307)

—The initiation and fulfillment of document change.

—Distribution of original documents and changes to documents.

—Removal of obsolete documents (or other method of ensuring that they don't get used).

—Maintenance of history files when required.

—Maintenance of Master Record Index or equivalent process.

• For computer-based systems, all of the above plus:

—Defined back up regime.

—Formal access controls.

—Methodology for the control of documents of external origin, including statutory and regulatory documentation, MSDS documentation, Standards, customer drawings, etc.

4.6 Purchasing

Defined methodologies and supporting records for:

• Review and selection of subcontractors.

• Identification of acceptable subcontractors.

• Monitoring and control of subcontractors.

• Removal from use of unacceptable subcontractors.

• Content and review of purchase orders.

• Review and release of product(s) at subcontractors' premises.

• Customer review and release of product(s) at subcontractors' premises—linkage to contract review process to identify the need.

4.7 Control of Customer Supplied Product

Defined processes and methodologies for the identification of customer supplied product addressing the following:

• Contractual identification of items (services) to be supplied.

• Clear definition of standards of acceptability to supplier and customer control and identification of any and all customer-supplied product records of loss, damage, or unsuitable deliveries/services.

(Continued on next page)

(Continued from page 308)

- Note the definition of 'product' in ISO 9001, paragraph 3.1, and attached notes. Also note that the 'product' may be for *"associated activities"* as well as for *"incorporation in the supplies."*

4.8 Identification and Traceability

- Definition of the applicability of the requirements for:

  —Identification of product from receipt to dispatch.

  —Any traceability requirements from design through to installation or beyond.

- Documented processes and methodologies for the tracking and control of the defined requirements.

- Maintenance of records appropriate for the defined requirements.

4.9 Process Control

Defined procedures and methodologies for:

- Production planning and control.

- Identification of the need for work instructions and provision of requirements.

- Identification of workmanship standards and provision of appropriate tools.

- Process approval.

- Process control monitoring.

- Identification of *"qualified processes"* equipment and personnel.

- Identification and fulfillment of environmental controls.

- Identification of any special handling needs.

- Equipment maintenance to meet process needs.

The collection, collation, storage and maintaining of appropriate records of all the above.

4.10 Inspection and Testing

Documented quality plans or procedures defining the requirements for:

- The inspection and testing of received materiel.

- In-process inspection and testing.

- Final inspection and testing.

(Continued on next page)

(Continued from page 309)

- Records of receiving and in-process inspections, and tests including the acceptability or otherwise of the results records of final inspections and tests, including verification of previously required records.

- Procedure for urgent release and positive recall, or statement prohibiting such action.

4.11 Inspection, Measuring, and Test Equipment

Documented procedures, methodologies and supporting records for:

- The identification of all test and inspection equipment used to verify product.

- Conformance identification of calibration methodologies, e.g.. either a defined process in-house or an approved external calibration facility traceability of calibration records to related equipment.

- Effective control of in-house test and inspection equipment to prevent invalida-tion of calibration.

- The basis for calibration when no external standard exists.

- Assessing the validity of historic results when equipment found to be out of calibration.

- The calibration status of each identified item.

4.12 Inspection and Test Status

- Documented procedures for the identification of the inspection and test status of all product from receipt through to dispatch and, when appropriate, installation, including the identification of any nonconforming product.

- Appropriate records.

4.13 Control of Nonconforming Products

Processes and methodologies for handling nonconforming products, including clearly defined responsibilities and authorities for:

- Isolating products that fail to meet specifications.

- Ensuring that nonconforming products are not shipped to customers.

- The review and sentencing of nonconforming materials.

- Identifying any contractual obligations which arise when nonconforming product occurs, including notifying customers producing records of review, disposition, waivers, repair, and rework

(Continued on next page)

(Continued from page 310)

4.14 Corrective and Preventive Actions

• Defined procedures for initiating, performing, and resolving nonconformances occurring with products (deliverables), processes, or systems.

• Data collection and analysis from all appropriate available sources to identify potential problems and initiate preventive actions.

• Reporting of preventive actions for management review.

• Records of corrective and preventive actions, including changes in procedures resulting from corrective and preventive action.

• Records of customer complaints.

4.15 Handling, Storage, Packaging, Preservation, and Delivery

Procedures addressing all aspects of:

• Product handling from receipt through to installation (where appropriate) designed to prevent damage or deterioration.

• Definitions of any special requirements (e.g., ESD, HAZCHEM).

• Storage which will protect the integrity of products at all stages.

• Packaging of all products to ensure safety in transit from receipt to customer.

• Any preservation techniques required to be applied at any stage in the process, from receipt to installation.

• The control of methods of delivery, i.e., use of approved transportation.

• Any special paperwork required, either by contract (link to contract review) or legal requirements (e.g., export documentation).

4.16 Quality Records

Definitions of quality records; this may be a single procedure or contained within the system documentation. The methodology must address these requirements:

• Who collects the data.

• Where data are stored.

• How data are stored.

• How long data are stored.

• How data are disposed of at the end of storage time.

(Continued on next page)

(Continued from page 311)

- How data are accessed.

- Are all data legible?

- Are data reasonably protected from damage or deterioration?

- Are there any contractual obligations for retention, access, etc.?

When records are stored on magnetic media, in addition to the above:

- Are provisions made for backup?

- Are provisions made for regular refreshing of long-term data?

- How is ongoing usability of magnetic data assured?

4.17 Internal Quality Audits

Procedures must provide for:

- A published schedule of audits based on the importance of the activity and the results of previous audits.

- The systematic and regular audit of the complete system.

- The timely review and closeout of audit findings.

- The use of trained auditors independent of the area under review. (see ISO 9001, 4.2.2)

- Records of audit results.

- Records of review of corrective actions for effectiveness.

- Use of audit reports to improve the system.

- Feedback of audit reports to management review.

4.18 Training

- A systematic and ongoing process for the identification of all training needs and the fulfillment of those needs.

- Records of the training provided.

4.19 Servicing

When contractually required, processes must be in place for:

- Identifying requirements.

- Maintaining records.

(Continued on next page)

(Continued from page 312)

- Linking back to 4.14, and providing for the review and analysis of warranty.

- Non-warranty service activity.

4.20 Statistical Techniques

- Declared responsibility for the identification of the need for statistical techniques to monitor and control processes and products, and the authority to introduce and implement such techniques.

- Documented procedures to implement and control statistical techniques wherever a work process or analytical process utilizes statistics.

- Link back to 4.18 and provide training in the use of statistical techniques whenever required.

Creating system documentation is often complex, because the requirements of ISO 9000 intermingle across functionality within an organization. Keeping track of compliance requires some form of logging system.

A schedule of conformity is a matrix which tracks the requirements of the relevant ISO 9000 model against the documentation produced within the system. Typically the requirements would form the 'X' axis and the procedures the 'Y' axis. A checkmark at an intersection indicates that the requirement is addressed by the appropriate procedure. Often individual procedures address several elements of the standard. Any 'not applicable' requirements can be indicated on the chart. This chart should be completely blocked out by the time that the implementation process is completed.

A sample schedule of conformity is shown in Table 8–1.

SYSTEM STRUCTURE

ISO 9000 places no restrictions on how systems should be structured. Indeed, in the introduction the authors state that *"it is not the purpose of these. . . Standards to enforce uniformity of quality systems. . . . The design . . . will be influenced by the varying needs of an organization . . ."* The only mandated item is a quality manual. Element 4.2 states that *"The supplier shall prepare a quality manual covering the requirements of this . . . Standard . . . [which] shall include or make reference to the quality-system procedures and outline the structure of the documentation used in the quality system."*

Table 8–1

Schedule of Conformity

Procedure Number & Name	ISO 9001—4.1 Management Responsibility					ISO 9001—4.2 Quality System		ISO 9001—4.3 Contract Review			ISO 9001—4.4 Design Control		
	4.1.1	4.1.2.1	4.1.2.2	4.1.2.3	4.1.3	4.2(a)	4.2(b)	4.3(a)	4.3(b)	4.3(c)	4.4.1	4.4.2	4.4.3
QUALITY MANUAL	X	X	X	X	X	X	X	X	X	X	X	X	X
MARKETING MK001 Order Intake								X					
MK002 Review of Quotations									X	X			
MK003 Pricing Policy													
PROCUREMENT PK001 Pricing Quotes									X	X			
PK002 Vendor Selection													

Many small organizations put the entire system into a single document, referencing any *"documents of external origin"* which may be appropriate and relevant. Larger organizations typically produce a tiered structure of documentation, because the single comprehensive binder approach is impractical.

The most common quality manual approach is the four-level structure, often illustrated using the pyramid diagram shown in Figure 8–1. The four levels consist of:

Level 1—The Quality Manual: sets out the general approach and policies of the company—the *why* of the system.

Level 2—Procedures: sets out the *what, when, where,* and by *whom* of the system, and provides the horizontal process linkages (internal customer-supplier links).

Level 3—Work Instructions: sets out the *how* of the operation and describes the activities performed and records created.

Level 4—Records: provides the *evidence of compliance* with the system.

System documentation should "cascade" from one level to the next, with any required traceability clearly defined within the documented structure.

Figure 8–1

Pyramid

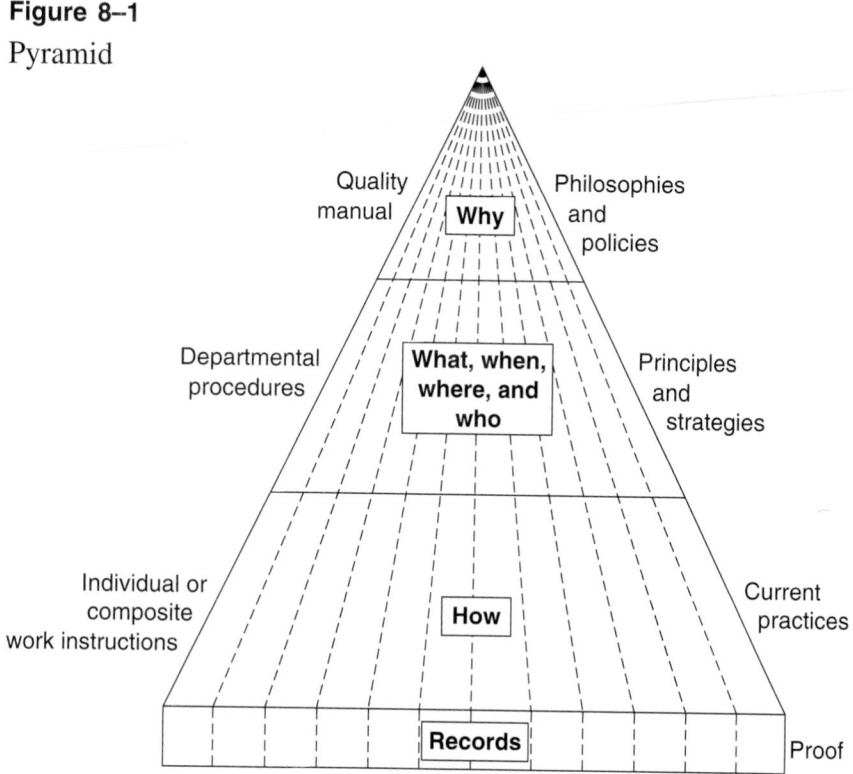

All documentation must meet the defined requirements for control as shown in ISO 9001, 4.5, and the system should be structured so that the ripple effects of any changes are downwards through the system, never upwards. For example, references between documents should only be by name or number, never revision status; level-two procedures, which may change with contracts or tasks, should only make generic references to level-three documents. These 'tricks of the trade' help to ensure that document control doesn't become a complete nightmare.

Level 1: The Quality Manual

A quality manual is a "document stating the quality policy and describing the quality system of an organization" (*ISO 8402, Quality management and quality assurance—Vocabulary*). Many organizations use their quality manual as a marketing tool, often printing it with glossy covers and photographs of products and people.

As such, the quality manual is often the first formal indication that a customer receives of the company's approach to the management of quality issues, and it should demonstrate the strength of the company's commitment to quality. It states *why* the company does the things it does. Typically the manual contains the company's quality policy statement and is often supported by a mission or vision statement.

Level 2: Company Operating Procedures

The second level of documentation is the operating procedures. These describe *how* the management system functions, demonstrate the linkages and interactions between functions and departments, and are often also used to define responsibilities and authorities. A procedure is defined as a "specified way to perform an activity" (ISO 8402, Quality management and quality assurance—Vocabulary).

The operating procedures should reflect the principles and practices defined in the quality manual. Their purpose is to define the way in which those principles and practices are transformed into management activities, and how these activities link with others within the company. A procedure describes what is to be done, when it is to be done, sometimes where it is to be done, and by whom it is to be done.

Level 3: Work Instructions

These are the *how-to* documents of the system and are typically localized to individual departments or company function. Work instructions provide the detailed information on how to perform a particular task; for example, operating instructions for a piece of equipment. Instructions should be produced when necessary to ensure consistent working methods and to achieve required levels of conformance and should be balanced against the training and experience of the personnel typically performing the task.

Box 2

Quality Plans

A documented quality system may also include quality plans. A quality plan can be defined as a "document setting out the specific quality practices, resources and sequence of activities relevant to a particular product, project or contract" (ISO 8402, Quality management and quality assurance—Vocabulary).

A note to this definition emphasizes that a quality plan usually references the applicable parts of the quality manual.

A quality plan is a stand-alone document that defines how quality will be managed on a specific project, contract, or product. The quality plan is based on corporate quality procedures, and the quality strategy for a particular application is developed from them. The quality plan should contain:

- A scope statement that defines the boundaries of the quality plan.

- The quality strategy, derived from the quality manual and any operating procedures or other core procedures.

- The methodology (a detailed definition of how quality will be managed), supported by flow charts, if necessary.

- Verification including objective evidence and methods that determine whether the planned controls and procedures have been implemented.

(See Chapter 4 for a discussion of the requirements for quality plans in ISO 9001.)

Remember that such documents should be adequate to enable another individual of similar education, experience, and background to perform the work in a competent manner. Work instructions must always reflect the requirements of the related departmental procedures, and as already stated the document control process should provide some level of traceability from the one to the other.

Level 4: Records

Although always included as part of the system hierarchy, the records produced by the management system are not really 'documents' in the sense in which the term is used within the ISO 9000 standard.

Documentation is subject to change and modification. Records are clearly permanent documents, but they are not subject to document control in the manner described in ISO 9001, element 4.5, because, once produced, a record remains for all time. If an activity is performed for a second time because the output—as demonstrated by the record—is inadequate, then a second record is produced.

Box 3

Typical Management System Records

Sales orders	Inspection reports
Receiving inspection data	Audit reports
Drawings	Maintenance records
Batch (route) cards	Nonconforming material records
Calibration data	Minutes of management reviews
Shipping documentation	Test and inspection records
Corrective action reports	Design validation records
Design review records	Purchase Orders
Design verification records	Training Records
Goods received notes	Preventive Actions
Vendor performance data	

A list of typical records from a management system are shown in Box 3. Records can include files, technical standards used in a design or production process, copies of relevant statutory regulations applicable at a given time or required to be maintained, drawings and specifications, as well as the data which document the output from the fulfillment of a procedure or work instruction. These forms and charts are typically referenced in the individual documentation that requires their use and must be designed to record all of the specified information.

The ultimate user of the system will benefit from well-planned naming and numbering conventions and an easy reference and location guide. For example, a company might choose to use the prefix "PUR" for the purchasing document or identify the level within the system using "OP" for a procedure or "WI" for a work instruction. These simple conventions can make the tracking and identification process much easier.

THE QUALITY MANUAL: STRUCTURE AND CONTENT

ISO 9001:1994, element 4.1.1, states that *"The supplier's management with executive responsibility shall define and document its policy for quality including objectives for quality and its commitment to quality."* Element 4.2.1 requires that the supplier prepare a quality manual. It is common practice for these two requirements to be joined together so the quality manual typically includes the quality policy statement and will either contain or reference corporate quality objectives. Often these objectives will exist at several levels:

- Global objectives engraved in stone and enshrined in the manual.

- Medium-term objectives described in business plans and tracked through management review meetings.

- Short-term objectives tracked through the operation of the management system itself.

The quality manual must also be the first part of the roadmap that runs through the entirety of the management system: Remember, a small organization may wish to keep the quality policy, operating procedures, and work instructions contained in a single document.

Many "off-the-shelf" ISO 9000 systems are offered on the market to help with this process, but proceed down this road with caution. No two companies operate in the same way: They have different client bases, management systems and styles, and different operational requirements. Some of these packaged solutions can help, but companies would be well advised to carefully weigh their costs and benefits against the benefits of a "made to measure system."

Quality Manual Content

A quality manual describes why a company does what it does in managing its processes. It should be a high-level document, simply written to describe the activities of the company, and preferably no more than 20 to 25 pages. Typically it will contain three discrete types of information—policy, system description, and procedural references, although the latter may also be included in the body of the manual, as in the sample manual shown in the Appendix at the end of this article.

Typically a quality manual is the document control information. This includes the controlled distribution, records of amendments, and the current issue status. Companies pursuing registration should remember to include their selected registrar on the formal distribution of the quality manual once the document is finalized and released.

Section 1: Introduction and Quality Policy

General statement.   Most manuals start with a general statement about the company, its background, its deliverables, and its operating philosophy. This section of the manual may also include the formal scope statement, which often also states why the company has decided to introduce a formal quality management system. It may be something about market pressures and the need to improve competitiveness; it may be more general, merely related to the growth and development of the company and the decision by management to formalize and document their commitment to quality products, processes or services.

Policy statement.   The formal declaration of the company's quality policy usually follows the introduction, often (and preferably) signed by the senior executive on the site

Box 4

Two Approaches to Writing a Quality Manual

Roderick Goult

There are two common approaches to writing a quality manual, and each has its advantages and disadvantages. The first approach is to write the manual following the layout and sequence of the standard. This has the advantage of making the manual very easy for the auditor to follow, but it will not follow the business flow too well. There is also a tendency for such manuals to simply parrot the standard and say nothing company-specific. If those traps can be avoided, this is the easiest approach to ensuring that nothing is left out of the quality manual.

The second approach is to write the manual in a way that follows the corporate business flow and then cross-map it to the standard. If the manual is going to be used to demonstrate conformance to more than one standard, i.e., to both ISO 9001 and Mil-Q-9858A (for example), then this is the better approach. For ease of use, it will need to be backed up by schedules of conformity that cross-reference the manual to each standard, but these are recommended anyway.

to which the policy and manual apply. In corporate structures, both a corporate quality policy and then a local quality policy statement is often included (see Box 5, Sample Policy Statement).

Good quality policy statements are short, to the point, and contain measurables. The policy should relate to the totality of the business, not merely part of it. For example, a commitment to compliant products may be all very well, but without a commitment to on-time delivery, many customers would remain unhappy!

The quality manual also should:

- Address issues of the company's organization and structure.
- Identify the management representative.
- Give details of management system responsibility and authority.

These preceding issues can either be addressed entirely early on in the manual, or the information can be included as the various topics arise throughout text. If, for example, the layout follows the appropriate ISO Standard, then responsibilities and authorities can be addressed under the response to element 4.1.

The organizational structure of the company will usually be addressed through a combination of organizational charts and text. Organization charts should only ever contain

Box 5

Sample Policy Statement

Example A

The company is committed to the delivery of products that comply with the agreed requirements, are delivered on time, and at the negotiated price.

All company staff are responsible for ensuring that the corporate goal is fulfilled.

Example B

The company is committed to implementing quality management programs which ensure comformance to contractual requirements on time and within budget. All staff are responsible for the quality of the company's deliverables.

Quality assurance programs are an integral part of the management methodology of the company and are designed to enhance and perpetuate the reputation of the company as a producer of reliable quality products.

Programs stress defect prevention aspects of quality, with the emphasis on building quality into the deliverable.

Management programs provide for control at all stages of the process from design through to installation and include subcontractors as partners in the process.

---

titles, never names, and should be regularly updated as things change. It is extremely important that the management representative is shown on the chart!

Section 2: Systems Outline

This section of the manual addresses the requirements of the standard, giving a brief description of what is done in each area of the company to control the processes that impact on the deliverables. Little detail should be given at this level (see Appendix 1, Sample Quality Manual). Not every 'shall' statement in the standard will be addressed in the quality manual, but it should be sufficient to reassure the reader that all relevant issues are being addressed.

Two common ways are used to format manuals. One follows the elements of the chosen standard and responds to them in the order in which they arise. The other method follows the process within the organization and describes that process in a logical manner. Both have advantages and disadvantages. The company must determine which method is appropriate. Whichever approach is used, make sure that the document allows for changes to be incorporated without forcing the complete reissue of the entire document. One way to achieve this is to have a single page for each subject, whether it is an element of the standard or a functional activity.

One of the issues the manual must address is where supporting documentation is held. It must either contain the relevant information or guide the reader or auditor to the person who is responsible for each procedure or group of procedures, and how these procedures are used, controlled, and updated.

Section 3: Procedure Index

An index of the company's procedures that support the systems outline.

PROCEDURE PLANNING AND DEVELOPMENT

Procedures are one of the most important parts of the management system. A procedure is a document which must accurately reflect the operation it describes, who is responsible for the activities, and any records which flow from those activities. Procedures define how the organization functions in each area activity and how the activities interrelate with each other—in short, managing the 'white space.' They can also be used as a benchmark for purposes of review and system improvement.

Procedures should be used to accomplish the following:

- Provide reference points for new personnel and reduce the need for verbal training.
- Define responsibilities and authorities.
- Help identify why errors have occurred.
- Enable the prevention of further error through modification of the procedure.
- Provide for the creation of records of the activities described.

Procedures are vital and useful tools; to ensure their credibility, some form of regular review is essential to enable dates and changes to be made when necessary to reflect current practice. This can be a discrete activity or an essential part of the internal audit process.

The development of the overall quality system is a management responsibility, as is the identification of those activities that require formal procedures. Careful planning and data collection are essential if the procedures are to be logical, well structured, and coherent.

Identify the Need

A procedure is necessary wherever its absence would adversely affect quality. A procedure exists to ensure that an activity is consistently performed and the need for a procedure may arise for several reasons:

- The overall system identifies a requirement for a process definition.

- An ISO 9000 requirement demands a procedure.

- Management requires to formalize significant activities.

- The person responsible for implementing activities requires clear definitions of the activities involved.

Once the need for a procedure is identified its creation should be authorized by the appropriate manager. The staff who are to develop and write the procedure should then be identified. Those actually involved in performing the task should be actively involved in the data collection and procedure writing process—ownership is important. Those who do a job are always the best source of data on the way a job should be performed. Every procedure writing activity should be put on a timeline and subject to some form of review, and these time lines should take account of the volume of work involved, including the following:

- Defining the scope.

- Collecting information relating to the activity.

- Preparing a draft procedure.

- Circulating the draft for comment.

- Getting agreement from those concerned.

- Obtaining authorization for issuing the procedure.

- Incorporating it into the document control system.

Identify Key Aspects of the Activity

- Start by establishing current practice through an activity review.

- Establish what documentation already exists for the activity.

- Establish and record the normal methods of performing the activity.

- Identify responsibilities and authorities for the activity.

- Determine the current levels of conformance, if any are defined, and how effectively they are being achieved.

- Identify aspects of the activity that significantly impact on conformance of the output to specified requirements.

Document Current Practice

Document the existing methods of performing the activity, review those methods, and reach agreement with all interested parties to ensure that the data are accurate before preparing the procedure. Ensure that the information collected clearly defines how each activity is carried out, how each step in the process is initiated, how it leads to the next step, and that any intervention points or decision points are clearly and accurately captured.

Flowcharting is a powerful tool for this process. Creating a flowchart forces the data to be presented in a sequential manner. Errors in the sequence of events are easily spotted and speedily corrected. Flowcharts also assist in identifying interfaces with other activities (and defining any required inputs and outputs). They enable parallel processes to be speedily identified, and make the process of corrective action when a 'no' decision is made at any point highly visible. This process of data collection may itself identify potential changes and improvement in working methods.

Review Current Practice

Once the flowchart or activity outline has been collated, a review of current practice should be held with the appropriate responsible personnel to evaluate the results and determine:

- If the specified objectives are being achieved by the process as defined.

- The method of achieving the required level of conformance if the specified objectives are not met.

- If those levels of conformance are adequate to support the rest of the process and the deliverable.

The review should seek out any gaps, duplications, or weaknesses in the process, especially with respect to cross-functional interfaces. Fixing any of these problems will lead to an improvement in the performance of the process.

WRITING PROCEDURES

Once all the data have been collected, collated, reviewed, refined, reexamined, and agreed on by all relevant parties, the process of actually writing the procedure can be undertaken. Remember that if a cross-functional issue is raised during the review, then a representative from the other function needs to be consulted. The author should ask a series of logical questions, extracting the answers from the data collected and formatting the output into a logical procedure. The format of the procedure should, of course, follow the "procedure for procedures" in the company system. Typically a "procedure for procedures" will define standard headings, typeface and size, title indents, line justification, contents sheets, and page layouts. Place the document under control as a 'draft' document as soon as it is complete. For document control activities, the author should refer to the company's document control procedure.

The actual writing of an effective procedure requires the application of some basic literary skills. The author should bear certain concepts and principles in mind when writing.

1. Only include data and directions specific to the activity.

2. Ensure that references to other procedures which address related activities are included when necessary.

3. If the procedure is getting lengthy, split the activities into two or more procedures. Good procedures will be typically between two and four pages in length. (Long test procedures (or similar) of a sequential nature may obviously run much longer than this, but such documents are used page by page, so that no one is trying to use many pages at once.)

4. Write for the usual end user—the degree of detail required in a procedure depends on skills and training provided for the personnel utilizing the document.

5. Involve everyone affected by the implementation of the procedure. This will create a sense of ownership and assist in the implementation of the procedure.

6. Ensure that there is a thorough review for usability.

7. When changes are made, go back round the review loop.

Ensure that every requirement is stated in a manner that makes it possible to audit. Every requirement from the procedure should be able to be converted into a question. Hence from the example given below, the statement that "All relevant parts of the form are to be completed with the information indicated" allows for the question "Have all parts of the form been completed as required?"

An example of the question-and-answer technique applied to the creation of a corrective action procedure follows.

Question One: What should this procedure achieve?

Define the scope and purpose of the document. Start with a statement of purpose. This should preferably be a single sentence which precisely summarizes why this document is being written.

Sanity check: If the purpose of this document cannot be summarized in a few short words, then ask the question "Why is this being written, is it needed?" The answer will probably be no!!

Scope—"This procedure defines the corrective action process." *Immediately the reader knows exactly what this procedure is trying to define and is led automatically to the next step. That step is to define the scope of the procedure.*

Is this to be a companywide procedure, for one particular department, or for even a subset of a department?

Purpose—"To ensure the effective tracking, reporting and control of all corrective action within the manufacturing unit of the company."

In this instance, it is clear that this procedure only applies to corrective action in one part of the company. Two key issues are resolved: applicability and whether or not a further procedure (or procedures) is required that describes the corrective action process in other parts of the company.

In the previous scenario it is logical to combine scope and purpose into a single statement. However, not all scope and purpose definitions will necessarily lend themselves to this approach. It often proves beneficial to keep the two statements separate even though in some cases that appears to create redundancy.

Question Two: Who should perform the task?

Once the scope and purpose are defined, consider the "who" of the task. The next section of the procedure should deal with responsibilities and define the responsibilities by precise job title. The need to define scope becomes even more obvious when reviewing of responsibilities. In the departmentally specific procedure example above, the corrective action process is likely to be allocated to a different individual than if the scope of the activity were companywide. Note that responsibilities can flow beyond a single department, and it is often from this careful review of "who" that other information becomes evident, such as the white space issues and who should approve the document.

Using this method of declaring responsibility and authority is an extremely effective method to ensure that the system meets the ISO 9000 requirement that "the responsibility, authority and interrelation of all personnel . . . shall be defined." Defining responsibilities within the body of procedures and work instructions makes those management roles highly visible, one of the declared intentions identified at the very beginning of the process.

Sanity check: If determining responsibilities proves difficult, then the process being described is probably inadequately defined, and it will be a worthwhile exercise to go back to the process flowchart and reexamine the whole subject.

See boxes 6 and 7.

Defining Responsibilities

1. The senior production engineer is responsible for the effective management of the corrective action procedure.

2. All other managers are responsible for ensuring that any CARs forwarded to them for investigation are responded to within the defined time lines.

Box 6

Suggestions for Writing Documentation

In writing the text, remember the KISS principle: *Keep It Short and Simple*. Use straightforward words and terms; do not use this documentation to demonstrate a vast and sophisticated command of the English language. Follow novelist George Orwell's suggestion to ask six questions before writing:

• What am I trying to say?

• What words will best express it?

• What image or idiom will make it more clear?

• Is this image fresh enough to have an effect?

• Could I have used fewer words?

• Have I said anything that is ugly?

More ground rules to follow include:

• Make the meaning very clear.

• Sift out all irrelevant material.

• Ensure that the text is grammatically correct; avoid the use of passive verbs and split infinitives.

• Search out errors in spelling and punctuation.

• Avoid jargon and "committee language."

• Use the simplest language to convey the thought.

• Use clear words and phrases.

• Use short sentences.

• Use punctuation thoughtfully and in a way to aid understanding.

• Separate ideas into individual sentences or paragraphs.

Remember that these are documents written to enable co-workers to perform their tasks more efficiently and consistently. Also remember that documentation users are intelligent; they just have different skills. Therefore:

• Do not write in an antagonistic manner.

• Do not write in a condescending manner.

• Do not write in a supercilious manner.

Box 7

Procedures as Bridge Builders

Roderick Goult

Procedures build bridges between departments—lines that are essential for the effective management of any organization. One of the fundamental problems with most business management structures is that they are organized in functional, vertical operational lines; this is the most efficient way to perform tasks. Satisfying customer needs, however, requires that a series of processes be fulfilled that essentially run horizontally across the organizational structures, trying to bridge the gaps between functional departments. These gaps are almost always where the problems lie. The task of the procedures is to ensure that the gaps are as seamless as possible.

Anyone who has worked in any company—even small companies—will know that "turf wars" between functional departments are common, and the manager who spends most of his or her time playing interdepartmental politics poses a problem. As a result, it is most important for procedures to clearly define responsibilities and authorities to ensure that work will be completed despite the management. It is quite rare to find problems within the vertical structures of an organization It is also quite rare to find those horizontal lines functioning unless a company is complying with ISO 9000 or a similar system. It is a case of managing the white space, and Figure 8–2 highlights the danger zones.

The process of achieving customer satisfaction can be thought of as building a vast jigsaw puzzle. Each employee in the company holds one piece of the puzzle, and the task is to complete the picture on time, every time, with every piece of the jigsaw puzzle correctly placed to create a masterpiece. Imagine trying to do that if the pieces are randomly distributed—which they usually are—and there is no clear picture. The system numbers each piece and tells the experienced jigsaw-puzzle solver how to complete the task.

3. Shop floor supervisors and operators are responsible for raising a corrective action report CAR (see boxes 8 and 9).

4. Whenever a production problem cannot be resolved on the shop floor, or where a problem is repetitive.

Question Three: What is the task?

"Outline and describe the task as simply as possible. Include when and where the task occurs, if appropriate." This section of the procedure is where the actual process is described. This description should be as concise as possible, basically highlighting key activities. There is no need to describe every action in minute detail—documentation

Box 8

Procedure for CAR

1. Any problem that cannot be resolved by shop supervision, or which is found to be repetitive, is to be reported on a CAR form. All relevant parts of the form are to be completed with the information indicated. The CAR will be signed by the individual who has raised the problem and passed to the shop supervisor (see Appendix 1 for the CAR form).

2. The shop supervisor will examine all CARs submitted and endeavor to identify a solution. Any CAR which is resolved at this stage will be signed off by the shop supervisor and passed on to production engineering for recording. CARs which cannot be resolved by shop supervision will be initialed as such and passed to production engineering.

2.1 The shop supervisor is responsible for determining whether or not the subject of the CAR is such that a production permit/waiver is required to enable production to continue during the investigation. If this is the case, the supervisor will notify quality accordingly and request the waiver, referencing the CAR.

3. Production engineering will log each CAR received and allocate a serial number from the CAR serial number list. The pink copy will be filed in the open file. CARs which have been closed by shop supervision will be filed in numeric order, and the green copy returned to the originator.

4. Open CARs will be passed to the senior production engineer, who will classify them as A, B, or C and determine which department is best equipped to resolve the problem. This will be noted on the form as indicated. The CAR will then be passed to the identified department or individual for resolution. Category A requires a response within 24 hours; category B within 72 hours; and category C within five working days.

5. The recipient of a CAR is responsible for ensuring that a response is generated within the designated time frame. The relevant sections of the form should be completed as indicated, and the form returned to the production engineering department.

5.1 Any CAR which requires longer than the designated time frame for a response should be annotated accordingly, and the yellow copy returned to production engineering. Production engineering will notify the originator accordingly.

5.2 Once the CAR is completed, the response will be approved by the appropriate departmental manager. If an engineering change is required to resolve the problem, the investigating engineer will initiate the change in accordance with the appropriate procedure and annotate the Request for Change form with the number of the CAR. The completed CAR will then be returned to the production engineering department.

(Continued on next page)

(Continued from page 329)

6. Production engineering will log the completed CAR, and the senior production engineer will review the response. Satisfactory responses will result in the CAR being closed out, filed in the closed file, and a copy sent to the originator. If production engineering does not accept the response, the CAR will be resubmitted to the relevant department with an explanation for the rejection. The resubmission will be logged in the open log, and a new category noted and advised to the recipient.

7. The senior production engineer is responsible for producing a monthly summary report on all currently open CARs and a quarterly analysis report. These reports will be circulated to all departmental managers.

should be aimed at the trained, skilled, experienced individual who normally performs the task, not written on the basis that recruitment is carried out by dragging someone off the street.

The objective is that the procedure highlights those actions which are unique to this company, or specifically important for this activity, e.g., the records to be kept. The task itself is typically performed within the professional training of the individual concerned.

Sanity check: If the procedure is running to more than two pages of letter-size paper using a 12-point, standard, san-serif typeface, then there is probably too much detail, or the procedure needs to become multiple procedures.

Question Four: Are there any related activities?

List any other documents directly integrated with the task.

Sanity check: List only related procedures which have been referenced in the text, or other referenced activities for which a specific procedure exists. In the example, the production permit/waiver procedure has been referenced as a directly related procedure, as is the Engineering Change Procedure.

Related documents:

1. Production Permit/Concession or Waiver Procedure
2. Engineering Change Request Procedure

Question Five: How is task fulfillment demonstrated?

"Detail the records created."

Box 9

CAR Records

1. Production engineering CAR log.

2. Production engineering open CAR file.

3. Production engineering closed CAR file.

4. Monthly summary report.

5. Quarterly analysis report.

The CAR log will be maintained on a rolling three-year basis. Closed CARs will be kept on active file on a rolling 12-month basis, and archived for two years prior to disposal. Reports will be retained for 12 months.

The procedure is now concluded. It has identified all the key players in the operation, the paperwork that is used, the methodology by which the records are to be kept, and how long they should be retained. Within the body of the text it has also identified who is responsible for those records—not directly, but by defining who performs the tasks concerned.

No further detail in a procedure is required than is illustrated by the example above. It is perfectly reasonable to assume that the personnel who will be involved in the responses to CAR in the case study know how to go about an investigation, since that is one of the factors determining the individual's suitability for employment.

Equally, no need exists to describe how to fill in the CAR form. Any good form is self-explanatory—not only that, but it is reasonable to assume that how to fill in a CAR will be part of the basic training that is provided to everyone who may be expected to use one. Never waste words on descriptions of activities which can be put down to training and basic common sense. Document the key points about the process without which individuals might not naturally act properly.

The five questions above focus and simplify the task of creating both procedures and work instructions. By applying this basic formula to the process, the documentation author is forced to think in clear, logical steps (see Figure 8–2).

Approve and Issue the Procedure

Once the procedure has been validated and agreed, it must be passed through the document control process for issue and distribution to all relevant staff. Another one of the responsibilities of the management is to ensure that relevant personnel not only receive but also effectively implement the procedure.

Figure 8–2

An ISO 9000 System Helps Manage the 'White Space'!

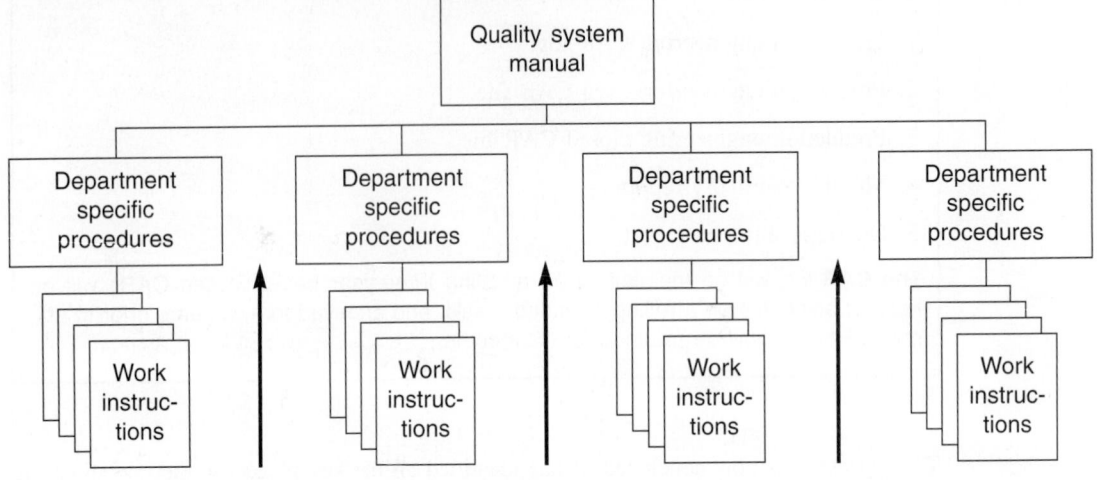

In summary, the evolution of a procedure from its inception to its withdrawal can be described as an eight-step process:

1. Identify the need.

2. Collect the data.

3. Prepare the document.

4. Review and validate.

5. Authorize and release.

6. Review and revise as required.

7. Rewrite whenever necessary.

8. Withdraw from the system when no longer applicable.

Procedure Structure and Format

Procedures should be consistent and easy to use. Uniformity and consistency are worth the effort. The six section format shown in the example has proved to be most effective (some organizations include a seventh section—definitions—in every procedure). The choice of format and layout is down to the individual company. A number of organizations use annotated flowcharts, with little or no formal text, and that works, too.

A consistent format can be achieved by using standard templates or a standard layout, which should be included in the "procedure for procedures" (see above). Templates on which to produce procedures should include the following:

- The company name, division, and logo.

- The document title.

- The document number.

- The document revision and date of issue. (No ISO 9000 requirement exists for the date of issue to be included on a document. The new ISO 14000 standard contains the requirement for environmental systems.)

Uniformity of document presentation can be obtained by defining document layout. Some consideration should be given to defining such issues as the columns on which the text should start and finish, the tabulation to be used for indents and the number of levels of indent to be acceptable. Consider also the typeface to be used and the size of that typeface. San-serif typefaces are more readable, and point sizes below 12 should be avoided. The appearance of headings in terms of upper- and lower-case type and the use of bolding is also worthy of consideration.

DOCUMENT CONTROL

Document control is an important aspect of the process and the methodology to be used should be determined early on the process. Key issues to be considered are:

- Defining responsibilities for documentation.

- Procedure numbering.

- Distribution.

- Revision methodologies.

Defining Responsibilities

Three areas need to be defined:

1. Who identifies the need, performs the reviews, and authorizes the issue of procedures?

2. Who is to prepare and write procedures?

3. Who is to control the documents?

Everyone should be encouraged to identify the need for a procedure; typically it is management who verifies the need. To eliminate conflicts within or between existing procedures, a review must take place of all procedures, especially when they are being added to an existing system. Review is important to ensure the procedure reflects current practice and provides adequate direction to the user and interfaces at both departmental and interdepartmental levels.

The approval of a procedure may require one or more 'signatures' (which can be electronic) depending on whether the procedure is a single department activity or a cross-functional procedure. Typically such signatory authority rests with department heads, managers, the quality assurance manager, or in exceptional cases, the CEO.

Again, preparation and writing of procedures must be done by those involved in their use. Everyone must be involved in the ISO 9000 management system.

Procedure Numbering

Administration and control of documents can either be centralized (not usually a good idea other than for cross-functional documents) or distributed to the departments who operate the procedures (generally a much better system). Management has to make this call. The control process must ensure that:

1. Procedural format is defined (see above).

2. There is a consistent numbering system.

3. The review and release process is defined, both for initial issue and subsequent change or withdrawal, is appropriate.

4. Master document storage, copy distribution, and copy disposal are systematic.

The concept of a numbering system that helps to identify the source and nature of a document has already been discussed. Whatever scheme is adopted, it is important to ensure that plenty of scope for expansion exists. The Standard also asks for a Master Record Index or equivalent. One may argue for a central register of documentation for *all* organizations, but enormous logistical difficulties exist both in compiling and then maintaining such a register. The devolved index is a more common practice, easier to maintain and more accessible to the user.

Distribution

The distribution of management system documentation can sometimes conflict with TQM concepts because controlled paper documents should only go out on a "need to know" basis. Those who merely *want* to know must seek out the documentation where it is needed and study it there! Scatter gun distribution creates enormous headaches for document controllers.

Controlled documentation means what it says—whoever is responsible for the document control process has to know where every single copy is, and to whom, or to which function within the company, it has been supplied. Examination of this process, both for effectiveness and for tracking the documents to assure they get where they are supposed to go, should be a regular part of the internal audit activity. The audit should review location and control of distributed documents. In particular the issue status of copies needs to be validated.

The document control procedure should prohibit unauthorized duplication of documents, because the uncontrolled copying of supposedly controlled procedures is one of the fastest ways a management system falls apart at the seams.

Revision Methodologies

A management system is, or should be, a living process. No company ever stands still. Processes and practices that were appropriate when the system was first created must keep up with the evolving business and new technologies. The process for revising a document should be substantively the same as that for its initial creation. Data collection, document review, and approval may be a much shorter process, but essentially the same activities should take place. Any change should be initiated by a formal request in much the same manner as a physical change to a product is typically initiated. The change will be triggered by an improved methodology, or as a result of the investigation of a problem, or because the existing process is obsolete—which is not necessarily the same as improved methodology.

Review and Approval

The Standard requires that the authority for the review and approval of changes to documents should be the same as for the original issue, *"unless specifically designated otherwise."* Careful thought about this wording leads to the inevitable conclusion that:

1. Signatory authorities should be described by position, not names.

2. If the authority for change is to be different from that for original issue, it must be declared up front.

Any request for change should provide details of the requested change, the reason for the change, the impact of the change, and the origin of the request.

Tracking Revision Status

Every document within the system must have a unique revision identifier. This is a basic requirement of the ISO 9000 Standards. It makes a great deal of sense to have that revision status on each page of the document, although the almost universal use of computerized systems for document production has resulted in page-dependent revisions being rapidly phased out. The system chosen for identifying the revision status of documents should be consistent throughout the organization.

Where practical, the standard asks that any revisions made within a document are indicated either within the document itself or in appropriate attachments. Within the document, text can be highlighted, marked in the margin, underlined, or italicized—whatever works. Alternatively, the nature of the change can be described in a change note distributed with each controlled copy for filing. The choice is up to the company.

Revision Records

Common practice is to include a record of the revisions at the front of each procedure or at the front of a manual containing a set of procedures. This helps to ensure that the user is in possession of the latest revision, although the final check is the revision status printed on the individual document itself, and cross-referencing the masterlist.

When revision record sheets are used, they should identify: the document name and/or number; the revision status (1, 1a, 1b, 2, etc.); the date of the revision; and a summary comment concerning the change made.

The master revision control sheet, or the change control document, should also carry the necessary authorizing signature. It should be reiterated that any change to any document within the system should find its way onto the Master Record Index—this, after all, is the oracle as far as the revision status of management system documentation is concerned.

Revision Issue

Many different ways exist to distribute revisions. Many companies now reissue entire procedures; some still work on page issue and re-issue only the changed pages. The latter course requires a far more complex document change process. Whatever the process, revisions must be issued to all holders of the procedure(s), and it is good practice to accompany the document with a document transmittal sheet. This may be simply for information, or it can be used for the recipient to acknowledge receipt back to the distribution source. Unfortunately this system rarely works effectively!

COMPUTERIZED DOCUMENTATION

The increasing use of computers for the storage and distribution of documents has rendered obsolete all the problems in these areas. Unfortunately, the removal of one set of problems produces another. One major problem is that information filed on computer is vulnerable to unauthorized changes, Another, potentially larger problem, is the risk of tampering by hackers from outside the company.

Some reasonable steps should be taken to protect against accidental interference—the use of passwords or structured access controls, and file protection by "read only" designations for open access documents. It is impossible to prevent deliberate abuse—the best that can be done is to assume that the users want to utilize the system properly and therefore take steps to prevent inadvertent tampering.

The other elements of review, authorization, and control at the front end and when changes are made remain the same. Soft masters also present issues about backup practices and the security of the data. Ideally, a master copy should be kept in a fireproof safe or at a separate location from the master file server.

IMPLEMENTING THE SYSTEM

Implementing ISO 9000 is discussed in detail in Chapter 7, however, it should be pointed out that in the grand scheme of things the design and documentation of the system is, comparatively, easy. Procedures and work instructions should be put into use as soon as they are created so that the "new" system gradually spreads across the organization.

The adequacy and effectiveness of the documented procedures should be carefully monitored throughout the implementation process. The audit activity is one way of doing this, but input is also needed from everyone using or affected by a document in the system. Encourage the most open and free feedback that can be achieved; encourage all employees to scrutinize the procedures and work instructions that affect them and provide constructive criticism. Not only does this activity eliminate any errors and omissions, but it also increases a sense of ownership in the user community.

Once the quality management system is up and running, plans should be made for a final, formal management review before the Registrar arrives. This review should methodically examine all the elements of the quality system and evaluate their compliance to the requirements of the ISO 9000 standard in use. No shortfalls should exist, particularly if the schedule of conformity has been maintained effectively. Any minor deficiencies and nonconformances which are found can be fed into the corrective action process to further enhance and improve the system.

CONCLUSION

Once completed, the quality management system should help the company perform more consistently and produce a lower cost of sales. It will enable both customers and external auditors to evaluate positive, proactive approaches to business management. The management system, however beneficial to the organization's customers, is primarily designed to benefit the organization itself. Achieving ISO 9000 registration will produce benefits, but they will always be secondary to the benefits produced by the effective operation of a value-added quality management system.

Box 10

How Much Documentation?

Question:

Clause 4.2.1 of ISO 9001 states: "The quality manual shall include or reference the documented procedures that form part of the quality system." Does this mean that all company procedures must be detailed or referenced in the manual? Describe an acceptable reference.

Answer:

The panel of experts agreed that it is not necessary or desirable to reproduce every work instruction in the quality manual, but the panel had differing interpretations on how far companies must go to satisfy the requirement.

Moderator Ian Durand, president of Service Process Consulting Inc., said that it is important to remember that employees are the primary customers for documentation. The structure, format, wording, and referencing of the quality system documentation must make it easy for employees to find and use those procedures and work instructions they need to implement the system as intended. "If they cannot, there is no hope for consistent deployment," Durand said. At a minimum, Durand said, the quality manual must reference every general procedure, and every work instruction must be referenced in the applicable general procedure.

Many companies rely on a three-level approach to documentation, he said, including a brief quality manual of 25 to 30 pages; more detailed but still general procedures for each of the requirements; and detailed work instructions where required to apply the general procedures to particular cases.

The quality manual frequently references several more detailed procedures. For example, it might describe the critical purchases a company makes and the criteria for approving suppliers. In some cases, there may be a further level of documentation, frequently called work instructions, which provide very detailed procedures. There might be a work instruction to describe the calibration procedures for a specific piece of laboratory measurement equipment, or a work instruction on how to clean and charge process equipment to produce processed materials.

"The standard merely requires that the quality manual outline the structure of the documentation covering the quality system requirements," Durand said. "In this case, the standards require that the quality manual reference every general procedure that forms part of the quality system." Similarly, Durand said, each general procedure must reference all applicable work instructions. He suggests that companies note where the procedures and work instructions reside.

(Continued on next page)

(Continued from page 338)

Many companies accomplish this, he said, by referencing each procedure by number in the body of the text of the quality manual section, followed by a list of all referenced procedures by number and title under a separate references section. Other quality manual sections are also cross-referenced as appropriate.

"Anything less than this makes understanding the requirements of the quality system a puzzle to employees and jeopardizes the requirement to effectively implement the documented procedures and the quality system contained in item 'b' of subclause 4.2.2," Durand said.

He added that a few smaller companies have made the quality manual a collection of general procedures. "In essence they opt for a two-level approach to their quality system documentation," he explained. "This approach is also consistent with the standards. In this case, in addition to explicitly referencing all related work instructions, there is usually more extensive cross-referencing among procedures."

An enlightened approach, according to Durand, is to include a cross-reference matrix showing every quality manual section and general procedure applicable to each job.

"This focuses the employees on those common and job-specific procedures and work instructions they need to know and follow to satisfy the requirements of the standard as it has been applied in their company," he said. "In addition, the requirement for quality planning requires a comprehensive description of those procedures and work instructions involved in meeting the specified requirements for products, projects, or contracts."

Allen Bailey, president of Compliance Quality International, said that many companies try to put too many work instructions in their quality manuals. That can be a mistake in the long run, he said.

"The quality manual as I look at it contains pointers that state where things are located," he said. "As a guide it should remain relatively unchanged with time, whereas your [work instructions] will change as you add and delete work processes."

Bailey agreed that it is necessary to reference every work instruction in the quality manual. "I would reference that the [work instructions] exist, but I wouldn't reference each one," he said. He suggests referencing the elements of ISO 9000 and noting where the procedures reside that satisfy each of the elements. According to Bailey, the reference should include the physical location of the document being referenced and possibly the title of the person with responsibility for it.

The quality manual outlines the company's overall policies in general terms and how it plans to respond to the requirements of the applicable ISO 9000 standard. That

(Continued on next page)

(Continued from page 339)

usually amounts to one or two pages covering each of the elements in the applicable external quality assurance standard.

Noel White, director of Advance Consulting International Ltd., said that smaller companies may want to include all procedures in the quality manual. "My view is that it's like any sensible structured information system," he explained. "You should be able to clearly and easily go from one level to another."

In some cases, White said, the quality manual may become unwieldy if all procedures are included in their entirety. Companies have the option of referencing procedures by document title, procedure title, and procedure number.

"It's better to reference downwards, so that if you are referencing a top-level clause in the quality manual it's clear where you need to go for further information about the detail of the procedure," he said. The reason for providing at least a complete reference for each procedure, he said, is to make the manual accessible to the reader.

"At the end of the day, it's a case of making the system easily understandable to anyone who wants to try to understand it," such as a new employee, White said. Theoretically "you could read the top-level manual, understand the system that's in place, and then [be directed] to the next activity in whatever area you are looking at." He suggested adding a section marked "references" under each subheading of the quality manual.

Denis McNamee, president of Quality Assurance Systems Inc. (Q.A. Systems), said it is not necessary to address all work instructions in the manual.

In some cases, he said, it would not be practical to list all work instructions. "For process control the list could be pretty extensive," he explained. "If you had 60 work instructions for process control, then you could say: 'There are approximately 60 work instructions that cover all essential areas of process control. These are some of the key ones.'"

McNamee defined key work instructions as those that are critical to the performance of the product, such as a testing work instruction.

"With computers it would be no big deal to include a [complete] list, but then the manual starts to become pretty extensive," he explained. "For the sake of brevity, it would be desirable not to print out every work instruction."

Appendix 1
Sample Quality Manual

Any Business, Inc.
123 Main Street
Anytown, USA

Phone: 700-555-2222
Fax: 700-555-2221
Quality Manual
Level 1 - POLICY
Controlled Copy No:

Authorized and Issued by:
Chris Smith

Chris Smith,
President and Chief Executive Officer

_____

Editor's note: ■ ■ ■ *indicates separate pages*

INDEX TO CONTENTS

ISO 9001

Page	Subject	
01	Cover sheet	
02	Index to Contents	
03	Introduction and Company Profile	
04	Register of Holders of Controlled Copies	
05	Amendment Record	
06	Policy Statement	
07	Management Responsibility	4.1
10	Quality System	4.2
11	Contract Review	4.3
12	Design Control	4.4
14	Document and Data Control	4.5
15	Purchasing	4.6
16	Control of Customer Supplied Product	4.7
16	Product Identification and Traceability	4.8
17	Process Control	4.9
17	Inspection and Testing	4.10
18	Control of Inspection, Measuring, and Test Equipment	4.11
19	Inspection and Test Status	4.12
19	Control of Nonconforming Product	4.13
20	Corrective and Preventive Action	4.14
21	Handling, Storage, Packaging, Preservation, and Delivery	4.15
22	Control of Quality Records	4.16
22	Internal Quality Audits	4.17
23	Training	4.18
23	Servicing	4.19
23	Statistical Techniques	4.20
24	Appendix—Content of Level 2 Procedures Manual	

■ ■ ■

INTRODUCTION AND COMPANY PROFILE

Any Business, Inc., was formed in 1980 by a group of people, the present directors, who shared many years of business management experience in the [specify] industry. Realizing that a niche existed in the market for an organization that combined technical expertise with the personal service available from the smaller concern, Any Business, Inc., was formed.

Proud of our reputation established in this industry, our total commitment is focused on continuing to offer customers across the spectrum of commerce, from traditional machines to the latest technology, an unparalleled service at competitive rates.

It is the ABI's endeavor to achieve and maintain Certification to ISO 9001: 1994, via the Accredited Registrar, Incorporated (ARI). We intend to display the ARI Certified Firm symbol on letterheads, advertising matter and Yellow Pages entries.

■ ■ ■

REGISTER OF HOLDERS OF CONTROLLED COPIES

00	**Master held on computer**	1
01	**President and Chief Executive Officer**	1
02	**Vice President of Quality**	1
03	**Assessment Authority**	1
04	**Quality Consultants**	1

AMENDMENT RECORD

Control of this manual is accomplished by updating the issue status of the manual itself, with amendment data shown in the record below. Changes are identified by a vertical bar in the right-hand margin.

■ ■ ■

QUALITY POLICY STATEMENT

This *Quality Assurance Manual* defines Any Business, Inc.'s policies and objectives regarding the application of the principles of controlled Quality Assurance to ensure that all services rendered by the company are of the required quality and comply fully with the customer's stated requirements and expectations.

The Quality System, as documented and implemented, is intended to comply fully with the requirements of ISO 9000 as applicable to the companyís operations.

The content of this *Quality Assurance Manual*, and supporting documents, is applicable to all company employees and shall be observed and implemented by all personnel as applicable to their activities. No deviation is permitted without the express permission of the undersigned.

The vice president of quality, Jennifer James, is hereby vested with full responsibility for the proper and timely implementation of the Quality System, together with the appropriate level of authority for ensuring its continuing effectiveness.

Chris Smith

Chris Smith, President and Chief Executive Officer
Any Business, Inc.

■ ■ ■

MANAGEMENT RESPONSIBILITY
ISO 9001 Clause: 4.1
4.1.1 Quality Policy

The company's Quality Policy, objectives, and commitment to quality are defined in the president and chief executive officer's Policy Statement (page 7 of this Manual). The Policy and the means of implementation are advised to all personnel by the publication of the Documented Quality System, communication on quality matters by the vice president of quality, and by the induction routines employed for new employees.

4.1.2 Organization
4.1.2.1 Responsibility and Authority

The Management Organization Chart (page 10 of this manual), defines the lines of responsibility of all personnel. The Schedule of Key Personnel Responsibilities and Authorities (page 9 of this manual), details job functions and levels of authority. Additionally, procedures clearly indicate responsibilities for their implementation.

4.1.2.2 Resources

Resources are demonstrated throughout the Documented Quality System and include Internal Quality Audit and Verification covering Administration, Customer Service, and all forms of Design and Process Control.

4.1.2.3 Management Representative

The Company's Management Representative, for all quality matters, both in-house and external, is Jennifer James, vice president of quality. Her responsibilities and authorities are defined in detail in the Schedule of Responsibilities (page nine of this manual). Her deputy is Terry Watson, vice president of technical services.

4.1.3 Management Review

The effectiveness of the Quality System is formally reviewed by management, chaired by the president and chief executive officer. The *Management Review Meeting* is scheduled annually, two months after the end of the company's financial year. The objective of the review is to examine demonstrated irregularities in the operation of the quality system itself and its contents. To this end, the vice president of quality maintains an ongoing analysis of corrective and preventive actions taken during the year and noncompliances identified by audit, appraisal by outside bodies, customer complaints, and nonconformity.

The totals are presented for the year and are compared by the meeting with data for earlier periods. As appropriate, the meeting determines any changes of policy, course or action. A formal record of Management Review is maintained, together with details of decisions taken.

Level 2 Procedure References

MP-01 Review of the Quality System
MP-02 Corrective Action
MP-20 Analysis of Review Input Data
MP-23 Preventive Action

■ ■ ■

Schedule of Key Personnel Responsibilities and Authorities

The company's management lies in the hands of the directors, who are the shareholders. As such, and in view of the fact that the company is relatively small, responsibility and authority are shared.

The *president and chief executive officer,* who is also the *accounts director,* has total responsibility for office administration matters, invoicing, credit control, and budgeting. He is authorized to purchase materials and services as required for the company's operations. He is also a focal point for customer contact and, as such, he effects contract review.

The *vice president of technical services* has responsibility and authority for the control of customer contact, contract review, technical matters, and control of day-to-day work. He is authorized to purchase materials and services as may be necessary for the implementation of the work. He is the Deputy Management Representative.

The *vice president of quality* is responsible for, and authorized to, set up and implement a quality system in accordance with the requirements of ISO 9001. She is the management representative for all quality matters. In addition, she works in parallel with the vice president, covering all the responsibilities and authorities assigned to him.

The *technical staff* carry out work at customers' premises and in the company's workshop. These activities include a sales element, with a vice president carrying out the contract review.

President and Chief Executive Officer
Chris Smith

Vice President of Quality
Jennifer James
Management Representative

Vice President of Technical Services
Terry Watson
Deputy Management Representiative

Foreman Technician

■ ■ ■

QUALITY SYSTEM
ISO 9001 Clause: 4.2
4.2.1 General
The company's documented Quality System is structured in two levels.

Level 1, the Quality Assurance Manual, defines policy, demonstrates scope and commitment and governs the implementation of the Level 2 procedures manual.

All quality related documentation is formally issue controlled, updated, and reissued by means of an issue and retrieval system designed to ensure withdrawal of outdated copy.

4.2.2 Quality System Procedures
The content of the Level 2 Procedures Manual is the company's norm for implementation of all activities. In the event that a customer stipulates requirements beyond the scope of, or differing from, the company's standards, a Project Quality Plan is prepared, listing the requirements, standards and specifications to be observed, acceptance criteria, and tests and inspections to be carried out. This Quality Plan shall take precedence over the company's documented quality system should any conflict arise.

Definitions used throughout the documented quality system are defined in ISO 8402 at the latest edition.

4.2.3 Quality Planning
The documented quality system is comprehensive and covers all aspects of both the company's activities and the requirements of ISO 9001 as applicable to those activities. Level 2 Procedures supporting this manual cover all requirements for Quality Planning, particularly the need for total review of the Quality System.

Level 2 Procedure References
MP-01 Review of the Quality System
MP-02 Corrective Action
MP-03 Control of Company Documentation
MP-29 Quality Planning

■ ■ ■

CONTRACT REVIEW
ISO 9001 Clause: 4.3
4.3.1 General
The Company maintains procedural instructions to govern Contract Review.

4.3.2 Review
Each inquiry, invitation to bid, or to supply, or perceived market opportunity is reviewed by the vice president of quality, with others as appropriate, in order to establish the acceptability of the inquiry content, ability of the company to supply, and completeness of data furnished. On receipt of an order, the response to the inquiry and the order are compared. In the event that any discrepancy is identified, the matter is referred to the customer to a point of agreement. Thereafter the order is accepted, acknowledged, and executed.

4.3.3 Amendment to Contract
In the event that the company or the customer should seek to vary the contract conditions or requirements, reversion shall be made to the details of Para. 4.3.2 above to a status of agreement with the customer.

4.3.4 Records
Formal records of all Contract Review activities are maintained.

Level 2 Procedure References:
MP-04 Contract Review

DESIGN CONTROL
ISO 9001 Clause: 4.4
4.4.1 General
The company maintains procedural documentation to control the design and development functions.

4.4.2 Design and Development Planning
Every contract accepted by the company that involves design and development activity results in a plan being created outlining the activities involved. Specific staff are assigned to each activity, and the estimated resources are identified for both personnel and equipment. As the design evolves the plan is updated to reflect progress.

4.4.3 Organizational and Technical Interfaces
The organizational and technical interfaces having input to the process are identified within the plan, and regular meetings of this group ensure that all necessary information is regularly reviewed. Minutes of these meetings are maintained, including all meetings with the client.

4.4.4 Design Input
All input information is documented within the activity, including relevant statutory and regulatory requirements, and the review process ensures that such inputs are adequate for the requirements of the design process. Any incomplete or ambiguous requirements will be identified in the minutes of review meetings and resolved with the appropriate parties before the design activity proceeds. Contract review is considered to be an essential part of this activity.

4.4.5 Design Output

The output from each stage of the design plan is documented as part of the design review activity and is stated in terms of measurable requirements, which are evaluated for:

- Compliance with the design input requirements.

- The appropriateness of the acceptance criteria either contained or referenced within the documentation.

- The proper and complete identification of any safety or performance critical aspects of the design.

No design documentation is permitted to be released prior to the successful completion of the necessary design review activity.

4.4.6 Design Review

The design plan identifies appropriate review stages for the project, and meetings involving all relevant personnel are held as required to ensure that the design and development activity is proceeding in an appropriate manner. Minutes of all such meetings are maintained by the design team leader.

4.4.7 Design Verification

Each stage of the development activity is subject to verification activities, which form part of the design review. These can take several forms, including:

- The use of alternative methods to validate design calculations.

- Comparison of the new design with an existing, proven design sufficiently similar in nature to provide a degree of assurance that the new design is adequate for the purpose for which it is intended.

- The performance of alpha and beta testing of prototypes.

4.4.8 Design Validation

At the conclusion of the design activity, and prior to the final design review and release to production, design validation is performed on all designs under conditions which replicate actual usage as closely as is feasible for the product involved. The results of this validation process are compared to the original design input requirements to ensure that the output from the design process meets the original intent of the customer.

4.4.9 Design Changes

Design change control is exercised from that stage of any project when the original designer of any stage of the product releases the design to other team members for evaluation, testing, or integration. A documented procedure controls the review, approval, and release of changes. Prior to such release, individual design engineers maintain designersí notebooks in which all changes to the design concept or realization are recorded.

Level 2 Procedure References

MP-05   Design Control
MP-06   Design Requirements Documents
MP-07   Design Verification and Review
MP-08   Design Acceptance Procedure
MP-09   Design Change

■ ■ ■

DOCUMENT AND DATA CONTROL
ISO 9001 Clause: 4.5
4.5.1 General
Quality manuals, procedures manuals, reference standards, formats, service manuals, and specifications are all subject to document control by procedure.

The President and Chief Executive Officer is the only person authorized to issue or reissue controlled manuals. Interim or handwritten changes are not permitted.

4.5.2 Document and Data Approval and Issue
All controlled documentation is reviewed by the vice president of quality prior to issue. A Master Document Control List is published to demonstrate issue currency. Obsolete documents are removed from circulation on a basis of one-for-one retrieval and reissue. All personnel have access to pertinent data, and outdated documents are retained for reference purposes.

4.5.3 Document and Data Changes
Only the vice president of quality may make changes to documents. The nature of changes is identified by means of a vertical bar in the right-hand margin opposite the change.

Level 2 Procedure References:

MP-03   Control of Company Documentation

MP-10   Reference Standards Control

MP-26   Document Transmittals

■ ■ ■

PURCHASING
ISO 9001 Clause: 4.6
4.6.1 General
The Company maintains procedures to ensure that purchased product conforms to specified requirements.

4.6.2 Evaluation of Subcontractors
All subcontractors and suppliers are assessed in accordance with documented criteria and, if acceptable, are placed on the Approved Suppliers List. Purchases are made only from organizations appearing on the list. The company maintains monitoring control as laid down by procedure. Subcontractor records are maintained.

4.6.3 Purchasing Data
Purchasing documents detail fully the data describing the product or service required, including reference to standards and codes where applicable. All such documents are reviewed for completeness of specified requirements prior to issue.

4.6.4 Verification of Purchased Product
4.6.4.1 Supplier Verification at Subcontractors' Premises
In the event that the company shall opt to verify purchased product at the subcontractor's premises, details of verification arrangements, and the means of product release are stated in the purchasing documents.

4.6.4.2 Customer Verification of Subcontracted Product
Where specified at Contract Review stage (Clause: 4.3), the customer, or agent, is afforded the right to verify product at the subcontractorís premises.

Level 2 Procedure References:
MP-12 Goods Receipt Routine
MP-13 Purchasing
MP-14 Maintenance of the Approved Suppliers List

CONTROL OF CUSTOMER-SUPPLIED PRODUCT
ISO 9001 Clause: 4.7
In this context, the receipt and handling of customer owned equipment, together with the free issue of any component, shall be viewed as Customer Supplied Product.

All product not purchased by the company and not the property of the company shall be subject to Goods Receipt Inspection. Company purchases shall also be identified by label or tag identifying ownership. Any discrepancy or damage shall be reported to the customer.

Level 2 Procedure References:
MP-11 Inspection Policy
MP-12 Goods Receipt Routine
MP-28 Control of Customer Supplied Product

PRODUCT IDENTIFICATION and TRACEABILITY
ISO 9001 Clause: 4.8
Identification and Traceability of Product, Work in Progress, and Stored Product is maintained by Goods Receipt Routines, accompanying documentation, and inspection records, respectively.

Level 2 Procedure References:
MP-11 Inspection Policy
MP-12 Goods Receipt Routine
MP-18 Handling, Storage, Packaging, Preservation, and Delivery

PROCESS CONTROL
ISO 9001 Clause: 4.9
All process activities are governed by documented procedures. The work is controlled by means of documented instructions and records of work carried out are maintained. No processes, the results of which cannot be fully verified by subsequent inspection or testing, are carried out.

Level 2 Procedure References:
MP-24 Servicing
MP-25 Calibration of Customer's Equipment
MP-27 Overhaul and Repair of Customer's Equipment

INSPECTION and TESTING
ISO 9001 Clause: 4.10
4.10.1 General
The Company maintains documented procedures for all forms of inspection and testing and for the records required to be maintained.

4.10.2 Receiving Inspection and Testing
Receiving Inspection and Testing is carried out on all goods received, whether purchased or customer supplied. No product may be used until goods receipt routine has been completed.

4.10.3 In-Process Inspection and Testing
Where applicable, in-process inspection and testing is implemented according to procedures or quality plans.

4.10.4 Final Inspection and Testing
Final Inspection is implemented, and recorded, on all services and product supplied.

4.10.5 Inspection and Test Records
Inspection and Test Records are maintained for all forms of inspection activity.

Level 2 Procedure References:
MP-11   Inspection Policy

MP-12   Goods Receipt Routine

CONTROL OF INSPECTION, MEASURING, and TEST EQUIPMENT
ISO 9001 Clause: 4.11
4.11.1 General
The company maintains procedures for all aspects of control of calibrated equipment. Control and calibration of instrumentation and calibration standards are in accordance with the requirements of ISO 10012, with tolerances amended to suit the practical usage of the equipment. All such equipment is uniquely identified, and each item is supported by a record demonstrating item identity, frequency of calibration, standards to be used, calibrations carried out, remedial actions, and state of accuracy prior to and after calibration.

4.11.2 Control Procedures
All calibrations are supported by certification referring traceability to National Standards. Full calibration records are maintained.

In the event that equipment demonstrates that it is out of tolerance prior to calibration, remedial retrospective action shall be implemented and the customer shall be advised in writing, if such discrepancy may have affected the accuracy of data supplied to the customer.

Level 2 Procedure References:
MP-02   Corrective Action

MP-16   Control and Calibration of Inspection, Measuring, and Test Equipment

MP-17   Noncompliances

MP-25   Calibration of Customer's Equipment

■ ■ ■

INSPECTION and TEST STATUS
ISO 9001 Clause: 4.12
Each inspection and test stage is recorded, with endorsement by the inspector and all equipment is labeled accordingly.

Level 2 Procedure References:
MP-11 Inspection Policy
MP-12 Goods Receipt Routine

CONTROL OF NONCONFORMING PRODUCT
ISO 9001 Clause: 4.13
4.13.1 General
The company maintains procedures for the control of nonconforming product. This includes those instances where any failure to provide service in accordance with agreed conditions is observed.

4.13.2. Nonconforming Product Review and Disposition
Product which is identified as nonconforming shall be reported to the vice president of quality for determination of disposition. Customer complaints are dealt with according to procedure, and observed noncompliances are similarly regularized by procedure.

Level 2 Procedures Reference:
MP-02 Corrective Action
MP-15 Customer Complaints
MP-17 Noncompliances

■ ■ ■

CORRECTIVE AND PREVENTIVE ACTION
ISO 9001 Clause: 4.14
4.14.1 General
The company maintains procedures for implementing both corrective and preventive action. Changes to the documented quality system resulting from these activities are implemented and recorded.

4.14.2 Corrective Action
Corrective action shall be instituted in all cases of failure factors.

These are:

- Nonconforming items and/or work in progress.

- Failures of personnel to observe the mandates of the Quality System.

- Recorded audit discrepancies.

- Recorded accreditation authority discrepancies.

- Customer complaints.

All the above are recorded in the Corrective Action Register and are subject to verified closeout and periodic review in order to identify trends and recurrences. An analysis of corrective actions is submitted to the Management Review, referred to under Clause 4.1, page 8 of this manual.

4.14.3 Preventive Action

Preventive Actions are implemented as a result of the identification of trends and recurrences identified by corrective action (4.14.2). Additionally, preventive action is applied to the ensurance of usability of equipment by periodic maintenance.

Level 2 Procedure References:

MP-02   Corrective action
MP-17   Noncompliances
MP-21   Internal Quality Audits
MP-23   Preventive Action

■ ■ ■

HANDLING, STORAGE, PACKAGING, PRESERVATION, AND DELIVERY
ISO 9001 Clause: 4.15
4.15.1 General

The Company maintains procedures for this subject.

4.15.2 Handling

Product is handled in order to preserve product integrity.

4.15.3 Storage

Storage of all product is secure and identified.

4.15.4 Packaging

Packaging is according to the nature of the product to ensure absence of damage in transit.

4.15.5 Preservation

Preservation is applied, as required by the nature of the product.

4.15.6 Delivery

Delivery or collection is as specified by the customer, or by company transport.

Level 2 Procedures Reference:

MP-18   Handling, Storage, Packaging, and Delivery

■ ■ ■

CONTROL OF QUALITY RECORDS
ISO 9001 Clause: 4.16

All job related records are maintained in dedicated files by customer name or contract reference, as applicable.

Company records pertaining to quality matters are held by the vice president of quality.

The responsibilities for preparation, maintenance and destruction, methods of destruction, filing and indexing systems, and periods of retention are all laid down by procedure.

Level 2 Procedure References:

MP-19   Preparation and Maintenance of Quality Records

INTERNAL QUALITY AUDITS
ISO 9001 Clause: 4.17

All areas of the Company, other than financial, health and safety, and fiscal are audited on a scheduled calendar basis such that all are covered at least once annually.

The audit system uses Serialized Reports and Corrective Action Requests, which are verified as closed out by the Internal Auditor when satisfied.

Objective evidence of all items, documents, activities, and personnel viewed is recorded within the report. Audit is implemented against a combination of the procedures listed on page 31 of this manual and the reference standard elements.

Level 2 Procedures Reference:
MP-02 Corrective Action
MP-21 Internal Quality Audit

TRAINING
ISO 9001 Clause: 4.18

Training requirements are planned annually in December of each year, using a formal Training Plan which highlights requirements for the coming year. Additionally, the Training Plan acts as a record of training and qualification achievements and is subject to periodic review and update as necessary.

All new employees receive Induction Training, and Quality Awareness Addresses are given initially and at two-year intervals.

The Vice President of Quality maintains individual personnel training records.

Level 2 Procedure References:
MP-22 Training

SERVICING
ISO 9001 Clause: 4.19

As a supplier of services, the company engages in servicing activities. These are, however, viewed as the company's normal process, addressed on page 14 of this Manual.

Level 2 Procedure References:
MP-24   Servicing
MP-25   Calibration of Customer's Equipment
MP-27   Overhaul and Repair of Customer's Equipment

STATISTICAL TECHNIQUES
ISO 9001 Clause: 4.20

As a supplier of service to the customer, the Company does not employ statistical techniques.

■ ■ ■

Level 2 Procedure References:
None

APPENDIX - Content of Level 2 Procedures Manual

Reference Clause	Contents	ISO 9001
MP-01	Review of the Quality System	4.1 / 4.2
MP-02	Corrective Action	4.14
MP-03	Control of Company Documentation	4.5
MP-04	Contract Review	4.3
MP-05	Design Control	4.4
MP-06	Design Requirements Documents	4.4
MP-07	Design Verification and Review	4.4
MP-08	Design Acceptance Procedure	4.4
MP-09	Design Change	4.4
MP-10	Reference Standards Control	4.5
MP-11	Inspection Policy	4.10
MP-12	Goods Receipt Routine	4.6 / 4.7 / 4.10
MP-13	Purchasing	4.6
MP-14	Maintenance of the Approved Suppliers List	4.6
MP-15	Customer Complaints	4.13 / 4.14
MP-16	Calibration and Control of Inspection, Measuring and Test Equipment	4.11
MP-17	Noncompliances	4.13
MP-18	Handling, Storage, Packaging, Preservation, and Delivery	4.15
MP-19	Preparation and Maintenance of Quality Records	4.16
MP-20	Analysis of Review Input Data	4.1
MP-21	Internal Quality Audits	4.17
MP-22	Training	4.18
MP-23	Preventive Action	4.14
MP-24	Servicing	4.19
MP-25	Calibration of Customer's Equipment	4.11
MP-26	Document Transmittals	4.5
MP-27	Overhaul and Repair of Customer's Equipment	4.9
MP-28	Control of Customer Supplied Product	4.7
MP-29	Quality Planning	4.2

Vice President of Quality   Jennifer James
Management Representative

President and Chief Executive Officer   Chris Smith

Vice President of Technical Services   Terry Watson
Deputy Management Representative

Details of amendment(s)

Appendix 2
Sample Procedures

These procedures are offered as examples of operating procedures that meet the requirements of ISO 9001. They are written with a presumption that the reader has some experience in ISO 9000 implemenentation and requires guidance rather than detailed direction.

These procedures exceed the basic requirements of ISO 9001 in several instances, thereby providing the user with opportunities for ongoing improvement. Note that the procedures address generic areas of the standard, which are the only areas where procedures can reasonably be applied universally.

SAMPLE PROCEDURE 1

Procedure Title   Management Review
Procedure Number   OP/MAG/01
Revision Status   01
Issue Date   01/01/96

PURPOSE

To ensure that the Executive Review of the management system is performed at regular, defined intervals and addresses all the relevant issues.

SCOPE

All Executive Management Reviews of the Quality Management System. (*If the site is one of several, then add the words "performed within the[insert here the name of the organization]" or some similar comment to clarify the scope*).

RELATED DOCUMENTATION

Identify here all procedures that produce summary reports for management which reflect upon the ability to review the effectiveness of the management system and address the items identified in the minimum agenda listed below. Examples would be: Summary Audit Reports (see 4.17); Summary Reports of Corrective and Preventive Actions Taken (see 4.14); Summary Customer Complaint Reports (see 4.14); Summary Reports on Process Capability Issues (see 4.9); Summary Reports on Test and Inspection Performance (see 4.10), etc.

RESPONSIBILITIES

(Job title of the most senior site executive)	To ensure that this procedure is followed. To ensure that any corrective actions determined as a result of the management review activity are carried out and validated for effectiveness.
All Managers	To attend all management review meetings as requested by the plant manager. To carry out any corrective action assigned from the meeting as effectively and quickly as is practicable.
All other company personnel	To cooperate in the Corrective Action process when asked.

PROCEDURE

1. The *(senior site executive)* is responsible for calling management review meetings to evaluate the working of the QMS at least *(here insert frequency of review meetings-suggestion, four times annually).*

2. The agenda for the management review meeting covers the minimum agenda listed below and any other matters which either the Plant Manager or meeting members may agree to include.

Minimum agenda:

   a. Review of status of corrective and preventive actions resulting from past reviews.

   b. Summary report of audit findings, corrective actions, and follow ups.

   c. Summary report of corrective and preventive actions taken, issues resolved, and issues still outstanding.

   d. Summary report of customer complaints.

   e. Summary reports of process compliance.

   f. Review of fulfillment of quality objectives.

   g. Identification of any changes in the system required to enhance any performance measures.

   h. Future requirements and their impact on the existing system.

3. Records of management reviews are maintained by *(insert title of manager responsible).*

RECORDS

The effective implementation of this procedure produces the following records:

1. Minutes of the management review meetings.

2. Corrective and preventive action plans.

It is the responsibility of the (plant manager) to maintain these records in an accessible and readily retrievable fashion. All records are maintained for a minimum of three years to enable long-term reviews of the effectiveness of corrective action to be performed when appropriate.

SAMPLE PROCEDURE 2

Procedure Title	Procedure for Procedures
Procedure Number	OP/DOC /01
Revision Status	01
Issue Date	01/01/96

PURPOSE

To ensure that all procedures created within the Quality Management System follow a consistent layout, content, and format.

SCOPE

All procedures created within the *(insert here the name of the organization)* QMS.

RELATED DOCUMENTATION

Document Control Procedure

RESPONSIBILITIES

All staff	To ensure that this procedure is followed at all times when writing or revising Quality Management System procedures.

PROCEDURE

1. Each procedure created within the Quality Management System is to be formatted in accordance with this procedure. As a minimum, the procedure will contain the following sections:

Purpose: Used to define why the procedure is being created.

Scope: Used to define the applicability of the document within the QMS.

Related Documentation: Used to identify any other documents related to the task described within the procedure or which contain technical information relevant to the activity described within the procedure *and to which the user should make reference.*

Responsibilities: Used to identify the key responsibilities that must be fulfilled for the procedure to be implemented effectively.

Procedure: Used to describe the activities involved in the fulfillment of the procedure at whatever level of detail is appropriate for the skills, experience, and training received in the task by the intended user of the document.

Records: Used to describe the records created by the use of the procedure, who is responsible for their identification and maintenance, and for how long they should be retained.

2. The master copy of each procedure is to have an attachment detailing the document control information and is signed either physically or electronically by the author and the approver.

3. Where necessary, a circulation list is to be attached to the document control master sheet for the procedure.

4. All procedures are to be produced on the identified documentation format sheet, and each page is to carry the name/number of the document, the revision status, the issue date, the page number, and the total number of pages in the document. The page count does not include the master sheet or the distribution list.

5. It is recommended that for maximum readability all documents be produced in a 13- or 14-point sans serif typeface. (This document is produced in 13pt Helvetica.)

RECORDS

The records which result from the effective implementation of this procedure are the QMS SOPs.

SAMPLE PROCEDURE 3

Procedure Title	Internal Audit
Procedure Number	OP/AUD/01
Revision Status	01
Issue Date	01/01/96

PURPOSE

To ensure that all internal audit activity is performed in a consistent manner by trained auditors and that results are adequately reported, followed up, and closed out as quickly as is reasonably practical.

SCOPE

All internal audit activity performed within *(insert here the name of the organization).*

RELATED DOCUMENTATION

Management Review Procedure
ISO 10011-2

RESPONSIBILITIES

Audit manager *(Put the title of whoever will be organizing the audits here)*	To ensure that this procedure is followed by all audit staff. To ensure that all audit findings are promptly reported to the manager of the area audited. To ensure that any adverse findings are followed up and closed out as quickly as possible. To prepare audit summary reports for review at the Management Review meeting at appropriate intervals.
All managers	To cooperate in scheduling and attending internal audits. To respond promptly to audit findings. To take steps to resolve any deficiencies reported in an effective and prompt fashion.
All other company personnel	To cooperate fully in the internal audit process and enable it to be a driver for the continuous improvement of the management system.
Executive management	To review audit summary reports at regular intervals as defined within the management review procedure. To require effective and prompt response from all managers to any adverse audit findings.

A. AUDIT INITIATION

1. The *(audit manager)* is responsible for developing an internal audit schedule each year and publishing it to all departmental managers at least one month prior to the first audit scheduled on the plan.

   The development of the plan is to take account of previous audit results when available and the impact of the area audited upon the quality of the final deliverable from the plant.

   At a minimum, the schedule contains the following information:

   a. Functional area of the company to be audited (planning, production, materials management, etc.).

   b. The week in which the audit is planned to take place.

   In addition, the audit plan may contain the names of the intended auditors.

2. The *(audit manager)* must ensure that auditors appointed to perform any of the audits are independent of the activities being audited, properly trained in accordance with the requirements of this procedure, and that no other potential conflicts of interest exist.

3. The auditors should be notified of their audit tasks at least one month prior to the scheduled date of the audit. The information to be provided to them includes the name of the manager of the area to be audited, details of the activities and/or procedures to be reviewed, any outstanding audit issues awaiting resolution, and the summary report from the immediate past audit when appropriate.

4. The appointed lead auditor is to contact the manager of the area to be audited and make appropriate arrangements for the timing of the audit, including any required opening and closing meetings.

5. When the audit team consists of more than one individual, the lead auditor is responsible for contacting the other team members to allocate tasks and provide them with the relevant information they require to perform the audit in accordance with the methods described during the training process.

B. AUDIT PRACTICE

1. Each audit should start with an opening meeting with the manager of the area being audited. This requirement may be waived at the discretion of the *(audit manager)* or at the request of the manager of the area under review.

2. Audit findings (opportunities) are recorded on an audit report form.

3. All findings are reported to the manager of the area being audited at the conclusion of the audit, and a provisional agreement is reached on the response time by which the manager will have a corrective action plan ready.

4. A summary report is issued by the *(audit manager)* within 10 working days of the audit.

5. Corrective action plans are reviewed by the *(audit manager)* once submitted by the area manager, and appropriate arrangements are made for the follow-up and close-out of the audit findings.

6. Corrective action plans and completed actions are posted forward to the next audit of the area to enable the auditor to validate that the corrective action has been effective.

C. AUDITOR TRAINING

1. All personnel performing internal audits must undergo a minimum of two days of formal audit training from an approved training provider.

   Approved training is either:

   a. In-house training approved by the *(audit manager)*.

   b. An IRCA registered Internal Auditor training course.

   c. An IRCA- or RAB-approved Lead Auditor training course.

2. Successful completion of an approved internal auditor training program is followed by a minimum of three audits in the company of a training auditor. This requirement may be waived when the training undertaken has been the Lead Auditor training program.

D. AUDIT REPORTING

1. A summary report of all audit findings is submitted for consideration at each Management Review.

RECORDS

The effective implementation of this procedure produces the following records:

1. An audit plan and statement of scope for each audit.

2. Individual audit findings (opportunities).

3. Corrective action plans, audit follow-up, and audit closeout records.

4. Summary audit reports.

5. Management Review audit reports.

6. Training records for individual auditors.

It is the responsibility of the *(audit manager)* to maintain these records in an accessible and readily retrievable fashion. Items 1, 2, and 3 are, as a minimum, retained until the next audit of the same area has been closed out. Items 4 and 5 are to be maintained for a minimum of three years to enable long-term reviews of the effectiveness of corrective action to be performed when appropriate.

ISO 9001—Issues addressed within system documentation.

The following list identifies those subjects that ISO 9001 requires be formally documented in the procedures which form the management system based upon this Standard. Linkages between elements are identified within the text, if experience has shown that they are often missed. The list is not to be considered exhaustive; there are also recommendations included in places which exceed the basic requirements of the Standard.

4.1 Management Responsibility

The quality policy, including the executive management commitment to quality.

The objectives for quality.

Records of any product quality problems.

Responsibility, authority, and interrelationship of all personnel within the management system scope.

Management representative responsibilities and authorities.

Management reviews, frequency of meetings, agenda, minutes

4.2 Quality System

System procedures that document the quality system and cover all aspects of the scope and all relevant parts of the Standard.

Outline structure of documentation.

Quality planning-either discrete quality plans or within the system documentation.

Identification of quality records.

4.3 Contract Review

Procedure for review of every contract.

Procedure for handling contract amendments and circulating necessary data to the appropriate departments (define "appropriate").

Records of the contract review activity.

4.4 Design Control

Process for assigning design tasks, responsibilities, and authorities

Definition of appropriate organizational and technical interfaces within and outside the design department.

Defined design process which includes methods for the formal handling of:

- Design input and output (link back to contract issues).

- Design reviews at predetermined stages in the process.

- Design evolution methodologies.

- Verification of design output.

- Validation of design output.

- Control of design change.

- Records of all significant design activities, including:

  —planning

  —formal inputs

  —formal outputs

  —regulatory/statutory compliance

  —reviews

  —verification

  —validation

  —changes.

4.5 Document and Data Control

Clearly defined structure and roadmap

Defined processes, responsibilities, and authorities for:

- The development, approval, and issue of documents.

- The initiation and fulfillment of document change.

- Distribution of original documents and changes to documents.

- Removal of obsolete documents (or other method of ensuring that they don't get used).

- Maintenance of history files when required.

- Maintenance of Master Record Index or equivalent process.

For computer-based systems, all of the above plus:

- Defined backup regime.

- Formal access controls.

Methodology for the control of documents of external origin, including statutory and regulatory documentation, MSDS documentation, Standards, customer drawings, etc.

4.6 Purchasing

Defined methodologies and supporting records for:

- Review and selection of subcontractors.

- Identification of acceptable subcontractors.

- Monitoring and control of subcontractors.

- Removal from use of unacceptable subcontractors.

- Content and review of purchase orders.

- Review and release of product(s) at subcontractors' premises.

- Customer review and release of product(s) at subcontractors' premises; linkage to contract review process to identify the need.

4.7 Control of Customer Supplied Product

Defined processes and methodologies for the identification of customer supplied product, addressing the following:

Contractual identification of items (services) to be supplied.

Clear definition of standards of acceptability to supplier and customer.

Control and identification of any and all customer supplied product.

Records of loss, damage, or unsuitable deliveries/services.

Note that the definition of "product" in ISO 9001, paragraph 3.1, and attached "notes." Also note that the 'product' may be for "*associated activities*" as well as for "*incorporation in the supplies.*"

4.8 Identification and Traceability

Definition of the applicability of the requirements for:

- Identification of product from receipt to dispatch.

- Any traceability requirements from design through to installation or beyond.

Documented processes and methodologies for the tracking and control of the defined requirements

Maintenance of records appropriate for the defined requirements.

4.9 Process Control

Defined procedures and methodologies for:

- Production planning and control.
- Identification of the need for work instructions and provision of requirements.
- Identification of workmanship standards and provision of appropriate tools.
- Process approval.
- Process control monitoring.
- Identification of *"qualified processes,"* equipment, and personnel.
- Identification and fulfillment of environmental controls.
- Identification of any special handling needs.
- Equipment maintenance to meet process needs.

The collection, collation, storage, and maintaining of appropriate records of all the above.

4.10 Inspection and Testing

Documented quality plans or procedures defining the requirements for:

- The inspection and test of received materiel.
- In-process inspection and testing.
- Final inspection and testing.
- Records of receiving and in-process inspections and tests, including the acceptability or otherwise of the results.
- Records of final inspections and tests, including verification of previously required records.

Procedure for urgent release and positive recall, or statement prohibiting such action.

4.11 Inspection, Measuring, and Test Equipment

Documented procedures, methodologies, and supporting records for:

- The identification of all test and inspection equipment used to verify product conformance.
- Identification of calibration methodologies, e.g., either a defined in-house process or an approved external calibration facility.
- Traceability of calibration records to related equipment.
- Effective control of in-house test and inspection equipment to prevent invalidation of calibration.

- The basis for calibration when no external standard exists.

- Assessing the validity of historic results when equipment is found to be out of calibration.

- The calibration status of each identified item.

4.12 Inspection and Test Status

Documented procedures for the identification of the inspection and test status of all product from receipt through to dispatch and, when appropriate, installation including the identification of any nonconforming product.

Appropriate records.

4.13 Control of Nonconforming Products

Processes and methodologies for handling nonconforming products, including clearly defined responsibilities and authorities for:

- Isolating products that fail to meet specifications.

- Ensuring that nonconforming products are not shipped to customers.

- The review and sentencing of nonconforming materiel.

- Identifying any contractual obligations that arise when nonconforming product occurs, including notifying customers.

- Producing records of review, disposition, waivers, repair, and rework.

4.14 Corrective and Preventive Actions

Defined procedures for initiating, performing, and resolving nonconformances occurring with products (deliverables), processes, or systems.

Data collection and analysis from all appropriate, available sources to identify potential problems and initiate preventive actions.

Reporting of preventive actions for management review.

Records of corrective and preventive actions, including changes in procedures resulting from corrective and preventive action.

Records of customer complaints.

4.15 Handling, Storage, Packaging, Preservation, and Delivery

Procedures addressing all aspects of:

- Product handling from receipt through to installation (where appropriate) designed to prevent damage or deterioration.

- Definitions of any special requirements (e.g., ESD, HAZCHEM).

- Storage that will protect the integrity of products at all stages.

- Packaging of all products to ensure safety in transit from receipt to customer.

- Any preservation techniques required to be applied at any stage in the process, from receipt to installation.

- The control of methods of delivery, i.e, use of approved transportation.

- Any special paperwork required, either by contract (link to contract review) or legal requirements (e.g., export documentation).

4.16 Quality Records

Definitions of Quality Records; this may be a single procedure or contained within the system documentation. The methodology must address these requirements:

- Who collects the data.

- Where data are stored.

- How data are stored.

- How long data are stored.

- How data are disposed of at the end of storage time.

- How data are accessed.

- Are all data legible?

- Are data reasonably protected from damage or deterioration?

- Are there any contractual obligations for retention, access, etc.?

When records are stored on magnetic media, in addition to the above,

- Are provisions made for backup?

- Are provisions made for regular refreshing of long-term data?

- How is ongoing usability of magnetic data assured?

4.17 Internal Quality Audits

Procedures must provide for:

- A published schedule of audits based on the importance of the activity.

- And the results of previous audits.

- The systematic and regular audit of the complete system.

- The timely review and closeout of audit findings.

- The use of trained auditors independent of the area under review

- Records of audit results.

- Records of review of corrective actions for effectiveness.

- Use of audit reports to improve the system.

- Feedback of audit reports to management review.

4.18 Training

A systematic, ongoing process for the identification of all training needs and the fulfillment of those needs.

Records of the training provided.

4.19 Servicing

When contractually required, processes must be in place for:

- identifying requirements

- maintaining records

Link back to 4.14 and provide for the review and analysis of warranty and nonwarranty service activity.

4.20 Statistical Techniques

Declared responsibility for the identification of the need for statistical techniques to monitor and control processes and products, and the authority to introduce and implement such techniques.

Documented procedures to implement and control statistical techniques wherever a work process or analytical process utilizes statistics.

Link back to 4.18 and provide training in the use of statistical techniques whenever required.

Appendix 3
ISO 10013:
Guidelines for Developing
Quality Manuals

The revised ISO 9001 standard requires a quality manual. In Subclause 4.2.1, a note refers companies to ISO 10013 for guidance on quality manuals.

The contents of ISO 10013 include the following sections:

1. Scope.

2. Normative reference.

3. Definitions.

4. Documentation of quality systems.

5. Process of preparing a quality manual.

6. Process of quality manual approval, issue, and control.

7. What to include in a quality manual.

The Annex to the standard includes a description of the typical quality system document hierarchy, examples of a procedure format, and a section of a quality manual. Following are some key points from the standard.

INTERNAL OR EXTERNAL PURPOSES

In *Clause 3, Definitions,* the standard makes a distinction between a quality management manual and a quality assurance manual. The former is designed for internal use only and may contain proprietary information, while the latter does not contain such information and may be used for customers and third-party assessors. The same issue is discussed in *Clause 4.3, Special applications of quality manuals,* where the standard points out that when internal and external manuals exist, they must describe the same quality system and not conflict.

PROCEDURAL SCOPE (4.1.1)

"Each documented procedure should cover a logically separable part of the quality system, such as a complete quality system element or part thereof, or a sequence of interrelated activities connected with more than one quality system element."

STRUCTURE AND FORMAT (4.2)

There is no required structure or format for quality manuals, so long as they convey the quality policy, objectives, and governing procedures of the organization clearly, accurately, and completely. A quality manual may:

- Be a direct compilation of quality system procedures.
- Be a grouping or section of quality system procedures.
- Be a series of procedures for specific facilities/applications.
- Be more than one document or level.
- Stand alone or otherwise.
- Be other numerous possible derivations based upon the particular organization's need.

PREPARING A QUALITY MANUAL (5)

Clause 5 suggests the following process for preparing a quality manual:

- List existing applicable policies, objectives, and procedures.
- Decide which quality system elements apply according to the selected standard.
- Gather information about the existing quality system.
- Obtain additional documentation or references from operational units.
- Determine the format and structure of the intended manual.
- Classify existing documents in accordance with intended format and structure.
- The actual writing of necessary procedures follows the above process.

WHAT TO INCLUDE IN A QUALITY MANUAL (7)

Clause 7 lists the suggested contents of a quality manual, including the following:

- Title, scope, and field of application.
- Table of contents.
- An introduction to the organization and the manual.
- Quality policies and objectives.

- Description of the organization, responsibilities, and authorities.

- Description of the quality system elements and/or references to quality system procedures.

- A definitions section, if appropriate.

- A guide to the quality manual, if appropriate.

- An Appendix that contains data necessary to support the quality manual.

BEYOND COMPLIANCE: MANAGING RECORDS FOR INCREASED PROTECTION

by Eugenia K. Brumm, Ph.D., CRM

Properly managing records is not just a smart way to conduct business. Appropriately structured, implemented and maintained records may keep a company out of costly legal proceedings.

Illustrating this point is the following case: *Telectron, Inc. v. Overhead Door Corp.*[1] Telectron brought suit against Overhead Door Corp. (OHD) for violation of antitrust laws. On the same day that Telectron's complaint and request for production of records were served on OHD, Mr. Arnold, its chief legal counsel (who was also the secretary of the company), ordered the destruction of massive amounts of relevant records. The court determined that ". . . Mr. Arnold ordered this destruction a willful and intentional attempt to place documentation, which he anticipated to be damaging to OHD's interests in this litigation forever beyond the reach of Telectron's counsel."[2] The destruction, also, was conducted in the absence of an established company records retention policy. The court entered a default judgment of liability on behalf of the plaintiff, and the absence of a records retention policy was the reason cited by the court as evidence of bad faith on the part of OHD.

Throughout the case, OHD claimed that it had destroyed records under a company a records retention and destruction program. During the testimony, however, a senior vice president of manufacturing stated that he was not aware of any records retention policy in the company. The court concluded that OHD's top management failed to provide any leadership or effective oversight regarding document retention. This was further evidenced by OHD's president, who claimed ignorance about the company's document retention policy. He stated that he believed such a policy had been established in 1966 or 1967, but he new none of the policy's provisions and he did not know if anybody was formally charged with implementing the policy. In other words, OHD claimed that it operated a records retention program but could not provide any written policies or procedures to that effect.

In this case, ample evidence revealed that OHD destroyed records willfully after the complaint was served. Eight months later, and after specifically ordering the destruction of records pertaining to the case, the legal counsel sent a memo to employees at OHD's Dallas headquarters informing them of a lawsuit with Telectron. The memo was too late, was limited in its distribution, did not provide enough information about the nature of the lawsuit, and did not direct the retention of records for the duration of the lawsuit. Because of the intentional and careless destruction of records, a default judgment was entered against the defendant. *The Wall Street Journal,* in September of 1987 estimated that the damage phase of the Telectron case could lead to a judgment against OHD as high as $69 million.[3] Because the final judgment is sealed, public disclosure of the actual amount that was awarded is not possible. A phone call to Telectron, however, revealed that Telectron did prevail in this case, and received a very large monetary settlement from OHD for an undisclosed amount.[4]

ISO 9000 RECORDS REQUIREMENTS AND LEGAL ISSUES

The ISO 9000 standards include both of the following records requirements:

1. Specific types of quality records must be created and maintained to support the quality operations in an organization.

2. Certain records management activities must be applied to those quality record types.

The requirement for specific quality records is interwoven throughout the standards and appears in overt statements, such as "Records of contract reviews shall be maintained."[5] Titles of sections, such as "Inspection and test records," also reveal requirements for specific record types.[6] Frequently, however, the requirements for specific types of quality records are implied in statements indicating that the organization must provide proof, without any mention being made of a record.

The second type of record requirement in the standards—for conducting records management activities—is clear and evident. Section 4.16, Control of quality records, includes requirements for the following records management activities:

- Retention schedule development and management.

- Disposition of records.

- Records protection program.

- Filing management.

- Development of access schemes.

- Design of indexing schemes.

- Forms management.

- Records creation management.

- Selection and management of equipment, media (micrographics, optical disk, software, hardware), and supplies.

Not only are these records management components required in ISO 9000, they are the same components that have resulted in default judgments and court-imposed sanctions against organizations because they do not exist, are inconsistently practiced, or are poorly conducted.

Another high-profile case that illustrates numerous records weaknesses is *Clara Carlucci, etc.* v. *Piper Aircraft Corporation.*[8] In 1976, three men perished in a crash of a Piper Cheyenne II at Shannon, Ireland. The pilot had taken off in low visibility conditions and soon afterwards, lost control of the plane and it crashed into the ground, killing all five aboard. The plaintiffs filed a complaint against Piper in 1978, alleging various design defects. For five years, the plaintiffs repeatedly requested records of the defendant and did not receive them.

The judge appointed a Special Master to examine why the requested records had not been handed over to the plaintiffs. In 1984, the judge entered a default judgment against the defendant in the amount of $10 million, based solely on Piper's misconduct with their records that had been requested during discovery. Like Telectron, Piper had claimed that it had already destroyed certain types of records that it should have produced during discovery, because it had a records retention policy and those records were destroyed in accordance with that policy. Piper, however, was unable to prove that a records retention policy actually existed in the organization. In his opinion, the judge stated:

> I am not holding that the good faith disposal of documents pursuant to a bona fide, consistent and reasonable document retention policy can not be a valid justification for failure to produce documents in discovery.

That issue never crystallized in this case because Piper has utterly failed to provide credible evidence that such a policy or practice existed."[9] Although the Director of Engineering, Calvin Wilson, Jr., testified that the records were destroyed based on the company's retention procedures, additional testimony led the judge to state the following:

> Piper presented no evidence to substantiate Mr. Wilson's claim that the document retention procedures are strictly complied with by Piper's employees. In fact, with the exception of improper destruction of the Product Condition Reports, . . . the defendant did not provide any evidence that the procedures are ever complied with by its personnel.
>
> This is particularly shocking in the light of my instruction . . . to Piper to detail for me its means of retaining and disposing of documents. Piper's absolute failure to provide any evidence on this issue must be construed as a tacit admission that the policy is a sham . . . Piper has utterly failed to demonstrate that its document retention policy is actually implemented in any consistent manner.[10]

The judgment in this case revolved solely around records issues, not the least of which was the selective destruction of records. Two flight test engineers, who were responsible for maintaining records in the flight test department, testified that they had received orders to destroy records that might be detrimental to Piper in a lawsuit. Initially, they destroyed "hundreds" of flight test documents, including those from other departments, and this practice continued after the beginning of the Carlucci lawsuit.

Because of this practice, suspicion surrounded the disappearance of several flight test reports from a notebook, since these particular reports pertained to characteristics of the aircraft that were in question. Because of this, the judge concluded:

> . . . that the defendant engaged in a practice of destroying engineering documents with the intention of preventing them from being introduced in law suits. Furthermore, I find that this practice continued after the commencement of this law suit and that documents relevant to this law suit were intentionally destroyed . . . Having determined that Piper intentionally destroyed documents to prevent their production, the entry of default is the appropriate sanction. Deliberate, willful and contumacious disregard of the judicial process and the rights of opposing parties justifies the most severe sanction . . . The policy of resolving lawsuits on their merits must yield when a party has intentionally prevented the fair adjudication of the case. By deliberately destroying documents the defendant, has eliminated the plaintiffs' right to have their cases decided on the merits. Accordingly, the entry of a default is the only means of effectively sanctioning the defendant and remedying the wrong.[11]

The case, therefore, was never decided on its merits, but was decided on the actions revolving around Piper's records. The same held true for the Telectron case. In both situations, the defendant organizations claimed that they had operated according to a company records retention and disposition policy. In both cases, however, the organizations could not provide any written policies or procedures to that effect and could not demonstrate that:

1. Such a policy was consistently implemented.

2. Employees in the organization were aware of such a policy.

3. Any records were kept identifying which company records were destroyed.

4. Records were officially authorized to be destroyed according to the company retention policy.

In both of these cases is the theme of retention policy and the question of its existence or non-existence as well as its inconsistent implementation. In both cases, as well, the judges questioned the credibility of the defendants regarding their inability to locate specific records that were repeatedly requested.

In cases such as Piper's, some of the most important records are quality records—production records, and records pertaining to quality control and testing. Of those, engineering design control and review records assume paramount importance in these cases, because they form the foundation of product liability records. Because activities conducted during the various design phases affect the performance of the resulting products, the importance of records pertaining to the design process cannot be overemphasized. Flaws and malfunctions often can be traced directly to an incomplete or weak design control process, and several studies provide evidence of this.[12]

The importance of the design control process itself is evident in the ISO 9001 standard, where it is one of the longest sections. It is a multi-faceted, complex activity that involves various phases, numerous reviews, and can result in tens of thousands of records, depending on the complexity of the product. Design records serve two basic purposes: (1) They provide evidence about the activities that have occurred, and (2) they collect data for the organization to use in pre- and post-production analysis. In the fullest sense, design records provide an accumulation of experiences pertaining to the product at hand, and they can serve as a mine of information for current and future designs.

Quality records, those required by the ISO 9000 standards—records related to design, testing, quality control, manufacturing, distribution, customer complaints and other aspects of the development and manufacturing process—are those that an organization should be able to produce if cited as a defendant in a product liability case. It is important to identify records that could be used to show how a product's design was developed and approved. Records that could pertain to product liability cases should depict a history of the product so that its design can be explained and the manufacturer can demonstrate that safety and quality were considered at each phase of the design, manufacturing, and marketing process.[13]

When a product is released for sale, a product history file should be created and retained for as long as necessary. In such a file, it is important to link information logically, if not physically, since many current products evolve from prior designs, and the current design often cannot be explained without references to records about the earlier design. Such records include, but are not limited to, the following:

- Design specifications.

- Engineering evaluations.

- Design reviews.

- Prototype testing.

- Design assurance testing.

- Final design selection.

- Quality control procedures.

- Inspection procedures.

- Product service history.

- Performance history.

These are the same records and procedures that are required by ISO 9000, with the first six record types being part of the design control paragraph in ISO 9001.[14] If manufacturers do not retain records that resulted from the design and manufacturing process, they could be questioned as to whether or not they actually conducted all of the performance and safety testing that they claimed.

STATUTE OF LIMITATIONS AND PRODUCT LIABILITY RECORDS

Quality records, such as those required in ISO 9000 and other quality standards, do become relevant in product liability cases. In determining retention times for such records, it is necessary to consider the statute of limitations for liability of your product. In some environments, such as nuclear and medical, government regulatory authorities impose retention periods for records. Some states have statute of limitations on product liability starting with the time that the product was first sold. In other states, the liability can exist for a long period of time—indefinitely.

Products can last for a long time. "In today's environment, product manufacturers cannot safely destroy records related to product development, testing, design and manufacturing until the product ceases to be used."[15] Most states, in fact, specify that the statute of limitations for personal injury begins at the time the injury occurs, and not when the product is sold. Manufacturers may be responsible throughout the life of the product for defects in design and manufacturing and for errors in instructions for using the product.

DEVELOPMENT OF A RECORDS RETENTION SCHEDULE

Retention times should not be assigned carelessly, but should be based on some legal, fiscal, operating, and/or historical need. Such a basis for determining retention times substantiates the retention schedule and makes it legally acceptable as an operating policy. If records retention times are not grounded in research and/or business need, they can be subject to dismissal in a court of law. As the following case illustrates, how the retention time itself was determined can become a point of question.

In *Lewy* v. *Remington Arms Co., Inc.,* the defendant, a rifle manufacturer, claimed that it routinely destroyed customer complaints and gun examination reports after three years, based on a routine retention and disposition program.[16] The case involved the accidental discharge of a rifle in the basement of Evelyn Lewy's home by her son, while he was unloading it. The bullet went through the ceiling and struck Mrs. Lewy in the leg, while she was seated in the living room. The U.S. Court of Appeals, Eighth Circuit rejected Remington's contention that they could not produce pertinent records because they were destroyed according to an established company retention policy. In this instance, the time for retention became the focal point. The court determined that ". . . a three-year retention policy may be sufficient for documents such as appointment books or telephone messages, but inadequate for documents such as customer complaints . . . the court should determine whether the document retention policy was instituted in bad faith."[17] In this case, the retention time itself for a particular category of documents, namely, customer complaints, was viewed with suspicion and it appeared as though the ". . . document retention policy [was] instituted in order to limit damaging evidence available to potential plaintiffs . . .".[18]

Organizations should not be establishing retention schedules that have shorter time frames than are expected by courts should those records become requested in litigation. In other words, records retention schedules should not be developed for the explicit purpose of eliminating records that may prove incriminating in a lawsuit. In addition, negligence and ignorance are not excuses in defending retention times that are assigned without any legal or operating basis. Organizations are expected to know the requirements for retention that govern the records for their line of business. In the case of *United States of America v. ABC Sales & Service, Inc., et al.,* the defendant claimed that it had discarded the requested files because the records were kept only as "space allows," which is usually not more than two years.[19] The court deemed this "retention schedule" to be improper, since ". . . three of the defendant's offices are in California and defendants admit in their answers . . . that debt collectors in that state are required to maintain collection files for a period of at least four years."[20] The court imposed a sanction against the defendant for failure to produce the requested records. This case illustrates that it is necessary to know the laws governing retention periods for records pertinent to your line of business.

To dispose of records when space dictates is not a retention policy, and ignorance of retention regulations is no excuse. Part of good business and operating practice is knowing what laws affect which record categories and being able to develop retention schedules based on those laws, as well as internal operating needs.

For these reasons, it is important to develop retention schedules methodically and systematically, by conducting research into the requirements and interviewing those who use and need the records. Developing a records retention schedule is a difficult, resource-consuming process, especially for those organizations that have been operating for several decades and have never addressed their records. Depending on the size, age, and complexity of the organization, the preparation of a retention schedule can take from six months to several years.

The following steps, listed in order, should be performed in order to develop a sound records retention schedule:

- Conduct a records inventory.
- Interview key personnel.
- Research records retention periods.
- Appraise the records.
- Prepare a draft schedule.
- Discuss the schedule with affected departmental managers and legal counsel.
- Obtain the necessary signatures.
- Duplicate and distribute the records retention schedule.

CONDUCTING A RECORDS INVENTORY

A records inventory serves as the basis for the entire records management program. A well-designed inventory collects information necessary not only to develop the retention schedule, but also to establish other aspects of records management required in the ISO standards: records protection programs, indexing schemes, retrievability, filing systems, and access schemes. It is logical to conduct a records inventory as the basis for a retention schedule, because it is impossible to define retention times for quality records if the organization does not know what records it maintains.

The more thoroughly the inventory is conducted, the less time will be spent diagnosing records problems and instituting other components of records management. Some elements of information that are needed for retention schedule development are also needed for other aspects of records management required in the standards. For example, developing a records protection program requires identifying the records, their volume, location, medium or format, and current equipment use. The same elements of information must be gathered about the records in order to develop a records retention schedule.

A physical inventory of quality records provides accurate information on which to base retention schedule decisions. The basic purpose of the inventory—to identify which quality records exist, where they are located and how many there are—is best served when the records management staff physically examines the records. It is not uncommon to find records in places that (1) are not obvious and (2) were not designed for records storage. Records are housed in coat closets; in basement rooms infested with insects and vermin; on the floor in empty offices; on top of file cabinets, chairs, and computer terminals; in grocery cartons stacked in warehouses; along the walls of hallways and offices; in mini metal storage units; and elsewhere. Without a records management program, most records that are so stored are unidentified, mixed with various records series, and often not protected from deterioration. They are also not readily retrievable. Both of these latter aspects are required by the ISO standards and both have become legal issues in court cases—protection from deterioration and ready retrievability.

In *Equal Employment Opportunity Commission* v. *C. W. Transport,* the defendant contended that it had been prejudiced ". . . by the loss of records . . . from natural disasters and similar circumstances.[21] In 1979 and 1980 severe flooding at the Chicago terminal destroyed the defendant's records stored in the basement. In 1985, another flood at Chicago caused by outside construction again destroyed records stored in the basement. A 1981 tornado destroyed the storage area of the Frankfurt, Illinois terminal. And in 1984 at the Richmond terminal, records were damaged when a pipe in a storage room froze and broke. Other causes of record loss were mislaying of records and employee pilfering. CWT maintains that it cannot catalog what records were destroyed, lost or misplaced."

The court had little sympathy for the defendant in this case, and rejected CWT's contention:

> Despite its claim that it cannot catalog precisely what records are missing, the company should know generally what records it maintained and, to some level of specificity, what records remain after the various natural disasters and other events. It does not seem unreasonable to require CWT to establish with more specificity what records are missing and why the loss of these records will materially prejudice its defense of this enforcement action.[22]

ACCESS TO RECORDS

The ISO 9000 standards require that records be "readily retrievable," a vague but tall order, since retrievability is the number one records problem in organizations, regardless of the organization and regardless of the record type or format. The issue of retrievability also has legal implications, since several court cases have viewed poor access to records as attempts by organizations to purposely "hide" the records and make them difficult if not impossible to retrieve. Poor retrievability can function in much the same way as intentional destruction of records, by making them unavailable to the plaintiffs during litigation. The following case illustrates this point. In a products liability case, the plaintiff, Brian Kozlowski, was severely burned when a pair of pajamas from Sears ignited.[23] After repeated requests, Sears failed to produce records of other related customer complaints, stating that it had ". . . a longstanding practice of indexing claims alphabetically by name of claimant, rather than by type of product."[24] Sears contended that there was ". . . no practical way for anyone to determine whether there have been any complaints similar to those alleged ". . . other than [by] going through all of the . . . claims in the Sears Index . . . which is the equivalent of an impossible task . . ."[25] The court stated the following:

> The defendant seeks to absolve itself of [its] responsibility by alleging the herculean effort which would be necessary to locate the document. The defendant may not excuse itself from compliance with Rule 34, Fed. R. Civ. P., by utilizing a system of record-keeping which conceals rather than discloses relevant records, or makes it unduly difficult to identify or locate them, thus rendering the production of documents an excessively burdensome and costly litigation. To allow a defendant whose business generates massive records to frustrate discovery by creating an inadequate filing system, and then claiming undue burden, would defeat the purposes of the discovery rules.[26]

In another case, *United States of America* v. *ABC Sales & Service, Inc., et al.* the defendants claimed that they had been making "diligent efforts to locate [395] files" that they had been asked to produce, but that they "were buried among several million files they maintain."[27] The court gave little weight to this reasons and stated: ". . . a business which generates millions of files cannot frustrate discovery by creating an inadequate filing system so that individual files cannot readily be located."[28] The inability to retrieve records because of poor records management practices and inefficient organizing, filing, and indexing schemes is not a valid excuse in a court of law. Organizations are expected to have access schemes for records that make sense for the record types that they are retaining.

INTERVIEW KEY PERSONNEL

Because it is not possible to gather all pertinent information by physically examining the records, it is common to interview users and keepers of the records. It is not necessary to interview all of the individuals in each department. In each organization there are usually key informants—those users and custodians of records who can provide the most information about them. Interviews are conducted for the following reasons:

- Procedural issues—to define the process and determine how specific records and series of records are used; to understand the interrelationships across records.

- Analytical issues—to understand actual or perceived problems with the records and actual or perceived barriers to productivity.

- Retrieval issues—to determine the characteristics of retrieval such as frequency, access points, active time frames, and so on.

- Historical issues—to understand when and why specific records procedures were developed.

- Retention issues—to determine the administrative need for the information.

RESEARCH RECORDS RETENTION PERIODS

NOTICE: There is no single source that provides an organization with all of the information about retention times for records. In fact, many records are not specified at all in any type of requirement. It is assumed, mistakenly, that (1) all requirements for records retention are clearly stated somewhere and (2) these requirements form the basis for most retention schedules. Both assumptions are false. There are multiple sources that provide information about records retention, and they must be unearthed.[29]

Federal rules and regulations exist; regulatory authorities publish their own requirements; state statutes can include retention requirements; municipal laws can affect records retention; and foreign governments' requirements can even enter the picture if the organization has operations in other countries. The basic federal rules and regulations are published in the *Federal Register* and in the *Code of Federal Regulations (CFR)*. In addition, some agencies and regulatory authorities may have jurisdiction over the type of business that the organization provides, and they have separate guidelines that must be consulted. State statutes and municipal laws must also be reviewed. Finally, the type of business an organization does or provides may be supervised by a self-regulatory organization, and their guidelines and rules can have the authority of law to member organizations.

APPRAISE THE RECORDS

After the inventory and interviews have been conducted, and the research into requirements has been completed, the records are appraised for their value—that is, whether they are being retained for legal reasons, to satisfy administrative or operating needs, or for historical reasons. A records appraisal is an examination of the data gathered through the records inventory, interviews, and research to determine the value of each record series.

PREPARE A DRAFT SCHEDULE

A draft copy of the schedule is then prepared and presented to the legal counsel and department heads for review. If the retention times have been based on cited requirements, a copy of the appropriate pages from the requirements with highlighted regulations should be attached to the recommended retention schedule to facilitate legal review. After the draft copy is reviewed, any necessary changes should be incorporated into the retention schedule.

DISCUSS THE SCHEDULE WITH AFFECTED DEPARTMENTAL MANAGERS AND LEGAL COUNSEL

The records manager should discuss the proposed retention schedule with departmental managers and with legal counsel to ensure understanding of the rationale behind the retention times that are being assigned, and to clarify any discrepancies.

OBTAIN THE NECESSARY SIGNATURES

When consensus is reached on retention times, signatures of the records manager, the department head, and the legal counsel should be affixed to the retention schedule or to a retention authorization form. At this point, the retention schedule is ready to be published and distributed. As several of the cases have emphasized, officials of the company should be aware of the retention schedule and employees should be knowledgeable about retention times for the records under their jurisdiction. Unless the retention policy is written, formalized and distributed, it is considered nonexistent.

DUPLICATE AND DISTRIBUTE THE RECORDS RETENTION SCHEDULE

There are two distribution methods that can be used, depending on the size and preference of the organization: (1) The entire retention schedule can be distributed to each affected unit and to the legal counsel, or (2) Only those portions that contain a department's records can be distributed to each department. At a minimum, the retention schedule should contain the following information:[30]

- The department name.
- The schedule number.
- The revision number.
- The effective date.
- The page number.
- The record series title and description.
- The office that holds the original copy.
- The media on which it is kept.
- The records retention period.
- The period in active use.
- The period in inactive use.

- The total retention period.

- It is also a good idea to include a column that provides the source or citation of the retention time.

The records retention schedule is not a fixed item. Because there are constant changes in the law and in the organization's operations, it is continuously open for revision. It is a dynamic document that must be updated regularly to incorporate any changes in record type, status, and retention periods.[31]

IMPLEMENTING THE RETENTION SCHEDULE

Just as important as its development is the active and attentive implementation of the retention schedule. Because the organization bears the ultimate responsibility, it is important that adherence to the authorized retention schedule be monitored. Part of the implementation process of a retention schedule is the periodic, scheduled destruction of records that have satisfied their retention periods. To maintain legal acceptability and demonstrate compliance, records must be destroyed according to schedule. If records are not destroyed according to schedule, but instead are destroyed whenever the organization gets around to it or runs out of storage room, the motives for destruction will be suspect. It can appear as though the organization has attempted to destroy unfavorable evidence, perhaps anticipating litigation or government investigation. This point is to be emphasized especially concerning quality-related records that provide evidence of product performance and product safety.

Courts routinely rule in favor of organizations that have records retention schedules in effect and follow them as the procedure dictates, since this does not amount to willful destruction of evidence with the intention of subverting litigation. Routine destruction of records is considered to be unintentional, since the records are being destroyed according to a procedure that was in place before any litigation materialized. An example of a court ruling pertaining to this is *Vick* v. *Texas Employment Commission.*[32] In this sex discrimination case, the court ruled that "... Texas Employment Commission (TEC) records on Vick were destroyed before the trial, apparently pursuant to Commission regulations governing disposal of inactive records ... There was indication here that the records were destroyed under routine procedures without bad faith and well in advance of Vick's service of interrogation. Certainly, there were sufficient grounds for the trial court to so conclude."[33]

The majority of courts have declined to punish evidence of destruction when a routine retention program is in place.[34] In such cases, however, there was ample evidence that records were destroyed during the normal course of business and that the retention schedule was adhered to not only for the records in question, but for others on the schedule as well. In *Moore* v. *General Motors*, the court ruled the following:

> There is no evidence that at the time defendant had any knowledge that it was facing litigation so that it was put on notice that it should not pursue its customary practice of destroy-

ing these records. Anyone knowledgeable of business practices and the cost of storing records in these times would find it reasonable and not smacking of fraud for the defendant, with no knowledge of pending litigation, to follow its customary practice."[35]

It is important for individuals who are responsible for implementing the retention program to be trained in the importance of implementation. Sporadic implementation gives the appearance of impropriety that can have adverse consequences. In *Capeluppo* v. *FMC Corp.* the records relevant to the suit were destroyed in manner that was inconsistent with the company's destruction procedures.[36] This was one basis for the court's decision ordering the company to pay double the plaintiff's fees and costs both in connection with the litigation and in connection with remedying the document destruction.

Scheduling records for destruction requires developing a timetable and preparing records that document the individuals involved and reveal that appropriate review processes have taken place. Under no circumstances should records be destroyed that have been requested for litigation or government investigation, regardless of the time indicated on the retention schedule.

RECORDS ON DEVELOPING A RETENTION SCHEDULE

The procedures and methods that were used in assigning retention times to records become important in litigation. Often, it is necessary to demonstrate the procedures that were followed in developing a retention schedule by providing evidence on the research that was conducted. The proper development and operation of a retention schedule become evidence that a systematic, methodical approach was followed.

If it becomes necessary to establish the existence of a valid records management program in court, your organization may be asked to produce evidence that proves the existence of such a program. Records managers and/or those responsible for developing the retention schedule and implementing it may be required to testify as witnesses. If such occurs, the following materials should be brought to the proceeding:[37]

1. Work papers and work sheets documenting the development of the records retention program, including legal research.

2. Signed approvals for the retention schedules and the entire program.

3. The description and the listing of all records destroyed over time, including the specific records dealing with the proceeding,.

4. The records retention manual, including modifications which existed at the time the particular records in question were destroyed.

The records manager or records custodian should explain the procedures followed in developing the records retention program and produce copies of the appropriate documentation if requested. The description and listing of all records destroyed is essential to prove that records have been destroyed in the regular course of business and that the records in question were treated like others that were destroyed under the program.

When records are destroyed routinely, based on a properly developed and administered retention schedule, their destruction is not considered to be in "bad faith"—in other words, purposely destroying records that may prove harmful to the organization. This is true if:

- The retention schedule is prepared prior to any litigation.

- The retention has been consistently implemented—i.e., records have been routinely destroyed based on their destruction due dates.

- It can be demonstrated that the retention schedule was based on one of the four values discussed earlier.

This is why it is important to retain a record of the methods used to determine retention times—so as to reveal that decisions were made for reasons that had nothing to do with the possibility of future litigation, even though that litigation might not be foreseeable.

SUSPENSION OF DESTRUCTION WHEN LITIGATION IS IMMINENT

Regardless of the soundness of a records retention program, document destruction must be suspended at the first sign of litigation. All units, departments, and individuals should be notified in writing that document destruction must cease when litigation is imminent. In *National Association of Radiation Survivors, et al.* v. *Thomas K. Turnage,* the court imposed monetary sanctions on the Veterans Administration for destroying relevant records after it had received a request for those records.[38] The court concluded:

> There is no question that relevant documents were destroyed and are now permanently lost . . . the defendant knew or should have known that these destroyed materials were relevant and discoverable . . . It is no defense to suggest, as the defendant attempts, that particular employees were not on notice. To hold otherwise would permit an agency, corporate officer, or legal department to shield itself from discovery obligations by keeping its employees ignorant. The obligation to retain discoverable materials is an affirmative one; it requires that the agency or corporate officers having notice of discovery obligations communicate those obligations to employees in possession of discoverable materials . . . The V.A.'s reckless and irresponsible abrogation of its responsibility to assure full compliance with discovery requests cannot be tolerated and excused, and is most assuredly sanctionable where it results in the wholesale destruction of potentially relevant material."[39]

The court further identified threats of retaliation against those V.A. employees who expressed concern over the continuation of records destruction after the agency was put on notice to produce the records.

The cessation or prevention of destruction has also been extended to include possibly "foreseeable" litigation. Risks inherent in routine evidence destruction when a lawsuit is imminent or reasonably foreseeable are well illustrated in cases involving destruction before a complaint has been filed. The reasoning in those cases suggests that the appropriateness of sanctions rests on whether litigation is foreseeable, rather than whether the destruction was routine. The foreseeability issue becomes most acute when a complaint has not yet been filed. This occurred in *Lewy v. Remington Arms Co.,* wherein the defendant had established a three-year retention policy for customer complaints.[40] The court concluded that ". . . certain documents should have been retained notwithstanding the policy. For example, if the corporation knew or should have known that the documents would become material at some point in the future then such documents should have been preserved."[41]

PERSONAL RECORDS VERSUS ORGANIZATION RECORDS

Furthermore, it is important that employees realize that they should not be creating their own individual files from records that are housed or maintained elsewhere, unless these records/files are clearly identified and brought into the records retention program. Often, employees feel that they have a need for records pertaining to their jobs, and they create copies for convenience's sake or because there is poor access to the records. Because the official records copy and authorized duplicates are part of the records retention program, they can be identified readily when destruction is scheduled. Personal employee convenience copies unknown to the records manager are not brought into the retention/destruction schedule. Such records, by their very existence, destroy the credibility and soundness of a retention schedule. They can also surface during discovery and cause untold problems for the organization. *The Wall Street Journal* reported on such an instance:

> A case in point was in the mid-1970s, when the Justice Department began a probe into allegations of price fixing in the folding-carton industry. After 18 months of investigations, prosecutors blew the case wide open with the discovery of two boxes of personal notes in the home of a retired Weyerhauser Co. administrative assistant. There . . . were details of years of illegal pricing agreements with officials from other companies. The material helped lead to price-fixing convictions of or no-contest pleas by 23 companies and 48 executives and to a huge $200 million civil settlement with customers.
>
> 'It wasn't a smoking gun—it was a nuclear warhead,' says Donald G. Kempf Jr., a Chicago attorney who represented several of the defendants. He said a proper document-destruction program might have eliminated the incriminating evidence in the normal course of business.[42]

Furthermore, employees should be aware that all records created in the organization, within the scope of its business, are considered to be company records that are open to subpoena should litigation occur. This includes, but is not limited to such items as calendars, appointment books, post-it notes, and so forth.

TERMINOLOGY

In the cases that have been discussed and in the ISO 9000 standards, the terms *record* and *document* are used interchangeably, and this clouds the issues and causes problems. There is a distinct difference between the terms records and documents in the standards, and the activities that control them. For purposes of the quality environment, the following definitions for the term record are equally valid:

1. The Association of Records Managers and Administrators International (ARMA), defines a record as "recorded information, regardless of medium or characteristics, made or received by an organization that is useful in the operations of the organization."[43]

2. *The Records Management Handbook* defines a record as "any information captured in reproducible form that is required for conducting business."[44]

The term *quality record* is well defined in ASME NQA-1-1994 as "a completed document that furnishes evidence of the quality of items and/or activities affecting quality."[45] Quality records often carry the distinction of having to be authenticated—bearing an authoritative signature.

Documents, on the other hand, as the term is used in quality activities delineated in the ISO 9000 standards, denotes procedures, policies, instructions, or other written or graphically depicted methods or ways of conducting oneself or the operations in a given organization. They explain what an organization plans to do and they instruct employees how to perform their tasks. They provide information about how the organization and its employees should operate. Unlike records, documents exist before the fact, providing guidelines, explanations, and instructions about how to operate. Records contain information about the activity and, thus, do not exist until after the activity has been performed. They come into being after the fact. In the legal cases discussed, the word document was used interchangeably with the word record to have the meaning of the word record. All of the legal cases that have been discussed, in other words, deal with various issues pertaining to records and not to documents, as they have been defined here.

Documentation, as it is used in the ISO standards, encompasses both records and documents. In some instances, the term is used when the standards discuss records requirements, and in other instances it is used when they discuss documented procedures.[46]

Document control is a system of managing, distributing, and keeping records on the documents that have been created by an organization as part of its overall quality system. Those who are unfamiliar with the profession often confuse records management and document control, because a complex series of records must be maintained about the document control system. Such records, like those in other quality functions, are the only way to prove that a tight document control system has been established and is operating.

For those complying with the standards, difficulty arises concerning controls that should be applied to quality documents under the heading "Quality Records Control" in Section 17.3 of ANSI/ISO/ASQC Q9004-1-1994, *Quality management and quality system elements—Guidelines.* The reverse also exists: the standards discuss quality records under the heading "Documentation of the Quality System" in Section 5.3 of ANSI/ISO/ASQC

Q9004-1-1994. The situation would be much clearer if Section 5.3, "Documentation of the Quality System," actually presented a list of records for the document control system, rather than entered into another category. Records are controlled in a completely different way than documents are. Records control includes the following activities:

- Controling the proliferation of records by instituting procedures to limit the number of copies that are made, reducing the length of records, and so forth.

- Purging to reduce the number of items that are retained as records. Often nonrecords are retained—for example, outdated notices about the Christmas party or thank-you notes from colleagues. In addition, multiple copies of a record need not be retained and filed.

- Determining the retention times of records—analyzing the records for retention based on legal, fiscal, administrative, and historical needs.

- Deciding between inactive and active records and moving the inactive records to less expensive storage.

- Developing logical, efficient access schemes so that records can be retrieved quickly.

- Disposing of records that have satisfied their retention requirements.

- Ensuring that blank records (forms) are available to those who need them.

- Protecting records from deterioration and destruction.

Document control is a different function, and it includes the following set of activities:

- Reviewing and approving all quality plans, procedures, and instructions before they are issued.

- Formally distributing documents on a need-to-know basis.

- Creating and maintaining records that reveal which individuals and/or functions have been issued which documents.

- Retrieving obsolete and superseded documents from individuals and functions.

- Ensuring that revised documents have gone through the same review and approval process that the original documents have.

Records about the document control process can include:

- Distribution lists for documents.
- Master lists of documents.
- Requests for documents.
- Periodic reviews of documents.
- Document change notices.

Records receive short shrift in most organizations. Some consider records to be a necessary evil, while some consider them to be only evil and unnecessary. At the very best, organizations tolerate records, albeit grudgingly, and devote token resources and personnel to their creation, maintenance, retrieval, and disposition. Realization of their innate importance to business operations surfaces only when they cannot be retrieved, when they need to be destroyed or conveniently lost to subvert evidence or the legal process, or when they are subpoenaed to be used in pending litigation.

The ISO 9000 standards require that organizations create and maintain records about all activities that affect quality, from the development of a quality plan to the design of a product and the processing of materials that comprise the product, through the manufacture/assembly of completed items, and even about conditions surrounding shipment of the product and its maintenance after it has been delivered. It should be common business practice to record information about the quality activities required in the standards, for internal productivity, for decision making and for sanity. The criticism that the standards require too many records is ill-founded. The fact that organizations balk at common-sense business requirements in the standards reveals the gap between common sense and common practice.

As has been discussed, many aspects of records management that are required in the ISO 9000 standards have become the subject of close scrutiny in courts of law. In numerous cases, organizations have suffered court-imposed sanctions because of poor, non-existent or inconsistent records practices. Retention times for records must be determined carefully, and should be based on analysis of the record types, research into legal requirements for retention and knowledge about the organization's operating and business requirements for those records. Furthermore, retention times should be part of a formally prepared, approved and distributed records retention schedule. Organizations can and do suffer adverse consequences from not properly developing sound records retention programs and not implementing those that they have developed. The records practices that have been addressed in court cases, however, extend beyond records retention schedule development into the arenas of retrievability, organizing schemes, and how well protected the records are. Organizations that claim an inability to access or retrieve records, those who claim to have misplaced records, those who hold that the records have been unintentionally destroyed and those that have sloppy retention procedures have suffered sanctions and have not had their cases decided on merit. Harsh financial penalties have been imposed on organizations that have failed to demonstrate sound records management practices and default judgments against such organizations are not uncommon.

While developing records management programs for ISO 9000 compliance, organizations can also protect themselves legally. Care and thought must be given to records management programs, especially the development of retention times. Used properly, a retention schedule can be a powerful tool in our highly litigious society. Perfunctory assignment of retention times is not tolerated by courts of law.

Endnotes

[1] *Telectron, Inc.* v. *Overhead Door Corp.,* 116 F.R.D. 107 (S.D. Fla. 1987).

[2] Ibid.

[3] Allen, "U.S. Companies Pay Increasing Attention to Destroying Files," *The Wall Street Journal,* September 2, 1987, p. 1. col. 1.

[4] Telephone conversation with Karen Veltri, Vice President, Telectron, December 1, 1995.

[5] ANSI/ISO/ASQC Q9001-1994. *Quality systems—Model for quality assurance in design, development, production, installation, and servicing,* p. 3.

[6] Ibid., p. 6.

[7] Ibid., pp. 8–9.

[8] *Carlucci* v. *Piper Aircraft Corp.,* 102 F.R.D. 472 (S.D. Fla. 1984).

[9] Ibid. at 485–486.

[10] Ibid. at 486.

[11] Ibid. at 485–486.

[12] See Eugenia K. Brumm, *Managing Records for ISO 9000 Compliance.* (Milwaukee, WI: ASQC Quality Press, 1995), p. 82.

[13] Jamie S. Gorelick, Stephen Marzen and Lawrence Solum. *Destruction of Evidence.* (New York: John Wiley & Sons, 1989), p. 343.

[14] ANSI/ISO/ASQC Q9001-1994, op. cit., pp. 3–4.

[15] Donald S. Skupsky, *Recordkeeping Requirements.* (Denver, CO: Information Requirements Clearinghouse, 1988), p. 93.

[16] *Lewy* v. *Remington Arms Co.,* 836 F.2d 1104 (8th Cir. 1988).

[17] Ibid. at 1112.

[18] Ibid.

[19] *United States* v. *ABC Sales & Service, Inc.,* 95 F.R.D. 316, 318 (D.Ariz. 1982).

[20] Ibid.

[21] *Equal Employment Opportunity Commission* v. *C. W. Transport, Inc.* 658 F.Supp. 1278 (W.D. Wisc. 1987).

[22] Ibid. at 1278, 1296.

[23] *Kozlowski* v. *Sears Roebuck & Co.,* 73 F.R.D. 73, 76 (D.Mass. 1976).

[24] Ibid.

[25] Ibid.

[26] Ibid. at 76

[27] *United States* v. *ABC Sales & Service, Inc.,* 95 F.R.D. 316, 318 (D.Ariz. 1982).

[28] Ibid.

[29] Donald S. Skupsky, JD, CRM has numerous publications devoted to legal requirements for records. He can be contacted at: Information Requirements Clearinghouse, 3801 East Florida Avenue, Suite 400, Denver, Colorado 80210, 303-691-3600.

[30] Association of Records Managers and Administrators (ARMA International). *Developing and Operating a Records Retention Program.* (Prairie Village, KS: ARMA International, 1986), p. 13.

[31] Ibid.

[32] *Vick* v. *Texas Employment Commission,* 514 F.2d 734 (5th Cir. 1975).

[33] Ibid.

[34] Gorelick, op. cit., p. 281.

[35] *Moore* v. *General Motors Corp.,* 558 S.W.2d 720, 735 (Mo.Ct.App. 1977).

[36] *Capeluppo* v. *FMC Corp.,* 126 F.R.D. 545 (D.Minn 1989).

[37] Association of Records Managers and Administrators, op. cit., p. 15.

[38] *National Association of Radiation Survivors* v. *Turnage,* 115 F.R.D. 543 (N.D.Cal 1987).

[39] Ibid.

[40] *Lewy* v. *Remington Arms Co.,* 836 F.2d 1104 (8th Cir. 1988).

[41] Ibid.

[42] Allen, op. cit, p. 1, col. 1.

[43] Association of Records Managers and Administrators (ARMA International). *Glossary of Records Management Terms* (Prairie Village, KS: ARMA International, 1989), p. 16.

[44] Ira A. Penn, Anne Morddel, Gail Pennix, and Kelvin Smith. *Records Management Handbook* (Brookfield, VT: Gower Publishing Co., 1994), p. 3.

[45] ASME NQA-1-1994, *Quality Assurance Requirements for Nuclear Facility Applications.* New York: The American Society of Mechanical Engineers, p. 7.

[46] See Brumm, op. cit., pp. 38–42 for clarification of this point.

A Quality System Checklist

9

by Robert W. Peach

The generic list presented in this article offers concise instruction and guidance for translating ISO 9000 requirements into a full-fledged quality system. Together with the previous detailed chapters on implementation and documentation, this "starter list" traces a logical path from the standards to system documentation.

INTRODUCTION

Early in the registration process, it is important to understand the ISO 9000 standard that you have selected for registration and develop quality system documentation.

Teams responsible for developing ISO 9000 documentation face the challenge of making the analysis and developing the documentation—a task they likely have never done before. Not only must they determine how to start the process, but also how to develop a schedule of activities and keep on it while remembering thoroughout how the documentation process should proceed.

The list that follows offers specific guidance for team members faced with translating the requirements contained in Clause 4 of the ISO 9000 standard into a comprehensive quality system. Together with documentation (quality manual, quality procedures, and operator instructions), it provides an orderly journey from the standards to system documentation.

Team members should recognize that these lists are generic. Team members should first review the content of the implementation lists and then modify them to meet their particular needs. Consider this a "starter list" to aid teams in their initial task of defining and guiding their assignment.

4.1 MANAGEMENT RESPONSIBILITY

1. Establish a quality policy.

   - Assign responsibility to an individual or team to develop the quality policy. It should include:

     —The organizations's quality objectives.

     —Management's commitment to quality.

     —Relevance to organizational goals.

     —Expectations and needs of customers.

   - A request for input from across the company to ensure "ownership" of the quality policy.

   - Establish comprehensive objectives.

   - Consider organizational goals.

2. Develop a plan to ensure that the policy is understood, implemented, and maintained at all levels.

   - Conduct an orientation for new employees.

   - Display copies of the policy.

   - Hold departmental meetings/discussions.

   - Reinforce and follow-up on the ideas in the policy.

   - Verify that awareness and understanding is uniform.

3. Define responsibility, authority, and how the assignments are interrelated.

   - Prepare organizational charts.

   - Review and expand job descriptions for those whose work affects quality and who have authority over the following:

     —Identifying problems.

     —Generating solutions.

     —Initiating corrective action to avoid recurrence of the problems.

     —Verifying implementation of the corrective action.

     —Controlling nonconforming product.

4. Identify resources requirements.

   • Provide resources and assign trained personnel for:

   • Management.

   • Work performance.

   • Verification activities.

5. Appoint a management representative who:

   • Ensures that the quality system is established and implemented.

   • Reports on the performance of the quality system.

6. Provide for management review of the quality system.

   • Assess quality audit results.

   • Ensure suitability and effectiveness in meeting policy and objectives.

   • Maintain records.

4.2 QUALITY SYSTEM

1. Determine the requirements of the standard, including both documentation and implementation.

2. Determine which ISO 9000 standard applies (9001, 9002, 9003).

3. Plan the structure of the documentation:

   • Quality manual—a typical outline includes:

      —Quality policy.

      —Organizational chart.

      —Quality assurance organization.

      —Statement of authority and responsibility.

      —Distribution list of controlled copies.

      —Quality system: clauses 4.1–4.20.

      —Procedures index.

      —Forms index.

   • Operating procedures.

   • Job instructions.

   • Records, forms, and specifications.

4. Establish existing company practices by using:

- Flowcharts.
- Procedures (written and unwritten).
- Work/job instructions.

5. Evaluate resources, present and needed:
   - Personnel.
   - Equipment and instrumentation.
   - Specifications and acceptance standards.
   - Quality records.

6. Establish a quality planning function to meet requirements for:
   - Products.
   - Projects.
   - Contracts.

7. Implement the quality system. Consider:
   - Quality plans.
   - Needed resources/time frames.
   - Updating procedures and instrumentation.
   - Identifying extreme measurement requirements.
   - Clarifying acceptance standards.
   - Compatible elements.
   - Quality records.

4.3 CONTRACT REVIEW

1. Document the customer's requirements.
2. Identify precontract practice.
3. Establish contract review procedures.
4. Verify the capability to meet requirements.
5. Internalize customer's requirements and resolve any differences.
6. Maintain control of customer purchase orders that are written under one contract.
7. Develop a plan for deployment.
8. Establish customer purchase order review procedures.

9. Obtain customer agreements.

10. Revise/improve procedures.

11. Evaluate revisions.

4.4 DESIGN CONTROL

1. Document all customer and other pertinent requirements (input).

2. Establish a plan for design control and assign responsibilities.

3. Assign qualified staff; provide adequate resources.

4. Obtain input from all cross-functional activities to establish interfaces.

5. Document the control procedures, with milestones required by the standard.

6. Design output to do the following:

    • Meet input requirements.

    • Contain reference data.

    • Meet regulations.

    • Consider safety.

    • Review documentation before release.

7. Provide output verification through the following:

    • Alternative calculations.

    • Comparison with proven design.

    • Qualification tests.

    • Review of documents before release.

8. Validate the design.

    • Ensure that design verification is successful.

    • Confirm that the final product meets user needs.

    • Assess the need for multiple validations.

9. Develop change control procedures.

    • Identification.

    • Documentation.

    • Review.

    • Approval.

4.5 DOCUMENT AND DATA CONTROL

1. List all documents.

2. Establish a plan to administer each category of document.

   • Document original procedures.

   • Verify review and approval of documents.

   • Consider the pros and cons of hard copy versus electronic media.

3. Investigate conformity to the plan.

4. Ensure accessibility at the work/job site. Accessibility includes:

   • A document master list or reference index of all documents

   • Organization of documents and data in a way that makes them available to those using them.

   • Formatting documents to provide real access to information.

   • Removal of obsolete documents.

5. Establish control over documents that become obsolete.

6. Establish/implement change control procedures.

   • Changes to documents are to be reviewed and approved by the same process followed for original documents.

   • Those functions approving changes are to have background information on which to base their review and approval.

   • The nature of the changes made should be identified on the document, as appropriate.

7. Investigate conformity to change procedures.

4.6 PURCHASING

1. Evaluate existing purchasing specifications and requirements.

   • Review the process for developing and approving specifications.

   • Update the procedures, if necessary.

2. Begin upgrading specifications as required.

   • Prioritize criticality in meeting requirements.

3. Prepare, review, and approve purchasing documents.

   • Refer to updated specifications.

4. Establish criteria for determining subcontractor acceptability.

   • Evaluate and select subcontractors based on their ability to meet requirements, including:

   —Product requirements (i.e., what's the subcontractor's product quality history?).

   —Delivery dependability.

   —Quality system capability (via quality audit/ISO 9000).

5. Develop a subcontractor classification system.

   • Start with a list of acceptable subcontractors.

   • Define the extent of control to be exercised over subcontractors based on:

   —Type of product.

   —Impact on final product quality.

   —Results of previous quality audits.

   —Previously demonstrated quality capability.

   • A qualified supplier list should include:

   —Raw materials.

   —Tooling.

   —Equipment.

   —Business service providers such as consultants and registrars (auditors).

6. Establish a record system. Keep records on the following:

   • Subcontractor's quality capability.

   • Established procedures for communicating requirements and performance with subcontractors.

   • Results of periodic subcontractor review.

   • Purchase contracts and supporting data.

   • Review and approval of purchasing data.

7. Deploy the plan through the following:

   • Develop a schedule.

   • Coordinate with receiving inspection.

   • Assign responsibility for administration.

8. Revise/improve procedures.

9. Evaluate revisions.

4.7 CONTROL OF CUSTOMER-SUPPLIED PRODUCT

1. Determine the existence of customer-supplied product (including test equipment).

2. Document the existing practice for:

    - Verification

    - Storage

    - Maintenance

3. Revise/improve procedures.

4. Evaluate revisions.

4.8 PRODUCT IDENTIFICATION AND TRACEABILITY

1. Establish customer and/or regulatory requirements.

2. Document existing traceability practices to include the following:

    - From your subcontractor

    - In your plant

    - To your customer

    - At/after installation

3. Revise/improve traceability procedures.

4. Consider types of traceability/identification:

    - Unit identification (serial number).

    - Lot identification.

    - Production date code.

5. Consider methods of identification:

    - Paper versus electronic

    - Labeling

    - Bar codes

6. Determine the following about the records to be kept:

    - Availability

    - Retention times

    - Responsibility

4.9 PROCESS CONTROL

1. Base process control on the quality plan.

2. Identify critical control points.

3. Define factors affecting key process controls (production, installation, and service):

   • Equipment

   • Work environment

   • Hazardous material control

4. Identify the following product requirements:

   • Specifications.

   • Workmanship standards.

   • Regulatory standards and codes.

5. Review existing monitoring techniques.

6. Develop control and approval procedures.

7. Develop work/job instructions.

8. Develop control equipment maintenance procedures.

9. Identify special processes.

10. Implement process change control.

11. Revise/improve procedures.

12. Evaluate revisions.

4.10 INSPECTION AND TESTING

1. Establish a separate plan or procedure for the following:

   • Receiving inspection and testing (consider the existing level of subcontractor control).

   • In-process inspection and testing.

   • Final inspection and testing.

2. Determine the policy; e.g., "Do not use until verified."

3. Identify categories of the product that are affected.

4. List all quality characteristics that are subject to inspection and test.

5. Ensure that the procedures for identifying specified requirements are available.

6. Provide for complete and current procedures at the point of inspection/test.

7. Provide for positive product identification/recall for urgent release.

8. Release product only when successful tests/records are complete.

9. Revise/improve procedures.

10. Evaluate revisions.

4.11 CONTROL OF INSPECTION, MEASURING, AND TEST EQUIPMENT

1. Identify all inspection and test requirements (Clause 4.10).

   • Measurements to be made.

   • Accuracy requirements.

2. List equipment and software available to conduct inspections/tests (fixed and portable).

   • Laboratory equipment.

   • Inspection and test equipment.

   • Production machinery.

   • Jigs, fixtures, templates.

   • Test software.

3. Identify recognized calibration requirements and verification procedures for each piece of equipment:

   • Both fixed and portable equipment

     —Required measurement capability

     —Known measurement uncertainty

   • Calibration schedules

4. Review and flowchart existing procedures and documentation for:

   • Measurements to be made.

   • Calibration procedures.

   • "Measurement uncertainty."

   • Identification of calibration status on equipment.

   • Out of calibration action.

   • Work environment control.

   • Handling and storage.

- Safeguarding against unauthorized adjustment.

- Rechecking intervals.

5. Revise/improve procedures.

6. Consider hard copy versus electronic.

7. Establish an effective record system.

8. Evaluate revisions.

4.12 INSPECTION AND TEST STATUS

1. Identify locations where inspection status is critical, such as:

- Receiving

- Production

- Post production

- Installation

- Servicing

2. Flowchart all processes.

3. Determine the means of identification/status:

- Marking, stamps.

- Tags, labels.

- Routing cards.

- Hard copy versus electronic records.

- Physical location.

4. Review positive release procedures and responsibility.

5. Revise/improve the quality plan or procedures.

6. Evaluate revisions.

4.13 CONTROL OF NONCONFORMING PRODUCT

1. Review and document your procedures for the following:

- Identification.

- Documentation.

- Segregation.

- Prevention of inadvertent use/installation.

2. Document the procedures for disposition, notification, and classification.

3. Assign authority for disposition approval.

4. Document the procedures for reinspection of repairs or rework.

5. Document the concession reporting and handling procedures.

6. Revise and approve your procedures.

7. Evaluate revisions.

4.14 CORRECTIVE AND PREVENTIVE ACTION

1. Separately identify procedures for corrective action (actual nonconformities) versus preventive action (potential nonconformities).

2. Carry out corrective action.

   - Assign responsibility to an individual or team.
   - Review the number and significance of complaints and returns. Evaluate their importance.
   - Prepare a flowchart of the present system.
   - Evaluate the effectiveness of present practices.
   - Provide resources:

     —Expertise.

     —Records, instruction procedures.

     —Defective product (for analysis).

   - Revise/improve procedures to:

     —Investigate the cause of nonconformities.

     —Analyze all processes.

     —Determine a final "fix" (i.e., an action plan).

     —Initiate action to prevent recurrence.

     —Apply new controls

   - Make permanent changes.
   - Evaluate revised procedures.

3. Carry out preventive action.

   - Assign responsibility to an individual or team.
   - Review existing preventive action activities.

- Prepare a flowchart of the present system.

- Evaluate the effectiveness of present practice.

- Identify appropriate sources of information:

  —Reports of purchased materials' quality

  —Processes

  —Waiver concessions

  —Audit results

  —Quality records

  —Service reports

  —Customer complaints

- Identify areas in which preventive action activities can be established or enhanced. Examples:

  —Product design

  —Process development

  —Process control

- Make use of preventive action tools such as Failure Mode and Effects Analysis (FMEA).

- Modify and improve procedures to:

  —Identify potential nonconformities

  —Initiate action to prevent occurrence

  —Apply new controls

- Report preventive actions that are taken.

- Evaluate revised procedures.

- Follow up on the effectiveness of the actions taken.

- Submit actions for management review.

4.15 HANDLING, STORAGE, PACKAGING, PRESERVATION, AND DELIVERY

1. Identify the critical points in the process.

2. Review available information, e.g., damage rates, shelf life.

3. Generate documentation for the following:

- Packaging designs.

- Unique customer packaging requirements.

- In-process handling procedures.

- Packaging, packing, and marking processes.

- Warehouse procedures.

- Inventory/stock management procedures.

- Transportation techniques/carrier selection.

- Storage, preservation, and segregation methods.

- Environmental impact.

4. Revise and improve procedures.

5. Evaluate revisions.

4.16 CONTROL OF QUALITY RECORDS

1. Review the list of documents (Clause 4.5).

2. For each category (function) of documents, review procedures to:

- Identify quality records.

- Collect quality records.

- Index quality records.

- Provide access to quality records.

- File quality records.

- Store quality records.

- Maintain quality records.

- Dispose of quality records.

3. Include a review of the following:

- Supplier records.

- Subcontractor records.

- Installation and servicing records.

4. For each document category, establish the following issues about the documents:

- Legibility.

- Identification with a product.

- Ability to be retained.

- Storage environment.

- Retention needs.

- Availability to customer.

5. Review requirements for quality records under the following clauses: 4.1.3, 4.2, 4.3.4, 4.4, 4.6.2, 4.7, 4.8, 4.9, 4.10, 4.11, 4.12, 4.13, 4.14, 4.17, and 4.18

4.17 INTERNAL QUALITY AUDITS

1. Identify the activities to be audited.

2. Establish the qualifications of audit personnel, including:

   - Experience

   - Training

   - Availability

   - Independence

3. Develop (or update) audit procedures to include:

   - Planning

   - Documentation

4. Conduct an initial (trial) quality audit.

   - Evaluate the adequacy of procedures.

   - Determine the effectiveness of procedures.

   - Verify compliance.

   - Determine the suitability of the working environment.

5. Establish a permanent quality audit program.

4.18 TRAINING

1. Identify training needs:

   - List all job functions.

   - Establish training requirements for each function.

   - Include the requirements in job descriptions.

2. Provide training based on:

   - Quality plan elements.

   - Process knowledge requirements: methods, equipment.

- Product knowledge requirements: specifications, workmanship standards.

- Cross-training.

- Extent of trainee's knowledge and skills.

- Other requirements: internal customer, delivery.

3. Establish and record personnel qualifications in individual personnel file to include:

- All required training completed.

- Education (initial, additional).

- Previous experience.

- Physical characteristics and limitations.

- Special training (safety, SPC).

- Medical records.

- Awards, rewards, promotions.

- Cross-training.

4. Develop and document a training plan (matrix) to include:

- Required training.

- Optional additional training.

- Qualifications of trainees.

- Periodic evaluation of effectiveness.

4.19 SERVICING

1. Identify customer service requirements.

2. Document the service requirements:

- Establish procedures.

- Perform the service.

- Report and verify that the requirements are met.

3. Revise/improve procedures.

4. Evaluate revisions.

4.20 STATISTICAL TECHNIQUES

1. Identify existing statistical applications and procedures.

2. Review status, correctness, and effectiveness of statistical technique applications such as in:

- Establishing process capability.

- Identifying potential problems.

- Verifying product characteristics.

3. Examine the quality plan for additional applications.

4. Provide for additional statistical applications.

5. Establish a training plan.

6. Select training personnel.

7. Conduct training sessions.

8. Evaluate the effectiveness and value of new applications.

Using ISO 9000 in Service Organizations

10

by Ian Durand, April Cormaci, and Robert Bowen

Services are the focus of virtually all business development today. The United States and other developed countries are clearly moving to an economy where service is the primary component. In fact, service organizations (including government, education, and healthcare) account for more than 80 percent of employment in the United States today.

The impact of services on daily life and global commerce has led to increased customer expectations for levels of quality in services. To survive and prosper, service organizations must couple an accurate, complete understanding of customer requirements with the organization, resources, and procedures to ensure that they continually meet the needs of their customers. In other words, they need effective, efficient quality systems.

Descriptions of the elements and structure of effective quality systems are readily available in the ISO 9000 family of standards. Producers of tangible goods are well aware of the value of these standards to business success and have been using them for many years. In fact, the number of companies that have achieved third-party quality system registration has been increasing exponentially over the last five years. Chapter 5 discusses quality system registration, and Chapter 19 discusses the use of ISO 9000 around the world.

Service organizations appear to be slower in recognizing the value of the standards and registration as a business tool. In addition, use is concentrated in those service categories that primarily support industrial companies. As a result, service organizations presently account for only about 12 percent of total quality system registrations in the United States. It appears that current, reported use is in response to specific customer requirements or corporate pressure. Yet the ISO 9000 standards are explicitly intended for application in service organizations, describe quality system models that clearly align with the business needs of services, and have proven their value.

The early sections of this chapter explore the characteristics of services and the current use of ISO 9001 and ISO 9002 by service organizations in the United States. The bulk of the chapter addresses the application of ISO 9001 and ISO 9002 in service organizations. ISO 9004, Part 2, Quality management and quality system elements—Guidelines for services, is discussed briefly, but the primary focus of the chapter is on the use of ISO 9001 and ISO 9002.

WHAT IS A SERVICE ORGANIZATION?

Service organizations can be recognized by the primary result of their interactions with customers. They either "transact" events on behalf of customers or they provide knowledge or information to customers. They rarely characterize themselves as designers and deliverers of tangible goods. A scan of the business and government listings in local telephone books shows the extent and diversity of organizations whose primary value is in providing services. Here are some examples:

Transaction Providers	Knowledge Providers
Banks	Architects
Equipment calibration services	Attorneys
Distributors of durable goods to industry	Consultants
Internal Revenue Service	Engineering firms
Long-distance companies	Physicians
Travel agencies	Schools
Utilities	Scientific research facilities

Box 1 lists more examples, and the rest of this chapter is liberally sprinkled with others.

A review of manufacturing organizations usually reveals many internal groups that provide services in support of manufacturing operations. These groups may, for example, process customer inquiries and orders, purchase raw materials, manage human resources, manage inventory, and administer computerized information management systems. In addition, the market offerings of many manufacturers often include services. For example, a manufacturer usually provides invoices to customers and arranges for delivery of products; it may also offer pre- or post-sales technical support. In fact, these kinds of services are becoming increasingly important to customers.

In reality, then, every company or organization provides services, and some provide tangible goods as well.

Characteristics of Services

For tangible goods, quality is defined largely in terms of physical characteristics and technical capabilities, for example, tolerances, size, weight, texture, chemical composition. For services, quality is defined in terms of time, events, and information.

Transaction-based services satisfy customers by fast and accurate processing of information or physical goods. For example, to address a customer claim, an insurance company enters customer-supplied data and compares the data to the terms and conditions of the insurance policy. Each of these actions can be thought of as a "transaction." The customer is interested in a fast and accurate response to the claim.

Box 1

Examples of Services*

Administration:	Personnel administration, computer system administration, office services.
Communications:	Airports and airlines; road, rail, and sea transport; telecommunications; mail movement; data transmission.
Financial:	Banks, insurance agencies, pension administration, property services, accountancy.
Health Services:	Medical staffs, doctors, hospitals, ambulances, medical laboratories, dentists, opticians.
Hospitality Services:	Catering, hotels, tourist offices, entertainment, radio, television, leisure activities.
Maintenance:	Electrical, mechanical, vehicles, heating systems, air conditioning, buildings, computers.
Professional:	Building design (architects), surveying, attorneys, law enforcement, security, engineering, project management, quality management, consultancy, training and education.
Purchasing:	Contracting, inventory management and distribution.
Scientific:	Research, development, studies, decision aids.
Technical:	Consultancy, photography, test laboratories.
Trading:	Wholesale sales, retail sales, distribution, marketing, packaging.
Utilities:	Cleaning services, trash collection, water supply, grounds maintenance, electricity, gas and energy supply, fire, police, public services.

*Adapted from Annex A of ISO 9004-2, Quality management and quality system elements—Part 2: Guidelines for services.

Quality in knowledge-based organizations focuses on effectiveness. For example, a physician works with a patient to produce "good health." The process of delivering a knowledge-based service is designed to place highly specialized knowledge in an active role in a customer's life.

Personal interaction is very often integral to the delivery of a service, although the interaction can take many forms. The world's most honored service profession, motherhood, is based on intensive physical contact. Building an office complex requires extensive, ongoing communication among owners, builders, architects, and the construction management team. The interactions between a client and a dry cleaner or delivery service may be short, sporadic exchanges of needs and information. For some services, highly automated equipment acts as surrogate for the service provider. Automated teller machines and interactive telephone systems for accessing information are examples.

Customers expect satisfying interpersonal experiences when dealing with employees of most services. In knowledge-based services, in particular, where the outcome often has an element of uncertainty (for example, a criminal trial), customers expect the interactions with the service provider to give them confidence in the effectiveness of the service. They expect the service provider to have a "good bedside manner," to borrow a phrase from the medical profession.

Immediate access to information is a crucial element for the success of most services. When customers interact with employees of transaction-based services (such as the reception staff at a hotel or the teller at the local bank), they expect employees to know the necessary procedures, to access customer records quickly, and to make things happen. When customers contract for knowledge-based services (such as medical or legal services, consultancy, financial management, or computer system administration and problem solving), they are clearly paying for experience, knowledge, and the most current information in the respective field.

The result (or output) of a service occurs at the interface with a customer. The customer's perception of satisfaction or dissatisfaction occurs immediately as well. Therefore, the traditional quality control approach used in manufacturing (produce, inspect, sort, and fix) has limited use.

For some transaction-based services, the traditional approach does have application. For example, an equipment calibration service or a distributor could use inspection or testing to verify accuracy and completeness. However, since traditional quality control has a potential impact on response time, it is not sufficient to assure quality.

For transaction-based services where the service is created and delivered on demand, the traditional approach is less useful.

Customers expect financial transactions at a bank, long-distance telephone calls and data transmission, or water supply to occur immediately and be satisfactory. For many knowledge-based services, in particular, this approach is simply not feasible. Consider the consequences of such an approach to heart surgery!

Because the traditional quality control approach is insufficient, impractical, or impossible, the primary way service organizations assure quality is by controlling the processes they use to provide their services. Briefly, controlling processes requires:

- Thoroughly understanding the activities involved and the extent to which the process is capable of producing the needed output.

- Tracking process performance; and taking action when performance is unsatisfactory.

Many services cannot be provided unless the customer supplies information or tangible goods that are used to produce, measure, or deliver the service. A long-distance company only places calls in response to a customer "request" made via a telephone operator or by dialing the destination number and other access numbers. An equipment calibration service must have customer equipment to calibrate. Accountants, attorneys, consultants, doctors, teachers, and private investigators must have customers (often called "clients") and extensive information about their situations and needs.

Some service organizations may be faced with shifts in external forces more often than most manufacturers. For example, public institutions (such as schools, not-for-profit groups, law enforcement agencies, and government agencies) must often adapt to rapid changes in funding, regulations, and laws. Computer support services are faced with the phenomenal rate of advancement in related technology.

WHY DO SERVICE ORGANIZATIONS USE THE ISO 9000 STANDARDS?

Surveys of registered companies in the United States and Canada by *Quality Systems Update* newsletter (published by Irwin Professional Publishing) have identified three primary reasons for pursuing quality system registration:

- Customer requirements or preferences.

- Competitive pressures.

- The need or desire for internal improvement.

No readily available data indicates any differences in motivation between service organizations and manufacturers. However, it is clear from the nature of the products offered by registered service organizations that these organizations primarily serve manufacturing companies. Only a small number serve commercial businesses, and almost none serve individual consumers. So, at the moment, it appears that customer requirements or competitive pressures are important motivators for service organizations with industrial or commercial customers.

Registration data clearly show that internal service organizations often seek registration independent of the rest of their larger organizations. The motivation for this independent action is not clear; however, analyses of survey data clearly indicate that, as a matter of corporate policy, several large, multisite companies have decreed that all organizational units shall be registered. In this era of staff reductions and work outsourcing, a reasonable surmise is that some internal organizations have used the ISO 9000 standards and registration to streamline and improve their operations as a matter of survival.

At least 20 percent of the service organizations with registered quality systems have multiple sites covered by registration. About half have quality systems that apply uniformly to multiple sites; the other half have chosen to register each site independently.

Survey data also show that implementing a quality system based on ISO 9001 or ISO 9002 in most organizations results in improvements that can range from increased productivity and decreased operating expenses to increased quality awareness and positive cultural change.

Other studies indicate that, typically, one-half to two-thirds of the work efforts in service organizations do not add value and result from poorly functioning processes and support systems. Data collected during registrations by British Standards Institution Quality Assurance (BSI QA) indicate that sampled service organizations saw reductions in the cost of rework, expediting, and other nonvalue-adding activities ranging from 15 to 60 percent. Several requirements in ISO 9001 and ISO 9002 facilitate improvements of this kind in process efficiency and effectiveness:

- Establishing adequate process control (Subclause 4.9).

- Understanding customer requirements as part of contract review (Subclause 4.3).

- Regular internal audits of quality-related operations (Subclause 4.17).

- Corrective and preventive action (Subclause 4.14).

- Regular management reviews of quality system effectiveness (Subclause 4.1.3).

Reported Users

Quality system registrations in Standard Industrial Classification (SIC) codes nominally used for services now account for approximately 12 percent of total registrations in the United States. In Canada, with its longer history of using third-party registration, more than 21 percent of registrations fall into these SIC codes. Service organizations in the United Kingdom account for an even larger, and growing, share of certifications (for our purposes, "certification" is equivalent to "registration").

Box 2 lists the SIC code categories for services to which significant numbers of registrations in the United States and Canada are ascribed to date. The first, second, and third largest categories, respectively, are:

- Distributors of durable goods for industry (5000).

- Business services (7300).

- Firms providing engineering, accounting, research, management, and related services (8700).

Surprisingly, about 10 percent of all registered service organizations are internal to larger companies. They provide support services such as processing of customer inquiries and orders, purchasing, testing laboratories, inventory management, and import/export management. These types of service organizations appear under SIC codes 5000, 7300, and 8700.

Box 2

SIC Categories for Services (with significant numbers of registrations to date in the United States and Canada)

4200—Motor Freight Transportation and Warehousing
   (A substantial number of registrations in the United States cover distributors.)
4400—Water Transportation
4700—Transportation Services
   (A significant number of registrations are not "transportation" per se, for example, inspection and test services and customs brokerage. In Canada, the numbers are somewhat skewed by the inclusion of several light manufacturing companies in this category.)
4800—Communications
   (In Canada, 5 percent of registered companies provide communications services, but 80 percent of these are branches of Northern Telecom or Ericsson Communications.)
5000—Wholesale Trade—Durable Goods
   (This category represents the largest number of registrations for service organizations.)
5100—Wholesale Trade—Nondurable Goods
7300—Business Services
   (This category includes an extensive and diverse variety of services. A significant component are companies that provide temporary staff. It should be noted that almost 25 percent of registrations ascribed to SIC code 7300 are for companies that develop and produce software; they are not included in the service organization percentages reported here.)
7600—Miscellaneous Repair Services
8700—Engineering, Accounting, Research, Management, and Related Services
   (The major specialties of registered firms are engineering/architecture, consulting and project management, testing laboratories, and calibration services.)
8900—Services Not Elsewhere Classified

HOW DOES A SERVICE ORGANIZATION USE ISO 9001 OR ISO 9002?

The ultimate objective of a quality system based on the ISO 9000 quality standards is to assure that customer requirements are consistently met through problem prevention at all stages of an organization's operations. This objective can apply to any organization, whether in the business of providing tangible goods, services, or both. In fact, the ISO 9000 standards specifically include service organizations as potential users. Box 3 lists the relevant definitions used in the standards that demonstrate this intent.

Box 3

ISO Definitions*

Product: Result of activities or processes.
                      Notes:

1. A product may include service, hardware, processed materials, software or a combination thereof.

2. A product can be tangible (e.g., assemblies or processed materials) or intangible (e.g., knowledge or concepts), or a combination thereof.

3. A product can be either intended (e.g., offering to customers) or unintended (e.g., pollutant or unwanted effects).

Supplier: Organization that provides a product to the customer.
Customer: Recipient of a product provided by a supplier.
                      Notes:

1. In a contractual situation, the customer is called the "purchaser."

2. The customer may be, for example, the ultimate consumer, user, beneficiary or purchaser.

3. The customer can be either external or internal to the organization.

Service: Result generated by activities at the interface between the supplier and the customer and by supplier internal activities to meet the customer needs.
                      Notes:

1. The supplier or the customer may be represented at the interface by personnel or equipment.

2. Customer activities at the interface with the supplier may be essential to the service delivery.

Service Delivery: Those supplier activities necessary to provide the service.

*From ISO 8402, Quality management and quality assurance—Vocabulary.

This section discusses application of ISO 9001 and ISO 9002 in service organizations. It assumes you are familiar with the structure of these standards and the specific requirements. (Chapter 4 reviews the requirements.)

Underlying Principles of the Standards

These five basic operating principles underlie the requirements in ISO 9001 and ISO 9002:

- Say what you do. In other words, document the way the organization ensures quality.

- Do what you say. In other words, consistently follow procedures. This reduces variability, prevents known problems, and provides the benefits of using proven techniques and procedures.

- Record what you did. Manage the business based on facts. Without a record of what occurred, made at the time it occurred, an organization is not likely to assess the level of quality effectively and less likely to be able to analyze performance, identify potential problems, or anticipate unsatisfactory results.

- Check on the results. Recording results is useless unless the results are analyzed and checked against specifications or criteria for quality.

- Act on the difference. If an organization does not take action when results are unsatisfactory or problems are indicated, consistent quality is difficult to maintain and quality cannot be improved. Lack of attention to this principle leads to decreasing quality, decreasing customer satisfaction, and negative impact on the business. Implementing this principle will lead to an ever more effective quality system.

In an effective quality system, these principles are reflected in ways that align with the basic needs and mission of the organization. For example, ISO 9001 and ISO 9002 require that an organization formally capture and maintain procedures ("Say what you do") for all quality system elements. This can seem overwhelming at first glance, but the standards do not intend that an organization create a bureaucracy or bury itself in paper. What they do intend is that the organization document to the extent that is effective and appropriate for its operation. Two major considerations are:

- The necessary documentation. Employees in different organizations or in different jobs in the same organization may need varying combinations of documented procedures, access to support data, training, and experience to do their jobs. Determining the appropriate "mix" is a decision each organization must make based on its unique characteristics and those of its employees. And the mix may vary considerably from one job to another, even in the same company.

- The vehicle(s) through which the documentation is provided. Documentation may be in any format and made available by any means that work for the employees who need to use it, for example, paper, electronic, flowcharts, or videotapes. Any of this documentation may be prepared in multiple languages.

Understanding the Language

A challenge for some service organizations is to "translate" terms used in ISO 9001 and ISO 9002 into a "language" that makes sense for services. The original issues of the standards were published in 1987 and use a structure and vocabulary drawn largely from manufacturing, particularly hardware manufacturing. Although the 1994 revisions maintain the structure of the earlier issues, they explicitly note the relevance and applicability of the standards to any combination of four generic product categories: hardware, software, processed materials, and services (as shown by the definition of "product" in Box 3). And they use a somewhat modified word choice to make the relevance more obvious. For example, hardware-oriented terms such as "parts," "material," and "batches" were changed to more generic terms such as "product," which, by definition, includes services. A key objective of future revisions is to make the standards more readily understandable for organizations with all types of products.

ISO 8402, Quality management and quality assurance—Vocabulary, can help service organizations understand quality-related concepts and terms used in the standards for services.

Examples of "translations" for four service environments are provided at the end of this chapter (see Appendixes 1 through 4). In each case, the basic principles underlying the quality system and the quality system elements remain intact, but the relationship among elements and the terms used to describe the meaning of each element are adjusted for the specific environment. The diagram at the beginning of each example summarizes primary stakeholders in each environment and primary processes and subprocesses.

Defining Your Quality System

The quality system elements described in Clause 4 of ISO 9001 and ISO 9002 provide a framework for an effective quality system. In practice, most transaction-based service organizations implement quality systems based on ISO 9002—that is, they do not include design control (Subclause 4.4)—and knowledge-based service organizations use ISO 9001.

In either case, operational quality systems include only those elements that are relevant to the service and the service delivery processes. For example, the requirements in Subclause 4.11, Control of inspection, measuring and test equipment, only apply to equipment that makes quantitative measurements (error rates, time intervals, etc.). If the quality measures of a service are all qualitative (gathered through test panels, customer surveys, etc.), the requirements in this subclause may not apply. Methods to ensure the validity of qualitative data might rely on staff skills and qualifications, which would be covered by Subclause 4.18, Training.

Conversely, the scope of a quality system may include aspects of service that are not specifically called out in the standards but are required by customers. For example, if accurate and timely invoicing is a customer requirement, then the accounting function of an organization must be included in the scope of the quality system. The organization would have to have process controls and verification procedures (subclauses 4.9 and 4.10) to assure that accurate invoices were generated in a timely manner. Training (Subclause 4.18) would be required for employees in the accounting function, and statistical analysis of errors (Subclause 4.20) might be required as the basis for tracking performance and initiating corrective action (Subclause 4.14).

In addition, any requirements for which the organization is accountable but that are fulfilled by subcontractors must be included in the organization's quality system. For example, an engineering firm might contract with a third party to inspect building modules at the builder's location. The engineering firm is accountable for effective inspection (required by Subclause 4.10, Inspection and testing), even though the verification is conducted by a subcontractor. The firm's documentation must indicate the way inspection and testing is carried out, and records of results must be available.

Applying the Requirements in a Service Environment

Over 1,000 service organizations in the United States and many thousands more in other countries have quality systems registered to ISO 9001 or ISO 9002. Applying an ISO 9000 quality system in a service organization is not difficult if the organization understands the intent underlying the requirements. Therefore, this section focuses on the intent of each of the quality system elements in the standards, as well as providing some insights and examples for service organizations. (See Box 4 in this chapter and Chapter 3 for ISO standards and other documents that are useful for service organizations.)

Management Responsibility (Subclause 4.1)

The standards intend that the quality system be integrated into daily business operations just as any other organizationwide management system would be. For this to happen, senior management must be actively involved in the implementation and maintenance of the quality system.

The requirements in Subclause 4.1 are the normal kinds of responsibilities that senior managers have, except in this case they are focused on the quality system. One of the most important vehicles for management involvement and continuous improvement of the quality system is periodic, formal, in-depth reviews of the suitability and effectiveness of the quality system required by Subclause 4.1.3. These reviews should consider a variety of information (for example, quality records, the results of internal quality audits, and subcontractor performance) as a means to evaluate:

- How well the quality system supports the quality policy, quality objectives, and strategic direction.

Box 4

ISO Standards and Other Documents Useful for Service Organizations

- **ISO 8402, Quality management and quality assurance—Vocabulary** lists definitions of terms used in the ISO 9000 family of standards. It can help in "translating" quality-related terms and hardware-related concepts for application in service organizations.
- **ISO 9001, Quality systems—Model for quality assurance in design, development, production, installation, and servicing, and ISO 9002, Quality systems Model for quality assurance in production, installation, and servicing** provide requirements for effective management systems used for quality assurance. They are being used by organizations producing all types of products to implement formal quality management systems, often as the basis for quality system registration.
- **ISO 9000, Part 1, Quality management and quality assurance standards— Guidelines for selection and use** discusses processes and networks of processes. This is a helpful guide for reading ISO 9004 Part 2.
- **ISO 9004, Part 2, Quality management and quality system elements—Guidelines for services** provides internal guidance for establishing and implementing a quality management system for mature service offerings and new products. It can be helpful in understanding the requirements in ISO 9001 and ISO 9002. It cannot be used as the basis for quality system registration.
- **ISO 9004, Part 4, Quality management and quality system elements—Guidelines for quality improvement** is a useful supplement to ISO 9004, Part 2 for users of the standards who are primarily interested in improving their competitive position through superior performance. It focuses on improving process performance.
- **ISO 10005, Quality management and quality system elements—Guidelines for quality plans** provides practical guidance for creating quality plans, which describe the way specific practices, resources, and sequences of activities combine to meet quality requirements for products, projects, or contracts. They are the result of the quality planning activity required by ISO 9001 and ISO 9002.
- **A Guide to the Interpretation and Application of the ISO 9000 Standards for Small Business**, while not specifically directed at service organizations, uses easily understandable language and examples that can be useful for managers with little experience in quality management systems.

- How well the quality system fulfills the requirements of the relevant standard.

- Opportunities for improvements in services and in the quality system.

Box 5 is an example of a quality policy developed by a mid-size building materials distributor. It ties together a work approach directed toward customer satisfaction, long-term positive change through continuous improvement, and personal gain through profit sharing. Box 6 shows examples of quality objectives and related subobjectives for an organization that tests for toxic chemicals in residential ground water.

Box 5

Sample Quality Policy—The Wolf Organization, Inc.

We want The Wolf Organization to be regarded by our customers, our suppliers, and our employees as the best and most profitable distributor of building materials in the markets we serve.

Total Customer Satisfaction

To this end, each member of The Wolf Organization is committed to the idea and practice of exceptional service that consistently meets or exceeds the expectations of our customers and our suppliers.

Outstanding Profitability and Profit Sharing

We are also committed to achieving outstanding levels of profit sharing by ensuring that each operating entity is the most efficient, the most productive, and, therefore, the lowest cost distributor of building materials in the markets we serve.

Continuous Quality Improvement

Each member of The Wolf Organization will continuously pursue quality improvement designed to provide every member of the organization with the knowledge and the skills needed to achieve the goals of our quality policy.

This Quality Policy is supported by an annual business planning process that results in measurable quality objectives and through monthly reviews of quality system operation by the senior management committee.

Quality System (Subclause 4.2)

The intent of Subclause 4.2 is to ensure that an organization has a formalized approach to assuring the quality of its services. This means the organization should:

- Perform quality planning for its services.

- Prepare documented procedures for work activities that impact the quality of services and service delivery processes and are consistent with the requirements of the relevant standard and the organization's quality policy.

- Prepare a quality manual that outlines the structure of documentation related to quality.

In transaction-based services, where the primary components of customer satisfaction are speed and accuracy, quality planning focuses on managing transaction processes. Examples of such processes are picking and shipping merchandise, processing insurance claims, collecting and disposing of trash, diagnosing problems in mechanical equipment, and preflight aircraft checks.

Box 6

Sample Quality Objectives

For a company that tests for toxic chemicals in residential ground water, **company-level** objectives might be:

- A 50 percent decrease in average time to report test results by the end of the current year.

- A two-day response to every customer request for testing by the end of the current quarter.

A subobjective for the sales office might be:

- 100 percent complete incoming samples by the end of the current year.

A subobjective for the analysis laboratory might be:

- 10 percent decrease in average interval between receipt of sample and delivery of test results to agent by the middle of the current year.

A subobjective of the personnel office might be:

- To hire and train five new lab technicians by the first quarter of the next year.

In knowledge-based services, where effectiveness is the primary component of customer satisfaction, quality planning often emphasizes controlling "front-end" processes, such as ensuring the qualifications of knowledge workers, and providing appropriate support (human and technological). Because knowledge-based services often deliver quite specialized services, it is also critical that the service organizations thoroughly understand customer needs, which may be stated, unstated, and often unrecognized by the customer. (The discussion on contract review in the next section expands on this last point.)

In some service organizations, "quality planning" may be a result of "strategic planning." For example, a large school district may identify a desired educational outcome (ranging from technology deployment to a level of student performance) over a four-year horizon.

In a slightly different way, a staffing organization might use a strategic planning approach as a means to perform quality planning.

To move toward the strategic goal of building competitive advantage, the staffing organization might develop a plan to assure quality of outcome for each of its major processes (for example, recruiting, database management, sales, etc.). Each plan might contain a flowchart of the process, objective measures that relate to the strategic goal of building competitive advantage, target values for the measurements, and the performance limits that would stimulate corrective action. All of these plans taken together could be considered a quality plan for the organization's major product: unique staffing solutions.

Contract Review (Subclause 4.3)

The intent of Subclause 4.3 is to ensure that an organization has confidence in its understanding of customer requirements and its ability to satisfy them. This requires that the organization thoroughly understand customer requirements before submitting a proposal, accepting a contract or order, or accepting a change to a contract or order, and verify that the organization can meet the requirements.

Customer requirements for all organizations, regardless of product, reflect the same four, general concerns:

1. Did I receive what I expected, and was it effective?

2. Was it complete?

3. Did I receive it when I asked to receive it and when I expected it?

4. Am I satisfied with the way I was treated during interactions with the organization?

These concerns are relatively easy to understand for transaction-based services. Customers primarily want their transactions processed quickly, accurately, and when they expect them to be processed. They also usually have some requirements related to the nature of the interaction with service providers.

For knowledge-based services, the four concerns require some thought. Understanding the aspects of the primary customer requirement of service "effectiveness," can be quite a challenge when the customer actually has an expectation that cannot be achieved or cannot be predicted.

For example, a government agency contracts with a research facility to evaluate the potential effects of a proposed drug, a company hires a consultant to support the implementation of an ISO 9000 quality system, or a defendant hires a lawyer to represent him in a criminal case. It is absolutely critical to the successful delivery of a knowledge-based service that the service provider fully understand what the customer expects and is able to meet that expectation. In these cases, the "contract review" process should focus on agreeing on possible outcomes, accepted approaches to use, ways to measure progress, and time frames.

In many cases, customers do not state all of their requirements, often because they are not aware of them. For example, most individuals who visit doctors or medical facilities want courteous treatment, clear explanations, and an opportunity to participate in decisions. Only a few of these same individuals, though, are likely to state these expectations. It is up to the service provider to discern or assume some of these expectations and make them characteristics of the service.

Contract review can pose a challenge where the customer is not well defined or more than one customer exists. For example, a medical group provides physicians on a rotating basis to monitor cardiac stress tests in a hospital. The "customers" of the medical service might be considered to be the person being tested, the hospital, or the insurance company that pays the bills. Each will have different requirements for that service.

For services more often than for tangible goods, customer requirements tend to be stated in subjective, qualitative terms. Customers might want "courteous reception" at the local veterinary office, "a great trip," "responsiveness," or "immediate availability." What do these phrases mean? How would an organization know if it were capable of meeting the requirements, and, in reviewing performance, if it did meet them?

One common approach is to rephrase subjective, qualitative requirements in a way that is meaningful for the service organization, based, of course, on exploration and verification with customers. Controllable requirements for "a great trip" may mean, for example that all trip documents are received by the customer five days before the trip, all hotel reservations are at four-star accommodations, and the specified rental car is available at the specified location and time. "Immediate availability" for electrical maintenance might mean an electrician arrives at the work site within one hour of an agreed-upon time.

A usual customer requirement in primary health care delivery is "to feel better." Contract review in this situation might take place when the client first enters the health care facility and might include a clear understanding of the symptoms and needs expressed by the client, a clear description of primary care service options and possible outcomes, and a recorded decision by the client regarding the options.

Added benefits of clarifying subjective requirements are that it makes it that much easier to design effective service outcomes and measure performance, gauge progress against objectives, keep records, and monitor the results of corrective and preventive action.

Box 7 illustrates contract review processes in two service environments.

Design Control (Subclause 4.4)

A basic premise of the standards is that most problems in quality can be traced back to incomplete or inaccurate specifications, problems with the design of the service, or problems with the design of the processes that produce and deliver the service.

Every organization does design. In service organizations, much of the design work is related to delivery processes. Examples in transaction-based services might include accomplishing overnight mail delivery, minimizing the occurrences of parts out of stock while controlling investment in inventory for distribution, scheduling aircraft flights, or designing scientific experiments.

An employment agency that customizes "staffing solutions" for clients not only needs a process for developing solutions, but must design the solution itself. Some transac-

Box 7

Examples of Formal Contract Review

In the order-taking process of an electronics distributor, a customer service representative:

- Answers a customer call.

- Enters customer information into the ordering database and reviews information (for example, name, address, credit card details) with customer.

- Enters order information (for example, item name and stock number, quantity, size) and reviews with customer.

- Accesses electronic inventory system to:

  —Determine availability or expected availability date.

  —Enter order to ship, or send notification to affected departments for out-of-stock items.

- Refers to chart for delivery options, discusses with customers, and enters agreed-upon delivery date and option into ordering database.

- Annotates entry with initials or code number for traceability.

- Handles any changes through electronic order change system that immediately alerts all affected departments.

In an elementary school prescreening process for a kindergarten student, a teacher:

- Enters biographical data from student application into student file.

- Administers an ESL (English as a second language) test and enters results into student file.

- Verifies home experience and/or disability of child with parents and enters data into student file.

- Refers to mainstream instruction or guidance and enters recommendations into student file.

Following this process:

- Guidance counselor develops an Individual Education Plan (IEP) based on student file.

- Parent signs IEP.

tion-based services also design tangible items such as financial statements, insurance policies, or report forms as the vehicle for documenting the service. Examples of processes in knowledge-based services might include hiring qualified personnel for a dental clinic, diagnosing an illness, designing a building, or developing a training course or an entire curriculum.

Whether to include design control in a quality system is a decision each organization must make. In general, if an organization has complete control over design and design is a major factor in ensuring quality of the service, then design control should be included (ISO 9001). For example, the design of learning experiences in an educational institution has a major effect on the outcome for the student. In addition, curriculum designs often must be changed in response to advances in instructional research, availability of new technology, or regulatory mandates. In contrast, if an organization provides a stable, predictable service that requires little, if any, ongoing design, design control is probably not an important element in the quality system. Distributors, for example, usually base quality systems on ISO 9002.

Document and Data Control (Subclause 4.5)

The standards intend that an organization create living, useful documentation and instructive data that supports continued use of best practices to achieve service quality. For documentation and data to be useful and consistently used, employees must be confident that they have accessible, accurate, complete, and current information.

A common characteristic of large service organizations, the "distributed work environment" emphasizes the necessity of document and data control. A distributor might have 45 separate warehouses throughout 12 states. To ensure that customers receive uniform service, employees doing the same job at all locations must have access to the same information and follow the same procedures, as well as having comparable competencies, as discussed in Training (Subclause 4.18) below.

Here are some examples of documents and data that impact the quality of their respective services and, therefore, would need to be controlled:

Service Organization	Examples of Controlled Documents and Data
Architectural firm	Drawings for physical construction, electrical wiring, heating, and plumbing
Commercial airline	Preflight checklists, navigation maps, aircraft maintenance procedures, approved subcontractor list
Hospital	Medical policies, administration procedures, setup procedures for emergency rooms, drug interaction reference material
Insurance company	Adjudication tables, policy terms and conditions (claims processing)

| School | Class schedules, tuition development procedures, admission procedures |
| Staffing organization | Candidate applications, employment availability databases |

Purchasing (Subclause 4.6)

The intent of Subclause 4.6 is to give an organization high confidence that vendors ("subcontractors" in the standards) will meet requirements for incoming goods and services that are critical to the quality of the organization's services and service delivery processes.

In service organizations today, with their increasing reliance on temporary staff and outsourcing, purchasing often breaks down into two, broad categories:

- Commercial purchasing (acquisition of equipment, supplies, and services).

- Resource acquisition (securing qualified assistance of external staff organizations).

A restaurant does a considerable amount of commercial purchasing. Its food products, delivery processes, and service environment are advertised to have certain characteristics. To provide those characteristics, the restaurant may need to purchase, for example, vegetables, fish, meats, breads, special condiments, and so on, as well as equipment maintenance and utility services. Using a subcontractor with a history of providing spoiled meat is soon going to spoil business!

For many other service organizations, the critical purchases are qualified resources. A health maintenance organization must establish a network of skilled providers who are not employees. A school district often, by law, must use nondistrict, off-site technical schools for certain students. Research and development organizations may outsource certain tests to highly specialized laboratories.

Control of Customer-Supplied Product (Subclause 4.7)

The intent of the requirements in this subclause is to prevent loss, misuse, or damage to anything customers provide that is used to produce, measure, or deliver a product. This chapter noted the necessity of customer-supplied product for many services earlier, and provided several illustrations. Here are others:

- Airlines must ensure that luggage is secured and protect it from damage or loss until it is returned to passengers.

- Financial institutions must protect deposits made to customer accounts.

- A laboratory must protect medical specimens taken during a physical examination for life insurance.

- A toxic waste disposal service must not damage waste containers while transporting them.

- A school must ensure that children are safe while on school property.

Product Identification and Traceability (Subclause 4.8)

The intent of this subclause is to ensure that an organization can identify and "recall" services and tangible goods related to services that do not meet customer requirements, and to provide the organization with information for preventive or corrective action.

Traceability refers to the ability of an organization to trace the history and delivery of a service or the components of the service at any point while the organization has responsibility. This may mean throughout the useful life of the service. The degree of traceability required is usually determined by regulatory requirements or by good business practice.

Identification refers to any suitable means that allows an organization to trace a service or component parts.

Some services are required to maintain identification and traceability, for example, toxic waste disposal services. However, many services use identification and traceability schemes as a matter of good business practice, that is, a proven method to assist them in service delivery that meets customer requirements. For example:

- Major package delivery services have schemes that allow them (and, sometimes, their customers) to trace the route of a package that has been delivered late or lost and to identify when a package was delivered and to whom.

- Airlines assign identification numbers to luggage to enable them to locate lost luggage or settle disputes.

- The first "treatment" many patients receive in a U.S. hospital is the affixing of an identification band around the wrist followed by the creation of a patient chart which details all subsequent treatment and records medical data.

- The judicial system assigns unique identifiers to cases.

- All insurance information can be traced by claim numbers.

- Universities use social security numbers or other means to identify student records.

- Schools have increasing responsibility to use and retrieve data and certificates supplied by parents.

- Staffing services must manage thousands of applicant documents such as work permits, training credentials, and birth certificates.

Some identification and traceability schemes must span organizations, some of which may be manufacturers and some service providers. For example, a firm that manufactures pacemakers will etch a unique serial number on each one (identification) and keep records of which medical facilities purchase which pacemakers (traceability). Each medical facility should keep records of which pacemaker was inserted into which patient (traceability).

For some new businesses, the service itself is identification and traceability. A dog or cat can now be identified by a uniquely coded microchip implanted under its skin. If a stray animal with such a microchip is found, a veterinarian can use a special instrument to read the code and access a database that can identify the owner. Similar but more sophisticated technology allows law enforcement agencies to trace stolen automobiles. Each runner in the 100th Boston marathon was fitted with an individual tracking chip!

In the event of a problem, however, most service organizations rarely have the option of "recalling" the service. While an airline may be able to recover lost luggage, a consulting session or an Internal Revenue Service audit cannot be undone. The primary importance, then, of identification and traceability for services is to furnish information that is the basis for corrective and preventive action.

Process Control (Subclause 4.9)

Underlying the ISO 9000 standards is the premise that all work is accomplished through processes. Every product, whether intermediate and delivered to an internal customer or final and delivered to an external customer, is the output of one or more processes. The importance of defining, controlling, and improving processes pervades the ISO 9000 family of standards, as well as being an explicit requirement in Subclause 4.9 of ISO 9001 and ISO 9002.

At a conceptual level, a process is a set of interrelated resources and activities that add value by transforming or changing specific inputs into specific outputs, as shown in Figure 10–1. The transformation usually requires a combination of information and tangible elements.

Every service organization has three essential processes:

- Designing the service in response to a perceived or stated customer need.
- Designing the processes that will deliver the service.
- Delivering the service.

Because the output of a service occurs at the interface with the customer, traditional quality control activities are largely inappropriate for ensuring service quality. Therefore, controlling the processes in service organizations becomes the primary way of maintaining consistent, satisfactory service performance.

Process control has a number of aspects:

- A defined process that is readily recognizable to people using the process. For example, maintenance personnel use the same series of maintenance checks when servicing any aircraft of a particular model and perform repair activities as necessary.
- Documented procedures that are consistent with the skill, experience, and knowledge of employees. An organization must make information available to

Figure 10–1

All Work Is Accomplished Through a Process

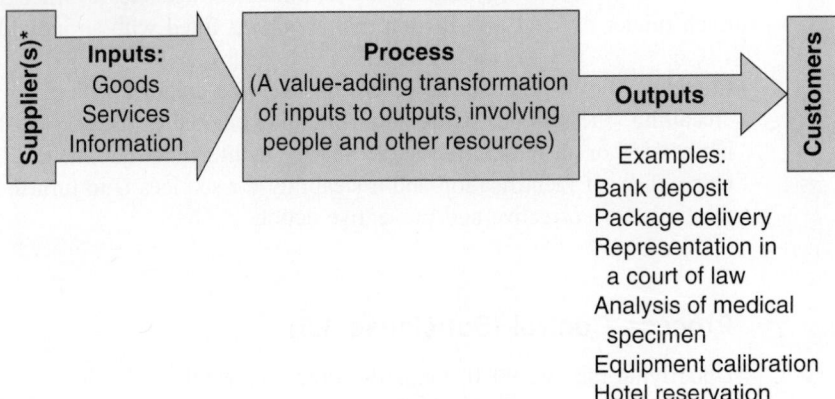

*Also called "vendor(s)." The standards use the term "subcontractor."

the extent that a lack of information has an adverse effect on quality. For example, a bank may train tellers in a set of basic transactions and then provide them job aids for less frequent types of transactions. A staffing service might only hire experienced interview and placement personnel. These employees might only need procedures for organization-specific processes, that is, the way the particular staffing service handles:

• Résumé review and entry into the candidate database.

• Assigning candidate identification numbers.

• Coding interviews.

• Management of candidate database.

• Maintenance of suitable equipment and a suitable working environment. Equipment can range from computer systems to hair dryers. Examples of work environment characteristics that might need to be maintained in a suitable way include noise level, air temperature, and exposure to hazardous substances.

• A system to monitor process and product characteristics that relate to customer requirements and to make changes as necessary to maintain consistency. This aspect of process control requires developing and instituting measures, analyzing collected measurements, and responding as necessary using normal process control techniques as well as corrective and preventive action procedures. Process measures are usually easier to develop and analyze for transaction-based services than for knowledge-based ones. For example, a distributor of building materials with a customer requirement for on-time delivery might institute a measure of the time from receipt of order to confirmation of delivery at a construction site. A security service might measure time from receipt of alarm signal to security force arrival at the site. A data transmission service might measure transmission network availability.

For knowledge-based services, it may be easy to determine what characteristic should be measured, but not necessarily as easy to define how to measure it. This results from the primary customer requirement for knowledge-based service: effectiveness. How would effectiveness be measured for services such as legal representation, research and development, or education?

The approach discussed under Contract Review (Subclause 4.3) for determining customer requirements can work for measures of process performance as well; that is, translating subjective, qualitative measurements into quantitative measurements. Often, measures relate more to the effectiveness of the outcome or result of the service than to the delivery process directly.

For example, a research organization might measure number and kind of errors in experiment design or implementation. A secondary school might evaluate Scholastic Aptitude Test scores or measure the number of graduates accepted by a university. A technical support service might measure the number of callbacks from the same customer for the same problem.

Inspection and Testing (Subclause 4.10)

The intent of the requirements in this subclause is simply to verify conformance to specifications as the service or its component parts move through the organization's processes. If, at any time, specifications are not met, procedures should be defined (see Inspection and Test Status, Subclause 4.12, below) to address the situation with minimal impact on the customer if possible. Inspection and test records are invaluable in improvement efforts.

In some service organizations, verification is performed automatically as part of providing the service, but not necessarily in a formal, documented manner. For example, tax accountants review figures as they prepare returns. Servers in restaurants confirm that the food delivered is the food that was ordered.

For other services, verification is a formal process step. Distributors often perform a check of package contents for completeness and accuracy and record the results on the packing slip. Schools frequently verify student performance (that is, that the school has effectively achieved transfer of knowledge) through the use of tests and homework.

Health insurance agencies verify claim information as part of on-line or manual processes. The *London Financial Times* reported that each insurance claim in most British agencies was rekeyed, on the average, 14 times for verification and processing. In the

Box 8

Examples of Inspection and Testing

For arranging travel:

Receiving: Verify information on client credit card to be used.

Verify availability of flights, cruises, rental cars, etc.

[Note: Client needs (destination, purpose, acceptable modes of transportation, time frame) would have been verified during contract review.]

In-process: Verify client has necessary documentation (passport, driver's license, visas).

Verify client meets health requirements.

Confirm all arrangements made via phone call or fax to client.

Final: Confirm that tickets/vouchers match agreements.

Verify client satisfaction with travel service.

At end of trip, verify with client that all arrangements worked as planned and expected.

For an outpatient surgical procedure:

Receiving: Review pre-admission test results.

Verify patient has fasted.

Ensure patient understands procedure.

Verify correct equipment in operating room.

Verify type of anesthetic with patient and doctor.

In-process: Measure respiration.

Monitor blood pressure.

Control bleeding.

Final: Verify cessation of bleeding.

Verify normal respiratory function.

Verify patient can tolerate food.

Review home care with patient.

Perform postoperative doctor visit.

Verify availability of escort to take patient home.

United States, the Health Care Financing Administration (Medicaid and Medicare) requires both final and in-process inspections to verify appropriate claim handling. Box 8 shows examples of inspection and testing approaches in two other types of service organizations.

Control of Inspection, Measuring, and Test Equipment (Subclause 4.11)

Process control and verification activities generate data about process and product characteristics that are critical for process control, as the basis for decisions on service acceptability, and as the basis for corrective and preventive action. However, in order to use the data in these ways, it must be trustworthy. Subclause 4.11 is the second longest subclause in ISO 9001 and ISO 9002, emphasizing the importance of ensuring that any equipment or device used to make quality-related measurements provides valid data.

The requirements only apply to equipment that makes quantitative measurements. Methods to ensure the validity of qualitative data should be covered by the skills and qualifications of the employees who collect and analyze qualitative data (see Training, Subclause 4.18, below).

Here are some suggestions for defining the terms used in the standards:

- *Equipment*—any device that generates data that can influence the quality of a service.

- *Calibrate*—to ensure that data generated by the device are trustworthy, that is, the data are a valid basis on which to make decisions.

- *Out of calibration*—not able to provide trustworthy data.

Many service organizations do not use equipment that would require calibration. However, here are some examples of equipment to which the requirements might apply:

- Software used to track time and expenses charged to specific jobs, such as in law practices or in consulting organizations.

- Gas chromatographs used in testing laboratories.

- Instruments used by land surveyors.

- Temperature gauges in food service.

This subclause can also find application in unexpected circumstances in service organizations. For example, a school might administer nationally recognized tests whose results are normalized and linked to national databases; this ensures valid test results locally as well as providing a national standard for comparison.

A staffing agency might use an industry-accepted candidate-screening exam for manual dexterity or visual acuity. In both examples, results must be traceable to a known standard of accuracy, even if the standard is only a local reference.

Inspection and Test Status (Subclause 4.12)

The intent of this subclause is to prevent the use, delivery, or installation of untested or "nonconforming" service (that is, services or their component parts that do not fulfill specified requirements). To this end, the standard requires that an organization have a system for identifying the status of a service relative to fitness for use throughout the organization's processes.

Here are some examples of identification of inspection and test status that might be helpful in understanding the intent:

Service Organization	Indication of Inspection and Test Status	Status
Commercial airline	Makes a notation in the aircraft's maintenance log and a central database of a minor maintenance problem that was found, and releases aircraft for use	Inspected and released for use
Dental office	Segregates instruments that have not been sterilized	Inspected, rejected, and awaiting disposition
School	Teacher annotates the file of a student whose homework assignment indicates a poor grasp of material	Inspected, rejected, and awaiting disposition
Security company	Blocks activation of newly installed alarm system until it can be checked	Not inspected

Control of Nonconforming Product (Subclause 4.13)

As the result of verification and process control activities, an organization may have identified a service or a component part as "nonconforming." The intent of the requirements in Subclause 4.13 is to prevent inadvertent use of nonconforming service components and to decide formally on their disposition (acceptance, rework, repair, regrade, scrap, or return to supplier). Instituting a systematic method to control nonconforming service components should prevent customer dissatisfaction.

This subclause is another one that can be difficult to apply in service organizations, especially where the service is a coming together of time, events, and people. In some cases, an organization will not be able to prevent use of the service. For example, a failure of cable television distribution is going to be obvious to the customer, as are interruptions of electric current or water supply.

In other cases, an organization may be able to shield the customer or minimize the effect of the problem. For example, a daily review of transactions in a bank may uncover an incorrect crediting of a deposit, which can be corrected so it has no effect on the customer's account. The chef in a restaurant may detect tainted food before cooking it. A teacher in a vocational school may discern a missed learning opportunity for a student.

Here are examples of nonconforming service disposition for these examples of nonconformance:

Nonconformance	**Disposition**	**Disposition Category**
Cable television distribution failure	Reinitialize connection	Repair
Incorrect crediting by bank	Credit correct account; debit incorrect account	Rework
Lost learning opportunity	Allow student to continue without the learning or Recreate opportunity	Acceptance Repair
Mismatched job and candidate	Conduct placement process again with different candidates or Determine that candidate matches another job	Rework Regrade
Recurrent adverse outcomes of medical treatment	Review medical history, symptoms, test results	Rework
Tainted food in restaurant	Return to supplier; dispose of in trash; submit to local health authorities	Scrap

The requirements in Subclause 4.13 require that an organization identify and suitably dispose of all service that does not meet requirements.

Corrective and Preventive Action (Subclause 4.14)

The requirements in Subclause 4.14 are intended to ensure that an organization identifies actual and potential nonconformances or other undesirable situations that pose a significant threat to the quality of services so that it can:

- Uncover and remove underlying causes of current problems.

- Resolve current and past problems.

- Prevent future problems.

Causes of nonconformance can be related to:

- Operations (for example, heavy background noise in an office may cause errors in order entry).

- Documentation (for example, procedures for developing training materials may be inadequate).

- Process control (for example, not enough service agents at airline check-in counters during peak travel times).

Embedded in the requirements for corrective action is the necessity to deal with identified problems in four time periods: the past, the present, the near term, and the future. For example, assume that elementary school teachers within a particular school district consistently report that the geography and social studies texts used are out of date due to rapid world changes.

Each occurrence of an incorrect textual element is documented as a "lost learning opportunity." The immediate remedial action (the present) might be to take the books out of circulation. In the near term, district policy might be changed to allow access to an all-news television station in classrooms, or teachers may develop lesson plans that compensate for the text shortcomings. Weekly quizzes verify the suitability of the short-term corrective actions. Identifying root causes and taking steps to prevent future problems will be the responsibility of the Strategic Planning Committee of the local Board of Education, ideally with the support of teachers. The Committee might decide to implement on-line learning via a database that is kept current. Scores on national history tests might be the vehicle for determining the effectiveness of this long-term solution. Dealing with past occurrences of lost learning may be quite difficult in this example. Conceptually, the school district could provide supplemental education to all students who used the texts since the information went out of date; practically, this is impossible.

Preventive action requirements focus on using appropriate information to identify potential problems before they result in nonconformances and take action to prevent the problems. For example, a review of records in a medical test laboratory indicates a high degree of accuracy in analysis—a valuable service characteristic from the point of view of doctors and patients. Other records show that, while the accuracy level is the best in the industry, the time to perform certain analyses differs widely among technicians. One approach to increase consistency and efficiency (and prevent potential overall deterioration) might be to observe several technicians. From records and observations, an organization might identify training needs for some technicians and techniques and approaches used by the most efficient and effective technicians ("best practices") that can be documented and communicated to others. Future records will confirm the effectiveness of implementing training and revising procedures to reflect best practices.

Handling, Storage, Packaging, Preservation, and Delivery (Subclause 4.15)

The intent of Subclause 4.15 is to preclude damage or deterioration of service-related components at any point during the service delivery process. Here the standard clearly emphasizes tangible products, although it is relatively easy to find application in many services. For example:

- An investment firm must protect the securities it holds until they are issued to investors.

- A lawyer must safeguard copies of legal documents such as wills.

• A mobile communications provider must preclude unauthorized use of the customer's number.

Note the close correlation between the requirements of this subclause and the requirements of Subclause 4.7, which requires the service organization to protect customer-supplied product from loss, damage, or misuse. Whether the customer or the service provides the "product" or service components, the service organization is responsible for assuring there is no harm done.

Control of Quality Records (Subclause 4.16)

A key principle underlying any effective quality system is management by fact, that is, decisions based on quantitative information. While documentation describes what must be done and how to do it, records capture the results of following procedures and instructive data. The intent of the requirements in this subclause is to ensure that the results collected and analyzed as the basis for action related to the quality system are legible, accessible, and protected from damage.

Records play critical roles in a quality system:

• They provide objective evidence that quality system elements are in place.

• They provide objective evidence that procedures are being followed.

• They hold the keys to understanding the source of problems by providing data for management reviews, internal quality audits, and corrective and preventive actions.

• They provide a vehicle for confirming the effectiveness of corrective and preventive actions.

Records are required by almost all of the quality system elements in ISO 9001 and ISO 9002, but, as is the case with documentation, the standards require that an organization create and maintain only those records that support the quality system.

"Paperwork" in some form is often the only physical manifestation of a delivered service. As such, it sometimes tends to increase with little regard for an overall record-keeping strategy. Implementing an ISO 9000 quality system is an opportunity to determine the records that are necessary to maintain the quality of the service and to review the effectiveness of current procedures for filing, establishing accessibility, and storage.

Knowledge-based services may be more challenged by the requirements in Subclause 4.16 than transaction-based services. Medical offices offer a prime example of a knowledge-based service that relies heavily on records. Patient records are a primary target for control. The information in them could be considered contract review records (agreements on treatments); customer-supplied product records (symptoms, past medical history, observations of the attending doctor); and nonconformance disposition records (action taken when treatment is ineffective).

In a multiphysician medical practice, different doctors may use different abbreviations and protocols for recording similar symptoms, observations, and treatments. The office itself may have no procedures regarding organization of file content, so that contents are not readily accessible. For the patient who may see more than one doctor, confusing or illegible records can severely impact service.

Medical offices should maintain most of the types of records required by the standards, as well, for example, records of training completed by all office personnel, results of periodic meetings of participating doctors to review how the office is doing in regard to caring for patients, process measures such as the average length of time a patient waits to see a doctor, and records of equipment maintenance.

Transaction-based services may more easily adapt or implement procedures to control records. For example, a software distributor, in order to operate, must keep records of orders (contract review) and is likely to have some process control measures in place, such as delivery intervals and returns. Of course, the distributor whose procedures require records of customer orders to be accessible for two years and who then stores them in cardboard boxes in a building with a leaky roof is not fulfilling the requirements of Sub-clause 4.16.

Internal Quality Audits (Subclause 4.17)

The intent of this subclause is to confirm that an organization's quality system:

- Fulfills the requirements of the relevant standard.

- Is suitable for the organization.

- Is effective in progressing toward quality objectives established by senior management.

The results of internal quality audits are a major input to regular management reviews of the quality system. Internal audits have three aspects:

- They are partly adequacy audits, determining whether the quality system as designed and documented is adequate to fulfill the requirements of the relevant ISO 9000 standard and the scope of the organization's quality system.

- They are partly compliance audits, determining whether the documentation is consistently followed throughout the organization.

- They are partly discovery audits, determining opportunities for improvement.

Over and above the mandated use, an organization can use internal audit in these significant ways:

- To assess operations prior to designing or enhancing a quality system to fulfill the requirements of ISO 9001 or ISO 9002.

- To act as a communication vehicle for identifying and extending the use of best practices throughout an organization.

- To build bridges across an organization.

Implementing internal audits in service organizations can be a challenge. For very small businesses, resources to perform the audits may be an issue. In knowledge-based services, historical hierarchies of jobs may pose interpersonal challenges—for example, paralegals auditing the management review process of partners in a law firm, or file clerks auditing purchasing procedures in a government agency.

Training (Subclause 4.18)

The intent of Subclause 4.18 is to ensure that all employees whose work affects the quality of services have the skills, knowledge, and experience (collectively called "competencies") to do their jobs in a way that supports the quality system and complements the level of detail in documented procedures.

Knowledgeable service providers are a cornerstone of success for service organizations. Depending on the type of job they are doing, they will need, to a greater or lesser extent:

- Training in the specific skills needed for the job, which may include interpersonal skills and skills in using computerized databases, as well as using documented procedures applicable to the job.

- Education in the specific discipline needed for the job, for example, civil engineering, gourmet cooking, architecture, electrical engineering, nursing, scientific disciplines, or law enforcement.

- Experience in the specific job skills, for example, piloting aircraft, performing surgery, conducting a case in a court of law, teaching.

- Training in the implementation of the quality system in the organization, for example, the way their job fits into the system.

- Training in skills needed to support the quality system, for example, internal auditing or analyzing process control data.

In general, individuals processing transactions will primarily need focused, job-specific skill training. For example, cashiers in retail stores by and large have relatively well-defined jobs with a minimal number of different activities; training should focus on the skills needed to do those activities, including interacting with customers and when and how to use supporting documentation.

A bank teller will probably need more training in completing many different banking transactions, as well as training in interpersonal skills, using electronic databases, and when and how to use supporting documentation. Note that many knowledge-based services also transact events for customers, for example, admitting patients to hospitals or preparing legal documents.

In general, individuals providing specialized knowledge or information will need extensive education, in a specific discipline, that emphasizes making independent, immediate decisions and interpersonal skills. Consider the educational background of a civil engineer, a teacher, registered nurse, or clinical psychologist.

In addition, most of the education of a "knowledge worker" will probably have taken place outside of the service organization and before the individual became a service provider. Note that many transaction-based services also use specialized knowledge and information, for example, a restaurant may employ a *cordon bleu* chef or a testing laboratory may employ a chemist. In addition to discipline-specific education, the knowledge worker may need, of course, organization-specific training for some elements of a job, for example, preparing reports or keeping records.

It is worth noting that typical "job descriptions" rarely capture needed competencies sufficiently. Even if they identify all job-specific competencies, they do not identify quality system-related responsibilities. Most service organizations will have to spend time developing and implementing systematic methods to determine job-specific competencies, assess the competencies of individuals, close the gaps, and keep records.

Regardless of the job, there are only three acceptable answers to the question, "How do you know what to do in your job?":

- "The procedures and work instructions are documented and I have a copy right here."

- "I have been qualified to do the job, and the records of that qualification are readily available."

- "I have a degree (or certificate, or have demonstrated competency), and records of that degree are available."

Servicing (Subclause 4.19)

Servicing is best thought of as "service support," which may be provided at any point in the delivery of a service. For example, a grounds maintenance organization may provide guidance to customers on specific services needed, run tests of soil acidity, or analyze plant specimens for particular diseases. An employee providing technical support to newly installed communications systems may diagnose problems on a customer's premises. Other examples of servicing are follow-up visits by doctors after surgery, repair of instruments by calibration services, or replacing a burned-out light bulb in a hotel room in response to a request from the room occupant.

An important component of customer satisfaction for a large insurance claims center may be the response of customer service personnel to requests for claim status while the claim is in process. To provide this customer satisfaction element, the claims center must have a procedure for handling inquiries, a procedure for recording results of interactions, and a method to determine customer satisfaction with responses.

Statistical Techniques (Subclause 4.20)

The intent of this subclause is to ensure that statistical techniques are considered and applied where they can be effective in establishing, controlling, and verifying process capability and service characteristics.

The most familiar use of statistical techniques is in monitoring and controlling processes. Given the criticality of process control to ensuring quality service, statistical techniques can be powerful tools for service organizations.

Many service and process measurements lend themselves to analysis using statistical techniques ranging from simple bar charts to sophisticated control charts. For example, a staffing organization might set a performance standard requiring that all customer inquiries receive a call back within 30 minutes. Simple run charts and bar charts, respectively, can immediately show percentages of late calls and extent of lateness. A Pareto diagram of causes for late responses can quickly highlight the critical few that deserve attention.

Histograms, Pareto diagrams, and cause-and-effect diagrams can help in the analysis of customer complaint data, identifying root causes of problems, and performing risk evaluations.

COMPARISONS WITH OTHER SERVICE INDUSTRY STANDARDS

Standards for service industries are few. Three are discussed briefly in this section: ISO 9004, Part 2, the U.S. Malcolm Baldrige National Quality Award Criteria, and the National Committee for Quality Assurance accreditation of managed healthcare facilities.

ISO 9004, Part 2, Quality management and quality system elements—Guidelines for services, provides guidance on establishing and implementing a quality management system for mature service offerings as well as new products. It uses language and concepts that are familiar to service organizations. This standard is intended for guidance within an organization only. It cannot be used as the basis for quality system registration, nor is it a guide for applying ISO 9001 or ISO 9002 to service organizations, although it can be helpful in understanding the requirements.

ISO 9004, Part 2, was published as an international standard in 1991 and recently adopted as a U.S. national standard. It may be considered a precursor of the next revision of the ISO 9000 core standards (ISO 9001, ISO 9002, ISO 9003, and ISO 9004, Part 1) because it focuses on managing business processes and using customer feedback as a means to improve overall business performance and customer satisfaction. The standard also introduces other total quality management principles into the ISO 9000 family of standards. Key for service organizations is the emphasis in this standard on the critical role of internal business processes in ensuring that both customer needs and internal business objectives are satisfied.

Topics covered in ISO 9004, Part 2, include:

- Managing customer relations.

- Measuring customer satisfaction.

- The critical need for process management, especially for processes that cross departments.

- The dependence of service quality on employee knowledge, skills, and motivation, in addition to the formal quality system.

- Employee involvement and teamwork.

- Quality planning and managing the life cycle of services.

- Problem prevention.

- The importance of continuous improvement based on quantitative measurements.

The U.S. Malcolm Baldrige National Quality Award (MBNQA) criteria describe the elements of one comprehensive, progressive quality management system model, and there is a specific award category for "Services." A quality system based on ISO 9001 or ISO 9002 would have essential building blocks for any more progressive quality management system, such as that described by the Baldrige criteria.

Many of the requirements of the ISO standards reflect progressive quality management, for example, quality planning, managing process control, and corrective and preventive action. In fact, many companies pursue registration to ISO 9001 or ISO 9002 as a milestone in their journeys toward "total quality management." (Chapter 13 compares ISO 9000, the Baldrige Award, Total Quality Management, the Deming approach, and continuous improvement.)

Two distinctions are important:

1. The requirements of ISO 9001 and ISO 9002 cover almost half of the Baldrige criteria. However, quite satisfactory compliance to ISO 9001 or ISO 9002 could result in only a modest Baldrige score. Conversely, a Baldrige winner may not pass a formal ISO 9000 assessment. This might occur because the requirements in the standards are much more focused and definitive than they are in the Baldrige criteria.

2. The current award categories of the MBNQA are limited to small business, manufacturing, and service, where service excludes government agencies, educational institutions, and healthcare systems.

The National Committee for Quality Assurance (NCQA) sponsors a system of third-party accreditation for managed healthcare systems. Several Fortune 500 companies now require their managed care providers to be "NCQA accredited." A quality system based on ISO 9001 can be a building block for an NCQA-accredited system. Here are important similarities and differences:

- The fundamental purpose is similar: ISO 9000 quality systems aim at preventing defects to achieve customer satisfaction. NCQA aims to achieve accurate, comparable health plan purchaser data and continuous improvement.

- Both pursue continuous improvement. ISO 9000 quality systems pursue continuing suitability, effectiveness, and improvement of the system. NCQA pursues constant improvement of clinical and operational effectiveness.

- Both rely on accurate information. ISO 9001 and ISO 9002 document and data control requirements demand 100 percent accurate, complete, authentic (prepared by a qualified person) information that is available wherever needed in the quality system. HEDIS "Report Cards" focus on the accuracy and completeness of clinical and operational data.

- Both systems rely on validation by an external third party, although the details differ somewhat. During on-site assessments, ISO 9000 auditors focus on comprehensive quality system operation, while NCQA relies on clinical experts to address medical management issues and focuses on achievement of specific outcomes. Both assessments rely on face-to-face interviews, documents, data, and records as evidence of quality improvement, but ISO 9000 audits assess all processes that impact the quality of products, while NCQA focuses on the core processes associated with healthcare management. Ongoing surveillance is also carried out differently. Continued effectiveness of ISO 9000 quality systems is usually assessed at six-month intervals, with a requalification required every three years. NCQA-accredited systems must be reassessed every three years.

Aside from standards, many services are subject to government and environmental regulations. The Securities and Exchange Commission regulates the securities industry. Architects must consider building codes and zoning board decisions. The Federal Aviation Administration regulates commercial airlines. Telecommunications services must secure tariffs to operate. Toxic waste disposal services must meet rigorous environmental regulations. Many services require operating licenses that are issued after evidence of competence is observed.

A PROVEN IMPLEMENTATION PROCESS FOR AN ISO 9000 QUALITY SYSTEM

For most organizations, implementing a quality system that satisfies the requirements of ISO 9001 or ISO 9002 in a manner that yields significant internal benefits will take at least 12 to 18 months. The cost and time involved qualify the effort as a major project.

The overall strategy and major activities in implementing an ISO 9000 quality system in a service organization are the same as for any other type of organization. Figure 10–2 is a brief overview of a proven, eight-phase implementation process. Chapter 7 describes this process more fully, discusses guidelines for successful implementation, and explores common quality system shortcomings.

This process assumes the establishment of a cross-functional Project Team to plan and guide the entire effort with demonstrated support from a Steering Committee composed of senior management. Smaller Element Teams composed of knowledgeable employees will actually design the quality system elements.

Phase 1 is perhaps the most critical phase: Pursuing implementation without support and demonstrated commitment throughout the effort from all levels of management is not likely to succeed. In Phase 2, the Project Team comes together, is educated on the requirements of the chosen standard, assesses the status of the current quality system against the standard, and develops a detailed plan of the way the project will be organized, staffed, and managed.

Once the initial work is underway, the Project Team moves into a project management role until implementation is complete. Phases 3 through 7 build on each other and progress logically, although movement back and forth among some of them may occur (indicated by the reverse flow arrows in the figure). In practice, activities should be done in parallel to the extent possible to minimize completion time.

This process was designed with medium-sized organizations in mind. Organizations with small staffs or very large staffs and multiple locations may need to make adjustments. Aside from the issue of size, every organization is unique and will have to tailor an implementation process to some extent.

The rest of this section briefly describes selected aspects of the implementation approach followed in two service organizations. Many of the activities in these case studies can be easily correlated to the process summarized in Figure 10–2. The case studies do illustrate the most important point about implementing an ISO 9001 or ISO 9002 quality system in any organization: Do only what makes good, long-term business sense for the organization.

A Case Study: ISO 9002 Implementation at a Staffing Services Company

On May 5, 1995, The Byrnes Group, located in York, Pennslyvania, became one of the first operational staffing companies to achieve ISO 9002 registration.

Why ISO 9000 in staffing services?

ISO 9002 was selected for implementation as a logical extension of continuous improvement initiatives already in place (such as working in teams, using problem-solving processes, and restructuring the organization). The primary business objectives for implementation were:

- To create a recognizable, operational system for a multisite business.

- To build a quality system that could be validated by independent assessment and recognized by customers.

Figure 10–2

Implementing an ISO 9000 Quality System (See Chapter 7 for a discussion of the phases.)

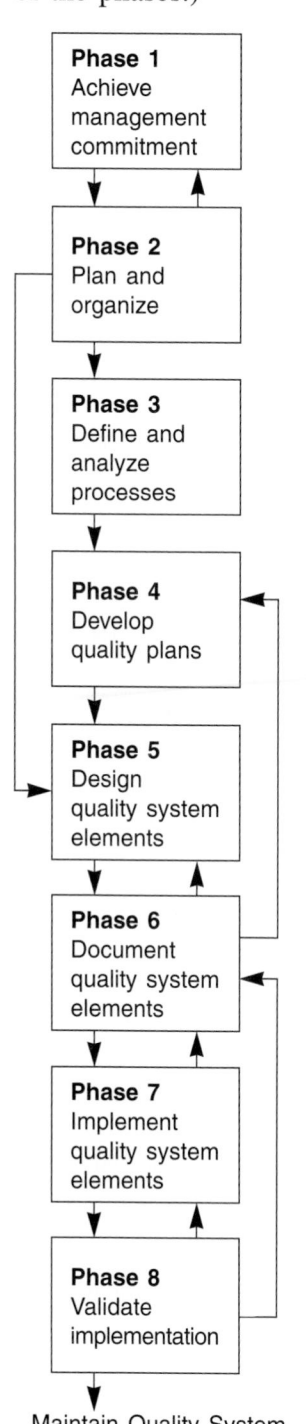

Objectives: To gain confidence that management will invest the necessary resources over the course of the implementation project, and to define the project.

Objectives: To assess the current quality system; to establish a structure, guidance, and processes that will support the implementation project; to begin work on quality system elements that impact the entire system; and to begin registrar selection (for registration).

Objectives: To describe the processes used to create and deliver services, and to measure process performance.

Objective: To create and document an integrated overview across all functions of the way specific practices, resources, and sequences of activities combine to meet customer requirements.

Objectives: To develop action plans for the design, documentation, and implementation of each quality system element, and to design/refine and validate relevant procedures.

Objective: To ensure that quality system documentation is appropriately developed, reviewed, and approved.

Objectives: To deploy new/refined procedures, to demonstrate their effectiveness, and to ensure consistent adherence to policies and procedures.

Objective: To provide high confidence that the quality system adequately covers the defined scope of the organization's quality system, and, if pursuing registration, that the system will be found in compliance during formal registration assessment.

Implementing ISO 9000

The primary obstacle to success was understanding how the standard affected day-to-day operational processes; that is, how to translate the basic intent of the standard meaningfully for the staffing business. "Product" became "successful candidate placement," and "nonconformance" or "scrap" became "unexpected events," "inaccurate information," "internal control failure," or "failure to place a correct job candidate."

The road map to implementation was customized based on the results of a baseline assessment conducted by external ISO 9000 experts and appropriate training. Following training, the organization developed a detailed chronology of events and critical paths called the "Project Plan."

Teamwork

The Byrnes Group used trained, cross-functional teams to document and streamline operational processes. Core processes were documented using flowcharts and simple text in such a way that participants in the process achieved a common understanding. An agreed-upon method to collect measurements for continuous improvement was established, and simple process monitors were put in place.

Procedures

Overall, 40 key procedures were written and approved, including procedures for:

- Management review meetings.
- Quality roles and responsibilities.
- Contract review.
- Design control for new services.
- Control of software affecting quality.
- Résumé process.
- Job ordering.
- Candidate numbering.
- Coding and interviewing.
- Controlled external complaints.

These procedures reflect the company's focus on improved process control and continuous improvement.

Teams implemented these procedures and similar ones in the Recruitment and Search, Contract and Technical, and Temporary Placement business units. Support for workers' compensation and payroll were included as necessary.

Quality System Registration

The culmination of the implementation effort was formal assessment and registration by Lloyd's Register Quality Assurance, Ltd.

A Case Study in Progress: An Educational Institution

Lancaster is the 10th largest school system in Pennsylvania, with steadily increasing enrollment. Currently, the district contains 19 schools, employs 1,200 faculty and staff members, and has an annual budget in excess of $75 million. The demographic mix is: 40 percent Caucasian, 38 percent Hispanic, 19 percent African American, and 3 percent Asian. The tax base is eroding, and urban poverty is increasing, with 58 percent of the students now qualified for free or reduced lunch programs.

ISO 9000 provides a management template for bringing together many initiatives for continuous improvement in the school district. It creates a systematic, well-organized platform to support implementation of strategic plans, technology, a decentralized, site-based management style, and active community participation.

Benefits for Education

As part of implementing an ISO 9000 quality system, the school district is examining the effect of each core work process on teaching and learning. It is expected that reductions in lost time and lost learning opportunities plus paperwork elimination will result in improved productivity and increased time available for teaching.

Further, an ISO 9000 quality system provides a vehicle to assess many activities throughout the school district, such as curriculum design and education delivery (including classroom activities and delivery of administration and support services, such as cafeteria services), and business processes such as purchasing.

Baseline Gap Analysis

Once the management of the school district was committed to the project, the first major activity was to conduct a baseline gap analysis comparing the requirements in ISO 9001 to the current state of activities in core teaching and learning processes throughout the district. The assessment concluded that fundamental building blocks for quality improvement were in place and working well, including a passionate commitment to children and dynamic leadership committed to continuous improvement in the teaching and learning processes.

Regarding the specific requirements of ISO 9001, the assessment team observed a lack of management review procedures to assess the health and effectiveness of the district's quality activities, and a variety of approaches to core processes. For example, "contract review" of customer needs (customers = students) was conducted in different ways at different levels throughout the district. At the time of the assessment, the district did not spell out a systematic approach to conducting such a needs analysis at key thresholds in the educational system.

Another example of diversity among approaches was identified in the area of "process control." The assessment team observed several successful approaches that fulfilled many of the process control requirements; however, they came together only through the initiatives of individual educators, not by following documented procedures. Consequently, many participants in the teaching and learning processes did not view themselves as part of an integrated, comprehensive process; rather, they saw themselves as a "series of professional islands."

Project Planning and Training

Current implementation activities include the development of a detailed timeline and preparing procedures for each element of an ISO 9001 quality system. Twelve cross-functional teams are preparing districtwide core process work instructions, which will be implemented at the building level in conjunction with site-based management concepts. Each team consists of 8 to 10 members representing professional staff and all administrative areas, as well as providing the opportunity for parent and student involvement.

The next phase of activity will be to train team members on preparing effective quality system documents and conducting internal assessment of the effectiveness of the documents.

Overall implementation is expected to take approximately 18 to 24 months.

A FINAL WORD

The experience of registered service organizations attests to the effectiveness of an ISO 9000 quality management system in service environments. This chapter has discussed the requirements of ISO 9001 and ISO 9002 as they might be applied in a service environment. Can any service provider interested in success and continuous improvement afford to ignore the potential place of the ISO 9000 standards in its future?

Appendix 1
Quality System Model for the Construction Industry

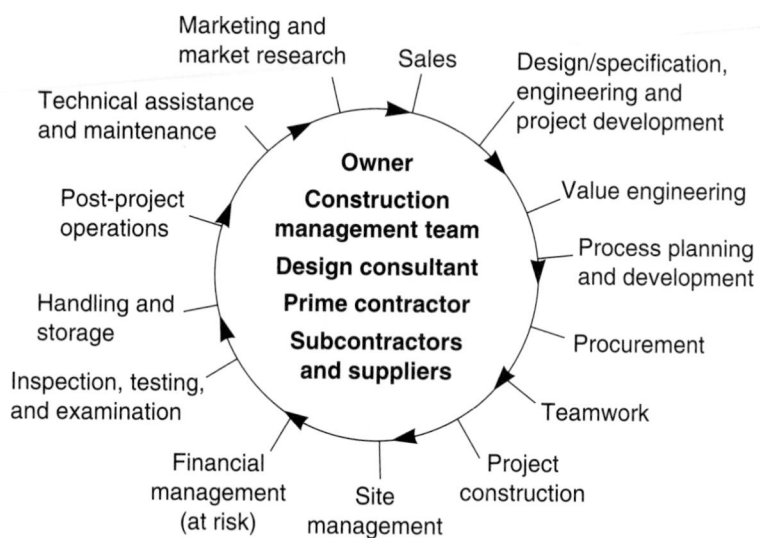

ISO 9001 Quality System Elements	For the Construction Industry
Management responsibility	Leadership policies, project plan
Quality system	Written quality manual
Contract review	Owner needs assessment process
Design control	Project design methodology; value engineering
Document and data control	Current, accurate, authentic information for everybody
Purchasing	Prime and subcontractor relationships; qualified suppliers
Control of customer-supplied product	Don't lose consigned items
Product identification and traceability	Material and information traceability; labeling rules
Process control	Site management; teamwork
Inspection and testing	Owner's checks; mandated inspections
Control of inspection, measuring and test equipment	Calibrate measuring instruments and process control equipment
Inspection and test status	Clearly identify nonconforming product or work
Control of nonconforming product	Prevent inadvertent use of unacceptable items or work
Corrective and preventive action	Organized method for continuous improvement
Handling, storage, packaging, preservation and delivery	Prevent damage or deterioration; protect owner-supplied items and information
Control of quality records	Keep only those records that are absolutely necessary
Internal quality audits	Comprehensive systems assessment; all personnel
Training	Job qualification process
Servicing	Check on complaints
Statistical techniques	Use acceptable, proven statistical methods

Courtesy of r. bowen international, inc.

Appendix 2
Quality System Model
for Education

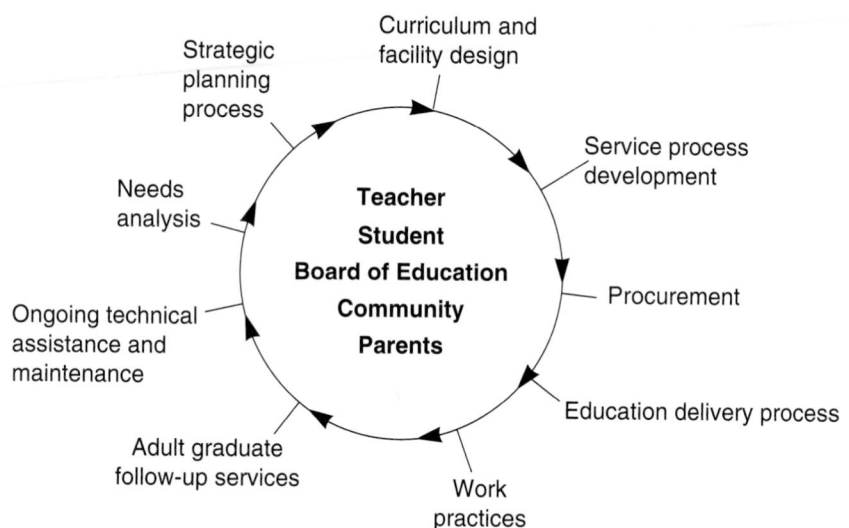

ISO 9001 Quality System Elements	For Education
Management responsibility	School Board policy; strategic plan
Quality system	Written quality manual
Contract review	Customer needs assessment process (teacher and learner)
Design control	Recognizable process to develop new curriculum
Document and data control	Current, accurate, authentic information for everybody
Purchasing	Commercial and instructional acquisition
Control of customer-supplied product	Don't lose documents parents provide
Product identification and traceability	Have readily retrievable student information
Process control	Instructional plans
Inspection and testing	Defined acceptance criteria
Control of inspection, measuring and test equipment	Calibrate, label equipment and national tests
Inspection and test status	Label all records
Control of nonconforming product	Keep track of lost learning opportunities
Corrective and preventive action	Organized method for continuous improvement
Handling, storage, packaging, preservation and delivery	Have defined information flows
Control of quality records	Keep only those records that are absolutely necessary
Internal quality audits	Comprehensive assessment; top-to-bottom; all personnel
Training	Everyone on staff needs a development plan
Servicing	Check on graduates
Statistical techniques	Use acceptable, proven statistical methods

Courtesy of r. bowen international, inc.

Appendix 3
Quality System Model
for Healthcare

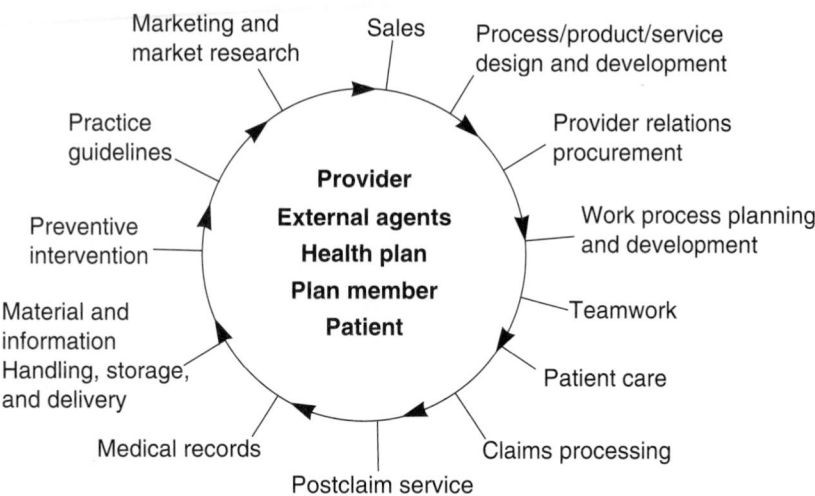

ISO 9001 Quality System Elements	For Healthcare
Management responsibility	Leadership policies; management review meetings
Quality system	Written quality manual; annual quality work plan
Contract review	Patient/member needs assessment process; encounter assessment
Design control	Product/service/process design methodology; value engineering
Document and data control	Current, accurate, authentic information for everybody
Purchasing	Provider relations; vendor management and selection
Control of customer-supplied product	Don't lose consigned information or samples
Product identification and traceability	Sample or information traceability; labeling rules
Process control	Healthcare delivery; claims processing; teamwork
Inspection and testing	Work process checking; mandated inspections
Control of inspection, measuring and test equipment	Calibrate measuring instruments and statistical techniques; verify primary credentials of clinical personnel
Inspection and test status	Clearly identify unacceptable work, information, or recurrent adverse outcomes
Control of nonconforming product	Prevent inadvertent use of unacceptable information or work
Corrective and preventive action	Organized method for continuous improvement; verify positive change
Handling, storage, packaging, preservation and delivery	Keep patients comfortable and safe; don't lose plan member or patient information or samples
Control of quality records	Keep only records that are absolutely necessary; everyone agrees on record-keeping protocols
Internal quality audits	Comprehensive systems assessment; all personnel; all functions
Training	Job qualification process; professional development plan
Servicing	Improve customer satisfaction; check on complaints and grievances
Statistical techniques	Use acceptable, proven statistical methods

Appendix 4
Quality System Model for the Staffing Industry

ISO 9001 Quality System Elements	For the Staffing Industry
Management responsibility	Leadership policies; management review meetings
Quality system	Written quality manual
Contract review	Customer needs assessment process
Design control	Method to design a staffing solution
Document and data control	Current, accurate, authentic information for everybody
Purchasing	Screen candidates
Control of customer-supplied product	Don't lose information or items given to you
Product identification and traceability	Traceability; file labeling rules
Process control	Finding and placing successful candidates; teamwork
Inspection and testing	Process gateways for each candidate
Control of inspection, measuring, and test equipment	Validate tests
Inspection and test status	Clearly identify unacceptable candidates, information, or work
Control of nonconforming product	Prevent inadvertent use of unacceptable candidates, information, or work
Corrective and preventive action	Organized method for continuous improvement
Handling, storage, packaging, preservation, and delivery	Don't lose information anyone gives you
Control of quality records	Keep only those records that are absolutely necessary
Internal quality audits	Comprehensive systems assessment; all personnel
Training	Job qualification process; personnel development plans
Servicing	Check on complaints and service results
Statistical techniques	Use acceptable, proven statistical methods

Courtesy of r. bowen international, inc.

APPLICATION OF ISO 9000 TO THE CONSTRUCTION INDUSTRY

by Paul A. Nee

The application of ISO 9000 to the construction industry makes both competitive and business sense in a global marketplace that increasingly rewards efficiency and superior customer service.

Interest in applying ISO 9000 to the construction industry has grown at a steady pace among North American-based companies (see Figure 10–1(a)). Some of that interest has been driven by countries that require ISO 9000 registration to bid on certain types of government and private sector projects. However, the adoption of ISO 9000 by an increasingly diverse stakeholder group from the chemical industry to the healthcare industry will continue to encourage the use of internationally accepted quality management practices.

THE DORMA EXPERIENCE

Dorma Door Controls, Inc. of Reamstown, Pennslyvania is a manufacturer of builders' hardware products for the construction industry. In January 1991, Dorma decided to pursue certification to ISO 9001 as a "stepping stone" to a Total Quality Management (TQM) philosophy. In April 1992, Dorma was registered. At the time around 300 registration certificates had been awarded in the United States.

The significance of Dorma's accomplishment was not fully appreciated by the marketplace, and there was little customer impact despite the fact that the effort strengthened Dorma's position as the quality leader in the builders' hardware industry.

However, Dorma's early success with registration put it in the unique position to become the industry ISO 9000 resource for training customers, competitors, and other related companies.

Dorma's market-leading experience has paid off. Many international and domestic customers are now requesting evidence of ISO 9000 registration as a requirement for doing business, a request that promises to be more common than not in the future.

Figure 10–1(a)

Construction-Related Registered Sites, 1990–1995

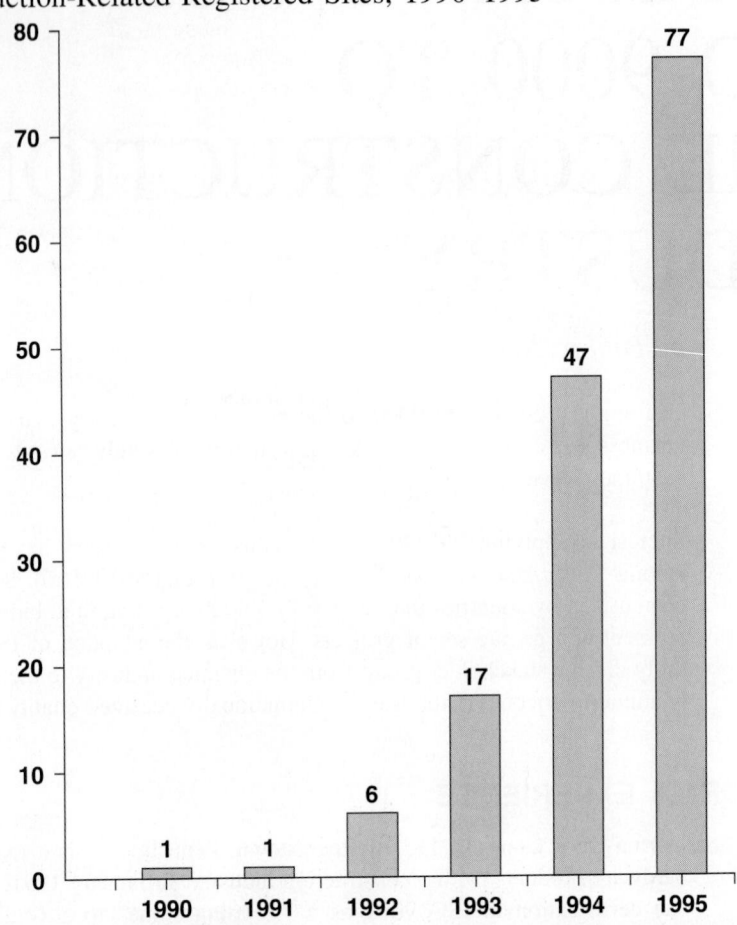

This chart represents an attempt by the editors to show the growth in construction-related industries using selected data from the *ISO 9000 Registered Company Directory*. It is not intended to be a complete listing of all construction-related industries using ISO 9000.

INDUSTRY CHANGES

The application of the ISO 9000 standard to the construction industry requires some interpretation, but the most difficult part of the process for most in this industry will be changing long established ways of doing business.

For example, many companies in the construction industry have poor instructions and inadequate process control. This results in lengthy "punch lists" that must be resolved before the project is complete. Excessive losses are experienced by contractors due to unsatisfactory work performance. ISO 9000 procedures will help avoid many of these costly, time consuming problems.

Editors Note: See the related chapter in this section, "Using ISO 9000 in Service Organizations" for more insight on the application of the standard to the nonmanufacturing sector. See also Appendix 1 of that chapter for a flowchart on the application of ISO 9000 to the construction industry.

HOW THE STANDARD RELATES TO CONSTRUCTION

The following is a clause-by-clause explanation of how the ISO 9001 standard can be applied to the construction industry. This is not a comprehensive implementation guide, but should be used as an overview of how a construction-related business approaches registration.

Subclause 4.1, Management Responsibility—The majority of construction-related companies do not usually involve executive management in setting and managing the company's quality system. Implementing this clause would improve communication throughout the organization. Responsibility, authority, and interrelationships would be identified, and accountability for quality encouraged.

Subclause 4.2, Quality System—Quality manuals and supporting procedures are documented and followed in most material and hardware manufacturers. On the other hand, many engineering and architectural organizations may not have a quality system in place. Most construction contractors do not even consider quality systems part of normal construction activities.

Subclause 4.3, Contract Review—Before a contract is signed, ISO 9000 requires that all differences are resolved during a documented contract review session. This clause is critical for engineering and architectural firms and should be an essential part of procedures adopted by all construction-related companies. Implementing ISO 9000 will assure careful review and confirmation of the customers' requirements, protect other organizations involved in the project, and reduce incidents of rework and replacement of products and/or service.

Subclause 4.4, Design Control—This clause is primarily the focus of engineering and architectural organizations. ISO 9000 requires that companies have design verification and validation procedures and adequate controls to assure an acceptable design.

Subclause 4.5, Document and Data Control—Document and data review during construction allows those overseeing a project to be informed of current status. Construction-related businesses are lacking in this area. Not only do such business practices create a more efficient business climate, but both customer and contractor are legally protected during the construction process.

Subclause 4.6, Purchasing—Most construction-related businesses use suppliers they trust, but ISO 9000 offers a more formal system for choosing suppliers. This clause has several advantages for the construction industry. It assures that an approved subcontractor's materials and services are incorporated into a construction project, and it draws project management into the process.

Subclause 4.7, Control of Customer-Supplied Product—Often construction projects involve customer-supplied materials. ISO 9000 registration formalizes this process in the construction industry. Ideally, controlling this aspect of the construction process is the responsibility of both builder and customer.

Subclause 4.8, Product Identification and Traceability—ISO 9000 assures that construction project materials are properly received, inspected, and identified prior to utilization. Current industry practices use an informal system to identify, trace, and control material entering a construction site. Product identifiers such as serial numbers, batch codes, or lot numbers are appropriate. Color coding, tags, labels, and physical markings are other adequate methods of identification. ISO formalizes this ad hoc system and insures that only acceptable quality products are utilized in the construction project.

Subclause 4.9, Process Control—Any construction project, no matter the size, must have its processes under control. For example, deciding what equipment to use, making sure that the project meets all applicable building codes, and setting criteria for workmanship should be standard procedure. ISO 9000 provides a framework for defining then ensuring that detailed planning is accomplished.

Subclause 4.10, Inspection and Testing—ISO 9000 requires a company to inspect incoming material, ensure process viability and then conduct a final inspection. It is hard to imagine a construction project that progresses much beyond the concept phase without some form of this clause in place. However, documented procedures for these activities and quality records are not normally an industry practice.

Subclause 4.11, Control of Inspection, Measuring and Test Equipment—Building any structure requires the use of numerous measuring devices, from a simple ruler to complex laser surveying equipment. However, ISO 9000 registration does not require that every level be calibrated unless it is used for verification of quality or acceptance to specification.

Subclause 4.12, Inspection and Test Status—Space is often limited and movement of material is difficult on many construction sites. ISO 9000 requires a company to identify materials that have been inspected and accepted by a responsible authority. Fortunately for the construction industry, the standard is flexible. Documented procedures allow for considerations, such as location of materials, to indicate their acceptability.

Subclause 4.13, Control of Nonconforming Product—Improperly placed concrete reinforcement rods can have a high financial liability cost. ISO 9000 requires a company to create a documented procedure to identify, document, evaluate, segregate, and eliminate nonconforming product.

Subclause 4.14, Corrective and Preventive Action—A common and recurring problem in the construction industry is failing to address symptomatic or recurring problems during construction. The same "minor detail" problems occur on every job, such as caulking, painting, and clean-up. Corrective and preventive action procedures would focus on the root cause of these problems. ISO 9000 helps ensure that past mistakes and inefficient practices are eliminated and continuous improvement is part of the quality system.

Subclause 4.15, Handling, Storage, Packaging, Preservation, and Delivery—Wet wallboard and warped paneling and studs do not contribute to a more attractive bottom line. Companies registered to ISO 9000 are required to document adequate handling, storage, packaging, preservation, and delivery procedures. This consistent approach not only eliminates waste, it also makes coordination of various construction activities manageable.

Subclause 4.16, Control of Quality Records—Final acceptance of a construction project often mandates documented procedures on how quality records are identified, collected, indexed, accessed, filed, stored, and maintained. ISO 9000 ensures that this takes place.

Subclause 4.17, Internal Quality Audits—A construction foreman is often the principal auditor for a construction project. For many companies, a documented procedure for quality system auditing is a new experience. However, these audits are a valuable quality tool for executive management.

Subclause 4.18, Training—Construction industry training is largely based on an apprentice system, with few formal requirements for many jobs. ISO 9000 requires an organization to assess employee training, provide adequate training where needed, and maintain appropriate training records. This will be difficult for many companies that often use unskilled temporary labor to keep costs to a minimum. Carefully documented training must be provided as needed.

Subclause 4.19, Servicing—Engineering and architectural firms, specification writers, or construction management organizations are likely to have a direct link to this clause. ISO 9000 requires that any organization with servicing activity document procedures and establish criteria for performance, verification, and service reporting. However, many construction organizations would not include this element in its audit.

Subclause 4.20, Statistical Techniques—Most construction organizations do not use statistical techniques. However, ISO 9000 requires that an organization identify the need or lack of need for statistical techniques. If statistical methods or techniques are used in the process, they must be defined in document procedures.

The construction industry could benefit from the wide acceptance of ISO 9000. Registered companies working toward the same goals often operate with a higher level of cooperation and customer satisfaction. Construction-related companies will experience improved performance and find fewer mistakes, delays, late penalties, and a shorter punch list for rework after the job is complete. For most companies, these efficiencies are payback enough for seeking registration.

THE ISO 9000 FAMILY AND RELATED STANDARDS

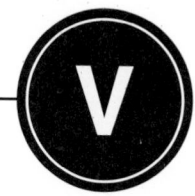

Chapter 11   The Future of the ISO 9000 Standards

Chapter 12   ISO 14000—The International Environmental
Management Standard

Chapter 13   Comparing ISO 9000, Malcolm Baldrige and
Total Quality Management

Chapter 14   Other Standards, Guidelines and Business Process Initiatives
Based on ISO 9000

Chapter 15   QS-9000 Quality System Requirements

The Future of the ISO 9000 Standards

11

by Ian Durand and April Cormaci

INTRODUCTION

While no one can reliably predict the future development and uses of the ISO 9000 family of standards, it is possible to make educated forecasts based on an understanding of past developments and of the current situation. This chapter provides a pragmatic view of likely developments into the early years of the 21st century that combines the lessons of history with marketplace forces evident in 1996.

The forecast goes beyond the standards themselves and considers their uses as well, since the two are inextricably linked it is the applications that shape and drive the development of useful standards.

THE DEVELOPMENT AND USE OF QUALITY STANDARDS

The development of the ISO 9000 standards and their predecessors, starting with MIL-Q-9858 in 1959, is summarized in Table 11–1.

Key Points

When coupled with information about usage, these key points from Table 11–1 will be useful in forecasting future developments and applications of the ISO 9000 standards:

Table 11–1

Summary of Significant Historical Developments in Quality System Standardization

Year	Standard	Uses
1959	MIL-Q-9858 Quality Program Requirements	Department of Defense procurement
1968	AQAP1 Requirements for an Industrial Quality Control System for a Quality Program	NATO procurement
1968	C-1 Specifications of General Requirements for a Quality Program (ANSI Z1.8-1971)	Quality assurance requirements for two-party contractual situations
1974–5	Canadian Standards Z299.1, .2, .3, and .4 quality assurance standards*	Primarily intended for two-party procurement purposes
1979	ANSI/ASME NQA-1 Quality Assurance Program Requirements for Nuclear Power Plants	Requirements and nonmandatory guidance for establishment and execution of quality assurance programs during design, construction, and operations of nuclear power plants
1979	Canada explores third-party registration and establishes QMI as a third-party registrar	Extend the use of Z299 standards to marketplace demonstration of appropriate quality system
1979	BS 5750 Parts 1, 2, and 3 Quality Assurance Requirements *	U.K. quality assurance system standards intended for industrial and commercial use
1979	ANSI/ASQC Z-1.15 Generic Guidelines for Quality Systems	Outlines basic elements for structuring and evaluating the quality systems for manufactured product
1979	ISO/TC 176 formed	Facilitate international trade by harmonizing national quality standards

*While BS 5750 is often cited as the prototype for ISO 9001, 9002, and 9003, the Canadian Z299 standards influenced the development of national standards in other countries and were also important to the development of international quality assurance standards. Not shown are numerous other national quality standards published in the 1970s.

- While formalized quality standards have been around since 1958, only those standards that were required by the marketplace, usually by customers but sometimes by regulation, came into widespread use. Salient examples are quality assurance standards required by the military and the nuclear power industry. Regulations developed for the food, pharmaceutical, and medical device industries often had elements of quality assurance systems embedded in them.

Table 11–1 (*continued*)

Year	Standard	Uses
1980	Netherlands establishes third-party certification of quality assurance systems	
1985	United Kingdom establishes third-party certification of quality assurance systems	
1987	ISO 9001, 9002, and 9003 Quality Systems—Models for quality assurance	Intended for two-party contractual situations
1987	ISO 9004 Quality Management and quality system elements—Guidelines	Development and implementation of a comprehensive and effective in-house quality system, with a view to ensuring customer satisfaction
1989	United States establishes third-party registration of quality assurance systems	
1989	Adoption of "New Approach" by European Community for essential requirements	Foster intra-Community trade
1994	Revision of ISO 9001, 9002, 9003, and 9004	Communicate application to all four generic product categories and recognize the use for third-party certification/registration

Stated another way, *the availability of standards in the absence of perceived need does not lead to usage.* If suitable standards are not readily available, customers or regulators will create suitable substitutes to meet pressing needs.

• This marketplace reality applied to the ISO 9000 standards as well. Much of the business world had not heard of the ISO 9000 standards until they were adopted by the European Community as a practical tool and given a great deal of international visibility around "EC '92."

• Even today, ISO 9000 means the ISO 9001 or ISO 9002 quality assurance standards to the vast majority of business users. Neither national nor international guidance standards for quality management have attracted significant interest or use to date.

THE IMPORTANCE OF THIRD–PARTY REGISTRATION

During the early 1990s, certification ("registration" in U.S. terms) of quality assurance systems by accredited third-party registrars began to spread throughout Europe and beyond. In part, this was fueled by publicity about "EC '92." But a more important factor was that large companies began to realize that registration could give them an edge in an increasingly competitive marketplace. Once large companies started becoming registered, they began to communicate the advantages to their suppliers. This sometimes was by persuasion, sometimes by decree.

Since 1991, the number of registrations worldwide has grown rapidly and shows no signs of saturation. For example, on a per capita basis, the United Kingdom has 29 times as many registrations as the United States, and the number is still growing at 25 percent per year. Australia and New Zealand have 13 times as many registrations as the United States, and Canada has twice as many. By mid-1996 the number of registrations worldwide is predicted to be 140,000. Based on surveys by Mobil Europe, average growth rates recently have been about 50 percent per year (see Chapter 19).

Higher-growth rates have been experienced in the rest of Europe, North America, and the Far East. By the year 2000, the number of worldwide registrations to ISO 9001, 9002, and derivative standards based on ISO 9001 could well exceed 500,000. It is not surprising that third-party certification/registration of quality systems is the dominant use by far of the ISO 9000 standards.

The pattern of industries making most use of the standards for registration is well established. In the United States, almost one-half of registrations are in three major industrial sectors, and only eight industrial sectors, all of them manufacturing oriented, account for almost 80 percent of the registrations (see Table 11–2).

Overall, only 12 percent of U.S. registrations are for services, and almost all of these companies serve industry. However, in Canada, 21 percent of the registrations are issued to service organizations. The two principal SIC codes for services in the United States are Business Services (7300) and Wholesale Distributors of Durable Goods (5000). Usage in Canada is similar.

Table 11–2

Primary Industrial Sectors Using Registration

SIC Code	Industrial Sector	Percentage
3600	Electronic and Other Electrical Equipment and Components, Except Computer Equipment	17.6%
2800	Chemicals and Allied Products	15.6%
3500	Industrial and Commercial machinery and Computer Equipment	13.4%
3800	Measuring, Analyzing and Controlling Instruments; Photographic, Medical and Optical Goods; Watches, and Clocks	8.9%
3400	Fabricated Metal Products, Except Machinery and Transportation Equipment	8.7%
3000	Rubber and Miscellaneous Plastics Products	5.5%
3300	Primary Metals Industries	4.8%
2600	Paper and Allied Products	4.4%

Registrations to ISO 9001 by U.S. companies that design and develop software account for a mere 2 1/2 percent of the total, and only one-third of these are under the TickIT scheme developed in the United Kingdom based on ISO 9000-3 (see Chapter 20).

Worldwide, 1095 TickIT certificates were issued,[1] approximately 0.7 percent of the total number of third-party certificates. Of these, more than three-fourths were issued in the United Kingdom, 7 percent were issued in the United States, and another 10 percent were issued to 7 more countries (France, India, Ireland, Japan, Netherlands, Norway, and Sweden). In only five countries do the TickIT certificates exceed 1 percent of the total certificates issued: India, Norway, United Kingdom, Sweden, and Ireland. (Note that Japan dropped plans for a scheme similar to TickIT in the face of pressure from U.S. software organizations.)

In the United States, ISO 9002 is used for more than 60 percent of the registrations, ISO 9001 for almost 40 percent, and ISO 9003 for less than 1 percent.

Small companies are reported to have difficulty using the ISO 9000 standards, although it is not clear whether this is difficulty in understanding and applying the standards or concern about the investment required to become registered. Nonetheless, about one-fourth of the registered companies in the United States have fewer than 150 employees, and only one-fourth have more than 1,500 employees.[2]

As a result of explosive growth of third-party registration, many thousands of quality professionals around the world have been attracted to the ISO 9000 phenomenon as auditors, consultants, and trainers. Globally, registration to ISO 9001 or 9002 has become a multibillion dollar industry, largely funded by industrial users.

Key Points

The patterns of growth and usage indicate that:

- *Customer needs play a dominate role in determining which economic sectors choose to use registration.* For example, there is almost no use by companies manufacturing, distributing, and selling consumer goods even though they could derive substantial internal benefits.

- *While hardware manufacturing industries dominate the use of registration, use is commonplace in processed material industries* such as chemicals, plastics, and paper.

- *The 1994 versions of ISO 9001 and 9002 are practical for use by service organizations.* The reasons for the low use by service industries in the United States relative to their numbers undoubtedly relate more to needs and customer pressures than difficulty in understanding and applying the standards in a service context.

- *Use of the existing ISO 9001 for companies designing and developing software is also feasible,* even if the absolute number of registrations is low.

- The relatively *low use of the TickIT scheme* based on ISO 9000-3 *calls into question the need for this supplementary standard* as a requirements document.

- *The large, embedded base of registered companies will likely resist changing ISO 9001 and 9002,* even if an incremental increase in value is likely.

- *The global industry of registrars, consultants, and training firms has a vested interest in stabilizing ISO 9001 and 9002 and their use* to protect employment security.

While the primary use of the ISO 9000 standards is for third-party registration, important uses in two-party relationships also exist. For example, the U.S. government, and the Department of Defense in particular, is rapidly transitioning from the use of internally developed quality standards to the use of national and/or international consensus standards. ISO 9001 and 9002 are assuming important roles in procurement by major government agencies.

DEVELOPMENTS IN THE ISO 9000 FAMILY

TC 176 has seen growth in a number of directions. For example, the number of participants at international meetings has grown so large that finding facilities large enough to host annual plenary meetings is challenging for the TC secretariat. The number of standards in the ISO 9000 family has grown correspondingly. Currently, 26 standards are either published or under active development, with more being proposed every year. In 1987, only 6 standards existed (see Chapter 2).

A contributing factor to the growth in the number of ISO 9000 standards has been the lack of market-based criteria for objective evaluation of proposed new work items. Nor is there a systematic mechanism to directly assess the needs and concerns of present or potential users. In this situation, TC 176 operated more in a "technology-push" than a "market-pull" mode.

Almost every new work item proposed had an underlying technical rationale. Those voting on these proposals both at the national and international levels are largely quality professionals drawn to technical completeness and detail. As a result, no proposed new work items were ever voted down. (The proposal that came closest to rejection was in 1992 for a guidance standard on Total Quality Management. The potential rejection was skillfully finessed by avoiding the use of the phrase "Total Quality Management.")

1994 Revisions to the Core Standards

The core standards, ISO 9000, 9001, 9002, 9003, and 9004 were revised and reissued in 1994, prompted in part by ISO/IEC Directives requiring a five-year review cycle. There were no major technical changes, although a number of useful refinements and enhancements were made (see Chapter 2).

Clause numbers and clause wording among the three quality assurance standards were aligned. The requirement for design control (Subclause 4.4) in ISO 9001 is the only difference between ISO 9001 and 9002. Under Subclause 4.4, ISO 9002 says simply that "The scope of this international standard does not include quality system requirements for design control." The use of ISO 9003 is small and shrinking, so the more substantive differences between ISO 9002 and 9003 are not of great import in the marketplace in general. The practical effect of this change is that there already is a *de facto* single quality assurance standard!

Just as important for forecasting the future is the fact that the relatively straightforward changes in the 1994 revisions took four years to accomplish. While complaints were voiced that the 1987 versions had a strong hardware manufacturing flavor, no strong marketplace forces were driving a speedy revision.

Growth of Derivative Standards

The embedding of ISO 9001 into industry-specific standards is another significant development. The most notable example to date is QS 9000, developed by the U.S. auto industry, but embedding has also happened within the petroleum industry, in ISO/TC 210 for Medical Devices, and in regulatory directives for medical devices. Note that QS 9000 is treated as a *de facto* international standard as use has spread to Europe and beyond.

Aside from persuasion, no effective mechanism exists to preclude more industries from following the same route. While intended to protect the integrity of the ISO 9000 quality assurance standards, Clause 6.6.4 of Part 2 of the ISO/IEC Directives specify criteria under which this can happen.

TC 176 Plan for the Year 2000

A dormant unease about the increasing number of standards in the ISO 9000 family erupted in vocal concern in early 1995 when a number of delegation leaders to TC 176, primarily from Europe, proposed substantial rationalization and simplification of the existing standards. Responses within TC 176 to this concern culminated in decisions by the plenary meeting in November 1995 to focus the ISO 9000 family on four primary standards, as shown in Table 11–3. This product line will be supported by additional documents as the need arises.

Underlying this vision for the Year 2000 ISO 9000 family is a belief that there is a clear need to simplify, consolidate, and integrate the product line, with minimal disruption of the established usefulness. A deliberate, staged development process was also adopted. It is envisioned that the coordinated development of ISO 9001 and ISO 9004 would be done by two subteams under the management of a single project leader.

Table 11-3

Year 2000 ISO 9000 Family*

Standard	Purpose
ISO 9000 Concepts & Terminology	Combines the former ISO 8402 and 9000-1
ISO 9001[†] Requirements for quality assurance *(ISO 9001 and 9004 are to be developed as a consistent pair of standards)*	The primary goal of ISO 9001 is to provide confidence, as a result of demonstration, in product conformance to established requirements
ISO 9004 Guidance for quality management systems	The primary goal of ISO 9004 is to achieve benefits for all stakeholder groups through sustained customer satisfaction
ISO 10011	Guidelines for auditing quality management systems

* From ISO/TC 176/N 243R.
[†]While a single quality assurance standard is anticipated, the numbers 9002 and 9003 will be available as placeholders, should the need arise for contractual requirements having a more focused scope than ISO 9001.

In response to feedback and input from different economic sectors, three ambitious objectives have been established for the simultaneous, coordinated revision of ISO 9001 and 9004:

- The two standards should complement each other without having significant overlap in content so that users could start with either and easily move to use of the other.

- Both standards should be readily understandable and usable for all four generic product categories without the need for supplementary guidance standards. To achieve this, it is planned to base the next version of the standards on a "business process model." The exact form of this model is yet to be defined.

- The standards, especially ISO 9001, should be harmonized with the ISO 14000 standards to facilitate use by those who choose to implement an integrated management system to fulfill the respective requirements.

Key Points

Here are some key points to consider:

- The growing number of derivative standards based on ISO 9001:1994 will impede change.

- Even simple revisions to ISO 9001 take a relatively long time because of the extensive investment in the current issue.

- The very ambitious objectives established by TC 176 for the next revision of the core standards can be expected to extend the time required to achieve consensus on any changes.

- ISO 9001 is already a *de facto* single quality assurance standard.

MARKETPLACE FORCES AT WORK TODAY

Seven significant forces at work in the marketplace today will impact the nature and pace of changes in the existing ISO 9000 standards through the early years of the 21st century. All the following forces impose significant restraints to varying degrees on changing ISO 9001. These forces are:

- *The desire for stability by the large and rapidly growing base of registered companies.* Companies already registered generally do not want to spend more money to make major changes to quality systems that, by and large, work reasonably well. Given that a wholesale change in ISO 9001 would create a new market for training and consulting services, industrial users would be suspicious of the motivations for significant changes if there was not a corresponding significant benefit to the existing users.

- *A growing number of industry-specific derivative standards explicitly based on ISO 9001:1994* with corresponding third-party registration or regulatory approval. The greater the number of these referenced uses of ISO 9001, the more difficult it will be to make significant changes of the type envisioned by TC 176 to ISO 9001. However, many in the U.S. auto industry would like the next revision of ISO 9001 to look very much like the existing QS 9000.

- *The vested interest of the global registration industry in maintaining and expanding the use of third-party registration.* Registrars and consultants generally do not want to have to learn a new or different approach as long as there are no signs of saturation in their client base.

- *The publication of the ISO 14000 standards raises the potential for integrated management systems* that can effectively respond to the needs for quality assurance, environmental management, and possibly other areas of management accountability. Some existing users of ISO 9001 and ISO 9002 report that this option is important to them. First achieving compatibility, then consistency, and finally harmonization of the ISO 9000 and ISO 14000 standards will constrain the types of fundamental changes possible in both sets of standards.

- *The demonstrated feasibility of using ISO 9001:1994 by companies providing services and developing software* when marketplace pressures require it. While there are still complaints that ISO 9001 requires interpretation for use in services and software development, currently there is not a clamor for significant revisions to better accommodate the needs of users in these product categories. In fact, standards developed by Subcommittee 7 of Joint Technical Committee 1 explicitly for software development could interface with the existing ISO 9001.

- *Each of the ambitious objectives established for the next revision pose major technical challenges.* Satisfactory responses to these challenges have yet to be developed, and they must then be supported by international consensus. This could take considerable time, notwithstanding the nominal schedules for standards development contained in the ISO/IEC Directives. While the kind of responses proposed to date would not significantly alter the intent of ISO 9001, they involve significant changes to the format of the standard. More subtle changes proposed in the past have triggered lengthy discussion and debate.

- *No new uses comparable in scope to third-party registration are foreseen that would drive rapid or significant changes in the content, coverage, or approach of the ISO 9000 standards.*

LIKELY IMPACT OF MARKETPLACE FORCES ON THE ISO 9000 STANDARDS

The marketplace forces described in the previous section will likely shape the use and revisions of the ISO 9000 standards as follows:

- The use of ISO 9001 and ISO 9002 will almost certainly grow through third-party registration. Other growth areas may include second-party relationships, which should expand as ISO 9001 replaces existing government standards. Other industries can be expected to make use of ISO 9001, whether or not third-party registration is involved. While government agencies and regulatory bodies are not likely to accept third-party registration, their endorsement of use of ISO 9001 by suppliers may grow.

- While various target dates from 1999 to 2001 have been proposed for the next issues of ISO 9001 and ISO 9004, all of the marketplace forces discussed previously will tend to delay formal publication of any revisions. However, the formal ISO/IEC process for gaining consensus on proposed standards and revisions will provide ample opportunity for comments and inputs on Committee Drafts (CDs) and Draft International Standards (DISs) well in advance of formal publication.

- Plans are to revise ISO 9001 and ISO 9004 as a coordinated pair to be published simultaneously. Many factors in draft preparation, testing applications, and gaining consensus will determine if this can be accomplished. Market pressures for faster revision of either ISO 9001 or ISO 9004 might develop and challenge the goal of simultaneous publication.

- Revision of ISO 10011, Guidelines for auditing quality systems, is targeted for 1997 or 1998. The revisions to the standard will undoubtedly adopt a number of features from the corresponding auditing standards in the ISO 14000 family. This harmonization will facilitate eventual harmonization of the systems standards in the two families.

- The proposed new ISO 9000, Concepts and terminology, incorporating the scope of existing ISO 8402, Vocabulary, and the concepts portion of ISO 9000-1, Guidelines for selection and use, is targeted for publication at the same time as the revised ISO 9001 and ISO 9004.

- ISO 9000-3, Guidelines for the application of ISO 9001 to the development, supply and maintenance of software, is targeted for reissuance as a guidance standard by 1997, following initial publication in 1991. Important changes include alignment of the clause structure with ISO 9001 and emphasis that the standard is guidance only.

- As part of the effort to minimize proliferation of standards, a number of the existing standards in the ISO 9000 family are planned to be combined into the reissued ISO 9001 and ISO 9004 standards. Some current development efforts may be released as technical reports rather than standards.

SUMMARY

As use of ISO 9001 grows, the desire to maintain the status quo and not change the content or format of this pivotal standard can be expected to be a significant factor. This interest may be balanced by the desirability of modifying ISO 9001 to be harmonized with ISO 9004 as it is revised to become a true foundation for TQM.

The impact of the marketplace will likely have a major effect on the decisions and timing of modifications to all of the standards in the ISO 9000 family.

Endnotes

[1] Data supplied by the DISC TickIT Office in London, United Kingdom as of May 1, 1996.

[2] From the *ISO 9000 Survey: 1996*, published by Irwin Professional Publishing.

ISO 14000—The International Environmental Management Standard

12

This chapter includes four articles describing ISO 14000 and its application and relationship to existing and developing international standards, including ISO 9000, Responsible Care® and other industry codes of practice. The chapter is intended to give the reader a solid understanding of ISO 14000 and related issues. Specifically, the chapter includes:

- An introductory article written by Glenn K. Nestel, of Roy. F. Weston, Inc., one of the worlds largest environmental consulting firms. Mr. Nestel, along with a team of authors from WESTON's Management Systems Division, introduces the ISO 14000 series standard and provides valuable information concerning the development and direction of the standard.

- An article by Suzan L. Jackson, of Excel Partnership, Inc., describing the integration of ISO 9000 and ISO 14000.

- An article by consultant Jack B. McVaugh, of Environmental Technology & Management, describing how to integrate ISO 14001 and the Chemical Manufacturers Association's Responsible Care program.

- Finally Marc E. Gold, Esq., of Manko, Gold & Katcher, an environmental law firm, explores the legal ramifications of the ISO 14001 standard.

AN INTRODUCTION TO ISO 14000

by Glenn Nestel

Private industry and regulatory agencies have sought for years to develop voluntary, consensus-based environmental management standards and codes of practice to solve concerns ranging from ozone depletion to deforestation and water pollution.

While some of these programs have been successful at rationalizing burdensome and often redundant country- and region-specific environmental compliance requirements, their overall effect is still not adequate to allow companies to easily conduct business on a global scale. ISO 14000 standards may provide a rallying point that will allow the international business community to get beyond technical, political, and geographical barriers and freely participate in the global marketplace.

To be fair, ISO 14000, at least in its early stages, will not produce this "level playing field" by providing companies with internationally accepted targets for effluents release levels and other technical requirements. As currently drafted, the standards exlude specific effluent and product standards. The ultimate promise of ISO 14000 is that wide acceptance of the standard will allow companies to prove a high level of environmental stewardship through the establishment of a consistent, documented environmental management system. This environmental management framework drives organizational development that makes environmental considerations part of every employees' job. Market forces are likely to drive the specific standard of excellence in specific-industry sectors.

REASONS BEHIND DEVELOPMENT

Creating a better internal management system, while a useful and worthy goal, is only part of industry's interest in ISO 14000.

Companies spend an estimated $500 billion globally on environmental protection. In the U.S. market alone, annual environment-related spending is about $250 billion. Experts predict at the current rate of increase that environmental spending will approach 3 percent of the U.S. Gross National Product by the year 2000.

This increasing cost of environmental compliance has particularly affected the process and natural resource industries. The Chemical Manufacturers' Association (CMA) estimates that environment compliance-related requirements, in addition to remediation spending, now average 15 percent of capital and 5–10 percent of operating expenses. Environmental compliance costs are similar for petroleum, petrochemicals, paper products, metals, glass, and other primary raw material industries.

Business Management Drivers

In addition to bottom-line concerns, companies are changing long-held environmental management axioms. Environmental compliance and remedial site clean-up is no longer viewed as just the "cost of doing business." Managing these costs has become an important part of a comprehensive, overall business strategy.

Efforts to minimize waste and redesign processes, products, and packaging to prevent pollution are all part of this new philosophy. Resource productivity, the measure of how efficient a company uses its capital, labor, energy, and raw materials resources, is an important competitive differentiation.

Companies whose products and processes make more efficient use of raw materials and energy produce less pollution, consequently resulting in pollution prevention.

Pollution prevention through substitution and process redesign, not end-of-the-pipe waste treatment, is now considered a more acceptable environmental management solution.

Companies that want to compete in the global marketplace must now integrate activities like planning, program implementation, and measurement into the same overall process used to manage all business activities. A growing number of companies realize that environmental activities cannot remain a staff responsibility, a lesson only recently learned by quality management professionals. Environmental protection, like quality system management, must be integrated into daily business operations.

Marketplace Advantages

Companies are also discovering that environmentally friendly products have greater marketplace appeal and help customers achieve their own environmental objectives.

For example, more than 90 percent of the total environmental burden associated with the use of cars, refrigerators, and washing machines is directly traceable to consumer use. Recyclable product components and other nonpolluting manufacturing systems that address the legitimate environmental concerns expressed by customers will have higher resource productivity and be competitively positioned as tomorrow's market leaders.

Stakeholders, including residents living near manufacturing facilities, customers and the investment community, are also more educated on environmental issues. Environmental policy can be influenced to a significant degree by this group of stakeholders. Implementing environmental management systems that protect the environment is an appropriate and ultimately cost-saving response.

Even McDonald's restaurant learned this lesson during the 1980s after it was criticized for the quantities of packaging used in its restaurants. McDonald's agreed to implement aggressive efforts to redesign its packaging as part of a legal settlement with the Environmental

Defense Fund. The fast food giant reduced by over 10,000,000 pounds per year the annual volume of solid waste it generated. The company identified another 35–40 million pounds of additional waste reduction opportunities. Not only did this campaign help create a positive environmental image for McDonald's, it also reduced the direct business costs.

ISO 14000, PART OF AN INTERNATIONAL TREND

Approximately 45 percent of governance requirements used by multinational companies are based on international standards. Europe has taken a leadership position in the development of environmental management standards. A voluntary environmental management standard, designated as BS 7750, was first published by the British Standards Institute (BSI) in 1991. Other industry groups have also responded to stakeholder pressures to improve environmental performance by developing industry codes of performance.

Even the international business and political community has adopted a code of responsible environmental management principles. In 1991, a United Nations–sponsored sustainable development conference called for internationally recognized environmental practices. The conference published a document known as the International Chamber of Commerce Sustainable Development Charter (ICC Sustainable Development Charter).

These and other environmental standards share many common elements and requirements. However, ISO 14000 may prove to be the only standard able to provide a flexible, internationally recognized framework for environmental management that will address the issue of costly and duplicative regulatory and customer requirements.

HOW ISO 14000 WAS DEVELOPED

In 1991, a panel of experts from ISO member countries formed the Strategic Advisory Group on the Environment (SAGE) to analyze global environmental trends. SAGE's task was to determine whether a set of common environmental management standards could serve to:

- Promote a common approach to environmental management similar to quality management.
- Enhance an company's abilities and measure its improvements in environmental performance.
- Facilitate trade and remove trade barriers.

SAGE's work was prompted in part by the success of the ISO 9000 quality management standards and by the national and international trends summarized in the proceding sections.

In the fall of 1992, SAGE recommended the development of a international set of environmental standards. A new ISO Technical Committee (TC 207) was formed in 1993 to develop these standards, now known as ISO 14000.

At the same time ISO was developing the standard that has become ISO 14000, the European Union (EU) had stated its desire to develop a single environmental management and auditing scheme for use among its member states. The EU scheme, known as the Eco-Management and Audit Scheme/Regulation (EMAS/EMAR) is still under development, but ISO 14000 is expected to be an acceptable standard to meet the EMS requirements of EMAS. ISO 14000 standard developers also hope that a single, internationally accepted standard will eliminate the need for country-specific environmental management standards such as the British Standard Institute's 7750 (BS-7750) and EMAS.

The ISO 14000 series of standards falls into two broad categories (see Figure 12–1). Three of the standards focus on a company's environmental and business management system. A second series of standards, life cycle assessment and environmental labeling, focus on the product development process.

STRUCTURE OF THE STANDARD

The ISO 14000 series, in a broad sense, is defined as a systematic approach to meeting environmental obligations. The standards are designed to assist a company manage and evaluate the environmental effectiveness of its activities, operations, products and services. The ISO 14000 series includes six main topical areas, described next.

Figure 12–1

Categories of ISO 14000 Development

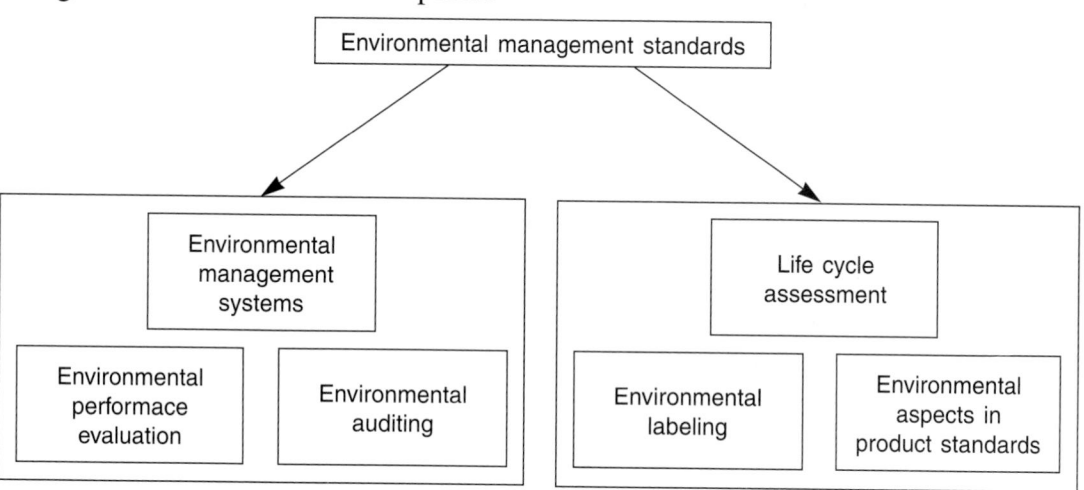

Table 12–1

Standard	Title	Scope
ISO14001	Environmental Management Systems Specification	Formulation of environmental policy and objectives based on legislative/regulatory requirements and environmental aspects and impacts
ISO 14004	General Guidelines on Principles, Systems, and Supporting Techniques management systems and principles	A guide for the development and implementation of environmental

Environmental Management Systems

The environmental management framework should identify resources for developing, implementing, measuring, and periodically reviewing a company's progress toward achieving the principles set forth in its mission statement and environmental policy. This framework should includes a company's:

- Organizational structure.
- Planning activities.
- Defined responsibilities, and.
- Standard practices, procedures, and processes.

See Table 12–1.

Environmental Auditing

Environmental auditing standards provide:

- General principles of environmental auditing.
- Guidelines for auditing environmental management systems.
- Qualification criteria for environmental auditors.

Audits are an essential element of an effective environmental management system. Qualified individuals should assess:

- Conformance with company and regulatory requirements.
- Progress toward achieving pollution prevention targets and objectives.

See Table 12–2.

Table 12–2

Standard	Title	Scope
ISO 14010	Guidelines for Environmental Auditing—General Guidelines on Principles, Systems, and Supporting Techniques	Provides general principles for all types of environmental auditing
ISO14011/1	Guidelines for Environmental Auditing Procedures—Auditing for Environmental Management Systems	Establishes audit procedures for planning and performing an environmental audit
ISO 14012	Guidelines for Environmental Auditing—Qualification Criteria for Environmental Auditors	Provides qualification criteria guidance for environmental auditors

Environmental Labeling

Environmental labeling provides guidance for three types of labels:

- Seal of approval.

- Single-claim labels.

- The environmental report card.

The standard seeks to harmonize labeling criteria and includes guidelines on environmental claims and label content. See Table 12–3.

Environmental Performance Evaluation

Environmental Performance Evaluation (EPE) seeks to measure, analyze, assess, and describe an organization's environmental performance. EPE is used primarily to assess progress toward meeting environmental objectives and targets, especially those associated with preventing pollution. EPE is an essential continuous improvement tool in ISO14001, a concept borrowed from TQM and other quality/continuous improvement standards. See Table 12–4.

Life Cycle Assessment

Life Cycle Assessment (LCA) is a tool for evaluating the environmental attributes, burdens, and impacts associated with a product, process, or service. LCA measures the environmental impact of a product from raw material extraction to final disposal. It includes the impact of manufacturing, distribution and transportation, product use, and recycling. LCA is used in the development of eco-Labeling criteria and in the development of environmentally superior products and services. See Table 12–5.

Table 12–3

Standard	Title	Scope
ISO14020	Goal and Principles of All Environmental Labeling	Provides guidance used in the label drafting process
ISO14021	Terms and Definitions for Self-Declaration Environmental Claims	Provides guidelines for environmental claims on goods and services
ISO14024	Environmental Labeling—Guiding Principles, Practices, and Criteria for Multiple Criteria-Based Practitioner Programs	Provides evaluation criteria for awarding labels on products and services

Table 12–4

Standard	Title	Scope
ISO 14031	Evaluation of the Environmental Performance of the Management Systems and Its Relationship to the Environment	Defines environmental management systems performance and provides implementation guidance
ISO 1403x	Evaluation of the Environmental Performance of the Operational Systems and Its Relationship to the Environment	Defines environmental operational performance aspects and provides implementation guidance

Table 12–5

Standard	Title	Scope
ISO 14040	Life Cycle Assessment—Principles and Guidelines	Provides a systematic set of procedures for compiling and examining environmental effects of a product or service, including materials and energy use
ISO 14041	Life Cycle Assessment—Goal and Definition/Scope and Inventory Analysis	Specifies guidelines and requirements to formulate life cycle assessment and inventory analysis
ISO 14042	Life Cycle Assessment—Impact Assessment	Proposes possible categories for an LCA impact assessment
ISO 14043	Life Cycle Assessment—Improvement Assessment	The basic content of this standard has still not been defined

Table 12–6

Standard	Title	Scope
ISO 14060	Guide for the Inclusion of Environmental Aspects in Product Standards	Establishes general product development criteria that will reduce environmental effects

Environmental Aspects in Product Standards

Environmental Aspects in Products and Standards is intended to raise awareness among standard writers and designers that product designs can affect the environment. The standard encourages companies to use recognized scientific methodologies that incorporate a products' life cycle in the design-making process. The standard also incorporates many current "design for the environment" methodologies. See Table 12–6.

STRUCTURE OF THE ISO 14001 STANDARD

The ISO 14001 environmental management standard is the management framework for systematically meeting environmental obligations (see Figure 12–2). All of the other standards relate to ISO 14001, including self-imposed company requirements. The standard also fits well with existing environmental management programs such as the CMA Responsible Care Principles and Codes of Practice. The ISO 14001 EMS standard is expected to be published as an international standard by the end of 1996.

ISO 14001 is a management framework for planning, developing, and implementing environmental strategies and related programs in an organization. This framework includes:

- A policy that states a commitment to a specified level of environmental performance.

- A planning process and strategy to meet this stated performance commitment.

- An organizational structure to execute the strategy.

- Specific objectives and targets.

- Specific implementation programs and related support tools to assist in meeting these stated objectives.

- Communications and training programs to execute the policy commitment.

- Measurement and review processes to monitor progress.

ISO 14001 can be used by any organization (regardless of size or business type) to develop and implement a formalized management process to improve environmental per-

Figure 12–2

How Standards Plug into an ISO 14001 Environmental
Management System

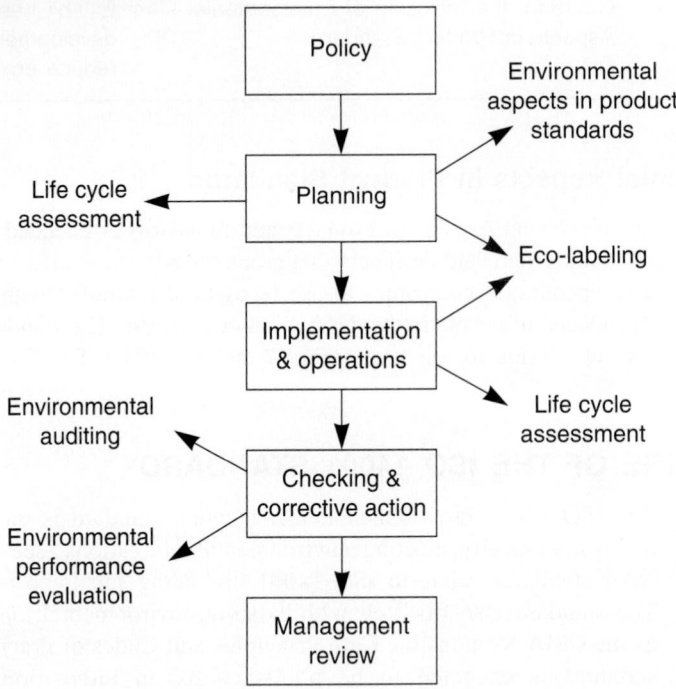

formance. ISO 14001 draws its core elements from proven management systems like the
ISO 9000 quality management series. It uses management concepts such as management
by objectives, organizational development models, and continuous improvement to measure, review, perform root-cause analysis, and take corrective action. A more detailed
comparison between ISO 14001 and the ISO 9000 series appears later in this article.

Elements of ISO 14001

Management Commitment and Environmental Policy

The most senior level of the organization must:

- Commit to a specific level of environmental performance appropriate to its activities, products, and services.

- Issue a policy statement that includes a commitment to pollution prevention and
  continuous improvement.

- Set objectives and targets within its policy statement that commit it to meeting
  environmental legislative and regulatory requirements.

Planning

The organization must establish and maintain processes to:

- Identify, evaluate, and manage environmental aspects and impacts of a company's operations, products, and services.

- Identify legal and other internally imposed requirements. Organizations usually fulfill this requirement when considering environmental aspects and objectives.

- Identify environmental performance objectives and targets for each relevant function within a company as stated in the environmental policy. Planning must also include a commitment to pollution prevention. When establishing environmental objectives, the organization should consider significant environmental aspects, relevant legal requirements, technological and financial aspects, business requirements, and stakeholders' views.

- Identify environmental management plans and programs to achieve a company's stated objectives and targets.

Implementation Operations

To effectively implement an environmental management program, an organization must:

- Define, document, and communicate the organizational structure as well as the roles, responsibilities, and authority of all participants. Adequate human, technological, and financial resources must also be committed to implementation. An appointed senior management representative must be assigned to oversee and review implementation progress.

- Ensure that all employees have the proper awareness and skills training to successfully execute their assigned roles in the implementation process.

- Establish a system to communicate internally and externally the requirements and expectations imposed by the environmental management system. The system should include a mechanism to receive and act on communications from outside parties.

- Document and record performance expectations and operating procedures as well as actual performance data. This information, along with legally required documentation, must be available for inspection.

- Establish and maintain a document control and archival system that meets both internal and legal requirements.

- Establish operation controls that identify, plan, implement, and maintain environmental requirements and procedures. The plan should integrate these activities into day-to-day business operations and expectations that are consistent with company environmental policies and objectives.

- Establish an adequate emergency preparedness and response program that is periodically tested for effectiveness.

Checking and Corrective Action

Continuous improvement process requirements in ISO14001 are met by measuring and evaluating implementation performance and effectiveness. The process is key to ensuring that a company is performing in accordance with stated policies and objectives. The key steps in this process are:

- Monitoring and measuring the effectiveness of environmental management activities.
- Correcting and preventing areas of nonconformance.
- Maintaining training, auditing, and review records.
- Performing environmental management system audits.

Management Review

Senior management must review periodically the environmental management system to ensure its adequacy and effectiveness. Any nonconformance must be corrected and preventive action taken.

The model for connecting the elements of ISO 14001 together into an organized framework is shown in Figure 12–3.

A successful ISO 14001 management framework links policy requirements and objectives in a company. If the policy states that an organization is committed to reducing its environmental impact then environmental performance targets, implementation programs, and measurement and monitoring must be in place to meet that stated goal.

DEVELOPING AN EMS SYSTEM

Initial Review

Document your current environmental management implementation processes relative to the elements of ISO 14001. The key questions are:

- Does the organization have an overall environmental management system in place, and how developed is that system?
- Is top management committed to compliance, pollution prevention, and continuous improvement?
- Are processes in place to track, interpret, and apply environmental statutory and regulatory requirements that apply to the organization?
- Are the management programs and implementation tools needed to achieve compliance and other environmental performance goals in place and operating effectively?

Figure 12–3

Evironmental Management System Framework

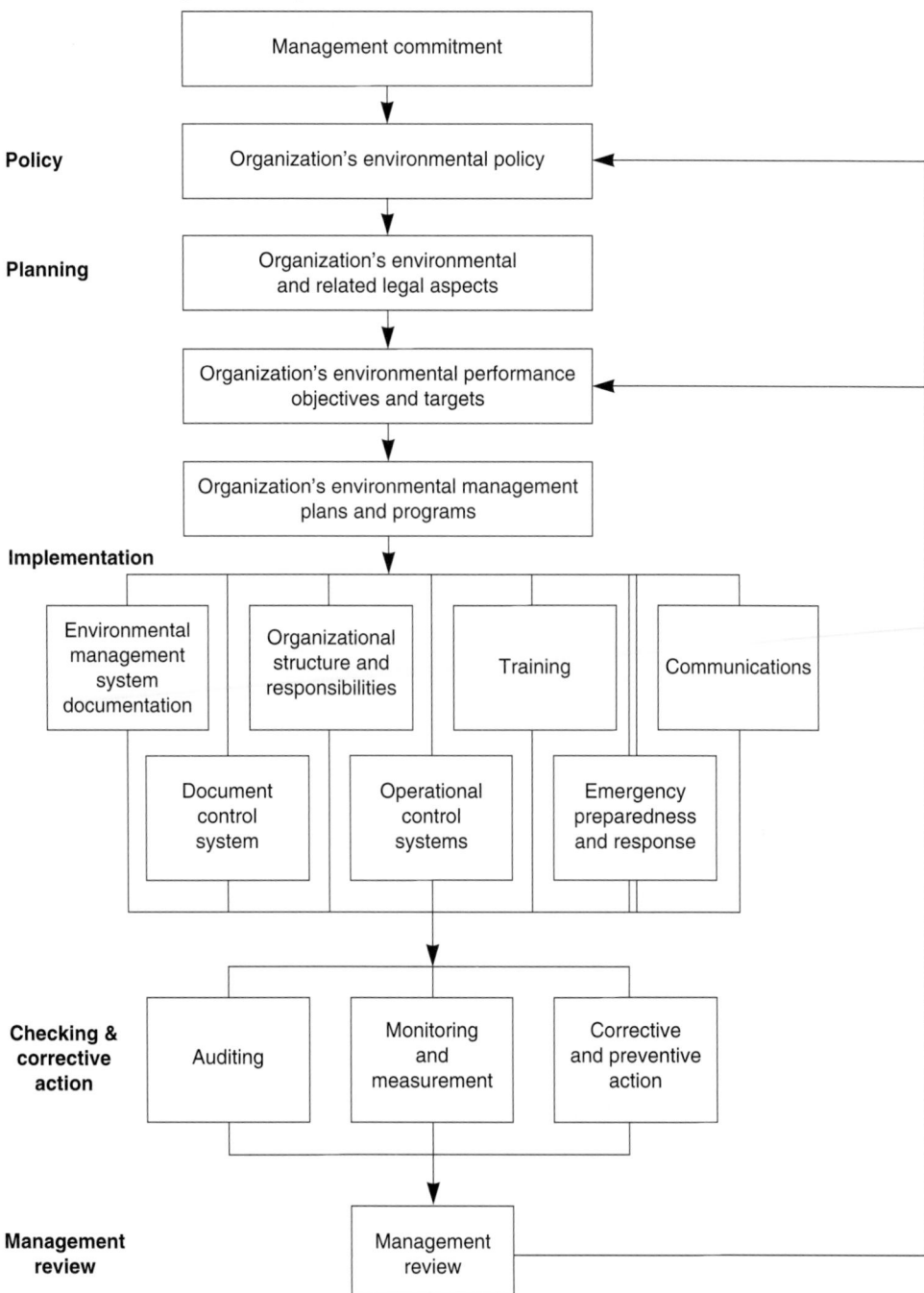

- Are organizational roles, responsibilities, and authority clearly defined and communicated throughout the organization?

- Are the necessary training and education programs in place, and how effective are they?

- Are the necessary environmental management documentation and operational controls in place to assure that policy commitments are carried out in day-to-day operations?

- Does the organization have an established environmental audit program and methods of measuring progress against improvement objectives and targets?

FORMULATION STRATEGY

How proactively an organization decides to pursue ISO 14000 depends on its business goals, expected rewards, and competitive considerations. A company's environmental strategy follows from its business strategy. Business and competitive considerations will also guide whether or not a company seeks formal registration of its environmental management system by an accredited registrar.

Here are some key questions to ask when developing and implementing an environmental management system:

Business Issues

- Is upper level management committed to the effort?

- Are available financial and technological resources consistent with your environmental strategy?

- Is your goal to achieve performance improvement or just to get registered?

- Should you consider a pilot implementation project to demonstrate the potential benefits to your organization and gain commitment to the effort?

Implementation and Certification Issues

- Do you need third-party registration?

- What documentation would a third-party registrar need to understand your system?

- Is the cost of achieving third-party certification worth the potential benefit in the marketplace?

- Are the legal risks of implementing an environmental management system offset by the benefits to your organization?

- Are you able to develop accurate environmental aspects and impact analyses of your organization, including legislative and regulatory requirements?

- Are available human, technological, and financial resources adequate to meet the requirements of your system?

Answering these questions will allow an organization to a make a sensible business decision. Here are general steps to use to develop and implement an environmental management system based on ISO 14000:

- Get senior management commitment.

- Conduct an initial audit of your organization's current environmental management as measured against ISO 14001 requirements.

- Develop an implementation plan that includes improvement objectives, and an implementation schedule and identifies necessary resources.

- Develop and/or update your corporate environmental policy.

- Identify and quantify the environmental aspects and impacts associated with your company's activities, products, and services.

- Identify and track legal and other environmental requirements associated with your company's activities, products, and services.

- Develop objectives that reduce the environmental impact and meet legal and other environmental regulatory and policy commitment.

- Develop environmental management and implementation plans, programs, and tools.

- Set up a program to perform ongoing implementation, monitoring, management reviews, and continuous improvement activities.

The Relationship to Other Environmental Management Systems Standards and Codes

ISO 14001, National Sanitation Foundation (NSF) 110, British Standards Institute's (BSI) 7750, Eco-Management and Audit Scheme (EMAS), the International Chamber of Commerce (ICC) Sustainable Development Principles, and the Global Environmental Management Initiative (GEMI) all emphasize implementing an effective environmental management system.

ISO 14001 is unique in its ability to provide a flexible framework that can incorporate major elements covered by NSF 110, BS 7750, EMAS, and GEMI and provides for an internationally recognized third-party registration. Table 12–7 illustrates the elements covered by each of these standards/principles.

Table 12–7

Summary of Different EMS Systems

Environmental Management System (EMS) Elements	ISO14001	BS 7750	EMAS	GEMI	NSF
Top management commitment	X	X	X	X	X
Initial review/assessment	X			X	
Organization mission/strategy					
Environmental policy	X	X	X	X	X
Organizational roles/ responsibilities	X	X	X	X	X
Environmental aspects/impacts	X	X	X	X	X
Objectives and targets	X	X	X	X	X
ID and track requirements	X	X	X		
Compliance requirements	X	X	X	X	X
Compliance guidance		X	X	X	
Compliance programs				X	
Environmental management programs	X	X	X	X	X
Implementation programs and tools	X	X	X	X	X
Training and communication	X	X	X	X	X
Environmental records and documentation	X	X	X	X	X
Environmental performance measurement	X	X	X	X	X
Financial performance measurement					
Environmental auditing	X	X	X	X	X
Management review	X	X	X	X	X
Continuous improvement process	X	X	X	X	X
Emergency preparedness and response	X			X	X
Third-party verification	X	X	X		

The Relationship to Responsible Care® and API STEP®

ISO 14001 principles are similar to such industry standards as the Chemical Manufacturing Association's (CMA) Responsible Care and American Petroleum Institute's STEP program. It is fully compatible with these standards. CMA and API standards consist of a series of explicit management principles and/or codes of practice. (See Jack McVaugh's article later in this chapter for a more detailed explanation of this issue.)

ISO 14001 is a broad framework for managing defined environmental commitments, objectives, and targets. ISO 14001, like other industry codes, requires a company to identify all environmental aspects and impacts of its operations, products, and services.

Unlike other industry codes, ISO 14001 requires a company to consider business-related risks in addition to environmental risks and legal requirements. ISO 14001 also requires formal management commitment, regular audits, and management reviews to measure performance and stimulate continuous improvement. Companies that have implemented any of these standards or codes are in a good position in begin meeting the requirements of an IS0 14001 environmental management system.

ISO 14000's Relationship to the ISO 9000 Standards

The ISO 14000 standards draw some of their core elements from the ISO 9001/9002 Quality Management Standards. These common core elements are:

- Continuous improvement management process.
- Resources and programs for effective implementation.
- Performance measurement.
- Periodic review.
- Root-cause analysis.
- Corrective action.

Section-by-section comparisons are difficult and can be misleading. Table 12–8 provides a side-by-side comparison of ISO 9001 and ISO 14001. (See Suzan Jackson's article on this subject later in this chapter.) ISO 9001 requirements are far more extensive and prescriptive. ISO 14001 is simpler and more flexible; however, key differences exists between the two standards. ISO 14001 is driven by a much larger group of formal stakeholders outside the traditional supplier/customer relationships that drives quality. Country, state, and region-specific environmental statutory and regulatory requirements must be tracked, interpreted, and implemented.

Other unique ISO 14001 requirements include:

- The initial review requirement.

Table 12–8

Side-by-Side comparison of ISO 14001 and ISO 9001

1.0	Scope	1.0	Scope
2.0	References	2.0	References
3.0	Definitions	3.0	Definitions
4.0	Environmental Management System Requirements	4.0	Quality Management System Requirements
4.1	General Requirements	4.1	Management Responsibility
4.2	Environmental Policy	4.1.1	Quality Policy
4.3	Planning	4.1.2	Organization
4.3.1	Environmental Aspects	4.1.3	Management Review
4.3.2	Legal and other Requirements	4.2	Quality System
4.3.3	Objectives and Targets	4.3	Contract Review
4.3.4	Environmental Management Programs	4.4	Design Control
		4.4.1	General
4.4	Implementation and Operation	4.4.2	Design and Development Planning
4.4.1	Structure and Responsibility	4.4.3	Design Input
4.4.2	Training, Awareness, and Competence	4.4.4	Design Output
		4.4.5	Design Verification
4.4.3	Communications	4.4.6	Design Changes
4.4.4	Environmental Management System Documentation	4.5	Document Control
		4.5.1	Document Approval and Issue
4.4.5	Document Control	4.5.2	Document Changes and Modifications
4.4.6	Operational Control		
4.4.7	Emergency Preparedness and Response	4.6	Purchasing
		4.6.1	General
4.5	Checking and Corrective Action	4.6.2	Assessment of Subcontractors
4.5.1	Monitoring and Measurement	4.6.3	Purchasing Data
4.5.2	Nonconformance and Corrective and Preventive Action	4.6.4	Verification of Purchased Product
		4.7	Purchaser Supplied Product
4.5.3	Records	4.8	Product Identification and Traceability
4.5.4	Environmental Management System Audits		
		4.9	Process Control
4.6	Management Review	4.9.1	General
		4.9.2	Special Processes
		4.10	Inspection and Testing
		4.10.1	Receiving Inpection and Testing

Table 12–8 (*concluded*)

	4.10.2 In Process Inspection and Testing
	4.10.3 Final Inspection and Testing
	1.10.4 Inspection and Test Records
	4.11 Inspection, Measuring, and Test Equipment
	4.12 Inspection and Test Status
	4.13 Control of Nonconforming Product
	4.13.1 Nonconformity Review and Disposition
	4.14 Corrective Action
	4.15 Handling, Storage, Packaging, and Delivery
	4.16 Quality Records
	4.17 Internal Quality Audits
	4.18 Training
	4.19 Servicing
	4.20 Statistical Techniques

- Environmental aspects and impacts identification and emergency planning and response.

- A broader group of stakeholders with changing requirements.

Nevertheless, for companies that have already invested in ISO 9001 or 9002 certification, it may still make sense to consider integrating common management elements from ISO 14001 (e.g., period management review and organizational development programs).

HOW TO PARTICIPATE IN THE ISO 14000 PROCESS

Participation in ISO 14000 development is open to any interested individual or organization. In the United States, the American Society for Testing and Materials (ASTM) administers the process of membership application. The cost to join the TC 207 committee is $250. For information on how to join, contact:

American Society for Testing and Materials
100 Barr Harbor Drive
West Conshohocken, PA 19428-2959
610-832-9725 Ext. 9721 (phone)
610-832-9555 (fax)
E-mail: service@local.astm.org
Administrator: Kathie Morgan

INTEGRATING ISO 9001 AND ISO 14001

by Suzan L. Jackson

Integrating ISO 9001 and ISO 14001 as part of a company's overall business management strategy makes sense in a global marketplace with a universal trading language increasingly based on international voluntary standards and codes of practice.

The ISO 9000 series of quality management standards has been an inspiration and model for much of the interest in internationally accepted management system standards. More than 100,000 sites worldwide have achieved ISO 9000 registration since the standards were published in 1987. They are also the most requested of the Geneva-based International Organization for Standardization's (ISO) more than 10,000 product and process standards.

ISO 9000 implementation has brought most of these companies new business opportunities and increased internal efficiencies. ISO 9001 registered companies contemplating ISO 14001 implementation have a new opportunity to squeeze even more value out of their system.

[**Editor's note:** Although references are made to ISO 9001 in the article, the benefits apply equally to ISO 9002.]

Many similarities exist between the two standards. In fact, all management systems, whether focused on quality, safety, or the environment, share certain core elements, including:

- Policy.
- Defined organization and responsibilities.
- Defined and documented standard practices.
- Control of critical operations.
- Document control (ensuring up-to-date documents are available where needed).
- Training.
- Records management.
- Internal audits.
- Corrective action.
- Management review for continual improvement.

ISO 9001 and ISO 14001 include similar although not identical requirements for all of these core elements. (See previous article for comparison of the main clauses in ISO 9001 and ISO 14001 and the degree of overlap between the two standards.)

DIFFERENCES BETWEEN ISO 9001 AND ISO 14001

Although quality and environmental systems are based on the same management system framework, some critical differences do exist. In addition to the core elements listed above, each management system also has some elements unique to its own particular focus. For example, a quality system includes an evaluation of suppliers and a review of customer contracts. An environmental system includes methods of evaluating environmental impacts and systems for responding to emergencies.

In addition to these specific clauses, a key underlying difference exists between a quality and environmental management system. In a quality management system, major system requirements are defined by customers. Although other stakeholder needs may be considered, customers are the only ones directly affected by the quality of the products and services. The goal of a quality system is generally clear—to meet or exceed customer requirements. The system becomes complicated only if customers have different requirements.

Environmental management systems do not offer such simplicity. A company's environmental impacts may still hold an interest for customers, but regulatory bodies, local communities, and environmental activist groups all have complex and often conflicting interests. International treaties, industry-specific programs, and codes of practice may also add additional layers of system requirements and increase stakeholder complexity.

Companies are challenged to meet these and other industry and regulatory requirements daily. Implementing ISO 9001 or ISO 14001 must make business sense. At the very least it must reduce the amount of resources devoted to managing these competing stakeholder requirements.

IMPLEMENTING ISO 14001

ISO 14001 is a model for a comprehensive environmental management system program. Here are some initial implementation guideposts.

1. Conduct a gap analysis to determine which system elements are in place and which need improvement.

2. Determine if the existing environmental and quality systems can be expanded to address key elements of the ISO 14001 standard. For example, many companies significantly improved their document control system after implementing ISO 9001. Ensuring that important environment-related documents are up to date and available may be possible using the existing ISO 9001 system.

3. Determine business priorities for improving existing systems and developing new systems using the results of the initial gap analysis.

Business needs and culture dictate the degree of ISO 9001/ ISO 14001 integration. Some businesses may completely integrate critical quality and environmental systems. Other businesses may utilize the same general control procedures, but carry them out separately for quality and environment. For example, an internal audit system for quality can also be expanded to include internal environmental systems auditing. Some companies may use the same audit procedures but different auditors for environmental and quality. Other companies may use the same (appropriately qualified) auditors for both systems, but still conduct the audits separately. Other companies may integrate auditing systems and conduct joint quality system and environmental management systems audits. Although the paths vary, the results are the same—efficient business management without redundancies.

CRITICAL EMS ELEMENTS

Despite the many similarities between the two systems, complete quality and environmental system integration is not possible due to some unique elements.

In an EMS, the most critical elements found are in the Planning section, which includes: environmental aspects; legal and other requirements; objectives and targets; and an environmental management program. The core framework for an environmental management system develops from this sequence of planning activities.

ISO 14001 requires an organization to identify how its products, services, and activities interact with the environment. The organization must define a method for determining which of its environmental aspects has or might have a significant impact on the environment. The definition of "significant" is left up to each organization. Satisfying this clause requires a thorough process for evaluating all business system elements.

While this process assignment may sound straightforward, determining significant impacts requires careful thought. A chemical plant may have excellent controls over its air emissions, water outflows, and waste management, but product transportation may present a considerable impact. ISO 14001 encourages companies to view environmental management systems in a broad sense, beyond just regulatory requirements.

ISO 14001 also requires a business to identify all other external requirements to which it subscribes, including industry-specific programs or codes of practice, international agreements, and voluntary initiatives. Programs such as Responsible Care for the chemical industry and chemical distributors, the International Chamber of Commerce (ICC) Charter for Sustainable Development, and the Public Environmental Reporting Initiative (PERI) are all part of this category of "other requirements." ISO 14001 requires a company to treat these voluntary initiatives just as it would legal or regulatory requirements.

The next step in setting up an ISO 14001 environmental management system is to identify specific environmental objectives and targets. Goals should be consistent with a company's business, financial, and operational policies and based on the company's environmental policy. In an integrated business system, the environmental objectives and targets fit within overall business goals; they don't describe a separate, stand-alone system.

Finally, the company must have a management program to help it meet each stated objective and target. The program must include defined responsibilities and timing.

For example, a business could have an objective to reduce its energy usage by 10 percent. The program for achieving this target could include detailed plans for improving operating efficiencies in the plant, working with the local utility company to reduce energy use during peak hours, or replacing equipment with newer, more energy efficient units. The end result of this integrated system is a framework for continuously improving the environmental system.

COMPARISON TO ISO 9001

ISO 9001 does not require exactly the same planning elements as detailed in ISO 14001. The standard does require a company to establish a quality policy and objectives and offer a plan to meet its quality requirements and objectives.

Some companies meet this requirement by creating a single, overall business policy that includes elements of both quality and environmental management. Some larger or more complex companies prefer to maintain individual quality and environmental policies. These policies should still be linked to an overall business policy and/or business mission. Integrating the process for setting and developing quality and environmental objectives may fit the needs of other organizations.

The exact format of these systems is not a critical concern. The key is to build a solid business system framework in which quality and environment are effectively managed as integral parts of the overall business philosophy and strategy. Once established, the management systems are focused on the needs of the business rather than on maintaining ISO standards compliance.

INTEGRATING ISO 14001 WITH THE CHEMICAL MANUFACTURERS ASSOCIATION'S RESPONSIBLE CARE®

by Jack McVaugh

A tendency exists among environmental professionals to see every new environmental initiative, voluntary or mandated, as adding just another onerous compliance burden. Admittedly, this is an understandable attitude, considering the continuous layering of regulations over the past 25 years.

The ISO 14001 standard, however, should be viewed as an opportunity to integrate an environmental management system (EMS) into the strategic business process—a goal that many companies seek, but few manage to achieve. Integrating management systems can create immediate benefits. For instance, one chemical plant environmental manager reports that combining various audits from different company functions reduced corporate auditing man hours from 6,000 to 2,000 per year.

ISO 14001 offers many opportunities for integration with existing activities because at least some management system elements are contained in, for example, EPA's Resource Conservation and Recovery Act, Pollution Prevention Act, and stormwater regulations, and OSHA's Process Safety Management regulations. Operating efficiencies can be realized by managing these programs together under one system.

Responsible Care, the chemical industry's voluntary management system initiative, focuses on the health, safety, and environmental aspects of the chemical product life cycle, from design and production through distribution and use. Integrating ISO 14001 and Responsible Care should be a natural partnership.

COMPARING ISO 14001 AND RESPONSIBLE CARE®

Each one of the six Responsible Care Codes of Management Practices describes a process for managing health, safety, and environmental aspects of a section of the chemical business, from manufacture through distribution and safe use. The Pollution Prevention Code, familiar to most environmental professionals in the industry, closely parallels ISO 14001. Fourteen Management Practices are required under this Code. Nine of the fourteen practices show a direct relationship with corresponding requirements in ISO 14001, and these linkages are described below.

Pollution Prevention Code of Management Practice 1

A clear commitment by senior management through policy, communications, and resources to ongoing reductions at each of the company's facilities in releases to the air, water, and land and in the generation of wastes.

- **ISO 14001 4.1, Environmental Policy** requires top management to commit to prevention of pollution, as well as to compliance and continuous improvement. The policy must be communicated and available to the public.

- **ISO 14001 4.3.1, Structure and Responsibility** requires top management to provide adequate resources, including technological, manpower and specialized skills, to operate the EMS and meet objectives and targets.

Pollution Prevention Code of Management Practice 3

Evaluation, sufficient to assist in establishing reduction priorities, of the potential impact of releases on the environment and the health and safety of employees and the public.

Implicit in **ISO 14001 4.2, Planning** is that environmental management priorities will be set using some kind of ranking process based on the severity of actual or potential adverse environmental impacts caused by the business.

Pollution Prevention Code of Management Practice 5

Establishment of priorities, goals, and plans for waste and release reduction, taking into account both community concerns and the potential health, safety, and environmental impacts as determined under Practices 3 and 4.

The ISO 14001 planning process (**ISO 14001 4.2, Planning**) includes provisions for determining the legal and regulatory compliance and other requirements, the environmental aspects of the business and their impacts on the environment, setting organizational objectives and targets that may go beyond regulatory compliance, and establishing action plans to meet these goals.

Moreover, **ISO 14001 4.2.3, Objectives and Targets** calls for these goals to be set taking into account the views of "interested parties," which may include employees and members of the public.

Pollution Prevention Code of Management Practice 6

Ongoing reduction of wastes and releases, giving preference first to source reduction, second to recycle/reuse, and third to treatment.

ISO 14001 requires commitment to pollution prevention in the environmental policy, and clear demonstration that this policy has been implemented. The involvement of top management to provide adequate resources to accomplish its stated objectives and targets ensures the use of the most technologically and financially effective methods and favors the hierarchy of actions stated in Management Practice 6. ISO 14001 also places emphasis on preventative measures instead of reacting to problems after they occur.

Pollution Prevention Code of Management Practice 7

Measurement of progress at each facility in reducing the generation of wastes and in reducing releases to the air, water, and land, by updating the quantitative inventory at least annually.

Measurement of progress in meeting policy commitments, as well as objectives and targets, is essential information for the management review process (ISO 14001 4.5, Management Review). Such information is used by top management to determine which parts of the EMS are working and which need adjustment. The informaton will also be used to set new objectives and targets during the next EMS cycle. Guidance on establishing a process for environmental performance evaluation is given elsewhere in the ISO 14000 Series (see ISO 14031).

Pollution Prevention Code of Management Practice 8

Ongoing dialogue with employees and members of the public regarding waste and release information, progress in achieving reductions, and future plans.

ISO 14001 4.2.3, Objectives and Targets, and 4.5 Management Review regard environmental management as a cyclical process. New objectives and targets may be set each time management reviews the system. This aspect of the standard creates an ongoing dialogue with employees, members of the public, and others because the views of interested parties must be taken into account each time objectives and targets are set.

Pollution Prevention Code of Management Practice 9

Inclusion of waste and release prevention objectives in research and in design of new or modified facilities, processes, and products.

ISO 14001 4.2.4, Environmental Management Programs requires that environmental management principles be applied on projects relating "to new developments and new or modified activities, products or services."

Pollution Prevention Code of Management Practice 11

Periodic evaluation of waste management practices associated with operations and equipment at each member company facility, taking into account community concerns and health, safety, and environmental impacts and implementation of ongoing improvements. Once again, a direct link to ISO 14001 4.5, Management Review.

Pollution Prevention Code of Management Practice 13

Implementation of engineering and operating controls at each member company facility to improve prevention and early detection of releases that may contaminate groundwater.

While this management practice is focused on groundwater contamination, similar provisions, but with a more general focus, are found in ISO 14001 4.3.6, Operational Control. In the Standard, operations and activities associated with all of the identified significant environmental aspects shall be planned to ensure that they are carried out under specified conditions.

CONCLUSION

Strong similarities exist both in structure and purpose between the ISO 14001 EMS and the Responsible Care Pollution Prevention Code. Both require senior management commitment; a planning process, including prioritization and goal setting; implementation and continuous improvement; measurement of progress; periodic evaluation of practices; and control of critical operations. And while both require input from interested parties, Responsible Care puts a great deal more emphasis on this aspect.

ISO 14001 will help companies accomplish some of the Responsible Care management practices that have presented difficulty or prevented the program from achieving its full potential. Responsible Care has not achieved the desired credibility with the public, according to some environmental groups, because no objective verification mechanism proves that companies are truly following its management practices. Although a company may self-declare conformance with the ISO 14001 Standard, independent, third-party certification of conformance is also an option.

Finally, Responsible Care's Management Practice 12 under the Pollution Prevention Code requires "implementation of a process for selecting, retaining, and reviewing contractors and toll manufacturers taking into account sound waste management practices that protect the environment and the health and safety of employees and the public." Often complying with this requirement meant costly and time consuming audits of suppliers, contractors, and toll manufacturers. ISO 14000 will allow many companies to implement this practice more simply and efficiently by making ISO 14001 registration a procurement contract requirement.

The ISO 14000 Series of Environmental Management Standards promises to be the most significant development in the history of environmental management. By breaking down ISO 14001, as well as existing regulatory and industry initiatives, into basic management system elements and looking for points of commonality between these elements, operating efficiency will be maximized.

History of Responsible Care®

In early 1985, the chemical industry was reeling from the news that a catastrophic release of chemicals had killed and injured thousands of people in Bhopal, India. In response to anxious questions from U.S. communities about the possibility of such an accident closer to home, the Chemical Manufacturers Association (CMA) launched the Community Awareness and Emergency Response (CAER) Program. This voluntary initiative called for chemical facilities to provide outreach to their respective communities and provide leadership in the development and/or improvement of local emergency plans.

(Continued on next page)

(Continued from page 503)

The spread of Total Quality initiatives among chemical companies in the mid-1980s, sent the CMA management to one of their "customers," the public, to see how well customer expectations were being met by the CAER program. The response was that the CAER Program was good, as far as it went, but did not address all of the concerns the public had about the chemical industry.

The public was concerned not only with emergency planning, but with wastes produced, the safety of chemical transportation and warehousing in their communities, the safety of manufacturing processes and worker health, and the safety of the products they purchased. When these concerns were tallied, a program which had been developed by the Canadian Chemical Producers Association, called Responsible Care, appeared to meet public and employee expectations and was adopted by CMA.

Recognition and response to community concerns about chemicals and chemical operations are listed first among the Guiding Principles of Responsible Care, and public input is a built-in feature throughout the process. Moreover, each of the six Codes of Management Practices addresses one area of public and employee concern about the chemical business and the life cycle of chemical products.

IMPORTANT LEGAL CONSIDERATIONS IN IMPLEMENTING ISO 14001

by Marc E. Gold, Esq.

Given the worldwide support for the ISO 9000 Quality Standards, many believe that standards for environmental management systems developed by the International Organization of Standardization through a similar process of global consensus will also enjoy the same favorable spotlight. While that result is possible, it is far from certain.

The market drivers, incentives, and potential advantages of ISO 9001 registration (e.g., meeting both customer specifications and objective product quality standards) are somewhat different from the motivators commonly linked to ISO 14001. One element of ISO 9000 that may prevent it from serving as a precursor model for ISO 14001 for industry in the United States is mandatory third-party registration, which serves as the cornerstone of the ISO 9000 Quality Standard and provides an essential measure of integrity to that process. This concern about third-party registration is not unfounded.

ISO 14001 provides for either third-party registration or self-declaration (ISO 14001, Section 1, Scope); in either event, based on periodic management system audits. Despite the two options expressly provided by the terms of ISO 14001, pressure by a variety of stakeholders for third-party registration is substantial and may be essential in order to provide the external program verification necessary to legitimize the ISO 14001 Environmental Management Systems Standard. This may be the result even though the current legal system in the United States does not easily accommodate such a process.

Domestic and multinational companies still may face pressure for registration to ISO 14001. For companies already registered to ISO 9000, the decision to seek ISO 14001 registration will most likely be initially market driven. However, whether reacting to customer demands, regulatory pressure, or the desire to improve the environment more aggressively than required by law, companies should proceed down the road to ISO 14001 registration with caution. The following points are key to making the registration decision:

1. Is the registration decision driven by customer or industry requirements?

2. Is registration likely to be driven in time by government regulatory or legal requirements?

3. Will registration improve the company's image even though seeing as ISO 14001 is not tied to product quality or specific environmental performance?

4. Will ISO 14001 registration be recognized in all markets served by the company?

The decision whether to register to ISO 14001 will not be clear-cut until industry, the regulators, and the public understand the potential value of the ISO 14001 process and its impact on environmental protection, cost efficiency, and legal compliance.

REASONS TO COMPLY WITH ISO 14001

Significant legal benefits can be achieved by developing and implementing an environmental management system like ISO 14001. Some of these benefits include:

1. Establishing a systematic program to achieve and maintain compliance with environmental laws and regulations.

2. Identifying future obligations under environmental laws and regulations.

3. Managing onsite and offsite environmental liabilities.

4. Facilitating compliance with related legal requirements, such as those administered by the Securities and Exchange Commission.

5. Positioning the company for more favorable treatment by regulatory and enforcement authorities.

The U.S. Sentencing Commission's Draft Sentencing Guidelines for Environmental Offenses is a good example of possible favorable treatment for ISO 14001 compliant companies. In the event of a criminal prosecution for violation of environmental laws, the Draft Sentencing Guidelines strongly favor the company with a program in place that is aimed at environmental compliance, awareness, and improvement—not merely minimum compliance but a comprehensive, documented, environmental management system founded on a clear commitment from management and supported by adequate training, follow-up, and disciplinary procedures for noncompliance. Thus, in the event of a criminal prosecution, some measure of protection is provided by a company's demonstrated, systematic, and documented commitment to environmental compliance and continuous improvement.

While ISO 14001 compliance can have a positive impact on a company's operations, several important legal issues are presented by the standard, principally relating to:

1. Collection, management, and dissemination of information about a company's environmental compliance status and its environmental aspects and impacts.

2. Integration of ISO 14001 into the current environmental legal system in the United States.

3. Potential liabilities caused by the mere existence of a global environmental management system standard.

COLLECTING SENSITIVE CORPORATE INFORMATION

Companies implementing ISO 14001 must demonstrate a commitment to compliance with legal requirements and any other voluntary standards to which the company subscribes (ISO 14001, Section 4.2(c)) and to "continual improvement and prevention of pollution" (ISO 14001, Section 4.2(b)). The company is free to identify the most important environmental aspects of its operations and their corresponding environmental impacts.

No specific environmental performance standards are established under ISO 14001. Instead, the company's environmental policy defines measurable objectives to address its environmental impacts. Senior management must participate in establishing the environmental policy, which reflects the company's size and an understanding of its potential environmental impacts (ISO 14001, Section 4.2(a)). The standard requires only that the environmental policy statement be communicated to employees and made available to the public (ISO 14001, Sections 4.2(e) and (f)).

Legal Requirements

The environmental aspects and impacts assessment is the critical first step in implementing the ISO 14001 Environmental Management Systems Standard. It provides the environmental and operational baseline upon which the company's environmental programs are developed and against which continual improvement is measured. Among the information to be gathered is the identification of legal requirements that must be met by the company and an evaluation of the company's current compliance status. How a company chooses to collect, review, distribute, and use this information will drive its need to develop confidentiality mechanisms. United States-based companies should pay particular attention to this requirement given the litigious nature of American society and the ability of regulatory agencies to obtain information.

Auditing Issues

Many companies already implement sophisticated environmental auditing programs to determine a facility's current environmental compliance status and assure compliance with future environmental requirements (e.g., periodic reporting deadlines, permit expiration dates, proposed legislation, and emerging regulatory programs). The scope of an ISO 14001 environmental management system and the initial assessment will extend much further than mere compliance with environmental laws and regulation and will likely include activities relating to best business practices and management systems that are not responsive to specific legal requirements.

Common-law legal privileges, such as the attorney-client privilege, the attorney work product privilege and, in some jurisdictions, the self-evaluative privilege, may be available to protect legal compliance information under specified circumstances, but only when certain procedures are established at the outset and steadfastly followed.

Those legal privileges may be more limited when applied to broader information collected and maintained under ISO 14001 that is unrelated to legal compliance or anticipated litigation. Audit or self-evaluative privilege legislation recently enacted in several states (and currently pending in the United States Congress and in other states) may provide an additional level of protection for information gathered under ISO 14001.

The scope of these laws differs from state to state, but the goal of encouraging a company to periodically evaluate and address environmental issues associated with its operations is a common theme. "The Environmental Protection Agency's Voluntary Environmental Self-Policing and Self-Disclosure Interim Policy Statement," 60 *Federal Register* 16875 (April 3, 1995), recognizes the importance of a process of self-evaluation; however, it rejects the notion of legal privilege.

Disclosure Questions

ISO 14001 avoids one area of potential corporate concern since it does *not* require public disclosure of a company's environmental aspects, impacts, or the details of its environmental management system. Instead, the company is required to make available to the public only the environmental policy document and merely to "consider" a procedure for external communication of "significant environmental aspects" and to document that decision (ISO 14001, Section 4.3).

Companies will respond to this element of ISO 14001 in different ways. The internal, corporate decision-making process may distinguish between external reporting to regulatory agencies or to local emergency response authorities as opposed to the public at large. However, the legal process in the United States may make it difficult to shield sensitive environmental information, and the public perception of a company attempting to do so may render moot the voluntary nature of public disclosure under ISO 14001. Requiring a company to "consider" more complete public disclosure and to document that decision may tilt the balance in favor of more detailed public disclosure, which may have been the underlying intent of the requirement all along.

Identifying Environmental Priorities

Ideally, a company will prioritize its environmental corrective actions to address first those conditions that present the most risk. In the United States, however, given the compartmentalized legal structure in which industry must operate, that flexibility is substantially limited. The operating conditions that may cause the most significant environmental impacts or pose the greatest risks often do not correlate to mandatory legal requirements or areas of potential noncompliance.

Perhaps greater environmental improvement can be achieved cost effectively by controlling the pollution from sources that are not regulated rather than those that are regulated. The relationship between current legal and regulatory requirements and the ability to establish priorities based upon risk is an issue that industry in the United States will have to work through as part of the ISO 14001 process.

ISO 14001 AND CURRENT LEGAL REQUIREMENTS

Integrating ISO 14001 with current environmental laws is one of the most challenging legal issues. Over the past 25 years, the environmental legal structure in the United States has developed on a media-by-media basis into a disjointed web of operational requirements, compliance obligations, and liabilities. In addition to regulatory compliance, the liability scheme set forth in the Comprehensive Environmental Response, Compensation and Liability Act (commonly known as Superfund) of strict, joint, and several liability has become the common denominator for government litigation and third-party claims relating to the remediation of contaminated sites. This comprehensive legal framework targeting future and past environmental liabilities creates an imposing domestic obstacle to the unconditional endorsement of a global environmental management system standard driven by priorities established by corporate policy and coordinated, multimedia risk evaluations.

Legal Incentives

Substantial environmental improvement can be achieved through the implementation of a comprehensive environmental management system, endorsed by senior corporate management and containing measurable goals and objectives. There is a fundamental need, however, to re-examine the manner in which environmental improvement will be accomplished over the next decade and to create sufficient legal incentives (and opportunities) to encourage industry to invest the additional sums necessary to continue the substantial progress that has been made.

The regulatory agencies have responded with a variety of pilot programs and experiments aimed at defining the benefits of comprehensive environmental management systems. For example, some regulatory agencies have attempted to implement facilitywide permitting programs, and several states and the Environmental Protection Agency have begun an active dialogue with industry and environmental groups in an effort to sort out the issues, impediments, and opportunities presented by ISO 14001. It remains to be seen whether and how the regulatory authorities will weave into current law and regulation (or support new legislation) legal mechanisms for industry to evaluate the environmental impacts of their operations and address those impacts on a priority basis, taking into account the risks posed by the conditions discovered.

INDUSTRYWIDE STANDARD OF CARE

A longer-term legal issue raised by the ISO 14001 Environmental Management Systems Standards is the potential for the standards to become the environmental performance benchmark against which all industry will be judged in the context of government enforcement, citizen suits, and third-party damage claims. Through the adoption of ISO 14001, the industry standard of care with respect to environmental performance could be redefined. Such a standard of care would then be used by those claiming injury or seeking damages from environmental exposures to strengthen their claims if industry defendants have not subscribed to the ISO 14001 standards.

Court Actions

Voluntary industry standards have been held by many state courts to be relevant factors in determining whether a company has acted negligently.[1] This result, while perhaps unintended, may already have occurred as trade associations establish core environmental principles intended to provide guidance to their members and serve as a public commitment to environmental protection and improvement. As a legal matter, when a voluntary code or standard is widely adopted, it becomes a bit more mandatory.

CONCLUSION

The movement toward environmental management systems may accelerate the process of achieving greater environmental improvement. Given the heightened awareness of the need for environmental progress through creative means, the time may be right for government, industry, and citizens to work together to forge a cooperative program which will build upon the principles of ISO 14001 to achieve continuous improvement in environmental protection and risk reduction.

Combining the financial incentives, business opportunities, and public sentiment supporting environmental programs worldwide, government authorities in the United States and domestic industry may well find themselves on the doorstep of new opportunities to build upon the environmental protection gains of the first 25 years of the modern environmental movement and to step into the next century with a renewed environmental spirit and a brighter environmental future. ISO 14001 could be a vehicle to fuel this new approach.

Endnote

[1]Consensus standards developed by the American National Standards Institute have been held to be admissible in defining an industry standard of care in at least two state courts. Other voluntary standards have also been held to have probative value in negligence cases.

Comparing ISO 9000, Malcolm Baldrige, and Total Quality Management

13

by Dr. Kenneth S. Stephens

INTRODUCTION

This chapter compares ISO 9000 to the Malcolm Baldrige National Quality Award guidelines and relates these disciplines to Total Quality. It provides an overview of the two quality initiatives, makes some comparisons, and emphasizes the importance of the complementarity of the two approaches to quality systems and the further pursuit of total quality.

With the 1994 revisions of the ISO 9000 Series, including their widespread adoption and implementation, these international/national standards continue to generate much attention. This is deserved, but needs understanding as to what the ISO 9000 Series is and what it is *not*. Such an understanding will contribute significantly to the correct implementation and use of these standards and guidelines on quality systems and management—to meet customer demands for a quality system *and* to establish a Total Quality system that has direct and long-term beneficial results for the enterprise itself. The ISO 9000 Series is not intended to be a standard for Total Quality. It is, however, a uniform, consistent set of system elements and requirements for quality assurance systems and management that can be applied universally within any total quality system. Its widespread adoption and implementation by companies and nations have brought about harmonization on an international scale and are supporting the growing impact of quality as a strategic factor in international, as well as national and local trade.

As detailed elsewhere in this *Handbook* the ISO 9000 Series consists of several compliance standards and guidelines for implementation and use. The most comprehensive compliance standards ISO 9001 (See Chapter 3). That consists of 20 broad-based requirements, that are expanded to 54 subsections that, in turn stipulate at least 135 more detailed more detailed requirements by way of "shall statements."

Particularly relevant to the topic of this chapter is the guideline ISO 9004-1 (see Chapter 3) consisting of 17 major guideline sections that are expanded to 142 subsections that, in turn, stipulate at least 261 recommendations by way of "should" statements.

Additional resources for total quality/management system elements, assessments, and criteria for designing, developing, and implementing such systems are the various quality awards. Among these are the Deming Prize, the Malcolm Baldrige National Quality Award (MBNQA), the European Quality Award, other national quality awards, a growing number of institutional and industry sector awards, and state quality awards. A major goal of the Malcolm Baldrige National Quality Award is to increase U.S. competitiveness worldwide. It recognizes U.S. companies for business excellence and encourages a proliferation of this excellence. "The Award promotes (1) understanding of the requirements for performance excellence and competitiveness improvement; and (2) sharing of information on successful performance strategies and the benefits derived from using these strategies." (As stated in the MBNQA *1996 Award Criteria*, page 1.) It consists of a set of criteria divided into seven major categories. These categories are further subdivided into 24 examination items and 52 areas to address. In assessing the categories and items a point score scale totaling 1,000 points is established. The criteria for the MBNQA address elements of total quality systems that are not included in the requirements of ISO 9000. These are examined subsequently.

Total Quality and its management counterpart, Total Quality Management (TQM), mean different things to different persons and organizations. These are merely handles (buzz words, if you like) for discussion, communication, developments, and applications. Programs with the TQM label have met with great success. Others with the same label have met with failure. Programs of similar content have also gone under many different labels (often because of the NIH factor—or "not invented here" syndrome). Many successful applications of Japanese origin (in particular) continue to use the term "Total Quality Control"—with the translation of "control" from Japanese to English implying "management" or "administration," rather than coercion or other negative connotations often associated with the term in Western culture. Labels such as IQM (integrated quality management), SQM (strategic quality management), TQ, reengineering, and whole system architecture are prevalent. What is significant here is that the nomenclature is not as important as what is being done. It isn't the terms that make the programs, but the activities being carried out, the philosophies, approaches, and attitudes being implemented, and the results being achieved—especially as they pertain to customer and employee satisfaction/retention; continuous improvement and innovation; and quality, productivity, and profitability enhancements.

The concept of Total Quality (or any suitable label), in it broadest aspect, is seen to go beyond the quality system prescribed by the ISO 9000 Series and, for that matter, beyond the criteria of the MBNQA, though the latter is perhaps more consistent with W. Edwards Deming's concerns about the competitive global marketplace. Deming (1986) warned that the western world needs a "transformation of the American style of management" and not merely a "reconstruction" or "revision." He further pointed out that this new way "requires a whole new structure, from foundation upward." Juran (1989) mentions a "quality revolution" to respond to the quality crisis. Of significant impact on the management of total quality are Deming's 14 points for management. We have not yet been successful in incorporating these philosophies into systems and programs such as the ISO 9000 Series and/or quality awards. These systems and programs help to point the way and provide partial mechanisms, but there is always the danger that their implementation may represent a "fix" rather than a "transformation"—merely plugging up holes in the dike rather than designing and building new and innovative dams.

Comparing the ISO 9000 Series requirements and guidelines, the MBNQA criteria, and Deming-based Total Quality Management (TQM) philosophies and proposed practices is a difficult task. To use a simple analogy, the ISO 9000 compliance standards and the ISO 9004-1 guidelines are like four starched, white business shirts—small, medium, large, and extra large—form-fitting but not expected to cover the whole body. MBNQA is like a giant, one-size-fits-all T-shirt with 24 pockets in which specific articles are to be placed. Deming-based TQM is like a whole change of wardrobe from which the user is expected to select appropriate apparel for his or her organization.

OVERVIEW OF THE TWO SYSTEMS

The ISO 9000 Series—Strengths and Limitations

A detailed technical content overview of the ISO 9000 Series is presented in Chapters 3 and 4. The overview presented here will look at the Series with a different perspective.

The ISO 9000 Series is a set of standards (and a complementary set of guidelines which are, themselves, standards in that they are prepared and promulgated by a standards' body). Thus it has both the advantages and disadvantages of standards, in general. As a set of standards it is subject to periodic review and revision. The first cycle of that process has been realized recently with the 1994 revisions—referred to as "phase I." More extensive, "phase II" revisions are already under consideration. See the papers by Tsiakals (1994), and Tsiakals and Cianfrani (1996). The standards need to be understood adequately in order to assure correct and beneficial implementation—together with other elements of a Total Quality System that are not a part of the Series. It is extremely important for enterprise/corporate managers and quality practitioners to understand that the ISO 9000 Series is not intended as a standard on Total Quality. A Total Quality System must go beyond ISO 9000. ISO 9000 is *a* quality system; it may not be *the* quality system best suited for any given enterprise. This is discussed subsequently in greater detail.

On the positive side, its strengths lie in the structure, which sets forth a uniform, consistent set of procedures, elements, and requirements that can be applied universally, albeit within limitations of interpretation and individual implementation. It provides a basis for designing, implementing, evaluating (assessing), specifying, and certifying (registering) a QA system. With widespread adoption (now a reality), it provides a common language for international trade with respect to the QA disciplines. It promotes (requires) a sound, well-documented contractual relationship between customer and supplier and a well-documented quality system by the supplier. Hence, it aims to establish a common understanding between these parties, based on agreed requirements and confidence to the customer that a quality system is in place that will have a positive impact on the continuing quality of supplies now and in the future.

For further understanding, the ISO 9000 Series is generic in two significant aspects. It is not *product or process specific*. It is a set of standards on the quality system, not on products/services or processes. Of course, the quality (QA) system (including the related quality manual—now a requirement) will, of necessity, have to contain specific subsystems related directly to the processes and the products or services to which it is being applied. Any quality system is developed and implemented, not in and of itself, but to assist in producing a product and/or offering a service. In this regard ISO 9001, itself, contains no less than 74 direct uses of the root word, "product(s)." There are another 8 uses of the term "production," and at least 31 uses of the root word "process." But in spite of these references to product and process, the ISO 9000 standards and guidelines do not set standards on product or process. Such standards must be established by the enterprises themselves to meet their internal and external customer requirements and made a part of the quality system under ISO 9000 that includes the documentation subsystem.

ISO 9000 is not even *quality system specific*. That is, it does not specify a fixed system (beyond the general requirements enumerated) for every enterprise (and this is, of course, not a criticism). It provides considerable flexibility to the enterprise to design and specify (document) its own system within the framework of the requirements and then directs attention to evaluating conformance to that system. This is, perhaps, a good place to dispel the common myth that the ISO 9000 standards only require that "You say what you do, and do what you say." As mentioned above, the standards, in particular ISO 9001, sets forth 20 broad-based requirements that are translated into at least 135 separate requirements via "shall" statements. Yes, you are required to say what you do (i.e., document) and do what you say (i.e., comply), but what you say and do had better meet the specific and implied requirements of the standards. You are not at complete liberty to say and do anything you want—if you wish to be found in compliance with ISO 9000. The reader may want to consult the paper by Marquardt (1995) for a further discussion of this point. ISO 9000 places considerable attention on documentation, which is a requirement in every single one of the 20 broad-based requirements, including document control (with the root word "document" occurring some 69 times throughout ISO 9001, and with only 16 of those occurrences in section 4.6 on document control). There are at least 10 occurrences of the root word "effective" in reference to the performance of the quality system designed and implemented.

ISO 9004-1-1994: Quality Management and Quality System Elements—Guidelines.

This standard (in the form of a guideline) plays a very special role with respect to the development and assessment of a quality system, especially in comparison with the Malcolm Baldrige National Quality Award. They both have in common an orientation toward self-development and/or self-assessment. ISO 9004-1 consists of a set of more than 90 quality system elements that should be considered when designing and implementing (and subsequently assessing) a quality system. It provides additional details for each of the broader categories of 18 quality system requirements referenced in ISO 9001 (2 of the 20 broad-based requirements are omitted, as discussed subsequently) and the other system (model) standards. A manufacturer needs to understand an operation in sufficient detail so that only the appropriate elements are selected for each step of the operation. The objective is to minimize the cost of the quality system while maximizing the benefits. ISO 9004-1 is intended to serve as a guideline for this task.

This standard is part of the noncontractual or guideline portion of the ISO 9000 Series. It is the largest document of the series, with 20 pages of text. (ISO 9001, with the most comprehensive set of 20 requirements, has only nine pages of substantive text). The purpose of ISO 9004-1 is presented well in ISO 9000-1 as,

> . . . reference should be made to ANSI/ISO/ASQC Q 9004-1-1994 by any organization intending to develop and implement a quality system. In order to meet its objectives, the organization should ensure that the technical, administrative, and human factors affecting the quality of its products will be under control, whether hardware, software, processed materials, or services. ANSI/ISO/ASQC Q 9004-1-1994 describes an extensive list of quality-system elements pertinent to all phases and activities in the life cycle of a product to assist an organization to select and apply elements appropriate to its needs.

ISO 9004-1 was based, heavily, on ANSI/ASQC Z1.15 (1979). *Generic Guidelines for Quality Systems.* Some of the criticisms leveled at the contractual quality assurance models (ISO 9001, 9002, and 9003), primarily in the form of omissions, are included in ISO 9004-1 as elements of quality systems. These include, among others, more on quality plans, quality improvement, quality costs, marketing quality, more on process quality, personnel and motivation, and product safety (reference to product liability has been deleted deliberately!). An even larger set of quality system elements is available from the author.

As mentioned earlier, the standard (guideline) consists of 17 major guideline sections, expanded to 142 subsections that in turn stipulate at least 261 recommendations by way of "should" statements. It is interesting to note that in its noncontractual category, two elements (requirements) of ISO 9001 are omitted from ISO 9004-1: contract review and customer-supplied product. Yet as a reservoir of potential elements for a quality system, these two elements could be important for inclusion in any quality system. This is to say that regardless of the category of noncontractual, ISO 9004-1 is a guideline on elements of a quality system. And even though it is intended for self-development/assessment, any quality system must be structured with "customers" in mind. Hence, contractual review and, for that matter, control of customer-supplied material are elements that must be given serious consideration for inclusion in a quality system.

On the limitation side, as for many standards passing through debate, review, negotiation, and consensus, the ISO 9000 Series represents a "least common denominator" in its coverage of the quality management/assurance/system disciplines. It would be good for everyone to understand this clearly, together with the fact that it is not a standard on total quality, and realize that the actual quality system that is optimum for a given enterprise or corporation may go well beyond the requirements, elements, and procedures of the ISO 9000 Series. See the remarks of Hayashi (1991) with respect to his concern, "that no TQC is anymore necessary." Many authors voice this caution (some as direct criticism). See, for example, Hutchins (1992), Kalinosky (1990), Kume (1992), Sayle (1988, 1992), Stephens (1993, 1994), and Zuckerman (1994a, 1994b, 1994c, and 1995), in the reference section. It *is* encouraging that many (perhaps most) companies implementing quality systems *do, in fact,* go beyond the requirements of the particular ISO 9000 standard used and incorporate other elements of total quality into their systems.

Some authors believe that there is too much emphasis on conformance rather than on adequacy and/or effectiveness. They believe that meeting the requirements is the principal concern (whereas many aspects of effective quality systems, especially beneficial to the enterprise itself, are not included as requirements). Others believe that short-term corrective action is emphasized over long-term improvement. Sayle (1988) addresses these and other concerns with respect to ISO 9000. The Series is also believed by many to contain a weak quality audit program. Sayle (1992) also discusses this aspect while drawing contrasts with management audits.

As alluded to above, there are many aspects of total quality systems not incorporated in the ISO 9000 Series. And, as also mentioned above, the name one gives to these systems is not as important as the content. The quality sciences have always been plagued by problems of semantics and the NIH syndrome. Young (and/or new) proponents of TQM, for example, often are ignorant of, or ignore, the fact that programs and systems with previous names as simple as quality control or total quality control (and existent as much as 25 to 35 years ago, including those developed in Japan) included such concepts and methodologies as project by project continuous improvement with a prevention orientation and quality teams; customer needs assessments and satisfaction programs; quality as a strategic business component, including its contribution to costs and cost reduction; design quality and innovation; and so on. This is not to say that important strides in refining and exposing these concepts to a wider audience have not been made in recent years; they have. But overzealous proponents of certain concepts have shown tendencies to idealize and ignore the conditions and necessities calling for a full range of tools (including statistical and others) for the total job of achieving quality and its related benefits.

With respect to understanding that the ISO 9000 Series is *not* a standard for total quality, in ISO 9000 the following are examples of concepts, principles, or methodologies that are either missing or less than adequately covered for a total quality system:

- Quality cost analysis and applications (other than as an element in ISO 9004-1 and a committee draft as ISO 10014).

- Top management involvement and leadership—top management-driven quality councils and quality as a business strategy encompassing planning, control, and improvement.

- Project-by-project improvement, pursued with revolutionary rates of improvement.

- *Joy and pride* in work—and employee participation, involvement, and empowerment via project teams and quality circles.

- Variation reduction, statistical process control, process capability, and process management.

- Production/inventory management systems such as JIT with TQC.

- The concept of single-sourcing, long-term cooperative supplier partnerships including product and quality system assistance (rather than mere assessment) based on trust and experience.

- Innovation of products and processes.

- Customer satisfaction including programs for surveying customer preferences and customer retention.

- Benchmarking of competitors as well as best in class products, processes, and systems.

- Deming's 14 points for management and the recognition of the necessity for transformation.

Juran (1994) lists the following exclusions from ISO 9000 as essentials to attain world-class quality:

- Personal leadership by upper managers.

- Training the hierarchy in managing for quality.

- Quality goals in the business plan.

- A revolutionary rate of quality improvement.

- Participation and empowerment of the workforce.

What has been said here must be properly interpreted and/or understood. It is not so much a criticism of ISO 9000 as an exhortation not to limit one's quality program/system to that of ISO 9000 alone. ISO 9000 is intended as a set of standards on QA systems, *not* for total quality systems!

In fact, alternative (or additional) resources for total quality/management systems assessments and criteria for designing and implementing such systems exist. Among these are the various quality awards such as the Deming Prize, the Malcolm Baldrige National Quality Award, and the European Quality Award. The latter, in particular, encourages self-appraisal and assigns 50 percent of the criteria to *results* in terms of people satisfaction (9 percent), customer satisfaction (20 percent), impact on society (6 percent), and business results (15 percent). See Conti (1991, 1995). See also EFQM (1992a, 1992b).

A further resource for quality system requirements is FDA's current good manufacturing practices (CGMPs). While these apply on a mandatory basis to applicable organizations, they are, nevertheless, available to any organization to study and use. Of significance here is the move toward the ISO 9000 series quality assurance system standards in the proposed rules published in the *Federal Register,* Vol. 58, No. 224 on November 23, 1993. Harmonization of CGMPs with ISO 9001 (for example) is being sought to promote competitiveness, improve the quality of manufacture, enhance safer and more effective medical devices, and promote a more comprehensive quality assurance system than presently offered by the CGMP requirements. Thus, an improved set of requirements is expected by merging the product, facility, and process specific requirements of the CGMPs with the more comprehensive quality assurance system requirements of ISO 9001.

As is now well known, the Automotive Industry Action Group (AIAG) published a set of requirements entitled *Quality System Requirements, QS-9000* in August 1994, and revised in February 1995. This document is an agreement between the three major American auto manufacturers and five truck manufacturers. QS-9000 contains all of ISO 9001, which is its foundation. Three additional clauses have been added, entitled Production Part Approval Process, Continuous Improvement, and Manufacturing Capabilities. Furthermore, an additional section lists some specific requirements of Chrysler, Ford, General Motors, and the truck manufacturers. While QS-9000 does not qualify, theoretically, as a "standard" because a consensus process was not used in its development, it does represent an alternate source of information and example for quality system requirements.

With consideration of these additional resources and incorporation of cogent elements, the resultant system should be better, more dynamic, more comprehensive, more effective, and more economical than that of ISO 9000 alone in its present form. This is illustrated by Figure 13–1, with ISO 9000 shown only as a basic foundation for a total quality system that for completeness includes other techniques and procedures as shown and mentioned above. Future revisions to the ISO 9000 Series, within the framework of its intended purpose, will address many of its inherent limitations. For example, the limitation of the "life-cycle model" versus a "process model" are being addressed by the phase II developments. With respect to its role within a total quality system, the phase II design specification prepared by the U.S. delegates to the ISO/TC 176 working group should be consulted, as given by Tsiakals (1994). See also Tsiakals and Cianfrani (1996). The papers by Peach (1994, 1996) should also be studied.

The Malcolm Baldrige National Quality Award

The global impact of national (and international) quality awards has become readily apparent from company/institutional advertisements, publicity, promotions, and the general literature. The criteria associated with these awards represents an important complement—to quality systems as significant alternatives an/or additional elements of

Figure 13–1

ISO 9000, Only a Minimum Foundation for Total Quality Systems

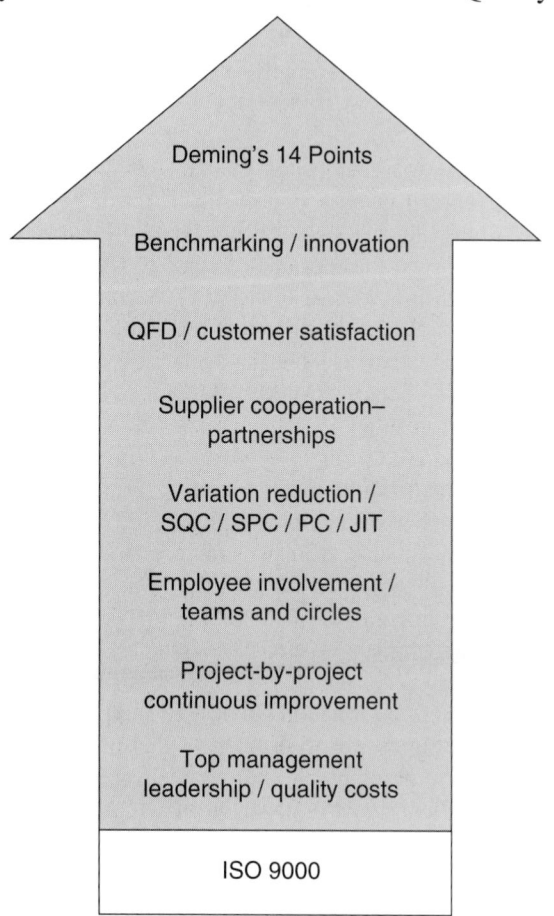

Deming's 14 Points

Benchmarking / innovation

QFD / customer satisfaction

Supplier cooperation–
partnerships

Variation reduction /
SQC / SPC / PC / JIT

Employee involvement /
teams and circles

Project-by-project
continuous improvement

Top management
leadership / quality costs

ISO 9000

quality systems otherwise promulgated by the ISO 9000 Series and its regional and national adoptions—as well as to quality management. Many important quality system elements not covered by the ISO 9000 Series are included in the criteria associated with the various quality awards. And while only a select few companies win these awards annually, it is the self-assessment, system development processes, associated with the use of the criteria at the corporation and enterprise levels, that contribute significantly to the spread of quality systems and management.

The three quality awards most sought after are the Deming Application Prize in Japan, the Malcolm Baldrige National Quality Award in the United States of America, and the European Quality Award. In the United States of America, in particular, a growing number of geographically local (institutional, state, city, etc.) quality awards are contributing significantly to the spread of quality concepts. Other developed countries have launched national quality awards, and a growing number of developing countries are also turning to these resources in formulating, establishing, and implementing National Quality Awards. A limited example is presented subsequently to illustrate the progress being made in this area of the quality disciplines, motivated by the U.S. Malcolm Baldrige National Quality Award (MBNQA).

The MBNQA was established in the United States in 1987 by way of the Malcolm Baldrige National Quality Improvement Act of 1987 that was signed by President Reagan on August 20, 1987 and became Public Law 100-107. The award was named after Malcolm Baldrige, the former Secretary of Commerce, who served from 1981 until his untimely death in 1987. Responsibility for the continuing management of the award is assigned to the National Institute of Standards and Technology (NIST, formerly the U.S. Bureau of Standards) which is an agency of the Department of Commerce. The American Society for Quality Control (ASQC) assists in administrating the award program under contract to NIST.

As stated in the MBNQA 1996 Award Criteria the award promotes "understanding of the requirements for performance excellence and competitiveness improvement; and sharing of information on successful performance strategies and the benefits derived from using these strategies." As a general goal the award promotes awareness of the importance of quality improvement to our nation's economy. The award's three eligibility categories are manufacturing companies, service companies, and small businesses. Two winners can be selected per category annually. Winners may publicize and advertise their awards and are expected to share information about their quality strategies with other U.S. organizations as an encouragement for the spread of the quality disciplines and a motivation to other organizations to develop and improve their operations to the extent of vying successfully for the award. For an early paper on MBNQA see Reimann and Haines (1990).

The award is based on a set of criteria for the evaluation of applicants, which in turn serves as the basis for self-development/assessment by companies wishing to improve their quality efforts, even when they have no intention to apply for the award. Purposes for the criteria are: (1) to serve as a working tool for planning, training, and assessment; (2) to help raise quality performance standards and expectations; and (3) to assist communications and sharing among and within organizations of important quality and operational performance requirements. The criteria are built around the following core values: customer-driven quality; leadership; continuous improvement; full participation, including employee participation and development; fast response; design quality and prevention; long-range outlook; management by fact; partnership development; and public responsibility in terms of corporate citizenship and responsibility.

The criteria framework for the award are further categorized into four major dynamic relationships and emphasis categories, as shown in Figure 13–2. As mentioned in the introduction, these categories are expanded into 24 examination items and 52 areas to address. The seven major categories are assigned point values adding to 1,000 to pro-

Figure 13–2

The Baldrige Award Criteria Framework—Dynamic Relationships

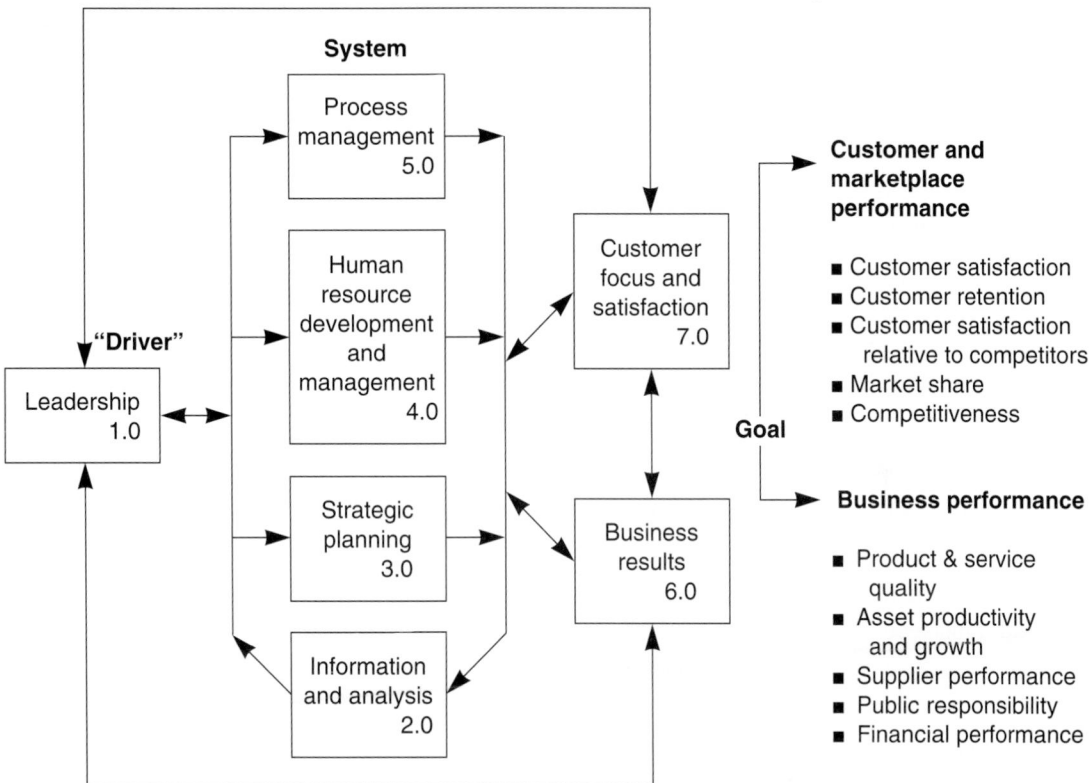

vide a numerical basis for degree of importance and assessment. The point values assigned to each category are further broken down to point values associated with the 24 examination items, as shown in Figure 13–3. It is instructive to note that two categories alone amount to 50 percent of the total point values, namely business results and customer focus and satisfaction (areas for which ISO 9000 is often criticized as lacking).

From 1988 through 1995 some 24 companies have received the award. The awardees are shown in Figure 13–4. As one of the purposes of the award is the sharing of information on quality programs, these companies may be contacted for such information. Application data over this same period is shown in Figure 13–5. The number of applicants reached a peak in 1991 and has declined since. One explanation relates to the awareness of companies, based on their self-assessment, that their programs are not yet ready for the high-performance standards expected. In the assessment of applicants both a paper evaluation (stage 1) and on-site visits (stage 2) are carried out, the latter stage executed for those applicants who score well in the first stage. Site visits data related to the applicants of Figure 13–5 are shown in Figure 13–6.

Figure 13–3

The Baldrige 1996 Award Criteria—Item Listing

1996 Categories/Items	Point Values
1.0 Leadership	**90**
1.1 Senior executive leadership	45
1.2 Leadership system and organization	25
1.3 Public responsibility and corporate citizenship	20
2.0 Information and Analysis	**75**
2.1 Management of information and data	20
2.2 Competitive comparisons and benchmarking	15
2.3 Analysis and use of company-level data	40
3.0 Strategic Planning	**55**
3.1 Strategy development	35
3.2 Strategy deployment	20
4.0 Human Resource Development and Management	**140**
4.1 Human resource planning and evaluation	20
4.2 High-performance work systems	45
4.3 Employee education, training, and development	50
4.4 Employee well-being and satisfaction	25
5.0 Process Management	**140**
5.1 Design and introduction of products and services	40
5.2 Process management: product and service production and delivery	40
5.3 Process management: support services	30
5.4 Management of supplier performance	30
6.0 Business Results	**250**
6.1 Product and service quality results	75
6.2 Company operational and financial results	110
6.3 Human resource results	35
6.4 Supplier performance results	30
7.0 Customer Focus and Satisfaction	**250**
7.1 Customer and market knowledge	30
7.2 Customer relationship management	30
7.3 Customer satisfaction determination	30
7.4 Customer satisfaction results	160
Total Points	1000

Figure 13–4

The Baldrige Award Winners

1988	■	Motorola
	■	Commercial Nuclear Fuel Division (Westinghouse)
	■	Globe Metallurgical
1989	■	Milliken & Co.
	■	Xerox Business Products and Systems
1990	■	Cadillac Motor Car Division (General Motors)
	■	IBM Rochester
	■	Federal Express Corporation
	■	Wallace Company, Inc.
1991	■	Solectron Corporation
	■	Zytec Corporation
	■	Marlow Industries
1992	■	AT&T Network Systems Group Transmission Systems Business Unit
	■	Texas Instruments Inc. Defense Systems & Electronics Group
	■	AT&T Universal Card Services
	■	The Ritz-Carlton Hotel Company Atlanta, Georgia
	■	Granite Rock Company
1993	■	Eastman Chemical Company
	■	Ames Rubber Corporation
1994	■	AT&T Consumer Communication Services
	■	GTE Directories Corporation
	■	Wainwright Industries, Inc.
1995	■	Armstrong Building Products Operations
	■	Corning Telecommunications Products Division

Upon reflection on the data of Figures 13–4 through 13–6 one might rightfully ask whether the effort involved in the award process is beneficial if so few companies participate from application to actual award. While this alone is viewed by the quality profession as beneficial in the establishment of a standard for quality excellence, of further benefit are the recognition of those who have achieved this standard, and the feedback from these companies as encouragement and example for future aspirants. A most significant additional benefit is the use of the criteria (standard) by thousands of companies carrying out self-developments and self-assessments for improving their quality efforts. While the full benefit in this regard may be one of those "unknowables," some indication is provided by Figure 13–7, which shows the number of application guidelines

Figure 13–5

The Baldrige Award Application Data, 1988–1995

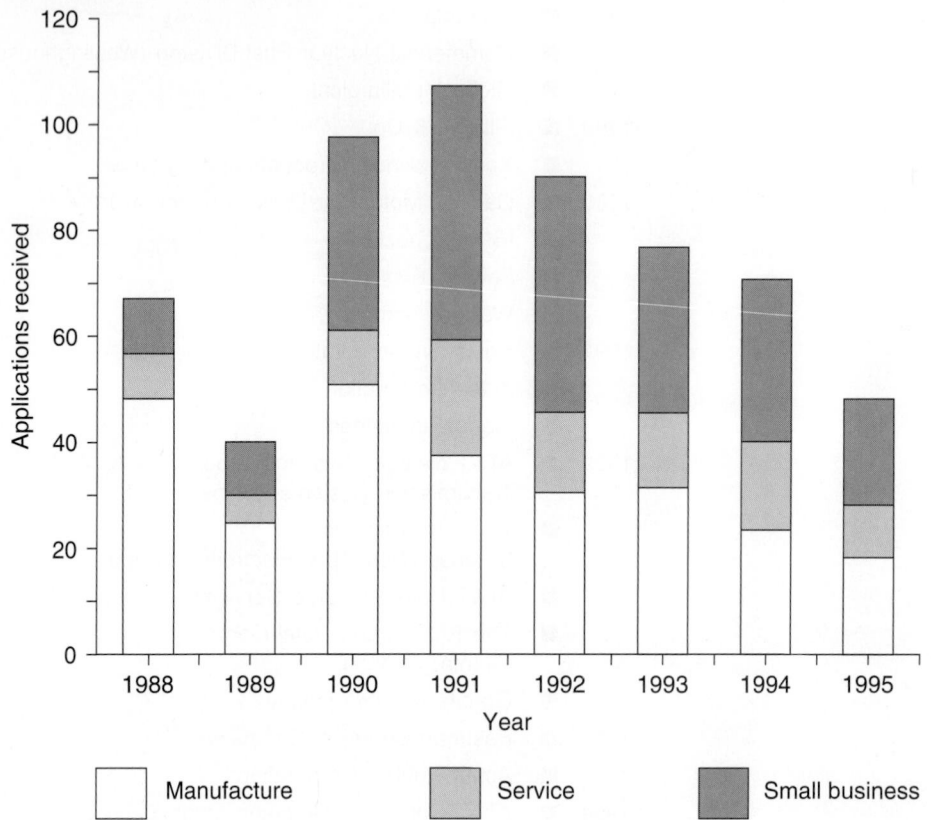

requested by year from 1988 through 1995. Even with declining applications, the guidelines continue to be in demand at levels well above 100,000 per year. Thus, while ISO 9000 stands as a standard (or set of standards) for a quality system, the criteria for the MBNQA stands, at least as a de facto standard, for Total Quality/Total Quality Management. In this respect the self-development/assessment of ISO 9004-1 and MBNQA have a lot in common. For a particularly cogent discussion of the importance of self-assessment see the paper by Conti (1995). In fact, recommended reading for the quality professional and management personnel is the entire Volume 5 on *The Best on Quality*, (International Academy for Quality, 1995), which is devoted to the subject of national and international quality awards. This volume contains numerous contributions on quality awards, globally—a literal handbook on the subject. The author's contribution to the referenced volume is principally in the developing country arena (Stephens, 1995a). A survey of national quality awards in developing countries was conducted and found to represent a significant development, contributing to quality systems and management globally. This is seen as an important development to complement quality systems and

Figure 13–6

The Baldrige Award Site Vistis Data, 1988–1995

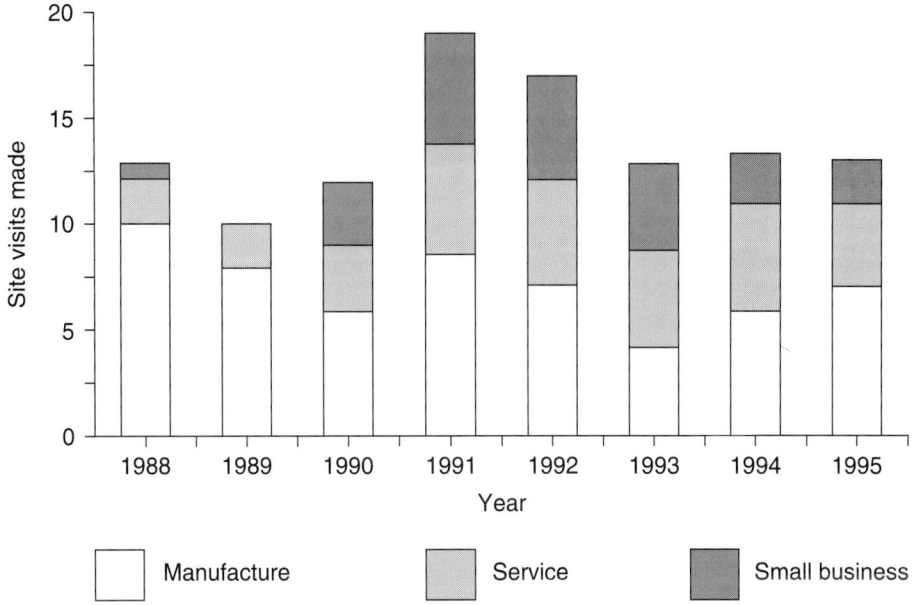

Manufacture   Service   Small business

management as promulgated by the ISO 9000 Series internationally. As an example the criteria for the Columbian National Quality Award is shown in Figure 13–8. Note that it is based to a large degree on the criteria of the MBNQA. However, there are some interesting additions, such as physical plant and installation and effects on environment (a harbinger of ISO 14000?).

It should be emphasized here that there is no guarantee of success in either the use of ISO 9000 or the MBNQA. In connection with the MBNQA, sharing success can be overwhelming and self-defeating, if not monitored and controlled carefully. This is true particularly for small companies. The Wallace Company (a 1990 recipient of the award, see Figure 13–4), with some 280 employees, received 60 to 70 requests for information per day after winning the award. In three months after winning the award they had 1,000 sign-ups for visits to their facilities. See Hart (1993). Shortly after winning the award the Wallace Company found itself in financial trouble. Some 70 employees were laid off. Company overhead increased by $2 million per year, and customers were unhappy with prices. According to CEO John Wallace, executives were giving speeches all over the United States and not getting sales. The award seemed to distract managers and took their attention away from some crucial parts of running the business. However, on the upside it was seen that employees were able to cope with the cutbacks after the financial problems because of their quality process participation and the openness it fostered. Their attitude of continuous improvement was retained.

Figure 13–7

The Baldrige Award Application Guidelines Data, 1988–1995

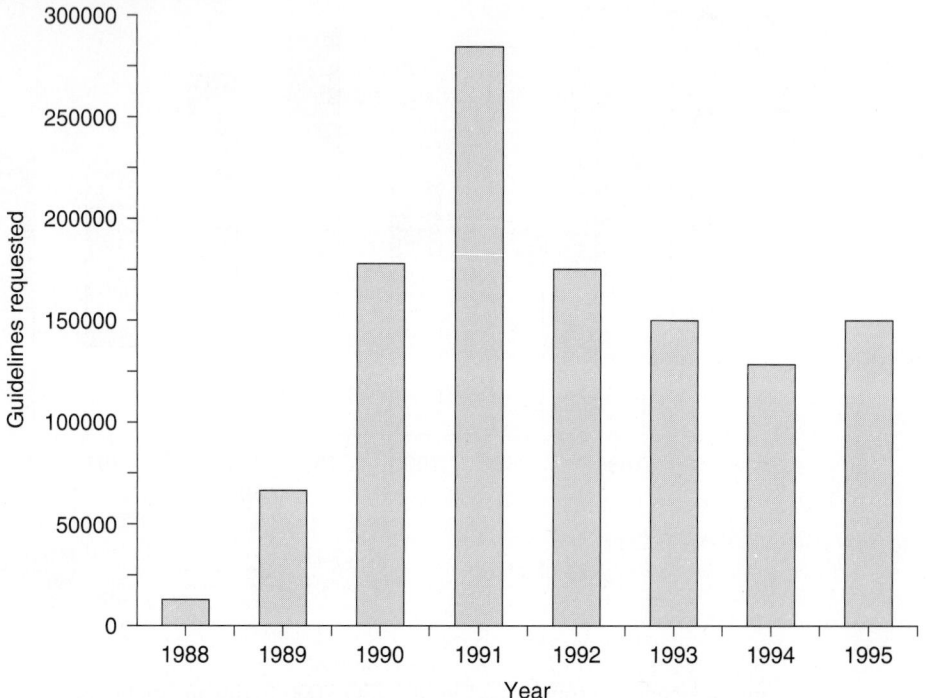

ISO 9000 COMPARED TO MBNQA

As indicated above, the ISO 9000 Series, especially with respect to the compliance standards, is intended to serve as models for quality assurance systems. The specific guideline, ISO 9004-1, has the somewhat broader role as a compendium of quality system elements and recommendations for quality management, primarily implemented on a self-development/assessment basis. In this respect it covers more items than the compliance standards, such as cost of quality, marketing, product safety, quality improvement, and personnel motivation. It does not cover contract review and customer-supplied product. The MBNQA is viewed as a higher level standard with respect to principles and tenets of Total Quality and Total Quality Management; with specific emphasis on customer management and satisfaction; on results that relate to product and service quality, operations and finances, human resources, and suppliers; on employee development and satisfaction; on higher level management involvement and leadership; on competitive benchmarking; and so on. It does not detail certain elements of quality systems, as found in the ISO 9000 compliance standards, such as documentation and document control, inspection, metrology, product traceability, contract review, customer-supplied product, control of nonconforming product, postproduction activities, and so on.

Figure 13–8

The Columbian National Quality Award Criteria

Areas and criteria	Maximum points
1.0 Customer Satisfaction	**180**

1.1 Management of customer relations	
1.2 Customer knowledge	
1.3 Feedback systems	
1.4 Results	

2.0 Leadership	**100**

2.1 Leadership by example	
2.2 Quality values	

3.0 Human Resources	**150**

3.1 Participation and involvement	
3.2 Training	
3.3 Evaluation and recognition	
3.4 Quality of life in the workplace	

4.0 Quality Strategy	**60**

4.1 Strategic planning	
4.2 Operations planning	

5.0 Quality Information	**90**

5.1 Data and sources	
5.2 Information analysis	

6.0 Quality Assurance and Improvement	**140**

6.1 Design and development of goods and services	
6.2 Control of operational process	
6.3 Control of administrative and support services	
6.4 Control of Measurement and test equipment	
6.5 Continuous improvement	
6.6 Documentation and recording of quality	
6.7 Audits or evaluations of the quality assurance system	

7.0 Supplier Relations	**60**

7.1 Quality in purchases	
7.2 Suppliers and subcontractors	

8.0 Physical Plant and Installation	**60**

8.1 Installation, cleaning, and maintenance	
8.2 Industrial security and environmental control	

9.0 Effects on Environment	**60**

9.1 Preservation of the ecosystems	
9.2 Promotion of quality culture in the community	

10.0 Achievements in Improvement	**100**

10.1 Improvement of products and services	
10.2 Improvement of support areas	
10.3 Comparison of results	
Total Points	**1000**

In an attempt to compare the two approaches Figure 13–9 identifies the extent to which the requirements of the compliance standards of the ISO 9000 Series align with the MBNQA guidelines/criteria. As seen by this interpretation, most of the line items in Figure 13–9 for each approach are aligned in multiple fashion, indicated by the large number of "highly aligned" symbols. An even larger number of "somewhat aligned" symbols exist, further indicating the multiple alignment of line items in the two approaches. No "at odds" symbols are shown on Figure 13–9, indicating reasonable harmony between the two approaches. For additional discussions of the relationships between ISO 9000 and MBNQA, see papers by Reimann and Hertz (1993), Pearl (1995), Majerczyk and DeRosa (1994), Peach (1994), and Stephens (1994, 1995b).

Comparison of Documentation and Control

The requirements for documentation and control are high for both ISO 9000 and MBNQA, though ISO 9000 is much more explicit, with respect to use of the root word "document" in every one of the 20 broad-based requirements (of ISO 9001) and a specific set of requirements for document and data control. Documentation and control in MBNQA is much more subtle and implied. An interesting analogy is the text by Feigenbaum (1991). If one looks in the index, the term "documentation" is not found. However, it would be a mistake to conclude that Feigenbaum doesn't include documentation in a total quality control system. Multiple references to documentation are salted throughout the text.

MBNQA has high requirements for results and results measurements, while ISO 9000 does not. The MBNQA requires measured process control; ISO 9000 requires documentation of process control where it exists, but does not necessarily require measured control.

Comparison of Degree of Prescriptiveness

The degree of prescriptiveness inherent in the two approaches to quality is defined by what and how work is to be done. The MBNQA guidelines spell out what must be done to attain a high score relative to the categories and items defined. It also spells out how each of the "areas to address" should be approached. Otherwise, there is a great deal of latitude within the bounds of the criteria for companies to express their individualities and uniqueness and management styles. There is no imposed stereotype. This can be seen by the diversity of the approaches taken by the 24 winners to date.

The ISO 9000 Series requirements for the compliance standards are generic, as mentioned earlier. They define specific areas to include in the quality system, essentially the "what" that is considered minimum for compliance. However, the requirements allow much latitude in "how" the system should be set up and how it should operate to accomplish the stated requirements. In this respect there is also no imposed stereotype. Companies with widely different quality systems have shown compliance and achieved certification. To a certain degree a lot is left to the interpretation of the management in

Figure 13–9

Extent to Which ISO 9000 Requirements Align with MBNQA Guidelines

How Does ISO 9001 Align with MBNQA?

Legend: ● Very well aligned ○ Somewhat aligned □ Little/no alignment ★ At odds

See previous page for a listing of MBNQA categories/items

ISO 9001 quality system standard	Leadership			Information and analysis			Strategic planning		Human resource development and management				Process management				Business results				Customer focus and satisfaction			
	1.1	1.2	1.3	2.1	2.2	2.3	3.1	3.2	4.1	4.2	4.3	4.4	5.1	5.2	5.3	5.4	6.1	6.2	6.3	6.4	7.1	7.2	7.3	7.4
4.1 Mgt.Resp.	●	●		○		●	●	●	●	●	●		●	●	●	●	●	●	●	●	●	○	●	○
4.2 Q System	●	●		●		●	○	●	●	●	●		●	●	●	○	●	●	●	●	●	○	●	○
4.3 Contract Rev.				○		○	●						●	●							●	○		
4.4 Design				○		○	○						●	●	○				○	○	●	○		
4.5 Docmnt/Data				○		○								●	●									
4.6 Purchasing				○	○	○		●						○	●	●	●		○	●				
4.7 Cust. Product				○		○		●						○	●	●	●			●				
4.8 Product ID				○		○								●	●	●	●			○				
4.9 Process				○		○					○		○	●	●	●	●		●	○				
4.10 Insp/Test				○		○			●	○				●	●	●	●			○				
4.11 Insp. Eqpmt.				○		○				○				●	●	●	●			○				
4.12 Insp. Status				○		○								●	●	●	●			○				
4.13 Nonconf. Prod.				○		○							●	●	●	●	●			○				
4.14 Corr. Prev.	●			●		●	○						○	●	●	●	●	●	●	○	●		●	○
4.15 Handling						○								●			●	●	●		●			
4.16 Q Records		○		●		●	●					○		●			●	●	●	○			●	●
4.17 Q Audits		●	●	●		●	●				●			●	●	●	●	●	●	○			○	
4.18 Training		●	●	○		●	○		○	●	○	○			●			●			○			
4.19 Servicing				●		●	○		○		●			●	●				●		●	●	●	●
4.20 Statistics	○	○		●		●	○		○		●	○		●	●	●	●	●	○	●	●		●	●
	1.1	1.2	1.3	2.1	2.2	2.3	3.1	3.2	4.1	4.2	4.3	4.4	5.1	5.2	5.3	5.4	6.1	6.2	6.3	6.4	7.1	7.2	7.3	7.4

Malcolm Baldrige National Quality Award (MBNQA) Criteria

setting up the system and the interpretation of the auditors in carrying out the certification process. Where these two interpretations agree sufficiently, registration is achieved. As indicated above, most companies recognize the benefits of inclusion of elements of quality systems not included in the specific requirements of ISO 9000.

ISO 9000, MBNQA, AND TOTAL QUALITY IN SUMMARY

As alluded to in the introduction and in the overview to ISO 9000, there is no standard established for total quality or total quality management, let alone agreement on the terms themselves. One definition of total quality is provided by the study group associated with the 1992 Total Quality Forum and included in Rampey and Roberts (1992) as:

> Total Quality is . . . a people-focused management system that aims at continual increase in customer satisfaction at continually lower real cost. TQ is a total system approach (not a separate area or program), and an integral part of high-level strategy. It works horizontally across functions and departments, involving all employees, top to bottom, and extends backwards and forwards to include the supply chain and the customer chain . . .

A definition for Total Quality Management used by the author is as follows:

Total Quality Management (TQM) is a process that integrates fundamental management art and techniques with the disciplines, principles, methodologies, activities, approaches, and techniques of total (strategic) quality to develop and implement successful business strategies throughout the organization (or any business entity).

Total (strategic) quality is viewed as:

- A clear focus on customer needs, wants, satisfaction, and delight.

- Continuous improvement and innovation of all processes, services, and products.

- Effective utilization, empowerment, and recognition of individuals under a team involvement/participation approach, including essential programs of education and training, shared values, decisions, and benefits.

- Sound planning for quality with fact-based decision making, variability reduction, defect prevention, and fast response systems that include recognition of the internal customer and the triple role of supplier/processor/customer at all processes.

- Integration of and mutual cooperation with suppliers based on longer-term partnership relations.

- Sensitivity to competitive comparisons via benchmarking, including "best in class" on noncompetitive but significant processes.

- Productivity, cost reduction, and profitability enhancements.

- Strong leadership by management at all levels with effective creation of policies, vision, mission, values, guiding principles, goals, and support communicated and implemented throughout the organization by example, with the establishment of a compatible corporate culture.

The ISO 9000 Series compliance standards achieve a great deal by offering a consistent set of requirements that can be applied anywhere with respect to elements of systems of quality assurance. The MBNQA guidelines and criteria likewise prescribe a set of de facto requirements more closely aligned with the concepts of total quality; in essence, building on the base established by ISO 9000. Hence, these two approaches are complementary and should be considered jointly in the development of a quality system. See for example the paper by Huyink (1996). However, there are elements, concepts, and philosophies of total quality not contained in either of these two approaches. Most of the differences lie in emphasis, approach, and philosophy, as embodied in Deming's 14 points for management and system of profound knowledge (see Deming, 1986, 1994; see also the paper by Puri, 1991). The term *Deming-based TQM* system is used in the following discussion.

Perhaps implied, but never required specifically, is the "transformation" emphasized by Deming. It would , of course, be difficult to measure and confirm. A Deming-based TQM initiative requires the substantial commitment of personal time and resources of senior managers in transforming the organization. Quality cannot be just another aspect of the business; it must become the way business is conducted. While ISO 9000 series registration does require support and involvement by senior management, it is not nearly so demanding as a Deming-based TQM transformation initiative. The MBNQA criteria for "Leadership" comes much closer to Deming's expectations.

Deming-based TQM is much more open than MBNQA or ISO 9000. It has no firm requirements other than to improve quality; to decrease costs by less rework, fewer mistakes, fewer delays, better use of machine-time and materials by the use of applied statistics; to improve productivity; to capture the market with better quality and lower price by meeting and/or exceeding customer needs through an understanding of the organization and the effects of current management practices; to remain in business; and to provide jobs and more jobs. It expects the senior managers of an organization to consider management style through a fact-based scientific examination of Deming's 14 points and then prove or disprove the implementation of those points as they apply to the organization. Deming expects senior managers to establish a controlled, customer-focused, continuously improving organization. That kind of organization has requirements, but practical application must define those requirements in a Deming-based TQM system. That is, form should follow function and necessity and/or potential benefit. Documentation showing how processes are to be accomplished is necessary, of course. But it is up to an organization to document processes to communicate effectively with those who need to know or might benefit from knowing. A Deming-based TQM system leaves the details entirely up to the organization. Deming-based TQM involves the most organizational involvement and organizational change of the three approaches.

Deming urges companies to drive out fear, eliminate slogans, exhortations, and management by objective, remove barriers that rob people of pride in workmanship, and institute education. These points are not addressed in the ISO 9000 Series and are not too explicit in the criteria of MBNQA.

ISO 9000 has several highly specific concerns about inspection and testing that could be at odds with Deming-based TQM. Deming emphasizes that an organization should "cease dependence on inspection to achieve quality." Deming says that companies should "eliminate the need for inspection on a mass basis by building quality into the product in the first place." ISO 9000 provides for cases in which in-process control makes later inspection unnecessary, but also emphasizes inspection by requirement. The ISO 9000 Series and Deming are in agreement that when inspection and testing are required, those doing the work should be trained and provided with appropriate equipment.

The ISO 9000 Series provides for the use of statistical techniques but does not emphasize them. If statistical process control methods are used, the ISO standard requires that procedures for SPC tools be documented and implemented as documented with appropriate training provided. The basis of Deming-based TQM is statistical understanding.

Deming-based TQM and MBNQA differ greatly in their approaches to benchmarking. Deming recommends that companies spend "time and effort focusing on what customers want and need, not what competitors are doing." Deming says that if you treat your customers right and continuously improve in what you provide to them, your competitors will be watching you, and you will always be ahead of those competitors, because you will be continuously improving while they are trying to catch up to you. Deming favors comparison of quality results and comparison of support-system quality results, but recommends that this comparison be made against the organization's previous record, not as a benchmarking device against competitors' results.

Deming is less concerned about measuring customer satisfaction and more concerned about developing a focused, continually improving relationship with customers and suppliers. While Deming supports organizational leaders helping to improve public quality awareness, it is not a requirement as expressed by MBNQA.

Deming-based TQM offers little guidance on either the nature of the work or how to set up a system. A Deming-based TQM system creates a learning organization. This learning tenet of TQM makes selecting a Deming-based approach to TQM even more persuasive. Deming provides depth of theory as well as application. Deming points out, "There is no learning without theory." He goes on to say: "All theory is wrong. Some is useful. Of course all theory is wrong! If it were right, it wouldn't be theory, now would it? It would be fact. But without theory, we cannot learn. Experience alone teaches nothing. If we have a theory, then experience can help us to learn."

After an organization has become effective through TQM transformation and has become a learning organization, the use of the MBNQA quality criteria can be an excellent means to improve an already effective organization, even if it doesn't actually apply for the award. However, an organization that applies for the Baldrige Award is likely to put more useful pressure on itself than one using the criteria solely on an internal basis.

It is reasonable for a company to spend some time internally with the MBNQA criteria. The MBNQA criteria are self-descriptive. Even if it is difficult to measure the number of points achieved when self-scoring, the practice is worthwhile. Winning the Baldrige award should not be a company's initial goal. The intent should be to improve. After sufficient improvement has been accomplished, then application should be made for potential recognition.

REFERENCES

ANSI/ASQC Z1.15 (1979), American National Standard, *Generic Guidelines for Quality Systems,* December 19, 1979, American National Standards Institute, Inc.

Conti, T. (1991), "Company Quality Assessments," *Total Quality Management,* June 1991, pp. 167–72, and August 1991, pp. 227–33.

Conti, T. (1995), "A Critical Review of Methods for Quality Awards and Self-Assessment," *The Best on Quality,* IAQ Book Series, Vol. 5, Milwaukee, Wisconsin: ASQC Quality Press. Chapter 17, pp. 211–33.

Deming, W. E. (1986), *Out of the Crisis,* Cambridge, Massachusetts: Massachusetts Institute of Technology, Center for Advanced Engineering Study.

Deming, W. E. (1994), *The New Economics for Industry, Government, and Education,* 2d ed., Cambridge, Massachusetts: Massachusetts Institute of Technology, Center for Advanced Engineering Study.

European Foundation for Quality Management (EFQM), (1992a), *Total quality management, The European model for self-appraisal, Guidelines for identifying and addressing total quality issues.* January 1992, Eindhoven, Netherlands.

European Foundation for Quality Management (EFQM), (1992b), *The European quality award.* June 1992, The European Quality Award Secretariat, Brussels, Belgium.

Feigenbaum, A.V. (1991), *Total Quality Control,* 3d ed., revised, New York: McGraw-Hill.

Hart, C. (1993), "What's Wrong and Right With the Baldrige Award," *Chief Executive,* No. 90 (Nov.-Dec. 1993), pp. 36–47.

Hayashi, A. (1991), "Japan's Policy Toward International Assessment and Registration System Using ISO Quality Assurance Standards," *Transactions, Seminar on Achieving*

Competitive Quality Through Standardization and Quality Control, MITI, JSA, UNIDO, SIRIM, October 29–31, 1991, Kuala Lumpur, Malaysia.

Hutchins, D. (1992), *Achieve Total Quality,* Cambridge, England: Director Books.

Huyink, D. S. (1996), "From ISO 9000 to Total Quality Management: How ISO 9000 Makes TQM Easier," *ASQC 50th Annual Quality Congress Proceedings,* May 13–15, 1996, Chicago, pp. 613–18.

International Academy for Quality (1995), edited by John Hromi, *The Best on Quality,* IAQ Book Series, Vol. 5. Milwaukee, Wisconsin: ASQC Quality Press.

Juran, J. M. (1989), *Juran on Leadership for Quality,* Milwaukee, Wisconsin: Quality Press, American Society for Quality Control.

Juran, J. M. (1994), "The Upcoming Century of Quality," *Keynote Address to the ASQC Quality Congress,* Milwaukee, Wisconsin: Quality Press, p. 15.

Kalinosky, I. S. (1990), "The Total Quality System—Going Beyond ISO 9000," *Quality Progress,* Vol. 23, No. 6 (June 1990), pp. 50–54.

Kume, H. (1992), "The Japanese Point of View on the ISO 9000 Standards," *Proceedings, 36th EOQ Annual Conference,* June 15–19, 1992, Brussels, Belgium, pp. 51–53.

Majerczyk, R. J. and DeRosa, D. A. (1994), "ISO Standards: The Building Blocks for TQM," *ASQC 48th Quality Congress Proceedings,* May 24–26, 1994, Las Vegas, pp. 642–50.

Marquardt, D. W. (1995), "The Missing Linkage in ISO 9001: What's Being Done About It?," *ASQC 49th Quality Congress Proceedings,* May 22–25, 1994, Cincinnati, pp. 1056–63.

Peach, R. W. (1994), "Planning the Journey From ISO 9000 to TQM," *ASQC 48th Quality Congress Proceedings,* May 24–26, 1994, Las Vegas, pp. 864–72.

Peach, R. W. (1996), "Putting the ISO 9000 Family of Standards to Use," *ASQC 50th Annual Quality Congress Proceedings,* May 13–15, 1996, Chicago, pp. 351–55.

Pearl, D. H. (1995), "Integrating ISO 9000 into an Internal Baldrige Audit," *ASQC 49th Quality Congress Proceedings,* May 22–25, 1994, Cincinnati, pp. 1018–25.

Puri, S. C. (1991), "Deming + ISO/9000, 'A Deadly Combination for Quality Revolution,'" *ASQC 45th Annual Quality Congress Transactions,* May 20–22, 1991, Milwaukee, Wisc., pp. 938–43.

Rampey, J. and Roberts, H. (1992), "Perspectives on Total Quality," *Proceedings of Total Quality Forum IV,* November 1992, Cincinnati, Ohio.

Reimann, C. W. and Haines, R.A. (1990), "The Malcolm Baldrige National Quality Award," *ASTM Standardization News* (October 1990), pp. 30–34.

Reimann, C. W. and Hertz, H.S. (1993), "The Malcolm Baldrige National Quality Award and ISO 9000 Registration," *ASTM Standardization News* (November 1993), pp. 42–53.

Sayle, A.J. (1988), "ISO 9000—Progression or Regression?," *EOQC Quality,* Vol. 1, pp. 9–13.

Sayle, A.J. (1992), "Audits—The Key to the Future," 1st Annual Quality Audit Conference, ASQC, February 27–28, St. Louis, Missouri, 1992.

Stephens, K. S. (1993), "Quality Systems and Certification—Some Observations and Thoughts," *EOQ Quality,* No. 1 (March 1993), pp. 5–12.

Stephens, K. S. (1994), "ISO 9000 and Total Quality," *Quality Management Journal,* Vol. 2, No. 1 (Fall 1994), pp. 57–71.

Stephens, K. S. (1995a), "National Quality Awards: A Developing Country Perspective," *The Best on Quality,* IAQ Book Series, Vol. 5, Milwaukee, Wisconsin: ASQC Quality Press. Chapter 16, pp. 179–210.

Stephens, K. S. (1995b), "Quality Systems and Total Quality," *Aseptic Pharmaceutical Manufacturing II,* edited by Michael J. Groves and Ram Murty, Buffalo Grove, IL: Interpharm Press, Inc. Chapter 3, pp. 41–79.

Stephens, K. S. (1996), "National Quality Awards—Complements to Quality Systems," ASQC 50th Annual Quality Congress Proceedings, May 13–15, 1996, Chicago, pp. 735–42.

Tsiakals, J. (1994), "Revision of the ISO 9000 Standards," *ASQC 48th Quality Congress Proceedings,* May 24–26, 1994, Las Vegas, pp. 873–81.

Tsiakals, J. and Cianfrani, C. (1996), "ISO 9001 in the Next Millennium: The Future Revision of ISO 9001, ISO 9002, and ISO 9003," ASQC 50th Annual Quality Congress Proceedings, May 13–15, 1996, Chicago, pp. 356–63.

Zuckerman, A. (1994a), "Mixed Reviews for ISO 9000," *Iron Age New Steel,* Vol.10, No. 2 (February 1994), pp. 40–43.

Zuckerman, A. (1994b), "EC Drops Ticking Time Bomb—It Could Prove Lethal to the ISO 9000 Community," *Industry Week* (May 16, 1994), pp. 44–51.

Zuckerman, A. (1994c)," ISO 9000 Skepticism," *Industry Week* (July 4, 1994), pp. 43–44.

Zuckerman, A. (1995), "One Size Doesn't Fit All," *Industry Week* (January 9, 1995), pp. 37–40.

Other Standards, Guidelines and Business Process Initiatives Based on ISO 9000

14

This chapter covers some potential new standards being developed based on ISO 9000, including a standard for occupational health and safety management, use of the ISO 9000 standard in the U.S. government, ISO 9000 and laboratory accreditation, and a short article about a program to allow companies reduce third-party audits under a program known as the Supplier Audit Confirmation.

AN ISO OCCUPATIONAL HEALTH AND SAFETY MANAGEMENT SYSTEM

Charles F. Redinger, MPA, CIH
Steven P. Levine, Ph.D., CIH

INTRODUCTION

This article addresses the development of a potential occupational health and safety management system (OHSMS) by the International Organization for Standardization (ISO). Formal consideration of an ISO OHSMS began in 1995 and was the central topic at an ISO OHSMS meeting in September 1996.

The application of the ISO "management systems" model to occupational health and safety (OHS) has intrigued many. This approach appears to offer the seeds of a new regulatory/management instrument that can address some of the inefficiencies of traditional command-control instruments in OHS control. Several theoretical and practical aspects of ISO's venture into social regulation are addressed in this article, including OHS control and management. While connections to ISO 14000 are made, the emphasis is on the development of an ISO OHSMS.

OHSMS OVERVIEW

Against this backdrop of ISO 9000 activity, a technically distinct but parallel effort to develop internationally accepted standards for corporate environmental practices has been proceeding. As presented in other portions of this book, the ISO Technical Management Board created Technical Committee 207 (TC 207) in 1992 to develop internationally recognized environmental management standards. Recently, TC 207 voted to advance draft standards for environmental management systems (EMS) and environmental auditing.

Concurrent with TC 207 efforts, published articles have posed questions regarding the applicability of the ISO 9000 quality system and its potential benefits to health and safety management systems.[1] Some organizations recognize a compatibility and have subsequently integrated occupational safety and health within the framework of their respective ISO 9000 system.[2,3] Irrespective of what standard or practice is employed, there appears to be an increasing trend within the industry to link quality, productivity, and occupational health with emphasis on sound managerial systems. Furthermore, businesses are increasingly managing health and safety not to simply comply with statutory requirements, but to achieve a marketplace advantage through gains in efficiency.[4]

RATIONALE FOR AN OHSMS

A novel approach to international organization and coordination was introduced with the management systems method embodied in ISO 9000. Many scholars have been intrigued with the evolution of the management systems approach. Some believe that the ISO management systems model may provide the necessary bridge between trade, environmental protection, and sustainable development.[5] By examining some of the basic tenets of the ISO management systems approach, an emerging theme is one that may be applicable not only to occupational health and safety, but also to other areas of social regulation where command-control instruments have been extensively used.

Numerous economists have been critical of the command-control approach to environmental, health, and safety regulation.[6] Arguments have been presented that show the inefficiencies of the command-control approach should be taken more seriously. Other instruments such as taxation, pollution credits, and market incentives also have shortcomings and are not always widely applicable because of ethical concerns. Conversely, the management systems approach appears to offer an instrument whereby negative externalities can be internalized through a system of modified market incentives; namely, within the ISO 9000 model, by tying environmental, health, and safety performance to market participation and procurement contract language.

An important component of the ISO quality management systems model is the manner in which a quality system is verified, that is, either through first-, second-, or third-party registration. The concept of third-party registration of management systems has been used throughout industry and commerce with increased frequency over the past decade.[7] The ISO 9000 and ISO 14000 body of standards developed by the International Organization for Standardization (ISO) relies on the use of accredited, independent third parties to verify an organization's conformance with a particular ISO standard. An organization achieves registration to a standard through an accredited registrar that audits the organization against the desired standard. In turn, the registrar is accredited by a national "competent body" established for that purpose.

Critics of the third-party registration process argue that the expense of the third-party's efforts is significant. A suggested alternative is to incorporate either a first- or second-party auditing process. Briefly, a first-party system, which would result in self-declaration of conformity to the ISO standard, essentially involves internal self-auditing within an organization; a second-party system involves a purchaser of goods and services conducting audits and contract qualifications of its suppliers. Critics of a third-party registration system have voiced support for first- or second-party auditing/qualification systems.[8]

Domestically, third-party, ISO-harmonized verification of environmental, safety, and health program adequacy has been promoted as a possible mechanism to assist the Environmental Protection Agency (EPA) and the Occupational Safety and Health Administration (OSHA) in achieving their goals. Third-party registration of environmental management systems is being considered by EPA, which "recognizes the development of and growing reliance on international voluntary management standards . . . [which] . . . if properly crafted and implemented, can provide a powerful tool for organizations to improve . . . compliance with [regulatory] requirements and move beyond compliance through [innovation]."[9]

The implementation of an ISO-harmonized third-party registration policy by OSHA, for instance, would represent a new policy approach that would augment and depart from the agency's command-control tradition. Such a policy shift would offer opportunities for encouraging innovation by focusing on occupational safety and health management systems.[10]

In 1993, the National Performance Review (NPR), conducted by the U.S. government to evaluate the efficiency and effectiveness of the federal bureaucracy, stated as a goal that OSHA would utilize third-party certification mechanisms to fulfill part of the Clinton administration's "reinvention" plan.[11] Since this NPR recommendation, OSHA, its stakeholders, and academics have been discussing its viability. Assistant Secretary of Labor for Occupational Safety and Health, Joe Dear, has stated on a number of occasions that OSHA is interested in pursuing the third-party certification concept as part of the agency's reinvention activities.[12]

To this end, the concept of third-party certification was discussed at an OSHA stakeholder meeting in Washington, D.C. on July 20–21, 1994. Some stakeholders voiced concern about the concept. In response to stakeholder feedback, OSHA is still considering a third-party certification policy;[13] however, the policy was not directly mentioned in the agency's May 16, 1995 reinvention plan.[14]

An OHSMS in the Context of Traditional Command-Control Approaches

During the late 1960s and early 1970s a wave of regulations affected American society in ways that previous regulation had not. Some scholars asserted that these regulations controlled Americans in ways that inappropriately attempted to modify behavior.[15] The Occupational Safety and Health Act was one of the regulations in the wave. Just as other regulations of this era affected specific segments of society, the act had a significant impact on the American workplace. It can be inferred that both the criticisms of scholars regarding agency capture and the response of the judiciary had an affect on the manner in which the social regulations of the late 1960s and early 1970s were implemented. While the command-control instruments used by OSHA have come under intense scrutiny over the past decade, based on the political and legal environment in their early years, one can see the origins of their use.[16]

Command-control regulations have been widely used throughout America to control environmental, health, and safety externalities. The basic mechanism involved with command-control regulation is that a governmental agency sets a standard that must be met by the producer of a good or service. Parties who are affected by command-control standards are required to meet the minimum requirements of the standards or face some sort of punitive action, typically in the form of monetary fines.

Compliance with command-control standards is usually enforced through mandatory self-monitoring, agency inspection, or third-party complaints. A potential problem with market-based approaches is that they must base their outcomes on a consumer's willingness to pay. With occupational health and safety, willingness to pay equates to willingness to diminish one's own health and well being.

Some economists suggest that willingness to pay considerations are the most efficient means to address occupational health and safety externalities.[17] In these cases, an employee can accept or decline a job if the wage does not fairly reflect the hazards. There are significant social and ethical problems with this rationale which include problems with asymmetry of information and ability to pay considerations.

The application of an ISO-harmonized first-, second-, and third-party registration system, when combined with the voluntary compliance programs presented in OSHA's reinvention plan, represents a potentially powerful market-based policy. This approach would reflect the differences in marginal costs among different work sites and would allow those most familiar with the site—the employer and employees—the opportunity to develop a site-specific occupational safety and health management system that would reflect the site-specific marginal costs. These programs, if ISO compatible, would in turn be certified by an accredited third-party registrar. In the case of employers who choose not to participate in such a market-based approach, the traditional OSHA command and control regulatory structure would be in place.[18]

EXPECTED BENEFITS OF AN OHSMS

As ISO begins formal consideration of an OHSMS, some of the benefits and open issues identified are summarized in this subsection.[19]

National/International Benefits

Some of the benefits of an OHSMS include the following:

- The utility of ISO OHSMS consensus standards could benefit domestic and international workers. The development of these standards could alleviate some of the inequity inherent in OHS regulations and governmental compliance activities that differ from country to country. Relevant national specification standards for OHSMS would be nested in the program requirements of such an ISO-type standard. Conformance to the ISO standard requires compliance with relevant national and local standards.

- The ISO process is well developed and has within it the 9000 and 10000 series protocols for standardized auditing, auditors, and registration procedures.

- Prevention-oriented OHSMS programs should be integrated with the design phases of industrial processes, not considered as separate entities. As such, an ISO OHSMS would be compatible with the scope of ISO 9001, with the net effect of minimizing the number of internal and external audits to which companies are subjected. By harmonizing an ISO 9000 quality management system and an ISO 14000 environmental management system, companies could address the logistical and financial barriers associated with multiple external evaluations.

- Multinational corporations may benefit from the evolution of complex intercountry philosophies to a singular health and safety approach. Exchanges of expertise (within the same company) in OHS, resulting in substantial cross-training, might be encouraged since OHS professionals would be using similar procedures to resolve similar problems. ISO 9000 does not specify how companies must design quality systems, nor would an ES&HMS. Therefore, innovation would be encouraged.

- There could also be other incentives built in the system for attaining registration. Contractual language could require trade partners to be OHSMS registered to be considered for major business contracts. This could also apply to U.S. federal contract awards. Also, corporate insurance premiums could possibly be reduced by participation in an ES&HMS.

- The language of the World Trade Organization (WTO) agreement supports creation of, and participation in, development of international conformity assessment standards. The agreement also suggests that developed countries, when requested, assist developing trade partners (less developed countries, or LDCUs) in their efforts to comply with technical standards and to give them special consideration with conformance.

If the spirit and intent of the WTO are applied to an ISO ES&HMS, LDCUs could be provided time and technical assistance, without fear of trade retaliation, to develop conformance strategies that are suitable to local social and political conditions. Finally, the WTO as currently drafted would not interfere with American public or private standards making activities.

Benefits to Industry and Government

By promoting the use of industry-driven OHS management systems, an OHSMS policy reaffirms the belief that industry, the main engine of sustained economic growth, should be unfettered by specification standards. Third-party registration is a market-based strategy for compliance which thereby creates flexibility and incentives for innovation.

There could also be benefits built into the system that would be favorable to small ISO OHSMS registrants. Several large U.S. firms are already showing a preference toward using suppliers who conform to the principles coupled with ISO 9000 registration. Small ISO registered firms, traditionally outside of EPA and OSHA oversight, could be brought into progressive, mainline ES&H management.

An ISO OHSMS model of independent third-party inspectors could potentially provide a value added service to federal OSHA and its state partners. First, the registration process would not interfere with OSHA enforcement activities. Second, firms that receive ISO OHSMS registration could potentially be removed from OSHA's programmed inspection schedule, thus reducing some of the work load from their compliance personnel. As a consequence, OSHA could focus its limited resources on the most immediate health threats to American workers and the environment.

This is a policy reform that fits into the category of the application of existing knowledge in pioneering attempts to effect institutional change to promote innovation. It is a shift in protection strategies from the "pollutant-by-pollutant, end-of-pipe, command-control approach to a prevention system-oriented approach." Provisions would be incorporated into new and existing regulations and programs that maximize flexibility for industry.

Many environmental and public health professionals may take for granted the significant contributions third-party certification has already made toward the protection of public safety, public health, and the environment during the last century. The reliance on third-party certification has been important as an adjunct to regulatory oversight in all developed countries, including the United States. In this paradigm, third-party accredited registrars and certified auditors would become important national and international business entities.

CURRENT STATUS OF DEVELOPMENT

As this book goes to press, ISO prepares to determine whether or not it will pursue the development of OHSMS standards in the near future. As such, the purpose of the ANSI meeting is to provide a forum for stakeholder groups and interested parties to express their viewpoints on the development of both an ANSI and ISO OHSMS.[20]

The five stakeholder groups identified by ANSI for participation in the workshop are: business and industry, labor, government, standards-developing organizations, and the insurance industry. While the response to an ANSI/ISO OHSMS set of standards is mixed, a trend appears to be evolving-as a greater understanding of the issues is gained, support for such standards increases. A brief discussion of the various stakeholder groups' interests and related issues follows.

Business and Industry

Support for an ANSI/ISO OHSMS in business and industry is mixed. Some observers would say that this stakeholder group is clearly against the development of an OHSMS standard. This observation is tempered by the fact that numerous firms, independent of the ANSI/ISO standards-making process, have already begun to restructure their OHS programs to conform with the ISO 9000 model. Preliminary information indicates that these firms are realizing greater efficiencies as well as reduced illness and injury rates. Organizations that support the development of an ANSI/ISO OHSMS view a potential standard as logical, efficient, and effective. Many of these firms have, or are in the process of obtaining, ISO 9000 certifications that include OHS programs. The organizations that are supportive and have begun to incorporate an ISO-compatible OHSMS include IBM, Raychem, Monsanto-Canada, and Kajima Construction.

A chief concern of industry is that an ANSI/ISO OHSMS will add another layer of expense and effort on top of existing OSHA regulations; that is, there will be more "outsiders" telling them what to do. Further, even though there is an understanding of harmonizing international OHS standards, there is doubt and suspicion whether or not overseas competitors will faithfully comply. Another concern is the third-party registration aspect of a potential ANSI/ISO OHSMS.

In addition, the industry is concerned about the effect an ANSI/ISO OHSMS would have on labor relations. Many of the concerns expressed during OSHA's efforts to develop a comprehensive Occupational Health and Safety Program Standard are germane to this point.

Another important issue associated with the ISO management standards is whether small- and medium-sized organizations with modest capital resources can afford to invest critical time and money to acquire registration. If the OHSMS is overly complex and registration excessively time consuming or expensive, many small or financially unstable organizations will undoubtedly question their ability to achieve registration. On the other hand, the potential for a negative impact on the health of workers and the attendant liability exposure may be of importance equal to or greater than the cost of registration.

Consequently, it is these firms that may achieve the greatest health and safety improvements by submitting the organization to a comprehensive health and safety management systems analysis. In recognition of the worldwide trend towards the formation of smaller organizations, accommodations for their needs should be considered prior to development of audit tools and procedures.

Labor

As with business and industry, labor support for an ANSI/ISO OHSMS is mixed. However, the observed trend is that as this stakeholder group gains a greater understanding of OHSMS concepts, support has followed. An evolving position within labor is that an ANSI/ISO OHSMS represents a possible means to improve occupational health and safety, since it appears that industry has effectively stalemated the development of specification standards. Further, if a reliable third-party mechanism is included in an OHSMS, it would allow OSHA compliance officers the expanded opportunity to focus on industries and organizations with the most serious compliance problems.

Labor's primary concern is that an ANSI/ISO OHSMS might in some way weaken or undermine past gains in worker health and safety standards and regulations. If implemented, the existence of an ANSI/ISO OHSMS might be used as an argument by industry to eliminate OSHA field compliance staff. Further, if employee input is not encouraged or allowed, organizations may effectively hire auditors who will not be objective.

Government

Over the past several years, OSHA has expressed interest in the principles embodied in a potential ISO OHSMS as the agency has considered a third-party registration policy. Specifically, at the American Industrial Hygiene Association Conference in May 1994, Assistant Secretary Joe Dear said " . . . if we were to have independent third-parties certify work places, who would do the certifications? What would be their qualifications? How would we ensure program integrity? What incentives could be developed to move in that direction?"[21]

On other occasions, Dear has stated that OSHA is interested in pursuing the third-party concept as part of the agency's reinvention activities. To this end, the concept of third-party certification was discussed at the OSHA stakeholders meeting in Washington, D.C. on July 20–21, 1994. Some stakeholders voiced concern about the concept. The discussion on a third-party registration policy at the stakeholders meeting was not connected to an ISO OHSMS. Since the stakeholders meeting, important distinctions between third-party registration and OHSMS have evolved. As these distinctions have been refined, a number of stakeholders who originally voiced opposition are currently supportive.

ISO OHSMS Development

In relation to the Clinton Administration's "reinvention" agenda, many of the core principles reflected in an ISO OHSMS are embodied in Vice President Gore's National Performance Review. Among its recommendations, the NPR addresses regulatory reform in a number of respects. Action steps which support a core principle—customer satisfaction—call for the elimination of so-called "regulatory overkill." One specific action item calls for "the Secretary of Labor [to] issue new regulations for work site safety and health, relying on private inspection companies or nonmanagement companies."[22]

In response to the NPR, OSHA developed an agency-specific reinvention plan. Two regulatory approaches are presented in the plan. The first is a choice of partnership with OSHA for firms with strong and effective health and safety programs. The second is traditional OSHA enforcement for firms that do not implement strong and effective health and safety programs. Organizations that have implemented strong health and safety programs will be "given the lowest priority of enforcement inspection, the highest priority for assistance, appropriate regulatory relief, and penalty reductions up to 100 percent."[23]

The plan's recommendations are based on six principles "for the protection of America's workers." The third principle states that "OSHA will initiate strategic, public-private partnerships to identify and encourage the spread of industry best practices to solve national problems." This principle correlates to the ISO quality management approach, which is based on management commitment and employee involvement.

Standards-Developing Organizations

Three standards-development organizations have been active within this stakeholder group. The American Industrial Hygiene Association (AIHA) has taken a lead role in developing a consensus on a potential ANSI/ISO OHSMS. Along with the American Society of Safety Engineers (ASSE) and the American Society for Quality Control (ASQC), the AIHA has observed the ISO 9000 and ISO 14000 developments to determine if these existing standards will facilitate improvement in working conditions. These organizations have subsequently concluded that a set of separate standards, one of which focuses on the unique aspects of occupational safety and health management systems, would be beneficial. Consequently, AIHA has led a collaborative effort with partner organizations, public and private, to develop an ISO 9000/14000-compatible Occupational Safety and Health Management System (OHSMS) standard. If ANSI elects to move forward with the development of an American OHSMS standard, the AIHA OHSMS could serve as the initial working document for use in the ANSI consensus process.

The Insurance Industry

The best way to summarize this stakeholder group position is that it is in favor of any standard/policy that reduces illness and injury. To the extent that an ANSI/ISO standard does this, the insurance industry will be supportive; to the extent that it does not, it will oppose it. This stakeholder group's customer base is industry. As such, it is anticipated that they will align with the positions of industry. Insurance is intrigued with the potential assessment tools and protocols that may be used in an ANSI/ISO OHSMS. Interest has been expressed in the use of universally accepted auditing tools, which could result in a database of validated leading indicators of workplace health and safety.

AN OHSMS AND ITS RELATION TO EXISTING STANDARDS/POLICIES WITHIN THE ISO FRAMEWORK

TC 207 structured the ISO 14000 standards so that they are, to the extent feasible, aligned with the ISO 9000 series.[24] This precedent, and the current development of an OHSMS congruent with ISO 9001 and ISO 14001, suggests that critical features of an OHSMS assessment instrument would likely parallel those found in ISO 9001 and ISO 14001.[25]

Traditional environmental health and safety audits are conducted to assess numerous endpoints, including regulatory compliance assurance, program effectiveness, training adequacy, liability identification, appropriate resource allocation.[26]

Alternately, ISO 9000 and ISO 14000 conformity assessments evaluate the extent to which an organization maintains and documents its management systems, not necessarily the endpoints. For example, an ISO 14001 audit would verify that the organization maintains a system that ensures environmental regulatory compliance. The audit would not verify compliance in and of itself.

A formal OHSMS is an orderly arrangement of interdependent activities and related procedures that drive an organization's occupational health performance. An OHSMS assessment instrument would need to evaluate these system features. As such, the assessment instrument would not require the capture, analysis, or evaluation of exposure samples.

An ISO 9000/14000-harmonized OHSMS standard would likely contain five major elements. These elements are presented below, accompanied by the implications for the assessment instrument. The assessment instrument would evaluate the following[27]:

1. *The presence of occupational safety and health policy and performance objectives.* An assessment instrument would have to contain elements that evaluate whether the safety and health policy legitimately addresses relevant site conditions and activities. Does the policy guide the setting of appropriate performance objectives? Is there a written commitment to comply with statutory requirements and industry practices?

2. *The adequacy of the occupational health management systems to achieve the policy objectives.* The management system review would examine factors such as planning and organizational procedures. The effective presence of these two factors should indicate whether the organization has deliberate, viable mechanisms to achieve health and safety policy objectives.

3. *The competency of individuals implementing the systems.* The best designed systems may be poorly implemented unless capable individuals ensure maximum performance. Assessment instrument contents would have to include an evaluation of personnel adequacy.

4. *Risk assessment, risk management, risk communication, and risk documentation.* The assessment instrument should evaluate the effectiveness of the company's efforts to assess environmental working conditions. The adequacy of the management system would be evaluated in light of policy objectives and statutory requirements. The assessment instrument would also contain instructions to auditors to verify that root causes of identified health and safety problems are methodically mitigated. Company communication efforts to both internal and external stakeholders would be assessed also. Finally, information management systems and support documentation would be examined.

5. *Organizational continuous review and improvement.* An important component of ISO 14000 currently absent in ISO 9000 is the continuous improvement feature. An ISO OHSMS would likely require health and safety continuous quality improvement. The OHSMS assessment instrument would evaluate the effectiveness of organizational efforts to continuously review and improve working conditions.

The assessment instrument would ideally contain auditor instructions to evaluate the company's compliance with applicable governmental specification standards, without listing the standards individually. Ironically, ISO 14001 does not mandate compliance with statutory requirements. Under the ISO 14000 model, companies are required to show "commitment" to compliance with governmental environmental regulations.[28] Compliance with applicable regulation is always viewed as the minimal requirement for conformance with an ISO-harmonized management system standard.

Comparison with Public and Private Assessment Instruments

Numerous public and private health and safety assessment instruments currently exist.[29] While some reflect a unique corporate philosophy and appear compliance driven, others such as the federal Voluntary Protection Programs (VPP) and the International Loss Control Institute's International Safety Rating System (ILCI-ISRS) merit closer inspection due to their systems-based, non-industry-specific assessment approach.

The ILCI-ISRS is a comprehensive health and safety assessment instrument that uses a numerical scoring system to rate organizational conformance to safety systems/practices ILCI considers important. The auditing tool contains 20 basic element areas, which cover issues including accident/incident investigations, communications, and management of change.

An ISO OHSMS assessment instrument would likely evaluate similar factors. The ISRS also attempts to measure specifics that fall outside the purview of an OHSMS assessment instrument. For example, the ISRS requires the presence of loss control bulletin boards, an off-the-job-safety program, and detailed safety and health training for senior management. The ISRS uses weighting factors for each auditable item and applies these factors and scores to all applications of the system. This practice is also not in accordance with an ISO-harmonized model.

Where the ILCI-ISRS is very detailed, the substance of the federal VPP is considerably more general and will likely be compatible with the forthcoming OHSMS. Under the VPP model, participant companies must adequately implement a comprehensive safety and health management system that contains six major areas of emphasis. These include: management commitment and planning; hazard prevention and control; worksite analysis; safety and health training; employee involvement in program planning; and annual evaluation of health and safety management systems.[30]

Additionally, companies must over the three years preceding the site inspection maintain an average of lost workdays and injury case rates at or below the rates of the most specific industry national average statistics published by the Bureau of Labor. Federal OSHA conducts on-site reviews every three years to ensure participant companies are committed to continuous health and safety improvement.

Although the VPP is broadly compatible with the OHSMS concept, site assessments would likely differ from OHSMS conformity assessment in several key areas. First, the OHSMS would probably not mandate employee participation in health and safety program planning. Second, the OHSMS would not prescribe acceptable, specific lost-workday rates. Third, some system elements required under VPP need to be implemented for at least a year prior to the site inspection. Specific time constraints would not be evaluated under an OHSMS model. Fourth, VPP site inspectors do not utilize a formal assessment instrument. Site evaluations are conducted by answering broad-based, open-ended questions that evaluate the six major program elements.[31]

Perhaps the most vexing questions about an OHSMS assessment instrument pertain to instrument validity and reliability. Researchers have been unable to identify published studies that evaluate the accuracy and repeatability of either publicly or privately held occupational safety and health assessment instruments. An appropriately designed assessment instrument should be reliable and valid. These attributes would facilitate parity among conformance evaluations, generate user confidence, and assist in outcomes research.[32]

The ISO 14010-12 auditing standards contain verbiage that promotes consistency and reliability. ISO suggests audits be conducted by "well-defined methodologies and systematic procedures." Furthermore it states "different audits may require different procedures." Such vague guidance coupled with the inherent difficulties associated with interpreting the intent of the standard may lead to uneven interpretation and place companies at "audit risk."

Audit risk refers to the depth of detail site assessors may require in their data collection efforts. For example, if a U.S. site auditor is assessing the adequacy of a company's respiratory protection program (CFR 1910.134), how much data should he reasonably review? A respiratory protection program should contain at minimum 10 basic components. Should the auditor review all 10 or terminate his actions after assessing the written operating procedures? (Recall, ISO 9000/14001 are management, not performance, standards) What if one auditor examines all 10 parts of the program while a second reviews only written operating procedures? The conclusions drawn from the differing approaches may not coincide.

Under an OHSMS standard, health and safety auditors would also have to rethink traditional approaches to site assessments. Many U.S. federal and state health and safety inspections and private-sector audits tend to be reactive and prescriptive. Alternately, an OHSMS conformity assessment would evaluate proactive management systems in an approach somewhat similar to the federal VPP.[33]

Another issue associated with the ISO OHSMS is achievability. Can organizations with modest resources invest critical time and money to acquire registration? If the OHSMS is overly complex and registration excessively time-consuming or expensive, many small or financially unstable organizations will undoubtedly question their ability to achieve registration. On the other hand, the potential for a negative impact on the health of workers and the attendant liability exposure may be of equal or greater importance than the cost of registration.

By nature small and medium-sized firms often do not have a large pool of resources dedicated to OHS issues. Consequently, these firms may achieve major health and safety improvements by submitting the organization to a comprehensive safety and health management systems analysis. Accommodations for small or poorly capitalized organizations should be considered prior to development of audit tools and audit procedures.

The European study, which suggests customers generally do not perceive a difference between ISO 9000 registered and nonregistered suppliers, raises an interesting question. If stakeholders do not perceive a benefit from conducting business with an OHSMS registered company, can the expense associated with acquiring registration be rationalized? If employee working conditions do not continuously improve to stakeholder satisfaction, the credibility of the entire process may fall suspect.[34]

Modifications to the Occupational Safety and Health Act of 1970 (i.e., OSHA reform) currently under consideration include proposals that exempt employers from routine OSHA inspections if the place of employment has received a workplace review provided by a "certified person."[35] If this amendment is approved as drafted, a well-designed, nationally accepted health and safety assessment instrument may be useful for both governmental and nongovernmental purposes.

Ultimately, development of an organizationwide management system assessment instrument might be practical. It would integrate all features of organizational performance. Elements such as accounting, personnel, environmental aspects, occupational health, information systems, and quality system considerations would necessarily be included in one seamless assessment instrument. This would reduce audit fatigue associated with multiple site assessments and place employee health and safety alongside business aspects as equals in organizational priorities.

The Occupational Safety and Health Administration

Amid intense debate, the Occupational Safety and Health Act was passed in 1970 as Public Law 91-596. The goal of the act was "to assure as far as possible every working man and woman in the Nation safe and healthful working conditions and to preserve our human resources . . . by providing for the development and promulgation of occupational safety and health standards."[36]

The act was one of numerous statutes that have been called the "fourth wave" of social regulations. This group of regulations has been criticized because of their far-reaching effect on American life.[37] As the federal deficit has steadily increased over the past several decades—some would say partially as a result of the social regulations of the 1960s and 1970s[38]—public administrators have been exploring ways to accomplish statutory mandates with fewer resources.

With respect to OSHA and the third-party policy, Assistant Secretary of Labor Joe Dear stated "to me, it's the resource question which is compelling us to think differently about how we're going to encourage voluntary cooperation efforts."[39] To this end, since 1980, OSHA's FTEs (full-time equivalents) have been reduced 40 percent, while the American workforce has increased by 21 percent.[40] Based on the initial signals of the 104th Congress, OSHA's future funding prospects are not encouraging.

On July 20–21, 1994, OSHA brought together a diverse group of stakeholders from industry, labor, unions, and different sectors of government to discuss OSHA's future. At the meeting, Dear said:

> I believe programs like OSHA are at a critical juncture. We know what our resource picture is like. It's totally inadequate for the job at hand, but we have to live with that. And there's great public doubt and lack of confidence in programs like ours, rightly or wrongly. Maybe we haven't done a good job of communicating our message, but failure to address that issue will further undermine confidence.[41]

It was against this backdrop that OSHA and its stakeholders discussed a number of reinvention ideas to support the agency's original statutory mandate. Two of the issues discussed included the third-party registration policy and a focused inspection program.

From a policy formulation perspective, the development of an ANSI/ISO OHSMS standard presents OSHA with an intriguing framework from which it can re-examine proposed and existing policy development activities in the agency. Two distinct aspects of the standards process discussed herein are the development of an OHSMS standard and the use third-party registrars as part of the OHSMS standard.

It is possible that an OHSMS standard could be developed that would not include the same third-party registration mechanism used in the ISO 9000 standards. However, if the third-party registration aspect is included, it will provide OSHA with a viable model in consideration of past third-party policy activities in the agency.

The development of an ISO-harmonized third-party registration policy, combined with the voluntary compliance programs presented in OSHA's reinvention plan, would represent a significant paradigm shift from the agency's traditional command-control regulatory approach. The use of third-party registrars who have been accredited through the ISO accreditation process would reflect a unique combination of privatization and deregulation. Such an approach would be consistent with the principles of the National Performance Review and could support OSHA in successfully meeting both its short- and long-term challenges.

It is important to recognize that an OSHA/ISO OHSMS-harmonized third-party registration policy would not represent a voluntary program unto itself, nor would it represent regulation by the least common denominator. When implemented in conjunction with voluntary compliance programs that are based on established standards, the third-party registration approach allows organizations to take advantage of: (1) health and safety quality management systems tailored to reflect both the firm's marginal abatement cost functions and its employee's social welfare functions; and (2) economy-of-scale savings based on preexisting health and safety quality management systems.

Third-party registration is a quasi-market-based/private-regulation strategy for compliance, which thereby creates flexibility and incentives. While the market-based strategy is sound, there is the potential that all stakeholder interests may not be maintained. Employee representation is crucial in a third-party registration policy. If properly crafted with the inclusion of employee participation and consideration of welfare functions, the values of both efficiency and equity can be fulfilled.

The American Industrial Hygiene Association OHSMS Guidance Document

In response to the interest and success of ISO 9000 and its "quality systems" approach, the American Industrial Hygiene Association (AIHA) has developed and is issuing an OHSMS guidance document to assist employers and employees in improving safe working conditions. It is AIHA's intention that the OHSMS may be used by practicing health and safety professionals in the United States and elsewhere as a basis for designing, implementing, and evaluating OHSMSs that may be established by organizations.

It is compatible with the ISO 9000 series standards, since ISO 9001:1994 was used as a template for its structure. The AIHA system is not meant to serve as either a U.S. national standard nor as an international consensus standard; however, it may be helpful in the consensus process. The main body of the AIHA OHSMS is contained in the quality systems requirements of section 4.0 (see Table 14–1).[42]

The AIHA document specifies the basic elements of an OHSMS for organizations wishing to implement a deliberate, documented approach to anticipation, recognition,

Table 14–1

AIHA Quality System Requirements

4.1 Occupational Health and Safety Management Responsibility

4.2 OHS Management Systems

4.3 OHS Compliance and Conformity Review

4.4 OHS Design Control

4.5 OHS Document and Data Control

4.6 Purchasing

4.7 OHS Communication Systems

4.8 OHS Hazard Identification and Traceability

4.9 Process Control for OHS

4.10 OHS Inspection and Evaluation

4.11 Control of OHS Inspection, Measuring, and Test Equipment

4.12 OHS Inspection and Evaluation Status

4.13 Control of Nonconforming Process or Device

4.14 OHS Corrective and Preventative Action

4.15 Handling, Storage, and Packaging of Hazardous Materials

4.16 Control of OHS Records

4.17 Internal OHS Management Systems Audit

4.18 OHS Training

4.19 Operations and Maintenance Services

4.20 Statistical Techniques

evaluation, prevention, and control of OHS hazards. The document does not state specific technical requirements, nor does it provide a prescriptive implementation regimen.[43]

CONCLUSION

The ISO management systems model presents a novel approach to meeting OHS goals. Some believe that in a broader context, ISO 14000 and an ISO OHSMS may provide the necessary bridge between competing interests and paradigms in trade, environmental protection, and sustainable development. The ISO approach appears to offer an instrument whereby negative externalities can be internalized through a system of modified market incentives; namely, by tying environmental, health, and safety performance to market participation and procurement language.

The OHSMS approach has the potential of leading to the most significant paradigm shift in domestic and international OHS since the inception of federal OSHA and formation of the International Labor Organization, respectively. Further, initial analysis indicates that this approach has the potential for reducing occupational illness and injury while improving economic efficiency—variables traditionally considered to be inversely correlated.[44]

If an effective OHSMS standard/policy is to be formulated and implemented, active debate is necessary. While the OHSMS approach does not represent a "cure all" to OHS problems, many believe that this approach would establish structures and procedures that overcome the inherent limitations of the traditional command-control approach by replacing the paradigm with the inherently inclusive, flexible, and self-correcting systems model that extends the range of what is possible.

Endnotes

[1]Dyjack, David, and Steven Levine, "Critical Features of an ISO 9001/14001-Harmonized Health and Safety Assessment Instrument." *In New Frontiers in Occupational Health and Safety: A Management Systems Approach and the ISO Model,* eds. Charles Redinger and Steven Levine. (Fairfax, VA: American Industrial Hygiene Association, 1996). 128.

[2]Raychem, "Raychem EHS Standard." (Menlo Park, CA: Raychem Corporation, 1995).

[3]American Industrial Hygiene Conference and Exhibition, Kansas City, Missouri, June 1995.

[4]Bell, C., "Managing for Business Advantage: Optimizing Health and Safety Performance at Tenneco" (paper presented at the Tenneco Risk Management Conference, Houston, Texas, November 1994).

[5]Dyjack, David, and Steven Levine, *In New Frontiers in Occupational Health and Safety: A Management Systems Approach and the ISO Model,* eds. Charles Redinger and Steven Levine. (Fairfax, VA: American Industrial Hygiene Association, 1996). i.

[6]Ibid.

[7]Redinger, Charles, "A Paradigm Shift at OSHA: ISO-Harmonized Third-Party Registration of Occupational Health and Safety Programs." *In New Frontiers in Occupational Health and Safety: A Management Systems Approach and the ISO Model*, eds. by Charles Redinger and Steven Levine. (Fairfax, VA: American Industrial Hygiene Association, 1996). 35.

[8]Workshop on International Standardization of Occupational Health and Safety Management Systems. American National Standards Institute, workshop held May 7–8, 1996 in Rosemont, Illinois.

[9]"Voluntary Environmental Self-Policing and Self-Disclosure Interim Policy Statement," *Federal Register,* 60:63 (April 3, 1995). 16877.

[10]Redinger, Charles, "A Paradigm Shift at OSHA: ISO-Harmonized Third-Party Registration of Occupational Health and Safety Programs." *In New Frontiers in Occupational Health and Safety: A Management Systems Approach and the ISO Model,* eds. by Charles Redinger and Steven Levine. (Fairfax, VA: American Industrial Hygiene Association, 1996). 68.

[11]National Performance Review Commission, Washington, D.C.: Government Printing Office, 1993. 6–7.

[12]Redinger, Charles, "A Paradigm Shift at OSHA: ISO-Harmonized Third-Party Registration of Occupational Health and Safety Programs." *In New Frontiers in Occupational Health and Safety: A Management Systems Approach and the ISO Model,* eds. by Charles Redinger and Steven Levine. (Fairfax, VA: American Industrial Hygiene Association, 1996). 36.

[13]Ibid.

[14]Occupational Safety and Health Administration, "The New OSHA—Reinventing Worker Safety and Health." Washington D.C.: Government Printing Office, 1995.

[15]Lowi, Theodore J.; *The End of Liberalism: The Second Republic of the United States*, 2nd ed. (New York: W.W. Norton & Company, 1979).

[16]Redinger, Charles; "A Paradigm Shift at OSHA: ISO-Harmonized Third-Party Registration of Occupational Health and Safety Programs." *In New Frontiers in Occupational Health and Safety: A Management Systems Approach and the ISO Model,* eds. Charles Redinger and Steven Levine. (Fairfax, VA: American Industrial Hygiene Association, 1996). 56.

[17]Kelman, Steven; "Occupational Safety and Health Administration." *The Politics of Regulation*, ed., Wilson, James Q. (New York: Basic Books, Inc., 1980). 264.

[18]Occupational Safety and Health Administration, "The New OSHA—Reinventing Worker Safety and Health." Washington D.C.: Government Printing Office, 1995.

[19]Levine, Steven and David Dyjack, "Development of an ISO 9000-Compatible Occupational Health Standard II: Defining the Potential Benefits and Open Issues." *In New Frontiers in Occupational Health and Safety: A Management Systems Approach and the ISO Model,* eds. Charles Redinger and Steven Levine. (Fairfax, VA: American Industrial Hygiene Association, 1996). 117–19.

[20]American National Standards Institute. "An Outline of Issues: International Standardization of Occupational Health and Safety Management Systems." ANSI, New York, New York, March 1996. Prepared for the Workshop on International Standardization of Occupational Health and Safety Management Systems. American National Standards Institute, May 7–8, 1996 in Rosemont Illinois.

[21]Dear, J., Keynote Address, American Industrial Hygiene Conference and Exhibition, Anaheim, California, May 26, 1994.

[22]National Performance Review Commission, Washington, D.C.: Government Printing Office, 1993. 62.

[23]Occupational Safety and Health Administration, "The New OSHA—Reinventing Worker Safety and Health." Washington D.C.: Government Printing Office, 1995. 8.

[24]Rule 15.9/14th ed CMS, "Other ISO rules en route," *Environment Today*, 1995.

[25]Dyjack, David, and Steven Levine. "Critical Features of an ISO 9001/14001-Harmonized Health and Safety Assessment Instrument." *In New Frontiers in Occupational Health and Safety: A Management Systems Approach and the ISO Model*, eds. Charles Redinger and Steven Levine. (Fairfax, VA: American Industrial Hygiene Association, 1996). 133.

[26]Liebowitz, A. J., ed. *Industrial Hygiene Auditing—A Manual for Practice.* The American Industrial Hygiene Association, 1994. 6–7.

[27]Dyjack, David and Steven Levine. "Critical Features of an ISO 9001/14001-Harmonized Health and Safety Assessment Instrument." *In New Frontiers in Occupational Health and Safety: A Management Systems Approach and the ISO Model,* eds. Charles Redinger and Steven Levine. (Fairfax, VA: American Industrial Hygiene Association, 1996). 134–135.

[28]International Organization for Standardization. Unofficial ISO Draft International Standard 14001. ISO Technical Committee 207, June 26, 1995, Geneva, Switzerland.

[29]Dyjack, David, and Steven Levine. "Critical Features of an ISO 9001/14001-Harmonized Health and Safety Assessment Instrument." *In New Frontiers in Occupational Health and Safety: A Management Systems Approach and the ISO Model,* eds. Charles Redinger and Steven Levine. (Fairfax, VA: American Industrial Hygiene Association, 1996). 137.

[30]"Voluntary Protection Programs to Supplement Enforcement and to Provide Safe and Healthful Working Conditions; Changes," *Federal Register*, 7/12/88, 53:133, pages, 26339–26348.

[31]Dyjack, David, and Steven Levine. "Critical Features of an ISO 9001/14001-Harmonized Health and Safety Assessment Instrument." *In New Frontiers in Occupational Health and Safety: A Management Systems Approach and the ISO Model*, eds. Charles Redinger and Steven Levine. (Fairfax, VA: American Industrial Hygiene Association, 1996). 138.

[32]Dyjack, David, and Steven Levine. "Critical Features of an ISO 9001/14001-Harmonized Health and Safety Assessment Instrument." *In New Frontiers in Occupational Health and Safety: A Management Systems Approach and the ISO Model,* eds. Charles Redinger and Steven Levine. (Fairfax, VA: American Industrial Hygiene Association, 1996). 139.

[33]Wenmonth, B. "Quality Systems and Environmental Management," *Environmental Health Review* (Australia), 23(2), 1994, 41-54.

[34]Dyjack, David, and Steven Levine. "Critical Features of an ISO 9001/14001-Harmonized Health and Safety Assessment Instrument." *In New Frontiers in Occupational Health and Safety: A Management Systems Approach and the ISO Model,* eds. Charles Redinger and Steven Levine. (Fairfax, VA: American Industrial Hygiene Association, 1996). 141.

[35]U.S. Congress. "Safety and Health Improvement and Regulatory Reform Act of 1995" (H.R 1834). Washington D.C.: U.S. Government Printing Office, 1995, 20–22.

[36]29 USCA, 651(b) . . . (9).

[37]Lowi, Theodore J. *The End of Liberalism: The Second Republic of the United States,* 2d (New York: W.W. Norton & Company, 1979).

[38]Gilpin, Robert. *The Political Economy of International Relations.* (New Jersey: Princeton University Press, 1987).

[39]*BNA Occupational Safety & Health Reporter,* June/1/94; The Bureau of National Affairs, 1231 25th Street, N.W. Washington, D.C. 20037. 3.

[40]*BNA Occupational Safety & Health Reporter,* July/27/94; The Bureau of National Affairs, 1231 25th Street, N.W. Washington, D.C. 20037. 421.

[41]*BNA Occupational Safety & Health Reporter,* July/27/94; The Bureau of National Affairs, 1231 25th Street, N.W. Washington, D.C. 20037. 404.

[42]American Industrial Hygiene Association. Occupational Health and Safety Management System (OHSMS) Guidance Document. Fairfax Virginia, May 1996.

[43]Ibid.

[44]Dyjack, David, and Steven Levine. "Critical Features of an ISO 9001/14001-Harmonized Health and Safety Assessment Instrument." *In New Frontiers in Occupational Health and Safety: A Management Systems Approach and the ISO Model,* eds. Charles Redinger and Steven Levine. (Fairfax, VA: American Industrial Hygiene Association, 1996). iii.

ISO 9000 IN U.S. GOVERNMENT AGENCIES

A number of U.S. government agencies are considering using ISO 9000 as a common standard guideline, or requiring registration. Many agencies see ISO 9000 as a viable option to replace and/or supplement their current quality standards.

The Government/Industry Quality Liaison Panel (GIQLP) was formed in 1994 to promote the use of a governmentwide quality standard for purchasing. GIQLP has held meetings with other governnment groups to promote its goals, including the Interagency Council on Standards Policy (ICSP). The ICSP has authority from the Office of Management and Budget to write government procument policy.

Use of international standards was part of an amendment to the National Technology Transfer and Advancement Act of 1995. The law requires federal agencies that do not use international standards to provide a written statement explaining why other standards were chosen.

In 1995, GIQLP met with officials from 12 federal agencies, including the Department of Defense (DOD) and the National Aeronautics and Space Administration (NASA), to discuss common use of the ISO 9000 standard. A memorandum of understanding was signed at the meeting that pledged the agencies to work toward a common quality requirement for government contractors based on ISO 9000. Both NASA and DoD have dropped government quality system standards in favor of ISO 9000.

The following list of U.S. government agencies considering ISO 9000 is current as of the publication of this *Handbook*. For the most up-to-date information, please call the contacts listed for each agency.

DEPARTMENT OF AGRICULTURE (USDA)

The USDA is approaching use of the ISO 9000 standards on many fronts. The standards are being considered for use in both internal and external applications.

Contact: Dan Glickman, 1400 Independence Avenue SW, Washington, DC 20250. Tel: 202-720-2791; fax: 202-720-5043.

DEPARTMENT OF COMMERCE (DOC)

International Trade Administration (ITA)

The Office of European Community Affairs (OECA) coordinates U.S. participation in the U.S./EU talks designed to establish one or more mutual recognition agreements (MRAs) covering testing and certification of EU-regulated goods and (where relevant) the approval/registration of the quality systems under which such goods are produced.

OECA also tracks references to the use of the ISO 9000 series (or their equivalent—the EN ISO 9000 series) in EU directives and assesses the potential impact on U.S.-EU trade.

Contact: Charles Ludolph, Office of EC Affairs, ITA/DOC, Herbert C. Hoover Bldg., Room 3036, Washington, DC 20230. Tel: 202-482-5276; fax: 202-482-2155.

NATIONAL INSTITUTE OF STANDARDS AND TECHNOLOGY (NIST)

In response to queries from customers of NIST's Office of Measurement Services, NIST managers have been holding discussions related to quality and the need for an internal quality policy in specific program areas, particularly with respect to the ISO 9000 series. These discussions indicate considerable support among both managers and staff for articulating such a policy. NIST is currently exploring its available options.

Contact: Dr. Carroll Brickenkamp, NIST, Room 232, Building 820, Gaithersburg, MD 20899. Tel: 301-975-2015; fax: 301-926-4751.

NATIONAL VOLUNTARY CONFORMITY ASSESSMENT SYSTEM EVALUATION (NVCASE)

NIST has proposed its own program for accrediting registrars where an acceptable accrediting program does not already exist-the National Voluntary Conformity Assessment Systems Evaluation (NVCASE) program. More than two years after it was first proposed, the NVCASE program went into effect in May 1994. The purpose of the program is to establish criteria and a system to evaluate and recognize specific conformity assessment activities.

Contact: New Contact, Standards Code and Information Program, NIST, Building 101, Room A629, Gaithersburg MD 20899, Tel. 301-975-4030, fax 301-926-2871.

NATIONAL VOLUNTARY LABORATORY ACCREDITATION PROGRAM (NVLAP)

The NIST National Voluntary Laboratory Accreditation Program (NVLAP) operated by the Office of Standards Services is doing the following:

Working towards internal compliance with ISO 9000 requirements

Planning to offer its accredited laboratories (as an option) quality system registration to ISO 9002.

The assessment processes would be combined to provide NIST-accredited laboratories with a cost-effective option for obtaining laboratory accreditation and quality system registration at the same time. Staff members are completing training in ISO 9000 requirements and how to conduct assessments.

Contact: Albert Tholen, NVLAP, NIST, TRF Bldg., Room 112, Gaithersburg MD 20899, 301-975-4017, 301-926-2884.

OFFICE OF STANDARDS SERVICES' WEIGHTS AND MEASURES PROGRAM

During 1992, the Office of Standards Services' Weights and Measures Program began upgrading the State Laboratory Accreditation Program in cooperation with the National Voluntary Laboratory Accreditation Program. Technical criteria are being developed based on ISO/IEC Guide 25, General requirements for the competence of calibration and testing laboratories. New requirements include a revision of laboratory quality manuals to meet ISO/IEC Guide 25 requirements. The Weights and Measures Program believes that once laboratories are accredited under the new criteria, they will fully meet customer ISO 9000 service requirements.

Contact: Georgia Harris, Weights and Measures Program, NIST, Building 101, Room A617, Gaithersburg, MD 20899. Tel: 301-975-4014; fax: 301-926-0647; e-mail gharris@nist.gov.

NATIONAL OCEANIC AND ATMOSPHERIC ADMINISTRATION (NOAA)

The unit responsible for electronic and digital charts is planning to become internally compliant with applicable ISO 9000 requirements.

Contact: Russell Kennedy, NOAA/NOS, N/CG 2232, SMC 3 Station 6558, 1315 East West Hwy, Silver Spring MD 20910, tel. 301-713-2719, fax: 301-713-4543.

DEPARTMENT OF DEFENSE (DOD)

The DOD, in conjunction with NASA, published *Guidance on the Application of ISO 9000-ASQC Q90 Series Quality System Standards* in February 1994. The document officially approved the use of the ISO 9000 series standards in contracts to eliminate unnecessary quality system requirements and to create a single system. The decision gives contractors the option of deciding if they want to use ISO 9000 to satisfy government contractual requirements for quality systems.

DOD program offices are authorized to use the ISO 9000 series standards in contracts for new programs instead of MIL-Q-9858-A and MIL-I-45208-A. (Applying ISO 9000 to existing contracts will be considered on a case-by-case basis.)

In addition, third-party registration will not be required nor will it be a substitute for government quality surveillance at the present time. Both DoD and NASA purchasing offices have the authority to require compliance with the international quality assurance documents, but the DOD does not intend to force ISO 9000 on contractors.

Contact: Frank Doherty, chief of industrial quality and productivity, Department of Defense, OASD (P&L) PR/IEQ, Pentagon (Suite 2A318), Washington, DC 20301. Tel: 703-695-7915; fax: 703-693-7038.

DEPARTMENT OF EDUCATION (DOED)

While DOEd has no plans at present for using the ISO 9000 standards within its programs, DOEd is collecting information on the possible applications of the ISO 9000 series standards within the education and training fields.

Contact: Ron Hunt, Special Assistant to the Director, US Dept. of Ed., 400 Maryland Ave SW, Room 3061, Washington DC 20202-3643, Tel. 202-401-1953.

DEPARTMENT OF ENERGY (DOE)

The DOE plans to endorse aspects of the ISO 9000 series standards, which will be referenced in their safety guide series. This series provides supplemental information regarding acceptable methods for implementing specific provisions of the DOE orders and rules.

The ISO 9000 series has already been used as one of the models for DOE's quality assurance order 5700.C, which spells out requirements for DOE personnel and contractors.

The department is also looking at the ISO 10000 series, which includes guidelines for auditing quality systems and quality assurance requirements for measuring equipment. In addition, the DOE is looking at other nongovernment standards developed by organizations such as the American Society of Mechanical Engineers and the American Society for Quality Control.

Contact: Gustave Danielson, Department of Energy, Group Code EH62, 19901 Germantown Road, Germantown, MD 20874. Tel: 301-903-2954.

DEPARTMENT OF HEALTH AND HUMAN SERVICES (DHHS)

Center for Disease Control (CDC) and National Institute for Occupational Safety and Health (NIOSH)

NIOSH is considering ISO 9000 quality standards for implementation in its certification programs.

Contact: Richard Metzler, M/S 1138, NIOSH, 944 Chestnut Ridge Road, Morgantown, WV 26505-2888. Tel: 304-284-5713; fax: 304-284-5877.

FOOD AND DRUG ADMINISTRATION (FDA) AND CENTER FOR DEVICES AND RADIOLOGICAL HEALTH (CDRH)

FDA is revising its medical device Good Manufacturing Practice (GMP) regulations to include requirements related to design control. The revised GMP was published in the *Federal Register* on November 23, 1993. The comment period has ended, and the FDA is making some changes to the proposed rule. The final rule should be published sometime in 1996, with an effective date 180 days after publication.

The GMP will be reorganized, and some of its language modified to harmonize it with ISO 9001. The revised GMP regulations will incorporate the requirements of ISO 9001 plus supplemental requirements specific to medical devices that are found in the present GMP regulations.

Global Harmonization

A Global Harmonization Task Force composed of government and industry representatives from the European Union, Canada, Japan, and the United States has been established to harmonize supplementary device requirements to ISO 9001 and 9002 and to examine the need for new documents and/or programs to encourage uniformity of inspections of quality systems. The first meeting was held in January 1993 in Brussels, where representatives discussed harmonizing FDA, Canadian, EU and Japanese GMPs and guidelines. They also formed three study groups:

1. Comparison of Regulatory Schemes.

2. Harmonization of FDA and EU GMP Requirements (contained in EN 46001).

3. Harmonization of Guidance Documents.

This third group's goal is to develop a generic guideline for applying EN 46001 to medical devices. It should reflect any differences between the proposed, revised GMP and EN 46001. Overall, the task force is developing a common GMP that will be used as a foundation for harmonization and to develop a guideline to ensure that the GMP is uniformly interpreted. If successful, these actions will form a solid foundation for a mutual recognition agreement.

Contact: Pam Wojtowicz, FDA/Center for Devices and Radiological Health, 12720 Twinbrook Parkway, Rm. T123, Rockville, MD 20857. Tel: 301-443-3426; fax: 301-443-4196.

DEPARTMENT OF INTERIOR (DOI)

Office of Acquisition and Property Management (PAM)

The Office of Acquisition and Property Management has been assigned responsibility for this issue since February 1993, though no specific activities other than information collection are underway. Preliminary discussions are currently being held with other DOI personnel involved in quality assurance and environmental matters to better assess the proper placement of standards-related responsibilities, including the ISO 9000 series standards, within the department.

Contact: Wiley Horsley, Automated Systems Division, DOI, MS-5512, 1849 C Street NW, Washington, DC 20240. Tel: 202-208-3347; fax: 202-208-6301.

DEPARTMENT OF LABOR (DOL)

Mine Safety and Health Administration (MSHA)

MSHA's Approval and Certification Center (A&CC) has examined the ISO 9000 standard for use within the agency and has provided training for a number of its staff members. The A&CC has also considered seeking accreditation by a third party to demonstrate compliance with the requirements of ISO/IEC Guide 25.

Contact: Ken Sproul or John Fain, MSHA/DOL, Industrial Park Road, RR 1 Box 251, Triadelphia, WV 26059. Tel: 304-547-0400; fax: 304-547-0400.

Occupational Safety and Health Administration (OSHA)

OSHA in its rulemakings and written documentation requirements is seeking consistency with quality management objectives. The documentation requirements of ISO 9000 have been used, in part, as a guide to the documentation requirements in the Process Safety Management Standard. The relationship between the ISO 9000 series and process safety is being addressed on a limited basis in outreach presentations on the OSHA Process Safety Management Standard.

Contact: Thomas Seymour, USDOL, OSHA, 200 Constitution Ave NW, Room N3605, Washington, DC 20210. Tel: 202-219-8061.

DEPARTMENT OF STATE

Office of International Communications and Information Policy

While the Office of International Communications and Information Policy has no plans at present for using the ISO 9000 standards within its activities, it is continuing to collect information on the requirements and applications of the ISO 9000 series standards.

Contact: Earl Barbely, State Department/CIP Bureau, Room 6317, 2201 C Street NW, Washington, DC 20520. Tel: 202-647-0197; fax: 202-647-7407.

FEDERAL TRADE COMMISSION (FTC)

The FTC is collecting information on the requirements/applications of the ISO 9000 series standards and the use of ISO 9000 registration claims in advertising and labeling.

Contact: Sidney Steinitz, FTC (H-200), 6th and Pennsylvania NW, Washington, DC 20580. Tel: 202-326-3282; fax: 202-326-2050.

GENERAL SERVICES ADMINISTRATION (GSA)

The Office of Business, Industry, and Governmental Affairs is coordinating GSA's ongoing study of ISO 9000 in collaboration with members of the GSA working group on quality management system registration. Included in the GSA action plan are discussions with trade associations, corporations registered to ISO 9000, the Small Business Administration (SBA) and colleagues in other federal agencies.

Contact: New Contact, GSA/Office of Business, Industry, and Governmental Affairs, 18th and F Streets NW, Washington, DC 20405. Tel: 202-501-4177; fax: 202-501-2806.

INTERNATIONAL TRADE COMMISSION (USITC)

While the USITC does not participate in any standards-developing groups or become involved in procurements requiring the use of such standards, it is interested in the impact of the development and application of international standards, such as the ISO 9000 series standards, on international trade and competitiveness.

Contact: David Rohr, USITC, 500 E Street NW, Washington, DC 20436. Tel: 202-205-3041; fax: 202-205-2338.

NATIONAL AERONAUTICS AND SPACE ADMINISTRATION (NASA)

NASA, in conjunction with the DOD, published *Guidance on the Application of ISO 9000-ASQC Q90 Series Quality System Standards* in February 1994. The document officially approved the use of the ISO 9000 series standards in contracts to eliminate unnecessary quality system requirements and create a single system. The decision gives contractors the option of deciding if they want to use ISO 9000 to satisfy government contractual requirements for quality systems.

NASA's current policy, as described in NHB 5300.4 (1B), "Quality Systems—Quality Program Provisions for Aeronautical and Space Systems Contractors" (dated April 1969), will be phased out. The ISO 9000 series, supplemented with the NASA Augmentation Requirements developed during the October 1993 Johnson Space Center ISO 9001 working group meeting, will become the basis for contractual quality assurance requirements. Applying ISO 9000 to existing contracts will be considered on a case-by-case basis.

Third-party registration requirements are optional for NASA applications. It remains the responsibility of the procuring NASA organization to assess suppliers' quality management systems.

Contact: Charles Harlan, Director of Safety Reliability and Quality Assurance, NASA, Lyndon B. Johnson Space Center, Houston, TX 77058. Tel: 713-483-3191.

NUCLEAR REGULATORY COMMISSION (NRC)

The NRC has not initiated programs or activities to formally review the ISO 9000 series standards. The focus of NRC activities in the area of quality assurance is on the requirements contained in 10 CFR 50, Appendix B, and the related NRC regulatory guides and industry standards.

Typically, a formal NRC review would be conducted in response to a request from a licensee (an organization licensed by NRC to operate a commercial nuclear power plant) who sought to use or rely on the ISO 9000 series in order to comply with NRC regulations and/or license commitments. To date, the NRC has not received a request from a licensee. Absent a request, the NRC staff is monitoring the development and implementation of the quality assurance standards in the ISO 9000 series standards.

Contact: John Craig, Deputy Director (or) Owen P. Gormley, Division of Engineering, Office of Nuclear Regulatory Research, NRC, Washington, DC 20555-0001. Tel: 301-492-3872; fax: 301-492-3696.

OFFICE OF MANAGEMENT AND BUDGET (OMB)

The Office of Management and Budget issued a revised version of OMB Circular A-119, "Federal Participation in the Development and Use of Voluntary Standards," to the heads of executive departments and agencies on June 25, 1993. Closing date for final review and comments was July 30, 1993. Section 7a(2) of the circular states the following:

International standards should be considered in procurement and regulatory applications in the interests of promoting trade and implementing the provisions of the Agreement on Technical Barriers to Trade and the Agreement on Government Procurement (commonly referred to as the 'Standards Code' and the 'Procurement Code,' respectively).

Annual OMB reporting requirements on the implementation of this circular by agency heads are also included in the revised circular.

Contact: Chris Jordan, Office of Federal Procurement Policy, OMB, Washington, DC 20503. Tel: 202-395-6812; fax: 202-395-5105.

U.S. POSTAL SERVICE

The Postal Service has recently initiated an effort to determine how ISO 9000 will/could impact U.S. postal activities.

Contact: Donald J. Burke, Manager of Quality Assurance (or) Mark Nepi, U.S. Postal Service, N.B. 4000, 475 L'Enfant Plaza, Washington, DC 20260. Tel: 202-268-4166 (Burke), 202-268-4642 (Nepi); fax: 202-268-4012.

GUIDE 25 AND LABORATORY ACCREDITATION

by Peter S. Unger

INTRODUCTION

Internationally, as well as in the United States, considerable debate and confusion exists about the similarities, differences, and relationship between laboratory accreditation (usually performed using ISO/IEC Guide 25, General requirements for the competence of calibration and testing laboratories) and quality system certification (or registration) to one of the three ISO 9000 series of quality system models, usually 9001, 9002, or 9003. Laboratories usually seek ISO 9002 registration.

Quality system registration certifies that a quality management system meets a defined model (ISO 9001, 9002, or 9003). It does not certify the quality of a particular product or service for compliance with specific technical specifications.

ISO/IEC Guide 25—1990

Unlike the ISO 9000 series, ISO/IEC Guide 25 was not established primarily as a contractual model for use between suppliers and their customers. Its aims are to:

- Provide a basis for use by accreditation bodies in assessing competence of laboratories.

- Establish general requirements for demonstrating laboratory compliance to carry out specific calibrations or tests.

- Assist in the development and implementation of a laboratory's quality system.

Historically, Guide 25 was developed within the framework of third-party accreditation bodies. Its early drafting was largely the work of participants in the International Laboratory Accreditation Conference (ILAC), and the latest edition was prepared in response to a request from ILAC in 1988.

To understand the significance and purpose of Guide 25 and its relationship to ISO 9002, it is essential that it be viewed in light of its development history, which was initially to assist the harmonization of criteria for laboratory accreditation. Guide 25 is now being used by laboratory accrediting bodies throughout the world and is the basis for mutual recognition agreements among accrediting bodies.

Laboratory accreditation is defined in ISO/IEC Guide 2 as "formal recognition that a testing laboratory is competent to carry out specific tests or specific types of tests." The key words in this definition are "competent" and "specific tests." Each accreditation recognizes a laboratory's technical capability (or competence) defined in terms of specific tests, measurements, or calibrations. In that sense, it should be recognized as a stand-alone form of specialized technical certification—distinct from a purely quality management system registration—as provided through the ISO 9000 framework.

Laboratory accreditation may also be viewed as a form of technical underpinning for a quality system in much the same way that product certification could be considered as another form of complementary underpinning for a certified quality management system.

SIMILARITIES AND DIFFERENCES

The ISO 9000 series and ISO/IEC Guide 25 are used as criteria by third-party certification bodies and both contain quality systems elements. The systems elements of ISO 9000 are generic; those of ISO/IEC Guide 25 are also generic but more specific to laboratory functions. The textual differences between ISO 9002 and Guide 25 are obvious; however, when interpreted in a laboratory context, it is generally accepted that the systems elements of the two documents are closely compatible. In the introduction to Guide 25 this compatibility is acknowledged: "Laboratories meeting the requirements of this Guide comply, for calibration and testing activities, with the relevant requirements of the ISO 9000 series of standards, including those of the model described in ISO 9002, when they are acting as suppliers producing calibration and test results."

It is not true, however, that laboratories meeting the requirements of ISO 9002 will thus meet the requirements or the intent of Guide 25. In addition to its system requirements (which are compatible with ISO 9002), Guide 25 emphasizes technical competence of personnel for their assigned functions, addresses ethical behavior of laboratory staff, and requires use of well-defined test and calibration procedures and participation in relevant proficiency testing programs. Guide 25 also provides more relevant equipment management and calibration requirements, including traceability to national and international standards for laboratory functions; identifies the role of reference materials in laboratory work; and provides specific guidance relevant to the output of laboratories—the content of test reports and certificates—together with the records requiring management within the laboratory.

Although Guide 25 contains a combination of systems requirements and those related to technical competence, for laboratory accreditation purposes the Guide is only used as a starting point. Guide 25 recognizes in the introduction that ". . . for laboratories engaged in specific fields of testing such as the chemical field . . . the requirements of this Guide will need amplification and interpretation . . . ."

The American Association for Laboratory Accreditation (A2LA) system of laboratory accreditation places these additional, technology-specific criteria in special program requirement documents such as the "Environmental Program Requirements."

In addition, technically specific requirements for individual test methods must be met, and a laboratory's competence must be publicly recognized. To summarize, the following is the hierarchy of criteria for laboratory accreditation purposes:

- ISO/IEC Guide 25.

- Any field-specific criteria.

- Technical requirements of specific test methods and procedures.

Fundamental Difference

Quality system registration (ISO 9000) asks:

- Have you defined your procedures?

- Are they documented?

- Are you following them?

Laboratory accreditation asks the same questions, but then goes on to ask:

- Are they the most appropriate test procedures to use in the circumstances?

- Will they produce accurate results?

- How have you validated the procedures to ensure their accuracy?

- Do you have effective quality control procedures to ensure ongoing accuracy?

- Do you understand the science behind the test procedures?

- Do you know the limitations of the procedures?

- Can you foresee and cope with any technical problems that may arise while using the procedures?

- Do you have all the correct equipment, consumables, and other resources necessary to perform these procedures?

The registration of a laboratory's quality management system is a component of laboratory accreditation—not a substitute. Quality system registration of a laboratory to ISO 9000 misses a key element—technical validity and competence.

Unfortunately, quality system registration of laboratories is already being seen as an easier route to some form of recognition for a laboratory than full accreditation.

OTHER CONSIDERATIONS

Apart from comparisons on the similarities and differences between the purposes of ISO 9000 and Guide 25 and their use for third-party conformity assessment purposes, it is important to examine the differences in skills and emphasis of assessors involved in the quality system registration and laboratory accreditation assessments.

Quality system registration places emphasis on the qualifications of the assessor. The systems assessor (often referred to as the Lead Assessor) is expected to have a thorough knowledge of the requirements of that standard. In current practice internationally, a quality system assessment team may or may not include personnel who have specific technical backgrounds or process familiarity relevant to the organizations being assessed.

For laboratory accreditation, the assessment team always involves a combination of personnel who have:

- Expert technical knowledge of the test or measurement methodology being evaluated for recognition in a specific laboratory.

- Specific knowledge of the policies and practices of the accreditation body and the general systems applicable to all accredited laboratories. Thus, the laboratory accreditation assessment includes a technical peer-review component plus a systems compliance component.

Some other key differences between ISO 9000 and laboratory certification assessment processes are:

- Laboratory accreditation involves appraisal of the competence of personnel as well as systems. Part of the evaluation of a laboratory includes evaluation of supervisory personnel, in many cases leading to a recognition of individuals as part of the laboratory accreditation.

- The technical competence and performance of laboratory operators may also be witnessed as part of the assessment process. The loss of key personnel may affect the continuing accreditation of the laboratory by the accrediting body. For example, A2LA recognizes key staff whose absence would reduce the laboratory's technical competence and may prompt a reassessment before it would be normally scheduled.

LABORATORY PRODUCT

The final product of a laboratory is test data. In many cases, laboratory accreditation assessments also include some practical testing of the laboratory through various forms of proficiency testing (interlaboratory comparisons or reference materials testing).

Quality system registration is not normally linked to nominated key personnel. The technical competence of managers and process operators is not a defined activity for quality system assessment teams. It is through the documented policies, job descriptions, procedures, work instructions, training requirements of organizations, and objective evidence of their implementation that quality system registrars appraise system personnel. Staff turnover is not an issue in maintaining registration status.

COMPLEMENTARY FUNCTIONS

Despite the differences in assessment and registration criteria, ISO 9000 and Guide 25 are complementary. Quality system registration for a laboratory should be viewed as a measure of a laboratory's capability to meet the quality expectations of its customers in terms of delivery of laboratory services within a management system model as defined in ISO 9002 or 9001—a "quality" job.

Furthermore, laboratory accreditation should be viewed by customers as an independent reassurance that a laboratory has the technical and managerial capability to perform specific tests, measurements, or calibrations—a "technically competent" job.

If validation is needed on both these characteristics, then a combination of quality system certification and laboratory accreditation may be appropriate. If a laboratory's function is purely for internal quality control purposes within an organization, it may be appropriate for the laboratory to operate within the overall ISO 9002 framework of the parent company. Nevertheless, such laboratories and their senior management may also benefit from the external, independent appraisal provided by the technical assessors used in laboratory accreditation.

However, if a laboratory issues certificates or reports certifying that products, materials, environmental conditions, or calibrations conform to specific requirements, they may need to demonstrate to their clients or the general community that they are technically competent to conduct such tasks. Laboratory accreditation provides the independent measure of that competence.

SCOPE OF ACCREDITATION/REGISTRATION

Organizations may be registered to a quality system standard within very broad industry or product categories. Naturally, organizations with a very narrow product range are registered in these terms.

Laboratories, on the other hand, are accredited for specific tests or measurements, usually within ranges of measurement, with associated information on uncertainty of measurement, and for particular products and test specifications.

Accreditation bodies encourage laboratories to endorse test reports in the name of the accreditation body to make a public statement that the particular test data presented has been produced by a laboratory that has demonstrated to a third party that it is competent to perform such tests.

The ISO 9000 series of standards are not intended to be used in this way. They address the quality system, not specific technical capability. The use of a quality system registration body's logo should not be used as a certification mark or endorsement as to the conformity of a particular product with its specified requirements. Similarly, it should not be used to endorse the competent performance of tests, calibrations, or measurements reported by laboratories. Only a logo or endorsement showing accreditation to Guide 25 or its equivalent for specific calibrations or tests denotes technical credibility and an expectation of valid results. Laboratories registered to ISO 9000 cannot make the same claim.

THE SPECIAL ROLE OF ACCREDITED CALIBRATION LABORATORIES

The ISO 9000 series requires that ". . . suppliers shall . . . calibrate . . . inspection, measuring and test equipment . . . . against certified equipment having a valid known relationship to nationally recognized standards."

Many calibration certificates presented to quality system auditors contain statements that the measurements or calibrations are "traceable to national standards." Some auditors also insist that suppliers' calibration documents provide cross-reference to the other reference standards used to calibrate their own devices providing a documented chain of traceability back to their own country's or international standards of measurement.

Since one of the fundamental requirements for accredited calibration laboratories is to have their own equipment traceable to national and international standards, both the interest and spirit of the ISO 9000 requirements are thus met when accredited calibration laboratories are used by suppliers. This principle has been recognized in ISO Standard 10012.1-1992. Clause 4.15, Traceability, states that ". . . the supplier may provide the documented evidence of traceability by obtaining his calibrations from a formally accredited source."

EUROPEAN POSITION

In April 1992, a statement issued by the European Organization for Testing and Certification (EOTC) included the following:

> . . . the only acceptable stand is to state that QS certification cannot be taken as an alternative to accreditation, when assessing the proficiency of testing laboratories. Not trying to underrate the QS certification procedure, it should nonetheless be underlined that, by being intended as a systematic approach to the assessment of an extremely broad scope of organizations and field of activity, it cannot include technical requirements specific to any given domain.

CONCLUSION

Before laboratories jump on the ISO 9000 bandwagon, they should understand whether this type of third-party recognition is really appropriate for the needs of their customers.

For the user of test data, the quality management systems approach to granting recognition to laboratories may be seen as deficient because it does not provide any assessment of the technical competence of personnel engaged, nor does it address the specific requirements of particular products or measurements. The ISO 9000 series state explicitly that they are complementary to—not alternatives to—specified technical requirements.

Users of test data should be concerned with both the potential for performing a quality job (quality system) and technical competence (ability to achieve a technical result). The best available method of achieving these two objectives is through laboratory accreditation bodies, operating according to best international practice, requiring laboratories to adopt best practices, and engaging assessors who are expert in the specific tests in which the customer is interested. Acceptance of test data, nationally or internationally, should therefore be based on the application of Guide 25 to assure the necessary confidence in the data's validity.

REFERENCES

ISO/IEC Guide 2—1993, General terms and their definitions concerning standardization and related activities.

International Laboratory Accreditation Conference, "Validity of Laboratory Test Data: The Application of ISO Guide 25 and ISO 9002 to Laboratories," June 1993.

Russell, Anthony J. 1994. "Laboratory Accreditation in a World-wide Perspective," Pittcon, (March 7).

International Laboratory Accreditation Conference Committee 1 on Commercial Applications, "Conformity Assessment: Testing, Quality Assurance, Certification, and Accreditation," February 1994.

European Organization for Testing and Certification (EOTC/AdvC/34/92), "Ascertaining the Competence of Test Laboratories in the Framework of EOTC Agreements Groups," April 15, 1992.

Bell, Malcolm. 1994. "Laboratory Accreditation," *TELARC Talk* (December).

Neumann, Lynne. 1995. "Validity of Test Data: The Application of ISO Guide 25 or ISO 9002," *ILAC Newsletter,* 8 (December).

THE SUPPLIER AUDIT CONFIRMATION

The Supplier Audit Confirmation (SAC) program is one of various industry lead efforts aimed at improving the effectiveness of the third-party ISO 9000 registration process. Under the SAC program, the results of a supplier's internal audits are taken into consideration by a third-party registrar. Supporters of the program assert that suppliers with mature quality systems can benefit from a system of more aggressive internal audits teamed with reduced external audits.

While the concept has been controversial since its introduction in 1994, it has slowly gained critical support among registered companies nationally and internationally by the technical organizations that support the registration and accreditation process. More than 50 companies, including some of the most recognizable names on the Fortune 500 list, have shown an interest in the concept. The program was first proposed by Hewlett-Packard and Motorola.

The Registrar Accreditation Board (RAB) and the American National Standards Institute have supported the concept of utilizing approaches such as the SAC program to allow reductions in audit activity to supplier systems that are in a mature state of implementation. A number of internationally recognized registrars also support the concept.

BASIC PRINCIPLES OF SAC

Here are some of the basic principles of SAC and how they apply to suppliers:

- Only suppliers that have "demonstrated effectiveness" of the management system can be allowed to use the SAC assessment methodology. A measure of demonstrated effectiveness is the degree to which the culture of the organization has embodied and implemented the concepts of management review, internal audit, and corrective and preventive actions over time. An organization that has practiced over a period of time (e.g. three years) and can show evidence of continuous application of these concepts should be considered to have demonstrated effectiveness.

- SAC assessment should be fully compliant with ISO/IEC Guides 61 and 62 as well as other relevant ISO/IEC guides.

- SAC assessments should not have restrictions based on size of the supplier organization, single or multisite registration, or industry sector of operation.

- SAC assessment should be based on third-party assessment methodologies that will result in an accredited registration by an accredited registrar. The resulting registration shall have equivalent integrity to any other accredited registration to the same standard, regardless of the assessment methodology utilized.

- SAC assessment should be based on a third-party assessment methodology that is reproducible, based on a set of principles and guidance, based on requirements relating to registration, and serves as a common framework from registrar to registrar.

- SAC assessment should be based on a nonprescriptive assessment methodology that does not add to or limit the interpretation of the requirements of the management system standard to a supplier. It shall allow for customization of the registration scope (e.g., multisite, business unit, company-level), as seen from the purchaser's perspective. The systems shall allow local site autonomy, providing a supplier organization with flexibility in design and implementation of its management system without defining a required organizational structure (e.g., central audit function need not be a requirement for a multisite registration).

- SAC assessment should allow flexibility in design and implementation of the assessment plan and schedule. For a supplier with demonstrated system effectiveness, the registrar may place reliance on the supplier's internal resources, under controlled and verified conditions.

- SAC assessment should minimize or eliminate non-value-added activities or processes that do not contribute value directly to the purchasers, suppliers, and governments of registration.

For the most current information on the program, visit the Hewlett Packard SAC home page at http:// www.corp.com/publish/ISO 9000/SAC or contact David Ling, Hewlett Packard Co., 1501 Page Mill Road, MS-5UL Palo Alto, California 94304 USA. Phone (415) 857-5057; e-mail: saciso9000@hpcea.ce.hp.com

QS-9000 Quality System Requirements

15

The Big Three automotive agreement on a common set of quality system requirements for its Tier-1 suppliers will have a tremendous impact on an industry that affects nearly every sector of the U.S. economy and has a significant impact on economies around the world.

Chapter 15 contains a basic explanation of these requirements by three experts from Alamo Learning Systems along with a history of QS-9000's development. In addition, 3-C Technologies, Inc. President Devan Capur explains how to implement QS-9000 as part of a useful business system that will ultimately affect how a company conducts business and manages its employees.

Finally, Chapter 15 offers a joint statement from the Big Three on why QS-9000 is important and its ultimate effect on industry nationally and internationally.

AN INTRODUCTION TO QS-9000

by Dave Fleischli, Bill Hooton, and Guy Hale

Quality System Requirements 9000 (QS-9000) is a fundamental set of quality management system requirements cooperatively developed and adopted by the Big Three automakers, Chrysler, Ford, and General Motors, along with certain truck manufacturers and other participating internal and external companies supplying production, service parts, and materials (see Box 1).

QS-9000 incorporates all 20 elements of the most comprehensive ISO 9001 standard and describes the minimum quality system requirements the Big Three auto manufacturers expect of its suppliers. QS-9000 compliant companies are required to register to either ISO 9001 (design and manufacturing) or ISO 9002 (manufacturing alone). QS-9000 requirements emphasize continuous improvement, defect prevention, consistency, and elimination of waste. (See Chapter 4 for additional requirements beyond ISO 9001 or 9002.)

The primary intent of the QS-9000 requirement for the automotive manufacturer according to Automotive Industry Action Group documents is "the development of fundamental quality systems that provide for continuous improvement, emphasizing defect prevention and the reduction of variation and waste in the supply chain." A QS-9000 compliant supplier will also benefit due to a reduced demand for second-party audits performed by the original equipment manufacturers (OEM). QS-9000 is not a product certification standard.

COMPANIES AFFECTED

Chrysler will require all Tier 1 production and service part suppliers to be third-party registered to QS-9000 by July 31, 1997. Ford does not require a QS-9000 audit by a third-party registrar; however, Ford did expect compliance with QS-9000 by June 1995. According to Ford, compliance means that the suppliers must:

1. Conduct a self-assessment.

2. Identify nonconformance issues.

3. Install a work plan to address these issues.

New suppliers to GM North America were to have been QS-9000 registered by January 1, 1996. All suppliers to GM Europe were to have been ISO 9000 registered by January 1, 1996. General Motors expects all of its existing production and service parts suppliers to be QS-9000 registered by December 31, 1997.

QS-9000 is not a replacement for any regulatory standard. Automobile and truck manufacturers may use QS-9000 in place of industry-specific standards, such as General Motors "Targets for Excellence" and Ford's Q-101 Quality System Standard.

REGISTRATION PROCESS

A number of internationally accredited ISO 9000 registrars hold QS-9000 accreditation. Registrars may seek accreditation to perform QS-9000 audits from at least 17 international bodies, including as of the July 1, 1996: the Registrar Accreditation Board (RAB) in the United States; the Dutch Council for Accreditation (RvA) in The Netherlands; and the United Kingdom Council for Accreditation Services (UKAS) in Great Britain. Reg-

Box 1

History of QS-9000

The history of the Big Three's QS-9000 requirements can be traced back to a June 1988 Supplier/OEM summer workshop sponsored by the American Society for Quality Control's (ASQC) Automotive Division.

At the meeting, concerns were raised over redundant documentation and audit requirements imposed individually by the Big Three. From this meeting the Supplier Quality Requirements Task Force was formed and charged with the task of harmonizing existing supplier documents and procedures. The group produced a series of documents beginning in 1990 beginning with the Initial Sample Warrant Form and the Measurement System Analysis Reference Manual. Over the next three years the task force produced four more documents, including:

- 1991—Fundamental Statistical Process Control Reference Manual.

- 1993—Production Part Approval Process (PPAP).

- 1993—Potential Failure Modes & Effects Analysis Reference Manual.

- 1994—Advanced Product Quality Planning & Control Plan Reference Manual.

In December 1992, the task force was given the green light from their Big Three managements to begin the process of developing a single document governing supplier evaluation and approval.

By September 1994, the task force released Quality System Requirements: QS-9000, which was based upon the 1994 revision of the ISO 9001 standard. A companion document, Quality System Assessment (QSA), was released at the same time. Representatives of the Truck Advisory Group (Freightliner Corporation, Mack Trucks, Inc., Navistar International Transportation Corporation, PACCAR, Inc., and Volvo GM Heavy Truck Corporation) participated in the development of QS-9000 and some adopted it in mid-1994.

More than 100 other auditable industry and individual automobile maker-specific requirements were added to the basic ISO 9001 standard. In February 1995 a second edition was released with additional third-party registration guidelines.

Interpretations

In early 1994, a working group consisting of the Big Three, registrars, accreditation bodies, and suppliers was formed to address common issues supporting the launch of QS-9000. That group is now known as the International Automotive Sector Group (IASG) and continues to meet semiregularly to address supplier questions concerning the process of assessment and the criteria of QS-9000. The IASG released the first officially sanctioned QS-9000 interpretations in March 1995. Updates to the interpretations are released several times a year, including two planned for 1997. These interpretations of the requirements are considered binding.

istrars are also admonished to follow a Code of Practice, in the registration process (Appendix B, QS-9000). (See Chapter 17 for more on accreditation of registrars.)

The registration contract is subject to reapproval every three years. To ensure that companies are following ISO 9000/QS-9000 requirements after certification is obtained, the registrar may conduct unannounced audits approximately twice a year.

THE QS-9000 AUDIT?

A QS-9000 audit process verifies implementation and compliance with the ISO 9000 standard plus additional industry and customer-specific requirements. The audit covers all elements of the supplier's quality system, even if these elements exceed QS-9000. Each on-site audit also includes a review of:

- Customer complaints and the supplier's response.

- Supplier internal audits.

- Management review results and actions.

- Progress made toward continuous improvement.

QUALITY SYSTEM ASSESSMENT (QSA)

The Quality System Assessment (QSA) is a list of general assessment instructions and guidelines used to determine QS-9000 conformance. The QSA may be used as a supplier's self-assessment or by a QS-9000 registrar. In addition, the QSA describes the assessment process and methodology.

QS-9000 compliant companies are subject to surveillance audits at approximately six-month intervals. A complete reassessment is not required. The surveillance audit ensures maintenance of the quality system. The re-registration provides for the long-term qualification of the company. QS-9000 currently applies to first-tier suppliers only, but may be extended to include second and third-tier suppliers as well at the customers' discretion.

In addition to requirements beyond ISO 9000, QS-9000 suppliers must meet a set of requirements specific to the auto industry and other requirements that are specific to individual manufacturers.

ADDITIONAL QS-9000 REQUIREMENTS

Chapter 4 of the *Handbook* discusses in detail additional requirements beyond the basic ISO 9001 standard. However, suppliers must meet both industry- and automaker-specific requirements beyond ISO 9000 registration. The following is a synopsis of sections II and III of QS-9000.

Section II of QS-9000 requires the following:

Box 2

QS-9000 Information Sources

QS-9000 Registered Company Directory

The American Society for Quality Control (ASQC) is the official Big Three sanctioned provider of the QS-9000 Worldwide Certified Company Directory. The directory includes all QS-9000 certified companies, all QS-9000 recognized accreditation bodies, and all QS-9000 qualified registrars. The directory is available on diskette, in printed form, and on the Internet. The database is updated monthly and printed quarterly.

ASQC

611 East Wisconsin Avenue, PO Box 3005, Milwaukee, WI 53201-3005
Telephone: (414) 272-8575
Fax: (414) 272-1734
E-mail: qic@asqc.org
QS-9000 Web Site: http://www/asqc.org/9000

ISAG Sanctioned QS-9000 Interpretations

In addition to the ISO 9001 standard, industry, and manufacturer-specific requirements, the Big Three published officially sanctioned interpretations of QS-9000. The interpretations are created by the IASG, which is made up of Big Three recognized Big Three accreditation bodies, QS-9000 qualified registrars (nominated from the Independent Association of Accredited Registrars (IAAR) and the Independent International Organization for Certification (IIOC), and Tier 1 automotive suppliers.

The interpretations are generated from questions submitted to the IASG by suppliers and other interested parties. The IASG meets quarterly.

Copies of the latest IASG-Sanctioned Interpretations are available from the ASQC. Companies wishing to submit questions to the IASG must fax those inquiries in English to 614-847-8556. Companies may also access interpretations from the ASQC QS-9000 Web Site at http://www.asqc.org/9000

Big Three Direct Contacts

Chrysler
Mr. Warren Norrid, CIMS: 484-08-02, 800 Chrysler Drive East, Auburn Hills, Michigan 48236-2757, Telephone: (810) 576-2701, Fax: (810) 576-2225.

Ford
Mr. Steve Walsh, Box 1517-A, NAAO, 17101 Rotunda, Dearborn, Michigan 48121
Telephone: (313) 845-8442, Fax: (313) 322-9778.

(Continued on next page)

(Continued from page 579)

General Motors
Mr. R. Dan Reid, GM-Powertrain Group, 895 South Joslyn Avenue, Pontiac, Michigan 48340-2920 (Stop 2R29), Telephone: (810) 857-0295, Fax: (810) 857-0301.

Organizational Contacts

Automotive Industry Action Group (AIAG)
The AIAG is a not-for-profit trade association organized to provide a forum to encourage communication, standardize business practices, and provide education for North American vehicle manufacturers and suppliers. Copies of the QS-9000 requirements can be purchased through the group. Contact AIAG, 26200 Lahser Road, Suite 200, Southfield, MI 48934; Order Telephone: (810) 358-3003, Other Inquiries: (810) 358-3253, Fax: (810) 358-3253.

Production Part Approval Process (1.0)

- Follow and comply with requirements in the *Production Part Approval Process* (PPAP) manual.

- Get approval for changes from original approval. May result in re-submittal.

- Verify that engineering changes are properly validated.

- Take responsibility for subcontractor materials and services.

- Get separate approval for "appearance items."

Continuous Improvement (2.0)

- Deploy a comprehensive continuous improvement philosophy that includes timing, delivery, price, and impact on all business and support activities, with specific action plans for processes most important to the customer. Construct an action plan.

- Identify opportunities for perfection of process methods as part of attribute characteristic. For variable processes, meet target values and reduce variation.

Opportunities for improvement include[1]:

- Machine downtime.

- Machine changeover times.

- Excessive cycle time.

[1]Adapted from *Memory Jogger 9000,* by Robert W. Peach, Diane Ritter, GOAL QPC, 1996.

- Scrap, rework, and repair.

- Use of floor space.

- Excessive variation.

- Less than 100 percent first-run capability.

- Process averages not centered on target values.

- Testing requirements not justified.

- Labor and materials waste.

- Excessive nonquality cost.

- Difficult assembly or installation.

- Excessive handling and storage.

- Re target values.

- Measurement system capability.

- Customer dissatisfaction.

Manufacturing Capabilities (3.0)

- Use a cross-functional team approach for the advanced quality planning process and evaluate current operations and processes. Consider plant layout, overall work plan, automation, ergonomics, line balance, production inventory levels, and labor content.

- Address potential sources of nonconforming units, using a methodology to prevent mistakes.

- Provide appropriate resources for tool and gauge design, fabrication, and full dimensional inspection, tracking and follow-up of subcontractors, and identification of customer-owned tools and equipment.

- Implement a system for tooling management, including maintenance and repair facilities and personnel, storage and recovery, set-up, and tool change programs for perishable tools.

Section III of QS-9000 contains requirements specific to Chrysler, Ford, General Motors, or truck manufacturers.

Chrysler-Specific Requirements

- Be knowledgeable of special requirements on parts identified with symbols:

  —shield (S) identifies safety characteristics;

—diamond (D) identifies special characteristics of a component, material, assembly, or vehicle assembly operation;

—pentagon (P) identifies special characteristics of fixtures, gages, developmental parts, and initial product parts.

- Provide a complete annual layout inspection to assure conformance.

- Conduct an internal quality system audit at least once a year, unless otherwise determined by the appropriate Chrysler representative.

- Perform design validation/production verification at least once per model year, unless a different frequency is specified.

- Submit a corrective action plan addressing all nonconformances.

- Be familiar with Chrysler packaging, shipping, and labeling.

- Perform a review of the supplier's process, using process sign-offs on new product by the advanced quality planning team.

Ford-Specific Requirements

- Be knowledgeable of special requirements on parts identified with an inverted delta ∇. These are control item products that have critical characteristics that may affect safe vehicle operation and/or compliance with government regulations.

- Control heat-treating processes according to Ford Manufacturing Standards and provide heat-treated steel components in conformance with Ford Engineering Material Specifications.

- Stop production shipments immediately, pending analysis of the process and corrective action, where Engineering Specification (ES) tests fail to meet performance requirements.

- Perform ongoing process and product monitoring on products for which SPC is in use and perform qualification of product characteristics on products not controlled with SPC and not included in a control plan.

- Implement Ford QOS standardized practices to achieve ever-increasing levels of customer satisfaction.

General Motors-Specific Requirements

- General Motors has published its specific requirements in separate guidelines. Questions on the content of these guidelines should be directed to the appropriate GM procuring division. Suppliers should verify at least annually that they are using the latest version of these documents.

Truck Manufacturers-Specific Requirements

- The truck manufacturers (Freightliner Corporation, Mack Trucks, Inc., Navistar International, Paccar Inc., and Volvo GM Heavy Trucks) have published their respective additional requirements separately. For information on these requirements, contact the truck manufacturer's Purchasing Department (the *QS-9000 Quality System Requirements Manual* itself does not list specific requirements).

Other Auto Industry-Specific Standards

The Big Three have discussed issuing other voluntary industry-specific standards, but none have been issued. However, requirements for the tooling and equipment industry, tentatively called TE supplement to QS-9000, were being discussed in mid-1996 when this *Handbook* went to press.

Box 3

Getting Started on QS-9000

Here is a simple 10-step program for QS-9000 implementation:

Step 1

Very early in the project one or more people, depending upon the size of the organization, must become experts in understanding the requirements. Identify these people and get them trained through courses available from the Automotive Action Industry Group (see Box 2 for contact information) and other sources. Obtain copies of the QS 9000 standard and related documents from the AIAG and study them. Begin selling the benefits of certification to upper management. Upper management can only become committed to certification when the benefits are understood. Buy-in from management can be a long-lead-time activity, so start early.

Step 2

Review your current business processes against the requirements in the standard. Develop a map to describe how, in general terms, the organization will address the requirements. This is your target system.

For example, purchasing (4.6) contains several unique requirements. The map should list how the organization plans to meet the element 4.6 requirements, regardless of whether the requirements are currently being met. Include in the map the name of the person responsible for meeting the requirement. As much as possible at this point collect opinions from the key process owners on how compliant each individual requirement is.

(Continued on next page)

(Continued from page 583)

The resulting outline will be a list of all the components that make up the quality system. These components can be policies, processes, procedures, forms, and records. At this stage these components can be formal or informal, informal meaning not documented or written. Once every component listed on the map is documented (as required) and implemented, and records are available to demonstrate compliance, the organization will be ready for the certification audit.

Step 3

Review the map and estimate the level of understanding of the standard and the level of compliance to the requirements. If the understanding and/or level of compliance is relatively low, perform a gap analysis to identify, by audit, specifics on where the organization is compliant and where there are gaps or long lead times. Use the results of the gap analysis to begin implementation of corrective actions. The gap analysis should ensure appropriate compliance to the three supplemental sections of the QS-9000 requirements:

- Section I—compliance to all 20 elements of ISO 9001 plus additional automotive requirements.

- Section II—specific quality system requirements common to all Big Three manufacturers.

- Section III—specific requirements outlined by individual manufacturers.

Step 4

Develop an implementation plan based on the map completed in step 2 and results of the gap analysis in step 3. All levels of management should be involved in crafting the implementation plan to ensure that executive management is committed to providing appropriate resources.

Establish a cross-functional implementation team. This is an essential element for success and reinforces ownership of the quality system. Make sure that the project manager, usually the quality management representative, is authorized to make important decisions and works well with all interests, including labor.

The detailed implementation plan should address the following:

- Systems or procedures to be developed.

- System objectives.

- Relevant QS-9000 sections.

- Personnel/teams responsibilities.

- Required approvals.

(Continued on next page)

(Continued from page 584)

- Required training.

- Required resources.

- Completion dates.

Most companies find that documentation systems are the weakest. However, quality and business planning, design control, purchasing, process control, and inspection and testing are often other problem areas.

Step 5

Using the information contained in the map, draft the quality manual or level 1 document. Since the map contains all the components that will comprise the target quality system, it should be an easy task to address each item through the standard. For example, it should be clear by this time whether servicing, 4.19, is a specified requirement for the organization. If not, the manual should state such. If it does apply, sufficient information about the owner, process, and general policies can be written.

Use several people to draft the manual. Involving people not only spreads the work but also increases ownership and buy-in for the manual and eventual quality system. The manual should not be perceived as a quality department manual.

Step 6

Select a registrar to perform the assessments of the quality system. As the Big Three deadlines for certification approach it is expected that registrar's backlog will increase and scheduling will become more difficult. The arrangement with the registrar will be long term so select carefully.

Step 7

Establish the internal quality audit program. Select individuals who can be objective and participate for an extended time period. Provide in-depth training for all members of the audit team. Once they are trained in the requirements and audit techniques, have the team plan and schedule the audit program for the first year. Conduct the first round of internal audits. Use the information obtained from the audit to take corrective actions and improve the quality system. Repeat the cycle if results indicate compliance gaps still exist.

Consider using the auditors as a team of quality experts to deploy knowledge of the quality system and QS 9000 requirements.

Step 8

Make sure employees understand QS 9000 requirements, the quality system, and the procedures and work instructions related to their work. Instruct employees on what to expect during the registrar's audit. Provide training as necessary.

(Continued on next page)

(Continued from page 585)

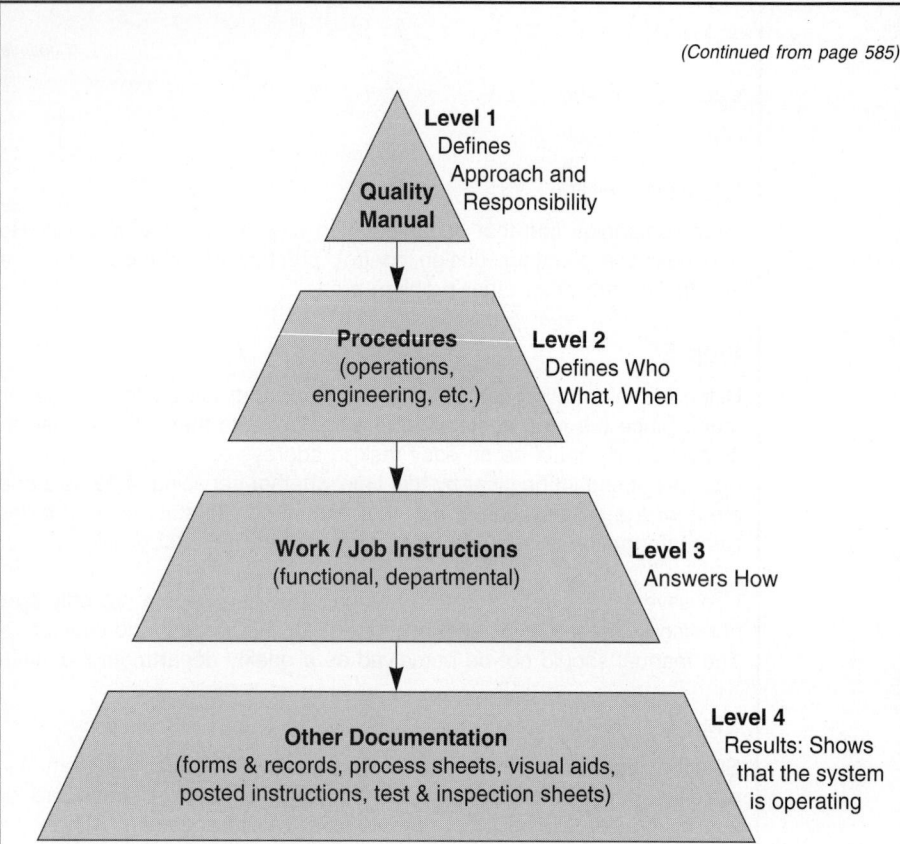

Level 1
Defines
Approach and
Responsibility
Quality Manual

Procedures
(operations,
engineering, etc.)
Level 2
Defines Who
What, When

Work / Job Instructions
(functional, departmental)
Level 3
Answers How

Other Documentation
(forms & records, process sheets, visual aids,
posted instructions, test & inspection sheets)
Level 4
Results: Shows
that the system
is operating

Step 9

Recheck everything. Confirm dates with the registrar. Review the initial implementation plan to ensure all activities are complete.

Initiate appropriate changes or corrective action to meet QS-9000 requirements. Use disciplined problem-solving methodology when a nonconformance to internal or external requirements is discovered. For example, QS-9000 requires that any parts returned from the customer (including parts from manufacturing plants, engineering facilities, and dealerships) must be analyzed and, where appropriate, corrective action and process changes made to prevent a recurrence.

Step 10

Conduct the registration audits. Consider whether a registrar's preassessment is needed. Conduct the final QS-9000 registration audit. Expect the audit to take several days, depending on company's size, complexity, and number of employees.

Expect an ongoing, six-month surveillance audit to ensure compliance to QS-9000 requirements. Initiate a continuous improvement program, with special attention given to customer-specific characteristics.

THE BENEFITS OF QS-9000

by R. Dan Reid, Steve Walsh, and Warren Norrid*

Since its 1994 launch in North America, many views have been offered as to the benefits of QS-9000.

The goal of QS-9000 is the development of fundamental quality systems that provide for continuous improvement, emphasizing defect prevention and the reduction of variation and waste in the supply chain. This goes beyond the basic quality control methodology provided for in ISO 9001, which was used as the foundation for QS-9000. The traditional quality control methodology is based largely on "inspection and testing," whereas QS-9000 is based more on prevention and process optimization.

A key aspect driving prevention is advanced quality planning (QS-9000, c. 4.2.3). QS-9000 requires the use of cross-functional teams, the use of a defined planning process (the Chrysler, Ford, GM Advanced Product Quality Planning and Control Plan reference manual) that generates outputs including a feasibility review, process failure mode and effects analysis (FMEAs) and control plans at prelaunch, and ongoing production phases. One of the early benefits identified by implementing QS-9000 is improved internal communications. The establishment of planning teams has encouraged various intracompany functions to work closely together toward a defined goal. The result has been better overall decision making as a result of greater inputs across the organization.

Also identified as a benefit is improved business processes. QS-9000 requires work that crosses more than one discipline to be documented. Many companies implementing QS-9000 are examining the methods by which work gets done and are taking this opportunity to reengineer the processes to simplify and streamline activities. Another result of the use of the cross-functional teams and improved documentation is that some companies have been able to increase management's span of responsibility and thus reduce the number of managers/supervisors required for employees. Savings associated with this can be substantial.

QS-9000 also requires the support functions (e.g., purchasing, engineering, marketing, shipping) to be considered in the quality system. Traditional quality systems may stop at the end of the production/assembly line; however, many problems for the customer can occur after this point (e.g., mislabeling of parts or containers). QS-9000 addresses these areas.

The elimination of defects leads directly to another identified QS-9000 benefit: improved customer satisfaction. For example, compliant companies:

- Benchmark inside and outside their industry to determine what "world class" is for their commodity or industry.

*Editor's Note: R. Dan Reid (General Motors), Steve Walsh (Ford), and Warren Norrid (Chrysler) are company representatives to the Chrysler, Ford, *and* General Motors Supplier Quality Requirements Task Force.

- Document performance trends and current quality levels and compare them with the "best."

- Track progress toward strategic business objectives and develop action plans for prompt solutions to customer related issues.

- Document process for determining customer satisfaction and dissatisfaction metrics and support with objective information and results.

The QS-9000 goal of reducing variation and waste in the supply chain leads to another benefit: improved quality. Advanced product quality planning is used to provide for quality. Subcontractor development provides for quality of purchased parts and materials. Statistical process control is used to monitor processes to ensure quality in production. Continuous improvement tools and methods then drive the improvement of quality.

Closely aligned with quality improvement is the related benefit of productivity improvement. QS-9000 requires continuous improvement in quality, service, and price. As scrap and rework are driven out of a company, the cost of quality as a percent of sales improves. This shows up on the bottom line immediately. As processes are streamlined, more value-added work is conducted, and more value can be passed on to the customers and stakeholders of a company.

This leads to another key benefit: cost savings. A 1996 survey of nearly 7,000 companies, based upon ISO 9000 registration, indicated one-time savings of up to $164,000 and annual ongoing savings of up to $291,000 per year as a result of registration.[1] Additionally, significant savings are also associated with elimination of redundant customer audits as a result of customer acceptance of QS-9000 registration as adequate assurance of a QS-9000 compliant quality system being in place.

In summary, the benefits of QS-9000 include improved business processes, enhanced management efficiency, improved customer satisfaction, tangible cost savings, as well as measurable quality and productivity improvements.

The worldwide launch of QS-9000 is still underway, and the full extent of its benefits may not yet be identified. Clearly, many companies have already proclaimed it as a worthwhile effort.

Endnote

[1]See Chapter 19 for more registered company statistics.

QS-9000 IMPLEMENTATION

by Devan R. Capur

Tier I automotive suppliers to the Big Three (General Motors, Ford, and Chrysler) have essentially little choice about whether or not to implement a QS-9000 compliant quality system. However, implementing a successful and maintainable quality system goes beyond simply meeting the customer mandate. It is important to understand that when you put together a quality system of this magnitude, it *must* add value. QS-9000, and ISO 9000 for that matter, adds value by bringing consistency, management systems integration, and quality to the customer.

In addition, it is important to recognize what QS-9000 *is* and what it is *not*. It is a quality process standard that drives a working, implemented, value added quality system. It is also a series of guidelines to help set up a living, changing set of cohesive business systems aimed at providing quality products and service to the customer. QS-9000 is not just a set of individual requirements, and it is not a series of documents and evidence books. A QS-9000 based quality system is not a system of documentation, but a *documented system*.

The critical focus of a QS-9000 implementation project should be on planning and assembling the quality management system. As implementers map their companies' roads to implementation, they must realize they will bring about major organizational change across all functional boundaries. Unlike quality programs in the past, the entire organization is affected by QS-9000—not just the quality assurance department.

However, QS-9000 is not a Total Quality Management (TQM) program. It is ISO 9000 plus supplemental requirements constructed principally of concepts from former Ford Q101, GM Targets for Excellence, and Chrysler Pentastar requirements. QS-9000 enhances ISO 9000 substantially with elements such as continuous improvement and production part approval and moves an organization toward TQM concepts.

Nevertheless, broad-based support for QS-9000 must be built right from the start. People must be prepared for changes and involved in making them. Remember, you can always document systems alone (although this is not recommended), but you can never implement alone. Educate, prepare, and involve key people in the implementation process.

BUILDING SUPPORT

To succeed, QS-9000 facilitators need good people skills and an understanding of the process of organizational change. The quality system being created cannot stand alone; it must become central to and be supported by the management process. This is no easy

Author's Note: Corey Alguire, of 3C Technologies, Inc., assisted in the preparation of this article.

task. A quality management system is focused on "core processes," but companies are organized around functional departments. Conflict is built into implementation.

The implementers' greatest challenge lies in creating a set of cross-functional processes in a functionally oriented world. Most people spend the bulk of their time focused on their individual responsibilities and not thinking of their position in or contribution to the grand scheme of things. QS-9000, however, focuses on integrated sets of activity that cross departmental boundaries and functional spheres of activity. Take the process of quoting (part of contract review), for example, as illustrated in Figure 15–1.

In most companies, work responsibilities are divided among separate spheres of activity connected to each other via various hand-offs and interactions. Employees are basically insulated from what goes on in other spheres of activity. This psychological and physical compartmentalization often makes it difficult for people to understand the systems perspective of QS-9000. Implementers must take care not to fall prey to this "functional mindset" and assign responsibility for entire elements or systems within QS-9000 to a single functional area.

Contract review, for example, should not be seen as strictly the responsibility of the sales department simply because they produce the foundation of quotes submitted to the customer. A robust, QS-9000 compliant system of contract review involves many functional areas working together to consider requirements ranging from manufacturing capacity and capability to formulating the exact definition of what is to be produced. A functioning system produces team results rather than a series of individual results.

POSITIONING FOR QS-9000 CHANGE

Because the changes that come with a QS-9000 implementation project are so far-reaching, understanding, and managing change is a key factor to implementation success. Change is a process, not an event—much like QS-9000 itself is a process and not an event. It is important for QS-9000 implementers to understand that they are working with people and teams, and that the people they work with will react to the changes that inevitably accompany QS-9000 implementation, especially when they perceive them as a direct threat. The following general tips should help prepare implementers for reactions to QS-9000 change.

- *Communication is critical.* Continually provide as much information as possible. Message repetition equals message retention, and genuine understanding of the situation will help alleviate a lot of anxiety about the unfamiliar. In many cases, the changes associated with a QS-9000 implementation project may be drastic, and in situations when changes are taking place, you cannot undercommunicate. Remember: everyone wants to know what is coming around the bend—let them know as often as possible and don't be afraid to repeat the same messages again and again.

- *Actively combat skepticism.* Skeptics will openly resist by pointing out similarities in current efforts to past failures with Q101, Targets for Excellence, and other

Figure 15–1

Functional/Organizational View of Quoting Activities

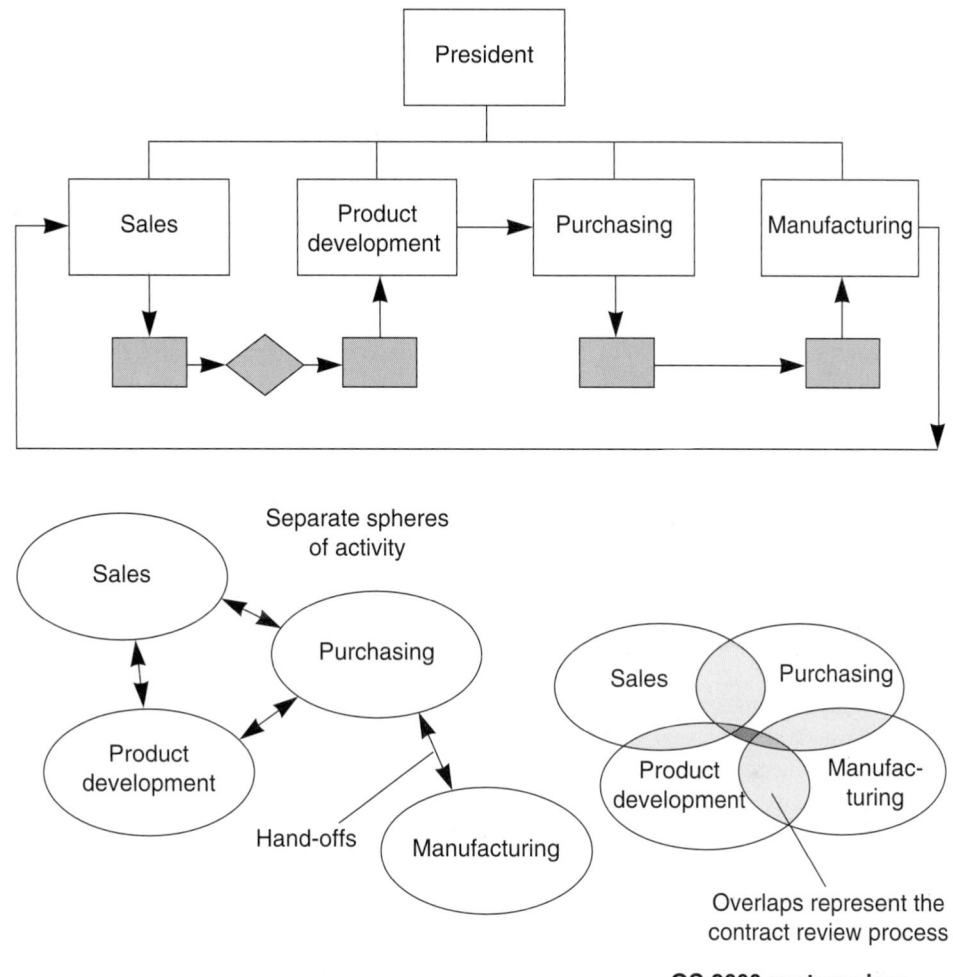

Functional view of contract review activities

QS 9000 system view of contract review

© 3C Technologies, Inc.

programs, or by widely communicating their opposition to change. Communication with skeptics (as with everyone else) must be consistent. They must see an overwhelming commitment to change in the organization. Demonstration of small, tangible successes will also help bring them around.

- *Involve people.* Help people understand the direct benefits QS-9000 will bring to them. Give them a personal stake in these changes, and they will become its defenders.

- *Empathize.* Implementers must be sensitive to people's concerns; listen to and address them. Implementers must also demonstrate their own commitment and sacrifice.

- *Don't wait for overt resistance.* Anticipate it and have a plan to deal with it in advance.

- *Manage the change process.* Change must be actively managed and *will* meet resistance. If resistance is not obvious in some form, it is either hidden . . . or nothing is really changing.

TEAMS AND QS-9000 IMPLEMENTATION

Implementing a QS-9000 quality management system requires efforts that cross all functional boundaries and involves nearly the entire organization. Planning for teams and reporting structures are worth careful consideration. Two types of teams that should be created for the QS-9000 implementation process. The first team is usually referred to as the steering committee. This is the group of people who guide the overall project, assign resources, and resolve cross-functional conflicts. In order to accomplish its tasks, it must consist of representatives from all key areas of the organization, and those representatives must have appropriate levels of authority. For example, Steering committee members could include figures such as vice presidents, upper-level managers, and union stewards.

The implementation team sets the implementation framework, defines the approach, and ensures the physical work of implementation (such as documentation, training, and systems integration). For the QS-9000 project process to succeed, the implementation team needs two key types of individuals:

- People with a good knowledge of current business processes and systems.

- People with real influence—formal or informal.

The first type of team member will be able to help identify where old systems meet the intent of QS-9000 and where new systems must be created or old systems revised. The second type of team member has the influence and respect to make things happen and get the job done. In addition, the implementation team must have a good mix of three fundamental skill types: people skills, system skills, and technical skills.

A subset of the implementation team is the core team. Implementation teams may be composed of people from all over the organization—some of whom provide input only into certain functional areas or systems and have limited involvement in defining the total project. The core team is a small group of individuals (organized formally or informally) dedicated to project definition and implementation. These are the people who spend the most time learning about QS-9000 requirements and what they really mean, defining project milestones and tracking progress toward the registration finish line. Many organizations elect to assign people to the core team as a full-time job. (See Table 15–1)

Table 15–1

Implementation Teams

Team	Composition	Function
Steering committee	• Upper management • All functional areas • Appropriate authority levels	• Guide overall project • Provide direction • Assign resources • Resolve conflicts
Implementation team	• People with influence (formal or informal) • People with system skills— knowledge of current systems and new requirements • People with people skills • People with technical skills	• Organize documentation efforts • Organize training • Coordinate ground-level implementation efforts
Core team	• Subset of implementation team dedicated to project	• Understand requirements • Define project and approach • Set timing, milestones, etc. • Track progress • Lead grassroots change

PATH TO IMPLEMENTATION

To successfully foster real teamwork and success when implementing an organization-wide quality system like QS-9000, it is important that everyone involved is held accountable for their project performance. Most, if not all, people working on the implementation project will be doing so in addition to their normal jobs. QS-9000 must therefore be made a clear priority, and performance should be rewarded.

The implementation team needs personal recognition. If people are doing a particularly good job on their QS-9000 responsibilities, make them visible heroes. Similarly, team members should not be penalized for their project involvement. There must be understanding among team members and their managers that they will be evaluated—formally and informally—based on their performance related to this project. QS-9000 implementation must become an integral part of team members' jobs; it can't be seen as just a side project to be done in their spare time.

Mapping the Implementation Path

Careful planning of the implementation process is the first step toward QS-9000 success. The project involves several phases, and although activities in separate phases may occur simultaneously, each stage has unique characteristics requiring different types of work. The first phase of the project involves understanding, planning, and positioning for QS-9000-related changes.

The next stage, mapping the implementation path, involves formalizing a plan to get to registration: selecting teams, mapping out project milestones and preparing for the implementation launch, designing the specifics of the quality system, organizing methods for making documented systems a reality, setting up a framework for evaluating the system's performance and fine tuning it, selecting a registrar, and planning for the registration audits. It is worth the effort to spend a lot of time preparing the implementation plan and developing methods of visibly tracking project timing and progress toward milestones (see Figure 15–2).

The first item to calculate when mapping out the project plan is how far the organization must travel to reach the registration finish line. Every company has some sort of management system, whether it is formal or not. However, the degree of formalization varies greatly from company to company and has a profound impact on the amount of work required to bring systems into compliance with QS-9000. Although each company starts down the path to QS-9000 with a different degree of readiness, all companies can be slotted into one of three basic categories to determine what approach they need to take toward implementation.

Figure 15–2

QS-9000 Project Process Diagram

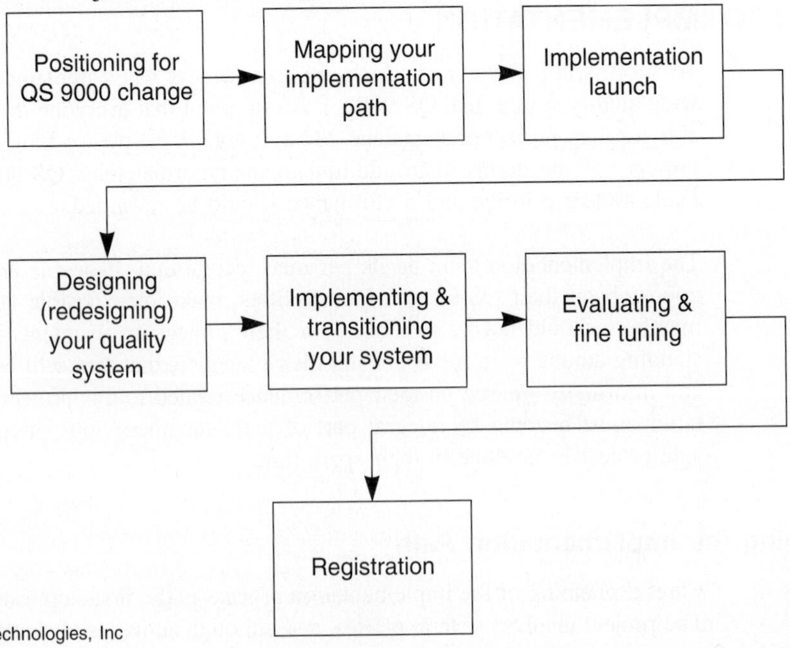

©3C Technologies, Inc

Path 1: Starting from Scratch

The companies that take this route have either few or no formalized quality management systems, or the formalized systems they use are not based on any particular management standard. This path involves *Greenfield implementation*—an industrial term that means starting with a blank sheet of paper, or at "ground zero."

A Greenfield implementation project will obviously involve more time and effort than the other paths since the process involves systems formulation, documentation, and implementation from the beginning stages. Implementers will face at least as much resistance as their colleagues in companies with some sort of quality system in place. On the other hand, starting from scratch has a distinct advantage over building on past efforts; people will probably have few or no preconceptions about quality systems, making them less likely to believe (falsely) that QS-9000 is strictly a quality program.

Path 2: Transition from Another Automotive Quality Standard

Companies with quality systems based on standards like Q101 and Targets for Excellence are already on their way to QS-9000 certification. Implementers on this path need to determine how current their Q1, TFE, and other documentation is and whether it reflects reality. Unlike the documentation required in the past, QS-9000 documents are intended to help people do their jobs. As a consequence, documents should no longer be written to please the customer, but to help *internal users*.

In addition, companies starting with established systems should build on the strengths of any existing quality planning processes and concentrate heavily on document control. Building or rebuilding an advanced product quality planning (APQP) process with a truly cross-functional team effort is a fundamental step toward a quality system that is actually effective rather than merely cosmetic.

APQP efforts in the past tended to be fragmented, project-specific, or driven by sample submission. A truly proactive quality planning process is essential to driving true defect and problem prevention and having a robust quality system. Finally, one of the biggest differences between standards like Q1 and QS-9000 is the switch from second-party audits (the customer audits the supplier) to third-party audits (audits by impartial registrars independent of both the customer and the supplier).

Path 3: Transition from ISO 9000

Because QS-9000 is based on the ISO series, companies that are already ISO 9000 registered have the shortest path to implementation. The biggest impact of the transition from ISO 9000 is the addition of the APQP process. APQP is a prescriptive quality planning process with set formats, stages, and interaction requirements that are more extensive and rigid than the requirements of ISO 9000. Use of measurement system analysis (MSA), more extensive calibration requirements, and Gage R&R studies are also additional requirements in QS-9000 that demand special attention and additional project lead-time.

QS-9000 Training and Communication

Another key factor to the success of a QS-9000 implementation project is training. For people to understand what they are expected to do, and for them to accept the new ways of working, they must be taught. Planning ahead for a comprehensive training program will ensure smoother implementation and an easier path to registration. Table 15–2 is a summary of the most important topics to consider for the QS-9000 training plan.

General QS-9000 training should be given to everyone. In addition, people are involved in different aspects of the system to different degrees. Topic-specific training should include the individuals and teams active in the new and changing systems. For training efforts to be successful, QS-9000 coordinators and trainers should follow these guidelines when presenting their subject matter to employees:

- Communicate to them what they must do and why it is important. Otherwise, they will have difficulty complying.

- Explain to them how to use the system. Be patient and detailed—new responsibilities are not immediately understood.

- Check regularly to ensure that the systems are being followed. Communicate and correct errors in both documentation and practices. Use internal auditing to validate what is working and uncover what is not.

Selecting Resources

Any project the magnitude of a QS-9000 implementation program requires extensive resources, and it pays to consider them during the planning stages. Aside from internal resources, numerous sources of information exist that can assist in the implementation

Table 15–2

QS 9000 Training Plan

Training Topic	Attendees
• Orientation training	• All employees—lasts about two hours
• Overview training	• Key people—lasts one day
• Documentation and implementation training	• Implementation team
• Internal quality audit training	• Subset of implementation team
• APQP training	• Cross-functional team
• Control plans	
• FMEAs	
• SPC training	• Cross-functional team
• PPAP training	• Cross-functional team
• QOS training	• Cross-functional team
• 8D training	• Cross-functional team

process. These include training and consulting services, computer bulletin board systems that focus on ISO/QS-9000 and the automotive industry, and scores of homepages on the World Wide Web.

Information is also available through Automotive Industry Action Group (AIAG), the American Society for Quality Control (ASQC), and the International Automotive Sector Group (IASG) which periodically publishes officially sanctioned interpretations.

REGISTRATION AND BEYOND

Successful completion of the registration audit is an event to be celebrated. Celebration and commemorations actually play a significant psychological role in the implementation process. Celebration is not merely for celebration's sake; it brings closure. QS-9000 registration is a perfect milestone for closure and provides an opportunity to reflect on the effort that went into the project and to experience the satisfaction of a major task completed. This is a chance to build pride and renew commitment to the company, energize employees, and fortify belief in the process and the organization.

Finally, don't overlook the possibilities registration to QS-9000 presents for the future. Registration is a significant, bold achievement and shows that the company has a world-class quality management system. But that system is a process—it must be perpetuated and evolved. Success in the future, maintaining registration, and improving the company's business all lie in enhancing and nurturing the system into a springboard to future excellence and new opportunities.

QS-9000 Registrars

Editor's Note: This list of QS-9000 registrars is current as of June 1996. For a more current list, contact the Registrar Accreditation Board, 800-248-1946, or Irwin Professional Publishing, 703/591-9008.

A.G.A. Quality, A Service of International Approval Services
8501 E. Pleasant Valley Road
Cleveland, OH 44131
216-524-4990
216-642-3463
Contact: Charles Russo
Accreditation: ANSI-RAB, RvA

(Continued on next page)

(Continued from page 597)

ABS Quality Evaluations, Inc.
16855 Northchase Drive
Houston, TX   77060
713-873-9400
713-874-9564
Contact: Patti L. Wigginton
Accreditation: ANSI-RAB, RvA

American Quality Assessors
1200 Main Street, Suite 107M
Columbia, SC 29201
803-779-8150
803-779-8109
Contact: Frank Degar
Accreditation: ANSI-RAB

AT&T Quality Registrar
650 Liberty Avenue
Union, NJ 07083
908-851-3058
908-851-3158
Contact: John Malinauskas
Accreditation: ANSI-RAB

British Standards Institution Quality Assurance
British Standards Institution, Inc.
8000 Towers Crescent Drive, Suite 1350
Vienna, VA 22182
703-760-7828
703-761-2770
Contact: Reg Blake
Accreditation: RvA

Bureau Veritas Quality International (North America), Inc.
North American Central Offices
509 N Main Street
Jamestown, NY 14701
716-484-9002
716-484-9004
Contact: Bryan Stansfield
Accreditation: ANSI-RAB, RvA

(Continued on next page)

(Continued from page 598)

Centerior Registration Services
Corporate Office
300 Madison Avenue
Edison Plaza
Toledo, OH 43652
419-249-5268
419-249-4126
Contact: William J. Niedzwiecki
Accreditation: ANSI-RAB

Davy Registrar Services, Inc.
One Oliver Plaza
Pittsburgh, PA 15222-2604
412-566-3086
412-566-3086
Contact: Leroy W. Pfennigwerth
Accreditation: ANSI-RAB

Det Norske Veritas DNV Certification, Inc.
16340 Park Ten Place
Suite 100
Houston, TX 77084
713-579-9003
713-579-1360
Contact: Steve Cumings
Accreditation: ANSI-RAB, RvA

Entela, Inc., Quality System Registration Division
2890 Madison Avenue SE
Grand Rapids, MI 49548-1206
616-247-0515 or 800-888-3787
616-247-7527
Contact: Brandon Kerkstra
Accreditation: ANSI-RAB, RvA

ITS Intertek Services
313 Speen Street, Suite 200
Natick, MA 01760
508-647-5147
508-647-6714
Contact: James P. O'Neil
Accreditation: ANSI-RAB, RvA

(Continued on next page)

(Continued from page 599)

KPMG Quality Registrar
150 John F. Kennedy Parkway
Short Hills, NJ 07078
800-716-5595
201-912-6050
Contact: Malcolm Appleby
Accreditation: ANSI-RAB, RvA

Lloyd's Register Quality Assurance Limited
33-41 Newark Street
Hoboken, NJ   07030
201-963-1111
201-963-3299
Contact: Bob Armstrong
Accreditation: ANSI-RAB , RvA

NSF International
3475 Plymouth Road, PO Box 130140
Ann Arbor, MI 48113-0140
313-769-6728
313-769-0109
Contact: Dilip Desai
Accreditation: RvA

OMNEX-Automotive Quality Systems Registrar
PO Box 15019
Ann Arbor, MI 48106
313-480-9940
313-480-9941
Contact: Dennis Hughey
Accreditation: ANSI-RAB

Perry Johnson Registrars, Inc
3000 Town Center, Suite 630
Southfield, MI 48075
800-800-7910
810-358-0882
Contact: Terry Boboige
Accreditation: ANSI-RAB

(Continued on next page)

(Continued from page 600)

Quality Management Institute
90 Burnhamthorpe Road W. Suite 800
Mississauga Ontario L5B 3C3
905-272-3920
905-272-3942
Contact: Michael J. Haycock
Accreditation: ANSI-RAB, RvA

Quality Systems Registrars, Inc.
13873 Park Center Road, Suite 217
Herndon, VA   22071-3279
703-478-0241
703-478-0645
Contact: Marshall Courtois
Accreditation: ANSI-RAB, RvA

SGS International Certification Services, Inc.
Meadows Office Complex, 301 Route 17 N
Rutherford, NJ 07070
800-747-9047
201-935-4555
Contact: Lois O'Brien
Accreditation: ANSI-RAB, UKAS

Smithers Quality Assessments, Inc.
425 W Market Street
Akron, OH 44303-2099
216-762-4231
216-762-7447
Contact: John Sedlak
Accreditation: ANSI-RAB, RvA

Steel Related Industries Quality System Registrars
SRI Quality System Registrar
2000 Corporate Drive, Suite 330
Wexford, PA 15090-7605
412-934-9000
412-935-6825
Contact: Peter Lake/James Bytnar
Accreditation: ANSI-RAB, RvA

(Continued on next page)

(Continued from page 601)

TRA Certification
700 East Beardsley Ave.
PO Box 1081
Elkhart, IN 46515
800-398-9282, or 219-264-0745
219-264-0740
Contact: Dean Hupp
Accreditation: ANSI-RAB

TUV America and TUV Product Service
TUV America, Inc., and TUV Product Service, Inc., (Headquarters)
5 Cherry Hill Drive
Danvers, MA   01923
508-777-7999
508-762-8414
Contact: Chris Stockwell
Accreditation: ANSI-RAB, RvA

TÜV ESSEN
2099 Gateway Plaza, Suite 200
San Jose, CA 95110
408-441-7888 or 800-TUV-4630
408-441-7111
Contact: Olga Rada
Accreditation: ANSI-RAB

TUV Rheinland of North America, Inc.
TUV Rheinland, North American Headquarters
12 Commerce Road
Newtown, CT 06470
203-426-0888
203-270-8883
Contact: Martin Langer
Accreditation: ANSI-RAB, RvA

Underwriters Laboratories, Inc.
1285 Walt Whitman Road
Melville, NY 11747-3081
516-271-6200
516-271-6242
Contact: Dan Keck
Accreditation: ANSI-RAB, RvA

CONFORMITY ASSESSMENT

Chapter 16   European Union and Conformity Assessment Requirements

Chapter 17   Registrar Accreditation

Chapter 18   Challenges Facing the ISO 9000 Industry

European Union and Conformity Assessment Requirements

This chapter provides the reader with a broad overview of conformity assessment by addressing the following topics:

- An explanation of conformity assessment.
- The European Union's single internal market.
- Conformity assessment's four major components.
- EU-wide directives.
- Harmonized standards.
- Consistent conformity assessment procedures.
- Competent certification and testing bodies.
- The European Union and other countries.
- Product Liability and Product Safety Directives.

This *Handbook* has focused on ISO 9000 and on the process of implementing the standard in companies. ISO 9000 is a global phenomenon. Thousands of companies around the world are registering to one of the ISO 9000 standards. For many companies the European Union's single internal market has been a key impetus for ISO 9000 registration. Although ISO 9000 registration is required only for some regulated products, the European Union is stressing ISO 9000 registration as an integral part of its overall goals. One of the European

Union's overall objectives is to develop a complex and comprehensive regulatory and product certification framework that may become a model for other regional groups of nations.

This chapter examines the European system of standards and product certification. It discusses the goals of the European Union and describes its conformity assessment procedures. For U.S. companies doing business in the European Union, it is important to understand the EU's efforts to establish Unionwide directives, standards and certification procedures.

THE EUROPEAN UNION AND CONFORMITY ASSESSMENT

by James Kolka

INTRODUCTION

Conformity assessment includes all activities that are intended to assure the conformity of products to a set of requirements. These activities can include the following:

- Approving product designs.

- Testing manufactured products.

- Registering a company's quality system.

- Accrediting organizations that perform testing and assessment procedures.

In general, conformity assessment includes all market access processes for a product or service that must be followed to bring that product or service to a market.

Governments use conformity assessment procedures to ensure that products sold in their countries meet their laws and regulations and to protect their citizens, public systems, and the environment from harm caused by products that enter their country.

Customers benefit from conformity assessment for the same reasons as do governments. Purchasers of products and services can use conformity assessment to identify suppliers whose products can be relied on to comply with critical requirements. Manufacturers, in turn, use conformity assessment procedures to demonstrate to their customers that their products comply with requirements.

THE EU'S SINGLE INTERNAL MARKET

The goal of the European Union's single internal market is to promote economic competitiveness and to become a powerful economic trading bloc by removing physical, technical, and fiscal barriers to trade. The free internal movement of goods, services, people, and capital from one member state to another is essential to economic growth.

Beginning in 1983 and anticipated in the 1985 document, *Completing the Internal Market, White Paper from the European Commission to the EuopeanCouncil*, the European Union has been developing a new approach to regulating products as one way to unify the European market. It has enlisted the aid of key European regional standards organizations to develop EU-wide, "harmonized" standards. The purpose of these standards is to eliminate the jumble of national standards of the individual 15 member states. The European Union identified 300 regulations (later reduced to 286) to implement the single internal market. Among these regulations were directives dealing with toy safety, machinery, electromagnetic capability, medical devices and others.

> Note: With the addition of Austria, Finland, and Sweden, the European Union increased its membership to 15 on January 1, 1995. In 1996 the EU formalized its assession plans to add 12 new member states in stages. These countries are Poland, Hungary, Czech Republic, Bulgaria, Romania, Slovakia, Estonia, Latvia, Lithuania, Slovenia, Malta, and Cyprus.
>
> These additions would increase the European Union's population to over 500 million. U.S. companies already exporting to these nations are finding "CE Marking" stipulated in contracts in anticipation of future EU membership.

Of most concern to U.S. companies wishing to do business with the European Union are technical barriers. These include different standards for products, duplication of testing and certification procedures for products, and differences in the laws of EU member states. These restrict the free movement of products within the European Union.

GOALS OF THE NEW SYSTEM

The European Union recognized that, as technical barriers were lowered, a new framework must replace them. The goal of this new framework is to create confidence among the member states in the following:

- Quality and safety of products sold in the European Union.

- Overall competence of manufacturers, including their quality procedures.

- Competence of the testing laboratories and certification bodies that assess the conformity of products.

The new framework involves EU-wide directives issued by the European Commission that will replace individual member state regulations. It also involves a comprehensive approach to conformity assessment.

This developing system encompasses all aspects of conformity assessment. The European Union refers to it as its Global Approach to Product Certification and Testing. The approach has four major components:

1. EU-wide directives.

2. Harmonized standards.

3. Consistent conformity assessment procedures.

4. Competent certification and testing bodies.

Each component is next discussed in detail.

EU-wide Directives

As noted in Chapter 2, products in the European Union are classified into two categories: Regulated products and nonregulated products.

Nonregulated Products

Most products sold in the European Union are nonregulated products such as paper and furniture. The European Union's strategy for removing technical barriers to nonregulated products is to rely on the principle of *mutual recognition* of national product standards established in 1979 by the European Court of Justice in the *Casis de Dijon* decision. According to the principle of mutual recognition, products that meet the requirements of one EU member state can freely circulate in other member states.

An EU purchaser of an unregulated product (such as cardboard boxes or manually operated hand tools) can continue to purchase U.S. products specified in terms of U.S. standards. Even if the product remains regulated at the national level and is not subject to harmonized standards, a U.S. product that meets the national requirements of one EU member state may enjoy free circulation throughout the entire European Union through mutual recognition.

Regulated Products

Only a small percentage of the total number of products sold in the European Union are regulated. Regulated products such as medical devices, pressure vessels, and personal protective equipment, however, make up approximately 50 percent of U.S. exports to the European Union, according to U.S. Department of Commerce estimates. They include those products that the European Union believes are associated with significant safety, environmental, or health concerns.

The EU Council of Ministers is working to remove technical trade barriers for regulated products by issuing Union-level directives. A *directive* is the official legislation promulgated by the European Commission and approved by the European Parliament and Council. It binds all members of the European Union, who are required to convert EU directives into national legislation and regulations. Existing laws and rules that conflict with the directive are invalid and are superseded by EU directives. After a transition period (in most instances), the regulated products must meet the requirements of the new directive.

Old-Approach Directives and New-Approach Directives

Prior to 1989, the European Union issued directives that are now known as old-approach directives. These directives were highly specific, detailing and defining all technical characteristics and requirements of a product. The problem with old-approach directives is that they are complicated, it is expensive to comply with them, and they are easily outdated due to technological advances. The term *old approach* also is confusing, because the European Union continues to adopt these types of directives. The 1992 automotive type approval directives, for example, are old-approach directives.

Nevertheless, these directives are binding on all manufacturers. If a company's product falls within the scope of an old-approach directive, it must meet the directive's requirements.

The European Union soon realized that the detailed blueprint it was drafting was slowing its progress in meeting the goals of a single internal market. To expedite the process, the European Union began issuing more "generic-type" directives, known as new-approach directives.

New-approach directives are based on the following four key elements:

- Essential environmental, health, and safety requirements.
- Presumption of conformity.
- Mutual recognition.
- Voluntary standards.

Essential Requirements.   New-approach directives stipulate the environmental, health, and safety requirements a product must meet to be considered safe for the marketplace. These requirements tell a company what must be done to comply with a directive. Technical standards-or the how-to specifications-are left to be spelled out by European regional standards organizations and by the member states themselves. The directives do not specifically list technical standards, but they do provide references for all appropriate supporting technical documentation. The number of standards per directive can be considerable. For example, 40 technical committees have drafted 650 technical standards for the Machinery Directive.

Presumption of Conformity.   If a product conforms to European harmonized technical standards that have been adopted into national law, it is assumed that the product conforms to the essential requirements contained in the applicable directive. For example, if a company declares that its product conforms to Committee for European Standardization (CEN) or European Committee for Electrotechnical Standardization (CENELEC) technical standards referenced in the directive that applies to the company's product, the product is presumed to conform with the applicable EU directive. (CEN and CENELEC are described in detail later in this chapter.)

Mutual Recognition.   Member states must accept products that are lawfully manufactured in any other member state, provided that the product meets EU-wide standards and/or the health, safety, and environmental concerns of the receiving state. The European Union is seeking to apply this principle not only to the acceptance of products, but also to test results and certification activities. Further, it intends to push for Mutual Recognition Agreements (MRAs) with non-EU nations to mutually accept test results and product certifications.

Voluntary Standards.   Each new-approach directive provides companies with various options to comply with the essential requirements of the directive. These include a range of conformity assessment procedures. To determine which conformity assessment procedure applies to a product, manufacturers should study the appropriate directive or directives, review the options for conformity assessment, and choose the preferred or acceptable option. Depending on the type of product and its potential safety risk, the choices can range from manufacturer self-certification to the implementation of a full quality assurance system. (The options are discussed below.)

Companies also have choices regarding the technical standard(s) to which their products can conform. These include the following:

- They can comply with technical standards appropriate to the directive. Conformance to the technical standard may involve a third-party evaluation; it depends on the specific directive and the procedure chosen for conformity assessment.

- They can conform to a non-European standard or to no standard at all. In this case, the company must demonstrate that its product meets the essential requirements of the directive, which may involve third-party approval of the demonstration.

In cases where there is a low safety risk, directives may allow a company to self-certify its product. However, most conformity assessment options require some third-party involvement in testing and certification.

Some manufacturers have abused the self-certification options of the Toy Safety Directive. This resulted in the approval of Council Regulation No. 339/93 on product safety conformity on February 8, 1993. It stipulates that all products must be properly marked and accompanied by documents that indicate product conformity with EU safety requirements. It most cases, this document will be a Declaration of Conformity for an EU new approach directive.

To be able to assess product compliance with EU legislation, it is important for manufacturers to identify relevant European standards and, where appropriate, third-party testing or certification entities authorized to assess product conformity to the requirements of specific directives. This information should appear on the Declaration of Conformity that accompanies a CE-marked product.

This information is published in the European Union's *Official Journal of the European Community*. For example, the first EU technical harmonization directive to be implemented, covering toy safety, came into effect on January 1, 1990. Reference harmonized standards were formally identified in the *Official Journal* on June 23, 1989. Lists of bodies authorized to carry out EU-type examination as referred to in the toy safety directive were first published over a period of months, beginning June 23, 1990. Currently these directives are republished with periodic updates in the *Official Journal* (see Appendix B for a list of key EU directives).

Transition Period.   In the case of some directives, provisions have been made for a transition period between the implementation of the directive and the date by which companies must comply. The purpose is to allow time for reference standards to be completed and sufficient testing facilities to be qualified and authorized.

For example, under the medical devices directives, transition periods have been established, ranging from two to four years from the date the directive is implemented. This allows manufacturers to continue meeting existing national standards during the transition period. However, they can market only to countries where their product complies with national standards. Under these circumstances, they cannot affix the CE mark and are not guaranteed free circulation for their products among all EU member states.

Manufacturers that meet the new EU-wide directives immediately will be able to sell medical devices throughout the European Union and European Free Trade Association (EFTA). Consequently a number of EU and U.S. medical device manufacturers are ignoring the transition period and are moving to complete certification as soon as possible. This gives them an edge over competitors, allows them to advertise compliance with new EU safety standards, establish an EU-wide marketing presence, and increase their market share.

With other directives, such as construction products and telecommunications terminal equipment, there was no transition period. Since the construction products directive covers a vast range of products and the system for product certification is not fully operational, the lack of a transition period has caused some difficulties in that industry. Interim procedures have been developed.

Requirements May Include Several Directives.   It is possible that a company must conform to more than one directive. For example, a commercial air-conditioning manufacturer would have to meet the requirements of five different directives-Machinery Safety,

Electromagnetic Compatibility, Low Voltage, Pressure Vessels, and Construction Products (which covers equipment installed in buildings, and building materials).

In another example, compressor-generators are covered by the requirements of the Machinery Directive. However, the air tank component of this equipment also must meet the requirements of the Simple Pressure Vessel Directive. The same is true for suppliers of air brakes that are incorporated into mobile industrial or construction equipment.

Manufacturers who supply components to other producers for incorporation into a product that is then exported to the European Union may find themselves expected to meet the technical requirements of European Union legislation and issue a Declaration of Incorporation to accompany the product in the EU. This type of declaration states that EU safety requirements have been statisfied, but that final CE marking responsibility belongs to the company incorporating the component into its machine.

The European Union also has issued directives that apply to all industry sectors. These include directives on product liability and product safety. Manufacturers will be required to comply with these as well. (See Appendix B for a list of major EU directives.)

Notified Bodies.   In addition to requirements, directives also list the appropriate government-appointed organizations, known as *notified bodies*, that are authorized to certify that a particular product conforms to the requirements of a directive. A notified body might be a testing organization, testing laboratory, the operator of a certification system, or even a government agency itself.

A notified body is designated by the competent authority of a member state from among the bodies under its jurisdiction. A *competent authority* is the national authority in each member country that has overall responsibility for the safety of products.

The phrase notified body derives from the fact that member states notify the EU Commission as to which bodies in their country are qualified to perform the specific evaluations stipulated in individual directives.

The duties of notified bodies are clearly spelled out in each directive, and lists of notified bodies vary depending on each directive. Every EU country must accept the results of conformity assessments by notified bodies in all other EU countries unless there is cause to believe the product was improperly tested. Notified bodies also are required to carry liability insurance.

The Competence of Notified Bodies.   Each member state must have confidence that its notified bodies are competent to declare conformity to a directive. In order to ensure members of the competence of notified bodies, the European Union has developed the Communitywide EN 45000 series standards for certification and testing. The European

Union developed a Council regulation to guide the creation of notified bodies and their compliance with the EN 45000 series. (Chapter 17 discusses the EN 45000 standards in more detail.)

Product Certification versus ISO 9000 Registration.   To satisfy the conformity assessment requirements of most EU new-approach directives and to affix the CE mark, a company must receive third-party approval from an EU notified body. Since this product certification approval is not the same as ISO 9000 registration by a registrar, it is critical to ascertain the quality assurance requirements of a directive and establish contact with an EU notified body to make certain that everything is in order. Some U.S. manufacturers have presented ISO 9000 registration to EU authorities and have been denied access to the marketplace because directive requirements and notified body approval have not been met.

It is possible that at least three elements will be necessary. For example, in most cases the new EU medical devices directives will require a registered ISO 9000 quality assurance system, augmented by compliance with the EN 46000 requirements (the application of ISO 9000 to medical devices) as guided by an EN medical devices guidance document. In addition, the guidance documents reference key clauses of ISO 9004-1.

A simple registration to ISO 9001 will not fulfill the requirements of the medical device directive. A company will require notified body certification of its product and the creation of a medical device vigilance system, in addition to the essential requirements set forth in Annex 1 of the directive. Similar essential requirements are set forth in each EU new-approach directive. In 1995, the European Commission published *The Guide to the Implementation of Community Harmonization Directives Based on the New Approach and the Global Approach*. It contains a chapter outlining the operating procedures and requirements for notified bodies.

The CE Mark.   The final result of the product certification process is the *CE mark*. A notified body is authorized to permit manufacturers to affix the CE mark, which signifies proof that a company has met essential health and safety requirements and the specific conformity assessment requirements to market its product in the European Union. If a directive calls for self-certification, (e.g., toy safety, machinery directives) the manufacturer takes responsibility for affixing the CE mark.

The CE designation, French for "Conformite Europeenne" (European Conformity), is required in order to sell any product manufactured or distributed under the new-approach directives. The CE mark will replace all national marks now used to show compliance with legislated requirements for regulated materials and products.

The requirements for affixing the CE mark are set forth in each directive (see Figure 16–1). Basically, four steps are needed to obtain the mark:

Figure 16–1

Except for Medical Devices, Information to Accompany CE Mark Manufacturers in European Union

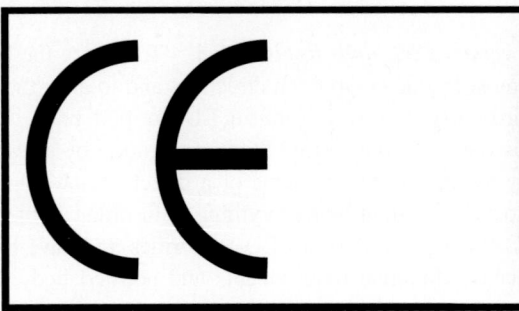

Should be listed in the Declaration of Conformity For Certain Medical Devices; the CE Mark should be accompanied by the identification number of the notified body.

• Conformance with the requirements of the appropriate EU directives

• Official registration with a notified body to the appropriate ISO 9000 standard (ISO 9001, ISO 9002, or ISO 9003), if quality system registration is required by the directive.

• Documentation of any test data required by the directives.

• Necessary certification by the appropriate notified bodies to verify compliance with the directive(s).

Each member state must allow products with the CE mark to be marketed as conforming to the requirements of the directive. The same rules apply regardless of the product's origin. Products that have been improperly certified will be refused entry or withdrawn from the market.

Old-Approach Directives.   Old-approach product safety directives, such as those for motor vehicles, tractors, and chemicals, contain detailed requirements for standards and test methods and specify required marks which must be applied to indicate conformance. Products subject to product safety requirements in these areas—as well as to new-approach directives—must bear both the CE mark and other marks required under EU legislation.

Additional Marks.   The CE mark alone is sufficient to market a product in the European Union. In some cases, however, customer acceptance of materials and products may hinge on the appearance of one or more additional certification and/or quality marks issued by bodies in that member state.

The CE mark does not preclude the continued existence of national quality or performance marks representing levels of quality, safety, or performance higher than those specified in EU legal requirements. Member states cannot require that products bear these marks as a condition of market access, but they can continue to exist on a voluntary basis and can be specified in private commercial contracts.

If a product incorporates a safety element subject to EU legislation requiring the CE mark, and national environmental control requirements (recycling, disposal, etc.), manufacturers could be required to obtain additional national environmental marks in order to market the product in a specific EU member state. Currently, Germany, the Netherlands, France, and other countries have disposal requirements for packaging waste. By June 30, 1999 EU-wide requirements will become law for recycling and the disposal of packaging waste that was adopted into national law under the EU Directive on Packaging and Packaging Waste 94/36/EC.

Harmonized Standards

The second key component of the European Union's conformity assessment system is harmonized standards. Harmonization refers to the process of creating uniform, EU-wide standards. The European Union believes that harmonized standards are essential to promote trade, not only in Europe but around the world. Standards have been steadily growing in importance and are becoming a strategic issue for business.

Ultimately, the aim of harmonization is a global system where manufacturers could produce to a single standard, be assessed by a single assessment or testing body, and have the resulting certificate accepted in every market. This is the goal of "make it once, test it once, sell it everywhere."

EU Standards and Regional Standardization Organizations

The essential safety requirements in the EU new-approach directives are broad guidelines only. In addition to issuing directives, the European Union is seeking to harmonize technical requirements by mandating the use of harmonized standards whenever possible. The task of developing specific technical standards to harmonize the many differing national standards of the EU countries into one set of common standards is carried out primarily by three European standard-setting organizations. These are the European Committee for European Committee for Standardization (CEN), the European Committee for Electrotechnical Standardization (CENELEC), and the European Telecommunications Standards Institute (ETSI). A fourth organization, the European Organization for Technical Approvals (EOTA), assesses the technical fitness of construction products for their intended use, even when no EU-wide harmonized standard or national standard exists for that product. (The four organizations are discussed in more detail below.)

These four organizations develop standards according to priorities set by the European Union and its member states. They also consult with existing national and international standardization organizations. CEN and CENELEC have negotiated agreements with the two international standards organizations, the International Organization for Standardization (ISO) and the International Electrotechnical Commission (IEC), to develop new standards. CEN and CENELEC will develop a new standard when:

- A standard does not already exist under ISO or IEC auspices.

- The standard cannot be developed at the international level.

- The standard cannot be developed at the international level within a specific time frame.

All member states must conform to each standard once it is formally adopted.

Committee for European Standardization (CEN)

CEN is the Committee for European Standardization (or Normalization, hence the "N"). This nonprofit organization is the world's largest regional standards group. It comprises delegates from 18 Western European countries—the 15 European Union nations plus three member nations of the European Free Trade Association (EFTA). CEN is composed of the national standardization institutes of these 18 countries. (The only nation missing from EFTA is Liechtenstein.)

CEN's main objective is to prepare a single set of European standards in place of numerous national standards. CEN works to remove any standardization differences among its 18 members.

Roles in Testing and Certification.   CEN promulgates standards when the European Union passes a directive. It also responds to EU requests to develop a standard when no directive has been issued. When necessary, CEN promulgates new standards which the member countries are obligated to adopt as their own national standards. CEN also creates and implements procedures for the mutual recognition of test results and certification schemes.

CEN and ISO.   CEN adopts ISO standards whenever possible and promotes the implementation of ISO and IEC standards. As far as possible, CEN avoids any duplication of work. CEN also works with ISO to draft new standards and has formal agreements with ISO for the exchange of information and for technical cooperation. Following the Vienna Agreement of 1991, CEN and ISO share common planning and have parallel votes during the development of standards and parallel votes for the joint adoption of technical standards.

Types of CEN Standards.   CEN publishes its standards in one of the following three ways.

European Standards (or European Norm, hence the EN designation) are totally harmonized, and the 18 member nations of CEN are obligated to adopt these standards as their own national standards. An EN must be implemented at the national level as a national standard and by withdrawing the conflicting national standard.

Manufacturer compliance with European standards is voluntary. But if an EN is met, it is presumed that this also fulfills the requirements of the directive that applies to the manufacturer's product. ENs for a new technology are prepared following specific requests from the European Union and EFTA.

Harmonization Documents (HD) allow for some national deviations in standards. The HD must be implemented at the national level, either by issuing the corresponding national standard or, as a minimum, by publicly announcing the HD number and title. In both cases, no conflicting national standard may continue to exist after a fixed date.

European Pre-standards (PrENs) are guidelines for expected ENs or HDs, or guidelines for rapidly developing industries. PrENs may be established as prospective standards in all technical fields where the innovation rate is high or where there is an urgent need for technical advice. CEN members are required to make the PrENs available at the national level in an appropriate form and to announce their existence in the same way as for ENs and HDs. However, any conflicting national standards may be kept in force until the PrEN is converted into an EN.

European Committee for Electrotechnical Standardization (CENELEC)

The European Committee for Electrotechnical Standardization (CENELEC) is CEN's sister organization and is also based in Brussels, Belgium. CENELEC is a nonprofit technical organization working to harmonize electrical standards among its 18 EU and EFTA member countries and is composed of delegates from those countries.

While CEN works closely with ISO to adopt standards on everything but technical issues, CENELEC works with its international counterpart, the International Electrotechnical Commission (IEC). CENELEC maintains an active working agreement with IEC, and 85 percent of the European standards adopted by CENELEC are IEC standards.

The procedures for the development of CENELEC standards are the same as those described for CEN. CENELEC publishes its standards in the same manner as CEN: as European Standards (EN), Harmonization Documents (HD), and European Pre-standards (ENV).

CENELEC Priorities.   CENELEC's priorities for developing standards are low-voltage areas, other electric equipment, and agreed-upon mandates for standardization from EU and EFTA countries. As mentioned earlier, manufacturers do not have to meet CENELEC standards if their products fulfill the essential requirements of the applicable directive(s). Products which meet CENELEC standards, however, are presumed also to fulfill the requirements of EU directives.

European Telecommunications Standardization Institute (ETSI)

The European Telecommunications Standardization Institute (ETSI) is the third sister organization in the CEN/CENELEC/ETSI regional triumvirate. It promotes European standards for a unified telecommunications system.

ETSI membership is open to all relevant organizations with an interest in telecommunication standardization that belong to a country within the European Confederation of Posts and Telecommunications Administrations. Users, research bodies, and others may participate directly in standardization work for Europe.

The process of publishing final ETSI standards is almost identical to the methods used by CEN and CENELEC. The three groups have formed a Joint Presidents' Group to handle common concerns regarding policy and management. The three groups also have signed a cooperation agreement to prevent overlapping assignments and work together as partners.

Non-European organizations interested in telecommunications are sometimes invited as observers to the technical work of ETSI. In addition, ETSI and the American National Standards Institute (ANSI), the U.S. standardization body, have agreed to an exchange of information concerning their respective work.

Consistent Conformity Assessment Procedures

The third component of the EU's overall conformity assessment approach is consistent conformity assessment procedures. The main principle is that, rather than adopting certification procedures on an *ad hoc* , directive-by-directive basis, as in the past, the European Union in the future will choose from a set of detailed conformity-assessment procedures. This plan is called the *modular approach.*

The Modular Approach

The modular approach to conformity assessment offers several procedures for a manufacturer to demonstrate compliance with directives. These range from a manufacturer's self-declaration of conformity to assessment of a quality system to type-testing of the product by a third party, depending on the health, safety, and environmental risks of the product. All future conformity assessment procedures will be based on one or more of these options.

The EU Council outlines the combination of procedures it considers appropriate for each directive and sets the conditions of application. The manufacturers themselves, however, have the final choice as to which of the procedures they will follow. The most recent statement of this approach is contained in Council Decision 93/465/EEC of July 22, 1993 entitled, *Concerning the Modules for the Various Phases of the Conformity Assessment Procedures and the Rules for the Affixing and Use of the CE Conformity Marking Which Are Intended to Be Used in the Technical Harmonization Directives.*

Figures 16-2 and 16-3 illustrate the certification options available to comply with the directives. If one or more directives is applicable to a particular product, the directives indicate whether a notified body must be involved, and if so, the extent of that involvement. Apart from Module A, the supplier has to involve a notified body for all other modules. The supplier is responsible for maintaining the conformity of its product to all relevant essential requirements.

As stated earlier, ISO 9000 registration alone, without product testing, is not sufficient to meet EU directive requirements. Quality system registration is a component of the conformity assessment requirements for some regulated products. The directives, however, require that products be tested to ensure compliance with the minimum requirements of the directive. In some directives, if a manufacturer has an ISO 9001-registered quality system, it can

Figure 16–2

Overview—EC 92 Conformity Assessment Procedures—The Modules.

	MODULES						
	A	B+C	B+D	B+E	B+F	G	H
Product Surveillance: Samples:	○	○			● OR		
Each Product:	○				●	●	
Q.A. Surveillance:			● EN ISO 9002	● EN ISO 9003			● EN ISO 9001
Type Testing:		●	●	●	●		○ Design
Technical Documentation:	①	≠	≠	≠	≠	≠	③
CE Mark Affixed by: Manufacturer:	CE	CE	CE★	CE★	CE★ OR		CE★
Third Party:					CE★	CE★	

○ Supplementary Requirements   ① Required to Be Available   CE CE Mark
● Action by Third Party   ≠ Required by Notified Body
   ③ Part of Quality System   CE★ CE Mark with the Notified Body Identification Symbol

Figure 16–3

Conformity Assessment Process.

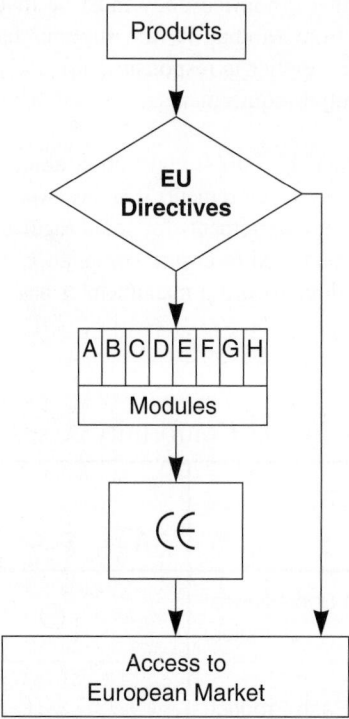

then self-declare conformity with the technical requirements of the directive, provided that all of the product safety essential requirements have been satisfied.

Description of Modules in Modular Approach

Two main phases in the modular approach exist: the design phase and the production phase. Both phases are covered by modules, which are further broken down into four types of examination:

- Internal control of production.
- Type-examination.
- Unit verification.
- Full quality assurance—EN ISO 9001.

These are examined in more detail below (see Figure 16-4).

Internal Control of Production (A).   Internal control of production allows manufacturers to self-declare conformity to the specific standard. Self-declaration is possible for

Figure 16–4

Confomity Assessment Procedure Modules

DESIGN	PRODUCTION
A. INTERNAL CONTROL OF PRODUCTION **Manufacturer** • Keeps technical documentation at the disposal of national authorities **Aa** • Intervention of notified body	**A. Manufacturer** • Declares conformity with essential requirements **Aa** • Tests on specific aspects of the product • Product checks at random intervals
B. TYPE EXAMINATION **Manufacturer submits to notified body** • Technical documentation • Type **Notified body** • Ascertains conformity with essential requirements • Carries out tests, if necessary • Issues EU type-examination certificate	**C. CONFORMITY TO TYPE** **Manufacturer** • Declares conformity with approved type • Affixes the CE-Mark **Notified body** • Tests on specific aspects of the product • Product checks at random intervals **D. PRODUCT QUALITY ASSURANCE** EN ISO 9002 • **Manufacturer** • Operates approved QS-production & testing • Declares conformity with approved type • Affixes the CE-Mark EN ISO 9002 • **Notified body** • Approves the QS • Carries out surveillance of the QS **E. PRODUCT QUALITY ASSURANCE** EN ISO 9003 • **Manufacturer** • Operates approved QS-production & testing • Declares conformity with approved type or essential requirements • Affixes CE-Mark EN ISO 9003 • **Notified body** • Approves the QS • Carries out surveillance of the QS **F. PRODUCT VERIFICATION** **Manufacturer** • Declares conformity with approved type or essential requirements • Affixes CE-Mark **Notified body** • Verifies conformity • Issues certificate of conformity
G. UNIT VERIFICATION **Manufacturer** • Submits technical doumentation	**Manufacturer** • Submits product • Declares conformity • Affixes the CE-Mark **Notified body** • Verifies conformity with essential requirements • Issues certificate of conformity
H. FULL QUALITY ASSURANCE **EN ISO 9001** • Operates an approved QS for design **EN ISO 9001** • Carries out surveillance of the QS • Verifies conformity of the design • Issues EU design examination certificate	**Manufacturer** • Operates an approved QS for production & testing • Declares conformity • Affixes the CE-Mark **Notified body** • Carries out surveillance of the QS

toys, electromagnetic compatibility, some weighing instruments for noncommercial use, and most types of machinery, as well as for some types of personal protective equipment, pressure vessels and equipment, recreational craft, and low-risk medical devices.

During the design phase, the manufacturer may carry out the procedure for conformity assessment itself. The manufacturer, however, must keep the technical documentation available in a technical construction file for review by the national authorities for at least 10 years after production of the product. This way, assessments and checks can be carried out to determine whether the product complies with the directive.

The producer has to provide insight into the design, manufacturing process, and performance of the product. The manufacturer must take all the steps necessary to ensure the manufacturing process guarantees that the product constantly complies with the essential requirements.

During the production phase, a notified body carries out testing on specific aspects of the product at random intervals. If allowed by this directive, this module can be used by a manufacturer who chooses to produce, not in accordance with the European standards, but directly in accordance with the essential requirements of the applicable directive, and can demonstrate that conformity to those requirements.

Type-Examination (B).   In type-examination, the design phase involves verification by a third party. In this module, the manufacturer has to present the technical documentation and one *product type* (typical example) to a testing organization of its choice. The testing organization assesses and draws up a type-examination certificate. This module must be supplemented by modules C, D, E, or F.

Conformity to Type (C).   The manufacturer can self-declare conformity to type with no quality system requirement. The manufacturer draws up a declaration of conformity for the approved type from module B, and keeps this for at least 10 years after manufacture of the last product.

Production Quality Assurance (D).   This module requires third-party certification to ISO 9002. ISO 9002 includes the entire production process, except for design. The manufacturer's quality system for production is approved by a testing organization. Then the manufacturer declares that his product matches the approved type.

Product Quality Assurance (E).   This module requires third-party certification to ISO 9003, for inspection and testing. Module E is the same as Module D, except that the quality system concerns only the end-production checks.

Product Verification (F).   This module requires testing the product and certifying conformity by a third party. The manufacturer ensures that the production process guarantees the product meets the requirements. Then the manufacturer declares conformity. An approved testing organization checks this conformity. This can take place sometimes by testing each product separately and sometimes by random testing. Finally, the testing organization issues a certificate of conformity.

Unit-Verification (G).   Unit-verification requires the manufacturer to submit technical documents for one-time projects, such as the construction of a power plant, to regulatory authorities. A notified body must certify the project/product by checking that the production process conforms with essential requirements.

Full Quality Assurance (H).   Full quality assurance requires the manufacturer to operate an approved quality system for design, production, and testing and to be certified to the European quality standard EN ISO 9001/ISO 9001 by a notified body. Manufacturers can avoid expensive, time-consuming product testing by instituting a full quality assurance system according to ISO 9001.

Degree of Complexity

As a general rule, the greater the safety risk associated with a product, the more complex the conformity assessment process. For example, in the EU Council *Directive on Personal Protective Equipment*, a manufacturer can probably choose to self-certify a product where the model is simple, the risks are minimal, and the user has time to identify those risks safely. Some examples are gardening gloves, gloves for mild detergent solutions, seasonal protective clothing, and gloves or aprons for moderate exposure to heat. Manufacturers, however, must choose either EC-type approval ISO 9002, production quality assurance, or EC type approval plus EC verification in cases where personal protective equipment is of a complex design, "intended to protect against mortal danger or against dangers that may seriously and irreversibly harm the health . . ." of individuals for example, surgical gloves.

Given a choice between EC-type approval plus EC verification or ISO 9002, most manufacturers will choose ISO 9002 registration. Under ISO 9002, the manufacturer submits the production quality system for approval, a preferable alternative to the more intrusive EU process of continuously submitting representative samples to a third party for screening.

Additional Requirements

In addition to using the basic framework of the ISO 9000 series, some EU directives have supplemental requirements. For example, to certify under the EU Construction Products Directive, a manufacturer also must comply with the additional requirements of the EN 45000 series of standards. These standards apply to laboratory, testing, and certification organizations.

Other product sectors for which additional guidelines have been developed are medical devices (*EN 46001, Particular requirements for the application of EN ISO 9001 (ISO*

9001) for medical devices) and aerospace products (EN 2000, EN 3042). Most likely, similar special requirements will be developed for other directives.

Competent Certification and Testing Bodies

The fourth and final component of the EU's comprehensive framework for conformity assessment is the role of certification and testing bodies. The goal of the European Union is to increase confidence of member nations in the work of these organizations so that the results of testing will be accepted throughout the European Union.

In its 1989 presentation entitled *The Global Approach to Certification and Testing*, the European Union outlined the following major elements of its program for certification and testing bodies:

- The credibility of the manufacturer must be reinforced. This can be achieved by promoting the use of quality assurance techniques.

- The credibility of and confidence in testing laboratories and certification bodies must be enhanced. This can be achieved by developing the EN 45000 series of standards to evaluate the competence of testing laboratories.

- The competence of laboratories and certification bodies is established through an accreditation process based on EN 45000 standards. This accreditation process involves a third-party evaluation and is discussed in more detail in Chapter 17.

According to this system, notified bodies must produce documentation proving they conform to the EN 45000 series. If the notified bodies are not formally accredited, the appropriate national authority in the member state where the notified body is located must produce documentary evidence that the notified body conforms to the relevant standards of the EN 45000 series.

Finally, there is a testing and certification organization at the European level called the European Organization for Testing and Certification (EOTC), which has the role of promoting mutual recognition agreements in the nonregulated sphere. The EOTC is discussed in detail in Chapter 17.

The European Union and Other Countries

As the European Union moves toward its goal of a unified market and maps out its comprehensive system for product regulation and certification, it also is defining its future relationship with other countries. One of the key components of this relationship is *nondiscrimination*. This means the same rules apply regardless of the product's origin. A corollary to this is acceptance of test reports or certificates of conformity from countries outside the European Union. These relationships, however, are still in a developmental phase. The issues discussed in this section include the following:

- Can notified bodies subcontract any of their activities to bodies outside the European Union? How much can be subcontracted?

- Are ISO 9000 registration certificates recognized throughout the European Union?

- Can a notified body be located outside the European Union?

Subcontracting.   Can notified bodies subcontract any of their activities to bodies outside the European Union? To some extent, yes. The European Union has proposed new rules for subcontracting and is moving to permit more extensive use of subcontracting. The European Union's general guidelines for subcontracting are stated below:

- EU notified bodies will need to hold subcontractors to the EN 45000 series of standards, including the requirements to maintain records.

- Subcontractors must contract with notified bodies and test to the same standards as the notified body.

- EU notified bodies "cannot subcontract assessment and appraisal activities".

- EU notified bodies remain responsible for any certification activity.

- Notified bodies can subcontract quality assessment audits, provided that they retain responsibility for the audit assessment.

The Guide to the Implementation of Community Harmonization Directives Based on the New Approach and the Global Approach (Brussels, 1995) discusses notified bodies, subcontracting, mutual recognition agreements, and so on.

Mutual Recognition.   Currently, U.S. companies that achieve ISO 9000 registration obtain whatever recognition that their accrediting entity has in the country in which the accreditation entity (i.e., United Kingdom Accreditation Services (UKAS) in the United Kingdom, RAB in the United States) is located. Other countries in the European Union and elsewhere can voluntarily choose to recognize the registration certificate. The registration certificates are not yet governed by EU legislation, and EU-wide recognition is not yet mandatory. Presently, a few national accreditation entities are negotiating mutual recognition agreements to recognize one another's ISO 9000 registrations.

Although the term *mutual recognition agreement* fairly describes what has been negotiated, it should not be confused with the EU legal term, *Mutual Recognition Agreement,* which will be governed by an EU legal document and will refer to product-sector Mutual Recognition Agreements negotiated between the European Union and third countries (United States, Canada, Japan, etc.). Issues of registrar accreditation and the recognition of registration certificates are discussed in more detail in Chapter 17.

Non-EU Notified Bodies.   Under the EU system, member states can designate only notified bodies from within the European Union. No subsidiaries or related enterprises located in a third country can perform full third-party product certification and quality system registration except under a legal Mutual Recognition Agreement (MRA) between the European Union and the government authorities of that country. In addition, their competence must be assessed by third parties, according to the provisions of the EN 45000 series.

MRAs will allow U.S. testing and certification organizations to act as notified bodies in the product areas covered by the MRA. In this way, non-EU notified bodies could award the CE mark for regulated products under the negotiated industry sector.

The U.S. government is actively negotiating MRAs with the European Union in areas where interest has been indicated on the part of the U.S. private sector. These include medical devices, pressure vessels, recreational craft, and electromagnetic compatibility.

THE EUROPEAN UNION AND U.S. CONFORMITY ASSESSMENT

Increasingly, the global economy is exercising an influence on the standards process. While both the European Union and the United States are major players in international conformity assessment talks, the European Union's influence now takes center stage as it works to harmonize standards and eliminate national barriers to trade. In November 1995 the European Union signed an agreement with MERCOSUR (the common market of Brazil, Argentina, Paraguay, and Uruguay) to collaborate on the development of technical standards.

Initially, the European Union directed its energies to critical health and safety issues in the 1985 EU white paper, *Completing the Internal Market, White Paper from the European Commission to the European Council.* More recently, however, EU standards activity is moving beyond the white paper to new areas of concern. In part, this process is being aided by various EU "green papers" that address topics from transportation to telecommunication. It is reasonable to expect this activity to continue and examine most every area of standards activity over the next 20 to 30 years.

The U.S. Response to the European Union

The critical issue for the United States is how to respond to these developments. On the one hand, the European Union and EFTA and their constituent member states have developed a standards structure which uses regional European quasi-legal standards organizations (CEN, CENELEC, ETSI, EOTC) to flesh out product standards. It is a process that fits comfortably into the framework of the EU and European legal code traditions of 17 EU and EFTA member states (only Ireland and the United Kingdom share the English legal system found in the United States). Furthermore, the remaining European nations and most Asian and Latin American nations share those same legal code traditions.

By way of contrast—except for areas of health and safety, where U.S. federal or state agencies have entered the conformity assessment process through regulatory procedures—most American products are manufactured in accordance with industry standards. Industry product sector associations, such as API, SAE, ASME, and NEMA have established voluntary quality standards that must be met and approved before a company is entitled to stamp that it has met association standards.

Quite clearly, the two systems do not provide an easy match. The EU system is quasi-legal in structure, and a significant part of the U.S. system is private and voluntary. The European Union has made it clear that it wants to interact directly with the U.S. government through a federal agency.

Currently, the National Voluntary Conformity Assessment Systems Evaluation (NVCASE) Program is being implemented by the National Institute of Standards and Technology (NIST). NVCASE is designed to serve as the U.S. governmental program that parallels its EU governmental counterparts and officially recognizes U.S. trade associations for product sector standards (see Chapter 17). This mechanism will allow trade associations in turn to recognize U.S. notified body equivalents that can participate in US/EU Mutual Recognition Agreements.

Whatever the decision, if a conformity assessment procedure is put forth as the U.S. procedure, it will have to be sanctioned by the United States. Likewise, only the European Union can sanction its conformity assessment procedures, government to government.

THE CRITICAL ROLE OF CONFORMITY ASSESSMENT

The critical role of conformity assessment procedures worldwide is just beginning to emerge. A company's decision to seek ISO 9000 registration should be part of an overall conformity assessment strategy. For example, a company's choice of an ISO 9000 series registrar might be limited by an EU directive or U.S. law. Approval of conformity assessment procedures for a specific product might also be required by the European Union.

In the United States, the regulatory process is guided by federal agencies such as the FDA, the Environmental Protection Agency, and the Consumer Product Safety Commission. These agencies drive much conformity assessment activity. Most U.S. products, however, are manufactured to private industry standards such as those adopted by the American Petroleum Institute and the Society of Automotive Engineers.

Despite the complexity of the issues, U.S. business must understand and prepare conformity assessment strategies for the products it sells. While ISO 9000 registration is a significant accomplishment for any company, it may not be sufficient to constitute a comprehensive, global market access program.

CONCLUSION

For those people reading about the European Union and its new structure for the first time, it probably seems like one of Bob Newhart's early comedy routines, in which he has Abner Doubleday call a company that sells party games and try to describe his new idea, a game called "baseball."

Naturally, the description of a baseball diamond, outfield, batter, pitcher, home plate, three bases, balls, bats, outs, innings, and nine players to a team sounded like total gibberish to the party game company. But what may have been gibberish in 1839 is now called our national pastime.

The European Union's new game may seem confusing and unsettled. From a U.S. perspective, it may not appear to be a level playing field. But it is becoming *the* playing field. Not understanding its dynamics could mean that a company's team is suited up for football only to discover that the other team is playing baseball.

For example, in 1995 Austrlia became the first nation to sign a MRA with the European Union, adopting new approach to medical device regulation. Other nations are expected to follow.

In a similar, but less formal fashion, India, Singapore, and Malaysia have adopted several of the European Union's new telecommunication standards. And, as noted earlier in this chapter, MERCOSUR, the common market of southern Latin America, signed a formal agreement to coordinate their technical standards developments.

To understand the EU system, it will be necessary to read the directives in depth. More than one directive or set of laws likely will be involved. Understanding the process will be crucial to becoming an effective competitor in what is the world's largest market. For those companies that get there early and establish a presence, the rewards could be substantial.

Editor's Note: European Union Conformity Assessment expert Alan Bailey provided valuable assistance and expertise in the preparation of this article and provided the directives referenced in Appendix B of this *Handbook*. Bailey is the author of a series of books on compliance with European Union Directives. See Appendixes A, C, and E for contact information.

Registrar Accreditation

by Joseph Tiratto

The role of the ISO 9000 registrar is to perform on-site audits of companies' quality systems and issue registration certificates. The accreditation of these registrars and the recognition of their certificates are critical matters. In fact, these two activities are the foundation of the ISO 9000 registration system.

The discussion focuses on registration and accreditation in Europe and the United States. It describes the developing relationship between the two and highlights the problems and differences that must be worked out along the road to mutual acceptance of registration and accreditation practices.

This chapter examines the following topics:

- Registrar accreditation in Europe, the United States, and Canada.
- Criteria for accrediting certified bodies—the EN 45000 series.
- Recognition of registration certificates.
- Auditor certification.
- QS-9000 and ISO 14000 Registrar Accreditation Issues.
- Interpretations of ISO 9000.

REGISTRAR ACCREDITATION

Companies seek registration for different reasons. Some companies seek registration to meet mandatory regulatory requirements. Others seek registration to gain strategic advantage in the marketplace, or simply as a means of improving internal quality systems.

The specific marketplace needs of a particular company often determine which registrar is chosen. For example, some registrars have more influence than others in certain regions of the world or in specific industries. Others may employ more appropriately qualified auditors for a particular industry. Yet, the primary difference between registrars is their accreditation—or lack of accreditation, in some cases. A registrar gains accredited status by meeting strict, internationally accepted guidelines and undergoing an appropriate audit from a handful of qualified accreditation bodies.

Not all issues that surround accreditation have been settled. Unaccredited registrars can be fully competent to conduct quality system audits and award ISO 9000 certificates. On the other hand, fully accredited registrars offer no guarantee of quality auditing services. Ultimately, it is the job of the accredited registrar to provide confidence to industry regarding their competence.

In addition, registrars can be accredited by more than one accrediting body. A registrar that offers multiple accreditations may allow a company to meet the requirements of its customers in different countries. For example, if a company does most of its business with customers in the United Kingdom, then seeking a registrar with United Kingdom Council for Accreditation Services (UKAS) accreditation makes sense. (See Chapter 5 for more on the process of selecting a registrar.)

ACCREDITATION BODIES IN EUROPE

In Europe, third-party assessors are regulated by governmental or quasigovernmental agencies. The Dutch Council for Accreditation (RvA) in The Netherlands and UKAS in the United Kingdom are two quasigovernmental bodies that certify organizations to perform third-party quality system audits.

Other accreditation bodies in Europe include Association franchise pour l'assurance de la Qualite (AFAQ) in France, UNICEI in Italy, and Asociacion Espanola de Normalizacion y Certificacion (AENOR) in Spain. The RvA, the first registrar accreditation body to be established, was until recently the only accreditation body in Europe that would accredit registration bodies outside its own country. The United Kingdom changed its policy in late 1995 and UKAS began offering accreditation to ISO 9000 registrars outside its borders.

ACCREDITATION AND REGISTRATION

The marketplace may be the most important registrar accreditation factor for companies whose products are not affected by a European Union (EU) directive. (Please refer to Chapter 16 for more information on regulated and nonregulated products.) Essentially, regulated products are those that affect health, safety, or the environment in a significant way.

If the company's product is regulated by the European Union, then often an accredited registrar with notified body status must be used for any quality system registration requirement of a directive. A registrar wishing to become an officially notified body must meet certain criteria and apply for such status. An appropriately accredited registrar is important in these strictly controlled venues because the results of its audit—the registration certificate—could be a company's ticket to do business in Europe.

ACCREDITATION IN THE UNITED STATES

The United States has developed its own accreditation system for several important reasons:

- To establish international credibility for U.S.-based registrars.

- To follow the precedent set by an EU resolution that calls for implementing accreditation bodies throughout the European Union.

- To keep pace with the worldwide move to establish accreditation bodies in each country.

In late 1989 The Registrar Accreditation Board (RAB) was established as an affiliate of the American Society for Quality Control (ASQC) to develop, along with its partner the American National Standards Institute (ANSI), a joint national accreditation program for ISO 9000 registrars. The program, called the ANSI/ RAB National Accreditation program, was intended to meet the challenge of ensuring the consistency of services offered by a rapidly expanding number of third-party U.S.-based registrars.

Registrar Accreditation Board

The RAB performs initial audits of registrars, issues certificates of accreditation, performs regular follow-up surveillance, and maintains a directory of accredited registrars. The RAB performs complete reassessments of accredited registrars every four years.

A group of qualified auditors perform the registrar audits, and an accreditation council evaluates the audit results to make final recommendations for the accreditation of a registrar. The RAB also includes an operations council, a board of directors, and an administrative office.

RAB Accreditation Criteria

The criteria used by RAB to accredit registrars are the same as those in Europe, which will enhance the mutual recognition of accreditation systems between the United States and other European countries. The confidence built by this exercise will eventually be spread to other U.S. trading partners. The final goal of these activities is to allow for the mutual recognition of registrar accreditations and the international acceptance of individual supplier quality system registrations. To this end, the RAB has incorporated the following international criteria into its own criteria:

- ISO 10011: Guidelines for auditing quality systems.

- ISO/IEC Guide 40: General requirements for the acceptance of certification bodies.

- ISO/IEC Guide 48: Guidelines for third-party assessment and registration of supplier's quality system.

- EN 45012: General criteria for certification bodies operating quality system certification.

RAB Recognition

RAB is seeking formal recognition of its registrar accreditation scheme in both the U.S. and European marketplace. Since the system to establish broad mutual recognition of accreditation bodies throughout Europe is still being debated, RAB has negotiated a bilaterial agreement with certain European countries to achieve mutual recognition of accredited registrars.

In August 1992, the RAB signed a Memorandum of Understanding (MOU) with the RvA that its supporters say will eventually lead to mutual acceptance of registrar accreditations performed on both sides of the Atlantic.

What Is an MOU?

James Kolka

A memorandum of understanding (MOU) is a formal agreement between two or more parties to confer a mutual exchange of benefits. Unlike a legal contract, which involves a specific exchange of goods and/or services for monetary consideration, an MOU follows no specific legal requirements.

For example, an MOU could be an agreement in which two accrediting institutions from two countries agree to develop formal protocols to recognize each other's accredited companies for a specific number of years. In effect, they agree to agree. Another example would be reciprocal arrangements between registration bodies that allow companies to obtain multiple or co-registrations with a single audit. The registration is recognized by both parties to the MOU.

While the shape of agreements and types of participating parties are not set, MOUs generally are used by organizations to confer mutual benefits. Naturally, the limits of an MOU are confined by the legal capabilities of the parties that sign the agreement.

NIST and the NVCASE Program

In setting up its EU-wide accreditation structure, the European Union prefers to work on a government-to-government basis. Since RAB is not a government organization, the U.S. Department of Commerce designated the National Institute of Standards and Technology (NIST) to establish criteria for conformity assessment in the United States.

NIST, through its office of Standards Services, offers on a fee-for-service basis a voluntary program to evaluate the recognized organization which support conformity assessment activities. The program, known as the National Voluntary Conformity Assessment Systems Evaluation (NVCASE) program, is designed to help U.S. manfucaturers satisfy applicable product requirements mandates by other countries through conformity assessment procedures conducted in the United States prior to export.

The program includes activities related to laboratory testing, product certification, and quality system registration. It does not include the actual registration of individual firms. After NVCASE evaluation, NIST provides recognition to qualified U.S. organizations that effectively demonstrate conformance with established criteria. Under the program, bodies such as RAB can apply for and gain government recognition through NIST. Quality system registrars, when directed by law or requested by another federal agency, may also gain recognition through the program.

NVCASE only responds to a specific industrial or technical need relative to a mandatory foreign technical requirement where there is no accreditation alternative available and where the absence of any alternative would result in significant public disadvantage. NIST operates its program at the recognition level only. It does not perform product testing or quality registration, nor does it accredit bodies to perform these conformity tasks. The NVCASE program is intended to recognize the activities of laboratory accreditors, testing organizations, or registrars.

NVCASE is organized as a separate office within NIST. It is fully funded by fees charged to accredit conformity assessment organizations. NVCASE maintains a listing of all recognized bodies and listings of qualified bodies that have been accredited or otherwise approved by a NVCASE recognized accreditor.

DESIGNATING U.S. NOTIFIED BODIES

The U.S. government has been actively negotiating with the European Union to develop a Mutual Recognition Agreement (MRA) concerning the acceptance of each other's conformity assessment results. The United States has proposed that the NVCASE recognition be one means of acceptance by the European Union of U.S. bodies as equivalent to EU notified bodies to approve products for entry to the EU marketplace.

At the time of publication these talks are active, and it is hoped that an agreement will be forthcoming in 1996. Names of all bodies considered competent to perform conformity assessment actions and acceptable to each side will be part of the agreement. (For more information on NVCASE, contact Robert L. Gladhill, Program Manager, 301-975-4273.)

Independent Association of Accredited Registrars

The Independent Association of Accredited Registrars (IAAR) is a nonprofit association of accredited management systems registrars operating in North America. The organization actively promotes the establishment and maintenance of effective management systems in North American companies through accredited registration. IAAR was formed in 1993.

IAAR works to encourage and facilitate consistent management systems registration through the use of accredited registrars. The association also promotes communication and professional practices among its members and works to educate companies and other organizations about the accredited registration process. In addition, IAAR promotes the integrity of the management systems registration process through member participation in a group forum and encourages its members to follow appropriate codes of professional practice.

IAAR works with other industry groups, government agencies, and other organizations to provide guidance in the appropriate use of international standards and the accreditation process. For example, the Big Three automakers consulted IAAR as it developed the QS-9000 requirements for auto industry suppliers.

All IAAR members are accredited to provide registration services in compliance with ISO Guide 48 and EN 45012.

ACCREDITATION BODIES IN CANADA

Standards Council of Canada (SCC) has been operating a registrar accreditation program since December 9, 1991. The program, known as National Accreditation Program for Registration Organizations (NAPRO), is voluntary and open to any registrar operating in Canada. The SCC is a federal Crown corporation with a mandate to foster and promote voluntary standardization. This connection gives the organization implicit government recognition of its programs.

Applicants for NAPRO accreditation are judged on a number of attributes including: organizational base, administrative practices, human resources, physical resources, documented policies and procedures, and independence of operation.

NAPRO is governed by guidelines contained in a Standards Council publication entitled *Criteria and Procedures for Accreditation Organizations Registering Quality System* (CAN-P-10). An Advisory Committee on Quality (ACQ), made up of experts in the field, oversees the accreditation program and provides guidance to the Standard Council on matters pertaining to quality.

Registrars that apply to the Standards Council for accreditation submit a written application describing their organization and resources, along with a fee as outlined in the published fee schedule. A team of assessors then carries out a five-step assessment process including:

- A preassessment meeting.

- An on-site assessment.

- Field observations of the applicant carrying out assessments of two clients.

- A final meeting to inform the applicant of the assessment team's findings, including any corrective action required.

- When the applicant meets all necessary requirements for accreditation, a report is made to the Standards Council's Registration Accreditation Sub-committee (RASC) and then to the full ACQ. A final recommendation is made by the ACQ to the Executive Committee of the SCC, which approves accreditations.

The entire procedure, from the completed application to the issuing of the accreditation certificate, takes approximately 11 months. Audits are done annually, and re-assessments every four years. SCC is working towards formal agreements with accreditation bodies in other countries that would provide official recognition of SCC-accredited registrars.

CRITERIA FOR REGISTRAR ACCREDITATION

The complexity of the European system, with its many unfamiliar acronyms, makes it difficult for a company to evaluate the competence of its registrar. The system used to evaluate ISO 9000 registrars is part of a larger framework established by the European Union for evaluating notified bodies.

The standard to evaluate registrars is known as the European Norm 45012 (see box). The standard is part of the EN 45000, which was adapted by two other EU organizations, the European Committee for Standardization and the European Committee for Electrotechnical Standardization (CEN/CENELEC). The guidelines are aimed at increasing the level of confidence in the certification, inspection, and testing bodies of the European Union.

EN 45012, *General Criteria for Certification Bodies Operating Quality System Certification,* is important to companies seeking ISO 9000 series standard registration. (See Chapter 5 for an outline of EN 45012 requirements and for more information on how to select a registrar.)

EN 45012: Criteria for Registrars

Organizations that perform quality system certification activities must be evaluated against the requirements of EN 45012. These criteria address the requirements for certified bodies at a national or European level. Implementation of this standard is the

The EN 45000 Series

The EN 45000 series consists of seven documents aimed at ensuring that declarations of conformity, test results, and product and quality system certificates from different national testing labs and certification bodies are equivalent. Many of these standards are modifications of ISO guides listed elsewhere in this chapter, which were written to provide general guidance to countries worldwide involved with product certification.

EN 45001, General Criteria for the Operation of Testing Laboratories details the issues that laboratories must address to demonstrate competence in product testing. These include test personnel, equipment, test methods, test reports, quality systems, and conflict of interest.

EN 45002, General Criteria for the Assessment of Testing Laboratories is designed for accreditation bodies that assess testing labs. EN 45002 incorporates the criteria of EN 45001. It also discusses other requirements for accreditation bodies, including a written accreditation process, published assessment methods, a minimum reassessment period, an opportunity for applicant laboratories to comment on the inspection report, a possible requirement for laboratories to participate in proficiency testing, and rules involving subcontracting of testing.

EN 45003, General Criteria for Laboratory Accreditation Bodies includes guidelines for organizations that want to become accreditation bodies. Requirements include, among others, free and accessible to applicants, independence of the accrediting body, the establishment of sectorial committees to advise the accreditation body and an appeals procedure.

EN 45011, General Criteria for Certification Bodies Operating Product Certification includes the criteria required for national or European recognition of a product certification body.

EN 45012, General Criteria for Certification Bodies Operating Quality System Certification looks specifically at the issue of quality system registration (certification). Its basic criteria parallels that of EN 45011.

EN 45013, General Criteria for Certification Bodies Operating Certification of Personnel applies to the certification of personnel, according to the same criteria as in EN 45011.

EN 45014, General Criteria for Declaration of Conformity goes into detail about the process of actually preparing the Declaration of Conformity to demonstrate conformance with a directive's requirements.

EN 45019, Guidance on Specific Aspects of Testing and Certification.

EN 45020, Definitions.

[To obtain the above standards, contact the American National Standards Institute, 212-642-4900 (ph), 212-302-1286 (fax).]

responsibility of the European Network for Quality System Assessment and Certification (EQNET).

Further guidance related to EN 45012 implementation is contained in the *ISO/IEC Guide 40, General Requirements for the Acceptance of Certification Bodies* and *ISO/IEC Guide 48, Guidelines for Third-Party Assessment and Registration of a Supplier's Quality System.*

Manufacturers or suppliers involved in ISO 9000 registration efforts should make sure that their quality system registrar has been accredited according to EN 45012.

When the framework for evaluating notified bodies in Europe is in place, each notified body will be assessed by a third party against the requirements of EN 45000 to ensure its notification within the Union. These standards will form the foundation for a system of mutual recognition within the European Union.

INTERPRETING EN 45000

The EN 45000 series addresses several controversial issues. Among the most significant are the definition of certification, the technical competence of assessors, the potential conflict of interest between consultancy and registration operations, and the peer review program.

In an effort to establish norms by which all European accreditation bodies should conduct their business, the European Accreditation of Certification (EAC) has published its interpretation of the EN 45000 series of standards.

Definition of Certification

Quality systems certification has a different meaning to customers, suppliers, and registrars within and outside the European Union. The EAC, realizing that mutual recognition of accreditation and quality systems certification requires that a certificate have similar meaning and value to all parties affected by it, has tried to harmonize these differences by defining the meaning of a quality systems certificate.

The EAC interprets EN 45012 *Clause 2, Certification of Conformity*, as follows:

> The certification should give the market confidence that the supplier is capable of systematically meeting agreed requirements for any product or service supplied within the field specified on the certificate, the 'scope of the supplier.'

This interpretation affirms that a supplier should give the market confidence that a product or service will meet the agreed-upon requirements. A customer who places an order with a certified supplier should expect consistency.

While this does not imply that the product or service meets specific technical requirements, both supplier and customer must understand that quality system certification and supplier conformance to agreed-upon customer requirements go hand in hand. Demonstrated product conformity to a technical standard is a separate issue.

Assessor's Technical Competence

The most effective way to ensure and promote market confidence in a certified supplier—and to increase industry belief in the value of the certificate—is for the certifying registrar to employ technically competent assessors.

The EAC interpretation of EN 45012 *Clause 7, Certification Personnel,* clarifies what is meant by competency. The EAC guidelines state:

> The assessment team needs a background which ensures that they understand the requirements relating to the system they are auditing. Each assessment team should have a general understanding and background in each technological and industrial sector in which it operates. It should be able to determine whether or not a particular quality system adequately covers the requirements of the standard in the area that it covered.

An assessor's technical competence should be demonstrated by an understanding of how the product is used, a general knowledge of the product's critical characteristics, the process by which the product is manufactured and tested, and familiarity with the appropriate product standards.

This competence can be demonstrated by a recognized industry certification, prior work experience, or technical society participation. The ability of a quality systems registrar to employ assessors with the necessary audit skills, technical depth, and product use knowledge ensures marketplace confidence in a certified supplier.

Consultancy

Consultancy and conflict of interest is an area of extreme importance to registrars, suppliers, and customers both in Europe and the United States. EAC guidelines state that if a registration body is owned by a person or holding company that also engages in consultancy, then that person or company is regarded as a consultant and the registration body must have an appropriate structure to prevent that entity from influencing certification.

In addition, the EAC guidelines prevent registration and consulting services from being marketed together to prevent even the perception of conflict of interest. These guidelines clarify that the registrar and its representatives must in no way imply that any business advantage is gained by using a registrar and consultant from the same organization.

Peer Review

The EAC's harmonized interpretation for applying the EN 45012 standard uses a peer review system assure that signatories of the EAC are complying with these guidelines. Peer review ensures that the certificates issued by an accreditation board and accredited registrar are valid within the EAC community. Valid may be loosely defined as a certificate that provides the necessary marketplace assurance.

RECOGNITION OF REGISTRATION CERTIFICATES

The international acceptance of registration certificates is a crucial issue. The ultimate goal is to develop a system that will allow a single ISO 9000 registration certificate to be recognized within an industry and in the rest of the world.

As mentioned earlier, in the private sector, recognition of the ISO 9000 certificate by a company's customers is the primary determinant of a certificate's acceptability. The accreditation status of the registrar and how this status is viewed by the marketplace is another matter.

At the member state level in the European Union, a registration obtained in one EU member state for a regulated product may not necessarily be accepted in other EU states on a bilateral basis. Other member states in the European Union can voluntarily choose to recognize the registration certificate.

Two organizations working on these issues in Europe are the European Network for Quality System Assessment and Certification (EQNET) and the European Committee for Quality System Assessment and Certification (EQS).

European Network for Quality System Assessment and Certification (EQNET)

EQNET was formed in 1990 by eight ISO 9000 registration bodies in Europe to promote cooperation among individual members based on bilateral sets of agreements. The intention is to further expand EQNET with the aid of multilateral contracts. The eight original members of the organization included: AFAQ (France), AIB-Vincotte (Belgium), BSI (Great Britain), DQS (Germany), DS (Denmark), N.V. KEMA (The Netherlands), SIS (Sweden), and SQS (Switzerland).

The main tasks of EQNET include the following:

- Cooperate to recognize the certificates issued by other members on the basis of existing contracts and to promote the recognition of certificates.

- Coordinate the certification of border-crossing groups of companies/organizations and the joint conduct of said certificates in a competent and efficient way.

- Issue several certificates at the same time on the basis of joint certification audits.

European Committee for Quality System Assessment and Certification (EQS)

EQS was formed to achieve the following primary goals:

- Harmonization of rules for quality system assessment and certification.

- Overall recognition of quality system certificates.

- Efforts to permit mutual recognition of the certificates of quality system certification bodies.

The group's ultimate aim is to avoid multiple assessment and certification of an organization's quality system and to develop confidence in quality system assessment and certification carried out by competent bodies.

ISO's Committee on Conformity Assessment (CASCO)

The work of recognition of registration certificates on an international level has been progressing largely through the efforts of ISO's Committee on Conformity Assessment (CASCO) and the International Accreditation Forum (IAF).

The goals of CASCO are the following:

- Study the means of assessing the conformity of products, processes, services, and quality systems to appropriate standards or other technical specifications.

ISO/IEC Guides Pertinent to Certification, Registration, and Accreditation

- Guide 2, General terms and definitions concerning standardization and related activities.

- Guide 7, Guidelines for drafting of standards suitable for use for conformity assessment.

- Guide 22, Information on manufacturer's declaration of conformity with standards or other technical specifications.

- Guide 23, Methods of indicating conformity with standards for third-party certification systems.

- Guide 25, General requirements for the competence of calibration and testing laboratories.

- Guide 27, Guidelines for corrective action to be taken by a certification body in the event of misuse of its mark of conformity.

(Continued on next page)

(Continued from page 640)

- Guide 28, General rules for a model third-party certification system for products.

- Guide 39, General requirements for the acceptance of inspection bodies.

- Guide 40, General requirements for the acceptance of certification bodies.

- Guide 42, Guidelines for a step-by-step approach to an international certification system.

- Guide 43, Development and operation of laboratory proficiency testing.

- Guide 44, General rules for ISO or IEC international third-party certification schemes for products.

- Guide 53, An approach to the utilization of a supplier's quality system in third-party product certification.

- Guide 56, An approach to the review by a certification body of its own internal quality system.

- Guide 57, Guidelines for the presentation of inspection results.

- Guide 58, Calibration and testing laboratory accreditation systems—general requirements for operation and recognition.

- Guide 59, Code of good practice for standardization.

- Guide 60, Code of good practice for conformity assessment.

- Guide 61, General requirements for assessment and accreditation of certification/registration bodies.

- Guide 62, General requirements for bodies operating assessment and certification/registration of quality systems.

- Prepare international guides relating to the testing, inspection, and certification of products, processes, and services and to the assessment of quality systems, testing laboratories, inspection bodies, and certification bodies, including their operation and acceptance.

- Promote mutual recognition and acceptance of national and regional conformity assessment systems.

- Promote the appropriate use of international standards for testing, inspection, certification, quality systems, and related purposes.

In 1992, CASCO recommended that a system be developed in which a single ISO 9000 registration would be recognized worldwide. As a result, OSP and the IEC formed a planning group on Quality Systems Assessment Recognition (QSAR). The group developed a proposal for a worldwide recognition system, which was accepted by ISO and IEC.

International Accreditation Forum (IAF)

The IAF was formed in January 1993 when 10 representatives of various international standards bodies and registrars met to begin a series of discussions on how international accreditation bodies could better cooperate in the effort to establish a complete network of mutual recognition agreements. The goal is a system whereby accreditations are recognized throughout the world.

At an April 30–May 1, 1993 meeting, the IAF considered a master list of 22 topics that members felt should be addressed and whittled it down to the following 9 primary areas of focus:

1. Requirements of complete reevaluation of a supplier's quality system, separate and distinct from surveillance.

2. Sphere of influence for accreditation bodies.

3. Structure of registrars, including role and makeup on the independent advisory board.

4. Use of satellite offices of an accredited registrar and the conditions for issuing registration certificate bearing the registration mark.

5. Mutual recognition of accreditation.

6. Minimum level of registration activity required before accreditation is granted.

7. Misleading ISO 9000 advertising.

8. Public announcement of applications for accreditation.

9. Allowing registrars to offer suppliers a choice of accreditation body marks and use of a quality mark system on packaging.

IAF has drafted a plan to form a Multilateral Recognition Agreement (MLA) group among member accreditation bodies. Both the QSAR and IAF plans require peer evaluation of accreditation bodies before being accepted for membership. In order to avoid duplication of effort and to simplify the QSAR program, IAF volunteered to assume the peer evaluation process for accreditation bodies. ISO and IEC councils have approved the recommendation that IAF should take on this role within the QSAR program.

The QSAR program will accept as a member any accreditation body that has demonstrated its competence by achieving membership in the IAF MLA group. Registrars will be eligible for QSAR membership by virtue of accreditation by a QSAR member accreditation body.

The ongoing work of CASCO, IAF and EQNET and EQS will eventually lead to a larger if not complete network of mutual recognition agreements.

THE EUROPEAN ACCREDITATION OF CERTIFICATION (EAC)

The European Accreditation of Certification is an association of accreditation bodies from the European Union (EU) and the European Free Trade Association (EFTA). The EAC's goal is the creation of a single European accreditation system covering products, quality systems and personnel. The EAC has published its own interpretations of the EN 45000 standards. (See below.) By harmonizing the definitions and interpretations of these rules, the EAC hopes to bring the business practices of all registrars into closer alignment.

THE EUROPEAN ORGANIZATION FOR TESTING AND CERTIFICATION (EOTC)

The European Organization for Testing and Certification (EOTC), proposed by the European Union for the purpose of dealing with conformity assessment issues, was created in April 1990. Its role is to promote mutual recognition of test results, certification procedures, and quality system assessments and registrations in nonregulated product areas throughout the European Union and EFTA. Its primary goal is to encourage equivalency of certificates and to avoid the duplication caused by multiple certifications.

The EOTC will also be responsible for providing technical assistance to the EU Commission in the implementation of some EU legislation, especially in the preparation of Mutual Recognition Agreements with non-EU countries.

EOTC Agreement Groups

EOTC aims to recognize the technical competence of the certification bodies (i.e., registrars and laboratories) in certain industry agreement groups. The purpose of an agreement group is to promote mutual recognition of test certificates by certification bodies throughout the EU, EFTA, and third world countries. These agreement groups assure mutual recognition of test reports and certificates by the certification bodies that participate in the agreement group. The EOTC also is considering extending the conformity assessment modular approach to old-approach directives and certain nonregulated products.

The EOTC has recognized 15 agreement groups. The industry sectors these groups cover vary from fire and security to information technology.

The EOTC's agreement groups bring value to the marketplace and benefit both the customers and agreement group participants by helping establish shared confidence in the test procedures and product and quality system evaluation procedures.

EOTC's Status

The EOTC previously was under the auspices of the European Commission; however, the EOTC signed an agreement with its 23 founding members to establish its independence as a private association. The founding members will continue to serve a leadership role in the nonregulated product conformity assessment scheme, but the association may help to organize agreement groups among notified bodies.

One of the EOTC's sectorial committees, the European Committee for Information Technology Testing and Certification (ECITC)—which oversees the agreement groups for information technology—is primarily responsible for this activity.

EOTC Published Agreement Groups

- WECC Calibration Laboratory Accreditation Systems.

- European Fire and Security Group (EFSG).

- International Instrumentation Evaluation Group.

- Open Systems Testing Consortium Recognition Arrangement (OSTC).

- Recognition Arrangement for European Testing of Electromagnetic Compatibility of Information Technology Products (EMCIT).

- Recognition Arrangement for European Testing for Certification for Office and Manufacturing Protocols (ETCOM).

- Short-Circuit Testing Liaison Agreement Group (STLA).

- Recognition Arrangement for Assessment and Certification of Quality Systems in the Information Technology Sector (ITQS).

- Low Voltage Agreement Group (LOVAG).

- Agreement Group for European Testing of Electromagnetic Compatibility in the Field of Electrotechnology (including electronics) (EMCEL).

- Agreement Group for Testing of Quality Label Schemes in the Aluminum Finishing Industry (QUALISURFAL).

- Certification and Testing of System Software (CATOSS).

- European Welding Foundation (EWF).

- Recreation Marine Agreement Group (RMAG).

- European Active Medical Devices Certification Agreement (EMEDCA).

AUDITOR CERTIFICATION PROGRAMS

Companies considering registration to ISO 9000 should also understand how auditors are qualified and what criteria are used. Several international guidelines are used. For example, *ISO 10011, Guidelines for auditing quality systems*, has been accepted by ISO as the qualifications criteria for auditors. It contains three parts (10011-1, *Auditing*, 10011-2, *Qualification criteria for quality systems auditors, and* 10011-3, *Management of audit programs*). The ISO 10011 series of guidelines has been widely accepted internationally, and while it is not a mandatory requirement it is an excellent resource for establishing consistent audit practice worldwide.

Recognition Arrangements for the Assessement and Certification of Quality Systems in the Information Technology Sector (ITQS)

ITQS, one of several agreement groups that EOTC has formally approved, is a good example of how an agreement group contributes to the harmonization process. Agreement groups are set up along industry lines and allow companies to rely on just one test or assessment that will be accepted throughout Europe.

The assessments carried out by the certification bodies (registrars) cooperating in ITQS are performed in a harmonized way, using common standards, techniques, and guidance material. ITQS membership is open to any organization in the world, provided it accepts and applies ITQS regulations, including an auditor guide.

The ITQS auditor guide is "unique because instead of being a guidance document for the certification applicant, it is mainly designed to be used by the certification bodies," according to Phillippe Caussin of AIB-Vincotte Inter. AIB-Vincotte Inter. operates the ITQS secretariat.

The auditor guide "tells the auditors what to look for when auditing an information technology firm, covers software and hardware development, and production and service activities," said Caussin. The purpose of an auditing guide is to ensure that information technology firms certified by an ITQS member are assessed on an equivalent basis. Quality managers at some U.S. information technology firms have cited the lack of experienced information technology auditors as one of the reasons the United States does not, as yet, have a strong, formalized system for information technology quality assurance.

Criteria for information technology auditors under the ITQS system include professional education or training or practical experience in information technology, as well as specific training in the understanding of quality control techniques applicable to information technology. "ITQS regulations also require that the auditors be evaluated on a regular basis by an evaluation panel, possibly from a professional society," according to Caussin.

While ISO 10011 guidelines cover all aspects of the conduct and management of an audit and the training of auditors, perhaps their most important requirement is that only certified lead assessors can lead the audit team and perform the assessment of a company's quality systems. (ISO 10011 is discussed in more detail in Chapter 6.)

The Institute for Quality Assurance

The Institute for Quality Assurance (IQA) in the United Kingdom was the first organization to govern and control the ISO 9000 auditor training and certification process. The program, known as the International Register of Certificated Auditors (IRCA), seeks to recognize the integrity and competence of quality systems auditors as measured against the criteria found in international standards.

Under the IRCA IQA scheme, auditor certification is reviewed every three years, with renewal dependent upon the ability of the registered auditor to meet criteria for assessment experience required at the time of the renewal.

Registrar Accreditation Board

In the United States, the Registrar Accreditation Board (RAB) has its own program for auditor assessor certification, The Certification Program for Auditors of Quality Systems. In addition, RAB has published a program to train ISO 9000 series auditors called *Requirements for the Recognition of an Auditor Training Course.*

This program is the final piece of a complete ISO 9000 series registration and accreditation scheme that includes the accreditation of registrars, certification of auditors and lead auditors, and the recognition of training courses.

Certification under RAB's Certification Program for Auditors of Quality Systems requires training in an approved course. Two types of courses are offered to meet RAB training requirements for RAB auditor certification:

1. A 36-hour Lead Auditor Training Course that includes ANSI/ISO/ASQC Q9000 (ISO 9000) series training, quality system audit training, and an examination that covers this information. An ability to communicate, both orally and in writing, and the requisite personal attributes and audit management capabilities are necessary to conduct and lead an audit.

2. A 16-hour Lead Auditor Training Course for ASQC Certified Quality Auditors that includes ANSI/ISO/ASQC Q9000 (ISO 9000) series training and an examination. The applicant's ability to communicate both orally and in writing will also be evaluated. The 16-hour course and an ASQC Certified Quality Auditor certificate meet general auditor training requirements for RAB certification. The Certified Quality Auditor (CQA) rating currently offered by the ASQC means that an applicant to the program has fulfilled requirements and has demonstrated knowledge of auditing.

The CQA rating does not exempt the candidate from the other experience required by RAB, including audit experience. A CQA rating, together with a 16-hour course, along with appropriate ISO 9000 training by the RAB-approved organizations, will exempt the candidate from the standard 36-hour course.

The CQA program will not be affected by RAB's quality system auditor certification program. RAB and ASQC are separate organizations with separate goals for auditor certification.

American Society for Quality Control

The ASQC has an auditor certification program that offers a CQA rating as described above. This rating signifies that an applicant has education and/or work experience in a specified field and demonstrates such knowledge through the successful completion of a written examination.

MUTUAL RECOGNITION OF AUDITOR CERTIFICATION

In March 1994, RAB and IQA reached an agreement that allows students who complete a lead assessor or lead auditor course sanctioned by one of the organizations to apply for certification through either RAB or IQA. The agreement also means that auditors who have been certified through one accreditation body are recognized by the other organization.

In September 1993, 16 nations met in Singapore to develop a plan to create a single, internationally accepted set of requirements for ISO 9000 series auditor certification and related course accreditation.

This process was further advanced in July 1995 when the International Auditor and Training Certification Association (IATCA) was created. IATCA is composed of 13 auditor certification and course accreditation organizations from 12 regions or countries. In addition to RAB representing the United States, organizations from the following regions or countries are involved with the IATCA program:

- Australia/New Zealand
- Brazil
- Canada
- China
- Chinese Taipei
- France
- Japan
- Korea

- Singapore

- Southern Africa

- United Kingdom

IATCA has developed a set of criteria governing auditor certifications and auditor training course approvals. These criteria are designed to eliminate the need for multiple certifications of auditors and multiple approvals of auditor training courses.

The key to recognition of an auditor certification organization is the IATCA peer evaluation process. When four participating organizations have successfully completed peer evaluations, they will sign an IATCA multilateral recognition agreement (MLA). Additional signatories will be added to the MLS as organizations qualify through peer evaluations. No auditor certifications or course accreditations may be graded under the IATCA criteria until at least four organizations sign the MLA.

RAB began to accept applications for certification as an IATCA grade in March 1996. Formal recognition of an auditor RAB IATCA grade certification will only occur, however, after the peer evaluation process is complete and the MLA is signed by RAB.

The current RAB auditor certification program will remain in place and, although current RAB certified auditors may choose to transfer to the RAB IATCA program, it will not be required to facilitate the transfer at the early stages. IATCA has set up a limited window of opportunity, until February 1, 1997, in which a certified auditor can make the transition to the new IATCA program by meeting certain criteria. After February 1, 1997, all applicants for IATCA certification, including auditors currently certified by RAB, must meet the full criteria rather than the simplified transitional requirements.

In addition to international acceptance of an auditor's certification to the new RAB IATCA program, an auditor training course providers' accreditation progam will also enjoy international acceptance. A course provider will now need to earn and maintain only a single accreditation for its auditor training course.

One of the important additions under the IATCA program is that each student must satisfy certain prerequisites before attending an ISO 9000 auditor training course. The course provider will administer pre-tests or a questionnaire to each prospective student to determine adequate knowledge of ISO 9000 standards. Only those students that demonstrate a basic understanding of the ISO 9000 standards will be accepted into the class.

ISO 14000 SERIES STANDARD FOR ENVIRONMENTAL MANAGEMENT SYSTEMS

ISO 14000 was published as an international standard in early fall 1996 (see Chapter 12).

Some registrars offered registration to the draft version of the standard, and the trend appears to be integration of the ISO 9000 and ISO 14000 systems. Registrars are planning for joining ISO 9000 and ISO 14001 accreditation audits.

As this Handbook went to press, ANSI and the RAB had begun the process of establishing a joint ISO 14000 accreditation program for registars and auditors. The new program will be called the ANSI/RAB National Accreditation Program (NAP) and represents an expansion of the existing ISO 9000 partnership between the two organizations.

QS-9000 QUALITY SYSTEM REQUIREMENTS

The QS-9000 Quality System Requirements were developed by Chrysler, Ford, and General Motors Supplier Quality Requirements Task Force. QS-9000, which was first published in August 1994 and revised in February 1995, incorporates the requirements of ISO 9001 plus sector-specific requirements and customer-specific requirements (see Chapter 15). The requirements were developed to standardize the quality systems requirements, documentation, and assessment tools for suppliers to automotive manufacturers and certain truck manufacturers.

Agreements have been reached with a number of accreditation bodies (including RAB) and their accredited third-party registrars to conduct QS-9000 registrations. QS-9000 auditors, in addition to being qualified ISO 9000 auditors, are required to take two one-day classes given by the Automotive Industry Action Group (AIAG) and pass an exam following the classes.

As of February 1, 1996, the following is a summary of activity of the QS-9000 program:

- 13 national accreditation programs have been approved to participate.
- 29 registrars have been qualified to issue certificates.
- 958 third-party auditors have been qualified to perform audits.

ISO 9001 INTERPRETATIONS

ISO Technical Committee 176 (TC176) agreed to take on the role of interpreting the standards following concerns that the ISO 9000 standards were not being implemented uniformly throughout the world. ISO TC 176 has assigned the task of providing the means for prompt interpretations of the original text to the Technical Committee Officers' Group (TCOG). These interpretations will be coordinated and communicated at an international level.

The interpretation process is still being developed within the ISO TC 176 Committee. Some questions, such as the roles of the individual member delegations and the method of distribution of the interpretations, need to be resolved.

In the meantime, a program for publishing interpretations of the ISO 9001 standard has been developed and implemented by the Canadian Technical Committee on Quality Management (CAC/ISO/TC 176). Under this program, the request for interpretations are reviewed and a proposed reply is prepared by the Interpretation Task Force. The proposed reply is then reviewed and voted on by the technical committee. If the proposed reply is approved, it becomes a formal Canadian position. The interpretation is then published, and ISO is notified of the interpretations. As of late 1995, 24 requests for interpretations had been answered. More requests are currently being processed.

QS-9000 INTERPRETATIONS

The automotive industry has developed and implemented a program for publishing interpretations of the QS-9000 requirements. The interpretations are developed by the International Automotive Sector Group (IASG), which consists of representatives from:

- Recognized ISO 9000 accreditation bodies.
- QS-9000 qualified registrars.
- Chrysler, Ford, and General Motors Supplier Quality Requirements Task Force.
- Tier 1 automotive suppliers.

Representatives from Chrysler, Ford, and GM must individually agree with interpretations before "agreed" status is achieved following a consensus of all members of the IASG. The interpretations are then incorporated into the "IASG Sanctions QS-9000 Interpretations" and are considered binding. As of late 1995 six sets of interpretations have been published.

CONCLUSION

Registrar accreditation and auditor certification issues are being addressed and refined, and considerable progress has been made in the past few years concerning international acceptance. As the registrar accreditation and auditor certification programs mature, remaining, unsettled questions will be addressed by the United States, European, and international groups.

Endnote

[1]For more information on the IQA scheme, write to: The Scheme Manager/The Registration Board for Assessors, PO Box 712, 61 Southwark Street, London, SW1W 0DQ, England; tel. 011-44-171-401-2988, fax 011-44-171-401-2725.

Editors Note: Thanks to Andrew J. Bergman who provided assistance in the preparation of this article. Mr. Bergman is vice president of the Houston-based management consulting firm Quality Specialist, Inc. QSI is an authorized distributor of European standards and assists companies with compliance. See QSI listing in Appendix A for contact information.

Figure 17–1

Accreditation Pyramid

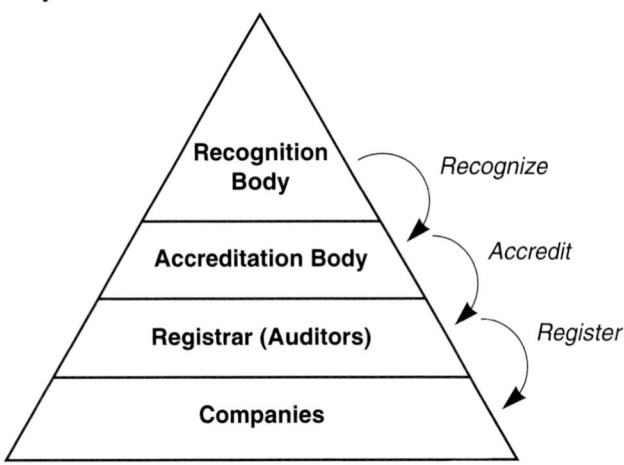

ANNEX 1

EN 45012 General Criteria for Certification Bodies Operating Quality System Certification— Outline of Requirements

1. Object and field of application—general criteria for a certification body operating quality system certification.

2. Definitions—definitions applicable to EN 45012.

3. General requirements—states that all suppliers have access to the services of the certification body and that the procedures under which the body operates shall be administered in a nondiscriminatory manner.

4. Administrative structure—requires the certification body to do the following:

   • Be impartial.

   • Choose the members of its governing board from among the interests involved in the process of certification without any single interest predominating.

   • Safeguard impartiality and enable participation from all parties concerned in certification system functions.

   • Have permanent personnel under the senior executive to carry out the day-to-day operations in a way that is free from control by those who have a direct commercial interest in the products or services concerned.

(Continued on next page)

(Continued from page 651)

5. Terms of reference of governing board—addresses the functions of the governing board of the certification body through the following:

   - Formulation of policy.

   - Overview of policy implementation.

   - Overview of finances.

   - Setting up committees as required.

6. Organizational structure—addresses the requirements for a certified body's organizational structure by calling for the following:

   - A chart showing the responsibility and reporting structure of the organization and in particular the relationship between the assessment and certification functions.

   - A description of the means by which the certified body obtains financial support.

   - A documented statement of its certification systems, including its rules and procedures for granting certification.

   - Documentation clearly identifying its legal status.

7. Certification personnel—requires the following:

   - Personnel be competent for the functions they undertake.

   - Information be maintained regarding relevant qualifications, training, and experience.

   - Records of training be kept up to date.

   - Personnel have available clear, documented instructions pertaining to their duties and responsibilities.

   - Personnel of subcontracted sources meet the requirements of EN 45012.

8. Documentation and change control—requires that:

   - The certified body maintain a system for controlling documentation related to the certification system.

   - Current issues of the appropriate documentation must be available at all relevant locations.

   - Changes to documents must be covered by the correct authorization and processed in a way that ensures direct and speedy action at the effective point.

(Continued on next page)

(Continued from page 652)

- Superseded documents be removed from use throughout the organization and its agencies.

- Certified suppliers be notified of changes that could be accomplished through direct mailing or by issue of a periodic publication.

9. Records—addresses the requirements for the following:

   - Maintaining a record system to demonstrate the way in which each certification procedure was applied, including assessment and surveillance.

   - Storing records for an adequate period.

   - Holding records secure and in confidence to the client, unless otherwise required by law.

10. Certification and surveillance procedures—addresses the following requirements:

    - That the certification body have documented procedures to enable the assessment, certification, and surveillance of quality systems.

    - That the certification body require the supplier to have a documented quality system.

    - That the certification body maintain regular surveillance of the supplier's quality system.

11. Certification and surveillance facilities—address the following requirements:

    - That the certification body have the required facilities in terms of certification personnel expertise and equipment to perform assessment, certification, and surveillance of the supplier's quality system.

    - That the certification body ensures that external bodies conform to the above requirement and that a properly documented agreement covering the arrangements, including confidentiality, be drawn up.

12. Quality manual—addresses the requirements that the certification body have a quality manual and documented procedures setting out the way in which it complies with the following criteria:

    - A quality policy statement.

    - A brief description of the legal status of the certification body.

    - A statement about the organization of the certification body, including details regarding the governing board, its constitution, terms of reference, and rules of procedure.

(Continued on next page)

(Continued from page 653)

- Names, qualifications, experience, and terms of reference of the senior executive and other certification personnel, both internal and external.

- Details of training arrangements for certification personnel.

- An organizational chart showing lines of authority, responsibility, and allocation functions stemming from the senior executive.

- Details of the documented procedures for assessing and auditing supplier quality systems.

- Details of documented procedures for surveillance of suppliers.

- A list of subcontractors and details of the documented procedures for assessing and monitoring their competence.

- Details of appeals procedures.

13. Confidentiality—addresses the requirements that the certification body ensure confidentiality of the information obtained in the course of its certification activities at all levels of its organization, including committees.

14. Publications—addresses the requirements that the certification body produce and update, as necessary, a list of certified suppliers with an outline of the scope of the certification of each supplier, and the requirement that the list be available to the public. This section further requires that a description of the certification system(s) be available in published form.

15. Appeals—addresses requirements that the certification body have procedures for considering appeals to its decisions.

16. Internal audit and periodic review—addresses the requirement that the certification body undertake internal audits and periodic reviews of its compliance with the criteria of EN 45012. The reviews are to be recorded and made available to persons having the right of access to this information.

17. Misuse of certificates—addresses requirements that the certification body exercise proper control on the use of its quality system certificates. This section also requires that incorrect references to the certification systems or misleading use of certificates found in advertisements, catalogs, etc., be dealt with by suitable actions. A further notation states that such actions include corrective action, publication of the transgression and, if necessary, legal action.

18. Complaints—addresses requirements that the certification body require the certified suppliers to keep a record of all complaints and remedial actions relative to the quality system.

(Continued on next page)

(Continued from page 654)

19. Withdrawal and cancellation of certificates—addresses the requirement that the certification body have documented procedures for withdrawal and cancellation of quality system certificates.

Challenges Facing the ISO 9000 Industry

18

by Donald W. Marquardt

"This chapter addresses challenges that will affect the credibility of the ISO 9000 standards and their implementation in the coming years. These challenges include documenting the scope of certification/registration; implementing the requirements for continuous improvements; implementing the requirements for statistical techniques; maintaining accuracy of translations and interpretations of the standards; considering alternate routes to registration; dealing with industry-specific adoptions of the ISO 9000 standards; and attaining compatibility with ISO standards for environmental management systems.

A three-part appendix to this chapter contains the detailed policy provisions of the Registrar Accreditation Board for documenting the scope of certification/registration; the first of the challenges summarized in the main body of this chapter."

ISO 9000 AS AN INDUSTRY

In a sense, the worldwide activities concerned with implementing the ISO 9000 standards are themselves an "industry." The registrar/certification bodies, accreditation bodies, auditor certification bodies, course provider organizations, consultants, auditors, publishers of trade magazines and books, journalists, and university and other academic activities constitute the infrastructure of a full-fledged industry.

Indeed, many organizations that engage in these activities operate internationally, with personnel in many countries. ISO 9000 now faces the types of challenges typical of a new, rapidly growing industry. Some of these challenges relate to the integrity and credibility of the industry and its component activities.

THE CHALLENGE OF CREDIBILITY OF REGISTRATION

Programs for third-party certification/registration of quality systems have grown worldwide, and the impact on the global economy has become enormous.

Existing international, regional, and national standards and guides for implementation of such third-party programs are providing the framework for programs worldwide. This framework includes requirements standards and guideline standards.

One focal point of activity is the ISO 9000 and ISO 10000 standards prepared by the ISO/TC176 committee. Another focal point is the implementation guides, such as ISO/IEC Guide 62, prepared by the ISO/IEC Committee on Conformity Assessment (CASCO). The elements of these various standards and guides have been adopted by regional bodies, such as the European Union, and by many nations worldwide.

In addition, many nations have set up implementation programs that reflect the three-level concept of certification, accreditation, and recognition.

- Quality systems *certification* bodies (registrars) evaluate suppliers' quality systems for conformity to the requirements standards.

- *Accreditation* bodies evaluate the management systems of certification bodies for conformity to international guides.

- *Recognition* bodies, usually government affiliated, provide national recognition of the quality systems certification (registration) and accreditation programs and thus facilitate the establishment of mutual recognition among nations in regulated areas.

The marketplace credibility of this network of international programs rests upon the elements of assurance and ethical principles incorporated in the various standards and guides and their implementation. The high degree of success to date in implementing registration programs worldwide indicates that the base of assurance elements and ethical principles is sound, at least in broad outline.

As the registration programs have become widespread among multiple industry/economic sectors and multiple cultures, new complexities are revealed.

The Registrar Accreditation Board (RAB), the U.S. accreditation body, has evaluated carefully the elements of assurance and ethical principles to ensure a credible program, giving due consideration to the complexities of the international system.

The three components of these elements are:

1. Defining the scope of registration of a supplier's quality system (defining the scope of registration of a supplier's quality system, i.e., specifying the standard, the geographical sites, the products, and the portions of the supply chain).

2. Policy, principles, and implementation regarding conflict of interest by registrars accredited under the joint ANSI/RAB program.

3. Code of conduct (for persons working on behalf of RAB).

All three components are considered essential to achieve the goal of marketplace credibility. The full text of the components is included in the chapter appendix. Some of the key points described in the first component, the Scope of Registration, are discussed below.

Scope of Registration of a Supplier's Quality System

In current practice worldwide there is great variability in the documented definitions of the scopes of supplier quality system registrations. This variability is observed from supplier to supplier for a given registrar, from registrar to registrar in a given nation, and from one nation to another. The term *registrar* is used in North America; the term *certification body* is used in Europe.

Greater consistency in defining and documenting this scope is an essential prerequisite to establish marketplace credibility of quality system registration, and to negotiate meaningful mutual recognition arrangements among nations.

There are other important benefits beyond the benefits of marketplace creditibility, registrars will benefit if the ground rules for defining scope are consistent for all parties. To adequately describe the scope of registration of a supplier's quality system, four basic questions must be answered.

Which Standard?

The statement of scope should identify the ISO 9000 series standard whose requirements have been the basis for evaluating conformity leading to the registration of the supplier's quality system (i.e., ISO 9001, ISO 9002, ISO 9003).

The customer must be confident that the selection of the appropriate standard jointly by the supplier and the registrar has taken adequately into consideration the amount and nature of product design activity that is involved in the products produced by the supplier, as well as the nature of the production processes through which the supplier adds value to the product.

Which Geographic Sites or Operating Units?

The statement of scope should identify the boundaries of the registered supplier quality system, in terms of the geographic location(s) of the facilities and activities or the operating units involved. An operating unit may be, for example, a ship.

The procedure should inform the customer whether the product the customer receives is produced within the registered quality system, even in situations where the supplier may have multiple sites or operating units dealing with the same product, not all of which may be registered.

Which Products?

The statement of scope should identify the boundaries of the registered supplier quality system in terms of the commercial product(s) that are processed. The term *products* includes all four generic product categories: hardware, software, processed materials, and services.

The customer should be informed whether the product the customer receives is processed within the registered system, even in situations where the supplier may deal with multiple products at the same site or operating unit, and not all of the products may be processed within the registered quality system.

Which Portions of the Supply Chain?

The procedure should inform the customer regarding:

- The starting points of the supplier's registered operations (e.g., the raw materials, parts, components, services, and intermediate products that are provided by sub-suppliers).

- The ending points of the supplier's registered operations (i.e., the remaining steps on the way to the ultimate consumer that are excluded from the supplier's registered operations).

- The nature of the value added that has been provided by the supplier's registered operations.

Where the registered quality system represents only a fraction of the supplier's operations or a fraction of the total value added to the product, this should be stated in registration documentation, and as a consequence customers may be aware of this fact.

The procedures should not invite suppliers who wish to be registered, but want to exclude portions of their operations from scrutiny by the registrar, to declare the excluded portions to be subcontractor operations. It does not matter whether the excluded portions are in another nation, elsewhere in the same nation, or simply another part of the same production site.

Procedures for this element of scope would apply also to support functions that are critical to product quality, such as a test laboratory, which may not be included in the supplier's quality system as registered to ISO 9001 or ISO 9002.

The Current Status

In practice today, most certificates of quality system registration are deficient because they do not provide definitive answers to all four questions. Part 1 of the chapter's appendix suggests ways that registrars can implement the above requirements. Four steps in this process are proposed:

1. Each accredited registrar provides information on the face of each registration certificate issued on the four elements discussed above.

2. All publishers of registration lists show the same information.

3. Each registered supplier maintains a mechanism to inform customers about the four elements, such as providing copies of the registration certificates when appropriate.

4. Each registrar establishes and maintains procedures and records regarding the above elements.

Part 2, policy, principles, and implementation regarding conflict of interest by ANSI/RAB-accredited registrars, and Part 3, code of conduct (for persons working on behalf of RAB) are discussed in this chapter's appendix.

RESPONSIBILITIES OF THE ORGANIZATIONS INVOLVED

ANSI/RAB and the U.S. program do not stand alone. The credibility of registration issues are of international concern.

The implementation of the ethical principles and elements of assurance incorporated in the various standards and guides is clearly important to the credibility of quality systems registration. Who should take responsibility for these? At a fundamental level, every person involved in any facet of quality system registration should take personal responsibility.

Specific organizations must have clear responsibilities for implementing the recommended procedures. For example, the code of conduct discussed in Part 3 is to be implemented at the accreditation body level, but corresponding codes of conduct are required of organizations operating at the registration level. The registrar conflict of interest policy and principles discussed in Part 2 is to be implemented at the registration level. The scope of registration definition discussed in Part 1 is to be implemented jointly by suppliers and registrars.

In each instance there are other organizational entities that have defined roles. For example, registrars audit supplier organizations for conformance; accreditation bodies audit registrars; and recognition bodies have formal processes to establish credibility for recognition.

Those who write international standards (e.g., ISO/TC176) and guides (e.g., ISO CASCO) have their own defined responsibilities. They set the ground rules by means of the standards and guides.

Editor's Note: Members of the RAB Board of Directors, in particular G. Q. Lofgren and R. W. Peach, have contributed to the RAB statements on the topic of this discussion of scope of registration. J. H. Hooper and others also provided comments and suggestions on Part 1.

THE CHALLENGE OF CONTINUOUS IMPROVEMENT

The philosophy of quality management and quality assurance has changed over the years. In the 1960s the virtually universal perspective of business managers was "If it ain't broke, don't fix it."

In that philosophical environment maintaining the status quo in quality achievement prevailed in most of industry and commerce. Today, as illustrated by the minimal use of ISO 9003, the philosophy of 30 years ago is giving way to a philosophy of continuous improvement. Continuous improvement is increasingly necessary for economic survival in the global economy and is becoming a widely pursued goal. It is the only reliable route to sustaining marketplace advantage for both customer and supplier.

The focus of quality assurance standardization is now on prevention of nonconformities. This requires a "process" focus, which is reflected in the required documentation and many other features.

In the first years of use since 1987 an unfortunate mind-set has been adopted by many registrars/certifiers and their auditors. This mind-set can be called the *status quo mind-set*. Reminiscent of the 1960s, it is characterized by the ditty, "Say what you do, do what you say."

This simple advice is correct as far as it goes, but is *far short of the requirements in the ISO 9000 standards of both 1987 and, especially, 1994.* It focuses only on adherence to established procedures. For example, the status quo mind-set ignores the *linked requirements* in ISO 9001 to demonstrate continuing adequacy of the quality system for business objectives and customer satisfaction.

Intrinsic to the 1994 revision of ISO 9001 is a reshaped mind-set, as summarized in Table 18–1. The reshaped mind-set is *a cycle of continuous improvement,* which can be depicted in terms of the classic plan-do-check-act management cycle.

Continuous improvement is a necessary consequence of implementing ISO 9001 (or ISO 9002). There are two groupings of linked clauses in ISO 9001 that work together to ensure continuous improvement.

The linkages among the clauses are really quite clear in ISO 9001, if your mind-set doesn't block them out.[1,2] The intention of the ISO 9000 standards has always been that the clauses are elements of an integrated quality system. When implementing any sys-

TABLE 18–1

The Reshaped Mind-Set

A cycle of continuous improvement (built into ISO 9001 by the linked requirements):

- *PLAN* your objectives for quality and the processes to achieve them.

- *DO* the appropriate resource allocation, implementation, training, and documentation.

- *CHECK* to see if

  —you are implementing as planned.

  —your quality system is effective.

  —you are meeting your objectives for quality.

- *ACT* to improve the system as needed.

---

tem, the interrelationships among the elements—that is, the linkages—are as important as the elements themselves.

The linkages among clauses in ISO 9001 can be recognized in three ways:

- Explicit cross references between linked clauses shown by parenthetic expressions "(see x.xx)" in the text of certain clauses.

- Use of key words or phrases in clauses that have linked requirements; for example, "objectives" for quality in clauses 4.1.1, 4.1.3.

- Content interrelationships which cause linkages among the activities required by two or more clauses; for example, clauses 4.1.1, 4.1.3, 4.17, 4.14.

Two groupings, a) and b), of linked clauses work together to ensure continuous improvement.

    a. Continuous Improvement via Objectives for Quality and Ensuring the Effectiveness of the Quality System

The 1994 revision of ISO 9001 expands and strengthens the requirements for executive management functions and links a number of clauses by requirements to define "objectives" for quality (clauses 4.1.1 and 4.1.3) and to "ensure the effectiveness of the quality system" (clauses 4.1.3, 4.2.2, 4.16, and 4.17).

    b. Continuous Improvement via Internal Audits and Management Review

The 1994 revision of ISO 9001 expands and strengthens the requirements of four clauses in the 1987 standard that link together for continuous improvement in the supplier organization. These are internal quality audits (clause 4.17), corrective and preventive action (clause 4.14), management representative (clause 4.1.2.3), and management review (clause 4.1.3).

The interplay of activities required by these four clauses provides a mechanism to institutionalize the pursuit of continuous improvement. In today's competitive economy, the objectives for quality must continually be more stringent in order to maintain a healthy business position. More stringent objectives for quality translate into the need for an increasingly effective quality system.

If the ISO 9000 implementation industry fails to embrace adequately the continuous improvement requirements in the standards, the competitiveness of the standards themselves will be seriously compromised in the global marketplace.

THE CHALLENGE OF STATISTICAL TECHNIQUES

From the earliest days of the quality movement, statistical techniques have been recognized as having an important role. In fact, during the 1940s and 1950s, statistical techniques were viewed as the predominant aspect of quality control. During succeeding decades the management system increasingly took center stage. In the 1987 version of ISO 9001, clause 4.20 on statistical techniques paid only lip service to its subject. The implementation of quality assurance standards worldwide has reinforced the deterioration of emphasis on statistical techniques.

It is critical that this situation be remedied and that statistical techniques and management systems each have their important place in quality. The 1994 version of ISO 9001 contains explicit, meaningful requirements relating to statistical techniques (clause 4.20), citing their relation to "process capability" and "product characteristics." There is direct linkage between ISO 9001 clause 4.20 and process control (clause 4.9, especially 4.9g), as well as indirect linkages to other clauses. These requirements, if conscientiously implemented, would guarantee that statistical techniques play important roles in quality under the ISO 9001 umbrella.

Unfortunately, most auditors, registrars, accreditation bodies, and their supporting consultants, and training course providers, have limited knowledge or experience in statistical techniques. Consequently, despite the requirements in clause 4.20 of ISO 9001:1994, the clause still is receiving too little emphasis.

This should be viewed as a challenge by supplier companies, registrars, auditors, consultants, and course providers. I hope many will seize these opportunities (which really are obligations) in the near future. Moreover, this places responsibility on accreditation bodies to ensure that the third-party registration system is implemented in conformance to the applicable international standards and guides. In particular, this includes ISO 9001 itself and clause 4.20.

In the United States, the Registrar Accreditation Board has taken initiatives. RAB has issued to all ANSI/RAB-accredited registrars a Bulletin stating RAB's intention to monitor the operations of registrars with respect to the linked requirements and the implementation of clause 4.20. (See also Marquardt 1995,1996.[1,2]) There are opportunities for

ISO/TC176, too. TC176 is cooperating with ISO/TC69 (Application of Statistical Methods) to provide guidance documentation on the use of statistical techniques when implementing the standards in the ISO 9000 family.

THE CHALLENGE OF STANDARDS INTERPRETATION

In actual application ISO standards are published by ISO in English and French (and often in Russian), which are the official ISO languages. ISO requires that "the texts in the different official language versions shall be technically equivalent and structurally identical" (ISO/IEC Directives, Part 1, 1.5, 1989). Sometimes ISO publishes standards in languages other than the official languages; then "each is regarded as an original-language version" (ISO/IEC Directives, Part 1, F.3, 1995). "However, only the terms and definitions given in the official languages can be considered as ISO terms and definitions" (ISO/IEC Directives, Part 3, B2.2, 1989).

When a nation "adopts" an ISO standard it is first *translated* by the national body into the national language and processed through the official national procedures for adoption. In the United States the translation issue is minimal, consisting, when deemed necessary, of replacement of British English (the ISO official English) with American English spellings or other stylistic editorial details. In the United States the adoption process follows American National Standards Institute procedures, which ensure the objectivity, fairness, and lack of bias that might favor any constituency; these are requirements for all American National Standards.

In situations where the national language is not one of the ISO official languages, ISO has, at present, no formal procedure for validating the accuracy of the national body translation. Translating from one language to another always presents challenges when great accuracy of meaning should be preserved. There are many ways in which the meaning may be altered in translation. These changes can be a troublesome source of nontariff trade barriers in international trade.

The challenge of interpretations of the ISO 9000 standards goes beyond problems of translation into languages other than the official ISO languages of English and French. In the global economy many *situations of use* are encountered, and the intended meaning of the standard is not always clear to those applying it. In such situations each member body of ISO is expected to set up interpretation procedures.

There will nevertheless be cases where an official ISO interpretation is required. ISO has, at present, no formal procedure for developing and promulgating such official interpretations. ISO/TC176 has taken an initiative with ISO Central Secretariat to establish an official procedure; ultimately ISO/TC176 should be the point of the final interpretation of the ISO 9000 standards, which it is responsible to prepare and maintain.

When the situation of use is a *two-party* contractual situation between the supplier organization and the customer organization, differences of interpretation should normally be revealed and mutually resolved at an early stage (e.g., during contract negotiation and contract review). Official international interpretations become more necessary in *third-party* certification/registration situations. In that environment interpretations must be applicable in a wide range of situations, and not prejudicial to maintaining a level playing field. Contract negotiations between supplier and customer are then free to focus on only those quality system requirements, if any, that go beyond the scope of the relevant ISO 9000 requirements standard, plus the technical specifications for the product.

THE CHALLENGE OF ALTERNATE ROUTES TO REGISTRATION

Organizations differ in regard to the status of their quality management efforts. Some are at an advanced state of implementation and effectiveness; others have hardly begun. Most are at some intermediate state. The ISO 9000 standards have as their primary purpose the facilitation of international trade. They are, therefore, positioned to ensure a level of effectiveness that is adequate for reducing nontariff trade barriers in international trade. This required level of effectiveness will tend to increase with the passage of time.

At any point in time there will be some organizations that have well-established, advanced quality management systems based on an approach that may go beyond the requirements of ISO 9001. For such organizations, the cost of registration/certification by the usual third-party route may be perceived to be high compared to the incremental value added to their quality management system.

In the United States a number of such companies that have major international presence, especially in the electronics and computer industry, have been working with organizations involved in the implementation of third-party certification/registration to devise an approach that would gain international acceptance. The approach would have to acknowledge their existing quality management system status and reduce the cost of certification/registration while supporting their international trade by providing the assurance conferred by accredited certification/registration. (More detail is provided in Chapter 14.)

THE CHALLENGE OF INDUSTRY-SPECIFIC ADOPTION
AND EXTENSION OF ISO 9000 STANDARDS

Some sectors of the global economy have witnessed industry-specific adoption and extension of the ISO 9000 standards. Such adoption and extension can be effective in a *very few industries* where there are special circumstances and appropriate ground rules can be developed and implemented consistently. These special circumstances are characterized by:

a. industries where the product's impact on health, safety, or environmental aspects is potentially severe, and as a consequence most nations have regulatory requirements regarding the quality management system of a supplier, or

b. industries that have had well-established, internationally deployed, industry-specific or supplier-specific quality system requirements documents prior to publication of the ISO 9000 standards.

Fortunately, in the very few instances so far, the operational nonproliferation criteria of the ISO/IEC Directives Part 2, clause 6.6.4 have been followed.

Circumstance a) relates to the medical device manufacturing industry. For example, in the United States the Food and Drug Administration (FDA) developed and promulgated the Good Manufacturing Practice regulations. The FDA operates under the legal imprimatur of the GMP regulations, which predate the ISO 9000 standards.

The FDA regularly inspects medical device manufacturers for compliance with the GMP requirements. Many of these are quality management system requirements that parallel the subsequently published ISO 9002:1987 requirements. Other GMP regulatory requirements relate more specifically to health, safety, or environmental aspects. Many other nations have similar regulatory requirements for medical device products.

In the United States the FDA is expected to promulgate during 1996 revised GMPs for medical devices that parallel closely the ISO 9001:1994 standard, plus specific regulatory requirements related to health, safety, or environment. The expansion of GMP scope to include quality system requirements related to product design reflects the recognition of the importance of product design and the greater maturity of quality management practices in the medical device industry worldwide. Similar trends are taking place in other nations, many of which are adopting ISO 9001 verbatim for their equivalent of the GMP regulations.

In ISO, a new technical committee, ISO/TC210 has been formed specifically for medical device systems. TC210 has developed standards that provide supplements to ISO 9001 clauses. These supplements primarily reflect the health, safety, and environmental aspects of medical devices and tend to parallel the regulatory requirements in various nations. These standards are in late stages of development and international approval at this time.

Circumstance b) relates to the automotive industry. In the years preceding publication of the 1987 ISO 9000 standards, various original equipment manufacturers (OEMs) in the automotive industry had developed company-specific, proprietary quality system requirements documents. These requirements were part of OEM contract arrangements when purchasing parts, materials, and subassemblies from the thousands of companies in their supply chain. The OEMs had large staffs of second-party auditors to verify that these OEM-specific requirements were being met.

Upon publication of ISO 9001:1994, the major U.S. OEMs began implementation of QS-9000, which is discussed in greater detail in Chapter 15. QS-9000 incorporates ISO 9001 verbatim, plus industry-specific supplementary requirements. Some of the supplementary requirements are really prescriptive approaches to some of the generic ISO 9001 requirements; others are additional quality system requirements which have been agreed by the major OEMs; a few are OEM-specific.

THE CHALLENGE OF COMPUTER SOFTWARE

The global economy has become permeated with electronic information technology (IT). The IT industry now plays a major role in shaping and driving the global economy. As with past major technological advances, the world seems fundamentally very different and, paradoxically, fundamentally the same. Computer software development occupies a central position in this paradox.

First, it should be noted that computer software development is both an industry and a *discipline.*

Second, many IT practitioners emphasize that computer software issues are complicated by the multiplicity of ways that computer software quality may be critical in a supplier organization's business. For example:

- The supplier's product may be complex software whose functional design requirements are specified by the customer.

- The supplier may actually write most of its software product, or may integrate off-the-shelf, packaged software from subsuppliers.

- The supplier may incorporate computer software/firmware into its product, which may be primarily hardware and/or services.

- The supplier may develop and/or purchase from subsupppliers software that will be used in the supplier's own design and/or production processes.

However, it is important to acknowledge that hardware, processed materials, and services often are involved in a supplier organization's business in these same, multiple, ways, too.

There is general consensus worldwide that:

- The generic quality management system activities and associated requirements in ISO 9001 are relevant to computer software, just as they are relevant in other generic product categories (hardware, other forms of software, processed materials, and services).

- Some things are *different* when applying ISO 9001 to computer software.

- There is at this time no worldwide consensus as to *which* things, if any, are different enough to make a difference and what to do about those that are.

ISO/TC176 developed and published ISO 9000-3:1991 as a means to deal with this important, paradoxical issue. ISO 9000-3 provides guidelines for applying ISO 9001 to the development, supply, and maintenance of (computer) software. It has been widely used. ISO 9000-3 offers guidance that goes beyond the requirements of ISO 9001.

In the United Kingdom a separate certification scheme (TickIT) for software development has been in operation for several years, using a combination of ISO 9001 and ISO 9000-3. The scheme has received both praise and criticism from various constituencies worldwide. Those who praise the scheme claim it:

- Addresses an important need in the economy to assure customer organizations that the requirements for quality in software they purchase (as a separate product or incorporated in a hardware product) will be satisfied.

- Includes explicit provisions beyond those for conventional certification to ISO 9001 to assure competency of software auditors, their training, and audit program administration by the certification body.

- Provides a separate certification scheme and logo to exhibit this status publicly.

Those who criticize the scheme claim it:

- Is inflexible and attempts to prescribe a particular life cycle approach to computer software development which is out of tune with current best practices in the field.

- Includes unrealistically stringent auditor qualifications in the technology aspects of software development; qualifications whose technical depth is not necessary for effective auditing of management systems for software development.

- Is almost totally redundant with conventional third-party certification/registration to ISO 9001, under which the certification body/registrar already is responsible for competency of auditors (and accreditation bodies verify competency as part of accreditation procedures).

- Adds substantial cost beyond conventional certification/registration to ISO 9001 and provides little added value to the supply chain.

In the United States a proposal to adopt a TickIT-like software scheme was presented to the ANSI/RAB accreditation program. The proposal was rejected, primarily on the basis that there was no consensus in the IT industry and the IT user community to support the proposal. Currently:

- ISO/TC176 is revising ISO 9000-3 to bring it up to date with ISO 9001:1994 and remedy some technical and architectural deficiencies.

- ISO/TC176 is planning the next revision of ISO 9001 with the long-term intention of incorporating quality assurance requirements stated in a way that will meet the needs of all four generic product categories without supplementary application guideline standards such as ISO 9000-3.

- Various national and international groups, conferences, and organizations are discussing whether there is a difference that makes enough of a difference to warrant a special program, and if so, what such a program should entail.

One thing that is clear is that currently no worldwide consensus exists. More detail on this subject is provided in Chapter 20.

THE CHALLENGE OF ENVIRONMENTAL MANAGEMENT SYSTEMS

Environmental management systems and their relation to quality management systems is a classic example of a challenge from the perspective of the ISO 9000 standards. Companies are likely to have to do business under both sets of requirements: the ISO 9000 standards from ISO/TC176 and the ISO 14000 standards from ISO/TC207. The opportunity for mutually beneficial consistency promises important benefits.

These benefits relate to the operational effectiveness of having one, consistent management approach in both areas of the business activities, and can translate also into cost benefits. The ISO Technical Management Board has mandated that TC176 and TC207 achieve compatibility of their standards.

In ISO standards development the compatibility of the ISO 9000 and ISO 14000 standards is one part of the standardization job. Compatibility of implementation requires that harmonization be established in each nation in the infrastructure of accreditation bodies, certification/registration bodies, and auditor certification bodies operating under internationally harmonized guidelines. At this time the ISO 14000 infrastructure is in its infancy. More detail is provided in Chapter 12.

Endnotes

[1]Donald W. Marquardt, "The Missing Linkage in ISO 9001: What's Being Done About It?," *Proceedings, ASQC 49th Annual Quality Conference,* American Society for Quality Control, Milwaukee, WI, 1995, pp. 1056–1061.

[2]Donald W. Marquardt, "The Importance of Linkages in ISO 9001," *ISO 9000 News,* Vol. 5, No. 1, January/February 1996, pp. 11–13.

Part 1: A Registrar Accreditation Board (RAB) White Paper—Defining the Scope of the Certification/Registration of a Supplier's Quality System

1. INTRODUCTION

Background

RAB observes that there is great variability in the documented definitions of scope of registration of suppliers' quality systems. This variability is observed from supplier to supplier for a given registrar, from registrar to registrar in a given nation, and from one nation to another. The term *registrar* is used in North America; the term *certification body* is used in Europe.

Benefits

The RAB believes that greater consistency in defining and documenting this scope is an essential prerequisite for:

- Establishing marketplace credibility of quality system certification/registration to ISO 9001, ISO 9002 or ISO 9003.

- Negotiating meaningful mutual recognition arrangements among nations.

Beyond the benefits of marketplace credibility, there are important benefits to registrars if the ground rules for defining scope are consistent for all parties.

2. PURPOSE OF THIS WHITE PAPER

The purpose of this white paper is to

- Analyze the concept of scope and define its elements.

- Describe for each element of scope a recommended approach, the current status in relation to the recommended approach, and the unresolved issues.

- Propose a principle upon which implementation procedures should be based.

- Propose implementation procedures for describing and publishing the scope of registration of a supplier's quality system.

- Provide guidelines for describing scope.

This white paper is intended to serve as a basis for discussion and consensus development on these issues by interested and affected parties in the United States and worldwide.

3. THE CONCEPT AND ELEMENTS OF SCOPE

To adequately describe the scope of certification/registration of a supplier's quality system, four questions must be asked:

Which standard?

Which geographic sites or operating units?

Which products?

Which portions of the supply chain?

In practice today most certificates of quality system registration are deficient because they do not provide definitive answers to all four questions.

The four basic questions lead to the four elements of scope.

3.1 Which Standard?

The statement of scope must identify the ISO 9000 standard whose requirements have been the basis for the registration of the supplier's quality system (i.e., ISO 9001, ISO 9002, ISO 9003).

3.2 Which Geographic Sites (or Operating Units)?

The statement of scope must identify the boundaries of the registered supplier quality system in terms of the geographic location(s) of the facilities and activities. The term *operating unit* applies in situations such as shipboard operations where the registered supplier quality system cannot be described in terms of a geographic location.

3.3 Which Products?

The statement of scope must identify the boundaries of the registered supplier quality system in terms of the commercial product(s) that are processed. The term *products* includes all four generic product categories: hardware, software, processed materials, and services.

3.4 Which Portions of the Supply Chain?

The statement of scope must identify the boundaries of the registered supplier quality system in terms of supply-chain criteria, specifically the starting and ending points of the registered quality system.

4. RECOMMENDED APPROACH, CURRENT STATUS, AND ISSUES

4.1 Selection of Standard

Approach

The procedure should provide confidence to the customer that the selection of the appropriate standard jointly by the supplier and the registrar has taken adequately into consideration the amount and nature of product design activity that is involved in the products produced by the supplier, as well as the nature of the production processes through which the supplier adds value to the product.

Status

The majority of quality systems today are being registered to ISO 9001 or ISO 9002. Most choices seem to be appropriate to the circumstances. In most cases the identification of the standard selected is visible to the customer.

Issue

In some cases ISO 9003 or ISO 9002 has been selected when it appears that a more comprehensive quality assurance model would be more appropriate. In some cases the mismatch may not readily be apparent to the customer.

An area where clarity is important is distributor operations. Many are registered to ISO 9003 under the rationale that the distributor does not produce the products themselves. However, a distributor's product is the services of acquiring, stocking, preserving, order fulfilling, and delivery. Hence ISO 9002 is appropriate to cover the production of these services. Distributors who design their service products should be registered to ISO 9001.

4.2 Specification of Boundaries in Terms of Geographic Location or Operating Unit

Approach

The procedure should inform the customer whether the product the customer receives is processed within the registered quality system, even in situations where the supplier may have multiple sites or operating units dealing with the same product, not all of which may be registered.

Status

Most customers today seem to be informed on this element of scope and are, when they deem it important, requiring suppliers to supply product processed at registered sites or operating units.

Issue

The lack of consistent procedures for scope description sets the stage for misrepresentation.

4.3 Specification of Boundaries in Terms of Product(s) Processed

Approach

The procedure should inform the customer whether the product the customer receives is processed within the registered quality system, even in situations where the supplier may deal with multiple products at the same site, and not all of the products may be processed within the registered quality system.

Status

Most customers today seem to be informed on this element of scope.

Issue

The lack of consistent procedures for scope description sets the stage for misrepresentation.

4.4 Specification of the Boundaries in Terms of Supply-Chain Criteria

Approach

The procedure should inform the customer regarding:

- The starting points of the supplier's registered operations (e.g., the raw materials, parts, components, services, and intermediate products provided by sub-suppliers).

- The ending points of the supplier's registered operations (i.e., the remaining steps on the way to the ultimate consumer that are excluded from the supplier's registered operations).

- The nature of the value added that has been provided by the supplier's registered operations.

Status

Specification of boundaries in terms of supply-chain criteria is not part of established procedures for describing scope. There appear to be many instances where the registered quality system represents only a small fraction of the supplier's operations, or a small fraction of the total value added in the product. This is not stated in registration documentation, and, as a consequence, customers may not be aware of this fact.

Issue

This situation invites suppliers who wish to be registered, but want to exclude portions of their operations from scrutiny by the registrar, to declare the excluded portions to be sub-contractor operations. It does not matter whether the excluded portions are in another nation, elsewhere in the same nation, or simply in another part of the same production site. It is possible that a supplier can be registered for the supply of an end product when only a small portion of the value added has been provided by the operations that have been subject to registrar audit and evaluation. The lack of consistent procedures for describing this element of scope appears to be providing a significant opportunity for misrepresentation by some suppliers in various nations.

This element of scope would apply also to support functions that are critical to product quality, such as a test laboratory, which may not be included in the supplier's quality system as registered to ISO 9001 or ISO 9002.

5. PRINCIPLE UPON WHICH IMPLEMENTATION PROCEDURES SHOULD BE BASED

5.1 Principle

There are many registrars, each registering many supplier quality systems. Each supplier is dealing with many customers. It is impractical to monitor adequately the operations of such a system solely by periodic audits.

Consequently, RAB supports the following principle regarding scope:

> *Primary reliance must be placed on the concept of "truth in labeling" by means of which every customer has routine, ready access to the information upon which to judge all four elements of the scope of a supplier's registered quality system.*

5.2 Relation to International Criteria

Generic requirements for the first three elements of scope are given in ISO/IEC Guide 62: 1996 (which supersedes Guide 48). Requirements for the fourth element (supply-chain boundaries) are not given explicitly in Guide 62 but are equally necessary.

Certificates of registration are the original records from which other records (e.g., lists of registered quality systems) are derived. Examination of samples of certificates and registers shows that even the first three elements are not universally or uniformly documented today.

6. FORMAT AND IMPLEMENTATION PROCEDURES FOR DESCRIBING SCOPE

To implement the "truth in labeling" principle, four types of requirements are proposed:

6.1 Each accredited registrar shall show on the face of each registration certificate it issues clear, specific, and sufficient information for *all four* elements of the quality system scope, as discussed in this document:

- Standard selected.

- Geographic sites or operating units.

- Products.

- Supply-chain boundaries.

6.2 All parties who publish registers that list certified quality systems shall show for each registered system, information on all four elements as required in item 6.1.

6.3 Each supplier with a registered quality system shall establish, maintain, and implement a mechanism to inform customers, as appropriate, about all four elements of its registered quality system. One such mechanism is to provide copies of all applicable certificates to a customer prior to consummation of a sales agreement.

6.4 Registrars and suppliers shall establish procedures and maintain records regarding implementation and notifications according to their respective requirements 6.1, 6.2, and 6.3.

7. IMPLEMENTATION GUIDELINES REGARDING ELEMENTS OF SCOPE

RAB proposes the following guidelines for "clear, specific and sufficient information" on each of the elements of scope. The information should be concise. Concise description of the boundaries for item 6.1 will often depend in part upon the interrelations of the boundaries defined for elements 3.1, 3.2, 3.3, and 3.4.

7.1 Selection of Standard (see 3.1 and 4.1)

Guideline

For the requirement in 6.1 simply show the number of the ISO 9000 series standard whose requirements were the basis for verifying conformity of the supplier's quality system (i.e., ISO 9001, ISO 9002, or ISO 9003).

7.2 Boundaries in Terms of Geographic Locations or Operating Units (see 3.2 and 4.2)

Guideline

For the requirement in 6.1 identify each of the facilities or portions of facilities, sites, and operating units in sufficient detail to distinguish those facilities associated with the certified/registered quality system from other facilities owned, operated, or controlled by the supplier and other parent, subsidiary, or affiliated organizations. In most cases a full street address is not necessary.

7.3 Boundaries in Terms of Product(s) (see 3.3 and 4.3)

Guideline

For the requirement in 6.1 identify the products and the generic product categories involved in sufficient detail to distinguish those products processed by the registered quality system from other products processed by the supplier and other parent, subsidiary, or affiliated organizations. In many cases a listing of generic product names, types, and/or trade names will be clear, specific, and sufficient.

7.4 Boundaries in Terms of Supply-Chain Criteria (see 3.4 and 4.4)

Guideline

For the requirement in 6.1 identify the processes that are provided by sub-suppliers and are not part of the supplier's registered operations. This is normally accomplished by specifying in sufficient detail which raw materials, parts, components, services, or inter-mediate products of any generic product category (hardware, software, processed materials, and services) are provided by sub-suppliers. Specify only those that are critical to the requirements for quality of the final product.

- If the starting and ending points in the supply-chain are different for the various products covered in a single registration certificate, the specifics should be made clear.

- If intermediate steps are done by a sub-supplier, creating multiple starting and ending points, the specifics should be made clear.

- If, from one shipment to another, the situation may change, the specifics should be made clear. An example is an intermediate product or component sometimes produced by the supplier and sometimes purchased from a sub-supplier.

In some cases the complexities of the supply-chain boundaries applicable for a given customer situation may be too voluminous to explain fully on a certificate and have to be made available to interested customers by means of supplemental documentation. For example, a certificate issued for supplier A could state the supply-chain aspects of scope as follows:

"Assembly of product T where approximately 40 percent of the value is represented by components manufactured in the registered quality system, including critical components X and Y, and the remaining materials and components are purchased according to Company A specifications and the requirements of ISO 9001, clause 4.6. Specific supplemental documentation is available to customers on request."

If the quality systems of some sub-suppliers are separately registered, then that information may be provided to customers but should not be shown on the supplier's certificate. This approach would apply also to a supplier's laboratory unit, which may be separately registered ("accredited") to ISO Guide 25 but not included in the ISO 9001/ISO 9002 registration.

8. EDUCATION OF CUSTOMERS

The "truth in labeling" principle will function most effectively when customers are educated in the issues and criteria. RAB recommends that educational materials be developed by accreditation bodies for distribution to customers. For example, the materials could suggest questions for customers to ask their potential suppliers, and discuss the scope issues related to those questions. Registrars could audit suppliers to verify that the information provided to customers is reasonable.

Part 2: Registrar Accreditation Board (RAB)— Policy, Principles, and Implementation Regarding Conflict of Interest by ANSI/RAB-Accredited Registrars

POLICY

An ANSI/RAB-accredited registrar, its employees, and agents shall not be involved in activities that constitute a conflict of interest, nor any activities that may reasonably be perceived to constitute a conflict of interest, with the registrar's registration services for any supplier.

PRINCIPLES

1. Consulting or the giving of advice to a supplier, or any of its employees or agents, by a registrar or any of its employees or agents (whether engaged as agents of the registrar, or as individual consultants) is considered a conflict of interest when the consulting or the giving of advice is provided under circumstances that are conducive to the fact or perception that the acceptance of the counsulting or advice enhances the likelihood of a favorable outcome of the registration services provided to the supplier.

2. Teaching activities may create or may be perceived to create the likelihood of concomitant consulting or the giving of advice when provided for a student group consisting exclusively or significantly of employees or agents of the supplier, or when provided on premises owned or arranged by the supplier, or when the supplier provides substantial funding or sponsorship.

3. Consulting or the giving of advice to a supplier may create, or may be perceived to create, the likelihood of conflict of interest, if any parent, subsidiary, or affiliate organization (or division thereof) of the registrar, or their employees or agents, also provide registration services to the same supplier, or a parent, subsidiary, or affiliate organization (or division thereof) of the supplier.

4. A pre-audit of a supplier by a registrar to assess readiness for a full audit that could lead to registration is not in itself a conflict of interest, if the results provided to the supplier consist only of audit findings in relation to the requirements of the applicable quality system standard and do not include consulting or advice on how to remedy any deficiencies encountered in the pre-audit.

5. If a registrar is itself a unit of a supplier and wishes to engage in first-party registration (i.e., to register other units of the supplier) or second-party registration (i.e., to register sub-suppliers of the supplier), such registration cannot be done under RAB accreditation.

6. A conflict of interest may exist, or be perceived to exist, where a registrar shares the same marketing organization, marketing procedures or facilities, and/or market plan with any parent, subsidiary, or affiliate organization that provides consulting.

7. A conflict of interest may exist, or be perceived to exist, where the organizational units, employees, or agents of the registrar organization have a significant financial ownership relation to the supplier organization or any of its parent, subsidiary, or affiliate organizations.

IMPLEMENTATION

1. Where a registrar seeking accreditation or re-accreditation by RAB does not have any potential conflicts of interest under Principles 1 through 7, the registrar shall do the following:

   • Attest to that fact in writing as part of the documentation supplied to RAB for accreditation or re-accreditation, and

   • Inform RAB whenever any change in this status may occur.

   This is subject to verification by RAB auditors on any audit of the registrar.

2. Where a registrar engages in consulting activities as described in Principles 1, 2, 3, or 6, the registrar must take special steps in addition to attestation (as in Implementation item 1), to assure the absence of conflict of interest.

   The following steps are required:

   a. *Option 1:* An ANSI/RAB-accredited registrar (including any of its parent, subsidiary, or affiliate organizations) shall not provide consulting to an organization and issue a certificate of quality system registration to a supplier unit that includes the organization for which consulting is provided, unless the time interval during which consulting is provided and the time interval during which registration services and subsequent surveillance are provided are separated by at least one year, and the registrar's procedures provide documentation and demonstration of that separation.

Option 2: The registrar's procedures, organizational assignments, and documentation thereof shall ensure and demonstrate that the employees or agents involved in the consulting have no part in, nor give nor receive information relating to, any of the activities leading to certification, registration, and subsequent surveillance for the supplier, nor is there any other communication regarding the supplier between those doing consulting and those doing certification and registration services, where the information, activities, or communication could reasonably be perceived to constitute a conflict of interest.

b. The registrar's advertising and other written materials for suppliers must disclose the nature of the dual consulting and registration services, and the nature of the organizational, procedural, and other steps the registrar has taken to ensure the absence of conflict of interest.

These provisions are subject to verification by RAB auditors in any audit of the registrar. RAB registration fees may reflect the additional effort required to audit items a and b.

c. The registrar shall, upon engaging the services of each employee or agent, and at least once per year thereafter, inform the employee or agent of the provisions in this policy and obtain the signature of the employee attesting to the absence of personal involvement in any conflict of interest under this policy, or documenting the circumstances if any such conflict or potential conflict has existed since the prior signature. These records shall be retained for RAB audit.

3. This statement shall be included by RAB in its "Procedures for Accreditation" and shall be supplied therewith to every registrar seeking accreditation or re-accreditation.

Part 3: Code of Conduct—Registrar Accreditation Board (RAB)

Scope: This Code of Conduct applies to members of the RAB Board of Directors, its Councils (e.g., on Operations and Accreditation), its employees, and auditors and others working on behalf of RAB.

Code of Ethics: RAB, being an affiliate company of the American Society for Quality Control (ASQC), adopts the ASQC Code of Ethics for all RAB personnel within the Scope of this Code of Conduct.

GENERAL STATEMENT OF THE RAB CODE OF CONDUCT

All persons within the Scope, in promoting high standards of ethical conduct, shall:

a) conduct themselves professionally, with truth, accuracy, fairness, and responsibility to their publics;

b) not misrepresent their qualifications, competence, or experience, nor undertake assignments beyond their qualifications;

c) treat as confidential all information gained in relation to RAB accreditation and auditor certification activities, unless authorized in writing by the RAB client and the RAB Board of Directors to disclose such information, and

—will not discuss such information with anyone outside those who have a need to know the information (e.g., for legitimate purposes of the accreditation, registration or auditor certification process);

—will not disclose any part of assessment findings, both during and after the assessment process;

d) not accept retainers, commissions, or valuable considerations from past, present, or potential registrars, or past, present, or potential training organizations or other interested parties, except where the person's other employment or consulting affiliations do not constitute a conflict of interest or likely perception of a conflict of interest, and either

   —these other employment or consulting affiliations are disclosed and available to interested and affected parties, or

   —if not disclosed and available, the person attests in writing that no such conflict of interest or likely perception of conflict of interest exists;

e) acknowledge that the provisions of clause d) apply to situations where the person may provide services to such registrars, training organizations, or other interested parties and also apply to situations where the person may provide confidential information or disclosures which may in any way influence an assessment process currently or in the future;

f) not serve any private or special interest in fulfillment of RAB duties; for example,

   —unduly influencing the registration or assessment of any supplier company, division, or business unit of which the person is employed or of which a consulting arrangement is in effect; or

   —unduly influencing actions concerning a primary competitor of any supplier company, division, or business unit of which the person is employed or of which a consulting arrangement is in effect;

g) not intentionally communicate false or misleading information which may compromise the integrity of the assessment and registration processes or decisions therein.

IMPLEMENTATION OF THE CODE OF CONDUCT

1. Any person within the Scope, upon perceiving or being informed competently of the existence of a conflict of interest under any of a) through g), or of the ASQC Code of Ethics shall self-exempt from deliberations and decisions of RAB that relate to the conflict of interest.

   For example, a conflict of interest or a likely perception of a conflict of interest within this Code exists when a person within the Scope accepts compensation from an organization for which the person has a direct influence as to whether the organization is accredited by, or accreditation is continued by, RAB as a registrar or as an auditor training organization. In particular, members of the RAB Councils and auditors have a direct influence on accreditation, whereas members of the RAB Board of Directors only have a direct influence in the event that an organization appealed a decision by the RAB Councils, which would then be decided by the Board of Directors.

In implementing provisions d), e), and f) a conflict of interest would be presumed not to exist for employment or consulting affiliations that have been terminated for at least one year.

2. All persons within the Scope shall annually sign a statement attesting to their understanding of this Code of Conduct and stating that either

—they have no instances of conflicts of interest since the last such statement they signed, or

—describing in sufficient detail each such instance and how each has been dealt with; these records will be retained in RAB files.

3. Annually, RAB shall conduct training and/or discussions of the Code of Conduct and its applications for all persons within the Scope and will retain records of the fact of such training and/or discussions.

ASQC CODE OF ETHICS

To uphold and advance the honor and dignity of the profession, and in keeping with high standards of ethical conduct, I acknowledge that I:

Fundamental Principles

I. Will be honest and impartial and will serve with devotion my employer, my clients, and the public.

II. Will strive to increase the competence and prestige of the profession.

III. Will use my knowledge and skill for the advancement of human welfare and in promoting the safety and reliability of products for public use.

IV. Will earnestly endeavor to aid the work of the Society.

Relations with the Public

1.1 Will do whatever I can to promote the reliability and safety of all products that come within my jurisdiction.

1.2 Will endeavor to extend public knowledge of the work of the Society and its members that relates to the public welfare.

1.3 Will be dignified and modest in explaining my work and merit.

1.4 Will preface any public statements that I may issue by clearly indicating on whose behalf they are made.

Relations with Employers and Clients

2.1 Will act in professional matters as a faithful agent or trustee for each employer or client.

2.2 Will inform each client or employer of any business connections, interests, or affiliations which might influence my judgment or impair the equitable character of my services.

2.3 Will indicate to my employer or client the adverse consequences to be expected if my professional judgment is overruled.

2.4 Will not disclose information concerning the business affairs or technical processes of any present or former employer or client without his consent.

2.5 Will not accept compensation from more than one party for the same service without the consent of all parties. If employed, I will engage in supplementary employment or consulting practice only with the consent of my employer.

Relations with Peers

3.1 Will take care that credit for the work of others is given to those to whom it is due.

3.2 Will endeavor to aid the professional development and advancement of those in my employ or under my supervision.

3.3 Will not compete unfairly with others; will extend my friendship and confidence to all associates and those with whom I have business relations.

Relations with Employers and Clients

Relations with Peers

ISO 9000 AROUND THE WORLD AND IN INDUSTRY

Chapter 19   ISO 9000 Registration Growth Around the World

Chapter 20   ISO 9000 in Various Industry Sectors

ISO 9000 Registration Growth Around the World

19

This chapter explains the growth of ISO 9000 registration in the United States and around the world in terms of the number of registrations since the publication of the ISO 9000 standard in 1987. It also describes how companies use the standard both to meet customer demands and foster continuous business process improvement.

Also in this chapter are results of two Irwin Professional Publishing surveys of U.S. and Canadian registered companies that shed some light on how companies view ISO 9000 registration and its external and internal benefits. The surveys were published in 1993 and 1996.

Worldwide numbers of registered companies are taken principally from *The Mobil Survey of ISO 9000 Certificates Awarded Worldwide (Fifth Cycle)*, published in December 1995. The survey results gathered by Dr. John Symonds, technical advisor, TQM, for Mobil Europe Limited, are the only currently existing statistics of worldwide registration.

Finally, Bohdan Dyczkowsky provides a bird's-eye view of registration activity in Canada and the marketplace forces driving registration in that country.

REGISTRATION IN THE UNITED STATES AND AROUND THE WORLD

by Mark Morrow

The number of ISO 9000 registered companies has grown from a mere handful in 1987, the year the series was first published by the International Organization for Standardization (ISO), to more than 127,000 sites (as of fall 1996, when this *Handbook* section went to press). In the United States alone, nearly 10,000 certificates have been issued (see Chapter 1 for a chart documenting U.S. and Canadian registration growth).

Although a worldwide database of registered companies is not available, it is clear that the influence of ISO 9000 is not limited to large industrialized nations. Some small independent nations in Europe, such as Monaco, claim at least five registered sites within their borders. Far East nations, such as Brunei and Fiji, also claim ISO 9000 registered sites, as do at least 19 nations in South America and 28 nations in Africa and West Africa.

Despite this interest in ISO 9000 (and in its related cousin, ISO 14000, for environmental management), quantitative and qualitative information about the benefits of registration was not available until *Quality Systems Update* newsletter (published by Irwin Professional Publishing) released in fall 1993 the first extensive survey of United States and Canadian registered companies. The 1993 survey, of 1,679 registered sites in North America at that time revealed that ISO 9000 registration brings companies bottom-line savings and significant internal process improvement. QSU's Survey Partner in 1993 was Deloitte & Touche Management Consultants.

A 1996 follow up survey of nearly 7,000 U.S. registered sites conducted by Irwin Professional Publishing and Dun and Bradstreet (Canada was not included in the follow up survey) revealed similar external and internal benefits for registered companies and bottom-line savings (see Figure 19–1).

Figure 19–1

Average One-Time Saving as a Result of ISO 9000 Registration

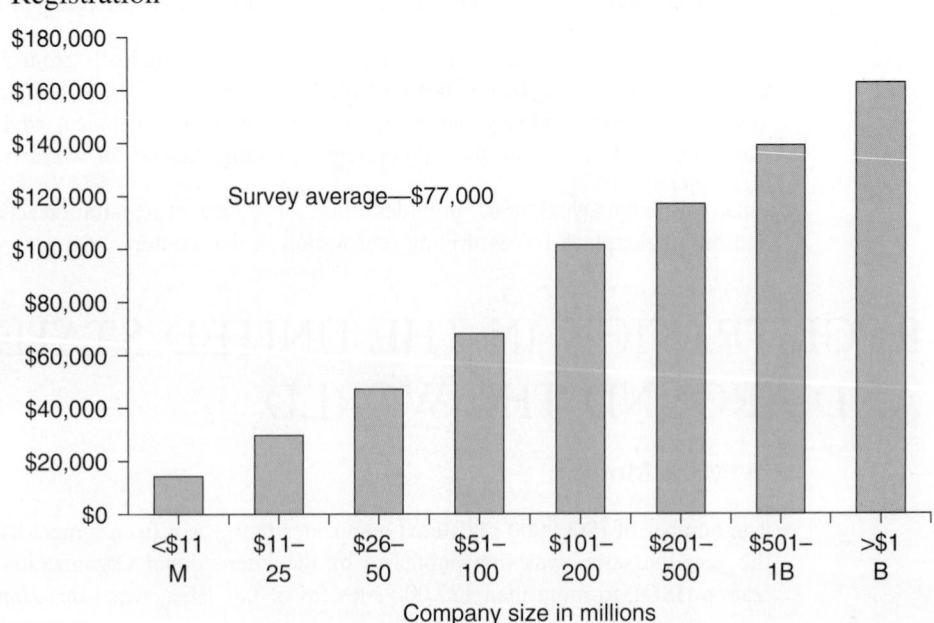

Company size in millions

REGISTRATION AROUND THE WORLD

A completely accurate assessment of worldwide ISO 9000 registration activity still awaits an organization to rise to the financial and logistical challenge of creating the reporting and databasing mechanisms to make this resource a reality. A few private and government affiliated organizations have plans to create such a database, but none have come to fruition.

The only resource available for registered company information around the world continues to be an accounting gathered by Dr. John Symonds of Mobil Europe, Ltd. Symonds has released *The Mobil Survey of ISO 9000 Certificates Awarded Worldwide* (now known as the Mobil Survey) since 1993. The first survey, released in January 1993 counted nearly 28,000 registered sites around the world. By September 1993, that number had increased to nearly 47,000.

Nine months later, the Mobil Survey reported that the number of registered sites had risen by nearly 24,000, sending the total number of registered sites worldwide to 70,364. In December 1995 (the last report available when this book went to press), the Mobil Survey reported that the number of registered company sites had risen to more than 127,000. Based on Irwin's own experience of tracking registered company activity in North American and the past two years of the Mobile Report, an estimate of at least 140,000 registered sites (as of June 1996) is probably close to the mark (see Figure 19–2).

BENEFITS FOR INDIVIDUAL COMPANIES—THE U.S. EXPERIENCE

Both the 1993 and the 1996 Irwin surveys show that companies find significant internal and external value in registration (see Table 19.1). It is also clear that the value companies ultimately find in the registration experience comes from a company's inter-

Figure 19–2

Worldwide Registration 1993–1996*

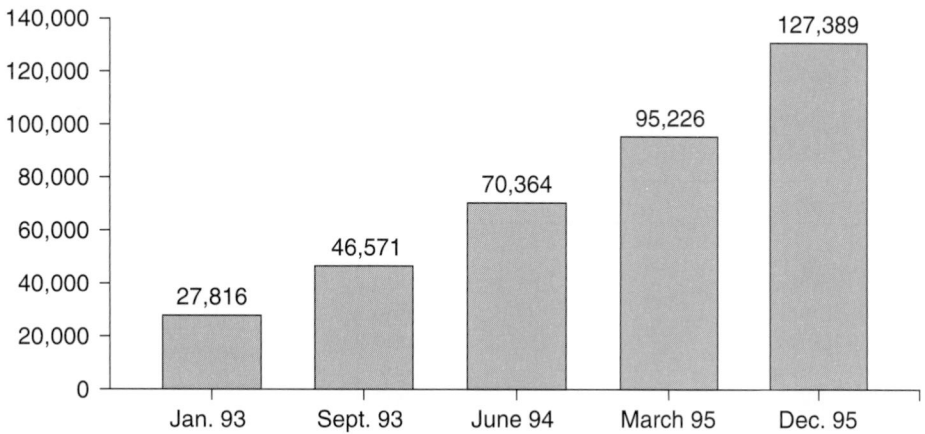

*From the Mobil Survey.
†Based on Mobil Survey and Irwin estimates.

Figure 19–3

ISO 14000 Implementation Plans

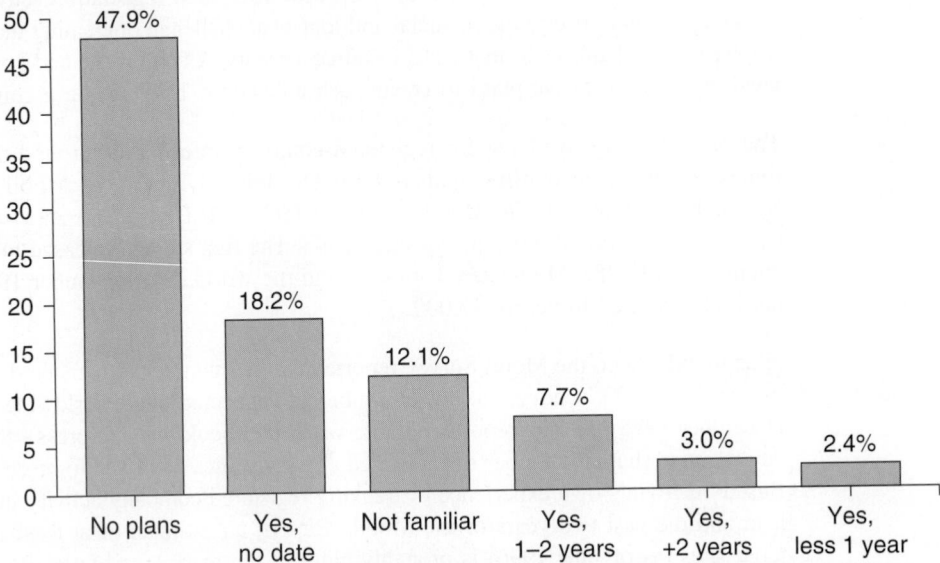

nal attitude about implementing the standard. Companies that approach registration as part of a continuous improvement program (even if registration is a customer requirement) appear to find the greatest value in registration.

While many of the questions asked in 1993 were repeated in 1996, developments such as the ISO 14000 series standard for environmental management (see Chapter 12) and QS-9000 requirements (see Chapter 15) necessitated adding new questions to the survey. The answers to those questions, reflected in Figure 19–3 and Tables 19–2 and 19–3, indicate that most companies do not expect to implement ISO 14000 or meet QS-9000 requirements until sometime beginning in 1997. The expectation for great numbers of ISO 14000 registration appears even further on the horizon.

Table 19–1

Average Ongoing Savings as a Result of ISO 9000 Registration

Company Size	1996
<$11 million	$23,000
$11–25 million	$57,000
$26–50 million	$72,000
$51–100 million	$102,000
$101–200 million	$167,000
$201–500 million	$123,000
$501–1 billion	$193,000
>$1 billion	$291,000
Survey average	**$117,000**

Table 19–2

QS-9000 Implementation Plans

• Implement before Nov. 1996	20%
• Implement in 1997	21.5%
• Implement, yes, but no target date	19.1%
• Implementation at least two years away	34%

Table 19–3

Why Companies are Considering ISO 14000

• Perceived internal benefits	70.2%
• Regulatory Pressure	67.4%
• Marketing/public relations	66%
• Customer pressure	40.8%

Table 19–4

Average Internal Cost for Registration

Company Size	1993	1996
<$11 million	$40,000	$51,000
$11–25 million	$91,000	$68,000
$26–50 million	$110,000	$87,000
$51–100 million	$107,700	$111,000
$101–200 million	$140,300	$166,000
$201–500 million	$242,400	$190,000
$501–1 billion	$239,100	$212,000
>$1 billion	$465,100	$321,000
Survey average	**$180,000**	**$138,000**

Other results from the 1996 Irwin survey show that on average, companies are spending less to become registered (see Tables 19–4, 19–5 and 19–6), but annual savings appear to be less than in 1993. Companies claiming to have an internal cost tracking system for ISO 9000 registration reported not only overall higher average cost for registration, but also higher annual and one-time savings as a result of registration.

Top benefits from registration did not change significantly between the 1993 and 1996 surveys. The top two external registration benefits—improved quality and competitive advantage—did not change between 1993 and 1996, and the top two internal benefits—improved documentation and quality awareness—also remained unchanged between surveys (see Tables 19–7 and 19–8).

Surprisingly, the number of respondents who believe that they will be able to reduce the number of audits conducted on suppliers was less in 1996 than in 1993—down to 74.9 percent from 83.7 percent (see Figure 19–4). Individually, 55.1 percent of the respondents claim to have significantly reduced the number of external audits; 38 percent of the respondents claim only a moderate decrease in audits.

Table 19–5

Average External Cost for Registration

Company Size	1993	1996
<$11 million	$13,300	$20,000
$11–25 million	$23,700	$34,000
$26–50 million	$19,500	$34,000
$51–100 million	$57,900	$46,000
$101–200 million	$42,200	$50,000
$201–500 million	$57,200	$82,000
$501–1 billion	$52,300	$62,000
>$1 billion	$85,900	$88,000
Survey average	**$42,900**	**$49,000**

Table 19–6

Average Registrar Costs for Registration

Company Size	1993	1996
<$11 million	$8,200	$11,500
$11–25 million	$11,600	$15,200
$26–50 million	$16,300	$18,000
$51–100 million	$17,800	$19,200
$101–200 million	$19,400	$23,100
$201–500 million	$23,400	$25,700
$501–1 billion	$32,800	$24,000
>$1 billion	$38,600	$27,400
Survey average	**$21,000**	**$19,800**

Table 19–7

External Benefits of Registration

1993		1996	
• Improved quality perception	76.3%	• Improved quality perception	83.3%
• Competitive edge	68.5%	• Competitive edge	69.6%
• Improved customer service	66.1%	• Reduced audits	56.1%
• Reduced audits	41.1%	• Improved customer service	N/A
• Increased market share	21.8%	• More product demand	28.6%
		• Increased market share	18.2%

Table 19–8

Internal Benefits of Registration

1993		1996	
• Improved documentation	72.7%	• Improved documentation	87.8%
• Greater quality awareness	74.7%	• Greater quality awareness	82.9%
• Positive culture change	47.3%	• Better communications	52.7%
• Better communications	38.7%	• Greater efficiencies	39.7%
• Greater efficiencies	32.3%	• Less scrap	19.1%
• Less scrap	20.8%	• More sales	4.6%
		• Positive culture change	N/A

Companies responding to the 1996 survey also appear to give ISO 9000 far less influence in the process of selecting suppliers. In the 1993 survey, 82.6 percent of the respondents said that ISO 9000 influences supplier selection (see Table 19–9). Only 62 percent of companies responding to the 1996 survey indicated that ISO 9000 registration status influences supplier selection. Finally, only 51 percent of the respondents in the 1996 survey said that a "positive difference" exists between a registered and nonregistered supplier (see Figure 19–1).

Despite the appearance of lessening influence of ISO 9000 as evidenced by the few datapoints represented in the survey statistics, companies in the United States and around the world still find value in the process of implementing ISO 9000. While the Mobil Survey does not list individual companies, the growth of registration, and expansion of its influence into new countries and marketplaces, provide ample evidence of the standard's importance and long-term viability.

Table 19–9

Influence in Supplier Selection

1993		1996	
• Influences selection	82.6%	• Influences selection	60.2%
• Encourage all	34.0%	• Encourage all	24.5%
• Encourage select	65.8%	• Encourage select	21.9%
		• May encourage	36.3%
		• Do not encourage	15.6%

Figure 19–4

Fewer Customer Audits Expected

Figure 19–5

Positive Differences
Between Registered/Nonregistered Suppliers

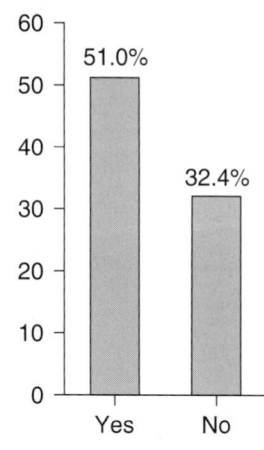

According to the March 1995 Mobile Survey, 86 countries had registration activity within their borders, up from 48 in January 1993. In the December 1995 survey, 99 countries claimed ISO 9000 registrations. While most of the registration activity is still concentrated in the United Kingdom (41.2 percent of the world share in December 1995) and the rest of Europe (31.4 percent in December 1995), registration is gaining significant footholds around the world. See Table 19–10 for a select list of registrations by country.

Copies of the latest Mobile survey may be obtained by contacting:

> Dr. John Symonds
> Technical Advisor, TQM
> Mobile Europe limited
> Mobile Court
> 44-(0)-171-412-4897 (telephone)
> 44-(0)-171-412-4152 (fax)

The information contained in the report offered by Mobil was gathered by Mobil personnel worldwide with the cooperation of a number of other organizations and certification

---

Where to Get More Information on the ISO 9000 Survey

The tables and charts on these pages represent the findings of the 1996 survey, but comparisons are made where appropriate to the 1993 survey. Copies of the complete 1996 survey are available from Irwin Professional Publishing. The *ISO 9000 Survey, Comprehensisve Data and Analysis of U.S. Registered Companies, 1996,* includes a copy of the survey instrument, the data reported back to Irwin from the responding companies, and an explanation of the survey methods and reponses. (Irwin Professional Publishing, 800/773-4607).

Table 19–10

ISO 9000 Registration Around the World*

Region/Country	January 93	September 93	June 94	March 95	December 95
Europe					
Austria	101	200	434	667	1,133
Denmark	326	608	916	1,183	1,314
Greece	18	46	90	162	248
Luxembourg	4	19	21	40	46
Rumania	0	0	6	15	42
Turkey	26	65	106	270	434
Far East					
China	10	35	150	285	507
Japan	165	434	1,060	1,827	3,762
Taiwan	43	96	337	1,060	1,354
Americas					
Argentina	3	9	23	37	86
Brazil	19	113	384	548	923
Mexico	16	24	85	145	215
Venezuela	1	1	2	56	81
Africa/W. Africa					
India	8	73	328	585	1,023
Israel	110	170	279	497	526
South Africa	933	1,146	1,367	1,369	1,454
Australia/ New Zealand					
Australia	1,668	2,695	3,710	5,299	8,834
New Zealand	194	489	918	1,180	1,692

*The Mobil Survey.

bodies. It is not intended to be a complete listing of every registration and is provided by the company to assist in the promotion of international quality and as a service to the quality community.

STATUS OF THE CANADIAN ISO 9000 REGISTRATION SYSTEM

Bohdan Dyczkowsky, P. Eng.

The number of registrations in Canada grew over 1,000 percent between late 1992 early 1996 (see Chapter 1 for registration growth chart). An analysis of these registrations reflects the distribution of manufacturing in Canada. Nearly half of all Canadian registrations are in the province of Ontario, the industrial leader in Canada. Significantly fewer registrations are recorded in Quebec and Alberta.

A further analysis of the SIC codes for Canadian registrations shows the largest number of registrations to be in SIC 3400, Fabricated Metal Products, except Machinery and Transportation Equipment. In second place is SIC 3500, Industrial and Commercial Machinery and Computer Equipment. SIC 3600, Electronic and Other Electric Equipment and Components, except Computer Equipment, is the third largest registered segment.

One expanding registered company SIC is the printing industry. An example of this is Quebecor Printing in Aurora, Ontario, one of the largest groups of printing companies in North America. For these companies, ISO 9000 registration has become a competitive tool that reaches beyond the purchase of new technology to improve quality. ISO 9000 helps these companies focus on the production process in order to be more competitive in an industry that often experiences steep hikes in the price of paper.

The service sector is another example of growing interest in registration beyond traditional manufacturing. Companies providing services such as project engineering and design engineering are seeking registration, as are temporary placement agencies (see Chapter 11 for more on registration in the service sector).

ANALYSIS OF CANADIAN REGISTRARS

The rapid growth of the ISO 9000 marketplace has resulted in an expansion of quality system registrars operating in Canada. Since the early 1990s, the number of registrars has grown from less than 5 to 14 registrars accredited by the Standards Council of Canada. The number ISO 9000 accreditations has probably peaked, as fewer registrars now seek multiple accreditations. The majority of registrars have one to three accreditations.

An analysis of registration by Canadian registrars shows that the majority of registrations recorded by these companies are in Canada. Some companies need multiple accreditations, and these options are also available in Canada. Another noticeable trend is that competition is resulting in reduced registrar costs. Registrars offer services focused on the customer, such as a review of a quality manual on the auditee's premises, free pre-assessment, and nonexpiring registration certificates.

TRENDS IN CANADIAN ISO 9000 REGISTRATIONS

The more competitive marketplace has empowered those seeking registration. Questions of price, credentials, and experience are standard practice. In addition, companies seeking a registrar are demanding a higher level of customer service from the registrars and will switch registrars if they are unhappy.

ISO 9000 in Various Industry Sectors

20

This chapter examines various uses of ISO 9000 among some key industries, including the software, chemical, and metals industry.

The first article, written by three software industry experts, Robert C. Bamford, Taz Daughtrey, and William J. Deibler, offers a detailed explanation of the application of ISO 9001 to the software development process. The article includes important background information along with a review of international committee work on standards development. The article also offers national and international information sources for other software initiatives including SPICE, Trillium, Bootstrap, and TickIT.

The next two articles on the chemical industry offer two perspectives on how ISO 9000 relates to the chemical process industry. James S. Bigelow and William Cox examine the application of ISO 9000 and QS-9000 to the chemical and process industries. Robert W. Belfit, Jr. examines the need for harmonization of process management initiatives including ISO 9000, ISO 14000, Responsible Care®, and other initiatives.

Finally, industry expert Peter B. Lake writes about the influence of ISO 9000 and QS-9000 on the metals industry and future directions for that industry.

THE APPLICATION OF ISO 9001 TO SOFTWARE DEVELOPMENT

by Robert C. Bamford, Taz Daughtrey, and William J. Deibler

INTRODUCTION

This article investigates the application of ISO 9001 to the software development process. It explores in detail a logical framework for applying the guidance in ISO 9000-3:1991 to the requirements of ISO 9001:1994. The article summarizes the current state of the changes proposed for ISO 9000-3 and concludes with an introduction to quality initiatives and other models related to software and ISO 9001, including TickIT, Bootstrap, Trillium, ISO SPICE, and the Carnegie Mellon University Software Engineering Institute Capability Maturity Model[SM].

Growing Acceptance

While ISO 9001[1] has achieved wide acceptance in a variety of industries, it is only recently that significant numbers of software development organizations have begun to investigate the requirements and benefits of ISO 9001.

Whether inspired by customer requirements, competitive pressure, or a desire to improve quality and efficiency, many of these software development organizations are beginning to explore ISO 9001's requirements for institutionalizing software engineering practices and for undergoing internal and external assessment. At the same time, there has been growing interest in the Carnegie Mellon University (CMU) Software Engineering Institute (SEI) Capability Maturity Model (CMM[SM]). Industry interest in the CMM has been raised by reports of qualitative and quantitative benefits realized by early adopters, primarily United States Department of Defense (DoD) software suppliers. Initial concerns about conflicts between ISO 9001 and the CMM have been allayed by an increasing understanding of the complementary nature of the two models.

To make the best use of this chapter, it is recommended that you have copies of ISO 9001[2] and ISO 9000-3[3] available for reference. Contact information for the International Organization for Standardization is included in Sources of Standards and Information, below.

APPLYING ISO 9001 TO SOFTWARE

In 1991, four years after the release of ISO 9001:1987,[4] the member nations of the International Organization for Standardization voted TC 176 on Quality Assurance to release ISO 9000-3[5] containing guidelines for the application of ISO 9001 to the development, supply, and maintenance of software. In practice, ISO 9000-3 is applicable to the development of all types of software, including commercial, off-the-shelf software (COTS), and software developed in a contractual situation (e.g., existing software customized for a specific customer or software developed to meet requirements defined in a purchase specification). ISO 9000-3 and ISO 9001 are equally applicable to stand-alone software, to software delivered as part of a system, and to software embedded in a system as firmware.

The ISO 9000 standards are divided into models and guidelines. The three models—ISO 9001, 9002, and 9003—define requirements. The remainder of the documents, including ISO 9000-3, contain nonbinding suggestions, clarification, and information. Although quality system registration for software product is *always* based on the current version of ISO 9001, it is still common to hear *incorrect* statements about a company being "registered to ISO 9000-3".

Applying ISO 9001 to software development requires an understanding of how the requirements defined in ISO 9001 relate to software development.

THE GUIDANCE IN ISO 9000-3

The logical source for implementors and auditors to turn to is ISO 9000-3. ISO 9000-3 divides its guidance into three parts:

- Framework (Part 4)

- Life-cycle activities (Part 5)

- Supporting activities (Part 6)

Figure 20–1 lists the headings that define the structure of Parts 4, 5, and 6 of ISO 9000-3.

While ISO 9000-3 contains useful guidance, significant analysis and interpretation are required before the guidance in ISO 9000-3 can be correlated with the requirements in ISO 9001. At least two of the reasons for the difficulty in correlating guidance and requirement are inherited from ISO 9001. First, ISO 9001 uses the language of manufacturing in ways that are, in places, difficult to generalize to other domains. Second, ISO 9001 incorporates familiar, fundamental terms that are not defined in any document associated with ISO 9000 standards (e.g., design and production).

ISO 9000-3:1991 compounds the problems inherited from ISO 9001 by:

- Perpetuating the use of undefined terms from ISO 9001.

- Implicitly redefining selected terms used in ISO 9001.

- Introducing new terms, some also without definition.

- Documenting the guidance in an order that, with a few exceptions, is completely independent of the outline structure of ISO 9001.

Figure 20–1

Clauses of ISO 9000-3

Clause of ISO 9000-3	
4	Framework
4.1	Management responsibility
4.2	Quality system
4.3	Internal quality system audits
4.4	Corrective action
5	Life-cycle activities
5.1	General
5.2	Contract review
5.3	Purchaser's requirements specification
5.4	Development planning
5.5	Quality planning
5.6	Design and Implementation
5.7	Testing and validation

Clause of ISO 9000-3	
5.8	Acceptance
5.9	Replication, delivery and installation
5.10	Maintenance
6	Supporting activities
6.1	Configuration management
6.2	Document control
6.3	Quality records
6.4	Measurement
6.5	Rules, practices, and conventions
6.6	Tools and techniques
6.7	Purchasing
6.8	Included software product
6.9	Training

To address the problems created by the inconsistency in organization, Annexes A and B were added to ISO 9000-3:1991 just prior to release. Unfortunately, the two annexes are, in places:

- Incorrect (e.g., ISO 9001 4.15 is mapped to ISO 9000-3 5.8; ISO 9000-3 6.5 is mapped to ISO 9001 4.11)

- Incomplete (e.g., ISO 9000-3 *General (life-cycle activities)* is not mapped in Annex A; ISO 9000-3 5.8 is not mapped to ISO 9001 4.4, 4.13, or 4.14)

- Inadequate—the annexes only define the relationship to the clause level (e.g., 4.1, 4.4) rather than the sub-clause level (e.g., 4.4.8).

A less significant complication arises from the fact that ISO 9000-3:1991 offers guidance on ISO 9001:1987, which was superseded by ISO 9001:1994. Because ISO 9001 remained relatively unchanged between the two revisions, much of the guidance in ISO 9000-3 can be readily applied to ISO 9001:1994.

A recent study by Stelzer, Mellis, and Herzwurm of 36 European software houses that had achieved ISO 9001 registration confirms the problems implicit in the current version of ISO 9000-3:

Surprisingly, also ISO 9000-3 is not of great help for software houses. Quality managers only read it when they begin to study the ISO 9000 family. When they realize that ISO 9000-3 is as difficult to read as all other parts of ISO 9000, they ignore it and use ISO 9001 instead.[6]

To address these problems, a revision of ISO 9000-3 is in development, scheduled for availability in late 1996 as a Draft International Standard (DIS) and eligible for release as an International Standard in mid-1997. (See below, Status of the Revisions to ISO 9000-3.)

To enhance the success of implementors and auditors in improving software quality and to suggest solutions for the problems with ISO 9000-3 described above, the next two sections of this article (Terms and definitions and Mapping ISO 9000-3:1991 to ISO 9001:1994) discuss and define:

- Terms used in ISO 9001:1994 and ISO 9000-3:1991, especially those which are not defined, as they relate to software development.

- The logic underlying the relationship between ISO 9001:1994 and ISO 9000-3:1991.

TERMS AND DEFINITIONS

The terms in ISO 9001 and ISO 9000-3 that present the greatest stumbling blocks to implementors and auditors in software development environments are:

- Design, stage, input, and output.
- Maintenance and servicing.
- Production.
- Verification and validation.

Design, Stage, Input, and Output

A key term not defined in any ISO 9000 standard is *design*. For software products (as defined in ISO 9001 3.1), the common interpretation is that design encompasses the complete software development process, including requirements analysis, design, and testing through acceptance (see ISO 12207[7] 3.8 for a definition of developer and 5.3 Development for further information about developer activities).

As referenced in ISO 9000-3:1991 5.6.3, *implementation* is, therefore, part of design, as are activities like *prototyping* and *coding*. In the language of ISO 9001 (see ISO 9001 *4.4 Design control*), the design process is divided into one or more stages. Each stage has *inputs* (see ISO 9001 4.4.4 and 4.4.6) and *outputs* (see ISO 9001 4.4.5 and 4.4.6). Design input may occur at multiple points in the complete software development process, and the outputs from one stage may become inputs to a subsequent stage.

Terminology in ISO 9000-3

ISO 9000-3 uses *stage* (ISO 9000-3 5.3.1 and 6.1.2 d), *phase* (ISO 9000-3 3.5, 5.4.6, 5.5.5, and 5.5.2), and *point* (ISO 9000-3 5.4.6), while defining only phase. For consistency with ISO 9001, read all three terms as *stage*.

ISO 9000-3 3.4 and 5.4 use *development* to describe the complete software development process, which is how *design* is used in ISO 9001. To compound the confusion, ISO 9000-3 5.6 uses *design* to describe activities that occur after the purchaser's requirements are specified and prior to implementation. While both of these represent common usage in software development, they are inconsistent with the usage in ISO 9001, which is the definitive document for determining compliance and for registration.

Maintenance and Servicing

For software product, *servicing* (ISO 9001 4.19) and *maintenance* (ISO 9000-3 5.10) are treated as equivalent terms, encompassing the activities related to the modification of software product due to a problem or a need for enhancement and extension as well as the migration and retirement of the software product (see ISO 12207 5.5).

In software organizations, the activities in the technical support function are included in *servicing*. To the extent that software development is required, maintenance activities are subject to the requirements of ISO 9001 *4.4 Design control*.

Production

Another key term not defined in any ISO 9000 standard is *production*. For software product, the common interpretation is that *production* includes the activities associated with design as well as the activities associated with software replication. This is consistent with the guidance in ISO 9000-3 and with the nature of software, which allows product (e.g., items that will ultimately be delivered to the customer) to be created early in the design process.

This interpretation affects those clauses in ISO 9001 that state their requirements apply throughout "all stages of production, installation, and servicing" (ISO 9001 4.8, 4.9, 4.10,[8] 4.11, 4.12, and 4.13). For software product, the requirements of these clauses apply throughout the design process. In addition, because maintenance for software feeds back to design, ISO 9001 *4.8 Product identification and traceability* is also applied to servicing activities. These extensions are illustrated in Figure 2.

Verification and Validation

ISO 9001 *4.4 Design control* contains three subclauses related to verification and validation:

Figure 20–2

Clauses Applied to Software Development and Maintenance Verification and Validation.[9]

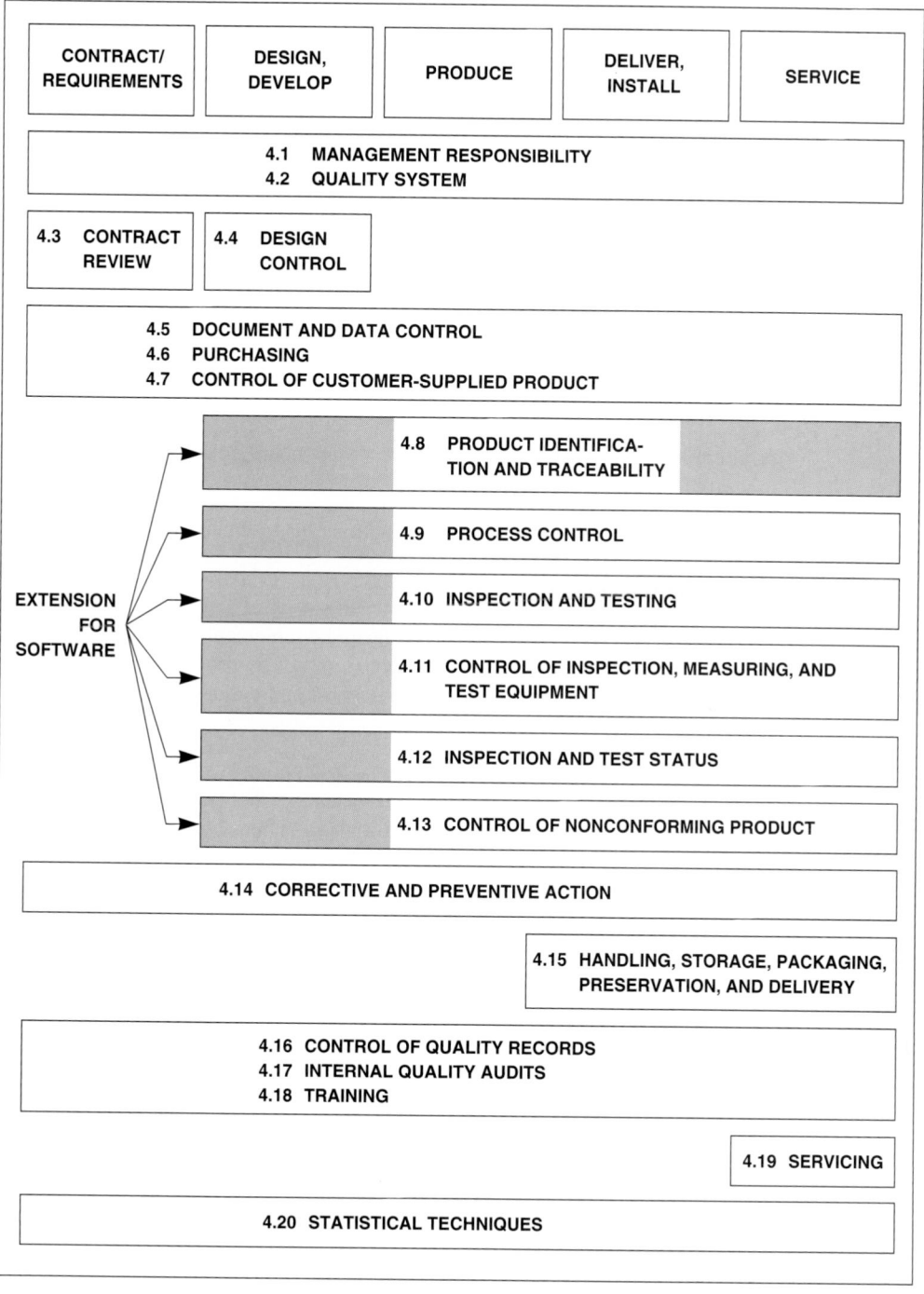

- ISO 9001 *4.4.6 Design review* specifies that reviews be "planned and conducted ... at appropriate stages of design."

- ISO 9001 *4.4.7 Design verification* specifies that verification be performed at appropriate stages of design to ensure that design stage outputs are correct and consistent with respect to the inputs to that stage. Design review is suggested by ISO 9001 as one appropriate method of verification. Examples of design-stage verification methods are reviews, tests, inspections, walk-throughs, and white-box testing.

- ISO 9001 *4.4.8 Design validation* specifies that validation be performed to ensure that product—the final design output—conforms to user needs and/or requirements. Examples of design-stage validation methods are functional tests, system tests, beta and acceptance tests.

- ISO 9001 *4.10 Inspection and testing* provides additional, valuable requirements:

ISO 9001 *4.10.2 Receiving inspection* defines requirements for:

—The verification of subcontracted work.

—The release of subcontracted software components prior to or concurrent with verification.

- ISO 9001 *4.10.3 In-process inspection and testing* supplements *4.4.6 Design verification* to specify that product be held (i.e., not be promoted to the next stage of design) until required verification has been completed.

- ISO 9001 *4.10.4 Final inspection and testing* supplements *4.4.7 Design validation* to specify that product not be dispatched (i.e., not be promoted for general distribution to customers) until required verification and validation has been completed.

- ISO 9001 *4.10.5 Inspection and test records* supplements *4.4.6 Design verification* and *4.4.7 Design validation* to require that records be kept of both verification and validation activities and that the release authority be identified.

MAPPING ISO 9000-3:1991 TO ISO 9001:1994

Annexes A and B in ISO 9000-3 are intended to provide a complete mapping between ISO 9000-3 and ISO 9001, but experience has proven that the annexes fall particularly short in the following areas:

- Customer-supplied product

- Configuration management

- Requirements definition

Each of these areas is discussed in the following paragraphs.

Customer-Supplied Product

Customer-supplied product, referred to as *included software product* in ISO 9000-3 6.8, refers to any product provided or specified by the customer for inclusion in the product or for use in developing the product. Typical examples of customer-supplied product are:

- Software components provided by the customer for inclusion in the product or for use in the development of the product.

- Commercial, off-the-shelf (COTS) products that the customer specifies be included in the product.

- Hardware platforms or components provided by the customer for use in development or in compatibility testing.

ISO 9001 specifies that the supplier—you—are responsible for controlling and ensuring the quality of customer-supplied product and for reporting any problems to the customer. For software product, the primary activity associated with customer-supplied product is appropriate verification and validation.

Configuration Management

Of all of the software engineering practices, software configuration management (SCM or CM) offers the greatest number of opportunities to address requirements found in ISO 9001. From a management perspective, the principles and practices of CM represent an accepted and understood foundation for implementing ISO compliant processes in software engineering organizations. In addition, the growing number of tools for automating CM practices are avenues for improving the efficiency and effectiveness of these processes.

Tracing the relationship between ISO 9001's requirements and CM practices begins with an examination of the guidance in ISO 9000-3.

ISO 9000-3 and Configuration Management

ISO 9000-3 contains two appendices, Annex A and Annex B, that provide cross-references between ISO 9001 and ISO 9000-3. According to Annex A, five clauses of ISO 9001 correlate to ISO 9000-3 *6.1 Configuration management:*

ISO 9001 4.4 Design control

ISO 9001 4.5 Document and data control

ISO 9001 4.8 Product identification and traceability

ISO 9001 4.12 Inspection and test status

ISO 9001 4.13 Control of nonconforming product

Each of these clauses of ISO 9001 contains a portion of the traditional CM process.

- ISO 9001 4.4 requires that design stage inputs and outputs be documented, reviewed, verified, controlled, approved, and modified according to documented procedures.

  In particular, ISO 9001 *4.4.9 Design changes,* in conjunction with ISO 9001 *4.14.2 Corrective action,* and *4.13 Control of nonconforming product,* requires that each change be traceable to an appropriate source and approval.

  For software product there should be a clear path between a change request spawned by a fault report or enhancement request and a change in a specific product component to correct the fault or implement the enhancement.

  An interested party should be able to pick up the path at any point and follow it forward to the released change and backward to the change request or to the fault report.

- ISO 9001 *4.5 Document and data control* addresses the identification, protection, approval, and availability of current issues of all pertinent product- and project-related documents and data, including designs, specifications, plans, schedules, and test and design data. Because a fundamental function of CM is making current configuration items available, the CM practices and tools can be applied to the control of product- and process-related documentation and data.

- ISO 9001 *4.8 Product identification and traceability* requires that each version of a configuration item be identified by some appropriate means.

- ISO 9001 *4.12 Inspection and test status* requires procedures to identify what verification steps and tests have been completed and what results have been achieved by the product or product components at each phase in the defined development life cycle.

- ISO 9001 *4.13 Control of nonconforming product* requires procedures to ensure that untested, defective, or incorrect versions (e.g., down level) of the product are not *inadvertently* used. This clause of ISO 9001 also requires procedures to determine the disposition of nonconforming product at all stages.

  For software, the bulk of the activity related to nonconforming product is in the correction of faults identified during all phases of development (e.g., during requirements definition, prototyping, integration testing, and beta testing) and after the product has been released (e.g., from customer reported faults).

Beyond ISO 9000-3

There are a significant number of additional areas of ISO 9001 that can be addressed through CM-related activities.

ISO 9001 4.1 Management responsibility

ISO 9001 4.2 Quality system

ISO 9001 4.4.2.2 Organizational and technical interfaces

ISO 9001 4.6 Purchasing

ISO 9001 4.7 Control of customer-supplied product

ISO 9001 4.9 Process control

ISO 9001 4.14 Corrective and preventive action

ISO 9001 4.15 Handling, storage, packaging, preservation, and delivery

ISO 9001 4.16 Control of quality records

ISO 9001 4.19 Servicing

ISO 9001 4.20 Statistical techniques

Each of these additional areas is addressed in the following paragraphs.

- ISO 9001 *4.1 Management responsibility*—Reports produced by the CM system on progress and exceptions support management review of the suitability and effectiveness of the development practices, as part of the quality system (ISO 9001 4.1.3).

- ISO 9001 *4.2 Quality system*—For a software engineering organization, the CM policies, procedures, and standards represent a significant portion of the quality system. Tools to support and automate the CM process support and enforce adherence to policies, procedures, and standards.

- ISO 9001 *4.4.2 Organizational and technical interfaces*—CM practices can go beyond control of configuration items to ensure that necessary information regarding status and change is communicated to affected individuals and organizations.

  Standard distribution lists and notification procedures linked to specific activities prevent significant missed communication. These procedures can also be automated through integrated workflow and groupware tools that increase the effectiveness and efficiency of the information exchange.

- ISO 9001 *4.6 Purchasing*—Intermediate or final product and all related documentation (e.g., specifications, plans, progress reports, test reports) received from the subcontractor can all be treated as if they were configuration items developed in-house. Applying CM practices to third-party development is of particular benefit for coordinating in-house integration and verification activities and for fault resolution.

- ISO 9001 *4.7 Customer-supplied product*—As described above, in software development, customer-supplied product includes software that is used in the development process or is included in the product to be delivered to the customer. This software is specified by the customer and supplied by the customer or by a third party; the software can be a standard, off-the-shelf (shrink-wrapped) product or one that is custom developed.

  Depending on how the included software is packaged and distributed, ISO requirements to verify, store, and maintain this software appropriately maybe met by considering the included software as a configuration item.

- ISO 9001 *4.9 Process control*—Process control contains requirements for documented procedures, suitable production equipment, monitoring and control of process and product characteristics, and approval of processes and equipment. In software engineering environments, the project management and CM processes combine to address the majority of these requirements.

By automating the product build process, a CM system contributes significantly to the effectiveness and efficiency of the software production process, both for intermediate versions of the product and for a released version. This becomes particularly significant when multiple versions of the product are being developed or maintained in parallel.

- ISO 9001 *4.14 Corrective and preventive action*—A significant portion of corrective action is creating the mechanism to ensure that customer-reported problems are resolved in an appropriate manner. The same requirements pertain to problems identified in the development process, starting from the point at which the software product or item comes under CM control.

Incidents must be tracked from report, through classification, and, if appropriate, to resulting changes in the product. CM practices, particularly those related to change management, product maintenance, and status accounting ensure that incidents that result in product change are always handled properly.

Significant opportunity exists for improving the efficiency of product support and software engineering organizations by minimizing the amount of manual intervention and effort in moving information between the problem tracking and CM systems.

- ISO 9001 *4.15 Handling, storage, packaging, preservation, and delivery*—For software product, the CM practices address all of the handling, storage, packaging, preservation, and delivery requirements at least to the point where responsibility for the product is turned over to software production.

ISO 9001 *4.15.2 Handling* specifies "methods of handling product that prevent damage or deterioration." For software product, this requirement is interpreted to include activities like virus checking if an outside replication vendor is used and off-site storage of product masters as a minimum level of disaster recovery.

At the end of the development process, automated support for the build process, included in most CM tools, reduces opportunities for error and can increase confidence in intermediate test results and final product integrity.

- ISO 9001 *4.16 Control of quality records*—In ISO 9001, quality records are the records that establish that processes were followed and quality requirements met. The definition of quality records includes records of product identification, nonconformity review and disposition, and all verification and validation activities, including design review minutes, test logs and records, and fault reports.

While these records are not documents (and are not subject to the requirements of ISO 9001 4.5), requirements for identification, collection, indexing, filing, storage, maintenance, and disposition of quality records can be addressed through procedures implemented as part of the CM system.

- ISO 9001 *4.19 Servicing*—ISO 9000-3 ties servicing to all aspects of software maintenance, including problem resolution, interface modification (e.g., support for additional or modified hardware components), functional expansion, or performance improvement.

  CM practices ensure that the product is maintained in an orderly manner and that each approved change can be prioritized, tracked, and managed to completion.

  Analysis of the data in the CM system related to all aspects of product maintenance can support systematic prioritization and planning for product and process enhancement. For instance, what modules change most often? What modules cause the most problems? Is the effectiveness of testing continuing to improve?

- ISO 9001 *4.20 Statistical techniques*—While ISO 9001 contains no specific requirements for statistical process control, as noted above, CM related activities generate a wealth of process and product data for analysis and comparison to plan: delivery dates, resources, benchmarking (e.g., lines of source code, executable size, performance), time to correct defects, and so on.

  If no "modern statistical techniques"[10] are implemented (e.g., Statistical Process Control, Design of Experiments), this data is considered input for ISO 9001 4.9d, which requires "monitoring and control of suitable process parameters and product characteristics."

  The data in the CM system is a primary input for problem analysis and the identification of root causes in products and processes.

SUMMARY: ISO 9001 AND CONFIGURATION MANAGEMENT

As described in the preceding paragraphs, of the 20 sections of ISO 9001 that define a supplier's capability to meet customer requirements, CM practices directly impact the requirements in 15 sections: Check marks in the table in Figure 20–3 indicate clauses of ISO 9001 that are addressed by CM practices.

Requirements Definition

Requirements definition is one of the areas in which ISO 9001 is least specific. This is, in part, because ISO 9001:1987 focused on two-party contractual situations, categorizing participants as *purchaser* and *supplier.* ISO 9001:1994 adopted language that supported the already common extension of the standard to the development of off-the-shelf, commodity products for which a contract was not a prerequisite. ISO 9001:1994 refers to *customers* and *suppliers.*

ISO 9001:1994 *4.3.2 Review* defines requirements for situations that do involve a contract. For software product, the key requirements are that the contract be reviewed to ensure that:

Figure 20–3

Clauses of ISO 9001 Addressed by Configuration Management Practices

Section of ISO 9001		
4.1	Management responsibility	✓
4.2	Quality system	✓
4.3	Contract review	
4.4	Design control	✓
4.5	Document and data control	✓
4.6	Purchasing	✓
4.7	Control of customer-supplied product	✓
4.8	Product identification and traceability	✓
4.9	Process control	✓
4.10	Inspection and testing	
4.11	Control of inspection, measuring, and test equipment	
4.12	Inspection and test status	✓
4.13	Control of nonconforming product	✓
4.14	Corrective and preventive action	✓
4.15	Handling, storage, packaging, preservation, and delivery	✓
4.16	Control of quality records	✓
4.17	Internal quality audits	
4.18	Training	
4.19	Servicing	✓
4.20	Statistical techniques	✓

- The requirements are adequately defined (ISO 9001 4.3.2a).

- The supplier has the capability to meet the requirements (ISO 9001 4.3.2c).

ISO 9000-3 *5.2 Contract review* and *5.3 Purchaser's requirements specification* offer suggestions for contracts related to software product. ISO 9000-3 5.3 addresses a circumstance frequently encountered in software development: if the customer is not capable of defining the requirements, the supplier should define and gain agreement with the customer on the requirements.

Although it is not listed in the annexes in ISO 9000-3, the last paragraph of ISO 9001:1994 *4.4.4 Design input* addresses the reality of changing requirements: "Design input shall take into consideration the results of any contract review activities."

For software products, the implication of this clause is that the software development organization and any supporting organizations impacted by contractual commitments— or technically competent representatives—should be involved or consulted in contract review and in the assessment of the impact and feasibility of requirements.

This involvement should continue from the initial contract through any changes. These clauses in ISO 9001 require that the supplier implement processes that minimize at least three common sources of risk and problems in software development projects:

- Other, uninformed organizations make commitments on behalf of software engineering.

- New or agreed-upon requirements are not communicated systematically to software engineering (e.g., *missed requirements* and *requirements creep*).

- Software engineering introduces changes into the product without consulting the customer (e.g., *gold-plating*).

STATUS OF THE REVISIONS TO ISO 9000-3

Several months after the approval of ISO 9001:1994, at the ISO Technical Committee 176 (TC176) Plenary in Toronto in September 1994, the decision was made to revise ISO 9000-3:1991. The work is the responsibility of Subcommittee 2 (SC2) of TC176. From 1994 to 1996, the document went through two drafts. With comments satisfactorily addressed, approval was given in June of 1996 to release the document as a Draft International Standard (DIS).

At the time this chapter of the *ISO 9000 Handbook* is being prepared, it is expected that the DIS will be available in late 1996. From the time the DIS is released, approximately seven months are allocated for voting, after which the document will be eligible for release as an International Standard.

One of the changes incorporated by SC2 is to restructure the document to conform to ISO 9001:1994. The guidance is now presented organized by clause of ISO 9001:1994.

A particular concern has been, and continues to be, the impact of the revisions on the software engineering community and on established programs, like TickIT (see below, *TickIT*). On one hand, can a significant revision of ISO 9000-3:1991 be justified, especially in light of the fact that a major overhaul of the ISO 9000 series is planned for the year 2000? On the other hand, can the worldwide software engineering community be asked to live with inadequate standards for a minimum of the next five years?

ISO 9001 AND THE CMM

The maturity framework, the foundation of the Capability Maturity Model, was published by the Carnegie Mellon University (CMU) Software Engineering Institute (SEI) in two technical reports.

The first technical report,[11] entitled *Characterizing the Software Process: A Maturity Framework* was republished as a 6-page article in *IEEE Software*.[12] The article described a five-level framework or model[13] for the engineering practices in a successful software development organization. The second was a 40-page report entitled "A Method for Assessing the Software Engineering Capability of Contractors."[14]

This technical report, sponsored by the United States Department of Defense for use as a procurement tool, contained a summary of the five-level framework, 101 questions, and a rating scheme to correlate the answers to the questions to the levels in the framework.

Together, these two reports presented the first version of what would become the SEI model for software engineering. In 1993, the SEI released two documents.[15] that described in detail 368 goals and practices associated with levels two through five of the framework. The framework, itself, remained unchanged from the version published in 1987. In 1994, a questionnaire corresponding to Version 1.1 was published, but without a publicly available rating scheme.

Two articles[16] and supporting papers, in particular, reached the same conclusions regarding the relationship between the achievement of an advanced CMM maturity level and obtaining ISO 9001 registration:

- No direct correlation exists between advanced maturity level and ISO registration. There are organizations that have achieved advanced maturity levels but which have not been recommended for ISO registration. There are ISO registered organizations that have been assessed at CMM Level 1—the beginning, default maturity level.

- The two models are completely compatible and mutually supportive. A software development organization can leverage the guidance in the CMM for achieving ISO registration; an organization can implement ISO 9001-compliant practices to enhance the success of its software engineering group, no matter what the maturity level.

These conclusions are further supported by the different scopes and sizes of the two models. In 450 pages, the CMM focuses exclusively and in great depth on contract software development, not addressing maintenance, support, or replication and distribution. In nine pages, ISO 9001 addresses from a broad perspective all aspects of product delivery: from contract review to development, to production, installation, and maintenance.

The relationship between the two models is represented graphically in Figure 20–4.

Contact information for the SEI is included in Sources of Standards and Information, at the end of this article.

Figure 20–4

The Relationship between ISO 9001 and the SEI CMM*

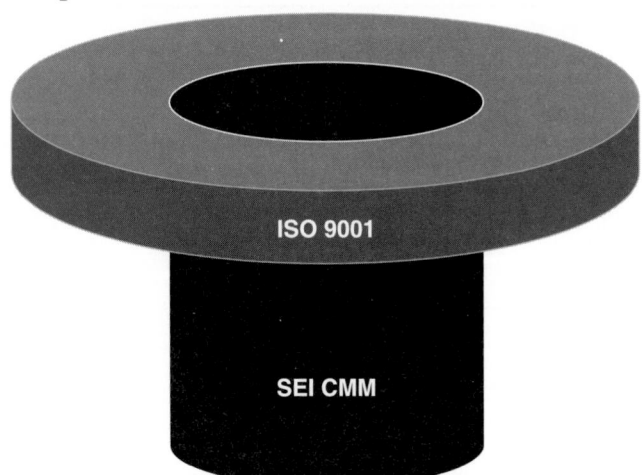

ISO 9001

SEI CMM

*Capability Maturity Model, published by Carnegie Mellon University Software Engineering Institute.

ISO 9000-3 AND THE IEEE SOFTWARE ENGINEERING STANDARDS

For nearly 15 years the Computer Society of the Institute of Electrical and Electronics Engineers (IEEE) has sponsored a growing family of standards related to software engineering, all under the umbrella of the *IEEE Standard for Software Quality Assurance Plans* (ANSI/IEEE Std 730). This standard requires developing a range of intermediate software products (software requirements specifications, software design descriptions, user manuals, etc.) and plans for various supporting activities (configuration management, verification and validation, testing, etc.).

Figure 20–5 contains a mapping of both the IEEE Software Quality Assurance standard and its derivative standards is shown below. These standards provide valuable ideas on how to satisfy the high-level requirements of ISO 9001 within the guidance of 9000-3.

Contact information for the IEEE is included in Sources of Standards and Information, at the end of this article.

OTHER PROGRAMS

A number of programs have been created based on ISO 9001 and the SEI CMM. Some of these programs are summarized in the following sections.

Figure 20–5

ISO 9000-3 and the IEEE Standards

ISO 9000-3:1991 Clause	IEEE 730 Section	IEEE Standards
4.1	3.6.2.1, 3.5.2.5	1028 Reviews and Audits
4.2	3.3	1002 Standards Taxonomy 1058.1 Project Management
4.3	3.6.2.7	1028 Reviews and Audits
4.4	3.8	
5.1		
5.2		
5.3	3.4.2.1	830 Requirements Specification
5.4	3.4.3	
5.5	3.3	1002 Standards Taxonomy 1058.1 Project Management
5.6	3.4.2.2	1016 Design Descriptions
5.7	3.7	829 Test Documentation 1008 Unit Testing
5.8		
5.9		
5.10	3.4.3	
6.1	3.4.2.6, 3.10	829 Test Documentation 1042 Configuration Management
6.2	3.13	
6.3	3.13	
6.4	3.5	
6.5	3.5	
6.6	3.9	
6.7	3.12	
6.8	3.12	
6.9	3.14	

TickIT

TickIT was introduced in 1991 in the United Kingdom by the National Accreditation Council of Certification Bodies (the NACCB).[17] In response to complaints from software development organizations regarding the quality and consistency of ISO 9001 registration assessments, the objective of the TickIT program is to help software organizations develop quality systems that add value to their businesses and that conform to ISO 9001.

TickIT is based on ISO 9001 and requires that the guidance in ISO 9000-3 be considered in assessing conformance to ISO 9001. TickIT also adds administrative requirements for the accredited registrar regarding auditor qualification and triennial, complete reassessment.

TickIT registrations, which numbered nearly 1,000 at the end of 1995, have shown rapid growth. About 350 new certificates were issued during 1995, of which about 25 percent were outside the United Kingdom, predominantly in Scandinavia, North America, and the Far East. To date, approximately 80 percent of the certificates are held by firms located in the United Kingdom, 7 percent in continental Europe, and another 7 percent in North America.

Until recently, accreditation to perform TickIT ISO 9001 registrations was limited to registrars who had a presence in the United Kingdom. A change in the charter now allows the United Kingdom Accreditation Services (UKAS) to function as an international accreditation agency, which allows any registrar to seek accreditation directly from UKAS.

The principal guidance document for TickIT is *A Guide to Software Quality System Construction and Certification Using ISO 9001:1994.* Issue 3.0 of the guide was published in October 1995.

SPICE

In January 1993, the International Organization for Standardization (ISO) and International Electrotechnical Commission (ISO/IEC) Joint Technical Committee 1 (JTC1) assigned responsibility for developing a suite of standards for software process assessment to Subcommittee 7 (SC7).[18] The standards development project is designated as Software Process Improvement and Capability Determination (SPICE).

As outlined in the initial requirements specification,[19] the purpose of the Software Process Assessment Standard (SPA) is to examine the processes used by an organization. The aim is to:

• Characterize current practices within the organization, identifying strengths, weaknesses, and the risks inherent in the process.

• Determine to what extent they are effective in achieving process goals,

• Determine to what extent they conform to a set of baseline practices.

Process assessment includes the determination of business needs, an assessment (measurement) of the processes used by the organization, and analysis of the current position. The analysis results are used to drive process improvement activities or determine the capability of an organization.

In late 1995, a proposal[20] was introduced for a change in the architecture of the SPA standard in response to U.S. Working Group concerns. The goal of the suggested changes is to ensure that SPICE serves as a framework for comparing models and assessment methods. The proposal addresses four perceived needs:

- For closer alignment of the SPA product suite with ISO 12207, including terminology.

- To make the standard less prescriptive.

- To simplify the model to address problems encountered during the first phase of trial assessments.

- To provide a clear route for migration/harmonization of existing models such as Bootstrap, Trillium, and the CMM.

Phase 2 trial assessments will be conducted from March 1996 until March 1997. To find out about becoming a participant in the Phase 2 trial assessments, download the following document: *www.esi.es/projects/spice/trials.html.* The regional coordinators for the trials are identified in the document and in Sources of Standards and Information, at the end of this article

Trillium

Trillium is a model for determining the capability of a supplier's processes for the development of telecommunications products. It was developed by Bell Canada, Northern Telecom, and Bell Northern Research. In use since 1991, the Trillium model, now at Release 3, is based on the SEI capability maturity model CMM version 1.1 and incorporates other standards, including ISO 9001 and ISO 9000-3, Bellcore TR-NWT-000179, the Malcolm Baldrige National Quality Award (MBNQA) criteria, and the IEEE Software Engineering Standards.

Trillium defines a set of best practices that reflect its specific telecommunications focus. Like the CMM, Trillium is intended for use as a self-assessment and improvement tool and as a supplier assessment tool. Trillium defines the benefits of higher capability in terms of responsiveness and minimum time to market, reduced costs, and increased ability to manage and meet commitments.

Bootstrap

A project funded by the European Commission as part of the ESPRIT program (*ESPRIT 5441 BOOTSTRAP: A European Assessment Method to Improve Software Development*), Bootstrap is administered and maintained by the Bootstrap Institute European Economic Interest Group (BI EEIG) in Milan, Italy. The focus of Bootstrap is to assess and improve the capability of a Software Producing Unit (SPU).

Bootstrap combines the ISO 9000 standards, European software engineering standards, and the SEI Capability Maturity Model CMM to create a basis for assessment and for consulting. The Bootstrap methodology includes both a diagnostic assessment of a software development process—including organization, methods and engineering capability, and tools and technology—and the creation of an action plan that defines the steps, implementation details, and time frames to improve the organization's ability to deliver quality products and services. The result of the assessment is a profile based on a Bootstrap assessment instrument that adds a second dimension to the CMM's levels: the process quality attribute.

> Bootstrap attempts to identify the attributes of a software development organization and a project and assigns all the questions in the questionnaire to process quality attributes as well as maturity levels.[21]

SOURCES OF STANDARDS AND INFORMATION

The following organizations provide information and source documents on the standards and programs described in this chapter. The documents and programs are listed alphabetically.

For Bootstrap
Bootstrap Institute
Via Adelaide Bono Cairoli 34
20127 Milan
ITALY
Tel: +39 2 261621 Fax: +39 2 2610927

For the CMM
Software Engineering Institute
Carnegie Mellon University
Pittsburgh, PA 15213-3890
USA
Tel: 412-268-7700, (412) 268-6815.
Fax: 412-268-5758
Internet:customer-relations@sei.cmu.edu
ftp address: ftp.sei.cmu.edu (for technical reports and documents, including [TR-24] and [TR-25])
http://www.sei.cmu.edu/SEI/HomePage.html

For the IEEE standards
IEEE
345 East 47th Street
New York NY 10017-2394
USA
IEEE Customer Service
445 Hoes Lane
PO Box 1331
Piscataway NJ 08855-1331
USA

Tel: 1-800-678-4333
Fax: 1-908-981-9667
http://www.computer.org
gopher://stdsbbs.ieee.org:/70/00/pub/ieeestds.htm

For ISO standards and standards information:

American National Standards Institute (ANSI),
11 West 42nd Street, 13th Floor
New York NY 10036
USA
Tel: 212-642-4900 and 202-639-4090
Fax: 212-302-1286 and 202-628-1886
http://hsdwww.res.utc.com

American Society for Quality Control (ASQC)
PO Box 3066
611 East Wisconsin Avenue
Milwaukee WI 53201-3066
USA
TEL: 414-272-8575, 800-248-1946
FAX: 414-765-8661
http://www.asqc.org

International Organization for Standardization
Central Secretariat
1, rue de Varembé
CH-1211 Genève 20
SWITZERLAND
TEL: 011-41-749-01-11
FAX: 011-41-22-733-34-30
Internet: ISO9000@ISOCS.ISO.CH
http://www.iso.ch/welcome.html (in English)
http://www.iso.ch/welcomef.html (in French)
gopher://sunny.stat-usa.gov:70/11/NTDB/Is

For SPICE

The following are the four regional contacts for SPICE.

Europe	Pacific Rim
Harry Barker	Terry Rout
Defense Research Agency	School of Computing
St. Andrews Road	Griffith University
Malvern, Worcs WR14 3PS	Queensland 4111

United Kingdom
Tel: +44-68-489-6106
Fax: +44-68-489-6246
habarker@dra.hmg.gb

USA
Mike Konrad
Software Engineering Institute
Carnegie Mellon University
Pittsburgh, PA 15213-3890
USA
Tel: +412-268-5813
Fax: +412-268-5758
mdk@sei.cmu.edu

Australia
Tel: +61-7-875-5046
Fax: +61-7-875-5051
terryr@kurango.cit.gu.edu.au

Canada
Jean-Normand Drouin
Bell SYGMA Telecom Solutions
700 La Gauchetiere (rm 13E1)
Montreal (Quebec) H3B 4L1
Canada
Tel: +514-870-0974 (tel)
Fax: +514-393-1391
jndrouin@qc.bell.ca

The following are the regional coordinators for the Phase 2 trial assessments.

Europe and South Africa
Phase 2 Trial Assessment
Coordinator
Bob Smith
SPICE Trials

European Software Institute
Parque Tecnológico de Zamudio #204
48016 Zamudio Vizcaya
Spain
Tel: +34.4.420 95 19
Fax: +34.4.420 94 20
e-mail: spice-trials@esi.es
www: http://www.esi.es/

**Canada and Central and
South America**
Phase 2 Trial Assessment
Coordinator
Jerome Pesant
1801 McGill College Avenue
Suite 800
Montreal (Quebec)
Canada
H3A 2N4
Tel: (+1) 514-398-8142
Fax: (+1) 514-398-1244
e-mail: pesant@crim.ca

Pacific Rim
Phase 2 Trial Assessment
Coordinator
Francis Suraweera
School of Computing and Information
Technology
Griffith University
Nathan Campus, Kessels Road
Nathan Queensland 4111
Australia
Tel: (+61) 7-3875-5018
Fax: (+61) 7-3875-5207
e-mail: F.Suraweera@cit.gu.edu.au
www: http://www-sqi.cit.gu.edu.au/spice/

USA

Phase 2 Trial Assessment
Coordinator
Dave Kitson
Software Engineering Institute
Carnegie Mellon University
4500 Fifth Avenue
Pittsburgh PA 15213-3890
USA
Tel: (+1) 412-268-7782
Fax: (+1) 412-268-5758
e-mail: dhk@sei.cmu.edu

India
Phase 2 Trial Assessment Coordinator
Srinivas Thummalapalli
8th Floor
Diamond Jubilee Commercial Complex
Hudson Circle
Bangalore 560 027
India
(Tel) (+91) 080.2211143
(Fax) (+91) 080.2211152
e-mail: tsrini@3se.soft.net

For TickIT
DISC TickIT Office
389 Chiswick High Road
London W4 4AL
United Kingdom
Tel: 44 71 602 8536
Fax: 44 71 602 8912

For Trillium
Acquisitions Technical Service
Bell Canada
2265 Roland Therrien
Longueuil, Quebec
Canada, J4N 1C5
Tel: +1 514 468 5523
Fax: +1 514 448 2090
e-mail: trillium@qc.bell.ca

Endnotes

1. ANSI/ISO/ASQC Q9001:1994, Quality Systems—Model for Quality Assurance in Design, Development, Production, Installation, and Servicing, American Society for Quality Control, Milwaukee, 1994. Because this chapter discusses multiple versions of the same standard, whenever there is a need to differentiate, the designation of a standard (e.g., ISO 9001) includes the year of issue (e.g., ISO 9001:1987 and ISO 9001:1994).

2. Ibid. See Sources of Information and Standards at the end of this article for information about obtaining copies of this standard.

3. ISO 9000-3, Quality management and quality assurance standards, Part 3: Guidelines for the application of ISO 9001 to the development, supply and maintenance of software, First edition, International Organization for Standardization, Geneva, Switzerland, 1991. See Sources of Information and Standards at the end of this article for information about obtaining copies of this standard.

4. ANSI/ASQC Q91:1987, Quality Systems—Model for Quality Assurance in Design/Development, Production, Installation, and Servicing, American Society for Quality Control, Milwaukee, 1987.

5. "ISO 9000-3" is read as "ISO 9000 part 3".

6. Dirk Stelzer, Werner Mellis, Georg Herzwurm, *Software Process Improvement via ISO 9000? Results of two surveys among European software houses*, University of Cologne, Chair of Business Computing, 50923 Cologne, Germany, as submitted for publication in Proceedings of the 29th Hawaii International Conference on System Sciences, January 3–6, 1996, Wailea, Hawaii, USA; this document is available on-line at:

www.informatik.uni-koeln.de/winfo/prof.mellis/publications/hicscopy.htm

7. ISO/IEC 12207, Information Technology—Software lifecycle processes, First edition, ISO/IEC, Geneva, Switzerland, 1995.

8. Note that ISO 9001 *4.10 Inspection and testing* and *ISO 9001 4.13 Control of nonconforming product* are not explicitly limited to "production, installation, and servicing". References in ISO 9001 4.10 to product and the explicit limitations in the closely associated clauses, ISO 9001 4.11, 4.12, and 4.13, support the common interpretation that ISO 9001 4.10 and 4.13 apply only to production, installation, and servicing.

9. R. C. Bamford and W. J. Deibler, Implementation as a Managed Process—A Software Perspective, Software Systems Quality Consulting, San Jose, Calif., 1993.

10. ANSI/ASQC Q9004-1-1994, Quality Management and Quality System Elements—Guidelines, American Society for Quality Control, Milwaukee, 1994. See 20.2.

11. Watts. S. Humphrey, *Characterizing the Software Process: A Maturity Framework,* CMU/SEI-87-TR-11, Software Engineering Institute, Carnegie Mellon University, Pittsburgh PA 15213, June 1987.

12. Watts. S. Humphrey, *Characterizing the Software Process: A Maturity Framework*, IEEE Software, March 1988, pages 73–79

13. This model is based on a model for maturity first published by Crosby (see Philip B. Crosby, *Quality Is Free,* McGraw-Hill Book Company, New York, 1979).

14. W. S. Humphrey, W. L. Sweet et al., A Method for Assessing the Software Engineering Capability of Contractors, CMU/SEI-87-TR-23, Software Engineering Institute, Carnegie Mellon University, Pittsburgh PA 15213, September 1987.

15. Mark C. Paulk et al., Capability Maturity Model for Software, Version 1.1, CMU/SEI-93-TR-24, Software Engineering Institute, Carnegie Mellon University, Pittsburgh PA 15213, February 1993; Mark C. Paulk et al., Key Practices of the Capability Maturity Model, Version 1.1, CMU/SEI-93-TR-25, Software Engineering Institute, Carnegie Mellon University, Pittsburgh PA 15213, March 1993.

16. R.C. Bamford and W. J. Deibler, *Comparing, contrasting ISO 9001 and the SEI Capability Maturity Model*, IEEE Computer, October 1993, Vol. 26, No. 10, IEEE Computer Society, page 68; R. C. Bamford and W. J. Deibler, Exploring the Relationship between ISO 9001 and the SEI Capability Maturity Model for Software Engineering Organizations: from 1987 to 1994, Software Systems Quality Consulting, San Jose, Calif., 1996; R. C. Bamford and W. J. Deibler, A Detailed Comparison of the SEI Software Maturity Levels and Technology Stages to the Requirements for ISO 9001 Registration, Software Systems Quality Consulting, San Jose, Calif., 1993; Mark C. Paulk, A Detailed Comparison of ISO 9001 and the Capability Maturity Model for Software, IEEE Software, January, 1995.

17. The NACCB merged with the National Measurement Accreditation Service (NAMAS) to form the United Kingdom Accreditation Services (UKAS). NAMAS accredited laboratories for specific tests or calibrations.

18. JTC1 SC7 is responsible for software engineering standards. TC176 is responsible for the ISO 9000 standards.

19. WG10/N017R Version 1.0, June 3, 1993.

20. WG10/N080.

21. Volkmar Hasse, Richard Messnarz, Robert M. Cachia, Software Process Improvement by Measurement, BOOTSTRAP/ESPRIT Project 5441. For additional information, see also Pasi Kuvaja et al., *Software Process Assessment & Improvement—The Bootstrap Approach,* Blackwell Publishers, Oxford, UK, ISBN 0-631-19663-3.

CHEMICAL INDUSTRY HARMONIZATION OF PROCESS MANAGEMENT INITIATIVES USING ISO 9000

by Robert W. Belfit, Jr.

The need for harmonization of process management initiatives within the chemical industry is clear. Most industry executives are faced with undertaking three or more initiatives; an exercise that leaves employees confused and fosters a lack of continuity among the programs. Both management and employees are now expressing a desire for simplification and/or harmonization.

David Hunter, editor in chief of *Chemical Weeks*, made this point in a January 10, 1996 editorial:

> Successful environmental management is falling prey to roadblocks at many of the 185 companies recently surveyed by Arthur D. Little, of Cambridge, Mass., with chemical firms being the biggest cohort in that study.
>
> Managers of environmental, health, and safety cite two key obstacles to improving their firms' environmental management: lack of cooperation between the environmental and the business issues in the company and their own failure to convince top management that environment is an important issue which must be integrated to business performance, rather than considered an outside operation to keep firms out of trouble.
>
> Pointing to a lack of resources, environmental, health, and safety managers complain that they often are not seen as equals to their business counterparts in many firms.

Mr. Hunter's comments are a familiar lament to those working in the industry. The solution may be found in the harmonization of ISO 9000 and other process management standards. Here are some key questions that must be addressed to move the harmonization process along.

What are some of the common elements or requirements of the ISO Quality Management System (ISO 9000), OSHA Process Safety Management (*Federal Register* C.F.R. 1910.119), Environmental Management System (ISO 14000), and the chemical industries' Responsible Care Guidelines?

Common requirements among these initiatives are training, document and data control, auditing and reviews, record keeping, product identification and traceability, process control, management of change, and establishing and maintaining documented procedures. The relationship of these initiatives to the basic ISO 9001 quality management system standard is depicted in Figures 1, 2 and 3.

- The management system of ISO 9001 has subfeatures that uniquely provide an umbrella of leadership in management, which formally enables the harmonization of these four initiatives. A principal element of interest is 4.1 Management responsibilities. The subelements of this clause are:

- Policy.
- Responsibility and authority.
- Resources.
- Management review.
- Executive management representative.

Implementation of the requirements, according to their fullest interpretation, blends all the requirements of the initiatives mentioned above into one process or system.

The policy statement is broadened to include process safety by reference to the regulations (29CFR OSHA 1910.119), Environmental Management (ISO 14001), Responsible Care® (guidelines) and ISO 9001 into one or two concise paragraphs.

The delegation of responsibilities and authority, according to section 4.1, should be expanded to include all of the elements of the initiatives identified above. Each individual should have responsibilities and authorities carefully delineated to include all of the requirements of the elements of the four initiatives. Additionally, each delegation of *responsibilities and authorities* must refer to a third element—*accountability*.

Likewise, the management representative can be one or more individuals depending on the company size and structure. The outcome will be the same as long as the total system covers all four initiatives.

Admittedly, attempting to meet all the initiatives simultaneously is complex and confusing to most employees and managers. However, harmonization through the ISO 9001 helps build a base to support this harmonization.

In addition, ISO 9000's acceptance around the world and its insistence on management involvement make it more suitable for this harmonization than Responsible Care®, Process Safety Management, and ISO 14000. ISO 9000 is generic, flexible, and permits interpretation. Individual companies decide the best implementation approach and whether it's best to combine all four activities or to keep them separate.

THE HEART OF THE SYSTEM

Any good management system has six basic elements. The first is management responsibility. It begins with a policy statement, a definition of responsibility and authority to all those who have responsibilities in the management system regardless whether it is safety, environmental, or ISO 9001.

Secondly, it requires that a management representative with executive responsibility ensures that the system(s) is developed, that it is maintained, and that it produces desirable results.

Figure 1

ISO 9001/0SHA 1910.119 Professional Safety Management

ISO 9001 Quality Management System	(a) Application	(b) Definition	(c) Employee Participation	(d) Process Safety Information	(e) Process Hazard Analysis	(f) Operating Procedures	(g) Training	(h) Contractors	(i) Prestartup Review	(j) Mechanical Integrity	(k) Hot Work Permit	(l) Management of Change	(m) Incident Investigation	(n) Emergency Planning	(o) Compliance Audits	(p) Trade Secrets
1.0 Scope & Field of Application	X															
2.0 Reference																
3.0 Definitions		X														
4.1 Management Responsibility			X									X	X			
4.2 Quality System																
4.3 Contract Review																
4.4 Design Control				X	X	X			X	X	X	X		X		
4.5 Document and Data Control				X								X				
4.6 Purchasing								X								
4.7 Purchasers Supplied Product																
4.8 Product ID & Traceability																
4.9 Process Control				X	X	X			X	X	X	X	X	X		X
4.10 Inspection & Testing																
4.11 Inspection, Measuring & Test Equip.																
4.12 Inspection & Test Equip.																
4.13 Control Nonconforming Product																
4.14 Corrective																
4.15 Handling, Storage, Packaging, Delivery																
4.16 Quality Records															X	
4.17 Internal Records																
4.18 Training							X									
4.19 Servicing																
4.20 Statistical Techniques																

Occupational Safety & Health Standard 1910.119 (column header group)

Additionally, the management representative has the responsibility to determine that the records of the internal audits and the corrective action are presented to the executive department for their review. Moreover, management must ensure appropriate resources are allocated so that systems are maintained. The executive management representative also has the responsibility to provide for internal audits and ensure that appropriate individuals receive records and data.

Figure 2

ISO 9001/ISO 14001 Environmental Management System

ISO 9001 Quality Management System		4.0 General	4.1 Environmental Policy	4.2 Planning	4.3 Implementation Operation	4.4 Checking and Corrective Action	4.5 Management Review									
1.0	Scope & Field of Application	X														
2.0	Reference															
3.0	Definitions															
4.1	Management Responsibility		X				X									
4.2	Quality System			X												
4.3	Contract Review															
4.4	Design Control															
4.5	Document and Data Control				X											
4.6	Purchasing															
4.7	Purchasers Supplied Product															
4.8	Product ID & Traceability															
4.9	Process Control				X											
4.10	Inspection & Testing					X										
4.11	Inspection, Measuring & Test Equip.					X										
4.12	Inspection & Test Equip.					X										
4.13	Control Nonconforming Product					X										
4.14	Corrective					X										
4.15	Handling, Storage, Packaging, Delivery															
4.16	Quality Records					X										
4.17	Internal Records					X	X									
4.18	Training				X											
4.19	Servicing															
4.20	Statistical Techniques															

A good management system also includes the development of the plans and all the elements covered by any of the standards or the initiatives. These plans are then concisely harmonized in an appropriate manual covering quality management, process safety management, or environmental management and responsible care. This single manual might be called a business system management manual and should also include a traditional business plan—both short and long term.

Figure 3

ISO 9001/Responsible Care

ISO 9001 Quality Management System		(1) Community Concerns Chemicals & Operations	(2) Chemicals, Manufactured, Transported, Disposed Safely	(3) Health, Safety, Environmental—A Priority	(4) To Report to Officials, Employees, Customers Chem-Health, Envir. Data	(5) Counsel Customer on Safe Use, Transportation and Disposal	(6) To Operate Facilities to Protect Envir., Health, Safety	(7) Conduct Research on Health, Safety & Envir. of Products and Waste	(8) To Work with Others on Past Problems	(9) Create Responsible Laws & Standards	(10) Promote Responsible Care by Sharing			
1.0	Scope & Field of Application													
2.0	Reference													
3.0	Definitions													
4.1	Management Responsibility	X		X						X	X			
4.2	Quality System													
4.3	Contract Review													
4.4	Design Control		X	X				X						
4.5	Document and Data Control													
4.6	Purchasing													
4.7	Purchasers Supplied Product													
4.8	Product ID & Traceability													
4.9	Process Control					X		X						
4.10	Inspection & Testing													
4.11	Inspection, Measuring & Test Equip.													
4.12	Inspection & Test Equip.													
4.13	Control Nonconforming Product													
4.14	Corrective													
4.15	Handling, Storage, Packaging, Delivery													
4.16	Quality Records													
4.17	Internal Records													
4.18	Training													
4.19	Servicing					X			X					
4.20	Statistical Techniques													

The third element of this basic system is the contract review, which covers the relationship with customers. This relationship not only covers government regulation, safety, and environment, but also includes the relationship with society and the community as well as the workers. Documented procedures ensure that all these expectations are met.

The fourth element of a successful management system involves purchasing. The cost of purchased products has been found to be a minimum of 60 percent of the cost to produce products, an important financial consideration. Additionally, the quality of the products purchased greatly impacts environmental and safety expectations.

The fifth element of an effective system is design control. In addition to the responsibility for properly designing the product, process engineering requires hazard analysis. Provisions for training, hazard analysis, and documentation of the procedures should be included. Adequate procedures must cover the 5 Ws and 1 H (what, why, where, who, when and how). Since the management system for ISO 14000 is modeled after ISO 9000, it readily becomes part of the umbrella organization.

The final element is the process control system. Not only should the equipment capability be evaluated, but also all those involved in operating the equipment. The safety of the workers and the community itself should be considered in the system. The operators and all those associated with an employees' activities must understand the hazards associated with the work and have appropriate training. It is part of the process safety management system to conduct hazard analysis, also known as Failure, Mode, Error Analysis (FMEA) in other industries. Essentially, FMEA is a process of considering "what happens if." This is a vital activity to ensure that all the activities meet the harmonized system's requirements.

The other fourteen elements of ISO 9000 are the ancillary ones—of great and necessary value as their purpose is to document and measure the results of the HEART elements. Thus the application of the ANCILLARY elements combined with the HEART ones provides a complete system whereby all of the initiatives can be implemented simultaneously.

ISO 9000 AND QS-9000 FOR THE CHEMICAL INDUSTRY

by James S. Bigelow and William E. Cox

INTRODUCTION

The chemical and process industry (CPI) is different from the manufactured goods industry. In particular, the CPI is often working with natural raw materials, automated process control is highly developed, and many of the processes occur at the molecular level. The ISO 9000 quality model is still highly applicable to the CPI, but some translation is necessary.

This article is will help workers, managers, and registrars understand the CPI issues which must be addressed. It will:

- Share our more than 16 years of experience with ISO 9000 in the CPI.

- Describe some of the changes necessary to convert from a parts-orientation to a process model.

- Compare QS-9000 to ISO 9001 and emphasize those areas specifically impacting the CPI.

The ISO 9000 and QS-9000 models both support good quality practices (blocking and tackling) and the concept of continuous improvement.

ISO 9004-3, Quality Management and Quality System Elements Part 3—Guidelines for Processed Materials, was not updated in 1994 along with other standards in the series. However, since ISO 9004-3 was issued in 1993, it still contains good advice that is specific to the CPI. The authors of this paper helped write the ASQC's recently published second edition of *ISO 9000 Guidelines for the Chemical and Process Industries.* This publication is the source for many of our comments on special issues for the CPI.

Several years ago, the Chemical Manufacturer's Association (CMA) suggested that the ISO 9000 model would be appropriate as a uniform model for use between suppliers and customers within the chemical industry. The CMA reasoned that once suppliers built confidence with their customers by using the model, less expense would be associated with auditing each other within the industry.

ISO 9000 AND THE CHEMICAL PROCESS INDUSTRIES

This section offers guidance on application of ISO 9000 in the CPI.

4.1 Management Responsibility

This section clarifies management's responsibilities and indicates that management must provide the resources necessary to implement and operate the quality system. Section 4.1 can be applied directly to CPI situations without interpretation. While it may seem somewhat prescriptive, we believe that the management review frequency should be at least annually, and preferably more often, if a genuine desire exists for continuous improvement. Customer complaints and internal statistical measures also ought to be part of these reviews, as well as supplier performance, results from internal and external audits, and corrective and preventive actions.

4.2 Quality System

A quality manual is a requirement, and the structure of the quality system must be included. In practice, those seeking registration have always been required by their registrars to develop a quality manual. In the chemical industry, organizations often refer to overriding standards in their quality manuals. These may include FDA Good Manufacturing Practices, the API Q1 standard, or other standard industry practices.

Quality planning activities are clearly required. A quality plan in the chemicals industry may differ from one in a hard goods industry. This may occur when the same processing equipment (our "hardware") is operated at different conditions (time, temperature, pressure, catalyst, etc.), or with different feedstocks (type and quantity) to produce different products. In some cases multiple grades are offered within a given product family.

The product quality plan called for in section 4.2 may be quite similar to the process control plan called for in section 4.9. Many chemical companies have published product slates, and customers order from the product sales specifications. Chemical products vary from near-commodity to high-specialty.

4.3 Contract Review

The critical issue is for suppliers to evaluate their ability to fulfill orders however they may be received. Section 4.3 includes tenders, contracts, and verbal orders. A focal issue for some chemical companies revolves around accepting blanket contracts for more than they can produce, much as airlines overbook seats on an airplane. "Capability" as used in this section is not necessarily a statistical process capability, but refers more to the overall ability of the supplier to fulfill the expectations of the customer.

Safety and other considerations require the chemical industry to define the point at which custody transfer takes place. Is the supplier or the customer responsible for the integrity of the product while it is on the way to the customer? Inspection and Test Status, section 4.10, defines conditions under which product may be "dispatched", and this is addressed further in 4.15.6, Delivery.

A footnote in section 4.3 refers to having established lines of communication and interface with the customer. This may be important in the CPI where shipment may involve lining up pipeline connections directly to the customer's tankage.

4.4 Design Control

In the CPI at least three kinds of design are common. Most CPI companies maintain fairly major distinctions between process, product, and plant engineering or design. All three types could come under an organization's scope of a quality system based on ISO 9001, although at a minimum, design of the product must be addressed.

In many cases, CPI organizations invent new products from their own initiatives, while seeking new ways to use a molecule. The customer's dollars are not at risk if the product finds no commercial success. In these cases, many organizations have followed the ISO 9002 model for production. ISO 9002 does not include design control as a requirement. However, other companies have found the discipline of design control as required by ISO 9001 to be beneficial to orderly product and process development.

It is often difficult to replicate results from lab-scale glassware in pilot plants and commercial plants. If the rules of similitude are not well developed in a given field, then the company may have no choice but to make full-scale trials in their commercial plants. If the design objectives have not been well defined in advance, this can be quite an expensive option.

4.5 Document and Data Control

Procedures for document control must be documented. Some of the CPI specific documents that should be controlled are specifications, formulas, recipes, standard and emergency operating procedures, laboratory and on-line test methods, sampling plans, calibration and measurement controls, and visual standards. Certain "documents of external origin," such as ASTM test methods, should be included if these documents are spelled out in specifications or contracts.

An issue in the CPI is how to have the current work instruction or procedure available at the point of use, which is often out in the unit and sometimes in areas that are exposed to the elements. Checklists and "cheat sheets" are sometimes used. The requirement for having written procedures at the point of use may not always be practical. In such cases, if the employee has adequate access to the procedures in a control room or other location visited during the performance of his duties, and if his level of training is such that constant reference to the documentation is not required, this should be adequate.

4.6 Purchasing

No CPI specific issues exist in following the ISO 9000 model for the purchase of conventional raw materials, catalysts, and chemicals. Special quality assurance plans should be considered in situations involving bulk shipments, contract terminals and laboratories, exchange agreements, toll processing, the operation of pipeline grids, and certain utilities such as process water. Careful consideration should be given to assure that all requirements of the customer are met.

Although ISO 9000 allows for the customer to include in the contract the right to inspect a subcontractors' operations, customers in the CPI seldom invoke this right.

4.7 Control of Customer-Supplied Product

Section 4.7 applies most directly to toll converters or in those cases where the customer provides the raw materials, packaging (including tote bins, rail cars, tank trucks, etc.), or other similar items that are used in your process. The supplier has the obligation to account for and preserve the condition of everything provided by the customer. In the CPI, these obligations are often spelled out in the contract in terms of loss factors.

4.8 Product Identification and Traceability

The principal requirement in section 4.8 is that you have procedures to identify all raw material, intermediates, and finished product throughout your processes. This provides flexibility in the means of identification, which is needed in the CPI.

While traceability is only a requirement if specifically stated in a customer's contract, the concept of providing full traceability is a good quality practice to assist in troubleshooting efforts. In addition, if an industry norm calls for traceability, such as FDA-GMP, this should alert the supplier to consider traceability a requirement even if not specifically referenced by the customer.

4.9 Process Control

Process control in the CPI often carries a different connotation than was intended by the authors of ISO 9000. The term is most often used to describe the systems of sensors, analyzers, and control loops employed in our processes. The term has broader meaning in ISO 9000 and is used to describe all the activities to control production, installation, and servicing.

Process control is typically an area of strength in the CPI. Automatic control strategies for normal and abnormal operations are usually well defined and documented. Training is often very thorough. The requirement for "suitable maintenance" is sufficiently vague as to explicitly require neither preventive maintenance nor written maintenance procedures. Typical preventive maintenance plans already used in the CPI should easily satisfy the maintenance subclause.

4.10 Inspection and Testing

High levels of in-process testing are a way of life in the CPI. All testing needed to meet the requirements of the quality plan should be covered either in this section or in the process control section. Close coupling of process units may prevent holding intermediates until testing is complete. In these cases, control plans should provide for the ability to contain the correct downstream inventory if adverse results are later reported. This is often accomplished with process computer data and time lag regression models.

Receiving inspection and testing covers both raw materials and customer-supplied materials. Storing bulk materials raises the potential for cross-contamination, and this should be addressed in the procedures. Some chemical raw materials, for example pipeline shipments, go directly into the process without ever going into inventory and this requires a high level of confidence in your suppliers. Inspection plans for raw materials should be appropriate to their criticality to the process.

ISO 9000 calls for final testing of the finished product to be complete before the product may be "dispatched." Carefully define these aspects of the transaction in the contract to answer questions of ownership and title transfer. The high cost of delays often allows CPI products to move down the supply chain before all the test results are available. This is acceptable practice provided that the supplier and customer have mutually agreed to the actual conditions.

As process capabilities improve, many in the CPI are correlating in-process test results with results on the finished product, thus eliminating some expensive testing. An example might be predicting a product's molecular weight based on reactor measurements, and correlating this by time lags to the finished product. Careful procedure design becomes even more important as this trend increases. Clear communications with the customers are a must.

4.11 Control of Inspection, Measuring, and Test Equipment

For the CPI, the scope of section 4.11 includes lab equipment, on-line analyzers, and equipment in the R&D lab that is used to verify product quality. In-process instrumentation is often used for process control and optimization purposes, rather than verification that product conforms to specified requirements. Many flow, temperature, pressure, and level instruments may not need to be included in this requirement.

While the focus is on calibration at prescribed intervals, it is generally accepted that statistical process control (SPC) may be used in the measurement processes. With this system, the equipment need only be taken out of service when SPC indicates it is out of statistical control. Before the equipment may be put back into service, the cause of the failure must be determined and eliminated. This process will likely include calibration as one of the final steps. It is important for the supplier to document the actual processes used.

National and international reference standards do not exist for many of the tests performed in the CPI. Industry standards are sometimes developed by sections of the industry (for example, ASTM ethylene standards for use with ASTM D3900). However, the supplier often must be responsible for developing, qualifying, and maintaining internal standards and reference materials.

Sampling and measurement system variation itself is often a large component of the total process variation experienced in the CPI. It is important, though not a requirement, for suppliers to do statistical studies of measurement capability.

4.12 Inspection and Test Status

Since the status of product from raw materials receipt through to the release of finished product must be known, this section is often related to the identification and traceability clause, section 4.8. Direct labeling of products in the CPI is often impossible, therefore the quality plans must identify other procedures to accomplish the same end. Special storage areas for solid products or hold tanks for liquids may be employed. A special case that requires definition is the one in which there is no ability to hold product. Examples are pipeline-linked units or direct shipments from the supplier's process unit to the customer's process.

4.13 Control of Nonconforming Product

Nonconforming product should be segregated into designated tanks, silos, tank cars, hopper cars, warehouse areas, and the like. Log notes and product classification systems become very important in tracking this storage.

When product flows are continuous, the absolute isolation of nonconforming product may not be possible. Supplier and customer should have plans to handle these contingencies. In some cases, the customer has the ability to accept the product, given prior notice that it is on the way, along with a description of what to expect. In contrast to the hard goods industry, where parts either fit or don't fit, CPI customers can often accept some deviation, depending on the following factors:

- Nature of the deviation.

- Duration and extent of the deviation.

- Ability to modify process conditions given sufficient advance notice.

- Expenses associated with shutting down and restarting.

- Price concession.

Section 4.13 provides several methods for disposing of nonconforming product, including reworking, acceptance by concession, regrading, and scrapping. To this list, the CPI adds blending and recycling, among others.

4.14 Corrective and Preventive Action

Three types of actions are to be considered:

- *Correction*—disposition of the problem product, perhaps including recall, rework or blending (per clause 4.13).

- *Corrective action*—finding and eliminating a root cause so that the problem does not *recur.*

- *Preventive Action*—eliminating potential causes before a problem ever *occurs.*

The concept of contingency planning is well developed in the CPI, given the long history of emphasis on safe operations, and this is analogous to taking preventive action.

Corrections, corrective actions, and preventive actions should all be included in the management review, not just preventive action as the standard literally requires.

4.15 Handling, Storage, Packaging, Preservation, and Delivery

From a pragmatic viewpoint, packaging, preservation, and delivery apply mostly to finished product. However, this clause also applies to internal distribution of raw materials as well as intermediate products. In the CPI, these situations are often addressed as part of process control. In any case, all of the points raised must be covered in one place or the other.

Issues for the CPI include avoiding cross-grade contamination and protecting products from deterioration other than normal aging. As many chemical products are unstable, shelf life and isolation of the products (such as nitrogen blanketing or inhibitor addition) must often be addressed. An issue that is sometimes missed in packaged chemicals is the useful life of the package itself.

Types of storage containers in the CPI are particularly varied and may include caverns, pipelines, ships, rail cars, piles, silos, bags, boxes, and crates. Issues that must be addressed include storage pressure and temperature, corrosion allowances, previous contents, and the characteristics of the product. Many of these same factors must also be considered for distribution and shipping.

4.16 Control of Quality Records

Several clauses of ISO 9001 specifically refer to this section indicating specific records that must be kept. Other clauses may encourage an organization to keep additional quality records, even though they are not specifically identified in the standard. Any type of storage medium is suitable to keep these records.

4.17 Internal Quality Audits

Many companies in the CPI have practiced internal safety audits for a long time. Internal quality audits are a logical extension of this practice. Internal audits are powerful in driving quality improvement. One benefit is working with the people in the process who are the most knowledgeable of actual or potential problems.

4.19 Servicing

Both ISO 9001 and 9002 include this requirement. In the CPI, servicing may occur in such areas as performance specifications, provision for technical service, emergency contacts, and other similar functions. It is not intended to include the typical goodwill or customer technical service work that is not billed separately and is considered to be part of the normal conduct of business. Therefore, this clause often does not apply to CPI firms.

4.20 Statistical Techniques

The proper use of statistical techniques is a requirement, not an option. The supplier is required to identify where statistical tools should be used and then provide procedures for their use.

Statistical techniques are widely used in the CPI to determine process and measurement capability, to set specification limits, design sampling plans, make conformance decisions, monitor processes, analyze causes of nonconformance, identify areas for improvement, analyze customer trends, and for many other applications.

SELECTING A REGISTRAR

Many organizations in the CPI decide to proceed beyond implementing an ISO 9000 based quality system to independent third-party registration. Because of the many special issues we have been discussing, it is important to select a registrar carefully. Your relationship with the registrar will likely last several years. Your registrar needs to be cognizant of and competent concerning special CPI issues. When you interview potential registrars you will want to ask for references in the CPI. The list below shows the IO registrars who are most active in the CPI (based on SIC codes 28, 29, and 30), according to the *Irwin Directory of ISO 9000 Registrations* (Feb., 1996).

ABS QE

BSI

BVQI

DnV

INTERTEK

LRQA

QMI

QSR

SGS ICS

UL

QS-9000 COMPARED TO ISO 9001

QS-9000 is a different model from ISO 9001. Some analysts attempt to minimize the differences, but these differences can be very significant. Depending on whether a company has already implemented some elements of TQM that go beyond the minimum requirements of ISO 9001, it could require as much as twice the amount of effort to implement and maintain a QS-9000 quality system. However, with its stronger emphasis on improvement, it may also generate significantly greater benefits.

Some of the additional QS requirements will have equal impact on CPI or non-CPI companies. Let's first consider the overall differences. Later, we will focus primarily on those differences that will have a particularly strong impact on businesses within the CPI.

Below are some data showing how different the two documents are:

Measurement	ISO 9001	QS-9000	% Increase
Number of primary elements (e.g., 4.x)	20	23+	15+
Number of secondary clauses (e.g., 4.x.x)	54	77	43
Number of "shalls"	137	250+	82+
Number of "shoulds"	0	34	infinite
Number of sanctioned U.S. interpretations	0	over 100	infinite
Number of pages in primary document	10	101	910

"+" indicates there are additional customer-specific requirements in Section III of QS-9000.

The data suggest that QS-9000 is much more comprehensive than ISO 9001. Obviously, it is difficult to simply quantify the differences. QS-9000 is designed to include clauses 4.1 through 4.20 in ISO 9001, plus additional requirements. These additional requirements include:

- Conduct benchmarking and competitive analyses.
- Develop short- and long-term business plans.
- Measure customer satisfaction and dissatisfaction.
- Use cross-functional teams.
- Exert stringent controls over your suppliers (subcontractors).
- Obtain written customer approval for product designs and control plans.
- Use data to trend performance in quality and operations.
- Use specific forms and methodologies.

- Make extensive use of statistical techniques in process control and measurement systems.

- Conduct improvement projects.

Some of these requirements are very similar to criteria of the Malcolm Baldrige National Quality Award. Others are more specific to the automotive industry. Use of some Baldrige elements is further evidence of a key difference. ISO 9000 was designed to be a minimum standard, achievable by any organization that put forth the effort required to manage quality. Baldrige was designed to define the very best. High scores and an award-winning level performance are achievable by only a limited number of companies that put forth a supreme effort evidenced by outstanding, quantifiable results.

ISO 9001 was designed to be a generic, non-industry specific standard. QS-9000 is very prescriptive by comparison, not only specifying *what* must be done, but in many cases *how* it should be done. Some of these prescriptive elements are couched as "shoulds" rather than "shalls." However, shoulds are really shalls, with a bit of latitude on how it is to be done.[1]

The biggest single difference for the CPI is the continual emphasis on parts. To a CPI reader, this can be confusing and frustrating. Dimensional tolerances are simply not important in manufacturing ethylene. It is also a bit difficult to inscribe a serial number on each atom comprising a molecule to provide the kind of ironclad part traceability envisioned for true part suppliers.

QS-9000 was intended to be applied to first-tier suppliers of parts directly to the Big 3 and others who also subscribe to QS-9000. The focus is on products that ultimately become a physical part of the finished automobile. However, there are two phenomena that make it applicable to chemicals, plastics, and other processed materials. The first is the inclusion of bulk materials within the control plan scope (4.2.3), and the second is the requirement (4.6.2) that first-tier suppliers must "develop" their suppliers using QS-9000, domino-style up the supply chain. This latter requirement is the one of crucial importance to the CPI, because it predicts that, while the CPI may not yet have felt much QS-9000 pressure, it will come in two to three years as the "development" begins to flow upstream. At this writing, there is some disagreement whether QS-9000 will be required for suppliers of "factory fills" such as lube oil, anti-freeze, greases, and hydraulic fluids.

A few conclusions may be made at this point:

- QS-9000 is much bigger than ISO 9000.

- QS-9000 is geared for parts, not for bulk processed materials.

- The CPI may not feel much QS-9000 pressure yet, but it will.

ISO 9000 Family of Documents	Additional QS-9000 Documents
ISO 9001, 2, 3, and Standards	QS-9000 the Requirements
ISO 9000, Selection	PPAP, Production Part Approval Process
ISO 9004, Guidance	APQP, Advanced Product Quality Planning
ISO 10011, Auditing Guidelines	QSA, Quality Systems Assessment
ISO 11012, Measurement	MSA, Measurement Systems Analysis
ISO 10013, Quality Manual	SPC, Statistical Process Control
ISO 8402, Definitions	FMEA, Failure Mode and Effects Analysis
ISO 9004-3, Processed Materials	

QS-9000 AND THE CHEMICAL PROCESSING INDUSTRY

The following elements of QS-9000 may have the biggest impact on the CPI (over and above ISO 9000):

- 4.2.3 Quality Planning and Advanced Product Quality Planning (APQP).

- 4.4 Design Control, the Production Part Approval Process (PPAP), the Scope of Your ISO Certificate, and ISO 9001 versus 9002.

- 4.9 Process Control.

- 4.10.2 Receiving Inspection.

- 4.11 Measurement System Variation Analysis.

- 4.15 Production Scheduling.

- 4.20 Statistical Techniques and II-2 Continuous Improvement.

4.2.3 Quality Planning

The first and perhaps biggest difference between QS-9000 and ISO 9000 and also current CPI practices is the use of the APQP—*Advanced Product Quality Planning and Control Plan* manual. This single requirement invokes another entire 100+ page manual. The APQP process greatly increases the customer's involvement in the product development process. In the CPI, where many companies are simultaneously customers, suppliers, and competitors, this will frequently be a very big change from current practice. The resulting product design and the associated control plan are in effect owned by the customer, insofar as the CPI supplier is not permitted to change anything without written customer approval. Thus, process operating conditions, raw material sources, and control methods come under the scrutiny and control of the customer.

During the various phases of product design, the supplier must identify special product characteristics that are critical to the product's performance in the customer's process. For these characteristics, Potential Failure Mode and Effects Analyses (FMEA) must be conducted at the design and production phases of the development process. Control plans must be prepared and approved by the customer for the prototype, prelaunch, and production phases of the development process. (In the CPI, more typical terminology is bench scale, pilot plant, and commercialization phases).

FMEA is similar to the Hazard and Operability (HAZOP) review frequently conducted in the CPI to manage safety related risks associated with processes. However, HAZOP analysis will not suffice for FMEA. FMEA is a formal, structured analysis of potential product performance problems and their likelihood of occurrence and detection. Preventive measures are emphasized to remove the cause of the potential quality problem.

4.4 Design Control, and
II-1 Production Part Approval Process

Much of the CPI has elected to register to ISO 9002 by limiting the scope of the certificate to manufacturing plants. Product development is often done at another location by a separate R&D organization, so this is a convenient separation of management systems. QS-9000 does not allow this separation.

The magic phrase in QS relative to element 4.4 is "design responsible." If your company (not site), at any of its locations is responsible for product design and development, then you must include this element. The CPI will typically follow this path, since the automotive customer almost never designs the bulk chemical product itself. In addition, the Big Three customers strongly prefer for the scope of the supplier's certificate to include all manufacturing locations of the automotive product as well as the related R&D function. The Big Three prefer for the scope of the supplier's certificate to include all manufacturing locations of the automotive product as well as the related R&D function. The Big Three prefer that their product is designed and manufactured under one unified quality system, rather than fragmented by location or function, as has often been practiced in ISO 9000 registered systems.

Design Control invokes the use of another 50-page manual called the Production Part Approval Process, or PPAP. This is referenced in clause 4.4.9 with respect to design changes, and also in clause II-1, the first of the three new automotive sector-specific requirements. Note that PPAP contains *requirements,* whereas the other five manuals are provided as *guidance.*

PPAP requires written customer approval of every key step and all changes to the design process. Additional requirements include analysis of measurement system variation, identification of special characteristics, process capability studies, and design and process FMEA.

While some flexibility is built into the process, all key decisions are subject to review and approval by the customer. For the CPI, this level of customer control is almost unprecedented and will either require waiver by an understanding customer, or a lot of adjustment on the part of the supplier. Existing products already being produced and sold into the auto-motive industry may have to be submitted to PPAP, if this has not already been done.

4.9 Process Control

Several important requirements have been added to this element, which is so critical to the process industries. Most large plants have extensive maintenance programs aimed at not only repair, but also preventive and predictive maintenance techniques. However, some smaller companies may be faced with a stronger requirement in QS-9000 for preventive mainte-nance than the relatively vague statement in ISO 9000 regarding "suitable" maintenance.

The Big Three added seven sub-clauses to this element. One strengthens the role of the control plan developed as part of the APQP process. However, the requirements around process capability will probably have the biggest impact on the CPI. For many parts of the CPI, the 1980s was the defining decade for SPC. QS-9000 makes SPC a permanent part of the system. Processes are required to have a P_{pk} value greater than 1.67 during the development phases and perhaps the initial period of commercialization. After the process has demonstrated stability, a C_{pk} value greater than 1.33 is required.

Whenever the process becomes unstable, the P_{pk} should again be greater than 1.67, and the product should be 100 percent inspected. The latter requirement for 100% inspection is frequently practiced in the CPI, where many processes are either unstable or incapable of meeting the 1.33 and 1.67 indices. This is often because specifications have been nar-rowly set without regard to statistical capability. Perhaps this will encourage the CPI to move away from unrealistically tight maximum or minimum specifications and require-ments, with frequent waivers, toward developing target values permitting an acceptable amount of variation around the target.

In clause 4.9.5, the parts concept adds to the confusion for application within the CPI. The term "job set-up" for the CPI is "plant, product, or grade start-up." Certainly the initial product produced at the end of a start-up period is tested to ensure it meets spec-ifications. However, comparison to the last product produced during the previous cam-paign has little significance for most of the CPI. So many variables exist, both controllable and uncontrollable, that current plant conditions are not necessarily compa-rable to the previous campaign.

CPI users should be cautious in developing the control plan and the possible undesirable constraints it may add to the control process. Any changes must be approved in writing by the customer.

4.10.2 Receiving Inspection and Testing

In the CPI, it is widely accepted practice to depend on a certificate of analysis (C of A) provided by the subcontractor for receiving inspection. This is especially true when a high level of confidence has been established in the subcontractor's manufacturing and testing processes. In other cases, some key properties may also be tested in the company's laboratory prior to use, but this is not always possible or necessary for all elements in the product specification. Often, the principal reason for testing inbound raw materials in the CPI is to make sure the material has not become contaminated while in transit. Pipeline shipments are often based on on-line analyzer results, perhaps backed up by laboratory testing and/or a composite sampling system.

QS-9000 specifies stringent requirements on receiving inspection, which in effect render the certificate of analysis practically useless. By contrast, if the subcontractor is certified to QS-9000 or ISO 9000, that in and of itself is sufficient! This would typically not be acceptable in the CPI, due to the high level of risk involved in using many raw materials.

Clause 4.10.2 specifies that if subcontractor certificates of analysis are used to accept or reject incoming product, they shall be *in addition* to another of the following methods:

- Receipt of statistical data (e.g., control charts).

- Testing in the supplier's laboratory.

- Second- or third-party audits of subcontractor locations.

- Product testing by a third-party accredited laboratory (e.g., ISO Guide 25).

Any one of these four alternatives is sufficient. Having the C of A does not add anything to the receiving inspection process according to QS-9000. The certificate of compliance (C of C) is explicitly ruled out, due to the requirement that any certificate must include actual test results.

4.11.4 Measurement System Analysis

QS-9000 relies more heavily on statistical techniques than ISO 9000. Variation of test equipment must be analyzed using techniques described in the guidance manual, *Measurement System Analysis*. Some CPI companies have felt comfortable accepting the test equipment manufacturer's statement for accuracy, without conducting variability studies including factors such as operators, equipment, and time. Unless waived, these types of studies will be required.

4.15.6 Production Scheduling

QS-9000 requires that production scheduling shall be order-driven. Such practices are not practical for some large-scale, continuous CPI processes. Ethylene manufacturing, for example, employs extensive networks of pipeline delivery systems used by multiple suppliers, and underground storage caverns. Such processes are run to long-term demand signals rather than for individual orders.

4.20 Statistical Techniques, and
II-2 Continuous Improvement

Many ISO 9000 users have complained about the weakness of the requirements in element 4.20. Whereas ISO 9000 requires that the need for statistical techniques be determined, it remains implied that anyone should actually have an understanding of what those techniques are and how they might be used. QS-9000 explicitly requires that concepts such as variation, stability, capability and overadjustment be understood throughout the supplier's organization "as appropriate." This requirement is further extended in clause II-2.3, where the supplier is required to demonstrate knowledge of 14 specific measures and methodologies.

ISO 14000 AND THE CHEMICAL PROCESS INDUSTRIES

ISO 9001 and ISO 14001 have many similarities. While ISO 9000 focuses on the on-purpose product, ISO 14001 addresses the byproducts. Many of the control systems called out in these standards are similar. How the CPI will accept ISO 14001 is still undetermined. An organization employing the ISO/QS 9000 and ISO 14000 models should use as many common processes as possible. Some of the common systems include:

- Development of policies and procedures.
- Definition of responsibilities.
- Document controls.
- Record keeping.
- Maintenance plans.
- Instrument controls.
- Internal auditing.
- Corrective and preventive action.
- Management review.

The CPI, and specifically the CMA, already have several environmental initiatives, such as Responsible Care®. The industry will have to reconcile these activities against the ISO 14001 standard and assess the relative value of compliance and/or certification to this new standard.

CONCLUSION

The Chemical and Process Industries are different from manufactured goods industries and require an appropriate translation of the ISO 9000 standards. Guidance has been given for application of ISO 9001 to the CPI to capture both good practices and continuous improvement. The ISO 9000 model is useful for CPI companies.

It is just a matter of time-—QS-9000 is coming to the CPI. When it does, companies will have some significant work to do to upgrade an ISO 9001 or 9002 certificate to QS-9000. Some of these requirements may seem unnecessary or bureaucratic, but remember: By keeping *improvement* in mind, the ISO 9000 and additional QS-9000 requirements can be used to drive true improvement in your organization.

Many have criticized ISO 9000 for its effectiveness to drive true quality improvement. QS-9000 is a giant stride to toward quality system requirements that require improvement rather than mere compliance with a minimum standard or to obtain a certificate. The CPI and other organizations should use this to their advantage.

Endnote

[1]According to AIAG training, Houston, TX, February 19, 1996.

ISO 9000 AND QS-9000 IN THE METALS INDUSTRY

by Peter B. Lake

INTRODUCTION

The metals industry continues to undergo changes in which ISO 9000 can play a strong supporting role. Essentially nearly all companies in the industry are seeking ISO 9000 registration, but they are not necessarily making it a requirement for their suppliers. Internal improvements engendered by ISO 9000 have been recognized. These positive attributes of ISO 9000 are expected to support Total Quality Management (TQM) efforts. The Big Three's Automotive QS-9000 requirements affect most in the metals industry and are a significant driver for registration. ISO 14000 environmental management standards will definitely impact the metals industry as usage of that standard increases.

THE METALS INDUSTRY CONTINUES TO UNDERGO CHANGE

The metals industry is often thought of in the context of producing, processing, and distributing a commodity class of products. Survival and success is predicated on how well an organization meets the constant pressures for cost containment, product quality, and reliability of delivery; all these goals are consistent with the effective implementation of ISO 9000.

Factors influencing the significant changes taking place within the industry include the challenges of globalization, the effects of ongoing downsizing, and an aging workforce and workplace. Adding to the competitive complexity are rapid advances in production technology.

In the meantime, customer requirements are reaching unprecedented levels of sophistication. One can no longer assume that market share is a function of production capacity! The age of highly tuned systems squeezing out every gram of efficiency is on its way to transforming an aging smokestack industry into a high-tech battleground.

ISO 9000 AND THE METALS INDUSTRY

The quality management systems of ISO 9000 (and QS-9000) require a management commitment to establish and maintain a well-organized, documented, and effectively implemented quality system. ISO 9000 also encourages standardization, consistency, and employee empowerment disciplines that are building blocks for better system performance and operational efficiencies. ISO 9000 disciplines are indeed "good medicine" for business.

Achieving registration usually involved the efforts of many cross-functional teams, the reissuing (or refining) of many policies, procedures, and work instructions, and establishing a real internal audit system to confirm the implementation of an effective quality system. After registration, as the systems mature, companies typically observe some important changes:

1. The internal champion changes from quality or sales to the head of operations, in whose departments the disciplines are making a measureable major positive impact on productivity and yield.

2. Employees are empowered—they do their jobs with less direct supervision and more consistency.

3. Systems efficiencies have improved because the process of formalizing and implementing procedures involved increased team work and adoption of a workable "best practice."

4. Internal audit system processes and their results become an important component of management review, and continuous improvement.

WHO'S GETTING ISO 9000 REGISTERED?

An informal survey done during Spring 1996 showed that, with few exceptions, all those in the metals industry were either registered or in the process of achieving ISO 9000. However, only a small percentage of these metals companies surveyed "required" ISO 9000 registration from their suppliers. Most companies chose to use the words "strongly suggest" with respect to registration for their key suppliers. This approach keeps ISO 9000 compliance a "supplier initiative for customer satisfaction" and can avoid the implied ownership in the supplier cost of compliance that the customer might have by requiring the supplier to get registered.

Some suppliers, when forced into ISO 9000 registration (not recognizing the internal benefits), have complained that the customer should somehow pay for the registration effort. In this industry still wearing the "commodity" label, there yet cries an argument based on the assertion that "they still buy on price," certainly a factor in for an industry enduring severe continuing cost pressures.

The same informal survey found that a large number of raw material and equipment suppliers to the metals industry are also obtaining registration, despite the absence of a customer requirement. Most segments of materials suppliers are in the process of achieving ISO 9000 registration, including coal and iron mines, various ferroalloy producers, limestone suppliers, and many nonferrous mining and smelting operations as well as scrap suppliers. Other segments seeking registration include refractory and graphite suppliers, equipment and various roll builders, engineering services, and other services such as outside processors, coil coaters, and heat treating operations.

INTERNAL IMPROVEMENTS ARE RECOGNIZED

The survey and other industry discussions show that, with rare exception, internal benefits obtained from the disciplines of ISO 9000 have been recognized. About 30 percent of the companies used the word "significant improvements." The majority of those surveyed used the phrase, "improving systems performance," a reasonable expectation considering the relatively short time (average one year) since most companies completed initial registration.

One nonferrous producer, who has been registered to ISO 9000 for several years, said, "The disciplines are a foundation for all of our quality goals. I cannot imagine being without ISO 9000 at this point."

Responses from those registered for at least a year often included an observation of the "empowerment" of employees. Internal consistency and documentation of expectations is appreciated at all levels in an industry that is experiencing ongoing change. In many cases, initial shopfloor concerns over ISO 9000 have been replaced with strong support.

ISO 9000—A GOOD FOUNDATION FOR TOTAL QUALITY MANAGEMENT (TQM)

Although the definition of TQM, whether Malcolm Baldrige, or others, remains loosely defined, a large majority of the metals producers surveyed claim that these efforts are "alive" in their organization! Most agree that the result of formally establishing and maintaining the ISO 9000 disciplines and registration will provide an improved foundation for their TQM efforts.

The regular surveillances required by ISO 9000 tend to enforce the ISO 9000 disciplines and strengthen hopes that company TQM will take on a similar constancy.

THE IMPACT OF THE AUTOMOTIVE INDUSTRY AND QS-9000

A majority of the metals industry companies are affected directly by QS-9000. Most metal producers are either Tier 1 suppliers or subcontractors to a Tier 1 supplier and are subject to the QS-9000 supplier development clauses.

For this reason QS-9000 plays a role in most major steel and aluminum metal producer's plans. Metal producers have already begun to require QS-9000 registration for some parts subcontractors or outside processors. Requiring QS-9000 registration of key subcontractors is one of the ways in which a Tier 1 supplier to the Big Three can meet the supplier development requirements.

This cascading of QS-9000 registration requirements to subcontractors has created a challenge for the principal Tier 1 suppliers and registrars in determining how a subcontractor can meet all the requirements of a QS-9000, which was designed for first-line auto part suppliers to the Big Three. Meeting this challenge requires good communication between registrar and subcontractor to assure common understanding of the scope and requirements of QS-9000.

The "good news" about QS-9000 for metal producers is that these requirements are nothing new. The "new" aspect of QS-9000 is a requirement to implement the system and show continuing effective compliance.

WHAT'S AHEAD FOR THE METALS INDUSTRY?

More ISO standards are coming in the years ahead, which will certainly affect the metals industry companies and their management systems. Experience with, and compliance to, ISO 9000 quality management system requirements will position them to extend their system disciplines to the areas of environment, health, and safety.

ISO 14001, the Environmental Management Systems standard, is expected to affect, one way or the other, every metal producer in the industry. Environmental compliance has always involved significant industry activity and effort, although many of the driving forces for, and benefits of, registration to ISO 14001 remain undefined and undetermined by the many standards and regulatory groups involved. These potential benefits to the industry will continue to be the subject of intense discussion and debate within the metals industry trade associations and companies.

Recent nationwide surveys have shown that most accredited registrars for ISO 9000 and QS-9000 in North America are also planning to provide registration services to these same companies to the EMS ISO 14001. Conversely, most companies intend using their ISO 9000 registrar for EMS registrations and other ISO 9000 based initiatives, such as occupational health and safety.

Contributors to the Third Edition of the ISO 9000 Handbook

CONTRIBUTORS

Allen R. Bailey, Canadian Standards Board Quality Auditor (ISO 9000), is a consultant and trainer in quality control and product compliance. As a consultant, he has assisted clients with implementing ISO 9000 Quality Systems, and with establishing material management programs, and revitalizing companies through effective implementation and training. He is registered with the International Register for Assessors of Quality Systems and TASA (Technical Advisory Service for Attorneys) and has taught Polymer Science and Material Engineering at Penn State and Harrisburg Area Community College.

He has written numerous articles and books on compliance. He also received several patents and served on the committee to provide technical expertise to the Pennsylvania High School Curriculum Committee. Mr. Bailey is a graduate of California State University, California, and holds two master's degrees, one from Marshall University and the other from Virginia Tech. He specialized in material engineering and physical chemistry. He served on the IEEE Committee on Sealants and Coating for IC, ASTM D4 on Wire Bonding, and SEMI Committee on Testing. He is an active member of the Western New York Consulting Association, American Chemical Society, Air and Waste Management Society, and ISHM.

Robert C. Bamford is a partner and founder of Software Systems Quality Consulting (SSQC), San Jose, California. Since 1990 SSQC has assisted software and hardware developers, manufacturers, and service providers in five related areas:

- Software engineering process and life-cycle definition.
- Education and training.
- Software quality assurance and testing.
- Business process reengineering and benchmarking.
- ISO 9000 registration.

Mr. Bamford holds a BS and MAT in mathematics and spent eight years teaching at the secondary and university levels. Since 1975, he has managed training development, technical publications, professional services, and third-party software development for National Semiconductor Corporation (Systems Division) and International Computers Limited (ICL). In 1987, he became involved in the implementation of a Crosby-based Total Quality Management system, facilitating quality courses, managing the education team, and serving on the corporate quality council.

In 1990, he participated in ICL's successful initiative to achieve ISO 9001 registration for its Santa

Clara hardware and software development center. He facilitated the teams, defining and documenting the company's complete processes. From 1990 until he left ICL, he managed teams of auditors conducting semiannual value-added internal audits of departments and projects.

During his affiliation with SSQC, Mr. Bamford has developed and published numerous courses, auditing tools, and articles on interpreting and applying the ISO 9000 standards and guidelines. His courses have been sponsored for their members by professional associations, including the ASQC, Semiconductor Equipment and Materials International (SEMI), and Software Engineering Forum for Training at California State University Long Beach, UC Berkeley, and UC Santa Cruz. He regularly performs independent assessments and surveys of organizations and works with companies to assist them in their engineering practices and business process definition. He was an active member of the Software Quality System Registration (SQSR) committee of the ANSI/ASQC Registrar Accreditation Board(RAB). This committee developed the U.S. equivalent of the UK TickIT program and guideline documentation. He was a principal author and editor of *A Guide to Software Quality System Registration under ISO 9001*. He is an active U.S. TAG member in the ISO/IEC JTC1 SC7—Software Engineering Standards subcommittee.

Dr. Robert W. Belfit, Jr., Ph.D., president and CEO of Omni Tech International, Ltd., who also leads Omni Tech's professional and technical quality efforts, has spent 29 years with the Dow Chemical Company in analytical product, process and quality functions. He shared responsibility for and was actively involved in the design, internal selling, and implementation of the quality systems used today by the Dow Chemical Company on a worldwide basis.

Dr. Belfit studied under Juran, Deming, and Crosby extensively and has had hands-on experience setting worldwide quality standards, implementing global quality management systems, setting up global methods of in-house testing, and coordinating company standards, specifications, test methods, databases, and service communications around the world.

Since 1986, in addition to serving as Omni Tech's CEO and developing the firm's reputation as a credible quality systems and services consulting or-

ganization, he has been deeply involved in national efforts to implement the new quality concepts throughout industry, such as those embodied in the ISO 9000 Series of Standards and QS-9000.

Dr. Belfit received his A.B. and A.M. in organic chemistry from Dartmouth College and his Ph.D. from Pennsylvania State University.

James S. Bigelow was, until July 1996, the quality advisor for Exxon Chemical Company in Houston, Texas. He has been an internal consultant for quality improvement since 1980. Prior to this assignment, he held positions at Exxon in engineering, marketing, manufacturing, new product development, and supply planning. Most of his work has been related to polymers. Mr. Bigelow received degrees in mechanical and electrical engineering from Lehigh University. He is a senior member of ASQC and a member of ASTM E11—Committee on Statistics. He is the lead U.S. delegate to ISO TC 176 Subcommittee 2 Working Group 12, which is charged with the revision of ISO 9004. Mr. Bigelow was an examiner for the Baldrige National Quality Award, served on the Board of Overseers of the Texas Quality Award and is currently serving as a judge for the Texas Quality Award. Jim retired from Exxon after 33 years of service. He is now associated with TQM consulting.

Robert D. Bowen is a principal in r. bowen international, inc. (rbi). Mr. Bowen is a certified quality engineer, certified quality auditor, and senior member of the American Society for Quality Control.

As principal of rbi, Mr. Bowen has developed a team of quality professionals and business associates with more direct experience implementing ISO 9000 than any other North American Resource. rbi has successfully implemented ISO 9000 and continuous business improvement in businesses responsible for over $11 billion in annual sales revenue. rbi has a 100 percent success rate at Initial Assessment by third-party registrars.

Before founding rbi, Mr. Bowen was senior technical resource for DuPont's ISO 9000 implementation. In that capacity, he presented ISO workshops to over 3,000 persons throughout the DuPont Corporation and its community of customers, suppliers, and business partners throughout the world. He was instrumental in achieving a $1,200 cost reduction associated with this implementation. Some "firsts" for Mr. Bowen include the following:

- First electronic certification in the United States.

- First distribution center certification in the United States.

- First worldwide ISO 9000 implementation.

- First to introduce ISO 9000 into many Pacific Rim countries.

- First Japanese certification of an electronics design facility.

- First staffing business to achieve ISO 9000 certification in the United States.

- First ISO 9000 assessment of a public school district.

- First United States application of ISO 9000 to an insurance and healthcare organization.

During his employment with DuPont Electronics, ISO 9000 was implemented in over 30 locations throughout the world. This process was the first global implementation of ISO 9000 and included over 8,000 persons in 15 countries.

Dr. Eugenia K. Brumm, Ph.D., CRM, president of the Quality Records Institute, is an independent consultant and trainer with over 17 years of experience in records management, document control, and quality systems. She provides consulting, training seminars, and on-site workshops in all areas of records management, specializing in compliance and regulatory records requirements. Through her international seminars and workshops, she has trained over 3,000 people in various aspects of records management techniques and principles, including records inventories, retention schedule development, records protection methods, filing systems, forms management, feasibility analysis, records costing issues, optical disk technology, micrographics, and inactive records storage.

Dr. Brumm is a former professor from the University of Texas at Austin, where she developed a graduate-level records management program. She has also worked for the nuclear industry, where she developed the records management program, the document control system, and the product certification process. She has designed and implemented successful records management programs within quality environments for private business and industry as well as for local and state government.

She is frequently requested as a speaker in ISO 9000 and records management.

Dr. Brumm holds a Ph.D. in library and information science from the University of Illinois as well as an M.S. in library science and an M.A. in Slavic languages from the same University. She is author of the book *Managing Records for ISO 9000 Compliance,* published by ASQC Quality Press. She has written for *ARMA Records Management Quarterly, Quality Progress, Records and Retrieval Report, Information Management Review, Document Management,* and other publications. She is a member of ASQC (American Society for Quality Control) and ARMA (Association for Record Managers and Administrators) and is chairperson on the ARMA International ISO Committee.

Devan R. Capur is the founder and president of 3C Technologies, Inc., a company dedicated to helping organizations achieve performance excellence by implementing customer focused, globally competitive change. 3C Technologies, Inc.'s key business thrusts are providing facilitation, training, and consulting services for ISO 9000, QS-9000, business process reengineering, and Total Quality Management. 3C is a Microsoft solution provider.

Mr. Capur's experience spans both the private and public sectors. He has been a featured speaker and presenter at national and international conferences. He has also contributed articles and written papers on ISO and QS-9000. He was recently invited to Europe and Asia to help companies with QS-9000 implementation. His customers include Registrars, OEMs, and small to mid-sized companies.

Mr. Capur holds an M.S. in industrial engineering and operations research. He is a QS-9000 lead assessor, is listed in the *Who's Who of Science and Technology,* and is a senior member of the Society of Manufacturing Engineers, Institute of Industrial Engineers, and ASQC.

3C Technologies, Inc. was rated number one in the 1995 worldwide Automotive Industry Action Group survey for QS-9000 trainers offering national and international training.

April A. Cormaci has spent more than 20 years as a writer and manager in the technical documentation field, including serving as a publications consultant for several books on quality management. She has more than six years' experience in customizing and delivering quality training to employees in a variety of services. Ms. Cormaci has consulted with interdepartmental teams managing and improving service processes and with executive management teams managing businesses. She has coauthored several articles and books, the most recent related to implementing a quality system based on the ISO 9000 quality assurance standards. She is currently the executive vice president of Service Process Consulting, Inc. Ms. Cormaci holds a B.A. in English from Duquesne University.

William E. (Bill) Cox is a widely recognized expert in the areas of ISO 9000 and Total Quality Management. He has assisted dozens of companies in the design and implementation of quality management systems based on the ISO 9000 series of standards. Many of his clients have successfully achieved certification, while others are at various stages in the process. His flexible approach is designed to tailor a system to fit a company's existing culture and quality improvement process. He is a quality systems auditor certified by the Registrar Accreditation Board and the ASQC. He is an author of the ASQC's *ISO 9000 Guidelines for the Chemical and Process Industries,* 2nd edition.

In addition to ISO 9000, Mr. Cox has an extensive background in developing and implementing broader Total Quality Management processes, including application of statistical process control. His expertise has been recognized by his selection to the 1993 and 1994 Board of Examiners for the Texas Quality Award, which is based on the Malcolm Baldrige National Quality Award. In 1975, Mr. Cox was awarded a B.S. degree, *magna cum laude,* in chemical engineering from the University of Tennessee. He held a variety of engineering and management positions during 16 years with Exxon Chemical Co., until he founded TQM Consulting in 1991. His professional activities have been devoted to quality management since 1985.

Taz Daughtrey is currently manager of educational services at the Naval Nuclear Fuel Division of Babcock & Wilcox in Lynchburg, Virginia. His previous work as a scientific programmer, a programming supervisor, and a software quality engineer focused on software-based systems in highly regulated environments.

Mr. Daughtrey has a dozen years' involvement in the development of software quality and engineer-

ing standards under the sponsorship of various professional societies. He has spoken and published extensively and has been a consultant to national laboratories, the U.S. Nuclear Regulatory Commission, the Canadian government, and the Japanese space agency.

A founding officer of the Software Division of the American Society for Quality Control, he served as its chair from 1990 to 1992. He has been active with both the Software Quality Systems Registration and Software Industry Quality Forum efforts. Currently review editor for the *Software Quality Journal,* he recently contributed a chapter on software for "Quality Planning, Control and Improvement in Research and Development."

William J. Deibler II is a partner and founder of Software Systems Quality Consulting (SSQC) San Jose California. Since 1990, SSQC has assisted software and hardware developers, manufacturers, and service providers in five related areas :

- Software engineering process and life-cycle definition.

- Education and training.

- Software quality assurance and testing.

- Business process reengineering and benchmarking.

- ISO 9000 registration.

Mr. Deibler has an M.Sc. in computer science and 18 years' experience in the computer industry, primarily in the areas of software and systems quality assurance, validation, and development. Since 1986, he has held engineering management positions at National Semiconductor Corporation (Systems Division) and International Computers Limited (ICL). As systems quality manager, he designed and implemented a quality system for ICL's Santa Clara Hardware and Software Development Center. The quality system was registered to ISO 9001 in 1990, one of the earliest registrations in the United States.

During his affiliation with SSQC, Mr. Deibler has developed and published numerous courses, auditing tools, and articles on interpreting and applying the ISO 9000 standards and guidelines. His courses have been sponsored for their members by professional associations, including the ASQC, Semiconductor Equipment and Materials International (SEMI), and Software Engineering Forum for

Training at California State University Long Beach, UC Berkeley, and UC Santa Cruz.

Bill regularly performs independent assessments and surveys of organizations and works with companies to assist them in their engineering practices and businesses process definition. He was an active member of the Software Quality System Registration (SQSR) committee of the ANSI/ASQC Registrar Accreditation Board (RAB). This committee developed the U.S. equivalent of the UK TickIT program and guideline documentation. He was principal author and editor of *A Guide to Software Quality System Registration under ISO 9001*. He is an active U.S. TAG member in the ISO/IEC JTC1 SC7-Software Engineering Standards subcommittee.

Ian Durand is president and principal consultant of Service Process Consulting, Inc., has been actively involved in ISO/TC 176 since 1986 and holds major leadership positions at the national and international levels. Mr. Durand has served as a senior examiner for the Malcolm Baldrige National Quality Award, specializing in service organizations. Prior to forming Service Process Consulting, Mr. Durand held a variety of technical and management positions during his 32 years at AT&T Bell Laboratories, including managing departments responsible for training in quality management and product realization, quality of long-distance services, and field engineering. While in the Quality Assurance Center, he developed a systematic methodology for managing and improving the performance of business processes that was standardized across AT&T.

Bohdan Dyczkowsky, director of MOYE Company Limited, is a native of Toronto, Ontario. He graduated from the University of Toronto with a BASC in mechanical engineering. He has broad manufacturing experience in construction, steel fabricating, nuclear piping, and electronics. Major career accomplishments include serving as the secretary of the Canadian Delegation to the ISO Technical Committee 176 during the development of the ISO 9000 series. Mr. Dyczkowsky is a registered professional engineer, a registered lead auditor with the IQA in England and the RAB in the United States, and an active member of the ASQC.

Dave Fleischli is the executive director of consulting services at Alamo Learning Systems and is responsible for development, delivery, and management of the quality system consulting practice. He refined and documented the FAST TRACK process to improve the level of service to customers. He successfully implemented ISO 9000 for three client companies in 11 months and managed the implementation of ISO and QS-9000 in 17 client companies. He has developed curriculum and delivered internal auditor and ISO implementation classes. He managed and trained Alamo Directors in providing consulting services. He has extensive experience implementing ISO and QS-9000 in a variety of industries including electronics, software development, automotive, defense and printing. His major strengths are in project management, process modeling and meeting facilitation. He has formal training in lead TickIT auditor, QS-9000 implementation, meeting facilitation and process modeling. He has an economics degree from California State University in San Jose, California and an A.A. in marketing from Foothill College in Los Altos, California.

Marc E. Gold is a founding partner of Manko, Gold & Katcher, a 22 lawyer firm, located in Bala Cynwyd, PA., that concentrates its practice exclusively in environmental law. Formerly, Mr. Gold served as a section chief in the legal branch of the United States Environmental Protection Agency, Region III, and was partner in the environmental department of a major Philadelphia law firm. In addition, Mr. Gold was an adjunct assistant professor at Temple University's School of Engineering, where he taught environmental regulation.

Mr. Gold has more than 20 years' experience in environmental law. His practice focuses on all aspects of environmental regulation and counseling covering hazardous waste, air and water pollution, and site remediation issues. Mr. Gold has been involved in several multifacility environmental audits and has participated in the development of corporate environmental policies and procedures, and has written and lectured extensively on environmental management systems including ISO 14000. In addition, Mr. Gold has handled the environmental aspects of major national and international corporate transactions. He has been listed in *The Best Lawyers in America* since 1989, and is also listed in *International Corporate Law's Guide to Environmental Law Experts*.

Mr. Gold currently serves on the board of directors of Big Brothers/Big Sisters of Philadelphia, and the Clean Air Council.

Mr. Gold is currently a member of the Natural Resources Section of the American Bar Association, and the Environmental, Mineral, and Natural Resources Section of the Pennsylvania Bar Association. In addition, he is a member of the Water Pollution Control Federation. Mr. Gold has served as vice chairman of the Zoning and Land Use Committee of the Pennsylvania Bar Association's Environmental Law Committee.

Mr. Gold receive his J.D. from Villanova University Law School and received his B.A. from the American University.

Rod Goult, president and CEO of the Victoria Group, has a broad range of experience in both manufacturing and service environments, although his specialties are software development and electronics design/manufacture. He is coauthor of the Internal Auditor and ISO 9000-3 course and principal author of the Lead Auditor program. Mr. Goult has worked with companies in the U.K. and Sweden on ISO 9001 certification and BABT approvals and has assisted many U.S. companies in achieving ISO 9001 certification as well. Since founding the Victoria Group in 1990, he has been a regular contributor to quality-related publications and a frequent speaker at conferences. He assisted with the RAB Auditor training as it evolved, served on the RAB-Software Quality Systems Certification Committee and is a member of the international conference on reciprocal auditor certification recognition. A senior member of the American Society for Quality Control, he is both an IRCA-registered lead assessor and an RAB-certified quality systems lead auditor.

Guy A. Hale is the founder and chairman of Alamo Learning Systems. He has developed, with the Alamo Staff, highly regarded management and employee training programs recognized nationally and internationally. In a recent study of the training industry's largest companies, Alamo was selected number one for product quality and customer service. It counts over 200 Fortune 500 companies as its clients.

Alamo's consulting practice is built around helping companies become registered in ISO 9000, QS-9000, and ISO 14000. Their Fast Track® ISO/QS 9000 Registration Program facilitates companies' registration in six months for ISO 9000/14000 and eight months for QS-9000. Alamo is the American

Electronics Association Preferred Provider. It is the largest company specializing in this technology.

Alamo is headquartered in Northern California and has offices in Detroit, New York City, Pittsburgh, St. Louis, Salt Lake City, Boston, and Chicago.

William Hooton is responsible for delivery of training and consulting services at Alamo Learning Systems. He works with clients implementing the Fast Track to ISO/QS-9000 program. Mr. Hooton served as a quality circle facilitator, product engineer, and draftsman at the Orscheln Company before becoming manager of training and development there. After several years in that position he worked as general manger of quality improvement. Before coming to Alamo, he served as director of quality at Dura Automotive Systems, Inc. He has more than 20 years of experience in various staff and management positions. His education and training includes Moberly Area Community College, the University of Missouri School of Engineering, Quality Circle Facilitator Training, Philip Crosby Quality College, ASQC Certified Quality Auditor, Missouri Quality Award Examiner, ISO 9000 Lead Assessor Training.

Greg Hutchins is a principal with Quality Plus Engineering, an engineering and management advisory firm in Portland, Oregon. He currently chairs ASQC's Education Division. He founded Greg's Outrageous Fortune Cookie Company. He wrote a column, "Hutchins on Jobs," for the *Oregonian* (480,000 circulation). He also wrote the following books:

> *ISO 9000 Registration in 10 Easy Steps,* John Wiley/Oliver Wright, 1994.
>
> *Taking Care of Business,* John Wiley/Oliver Wright, 1994.
>
> *ISO 9000,* John Wiley/Oliver Wright, 1993.
>
> *Quality Auditing,* Prentice Hall, 1992
>
> *Strategies for Purchasing Quality,* Dow Jones Irwin, 1991.
>
> *Introduction to Quality: Management, Assurance, and Control,* Macmillan, 1991.

Suzan Jackson is business development manager of Environmental Services for Excel Partnership, Inc., which provides management systems training and consulting services. Prior to joining Excel, Sue worked for DuPont for eight years in a variety of positions related to quality and environmental management. She is a member of the U.S. Technical Advisory Group (TAG) to TC 207, the ISO committee which is developing the 14000 series. Sue has authored numerous articles on ISO 9000 and ISO 14000 and has recently completed a book, *ISO 14001 Implementation Guide: Creating an Integrated Management System,* which will be published in Nov. 1996 by John Wiley & Sons.

James W. Kolka is an international legal consultant. He has a Ph.D. in political science/international affairs from the University of Kansas and a J.D. with a background in product liability and environmental law from the University of Wisconsin-Madison. He has published numerous articles and books, served as a full professor at several universities, served as vice president at Drake University and in the University System of Georgia, and held senior management positions in the University of Wisconsin System. He developed Wisconsin's statewide Adult Extended Degree Program and was instrumental in the creation of the Center for International Standards and Quality at Georgia Tech. He serves as executive in residence at the Ivan Allen College of Management and International Affairs at Georgia Tech.

Dr. Kolka has conducted seminars, consulted on product, services and environmental liability, product safety, EU product certification, strategic planning, and the integration of ISO 9000 quality assurance systems and ISO 14000 Environmental Management Systems with preventive law programs for the U.S. and the EU markets. Fluent in Spanish and conversant in German, Portuguese, and Italian, he has worked throughout the United States, Canada, Europe, and Latin America.

His most recent books are *European Community Product Liability and Product Safety Directives,* 1992, written with Gregory Scott; *EC Medical Directives Report,* 1992, written with Gregory Scott and David Link; two chapters of the *ISO 9000 Handbook,* 2nd edition, 1994; *The European Union Machinery Directive: Compliance Manual For Trade,* 1995, written with Bruce McIntosh; and *ISO 9000 for Lawyers,* 1996.

Brian P. Kujawa is a research librarian with extensive experience in legal and technical searches. He is an expert with international internet databases and is familiar with the European and FDA regulation process and the interrelationship among the regulations.

Mr. Kujawa has a bachelor's and master of library science from SUNY at Buffalo. He is also a licensed paralegal in the state of New York (Canisus College). He has received numerous awards and written many articles in the area of library research.

Peter Lake is president of the SRI Quality System Registrar Inc., a registrar accredited by RAB and RvA to certify companies to the international ISO 9000 quality system standards and to the automotive requirements, QS-9000. SRI was among the first five accredited registrars in North America.

Dr. Lake serves as vice president of the Independent Association of Accredited Registrars, IAAR, the North American trade association consisting of accredited registrars.

He is also the IAAR chairman—Auto Sector Group and is a member of, and public contact for, the International Automotive Sector Group, IASG, which develops and releases binding interpretation and changes to the QS-9000 automotive requirements of the Big Three. He is a metals industry representative to the USA TC 176 ISO Team.

In addition, Dr. Lake:

- Served as a Malcolm Baldrige senior examiner in 1991 and 1992.

- Was employed as National Steel's Corporate quality manager and earlier as director of National Steel's Detroit-based Product Application Center.

- Chaired the American Iron and Steel Institute's Committee on Quality.

- Chaired/directed the "Steel Industry Shared Audit Program", SISAP.

He is in his third year as a judge for the Greater Pittsburgh Total Quality Awards. Dr. Lake is also an RAB—certified lead auditor.

For the past 27 years he has served on the board of directors of Gorman-Rupp Company and ASE company located in Ohio.

Steven Levine, Ph.D., CIH, QSPA, began his career in industry where worked for eight years. He worked for Stauffer Chemical Company, Ford Motor Company, and Oil and Hazardous Materials (OHM) Company. He has been at the University of Michigan for 13 years and currently is professor of occupational and environmental health. In addition to his numerous publications, he is coeditor of the book *New Frontiers in Occupational Health: A Management Systams Approach and the ISO Model*. He is also a major contributor to the development of the American Industrial Hygiene Association's Occupational Health and Safety Management System guidance document. He is a member of the ISO 9000 and 14000 advisory councils of NSF International Strategic Registration Ltd., was a member of the U.S. Technical Advisory Group for ISO TC-207, was a member of the AIHA-ACGIH-AAIH-ABIH Unification Task Force, and is the director of the University of Michigan WHO Collaborating Center for Occupational Health. He has been honored with the AIHA-ACGIH Drum Buster Award, and the AIHA John White Award. He is certified by the ABIH in both comprehensive and chemical practice, and by the IRCA against the ISO 10011.2 requirements.

Donald W. Marquardt is the leader of the U.S. delegation to TC176 and is chairman of the U.S. Technical Advisory Group to TC176. He has been active in TC176 since its initial meeting in 1980. He has served in a variety of international leadership roles in TC176, currently as a member of the Chairman's Advisory Group. He has been convener of several working groups, including those that prepared ISO 9000:1987, ISO 9000-1:1994, and ISO 9000-2:1993. Mr. Marquardt was chair of the international Task Force that prepared the Vision 2000 report establishing the strategic intent for implementation and revision of the ISO 9000 standards in the global marketplace of the 1990s. He also chaired the TC176 Strategic Planning Advisory Group subsequent to the Vision 2000 Task Force.

Mr. Marquardt has served since 1990 on the board of directors of the Registrar Accreditation Board, which operates in partnership with ANSI, the American National Accreditation Program for Registrars of Quality Systems.

He is president of Donald W. Marquardt and Associates, established in 1991, which provides consulting and training in the fields of quality management, quality assurance, ISO 9000 standards, strategic planning, applied statistics, and organizational change. Prior to his current business, his career of 39 years with the DuPont Company included assignments in research, management of engineering services, and leadership of the DuPont Quality Management and Technology Center, which he organized.

He is a former president of the American Statistical Association (ASA) and a former senior examiner for the Malcolm Baldrige Award. He is a fellow of the American Association for the Advancement of Science (AAAS), the ASA, and the ASQC and is an elected member of the International Statistical Institute. His awards include the ANSI Meritorious Service Award, the ASA Founders Award, and the ASQC Shewhart Medal.

Jack McVaugh is a chemical engineer with a master's degree in environmental engineering from the University of Delaware, a professional engineer, and a registered environmental manager. After graduate school, he provided process design, start-up, and troubleshooting services while solving industrial wastewater treatment problems on behalf of Envirex, an equipment manufacturer. At Miller Brewing Co., he was introduced to compliance management issues in air permitting and the emerging RCRA hazardous waste regulations.

The combination of process engineering and regulatory experience served him well as manager of environmental affairs for Akzo Chemicals Inc. There, he managed a group of in-house consultants who, among other accomplishments, developed a highly successful waste elimination program, carried on a vigorous environmental auditing activity, managed environmental control projects in the United States, Canada, and Brazil, and obtained the first RCRA Part B Incinerator Permit in EPA Region V. Later at Akzo, he ran a small polymers marketing business, purchased chemical raw materials, and managed purchasing, logistics, quality assurance and customer services functions for the Catalyst Business Unit. In 1994, he formed Environmental Technology & Management, offering environmental management consulting services to industry. The firm's mission is to use industrial experience in environmental, quality, and operations management to help companies gain control of their environmental affairs. He is also involved with the development of the ISO 14000 Series of International Environmental Standards through membership on the U.S. Technical Advisory Group to ISO Technical Committee 207.

Mark Morrow is executive editor at Irwin Professional Publishing in Fairfax, VA. In 1991, he founded the newsletter *Quality Systems Update*, now one of the most respected publications in the industry, covering ISO 9000 and developments in related standards, including ISO 14000 and QS-9000.

Mr. Morrow has more than 20 years' reporting and editing experience, which includes assignments covering Capitol Hill, the Department of Defense, and other federal agencies. He is a member of the American Society for Quality Control Education Division and serves on the ASQC Education and Training Board. He also serves on the advisory board of the EC 92 Institute. Mr. Morrow has also worked as a professional photographer, completing assignments for magazines such as *Esquire, People,* and *Fortune.*

In 1987, Mr. Morrow published a book of portraits and essays of southern writers (*Images of the Souther Writer,* University of Georgia Press), which received widespread national critical acclaim.

David Middleton is president of Excel Partnership, Inc., an international training and consulting organization. He is responsible for managing EXCEL's activities in the United States. Mr. Middleton's experience in quality management consultancy and in the application of ISO 9000 extends over 12 years, particularly in the textile, packaging, apparel, transport, automotive, and process industries. He guided a wide cross section of industrial and service companies to registration including a number of organizations first in their field. His approach to implementing quality systems is supported by earlier experience in production and commercial management in the Middle East, Far East, the United States and Europe.

Paul A. Nee is the manager of quality at DORMA Door Controls Inc. of Reamstown, Pennsylvania. He is a graduate of Williamsport Technical Institute in tool design technology and Elizabethtown College in business management. His work experience includes 30 years in inspection, quality control, quality assurance, and quality management. Eighteen years of this experience has been at DORMA Door Controls Inc. For the past five years he has been responsible for the development, implementation, and maintenance of the ISO 9001 Quality Assurance System. In addition, he is the facilitator for total quality management at DORMA. During the past two years, he has presented 35 ISO seminars to construction-related industries and other professional groups.

Articles Published :

"The International Expectation—ISO 9000." *Doors and Hardware,* March 1993.

"Making the Grade." *The Construction Specifier,* June 1993.

"Commitment Makes the Difference." *Doors and Hardware,* Oct. 1993.

"ISO 9000 Revisited." *The Construction Specifier,* May 1994.

"Direct Route to Quality." *Doors and Hardware,* Sept. 1994.

"Heading in the Right Direction." *Doors and Hardware,* Feb. 1995.

"Accelerating Quality." *Doors and Hardware,* May 1995

Books published :

ISO 9000 In Construction, John Wiley and Sons, April 1996.

Glenn K. Nestel is a vice president and senior manager in the Management Systems Division of Roy F. Weston, Inc., where he is responsible for leading business development and overseeing the management consulting practice. Mr. Nestel brings 27 years of environmental, health, and safety management and business management experience to this assignment. He has held a variety of senior management positions in several multinational companies, with corporatewide responsibilities for developing policies, requirements, programs, management practices, and tools to support company business operations and auditing, as well as performance measurement activities. This experience includes several assignments for a Fortune 100 company as corporate occupational safety and health manager and manager of a newly formed business unit set up to manage corporatewide site remediation liabilities.

Mr. Nestel's consulting practice is focused on helping clients develop and implement management systems that integrate environmental protection and occupational safety and health into business operations. He has pioneered the development of assessment and benchmarking methodologies for evaluating an organization's environmental, health, and safety management systems and for developing and implementing long-term improvement strategies.

He has extensive experience developing the programmatic and performance support tools for implementing these strategies, including education/training and information management systems. He also has extensive experience in ISO 9000, Total Quality Management, and continuous improvement processes in both business and environmental applications. Mr. Nestel has extensive experience applying environmental management systems and codes, including the emerging ISO 14000 standards. He is currently working with the water utility industry as lead investigator to develop a fully realized accreditation program similar to that used by the hospital/healthcare industry. Malcolm Baldrige elements have been incorporated into this performance model.

James P. O'Neil is president of Inchcape Systems Registration/Intertek Services, one of the leading internationally recognized registrars certifying quality (ISO 9000), automotive (QS-9000), and environmental (ISO 14001) management systems. Prior to this, Mr. O'Neil founded and was president of National Quality Assurance, USA and previously held the position of corporate director of quality for National Technical Systems, one of the largest independent testing facilities in the United States. Mr. O'Neil has experience in nuclear power regulatory interfaces, manufacturing assessments, design analysis, and compliances issues, and lectures throughout the world on quality, world trade, and management issues. He holds IRCA registered lead assessor certification and is on the board of directors for the International Standards Institution.

Elizabeth A. Potts is a chemical engineering graduate from the University of Illinois and holds a master of business administration from Ashland University. She is currently director of quality for Ashland Chemical Company. She is the former president of ABS Quality Evaluations, an accredited third-party certifier of management systems to ISO 9000, QS-9000 and ISO 14000. Previously, she was employed as the quality manager for the American Gas Association Laboratories where she was responsible for product certification inspections, ISO 9000 registration and internal quality program development and implementation. She has also been employed by Babcock & Wilcox as a quality control manager responsible for development and implementation of quality programs for nuclear reactor components and other defense related components.

Ms. Potts is a member of the U.S. Technical Advisory Group to ISO Technical Committee 176 on quality assurance and quality management, and is a member of the board of directors of the Registrar Accreditation Board (RAB). She is also a registered ASQC certified quality auditor and an IQA registered assessor.

Roger C. Pratt is manager of the quality training Office at Pacific Northwest Laboratory in Richland, Washington. The Laboratory is operated by the Battelle Memorial Institute for the U.S. Department of Energy.

Charles Redinger, MPA, CIH is a University of Michigan doctoral student who is conducting research in public policy issues related to occupational health and safety management systems. He has a master's degree in public policy from the University of Colorado, is a member of the public policy honor society Phi Alpha Alpha, and is a Kemper Fellow in Public Health.

He is coeditor of the book *New Frontiers in Occupational Health: A Management Systems Approach and the ISO Model.* He is also a contributor to the American Industrial Hygiene Association's Occupational Health and Safety Management System guidance document.

As a cofounder and officer of a national consulting firm, he has over 12 years of industrial hygiene and environmental consulting experience in both the public and private sectors. He sat for two years on the governor of California's Asbestos Task Force and is currently assisting Federal OSHA in the analysis of several policy approaches related to occupational health and safety management systems. He is certified by the ABIH in comprehensive practice and is a registered environmental assessor in California.

Kenneth Snowden Stephens, Ph.D., is associate professor of industrial engineering technology at the Southern College of Technology in Marietta, Georgia. Prior to that he served UNIDO for 18 years in such positions as senior industrial development officer, quality, country director and senior industrial development field adviser in China and Pakistan, and as standardization, certification and quality expert, consultant and chief technical adviser on assignments in Ethiopia, Mauritius, Turkey, Nigeria and Thailand. He is a fellow of ASQC, an academician of IAQ, a member of EOQ, ASTM, and full member of Sigma Xi. Some of his recognitions and honors include: certificate in quality engineering from ASQC, academician of the International Academy for Quality, registered professional engineer—state of Pennsylvania. He received ASQC's E.L. Grant Award, AIIE's Book-of-the-Year Award (with Wadsworth and Godfrey) for *Modern Methods for Quality Control and Improvement,* John Wiley & Sons, Publishers, 1986, Annual AIIE Conference, Washington, DC, June 1987. He also is the recipient of the ASQC Metropolitan Section's Ellis R. Ott Award to recognize his outstanding contribution in joining quality control technology and management in such a way that each supports and enhances the other, 1990 Metropolitan Section Annual All Day Conference on Quality Control and Statistics in Industry, September 14, 1990. He also received ASQC's E. Jack Lancaster Award, 1994 Annual Quality Congress, Las Vegas, May, 1994. Dr. Stephens received a B.S. in industrial science from LeTourneau Technical Institute. He earned an M.S. and a Ph.D. in applied and mathematical statistics from Rutgers University, the State University of New Jersey (under direction of Harold F. Dodge & Ellis R. Ott, respectively).

Joseph Tiratto (P.E.) is an international consultant on quality systems, a registered lead auditor with IQA (IRCA) and RAB, and an approved lead auditor Instructor. He holds bachelor degrees in naval architecture and marine engineering and mechanical engineering, and holds a master's degree in quality management. He is a registered professional engineer (PE), a licensed marine engineer, a chartered engineer (CENG) (United Kingdom) and a registered European engineer (EUR.ING)(EC & EFTA countries). He is a member of The American Society for Quality Control, The American Society of Mechanical Engineers, The Society of Naval Architects & Marine Engineers, and The Institute of Marine Engineers (U.K.). He is a member of the U.S. Technical Advisory Groups (TAGs) to ISO/TC 176, ISO/TC 69, IEC/TC 56, and ASQC/ASC Z-1 committee on Quality Assurance, He is a past member and secretary of the board of directors of the Registrar Accreditation Board (RAB).

Mr. Tiratto has over 35 years of management experience in engineering, production, operation, quality control, quality assurance, and auditing. He has developed and implemented quality systems and has conducted audits to ISO 9000 quality assurance standards in 17 countries worldwide. He directed the development of quality systems for a registrar that met the requirements of the RAB in the United States and the RvA in Netherlands. He has published

papers, written chapters in handbooks, lectured on ISO 9000 standards and quality system registration, and teaches ISO 9000 lead auditor courses.

Peter S. Unger is president of the American Association for Laboratory Accreditation (A2LA). He served as vice president of the Association for 10 years. A2LA is a nonprofit, membership society administering a broad spectrum, nationwide laboratory accreditation system.

Prior to that, Mr. Unger served as associate manager of laboratory accreditation at the National Bureau of Standards (now the National Institute of Standards and Technology). He has been involved with laboratory accreditation on the national level since 1978. He began his career with the U.S. General Services Administration on various assignments involving standards writing, testing, quality control, and procurement research.

He is chairman of ASTM E-36 on Conformity Assessment and the convener of the working group of the International Laboratory Accreditation Co-operation (ILAC) on Assessor Qualifications & Competencies.

Mr. Unger has a bachelor's degree in systems engineering from Princeton University and a master's in environmental management from George Washington University. He is an RAB-certified lead auditor (QO2335) and IRCA certified lead assessor (A005890).

Bud Weightman is president of Qualified Specialists, Inc. (QSI) and an international consultant who has extensive experience designing, implementing and assessing quality management systems, particularly ISO 9000 and similar standards. Mr. Weightman is a registered lead assessor (UK), IRCA/RAB lead assessor, Texas quality award examiner (1993), Houston Awards for Quality Assessor (1993), and an ASQC CQA. He has personally assisted more than 50 companies worldwide achieve third party quality system/product registration/licensing. In addition, he is an active member of the U.S. Technical Advisory Group to TC 176. Mr. Weightman has had several of his articles on ISO 9000 published in national and international trade magazines. He has over 23 years' experience in quality systems, and he engages in public speaking and training on the subject of auditing, quality, ISO 9000 and registration.

Standards and Directives

This appendix contains sources for information about EU standards and directives, including the following information:

- ISO 9000 and ANSI/ISO/ASQC Q9000 standards
- EU directives
- Additional EU standards
- EC 1992 single market information
- Electronic information

ISO 9000 STANDARDS

To order the ISO 9000 Standards, call ASQC or ANSI at the addresses and phone numbers listed below:

American Society for Quality Control (ASQC)

611 East Wisconsin Avenue, P.O. Box 3005
Milwaukee, WI 53201-3005, U.S.A.
Tel: 800-248-1946; Fax: 414-272-8575
E-mail Address: http://www.asqc.org

The ANSI/ISO/ASQC Q9000 Series *Complete Set*
    List Price: $73.00, Member Price: $66.00
    Item: T3000

ANSI/ISO/ASQC Q9000-1-1994 *Quality Management
    and Quality Assurance Standard—Guidelines for
    Selection and Use*
    List price: $16.50, Member Price: $14.75
    Item: T9000

ANSI/ISO/ASQC Q9001-1994 *Quality Systems—
    Model for Quality Assurance in Design,
    Development, Production, Installation, and
    Servicing*
    List Price: $20.00, Member Price: $18.00
    Item: T9001

ANSI/ISO/ASQC Q9002-1994 *Quality Systems—
    Model for Quality Assurance in Production, Instal-
    lation and Servicing*
    List Price: $17.50, Member Price: $15.75
    Item: T9002

ANSI/ISO/ASQC Q9003-1994 *Quality Systems—
    Model for Quality Assurance in Final Inspection
    and Test*
    List Price: $15.00, Member Price: $13.50
    Item: T9003

ANSI/ISO/ASQC Q9004-1-1994 *Quality Management
    and Quality System Elements—Guidelines*
    List Price: $27.50, Member Price: $24.75
    Item: T9004

ANSI/ISO/ASQC Q9004-2-1991 *Quality Management
    and Quality System Elements—Guidelines for
    Services*
    List Price: $30.00, Member Price: $27.00
    Item: T202

ANSI/ISO/ASQC Q9004-3-1993 *Quality Management
    and Quality System Elements—Guidelines for Pro-
    cessed Materials*
    List Price: $30.00, Member Price: $27.00
    Item: T200

ANSI/ISO/ASQC Q9004-4-1993 *Quality Management
    and Quality System Elements—Guidelines for
    Quality Improvement*
    List Price: $30.00, Member Price: $27.00
    Item: T204

ANSI/ISO/ASQC A8402-1994 *Quality Management
    and Quality Assurance—Vocabulary* (Revision and
    Redesignation of ANSI/ASQC A3-1987)

    List Price: $25.00, Member Price: $22.50
    Item: T54

ANSI/ISO/ASQC Q9000-2-1993 *Quality Management
    and Quality Assurance Standards—Generic Guide-
    lines for the Application of ANSI/ISO/ASQC 9001,
    9002, and 9003*
    List Price: $30.00 Member Price: $27.00
    Item: T201

ANSI/ISO/ASQC Q9000-3-1991 *Quality Management
    and Quality Assurance Standards—Guidelines for
    the Application of ANSI/ISO/ASQC 9001 to the De-
    velopment, Supply, and Maintenance of Software*
    List Price: $28.00, Member Price: $25.00
    Item: T203

ANSI/ISO/ASQC Q10007-1995 *Quality Manage-
    ment—Guidelines for Configuration Management*
    List Price: $32.00, Member Price: $29.00
    Item: T58

ANSI/ISO/ASQC Q10013-1995 *Guidelines for Quality
    Manuals*
    List Price: $32.00, Member Price: $29.00
    Item: T205

ANSI/ISO/ASQC Q10011-1994 *Series (American Ver-
    sion of the ISO 10011 series) Guidelines for Audit-
    ing Quality Systems*
    List Price: $35.00, Member Price: $31.50
    Item: T10011

ISO 10012-1-1992 *Quality Assurance Requirements for
    Measuring Equipment—Part 1: Metrological Con-
    firmation System for Measuring Equipment*
    Price: $48.00
    Item: T211

ANSI/ASQC D1160-1995 *Formal Design Review*
    This standard can provide additional assistance in
    satisfying clause 4.4.6 of the ANSI/ISO ASQC
    Q9001-1994.
    List Price: $17.00, Member Price: $15.00
    Item: T218

ANSI/ASQC E4-1994 *Specifications and Guidelines
    for Quality Systems for Environmental Data Col-
    lection and Environmental Technology Programs*
    List Price: $25.00, Member Price: $22.00
    Item: T55

IEC/ISO Guide 25-1990 *General Requirements
    for the Competence of Calibration and Testing
    Laboratories*
    Price: $33.00
    Item: T25

ANSI/ASQC Q2-1991 *Quality Management System and Elements for Laboratories—Guidelines*
List Price: $12.95, Member Price: $11.95
Item: T48

American National Standards Institute (ANSI)

Attention: Customer Services
11 West 42nd Street
New York, NY 10036, U.S.A.
Tel: 212-642-4900; Fax: 212-302-1286

ANSI Products:

ISO 9000 Compendium (6th Edition) *International Standards for Quality Management*
(Contains: ISO 8402, 9000-1, 9000-2, 9000-3, 9000-4, 9001, 9002, 9003, 9004-1, 9004-2, 9004-3, 9004-4, 10005, 10007, 10011-1, 10011-2, 10011-3, 10012-1, 10013, and DIS 10012-2). Replaces ISO 9000 Compendium (5th Edition)
Price: $250.00

ISO 9000-1:1994 *Quality Management and Quality Assurance Standards—Part 1: Guidelines for Selection and Use.* Replaces ISO 9000:1987
Price: $56.00

ISO 9001:1994 *Quality Systems—Model for Quality Assurance in Design, Development, Production, Installation, and Servicing.* Replaces ISO 9001:1987
Price: $44.00

ISO 9002:1994 *Quality Systems—Model for Quality Assurance in Production, Installation, and Servicing.* Replaces ISO 9002:1987
Price: $40.00

ISO 9003:1994 *Quality Systems—Model for Quality Assurance in Final Inspection and Test.* Replaces ISO 9003: 1987
Price: $37.00

ISO 9004:1994 *Quality Management and Quality System Elements—Part 1: Guidelines.* Replaces ISO 9004:1987
Price: $65.00

ISO 9000-3: 1991 *Quality Management and Quality Assurance Standards—Part 3: Guidelines for the Application of ISO 9001 to the Development, Supply, and Maintenance of Software*
Price: $52.00

ISO 9004-2: 1991 *Quality Management and Quality System Elements—Part 2: Guidelines for Services*
Price: $56.00

ISO DIS 9004-2 *Quality Management and Quality System Elements—Part 2: Guidelines for Services*
Price: $70.00

ISO 9001:1994 *Quality Systems—Model for Quality Assurance in Design, Development, Production, Installation, and Servicing.* Replaces ISO 9001:1987
Price: $ 44.00

ISO 10012-1: 1992 *Quality Assurance Requirements for Measuring Equipment—Part 1: Metrological Confirmation System for Measuring Equipment.* Replaces ISO DIS 10012-1:1990
Price: $48.00

ISO 10011-1:1990 *Guidelines for Auditing Quality Systems—Part 1: Auditing*
Price: $37.00

ISO 10011-2: 1991 *Guidelines for Auditing Quality Systems—Part 2: Qualification Criteria for Quality Systems Auditors*
Price: $34.00

ISO 10011-3:1991 *Guidelines for Auditing Quality Systems—Part 3: Management of Audit Programmes*
Price: $28.00

Where to Get EU Directives

European Union Depository Libraries in the United States

The European Union has established a network of libraries to give Americans access to all its official publications. All depositories automatically receive, free of charge, one copy of most EU periodical and monograph publications. The collections are available to the public during the library's regular working hours, free of charge and without any conditions. Many of the libraries also offer interlibrary loan services. The official publications contain information ranging from policy developments in all sectors of EU activity to statistics on trade demographics, agriculture, transport, energy, and economic indicators.

Of particular importance is the *Official Journal of the European Communities,* the equivalent of the Federal Register. The "C" section of the Official Journal includes proposed legislation and other important notices; the "L" section has the final texts of legislation. Annual indices and the *Directory of Community Legislation in Force* provide subject and numeric reference access to legislation. Depository collections also contain the legislative proposals and communications of the Commission in their original "COM" document form, as well as reports and debates of the European Parliament, opinions of the Economic and Social Committee, and decisions of the Court of Justice.

Subscriptions to the *Official Journal* are available in the United States and Canada through UNIPUB, the North American representative of the European Union for all publications. The 1996 annual subscription cost is $1,165.00.

Table B.1   EU Product Directives

Referencing ISO 9000 (EN ISO 9000) Standards as a Component of the Product Certification Process	
ISO 9000 (EN ISO 9000) Reference	
Adopted Directives	
Construction Products	9002 or 9003
Gas Appliances	9002
Personal Protective Equipment	9002 or 9003
Non-Automatic Weighing Instruments	9002
Active Implantable Medical Devices	9001 or 9002
Telecommunications Terminal Equipment	9001 or 9002
Pressure Equipment	9001 or 9002
Non-Automatic Weighting Instruments	9001 or 9002
Medical Devices	9001 or 9002
Elevators (Lifts)	9001 or 9002
Equipment for Use in Potentially Explosive Atmospheres (mines, surface extractions)	9001 or 9002
Recreational Craft	9001 or 9002
Planned Directives	
Cable Ways Equipment	9001
Amusement Park and Fairground Equipment	9003
Fasteners	9002

Table B.2 EU Legal Requirements for Industrial Equipment and Consumer Goods

Directive	Citation Number	Official Journal	Date of OJ	Current Status	Date of Implementation
(List Current through 15 February 1996) Bibliography Prepared By: Brian P. Kujawa, Research Librarian, International Quality Press					
General Extension of information procedures on standards and technical rules	83/189/EEC	L 100	4/26/83	Adopted	
(1st Amendment of Directive 83/189/EEC)	88/182/EEC	L 81	3/26/88	Adopted	1/1/89
(2nd Amendment of Directive 83/189/EEC)	94/10/EC	L 100	4/19/94	Adopted	
(Amendment— Updating List of National Standards Bodies)	Decision 96/139/EC	L 32	2/10/96	Adopted	Immed.
Modules of Conformity Assessment	Decision 93/465/EEC	L 220	8/30/93	Adopted	Immed.
Good Laboratory Practice	88/320/EEC	L 320	6/9/88	Adopted	Immed.
(1st Amendment of 88/320/EEC)	87/18/EEC	L 18	12/18/86		
(2nd Amendment of 88/320/EEC)	90/18/EC	L 18	12/18/89		
Appliances Appliances burning gaseous fuels	90/396/EEC	L 196	7/7/90	Adopted	1/1/89
(Amendment— CE Marking)	93/68/EEC	L 220	8/30/93	Adopted	1/1/95
Notified Bodies:	95/C 280/01	C 280	10/25/95	Communication	
Standards:	94/C 334/08	C 334	11/30/94	Communications	
	95/C 187	C 187	7/27/95		
Civil Aviation Council Regulation on technical requirements in civil aviation	No. (EEC) 3922/91	L 373	12/31/91	Adopted	1/1/92
Construction Products	89/106/EEC	L 40	2/11/89	Adopted	6/28/91
(Amendment— CE Marking)	93/68/EEC	L 220	8/30/93	Adopted	1/1/95
Notified Bodies /Interpretative Documents for the implementation of Directive 89/106/EEC	95/C 280/01	C 280	10/25/95	Communication	
	94/C 62/01	C 62	2/28/94	Communication	

Table B.2   *(continued)*

Decision Implementing Art. 20 of Directive 89/106/EEC	Decision 94/611/EC	L 241	9/16/94	Adopted	
Decision Implementing Art. 20 (2) of Directive 89/106/EEC	Decision 95/204/EC	L 129	6/2/95	Adopted	
(Corrigendum)	—	L 217	9/11/95		
Common Procedural Rules for European Technical Approval	Decision 94/23/EC	L 107	1/20/94	Adopted	
Notified Bodies:	94/C 206/04	C 206	7/26/94	Communication	
	95/C 211	C 211	8/15/95		
Electrical Equipment					
Electrical equipment for use in potentially explosive atmospheres (supersedes Directives 76/117/EEC, 79/196/EEC, and 82/130/EEC)	94/9/EC	L 100	4/19/94	Adopted	7/1/2003
Notified Bodies:					
Directive 76/117/EEC	94/C 80/06	C 80	3/17/94	Communications	
	95/C 215/02	C 215	8/19/95		
Standards:	Commission				
(stated as Annexes to	Directives				
Directives 79/196/EEC				Adopted	
and 82/130/EEC as	94/26/EC	L 157	6/24/94		7/14/94
amended to adapt for technical progress)	94/44/EC	L 248	9/23/94		10/13/94
Low Voltage Equipment	73/23/EEC	L 77	3/26/73	Adopted	8/75
Electrical appliances: standards					
(Amendment— CE Marking)	93/68/EEC	L 220	8/30/93	Adopted	1/1/95
Notified Bodies and Symbols:					
Revised Lists:	92/C 210/01	C 210	8/15/92	Communications	
Amendment—	95/C 214/02	C 214	8/18/95		
EN Standards and	92/C 210/01	C 210	8/15/92		
HDs:	93/C 18/04	C 18	1/23/93		
	93/C 319/02	C 319	11/26/93	Communications	
	94/C 169/04	C 169	6/22/94		
	94/C 199/03	C 199	7/21/94		

Table B.2 *(continued)*

Electromagnetic Compatibility Radio Interference (L 127)	**89/336/EEC**	**L 139**	**5/23/89**	**Adopted**	**7/1/88**
(Amendment—Transition Period)	92/31/EEC	L 126	5/12/92	Adopted	10/28/92
(Amendment—CE Marking)	93/68/EEC	L 220	8/30/93	Adopted	1/1/95
Notified Bodies:	95/C 280/01	C 280	10/25/95	Communication	
Standards:	92/C 44/10	C 44	2/19/92	Communications	
	92/C 92/02	C 92	4/10/92		
	94/C 49/03	C 49	2/17/94		
	95/C 241/	C 241	9/16/95		
	95/C 325/05	C 325	12/6/95		
Instruments Non-Automatic Weighing Machines	90/384/EEC	L 189	7/20/90	Adopted	7/1/92
(Amendment—CE Marking)	93/68/EEC	L 220	8/30/93	Adopted	1/1/95
Notified Bodies:	95/C 280/01	C 280	10/25/95	Communication	
Standards:	93/C 104/04	C 104	5/7/93	Communications	
	94/C 153/09	C 153	6/4/94		
Measuring Instruments and means of metrologic control	**71/316/EEC**	**L 202 [Eng. Spec. Ed.—1971 (II)]**	**9/6/71** (Dec. 1972)	**Adopted**	**3/26/73**
(Amendments—Ascension of United Kingdom and Ireland)	72/427/EEC	L 332 [Eng. Spec. Ed— 1972 (28-30 Dec.)]	12/28/72	Adopted	1/1/73
(Amendment—Technical Update)	83/575/EEC	L 332	11/28/83	Adopted	1/1/85
(Amendment—Update of Ireland Symbol)	87/354/EEC	L 192	7/11/87	Adopted	12/31/87
(Amendment—Ascension of Greece, Spain and Portugal)	87/355/EEC	L 192	7/11/87	Adopted	12/31/87
(Amendment—Technical Updates)	88/665/EEC	L 382	12/31/88	Adopted	12/31/88
(Amendment—EFTA Nations' Symbols)	94/C 320/06	C 320	11/17/94	Adopted	Immed.

Table B.2 *(continued)*

Units of Measure	80/181/EEC	L 39	2/15/80	Adopted	10/1/81
(Amendment—Update)	85/1/EEC	L 2	1/3/85	Adopted	1/1/85
(Amendment—Update)	89/617/EEC	L 357	12/7/89	Adopted	12/31/89
(Proposed Update)	91/C 185/06	C 185	7/11/91	Proposal	
Lifts					
Standards applied to electrically operated lifts	90/486/EEC	L 270	10/2/90	Adopted	
Safety requirements for lifting appliances for persons	95/16/EC	L 213	9/6/95	Adopted	
Medical Devices					
Electromedical Equipment	84/539/EEC	L 300	11/19/84	Adopted	
(Amendment—	93/42/EEC	L 169	6/18/93	Adopted	1/1/96
Medical Devices)	90/385/EEC	L 189	7/20/90	Adopted	7/1/92
Active Implantable Medical Devices					
(Amendment— CE Marking):	93/68/EEC	L 220	8/30/93	Adopted	1/1/95
Notified Bodies:	95/C 280/01	C 280	10/25/95	Communication	
Standards:	94/C 277	C 277	10/4/94	Communications	
	95/C 204	C 204	8/9/95		
	95/C 307/09	C 307	11/18/95		
Medical Devices	93/42/EEC	L 169	6/18/93	Adopted	1/1/96
Notified Bodies:	95/C 280/01	C 280	10/25/95	Communication	
Standards:	94/C 277	C 277	10/4/94	Communications	
	95/C 204	C 204	8/9/95		
	95/C 307/09	C 307	11/18/95		
	95/C 307/10	C 307	11/18/95		
In-Vitro Diagnostic Devices	95/C 172/02	C 192	7/7/95	Proposal	
(Economic & Social Committee Opinion)	96/C 18/02	C 18	1/22/96		
Machinery	89/392/EEC	L 183	6/29/89	Adopted	12/31/92
(Amendment— Moving Machines [FOPs, ROPs])	91/368/EEC	L 198	7/22/91	Adopted	1/31/96
(Amendment— Lifting and Loading)	93/44/EEC	L 175	7/19/93	Adopted	1/1/97
(Amendment— CE Marking)	93/68/EEC	L 220	8/30/93	Adopted	1/1/95

Table B.2 *(continued)*

Notified Bodies:	95/C 280/01	C 280	10/25/95	Communication	
Standards:	92/C 154/03	C 154	6/24/92	Communications	
	93/C 229/03	C 229	8/25/93		
	94/C 207/03	C 207	7/27/94		
	94/C 377/08	C 377	12/31/94		
	95/C 165/03	C 165	6/30/95		
	96/C 42/05	C 42	2/14/96		
Telecommunications Terminal Equipment					
Approximation of member states laws concerning telecommunications terminal equipment, including mutual recognition of their conformity	91/263/EEC	L 128	5/23/91	Adopted	11/6/92
(Amendment— CE Marking)	93/68/EEC	L 220	8/30/93	Adopted	1/1/95
Notified Bodies:	95/C 280/01	C 280	10/25/95	Communication	
Standards [NETs]	89/C 210/02	C 210	8/16/89	Communications	
(under its predecessor,	92/C 143/03	C 143	6/5/92		
Directive 86/361/EEC):	93/C 53/05	C 53	2/24/93		
(Supplement— satellite earth stations)	93/97/EEC	L 290	11/24/93	Adopted	
(Implementation— general attachment requirements for public pan-European cellular digital land-based mobile communications)	Decision 94/11/EC	L 8	1/12/94	Adopted	
(Implementation— telephone application requirements for public pan-European cellular digital land-based mobile communications)	Decision 94/12/EC	L 8	1/12/94	Adopted	
(Implementation— attachment requirements for terminal equipment interface for ONP 2,048 kbits/s digital unstruc-tured leased lines)	Decision 94/470/EC	L 194	7/29/94	Adopted	

Table B.2 *(continued)*

(Implementation—general terminal attachment requirements for Digital European Cordless Telecommunications [DECT])	Decision 94/471/EC	L 194	7/29/94	Adopted	
(Implementation—telephony application requirements for Digital European Cordless Telecommunications [DECT])	Decision 94/472/EC	L 194	7/29/94	Adopted	
(Implementation—attachment requirements for terminal equipment interface for ONP 64 kbits/s digital unstructured leased lines)	Decision 94/821/EC	L 339	12/29/94	Adopted	
(Implementation—general terminal attachment requirements for Digital European Cordless Telecommunications [DECT], public access profile [PAP] applications)	Decision 95/525/EC	L 300	12/13/95	Adopted	
(Implementation—attachment requirements for handset terminals for ISDN; telephone 3.1 kHz teleservices)	Decision 95/526/EC	L 300	12/13/95	Adopted	
Personal Protection Equipment (PPE)	89/686/EEC	L 399	12/30/89	Adopted	12/1/91
(Amendment—CE Marking)	93/68/EEC	L 220	8/30/93	Adopted	1/1/95
(Amendment—Two-wheeled motor vehicle helmets and transition period)	93/95/EEC	L 276	11/9/93	Adopted	6/30/95

Table B.2 *(continued)*

(Proposed Amendment— Repeal of year CE Marking affixed requirement)	96/C 23/07	C 23	1/27/96	Proposal	
Notified Bodies Standards	95/C 280/01	C 280	10/25/95	Communication	
	92/C 44/10	C 44	2/19/92	Communications	
	92/C 240/05	C 240	9/19/92		
	93/C 345/05	C 345	12/23/93		
	94/C 359/06	C 359	12/16/94		
	95/ C	C 224	8/30/95		
	224/03	C 7	1/8/96		
	96/C 7/09				
Health and safety requirements for the use by workers of personal protective equipment at the workplace (3rd Directive within the meaning of Art. 16 (1) of Directive 89/391/EEC)	89/656/EEC	L 393	12/30/89	Adopted	
(Implementation):	89/C 328/02	C 328	12/30/89	Communication	
Toys approximation of the laws of Member States on the safety of toys	88/378/EEC	L 187	7/16/88	Adopted	6/30/95
(Amendment— CE Marking)	93/68/EEC	L 220	8/30/93	Adopted	1/1/95
Notified Bodies Standards	95/C 280/01	C 280	10/25/95	Communication	
	89/C 155/02	C 155	6/23/90	Communications	
	90/C 154/03	C 154	6/23/89		
	93/C 237/02	C 237	8/30/93		
	94/C 129/13	C 129	5/11/94		
	95/C 156/04	C 156	6/22/95		
	95/C 265/05	C 265	10/12/95		
Simple Pressure Vessels	87/404/EEC	L 220	8/8/87	Approved	7/1/90
(Amendment— Transition Period)	90/488/EEC	L 270	10/2/90	Approved	7/1/92
(Amendment— CE Marking)	93/68/EEC	L 220	8/30/93	Approved	1/1/95

Table B.2   *(continued)*

Notified Bodies	95/C 280/01	C 280	10/25/95	Communication	
Standards	92/C 104/04	C 104	4/24/92	Communications	
	95/C 162/04	C 162	6/27/95		
Product Liability	85/374/EEC	L 210	8/7/85	Adopted	7/1/88
Access of Consumers	COM	N/A	11/16/93	Commission	
to Justice and the	(93) 576			Green	
Settlement of	final			Paper	
Consumer Disputes in					
the Single Market					
(Economic and Social	94/C 295/01	C 295	10/22/94		
Committee Opinion)					
General Product Safety	92/59/EEC	L 228	8/11/92	Adopted	7/1/95
Protection of individuals	95/46/EC	L 281	11/23/95	Adopted	11/23/98
with regard to the					
processing of personal					
data and of the free					
movement of such data					
Worker Safety	89/391/EEC	L 183	6/29/89	Adopted	
Minimum safety and	89/654/EEC	L 393	12/30/89	Adopted	12/31/92
health for work					
equipment used by					
workers at work					
(1st Directive within the					
meaning of Art. 16 (1)					
of Directive					
89/391/EEC)					
Minimum safety and	89/655/EEC	L 393	12/30/89	Adopted	12/31/92
health for work		L 59		Corrigendum	
equipment used by					
workers in the					
workplace (2nd					
Directive within the					
meaning of Art. 16 (1)					
of Directive					
89/391/EEC)					
(Amendment—	92/57/EEC	L 245	8/26/92	Adopted	
(Amendment—	94/C 104	C 104	4/12/94	Proposal	
	95/C 246	C 246	9/22/95	Amended Prop.	
Manual handling of	90/269/EEC	L 156	6/21/90	Adopted	12/31/92
loads where there is a					
risk particularly of back					
injury (4th Directive					
within the meaning of					
Art. 16 (1) of Directive					
89/391/EEC)					

Table B.2 *(continued)*

Visual Display Units [including workstations] (5th Directive within the meaning of Art. 16 (1) of 89/391/EEC)	90/270/EEC	L 156	6/21/90	Adopted	12/31/92
Risks related to exposure to carcinogens at work (6th Directive within the meaning of Art. 16 (1) of Directive 89/391/EEC)	90/394/EEC	L 197	7/26/90	Adopted	
Risks related to exposure to biological agents at work (7th Directive within the meaning of Art. 16 (1) of Directive 89/391/EEC)	90/679/EEC	L 374	12/31/90	Adopted	11/09/93
(Amendment—genetically modified biological agents)	93/88/EEC	L 268	10/29/93	Adopted	4/30/94
(Amendment—	95/	L 155	6/30/95	Adopted	
Temporary or mobile construction sites (8th Directive within the meaning of Art. 16 (1) of Directive 89/391/EEC)	92/27/EEC	L 245	8/26/92	Adopted	
Provision of safety and/or health signs at work (9th Directive within the meaning of Art. 16 (1) of Directive 89/391/EEC)	92/58/EEC	L 245	8/26/92	Adopted	
Protection of workers in surface and underground mineral-extracting industries (12th Directive within the meaning of Art. 16 (1) of Directive 89/391/EEC)	92/104/EEC	L 404	12/31/92	Adopted	
Risks arising from physical agents	93/C 77/02	C 77	3/18/93	Proposal	

Table B.2   *(concluded)*

Safety of Recreational Craft	94/25/EC	L 164	6/30/94	Adopted	
Standards:	95/C 255/	C 255	9/30/95	Communication	
Energy Labelling Directive on the indication by labelling and standard product information of the consumption of energy and other resources by household appliances	92/75/EEC	L 297	10/13/92	Adopted	
(Implementation—electric ovens):	79/531/EEC	L 531	6/13/79	Adopted	6/1981
(Implementation—household electric refrigerators, freezers and their combinations):	Commission Directive 94/2/EC	L 45	2/17/94	Adopted	3/9/94
(Proposed Directive—energy efficiency requirements for household electric refrigerators, freezers and their combinations):	94/C 390 96/C 49/07	C 390 C 49	12/31/94 2/20/96	Proposal Amended Prop.	
(Implementation—energy efficiency requirements for household electric washing machines):	Commission Directive 95/12/EC	L 136	6/8/95	Adopted	6/28/95
Standards:	95/C 312/05	C 312	11/23/95	Communication	
(Implementation—energy efficiency requirements for household electric clothes dryers)	Commission Directive 95/13/EC	L 136	6/8/95	Adopted	6/28/95
Standards:	95/C 312/06	C 312	11/23/95	Communication	

Editors's Note: Almost every Directive above was also amended by the "Treaty of the European Union" (officially known as *The Agreement on the European Economic Area [EEA] between the European Communities, their Member States...*) [OJ No. L 1, 3.1.1994].

UNIPUB
4611-F Assembly Drive
Lanham, MD 20706-4391,
U.S.A.
Tel: 301-459-7666; Fax: 301-459-0056

European Union Depository Libraries

University of Arizona
International Documents
University Library
Tucson, AZ 85721, U.S.A.
Tel: 520-621-6441

University of Arkansas
Documents Department
UALR Library
2801 S. University Avenue
Little Rock, AR 72204, U.S.A.
Tel: 501-569-3000

University of California
Documents Department
General Library
Berkeley, CA 94720, U.S.A.
Tel: 510-642-6000

University of California
International Documents
Public Affairs Service
Research Library
Los Angeles, CA 90024, U.S.A.
Tel: 310-825-4732

University of California
Social Sciences and Humanities
Library
La Jolla, CA 92093, U.S.A.
Tel: 619-534-2230

University of Southern California
International Documents
Von KleinSmid Library
Los Angeles, CA 90089, U.S.A.
Tel: 213-740-1767

Stanford University
Western European Collection
The Hoover Institution
Stanford, CA 94305, U.S.A.
Tel: 415-723-2300

University of Colorado
Government Publications
University Library
Boulder, CO 80309-0184, U.S.A.
Tel: 303-492-7477

Yale University
Government Documents Center
Seeley G. Mudd Library
38 Mansfield
New Haven, CT 06520, U.S.A.
Tel: 203-432-3203

American University
Law Library
4400 Massachusetts Avenue, NW
Washington, DC 20016, U.S.A.
Tel: 202-885-1000

Library of Congress
Serial Division
Madison Building
10 First Street, SE
Washington, DC 20540, U.S.A.
Tel: 202-707-5000

University of Florida
Documents Department
Libraries West
Gainesville, FL 32611, U.S.A.
Tel: 352-392-0345

Emory University
Law Library
School of Law
Atlanta, GA 30322, U.S.A.
Tel: 404-727-6824

University of Georgia
Law Library
Law School
Athens, GA 30602, U.S.A.
Tel: 706-542-1922

University of Hawaii
Government Documents
University Library
2550 The Mall
Honolulu, HI 96822, U.S.A.
Tel: 808-956-7204

University of Chicago
Government Documents
Regenstein Library
1100 E. 57th Street
Chicago, IL 60637, U.S.A.
Tel: 312-702-4685

University of Illinois
Law Library
School of Law
504 E. Pennsylvania Avenue
Champaign, IL 61820, U.S.A.
Tel: 217-333-2290

Illinois Institute of Technology
Law Library
565 W. Adams Street
Chicago, IL 60661, U.S.A.
Tel: 312-567-3000

Northwestern University
Government Publications
University Library
Evanston, IL 60208, U.S.A.
Tel: 847-491-3741

Indiana University
Government Documents
University Library
Bloomington, IN 47405, U.S.A.
Tel: 812-855-0100

University of Notre Dame
Document Center
Hesburgh Library
Notre Dame, IN 46556, U.S.A.
Tel: 219-631-5000

University of Iowa
Government Publications Library
Iowa City, IA 52242, U.S.A.
Tel: 319-335-5927

University of Kansas
Government Documents and Maps
University Library
6001 Malott Hall
Lawrence, KS 66045, U.S.A.
Tel: 913-864-3956

University of Kentucky
Government Publications
Margaret I. King Library
Lexington, KY 40506, U.S.A.
Tel: 606-257-8396

University of New Orleans
Business Reference
Earl K. Long Library
New Orleans, LA 70148, U.S.A.
Tel: 504-286-6000

University of Maine
Law Library
246 Deering Avenue
Portland, ME 04102, U.S.A.
Tel: 207-780-4350

Harvard University
Law School Library
Langdell Hall

Cambridge, MA 02138, U.S.A.
Tel: 617-495-3455

University of Michigan
Law Library
Ann Arbor, MI 48109-1210,
U.S.A.
Tel: 313-764-9324

Michigan State University
Documents Department
University Library
East Lansing, MI 48824-1048,
U.S.A.
Tel: 517-353-8740

University of Minnesota
Government Publications
Wilson Library
Minneapolis, MN 55455, U.S.A.
Tel: 612-624-0303

Washington University
John M. Olin Library
1 Brookings Drive
St. Louis, MO 63130, U.S.A.
Tel: 314-935-5489

University of Nebraska
Acquisitions Division
University Libraries
Lincoln, NE 68588-0410, U.S.A.
Tel: 402-472-7211

Princeton University
Documents Division
Firestone Library
Princeton, NJ 08544, U.S.A.
Tel: 609-258-3000

University of New Mexico
Social Science Coll. Dev.
Zimmerman Library
Albuquerque, NM 87131, U.S.A.
Tel: 505-277-0111

Council on Foreign Relations Library
58 E. 68th Street
New York, NY 10021, U.S.A.
Tel: 212-734-0400

New York Public Library
Research Library, Ecn. and Pub.
New York, NY 10017, U.S.A.
Tel: 212-930-8800

New York University
Law Library
School of Law
40 Washington Square South
New York, NY 10012, U.S.A.
Tel: 212-998-6300

State University of New York
Government Publications
Library
1400 Washington Avenue
Albany, NY 12222, U.S.A.
Tel: 518-442-3300

State University of New York
Government Documents
Lockwood Library
Buffalo, NY 14260, U.S.A.
Tel: 716-645-2814

Duke University
Public Documents Department
University Library
Durham, NC 27708, U.S.A.
Tel: 919-660-5851

Ohio State University
Information Services Dept.
University Library
1858 Neil Avenue Mall
Columbus, OH 43210, U.S.A.
Tel: 614-292-6175

University of Oklahoma
Government Documents
Bizzell Memorial Library
401 W. Brooks
Norman, OK 73019, U.S.A.
Tel: 405-325-3341

University of Oregon
Documents Section
University Library
Eugene, OR 97403, U.S.A.
Tel: 541-346-3111

University of Pennsylvania
Serials Department
Van Pelt Library
Philadelphia, PA 19104 , U.S.A.
Tel: 215-898-7555

Pennsylvania State University
Documents Section
University Library

University Park, PA 16802, U.S.A.
Tel: 814-865-4700

University of Pittsburgh
Government Documents
Hillman Library
Pittsburgh , PA 15260, U.S.A.
Tel: 412-648-7800

University of Puerto Rico
Law Library
Rio Piedras, PR 00931, U.S.A.
Tel: 809-764-0000

University of South Carolina
Documents/Microforms
Thomas Cooper Library
Columbia, SC 29208, U.S.A.
Tel: 803-777-4866

University of Texas
Law Library
School of Law
727 E. 26th Street
Austin, TX 78705, U.S.A.
Tel: 512-471-7726

University of Utah
International Documents
Marriott Library
Salt Lake City, UT 84112, U.S.A.
Tel: 801-581-6273

University of Virginia
Government Documents
Alderman Library
Charlottesville, VA 22903, U.S.A.
Tel: 804-924-3021

George Mason University
Center for European Studies
4001 N. Fairfax Drive
Arlington, VA 22203, U.S.A.
Tel: 703-993-8200

University of Washington
Government Publications
University of Washington Libraries
Seattle, WA 98195-2900, U.S.A.
Tel: 206-543-2100

University of Wisconsin
Documents Department
Memorial Library
728 State Street
Madison, WI 53706, U.S.A.
Tel: 608-262-3193

ADDITIONAL EU STANDARDS

Information Sources

To find out whether your product is covered by harmonized (EU-wide) standards, first call:

Office of European Union and Regional Affairs
International Trade Administration
Department of Commerce
14th Street & Constitution Avenue, NW Room 3036
Washington, DC 20230, U.S.A.
Tel: 202-482-5276; Fax: 202-482-2155

Charles Ludolph, Director, Office of European Union and Regional Affairs
Ken Nichols, Director of Commercial Policy Division; Tel: 202-482-3187
Lori Cooper, Director of Trade Policy Division; Tel: 202-482-5279

This office will send the standard to you if it is in the EU office files. If the standard is in the "proposal and commentary stage" it will be necessary to examine *Information and Notices,* a frequent publication of the Official Journal of the European Community. This publication can be found in any state's EU full depository library or any partial depository library. The OJEC can also be found in many law libraries.

National Center for Standards and Certification Information (NCSCI)
National Institute of Standards and Technology (NIST)
Department of Commerce
Building 820
Room 164
Gaithersburg, MD 20899-0001, U.S.A.
Tel: 301-975-4040; Fax: 301-926-1559

Office of the U.S. Trade Representative
WTO & Multilateral Affairs
Technical Barriers to Trade Winder Building
600 17th Street, NW
Room 513

Washington, DC 20506, U.S.A.
Tel: 202-395-3063

Delegation of the European Commission
2300 M Street, NW
Washington, DC 20037, U.S.A.
Tel: 202-862-9500; Fax: 202-429-1766

Delegation of the European Commission
3 Dag Hammarskjold Plaza
305 E. 47th Street
New York, NY 10017-2301, U.S.A.
Tel: 212-371-3804; Fax: 212-688-1013

United States and Canadian Contacts

For information on various aspects of EU activities related to standardization, contact the Office of European Union Regional Affairs at the Department of Commerce above, or:

American National Standards Institute
11 W. 42nd Street
13th Floor
New York, NY 10036, U.S.A.
Tel: 212-642-4900; Fax: 212-398-0023

The American National Standards Institute
Washington Office
655 15th Street, NW
Suite 300
Washington, DC 20005, U.S.A.
Tel: 202-639-4090; Fax: 202-628-1886

American Society for Quality Control (ASQC)
611 E. Wisconsin Avenue
P.O. Box 3005

Milwaukee, WI 53201-3005, U.S.A.
Tel: 414-272-8575; Fax: 414-272-1734

British American Chamber of Commerce
41 Sutter Street
Suite 303
San Francisco, CA 94104, U.S.A.
Tel: 415-296-8645; Fax: 415-296-9649

Canadian Standards Association
178 Rexdale Boulevard
Etobicoke, ON M9W 1R3, Canada
Tel: 416-747-4000; Fax: 416-747-2475

Compliance Engineering
1 Tech Drive
Andover, MA 01810, U.S.A.
Tel: 508-681-6600; Fax: 508-681-6637

Global Engineering Documents
Suite 407
7730 Carondelet Avenue
Clayton, MO 63105, U.S.A.
Tel: 800-854-7179; Fax: 314-726-6418

**Information Handling Services
(IHS)**
15 Inverness Way East
Englewood, CO 80112, U.S.A.
Tel: 303-790-0600; Fax: 303-397-2599

Document Center
1504 Industrial Way
Unit 9
Belmont , CA 94002, U.S.A.
Tel: 415-591-7600; Fax: 415-591-7617

**General Services Administration
(GSA)**
Federal Supply Service
490 L'Enfant Plaza
Suite 8214
Washington, DC 20407, U.S.A.
Tel: 202-755-0300; Fax: 202-755-0290

Intertek Technical Services
9900 Main Street
Suite 500
Fairfax, VA 22031, U.S.A.
Tel: 703-591-1320; Fax: 703-273-4124

Technical Standards Services
4024 Mount Royal Boulevard
Allison Park, PA 15101, U.S.A.
Tel: 412-487-7007; Fax: 412-487-6718

European Union Contacts

European Commission
Rue de la Loi 200
B-1049 Brussels, Belgium
Tel: 32-2-299-1111; Fax: 32-2-295-0138

Council of the European Union
Rue de la Loi 170
B-1048 Brussels, Belgium
Tel: 32-2-285-6111; Fax: 32-2-285-7397

**European Committee for Standardization
(CEN)**
**European Committee for Electrotechnical
Standardization (CENELEC)**
**European Organization for Testing and
Certification (EOTC)**
Secretary General's Office
Rue de Stassart 36
B-1050 Brussels, Belgium
Tel: 32-2-550-0811; Fax: 32-2-550-0819

**European Telecommunications Standards
Institute (ETSI)**
650 Route de Lucioles
F-06921 Sophia-Antipolis Cedex, France
Tel: 33-92-94-4200; Fax: 33-93-65-4716

**International Electrotechnical Commission
(IEC)**
Rue de Varembe 3
P.O. Box 131
CH-1211 Geneva 20, Switzerland
Tel: 41-22-919-0211; Fax: 41-22-919-0300

**International Standards Organization (ISO)
Rue de Varembe 1**
CH-1211 Geneva 20, Switzerland
Tel: 41-22-749-0111; Fax: 41-22-733-3430

**EC Committee of the American Chamber
of Commerce**
Avenue des Arts 50, BTE 5
B-1040 Brussels, Belgium
Tel: 32-2-513-6892; Fax: 32-2-513-7928

European Free Trade Association
9-11 Rue de Varembe
CH-1211 Geneva 20, Switzerland
Tel: 41-22-749-1111; Fax: 41-22-733-9291

European Parliament
Press and Information Office
Rue Belliard 97
B-1047 Brussels, Belgium
Tel: 32-2-284-2111; Fax: 32-2-23-6933

European Parliament
Centre Europeen
Plateau du Kirchberg
L-2929 Luxembourg
Tel: 352-43-001; Fax: 352-4300-4842

GATT
Centre Wiliam Rappard
154 Rue de LaU.S.A.nne
1211 Geneva 21, Switzerland
Tel: 41-22-739-5111; Fax: 41-22-731-4206

SINGLE INTERNAL MARKET INFORMATION

For information on the 1992 Single Market program, background information on the European Union, or assistance regarding specific opportunities or potential problems, contact the Office of European Union Community Affairs Single Internal Market Information Service at 202-482-5276. Specific industry offices are listed below.

Department of Commerce
Office of European Union Community Affairs
14th Street and Constitution Avenue, NW
Room 3042
Washington, DC 20230, U.S.A.

Aerospace, Sally Bath, Office of Aerospace, Room H2128; Tel: 202-482-4222

Autos, Henry Misisco, Office of Automotive Affairs, Room 4036; Tel: 202-482-0554

Basic Industries, Michael Copps, Office of Basic Industries, Room 4039; Tel: 202-482-0614

Chemicals Division, Stuart Keitz, Office of Basic Industries, Room 4039; Tel: 202-482-0128

Computer Equipment, Tim Miles, Office of Computer Equipment; Tel: 202-482-2990

Consumer Goods, John Frielick, Office of Consumer Goods; Tel: 202-482-0337

Forest Products and Building Materials, Gary Stanley; Tel: 202-482-0376; Charles Pitcher; Tel: 202-482-0385

Industrial Trade Staff, Heather West, Room 3814A; Tel: 202-482-2831

Microelectronics, Medical Equipment & Instrumentation, Marge Donnelly, Room 1015; Tel: 202-482-5466

Service Industries, Bruce Harsh, Office of Service Industries, Room 1114; Tel: 202-482-3575

Telecommunications, Myles Denny-Brown, Office of Telecommunications, Room 4324; Tel: 202-482-4466

Textiles and Apparel, Troy Cribb, Office of Textiles, Apparel, and Consumer Goods Industries, Room 3001A; Tel: 202-482-3737

U.S. & Foreign Commercial Service, Room 3802; Tel: 202-482-5777

ADDITIONAL U.S. GOVERNMENT CONTACTS

Additional U.S. government contacts for information on European Union matters include:

European Commission
Delegation of the European Commission
2300 M Street, NW
Washington, DC 20037, U.S.A.
Tel: 202-862-9500; Fax: 202-429-1766

Delegation of the European Commission
3 Dag Hammarskjold Plaza
305 E. 47th Street
New York, NY 10017, U.S.A.
Tel: 212-371-3804; Fax: 212-688-1013

Department of Agriculture, Foreign Service Agency
AG 0574, P.O. Box 2415
Washington, DC 20013, U.S.A.
Tel: 202-720-5267

Small Business Administration, Office of International Trade
409 Third Street, SW 6th Floor
Washington, DC 20416, U.S.A.
Tel: 202-205-6720

Department of State, Office of European Community and Regional Affairs
Room 6519
Washington, DC 20520, U.S.A.
Tel: 202-647-2395

Department of State, Office of Commercial and Business Affairs
Room 2318
Washington, DC 20520, U.S.A.
Tel: 202-647-1625

U.S. Export-Import Bank
811 Vermont Avenue, NW
Washington, DC 20571, U.S.A.
Tel: 202-565-3946; Fax: 202-565-3380

For advice or information about any aspect of exporting to the European Union, contact a local International Trade Administration (ITA) District Office or speak to the appropriate desk officer at the U.S. Department of Commerce's International Trade Administration:

Belgium, Luxembourg, Netherlands: 202-482-2905

Denmark: 202-482-3254

France: 202-482-6008

Germany: 202-482-2434

Greece: 202-482-2177

Ireland: 202-482-3748

Italy: 202-482-0010

Portugal: 202-482-4945

Spain: 202-482-3945

United Kingdom: 202-482-3748

Other Business Information Resources

World Trade Center
401 E. Pratt Street
Suite 2432
Baltimore, MD 21202, U.S.A.
Tel: 410-962-4539

American Association of Exporters and Importers
11 W. 42nd Street
30th Floor
New York, NY 10036, U.S.A.
Tel: 212-944-2230; Fax: 212-382-2606

Federation of International Trade Associations
1851 Alexander Bell Drive
Suite 400
Reston, VA 22091, U.S.A.
Tel: 703-620-1588; Fax: 703-391-0159

National Council on International Trade and Documentation
818 Connecticut Avenue, NW
12th Floor
Washington, DC 20006, U.S.A.
Tel: 202-331-4328; Fax: 202-872-8696

U.S. Chamber of Commerce
1615 H Street, NW
Washington, DC 20062, U.S.A.
Tel: 202-659-6000; Fax: 202-463-5836

U.S. Council for International Business
1212 Avenue of the Americas, 21st Floor
New York, NY 10036, U.S.A.
Tel: 212-354-4480; fax: 212-575-0327

HOTLINES

EU Hotline
301-921-4164

This hotline reports on draft standards of the European Committee on Standardization (CEN), the European Committee for Electrotechnical Standardization (CENELEC) and the European Telecommunications Standards Institute (ETSI). It also provides information on selected EU directives. The recorded message is updated weekly and gives the product, document number, and closing date for comments.

GATT Hotline
301-975-4041

This hotline provides current information, received from the GATT Secretariat in Geneva, Switzerland, on proposed foreign regulations which may significantly affect trade. The recorded message is updated weekly and gives the product, country, closing date for comments (if any) and Technical Barriers to Trade (TBT) notification number.

The Export Hotline
800-U.S.A.-XPORT (800-872-9767)

Supported by the U.S. Department of Commerce, this privately sponsored fax retrieval system can provide reports on country markets, trade fairs, trade and investment issues, key contacts, shipping requirements, and news/risks/advisories.

The Trade Information Center
800-U.S.A.-TRADE (800-872-8723)

Operated by the Deparment of Commerce, this is a focal point for information on federal export assistance programs.

Consultants/Training Appendix

ISO 9000 CONSULTANTS AND TRAINING SERVICES

This section profiles nearly 200 companies that provide ISO 9000 consulting and training services. While it is not a comprehensive or official list, it gives a sampling of the ISO 9000 services available to companies seeking information about the registration process. The section is divided into the following three parts:

- Companies providing ISO 9000 consulting

- Companies providing ISO 9000 training

- Companies providing both services.

For information concerning the specific services offered by these companies, contact the companies directly through the contact information provided. Acronyms used by consultants and trainers are listed on the following page.

Editor's Note: *This listing is provided for your information and consideration. IRWIN Professional Publishing does not endorse any of the consultants or trainers by including them in this Handbook. IRWIN Professional Publishing has made every effort to verify the accuracy of the information contained in this section. Please notify IRWIN Professional Publishing of any errors.*

ACRONYMS USED BY CONSULTANTS AND TRAINERS

AIAG	Automotive Industry Action Group		IIE	Institute of Industrial Engineers
ANSI	American National Standards Institute		IQA	Institute for Quality Assurance
API	American Petroleum Institute		IRCA	International Registrar of Certified Auditors
ASME	American Society of Mechanical Engineers		ISO	International Organization for Standardization
ASQC	American Society for Quality Control			
ASTM	American Society for Testing and Materials		MBNQA	Malcolm Baldrige National Quality Award
CMC	Certified Management Consultant		NIST	National Institute of Standards and Technology
CQA	Certified Quality Auditor		PE	Professional Engineer
CQE	Certified Quality Engineer		PQA	Preproduction Quality Assurance
CRE	Certified Reliability Engineers		QA	Quality Assurance
CSA	Canadian Standards Association		QMS	Quality Management System
DoD	Department of Defense		RAB	Registrar Accreditation Board
DOE	Department of Energy		RBA	Registration Board for Assessors
EQA	Effective Quality Workshop		RVA	Dutch Council for Accreditation
FCC	Federal Communications Commission		SPC	Statistical Process Control
GMP	Good Manufacturing Practices		TAG	Technical Advisory Group
IAAR	Independent Association of Accredited Registrars		TC	Technical Committee
IAEI	International Association of Electrical Inspection		TQM	Total Quality Management
IEEE	Institute of Electrical and Electronic Engineers		UKAS	United Kingdom Accreditation Service

COMPANIES PROVIDING ISO 9000 CONSULTING

ASSOCIATED QUALITY CONSULTANTS

P.O. Box 1412
Herndon, VA 22070-1412, U.S.A.
Tel: 800-604-6149; Fax: 703-716-0479
E-mail Address: help@quality.org
Home Page Address: http://www.quality.org/qc

Contact: Bill Casti, President

In Business: Two years

Clients: Penfield Gill Inc. (Boston, MA); private individuals; Federal Emergency Management Agency (FEMA)

Qualifications/Certifications: Senior Internet Administrator, Bell Atlantic Network Services; ASQC Certified Quality Auditor (CQA); Chairman for Electronic Media, ASQC Section 0511 (Northern Virginia); over 18 years' experience in Quality field.

Additional Information:

E-mail Address 2: bill@casti.com
E-mail Address 3: e_media@cais.com
Home Page Address 2:
http://www.casti.com/casti/Bill.html

BREWER AND ASSOCIATES, INC.

2505 Locksley Drive
Grand Prairie, TX 75050, U.S.A.
Tel: 214-641-8020; Fax: 214-641-1327
E-mail Address: BAIQUALITY@AOL.COM
Home Page Address: http://www.BAI-BK.com

Contact: Ronald D. Kurtzman, Senior Vice President

Contact 2: Laurie B. Miller, Vice President, Operations/Finance

In Business: 13 years

Clients: DuPont, American Airlines, Mobil, Goodyear, Sandoz Chemical, Occidental Chemical, ALCOA, Amoco Chemical, AT&T, Federal Express, U.S. Postal Service, IBM, Nabisco, Johnson & Johnson, Department of Defense, Fort Howard Paper Company, and hundreds more. B-K Education Services has trained over 10,000 people in 10 countries.

Qualifications/Certifications: BAI's associates are primarily Lead Auditors and include two Baldrige Examiners, an Air Force Quality Award Judge, members of the ASQC National Ethics Committee and Boards of Governors/Directors for quality registrars, publications and universities, and a member of the TAG 207 Task Force for Management and Audit Subcommittees (ISO 14000). BAI joined the U.S. TAG to ISO TC 207 for Environmental Management in 1994.

DEVON HUNTER CONSULTING

3-202 Second Avenue
Ottawa, Ontario K1S 2H7, Canada
Tel: 613-233-9319; Fax: 613-233-9319
E-mail Address: mhunter@fox.nstn.ca

Contact: Murray Hunter

In Business: 6 years

Clients: Public and private sector, services, nonprofit organizations

Government of Canada: Treasury Board, Health Canada, Bank of Canada

Private Sector: Bell Northern Research, Digital Equipment, Stentor

Qualifications/Certifications:

- Two books written, based on the experience of companies with ISO 9000: *ISO 9000: Making Quality Happen* and *Quality in Geomatics/ A Practical Guide to ISO 9000.*

- Lead Assessor Training.

THE FARRELL GROUP

21311 Hawthorne Boulevard
Suite 230
Torrance, CA 90503, U.S.A.
Tel: 310-316-4420; Fax: 310-543-2194
E-mail Address: farrellgrp@aol.com
Home Page Address:
http://www.exit.109.com/~leebee

Contact: Thomas E. Pellegrini, Consultant

In Business: Five years

Clients: In excess of 100 clients. The client size ranges from small organizations to Fortune Fifty companies spanning a wide variety of industries.

Qualifications/Certifications: The partners and consultants are former Malcom Baldrige Senior Examiners and State Quality Award Examiners. The work experience spans numerous industries including manufacturing, service, and healthcare.

GEORGE S. MAY INTERNATIONAL COMPANY

303 South Northwest Highway
Park Ridge, IL 60068, U.S.A.
Tel: 800-999-3020 x416; Fax: 847-825-7937

Contact: Ed Culhane

In Business: 71 years

Clients: Over 370,000 clients in 3,000 industries

Qualifications/Certifications: All consultants have a college degree in business or a technical discipline plus a minimum of 10 years experience. In addition, some of our staff are ISO 9000 certified Lead or Professional Auditors.

GTW ASSOCIATES

1012 Parrs Ridge Drive
Spencerville, MD 20868, U.S.A.
Tel: 301-421-4138; Fax: 301-421-4138
E-mail Address: GWILLINGMY@aol.com

Contact: George T. Willingmyre, P.E., President

In Business: One year

Clients: Microsoft Corporation

National Mining Association

Services are intended for businesses with U.S. or international standards, certification or regulatory issues; trade associatons and standards organizations; and government agencies.

Qualifications/Certifications: George T. Willingmyre, P.E. is the president, principal, and founder of GTW Associates. The company assembles project teams of experts tailored to the individual client needs. Mr. Willingmyre was vice president and headed Washington Operations for the American National Standards Institute (ANSI) from 1989 to 1995. Mr. Willingmyre helped resolve the United States/Japan dispute over Japan's plan to require unique quality system requirements for computer software, directed ANSI's role in the ANSI-RAB American National Accreditation Program for Quality System Registrars now the ANSI/RAB National Accreditation Program and served as the founding secretary of the International Accreditation Forum. He was responsible for public policy and government relations, ANSI's Government, Organizational and Company Member Councils and national programs for testing and certification to standards.

INQC CONSULTING

50-B Bearfoot Road
Northborough, MA 01532, U.S.A.
Tel: 508-393-1303; Fax: 508-393-1874
E-mail Address: pernix01@AOL.COM

Contact: David Manalan, Principal

In Business: Five years

Clients: Stat-a-Matrix, Expertech Associates, National Medical Care-DSD, Tomlin Associates, Pall-Filtron, Garland, MFG., Remington Products, RA-QA Associates, Infusaid-Strato Medical, Genesis Orthopedics

Qualifications/Certifications: RAB Certified Quality Systems Auditor, ASQC Certified Quality Auditor, Massachusetts Registered Toxic Use Reduction Planner, General Practice Examiner, Massachusetts Quality Award 1993, Boston Section-ASQC-Instructor, Co-Chair, ASQC-Boston Pollution Prevention Subcommittee Member-IPSE, AAMI, AIM, PDA, ASQC, Turpa, B.S. Chemical Engineering (MIT), Reviewer-ASQC Multimedia Review Board.

Additional Information:

E-mail Address 2: 73367.1725@compuserv.com

INSTITUTE OF CERTIFIED MANAGEMENT CONSULTANTS OF CANADA

BCE Place, 181 Bay Street
P.O. Box 835
Heritage Building, Galleria, Floor 2B
Toronto, Ontario M5J 2T3, Canada
Tel: 416-869-3001; Fax: 416-869-3037
E-mail Address: Compuserv-102774,1361

Contact: Heather Osler, Executive Director

In Business: ISO 9000 Initiative started in 1994.

Qualifications/Certifications: Listed by the ISO 9000 Forum as the registrar of quality consultants in Canada.

ISO COMMUNICATIONS

P.O. Box 1110
Amherst, VA 24521, U.S.A.
Tel: 804-946-2211; Fax: 804-946-2411
E-mail Address: ISOCOM@AOL.COM

Contact: Louise C. Rozene, President

Contact 2: James B. Kohler II, Director of Marketing

Contact 3: Karen Lupton, Art Director

In Business: Four years

Clients: DNV Certification Inc., Ericsson Inc., Framatome, GE Drive Systems, Limitorque Corp., National Quality Assurance, USA, Pilot Software, Progress Printing, TUV America, Wiley & Wilson, Xyplex

Qualifications/Certifications: Principals in the company have more than 15 years of experience each in journalism, public relations, graphic design, and marketing, the last four of which have been spent solely in quality communications in business.

MILLER COMPANY

3331 Cochran Drive
Lancaster, PA 17601, U.S.A.
Tel: 717-898-7971; Fax: 717-290-8653

Contact: Ted Miller, Consultant

In Business: Six years

Clients: Predominantly private sector commercial and industrial, ranging from Fortune 500 to small and medium sized enterprises. Specific list available upon request. Geographic areas include domestic and international.

Qualifications/Certifications: Lead assessor training for ISO 9000 certified by the Assessor Registration Board for Assessors of the United Kingdom.

SIMCOM, INC.

P.O. Box 420511
Atlanta, GA 30342, U.S.A.
Tel: 404-303-7799 ; Fax: 404-303-0192
E-mail Address: simcom@mindspring.com
Home Page Address: http://www.eurocom.com

Contact: Michael Shealy

In Business: Since 1992.

Clients: Allen-Bradley Co., AMP, AT&T, Bay Networks, Cincinnati Milacron, Compaq Computer, Deere and Co., Digital Equipment Corp., Eastman Kodak, General Electric, Graco, Inc., Hewlett Packard Co., Honeywell, IBM Corp., Ingersoll-Rand, KPMG Peat Marwick, Silicon Valley Group, Inc., Motorola, Northrup Grumman, Nynex, Philips Medical, Samsung, Siemens, Sony, Sun Microsystems, Texas Instruments, USF&G

Qualifications/Certifications: Simcom is an international education and information services company. It is structured as a consortium of over 50 individuals specializing in the technical aspects of international trade. These experts come from both the public and private sectors, including testing, certification, and registration entities. Simcom serves as a comprehensive, technical, unbiased clearinghouse to meet diverse yet pertinent compliance needs.

SOFTWARE ENGINEERING PROCESS TECHNOLOGY

2725 NW Pine Cone Drive
Issaquah, WA 98027, U.S.A.
Tel: 206-451-1051; Fax: 206-557-9419
E-mail Address:
73211.2144@COMPUSERVE.COM

Contact: Stan Magee, President

In Business: Four years

Clients: Client list is available

Qualifications/Certifications: Stan Magee is president of Software Engineering Process Technology Company specializing in the implementation of software process technology for U.S. and international corporations and organizations. Mr. Magee is vice chair of the United States SC-7 TAG (Technical Advisory Group) which determines the U.S. position on international (ISO) Software Engineering Standards. He has been a U.S. delegate to the International Plenary meetings since 1986. He was the lead U.S. delegate on the ISO 9000-3 update meeting in Durban, South Africa (1995). Mr. Magee is coauthor of the book *Software Engineering Standards and Specifications,* Global Publications, 1994. Mr. Magee is also a Washington State quality examiner.

THOMAS F. BRANDT ASSOCIATES
Innovative Technology Management

P.O. Box 24765
5232 Kellog Avenue South
Edina, MN 55424, U.S.A.
Tel: 612-926-9222; Fax: 612-925-2278
E-mail Address:
71441.2051@COMPUSERVE.COM

Contact: Thomas F. Brandt, Principal

In Business: Seven years

Clients: CyberOptics Corp., Remmell Engineering, Inc., KMA Associates International, Inc., APG

Cash Drawer Company, Despatch Industries, Inc., Bureau Veritas Quality International (NA), Teltech

Qualifications/Certifications: Lead Auditor (Registrar Accreditation Board & International Register of Certified Auditors), Qualified for QS-9000, Senior Examiner, Minnesota Quality Award 1992, 1993, 1994, Member ASQC, Member Minnesota Council for Quality; Life Member, Institute of Electrical and Electronic Engineers; 40 years experience in manufacturing as design engineer, technical manager, senior executive, and consultant. Business experience in Europe and Asia.

WC TERRY AND ASSOCIATES

P.O. Box 589
Ossian, IN 46777, U.S.A.
Tel: 219-622-7950; Fax: 219-622-7950
E-mail Address: wcterry@cris.com
Home Page Address:
http://www.cris.com/~wcterry/

Contact: William C. Terry, Principal

In Business: Eight years

Clients: First tier and second tier automotive suppliers. References furnished upon request.

Qualifications/Certifications: WC Terry and Associates has been trained by Eastern Michigan University in QS-9000, achieving preferred vendor status. Mr. William Terry, Principal, has over 20 years of experience in the Quality Assurance field. As a Quality Engineer, he has developed and submitted FMEA's, AQP's to the big three.

COMPANIES PROVIDING ISO 9000 TRAINING

ACCADEMIA QUALITAS

2297 Guenette
Ville St. Laurent, Quebec H4R 2E9, Canada
Tel: 800-263-0128/514-333-5767
Fax: 514-333-5768
E-mail Address: assist@ACCADEMIA.COM

Contact: Isabelle Cerutti, Seminar Coordinator

Contact 2: Nicole Gagné, Quality Representative

In Business: Six years

Clients: John Deere, IBM, IST, ICI, CGI, ABB, SNC Lavalin, SNC Industrial Technologies, Alcan, Bell Canada, Northern Telecom, Teleglobe Canada, United Westburne, Petro-Canada, YKK, CE Zinc, Agriculture Canada, Industry Canada, Health and

Protection Branch, Montell, Liquid Air, JWI Johnson, GTE Control Devices, Cascades, Pratt & Whitney, Harris Farinon, Stone Consolidated, Weyer Hauser, Seagram, C-MAC, Imperial Oil, EKA Nobel, Inco, Camco, Pirelli, Sidbec Dosco, Stelco McMaster, SGL Canada, Bell Sigma, American Iron & Metal, Lauralco, Bridgestone/Firestone, Dominion Textile, Abbott Laboratory

Qualifications/Certifications: Registration to ISO 9001:94 in process.

Additional Information:

E-mail Address 2: denis@accademia.com

AIB REGISTRATION SERVICES

1213 Bakers Way
Manhattan, KS 66502, U.S.A.
Tel: 913-537-4750; Fax: 913-537-1493

Contact: William Pursley, Director

Contact 2: Leonard Steed, Lead Assessor

Contact 3: Janet Bloomfield, Administrative Assistant

In Business: AIB—77 years; AIBRS—Three years

Clients: The AIB is a not-for-profit institution with over 750 members and over 5,000 participating companies representing the food and allied industries. Major food processeors, retailers, distributors, and allied members support the activities of the AIB and its services. Members are representatives of all major food groups and processors. The AIB also provides its services internationally in over 68 countries in 1995.

Qualifications/Certifications: AIBRS is undergoing the accreditation process with the Registrar Accreditation Board (RAB). Certified lead assessors and team auditors performing training and registration audits will have the appropriate food industry background. The advisory board for AIBRS is only comprised of personnel working in or servicing the food industry. AIB is accredited to the North Central Association of Colleges and Schools and issues continuing educational credits on all training seminars. AIBRS is currently the only registrar totally dedicated to the food industry and its suppliers.

AMERICAN ASSOCIATION FOR LABORATORY ACCREDITATION (A2LA)

656 Quince Orchard Road
Suite 620
Gaithersburg, MD 20878, U.S.A.
Tel: 301-670-1377; Fax: 301-869-1495
E-mail Address: a2la@aol.com
Home Page Address:
http://users.aol.com/a2la/a2la.htm

Contact: Peter S. Unger, President

Contact 2: Roxanne M. Robinson, Vice President

In Business: More than 18 years

Clients: A2LA has accredited over 800 testing laboratories to ISO/IEC Guide 25, registered a number of laboratories and lab suppliers to ISO 9000 and certified over 200 lots of reference materials.

Qualifications/Certifications: The A2LA quality system registration program for labs and laboratory related suppliers has been accredited by the Registrar Accreditation Board (RAB). A2LA's programs are also recognized via mutual recognition agreements (MRAs) in Australia, Canada, Hong Kong, New Zealand and Singapore and over 30 federal, state, and local government agencies, companies, and associations. These partners recognize the competence, quality of service, and cost effectiveness of A2LA.

AMERICAN ELECTRONICS ASSOCIATION

5285 S.W. Meadows Road
Suite 224
Lake Oswego, OR 97035, U.S.A.
Tel: 503-624-6050; Fax: 503-624-9354
E-mail Address: Jennifer_O'Donnell@aeanet.org
Home Page Address: http://www.aeanet.org

Contact: Jennifer O'Donnell, Field Relations Manager

In Business: 50+years

Clients: High-technology industry

AMERICAN QUALITY RESOURCES, INC.

P.O. Box 817
1141 Main Street
Coventry, CT 06238, U.S.A.
Tel: 800-8USA-ISO/800-887-2476/860-742-5582
Fax: 860-742-1232

Contact: Michael P. Enders

In Business: Three years

Clients: Small and medium-sized manufacturers of air-craft engines, chemicals, electronics, communications, automotive, plastics molding, cosmetics, medical devices, general manufacturing and primary metals, on-going work in Hungary with the state-owned electric utilities for the past two years using AQR custom ISO 9000 training classes. Federally-funded training provided under JTPA for displaced workers resulting in 80 percent placement rate.

Qualifications/Certifications: The Principals of AQR have successfully led more than seventy companies to ISO 9000 registration. The AQR Management team has 110 years combined of experience in quality assurance across a wide range of industries. AQR has trained hundreds of attendees in our ISO 9000 classes.

ASQC

611 E Wisconsin Avenue
P.O. Box 3005
Milwaukee, WI 53201-3005, U.S.A.
Tel: 800-248-1946/ 414-272-8575
Fax: 414-272-1734
E-mail Address: asqc@asqc.org
Home Page Address: http://www.asqc.org

Contact: Monica J. Wickert, Manager, Education Development and Promotion

In Business: 50 years

Clients: Xerox, Pacific Bell, Motorola, Argonne National Laboratory, Johnson Controls, FDA, etc.

Qualifications/Certifications: ASQC is the leading quality improvement organization in the United States, with more than 130,000 individual and 1,000 sustaining members worldwide.

CENTER FOR MANUFACTURING SYSTEMS

New Jersey Institute of Technology

University Heights
Newark, NJ 07102, U.S.A.
Tel: 201-596-2876; Fax: 201-596-6438
E-mail Address: gold@admin.njit.edu

Contact: Art Gold

In Business: Seven years

Clients: New Jersey Manufacturing Companies

Qualifications/Certifications: NIST MEP, Advanced Technology Center of NJ Commission on Science and Technology, NJ State University.

COLUMBIA AUDIT RESOURCES

6951 West Grandridge Boulevard
Kennewick, WA 99336, U.S.A.
Tel: 509-783-0377; Fax: 509-783-1115
E-mail Address: Arter@quality.org
Home Page Address:
www.asqc.org/membinfo/divisions/qad.html

Contact: Dennis R. Arter, Owner

In Business: Twelve years

Clients: Exxon Chem, FDA, Boeing, Nortel, Union Carbide, Campbell Soup

Qualifications/Certifications: Author of *Quality Audits for Improved Performance* and active in ASQC Quality Audit Division. Frequent contributor to ISO 9000 internet discussion list.

COMPLIANCE QUALITY INTERNATIONAL

#5 Hidden Trail
Lancaster, NY 14086-9685, U.S.A.
Tel: 716-685-0534; Fax: 716-685-0048

Contact: Allen R. Bailey, President

Contact 2: George Reed, Partner

In Business: Three years

Clients: Compliance Quality International specializes in small to medium-size businesses that export to Europe in the electronic and electrical equipment, medical devices and service.

Qualifications/Certifications: Compliance Quality International's staff consists of National Registration Scheme and/or Canadian trained assessors, and subject matter experts from a wide range of technical backgrounds and market sectors who are all quality professionals. Mr. Bailey is the author of a series of books on E.U. conformity assessment published by St. Lucie Press, Delray Beach, FL.

ENTELA INC.

Quality System Registration Division

2890 Madison Avenue
Grand Rapids, MI 49548, U.S.A.
Tel: 800-888-3787/ 616-247-0515
Fax: 616-248-9690
E-mail Address: sdemarco@entela.com

Contact: William Vosburg, General Manager

Contact 2: Brandon Kerkstra, Account Manager

Contact 3: Mary Poe, QSRD Coordinator

In Business: Since 1993.

Clients: More than 500 company sites under contract for registration services. Clients represent a wide range of designers and manufacturers from the automotive, rubber, plastics, electronics, and metals industries. A comprehensive client list of registered companies is available by calling 800-888-3787.

Qualifications/Certifications: Dutch Council for Accreditation (RVA), Registration Accreditation Board (RAB), Member of Independent Association of Accredited Registrars (IAAR)

Additional Information:

E-mail Address 2: bvosberg@entela.com
E-mail Address 3: bkerkstra@entela.com

EXECUTIVE IMPROVEMENT SOLUTIONS

744 Pacheco Street
San Francisco, CA 94116, U.S.A.
Tel: 800-810-EXIS; Fax: 801-467-8651

Contact: James Davies

Contact 2: Paul Webb

GEORGE MASON UNIVERSITY

Executive Programs

Corporate and International Training—CPD

Executive Programs, 2214 Rock Hill Road
Suite 400
Herndon, VA 22070, U.S.A.
Tel: 703-733-2804; Fax: 703-733-2820
E-mail Address: cgibbs1@gmu.edu
Home Page Address: www.gmu.edu

Contact: Christine Gibbs, Assistant Program Director— Contract Training
Tel 2: 703-733-2802

Contact 2: Janet Niblock, Director
Tel 3: 703-733-2800

Contact 3: Mary Salmon, Open-Enrollment Programs

In Business: Since 1988

Clients: Managerial, technical, and professional employees from various small to large size companies, nonprofit organizations and government agencies.

Qualifications/Certifications: Established in 1988, the mission of the Executive Programs is to provide a broad range of high-quality, noncredit professional development seminars to meet the needs of private sector, as well as local, state, and federal government organizations both domestically and internationally through open enrollment and in-house contractual professional development courses.

Additional Information:

E-mail Address 2: jniblock@gmu.edu
WWW Address: www.gmu.edu

GEORGE WASHINGTON UNIVERSITY

Continuing Engineering Education
2029 K Street, NW
Suite 200
Washington, DC 20006, U.S.A.
Tel: 800-424-9773/ 202-496-8444
Fax: 202-872-0645
E-mail Address: ceepinfo@ceep.gwu.edu
Home Page Address: http://www.gwu.edu/~ceep

Contact: Pat Murphree, Director

Contact 2: Sue Zimmerman, Deputy Director

In Business: 28 years

Qualifications/Certifications: Currently, GWU's courses are being taught by a team of international standards professionals including Anton Camarota, Eugenia K. Brumm, Lewis Farina, and Peter Corradi.

GLOBAL ENVIRONMENTAL MANAGEMENT SYSTEMS

W348 N6069 California Avenue.
Oconomowoc, WI 53066 , U.S.A.
Tel: 414-569-1890; Fax: 414-569-1890
E-mail Address: Dburd52011@aol.com

Contact: David Burdick, P.E.

In Business: Two years

Clients: All sizes of manufacturing and service businesses. A comprehensive list is available on request.

Qualifications/Certifications: Global Environmental Management Systems is dedicated to environmental management and system implementation and assessment. Thirteen years experience reflects a

blend of international business experience, manufacturing expertise, and technical environmental knowledge. Practical knowledge and skills in: environmental and quality management system assessment and registration, environmental regulation, electrical manufacturing and assembly design, production packaging, food processing, and automotive fabrication. Languages: English, Danish, and some casual fluency in Japanese and German.

OKLAHOMA STATE UNIVERSITY— OKLAHOMA CITY

Center for Organizational Improvement

900 North Portland Avenue
Oklahoma City, OK 73107-6195, U.S.A.
Tel: 405-945-3278; Fax: 405-945-3397

Contact: William J. Nelson, Director

Contact 2: Judy Determan, Coordiantor

In Business: 11 years

Clients: Tinker Air Force Base, Hollytex Carpet Mills, Fred Jones Manufacturing, Blue Cross/Blue Shield, Integris Health Services, AT&T, Zebco Inc., JC Penney Headquarters, Kerr-McGee Chemical Corp., Oklahoma Gas & Electric Services, KF Industries, ELF ATOCHEM Corp., Vance Air Force Base, John Crane Lemco Corp., Oklahoma Tax Commission, and General Motors

Qualifications/Certifications: COI has been instrumental in helping many companies attain their ISO 9000 certification and has met the needs of manufacturing, government, and service companies in the field of Total Quality Management. COI's ISO 9000 instructor is registered as a Lead Assessor in both the United Kingdom (RBA/IQA) and the United States (RBA) and has performed nearly 100 audits in various industries. He is also lead tutor for two Registered Lead Auditor/Assessor courses.

QUALITY MANAGEMENT INSTITUTE

Sussex Center
90 Burnhamthorpe Road West, #300
Mississauga, Ontario L5B 3C3, Canada

Tel: 800-465-3717/905-272-3920
Fax: 905-272-3942
E-mail Address: qmi@inforamp.net
Home Page Address:
http://www.pic.nt/qmi/index.html

Contact: Catherine Neville, President

Contact 2: Diana Pryde, Marketing & Communications Manager

Contact 3: Michael Haycock, Business Development Manager

In Business: Since 1979

Clients: Customers from industries as diverse as petro-chemicals, machinery, and transportation, metal fabrication, electrical, wholesale/retail, food, wood, pulp and paper, and increasingly the automotive and service sectors. Over 1900 registrations completed. QMI Directory of Registered Companies available upon request.

Qualifications/Certifications: QMI is accredited by SCC (Standards Council of Canada), RAB (Registrar Accreditation Board) USA and RvA (Dutch Council for Accreditation), and offers registration to ISO 9000, QS-9000, and ISO 14000.

RSG ENTERPRISES

936 Hyacinth Drive
Delray Beach, FL 33483, U.S.A.
Tel: 407-278-5406; Fax: 407-278-5406

Contact: Ralph S. Gootner

Contact 2: N. Hout-Cooper

In Business: Five years

Clients: Maritime Products; Medical—Plasma Collection; Packaging—Medical Containers; Plastic Injection Molding—Maritime, Medical, Automotive; Industrial Wrapping Machines; Thin-Film Plastic Extruders; Air Freight Cargo Consolidators; Industrial Container Manufacturing—Chemical; Printed Circuit Board Manufacturing; Distribution—Nuts/Bolts/Hardware; Service—Real Estate International; Electronics—Safety Detection Equipment; Computers—CFR-50; Floral Distribution—Bouquets/Arrangements

Qualifications/Certifications: Qualtec ISO-9000 Implementation, BSI lead auditor, BSI 1994 revisions, FIU—NASA Funded ISO 9000 Training and Services.

SGS INTERNATIONAL CERTIFICATION SERVICES CANADA INC.

5925 Airport Road
Suite 300
Mississauga, Ontario L4V 1W1, Canada
Tel: 800-636-0847/ 905-676-9595
Fax: 905-676-9519

Contact: Client Services Department

Surveillance Policy: SGS ICS Canada, Inc., registration certificates are valid for three years. Surveillance audits are conducted twice per year. Certificates may be offered on a continuous registration basis.

In Business: SGS ICS has been providing quality systems registration services since 1985. SGS ICS Canada, Inc. officially opened for business in Canada in January, 1993.

Clients: Accreditations include:

SCC-Canada, ANSI-RAB-USA, UKAS-UK, JAS ANZ-Australia/New Zealand, RvA-Netherlands, DAR-Germany, Sincert-Italy, S.A.S.-Switzerland, Belcert-Belgium, JAB-Japan, INMETRO-Brazil

Qualifications/Certifications: SGS ICS Canada, Inc., currently holds accreditation to 35 of the 39 NACE Codes, and SGS ICS Inc., has trained auditors covering 234 industry SIC Codes worldwide.

All SGS ICS Inc. lead auditors are registered with the International Register of Certificated auditors (IRCA). The ability of lead auditors to participate in Specific industry sector audits are to qualified SIC or NACE codes. Each audit team will have, as a member, a qualified auditor as related to the specific industry sector in addition to a qualified lead auditor

SGS INTERNATIONAL CERTIFICATION SERVICES, INC.

Meadows Office Complex
301 Route 17 North
Rutherford, NJ 07070, U.S.A.

Tel: 800-747-9047/ 201-935-1500
Fax: 201-935-4555

Contact: Lois O'Brien, Sales Manager

In Business: Seven years

Clients: SGS ICS has registered over 9500 facilities in a broad range of the manufacturing and service

sectors. A comprehensive list is available upon request.

Qualifications/Certifications: Accreditations: RAB (United States), JASANZ (Australia/New Zealand), Belcert (Belgium), INMETRO (Brazil), SCC (Canada), INN (Chile), DAR (Germany), Sincert (Italy), JAB (Japan), RvA (Netherlands), SAS (Switzerland), UKAS (United Kingdom)

COMPANIES PROVIDING ISO 9000 TRAINING & CONSULTING

3C TECHNOLOGIES, INC.

Suite 202
6834 Spring Valley Drive
Holland, OH 43528, U.S.A.
Tel: 800-3C SOLVE/ 419-868-8999
Fax: 419-868-8993
E-mail Address: 3c@3ctech.com
Home Page Address: http://www.3ctech.com

Contact: Devan R. Capur, President & General Manager

In Business: Seven years

Clients: Mack Trucks, Mazda, Honda, Ford, GM, Arvin Automotive Industries, Milliken & Co., Underwriters LaB.S. (UL), Cooper Tire, Amcast, Lenawee Stamping, Standard Products, Defiance Metal Products.

Qualifications/Certifications: 3C Technologies, Inc., has an exceptional track record of customer registrations in ISO 9001, 9002, and QS-9000 all on the first try. Rated #1 in training in the 1995 worldwide AIAG survey for QS-9000 trainers offering national and international training. 3C has trained over 5,000 people in the United States, Europe, and Asia, including registrars. Serving diverse industries worldwide, such as automotive, plastics, metal, rubber, textile and software, our team has seasoned facilitators who are RAB and IQA certified. 3C Technologies, Inc., is a Microsoft Solution Provider.

ADVENT MANAGEMENT INTERNATIONAL, LTD.

P.O. Box 1717
West Chester, PA 19380-0057, U.S.A.
Tel: 610-431-2196; Fax: 610-431-2641
E-mail Address: 102463.430@Compuserv.com

Contact: John J. Reddish, President

In Business: 18 years

Clients: Osram/Sylvania, Johnson Matthey, Courtney Division Geo Chemicals, Connecticut Steel, SHW, Inc., Durham Industries, Dri-Print, Schenkers International Forwarders, Omni North America, BDP, Inc., Gafbarre, Inc., Pegasus, Action Technologies, Union Transport, Unifiber Corporation, Cosmair, Inc.

Qualifications/Certifications: Advent consultants come from a broad range of backgrounds. Credentials include: PE, IQA Lead Auditor, CQA Lead Auditor and Certified Management Consultant (CMC). Each ISO engagement is led by a trained lead auditor and supported by trained auditors and supporting specialists with a variety of credentials. Staff includes 10 lead auditors, 16 internal auditors and more than 20 consultants.

ALAMO LEARNING SYSTEMS

3160 Crow Canyon Road
Suite 335
San Ramon, CA 94583, U.S.A.
Tel: 800-829-8081/510-277-1818
Fax: 510-277-1919
E-mail Address: alamols@aol.com

Contact: Lance Hale

In Business: 20 years

Clients: Services for all sizes and types of companies. Over 150 Fortune 500 serviced.

Qualifications/Certifications: General Services Administration—Federal Supply Schedule; "Preferred Provider" American Electronics Association;

Members of American Society of Training and Development, Automotive Industry Action Group, American Electronics Association, and California Manufacturers Association.

ANOKA-HENNEPIN TECHNICAL COLLEGE

1355 W. Highway 10
Anoka, MN 55303-1590, U.S.A.
Tel: 612-576-4769; Fax: 612-576-4802
E-mail Address: C.Skotterud@ank.tec.mn.us

Contact: Clo Skotterud, Program Director

In Business: 20+ years

Clients: Small to mid-sized manufacturing and services organizations.

Qualifications/Certifications: AHTC has worked with numerous companies in the areas of quality and other technical training. This includes assisting companies with meeting certification standards as well as ongoing monitoring to maintain the certifications.

APPLIED QUALITY SYSTEMS

2595 Hamline Avenue N
St. Paul, MN 55113, U.S.A.
Tel: 612-633-7902/ 800-633-2588;
Fax: 612-633-7903
E-mail Address: AQSYSTEMS@AOL.COM

Contact: Mark Ames, President

In Business: Six years

Clients: Seagate Technology, U.S. Navy, U.S. Steel, Onan, Ford Motor Co., Rubbermaid, JI Case, Martin Marietta, Southern California Edison, Phoenix, International, Ramsey Technology, Viking Press (Banta Corporation), Alcon Surgical, Deluxe Check, Anchor Hocking, Caterpillar, Cardinal Glass, Matsushita, Wheelabrator.

Qualifications/Certifications: Applied Quality Systems' instructors are professional and experienced practitioners. Their backgrounds include extensive experience in the manufacturing and service sectors.

Certifications include: CQE, CQA, and RAB Lead Assessor. All staff members are ASQC members.

ARCH ASSOCIATES

ISO 9000 Series Training & Support Services
41620 Six Mile Road
Suite 102
Northville, MI 48167-2399, U.S.A.
Tel: 810-449-5433; Fax: 810-449-5434

Contact: William M. Harral, Director

In Business: 13 years

Clients: Clients are primarily in the following industries: Automotive, general manufacturing, basic metals, metal fabricating or forming, machinery/tools design and manufacture, electromechanical, plastic, microbiology, contracted design and other service industries. Client size ranges from small organizations such as Continental Carbide and McQuade Industries that have been featured in Detroit Business and Nation's Business.

Qualifications/Certifications: William M. Harral, QS-LA, CQE, CQA, CRE, PEIT, has 20 years' experience in various engineering, manufacturing, planning, and quality management positions at Ford Motor Company prior to founding Arch Associates in 1983. Mr. Harral is author of numerous articles, books, and handbooks on quality. He is an active ASQC fellow, past-chair of the Greater Detroit section, Region 10 section's deputy director, counselor for Automotive and Audit Divisions, Audit Division standards committee, Quality Management Division Technology Group chair and executive secretary of the General Technical Council. At Automotive Industry Action Group (AIAG), he has served on the Continuous Quality Improvement Project Team since 1984. He currently chairs the Standards Committee and serves as ISO 9000 series technical advisor to the Truck Advisory Group.

B-K EDUCATION SERVICES

2505 Locksley Drive
Grand Prairie, TX 75050, U.S.A.
Tel: 214-660-4575; Fax: 214-641-1327
E-mail Address: BAIQUALITY@AOL.COM
Home Page Address: http://www.bai-bk.com

Contact: Laurie Miller, Vice President

In Business: 13 years

Clients: DuPont, American Airlines, Mobil, Goodyear, Sandoz Chemical, Occidental Chemical, ALCOA, Amoco Chemical, AT&T, Federal Express, U.S. Postal Service, IBM, Nabisco, Johnson & Johnson, Department of Defense, AlliedSignal, Weyerhaeuser, Fort Howard Paper Company, Hoechst Celanese Corporation, and many others. B-K has trained over 10,000 individuals in 10 countries in its quality training courses.

Qualifications/Certifications: The course authors, Clyde Brewer, Dick Kleckner, and Claud Westbrook, have over 90 years' combined experience in quality systems management, including 45 years encompassing all aspects of the certification/accreditation process. They have conducted hundreds of audits for qualification and certification of quality systems to worldwide recognized standards, including the ISO series standards.

BESTERFIELD AND ASSOCIATES
Continuous Quality Improvement Consultants

2588 Westrick Drive
St. Louis, MO 63043-4143, U.S.A.
Tel: 314-878-4488; Fax: 314-878-4488
Home Page Address: DBSTRFIELD@AOL.COM

Contact: Dale H. Besterfield, Principal

In Business: 15 years

Clients: All sizes of manufacturing and service businesses.

Qualifications/Certifications: Dr. Besterfield's services have been used by Fortune 500 companies, universities, and small companies. He has conducted public seminars on ISO/QS-9000, SPC/SQC, TQM, and DOE. In addition he was an examiner for the Baldrige Award. Dr. Besterfield has two best-selling textbooks, *Quality Control, 4e,* 1994 and *Total Quality Management,* 1995. He organized and participated in a six-day seminar on quality assurance at Damascus, Syria, and was a visting research professor in Taiwan.

BTI SERVICES, INC.

1603 Hill Top Drive
Lawrenceberg, TN 38464, U.S.A.
Tel: 800-881-9917/615-766-5062
Fax: 615-766-5067

Contact: Massey Ghatavi, President
375 Finley Avenue, Suite 203
Ajax, Ontario L1S2E2, Canada
Tel 2: 800-881-9917/905-686-0101
Fax 2: 905-686-8101

In Business: Three years

Clients: Society for the Plastics Industry of Canada, EnviroCare, Smith & Wesson, Murray Inc., PCO, KNT Plastics, Horizon Plastics with many more in both the service and manufacturing industries.

Qualifications/Certifications: BTI Services, Inc. was established to focus on assisting organizations make the transition into a competitive market environment across many sectors of industry and commerce in North America. The principals of BTI have extensive experience in providing organizations, both in the service and manufacturing industries, the required framework and discipline to achieve their business goals and objectives. The senior principals of BTI have experience in industry and as management consultants working with major international organizations with whom BTI Services still maintains an associate relationship. Qualifications include:

- IQA Certified ISO 9000 lead assessor.
- RAB certified ISO quality systems lead auditors.
- American Society for Quality Control (ASQC).
- Association for Manufacturing Excellence (AME).
- AIAG qualified auditors specializing in the automotive industry.

BURNHAM ASSOCIATES INTERNATIONAL, INC.

3110 Trenholm Drive
Oakton, VA 22124-1328, U.S.A.
Tel: 703-620-0093; Fax: 703-620-0094

Contact: Howard Mappen, Vice President Consulting

Contact 2: Charles Green, Internal Auditor Training
Leader
La Paz, #35, Fracc. Valle Ceylan
Tiainepantia, Edo 54150, Mexico
Tel 2: 525-388-9901; Fax 2: 525-753-2325

Contact 3: Alejandro Sanchez, Manager, Latin American Market Development

In Business: Five years

Clients: ISO 9000 Clients include: Life Technologies, Inc., Veeder-Root Corporation, Pacific Communications Sciences, Inc., Telecommunications Techniques, Inc., All-Controls, Inc., BWI-KartridgPak, Inc., Hoppmann Corporation, KNORR Brake Corporation, Muller Manufacturing, Ltd., MICROLOG Corporation, Noise Cancellation Technologies, Inc., Metal-Tech, Inc., NIR Systems, Inc., MUSCO Lighting, Inductotherm, Inc., W.L. Gore and Associates, Mateer-Burt Company, Inc., Universal Dynamics, Inc., Sprint International.

Training clients include: U.S. Coast Guard Yard, Sprint International, Hoppmann Corporation.

Qualifications/Certifications: All consultants are IQA qualified and have a minimum of 35 years of experience. Consultants are degreed industrial engineers.

BYWATER, INC.

Two Stamford Landing
68 Southfield Avenue
Stamford, CT 06902, U.S.A.
Tel: 203-973-0344; Fax: 203-973-0345

Contact: George Russel, President

In Business: 15 years

Clients: Cincinnati Bell, ICI (C & P), Monsanto, Philips USA, Kodak, Shell International, AM General, LTV Steel, RAM Mobile Data, Pioneer, Connaught Laboratories.

Qualifications/Certifications: Bywater's consultants and trainers are permanent salaried staff, operating from bases throughout the United States, United Kingdom, Europe, Middle East, North America,

and Australia. They are all experienced, professionally qualified practitioners with substantial hands-on experience turning quality management principles into practice. In addition, many are registered lead assessors with specialist knowledge in the automotive, chemical, electronics, food, pharmaceutical, plastics, telecommunications, healthcare, services, and software industries.

C.L. CARTER, JR. AND ASSOCIATES, INC.

1211 Glen Cove Drive
P.O. Box 5001
Richardson, TX 75080, U.S.A.
Tel: 214-234-3296; Fax: 214-234-3296

Contact: C.L. (Chuck) Carter, V.P., Executive Consultant

Contact 2: G.M. Carter, President

In Business: 32 years

Clients: All sizes of manufacturing and service firms.

Qualifications/Certifications: Registered Lead Auditor by RAB and IRCA; Registered Professional Quality Engineer; Certified by ASQC as CQA, CQE, CRE; Certified by Society of Manufacturing Engineers as C.Mfg.E.; Fellow of ASQC; Certified Management Consultant by Institute of Management Consultants, Inc. N.Y.C., NY; 45+ years of Quality Assurance Experience.

CARPENTER CONSULTING, INC.

39487 Village Run Drive
Northville, MI 48167, U.S.A.
Tel: 313-420-3582; Fax: 313-420-1103

Contact: David Carpenter, President

In Business: Three years

Clients: Diamond Chrome Plating, Duncan Electronics, Fisher Corp., Globe Precision, Hatch Stamping, Precision Stamping, Select Tool and Gage, Sigma Stamping, Valeo Thermal Systems, Wedge Mill Tool, Washtenaw Community College.

Qualifications/Certifications: Mr. Carpenter is an ASQC, CQA, and CQE. He is certified by the RAB and AIAG for ISO and QS-9000 auditing. He has 40 years of experience in industrial quality assurance and related fields.

CEEM INC.

10521 Braddock Road
Fairfax, VA 22032-2236, U.S.A.
Tel: 800-745-5565/ 703-250-5900
Fax: 703-250-5313

Contact: Jeff Henriksen, Vice President Marketing

Contact 2: Brooks Cook, President

Contact 3: Jack Sweeney, Vice President Training Services

In Business: 17 years

Clients: CEEM's clients consist of manufacturing, service, and process companies of all sizes that are concerned with environmental management systems and with quality.

Qualifications/Certifications: CEEM is an employee-owned company that has a staff with extensive experience and expertise in conference management, journalism, marketing, and training programs. CEEM is in the process of ISO 9001 certification.

CHARRO PUBLISHERS, INC.

The Trainers Workshop

Box 3442 Jayhawk Station
Lawrence, KS 66046-0442, U.S.A.
Tel: 913-865-4306/800-598-9009
Fax: 913-865-4311
E-mail Address: CHARROPubs@AOL.COM

Contact: C.W. Russ Russo, Managing Director

In Business: Eight years

Clients: Trainers and educators responsible for training in commercial organizations, including service industries and manufacturers, as well as nonprofit organizations, including government, academic and charitable organizations.

Qualifications/Certifications: CHARRO/The Trainers Workshop have focused their services for

experienced corporate trainers involved in quality change projects within their corporate training function as well as educational leaders embarking on a quality management process in their educational institutions.

CONSULTANTS IN QUALITY INC.

319 Friendship Street
Iowa City, IA 52245-3916, U.S.A.
Tel: 319-337-8283; Fax: 319-351-3140

Contact: Richard B. Stump, Management Consultant

In Business: Six years in current company, with five years' prior consulting with Rath & Strong, Lexington, MA

Clients: Motorola, Inc., Dana Corp., Intel Corp., CSX Corp. (Commercial Services), Baxter Healthcare, U.S. Department of Education, and a variety of smaller businesses, working in such areas as electronics, plastics, hoists, construction, forgings, electric ballasts, and construction engineering.

Qualifications/Certifications: Dick Stump is both an ASQC Certified Quality Engineer and Quality Auditor, a Certified Quality Action Team Facilitator and a BSI-trained Lead Auditor for ISO 9000 quality systems. Mr. Stump is a technical expert for the NIST-sponsored NVLAP Fasteners and Metals program. He served two years with NIST's Malcolm Baldrige National Quality Award (for companies exhibiting a Total Quality Management System).

CONTEMPORARY CONSULTANTS CO.

15668 Irene Street
Southgate, MI 48195, U.S.A.
Tel: 313-281-9182; Fax: 313-281-4023

Contact: Dean H. Stamatis

In Business: 12 years

Clients: Ford Motor Co., IBM, Hewlett Packard, Life Scan, Dell Computers, Hermes Automotive

Qualifications/Certifications: Contemporary Consultants have an average of 15 years' experience in the quality field and all have graduate degrees in either engineering and/or management education. Contemporary Consultants have published many books and articles in the field of quality management. Most of the staff are certified auditors, lead auditors, CQE, certified manufacturing engineers and/or CQA.

DBS QUALITY MANAGEMENT INTERNATIONAL

894 Carriage Way
Lansdale, PA 19446, U.S.A.
Tel: 215-368-6266; Fax: 215-393-4873
E-mail Address: MIKEDOCRT@AOL.COM

Contact: Michael J. Dougherty, RAB QS-LA Principal Consultant

In Business: Five years

Clients: Industries Served include: Aerospace, Defense, Automotive, Medical Device, Medical Packaging, Computer, Telecommunications, Electronics, and Service Sectors.

Qualifications/Certifications: Since 1991 when DBS Quality Management International was established, the company aided over 50 clients through the certification process. DBS has also provided implementation training services for hundreds of attendees at public and private on-site training programs.

DELAWARE COUNTY COMMUNITY COLLEGE

Center for Quality and Productivity

901 S Media Line Road
Media, PA 19063-1094, U.S.A.
Tel: 610-359-5288; Fax: 610-359-7334
E-mail Address:
PLMCQUAY@DCCCNET.DCCC.EDU

Contact: Paul L. McQuay, Executive Director, International Studies
Tel 2: 215-359-7344 ; Fax 2: 610-359-7393

Contact 2: Donald K. Entner, Director
Tel 3: 215-359-5035; Fax 3: 610-359-7393

Contact 3: Angela Kitson, Marketing Manager

In Business: 28 years

Clients: Defense Personnel Support Center, Life Sciences International, Mars Electronics International, Matheson Gas Products, North American Drager, Tecot Electric Supply Company, Inc., Teleflex, Inc., U.S. Coast Guard, U.S. Steel-Fairless Plant, Speakman Company.

Qualifications/Certifications: Delaware County Community College provides training and consulting services to organizations implementing Total Quality Management and to companies seeking ISO 9000 certification. The College has received the approval of the Federal Quality Institute to provide assistance to federal agencies implementing TQM and considering the implications of ISO certification in their procurement process. The College's ISO trainers and consultants are certified lead auditors who have worked with manufacturing, service, and government agencies.

Additional Information:

E-mail Address 2:
DEntner@DCCCNET.DCCC.EDU

E-mail Address 3:
AKitson@DCCC.NET.DCCC.EDU

DEMMA AND DAVIS CONSULTING

5208 Fox Hills Drive
Fort Collins, CO 80526, U.S.A.
Tel: 970-229-1402; Fax: 970-229-1402

Contact: Michael Demma, Partner
Tel 2: 303-682-4010

Contact 2: Ron Davis, Partner

In Business: Six years

Clients: Coca-Cola, U.S.A., Hughes Electronics, Bell of Canada, Allied-Signal, Domino Sugar, Pittsburgh Corning, New York City Transit Authority, Bridgestone/Firestone, Delphi Harrison Thermal Systems, Delphi Saginaw Steering Systems, Allison Transmission, Delco Electronics, Delphi Energy and Engine Management Systems, PACCAR-Kenworth & Peterbilt Trucks, Mitchell Corporation, Southco, PPG Industries-Flat Glass Group, AlliedSignal, Environmental Catalyst, Alumnax Mill Products, The Budd Company, Abbott Laboratories, Becton Dickinson, U.S. Steel, Ohmeda/BOC Healthcare, Stryker, J&J Medical, Bayer Diagnostics, IBM

Corporation, Colorado Memory, Hewlett-Packard, J.D. Edwards, Cook, Composites and Polymers, Norgen/IMI, Seagate Technology.

Qualifications/Certifications: For more than six years, Demma and Davis Consulting has helped manufacturing, service, and educational institutions with the development and implementation of quality systems.

The staff are registered lead assessors and specialists in many different industries. Some of these industries would include: automotive, computer hardware, software, medical device, steel, education, pumps, paints/coatings, transportation, and plastics.

DET NORSKE VERITAS, INC.

4546 Atlanta Highway
Loganville, GA 30249-2637, U.S.A.
Tel: 800-486-4524 x204/ 770-466-2208
Fax: 770-466-4318

Contact: Sam Stockton

In Business: Since 1864

Clients: ABB, AT&T, Baldwin Filters, Caterpillar, Chevron, Dana Corp., Dowell Schumberger, Eaton Corp., Exxon, FMC, Hoechst Celeanese, Kerr-Mcgee, Martin-Marietta, Mitsubishi Consumer Electronics, Monsato, Phelps Dodge, Polaroid Corp., Texas Instruments.

Qualifications/Certifications: Det Norske Veritas, Inc., consultants and instructors are among the most knowledgeable in the industry. Each brings a wide experience from a variety of industries and many have advanced degrees and registrar experience.

DIEHL AND ASSOCIATES, INC.

7164 Beechmont Avenue
Suite 205
Cincinnati, OH 45230, U.S.A.
Tel: 513-232-1781; Fax: 513-231-6753

Contact: Thomas P. Diehl, President

Contact 2: Geoffrey Downer, Vice President

In Business: Seven years

Clients: Toyota, Stevens International, Parker & Amchem, Senco Products Inc., Henkel Corporation, Cross Pointe Paper, University of Kentucky, Badger Paper, ITT, Pilot Chemical, DHL Air, APS Materials, United Lubricants, Institute of Advanced Manufacturing Sciences, ITW Vortec, Shelbyville Chamber of Commerce, ITW Angleboard, Crane Connectors, Western Kentucky University, R-K Electronics, The Neatsfoot Oil Refineries, Ahaus Tool and Engineering, H.B. Fuller Co., H.B. Fuller Automotive, Parkway Products, Imperial Adhesives, Millers Falls Tool Co., S+S Fasteners, Pentaflex, Applied Intelligent Systems, Advanced Products Technology, Inc., Color & Composite Technologies, Inc.

Qualifications/Certifications: All consultants are Registered Lead Assessors and Tutors. They have a combined experience of 52 years working with the ISO 9000/BS 5750 series of standards and have undertaken in excess of 1,300 external audits against the Standard. In addition all consultants are certified as QSA lead auditors to lead the Third-Party Certification Body's audit team in assessing companies against the Automotive QS-9000 Standard.

Our professional consultants have taught the lead auditor course to over 3,200 delegates over the past 12 years.

DILLINGHAM QUALITY CONSULTING, INC.

37264 Thinbark Road
Wayne, MI 48184, U.S.A.
Tel: 313-721-5606; Fax: 313-721-8271

Contact: David R. Dillingham, President

In Business: Eight years, plus 40 years working on quality systems.

Clients: Automotive suppliers, general manufacturing, basic metals, metal fabricating, machine tool, plastic molding. Examples include: Truesdell Enterprises, Budd Corp., Hydro Aluminum-Bohn, ITT Teves, Molded Materials, Hatch Stamping, Valeo Thermal Systems, Sigma Stamping, U.S. Army Automotive-Tank Command, Duncan Electronics, Koppy Corporation.

Qualifications/Certifications: David Dillingham was employed at General Motors for 37 years. He had a number of management positions in purchasing, reliability, and quality control. He is a past chairman of the Greater Detroit Section of ASQC and a fellow in the national organization. He is a Certified ASQC Reliability Engineer, Quality Engineer, Quality Auditor, Certified RAB Quality Systems Lead Auditor, and an examiner for the Michigan Quality Council.

DISTRIBUTION SOLUTIONS, INC.

349 Wooster Way
Lancaster, OH 43130, U.S.A.
Tel: 614-833-0150; Fax: 614-833-0150
E-mail Address: ROGERGARVER@MSN.COM

Contact: Roger Garver, Principal

In Business: Three years

Clients: Borden, Columbus Coated Fabric Division, Consolidation Services, Inc., Interamerican Logistic, and MARK VII Transportation, Inc.

Qualifications/Certifications: Roger Garver, principal, has more than 10 years of management consulting experience in both manufacturing and transportation companies. His Certifications include: APICS Certification in Production and Inventory Management; ASQC Certified Quality Manager; certified in MTM 1,2,3, and applied ergonomics; Certified ISO 9000 Lead Auditor; certified in value analysis and value engineering.

DONALD W. MARQUARDT AND ASSOCIATES

1415 Athens Road
Wilmington, DE 19803, U.S.A.
Tel: 302-478-6695; Fax: 302-478-9329

Contact: Donald W. Marquardt, President

In Business: Five years

Clients: Industrial, Commercial and Service companies, and trade associations in the United States and worldwide.

Qualifications/Certifications: Mr. Marquardt was with the DuPont Company for 39 years. His assignments included management of engineering services and consulting services. He organized and managed the DuPont Quality Management and Technology Center. Mr. Marquardt is chairman of the U.S. Technical Advisory Group and leader of the U.S. delegation to the ISO/TC176 committee, which developed the ISO 9000 standards. He was chair of the ISO/TC176 international task force that prepared the Vision 2000 report establishing the strategic intent for implementing and revising the ISO 9000 series in the 1990s.

EAST CONCORD ASSOCIATES

5512 Concord Avenue
Edina, MN 55424, U.S.A.
Tel. 612-927-0860; Fax: 612-927-5313
E-mail Address: EASTCON@AOL

Contact: Peter Malkovich

In Business: Nine years

Clients: Liquid Carbonic, Butler Manufacturing Company, Engineering Design Group, Canadian Trucking Association, Reality Interactive, Safetran Systems, St. Paul Brass & Aluminum Foundry, Atlantis Transportation Services, Universal Fasteners Inc., Alusuisse, Datalink, Eleventh Mines, Illbruck, IMED Corporation, Marine Associates, Omnetics Connector Corporation, Oglebay Norton Company, Hendrie, Process Management International, rms, TECo Enterprises, Pottter Production Corporation, Service Quality Associates, Zytec, Startex, AMCOM Corporation, Gaskatape Inc., White Oak Transport.

Qualifications/Certifications: A principal, Pete Malkovich, coauthored two ISO 9000 books:

* *Achieving ISO 9000 Registration*
* *Understanding the Value of ISO Registration*

East Concord Associates has assisted manufacturing, process, and service industry organizations with successful and effective registration.

Additional Information:

E-mail Address 2:
PETE@LEARNSTAR.TECOINC.COM

EASTERN MICHIGAN UNIVERSITY

Center for Quality
34 North Washington Street
Ypsilanti, MI 48197, U.S.A.
Tel: 800-932-8689// 313-487-2259;
Fax: 313-481-0509
E-mail Address: Kathy.Trent@emich.edu
Home Page Address:
http://www.emich.edu/public/cq/cq.html

Contact: Kathy Trent, Program Coordinator

Contact 2: Terry Carew, Accounts Manager

In Business: 15 years

Clients: Allegro Microsystems, Allied Signal, American Yazaki, Ervin Industries, Ford Motor Company, General Motors Corporation, GNB Incorporated, Harman Electronics, Imperial Oil, Johnson Controls, Mazda, Motorola, Philips Display, Raychem, Rouge Steel, Tilden Mines, TRW, W.R. Grace.

Qualifications/Certifications: The Center for Quality's instructors are seasoned trainers with extensive industrial experience. Over 20,000 individuals from over 4,000 companies have participated in EMU sponsored training programs.

Additional Information:

E-mail Address 2: Terry.Carew@emich.edu

ELLIS AND ASSOCIATES, INC.

Quality Engineering Consultants
430 Craigs Creek
Versailles, KY 40383, U.S.A.
Tel: 606-873-7459; Fax: 606-873-1353
E-mail Address: Jimikellis.@AOL.COM

Contact: James Ellis, Principal Consultant

Contact 2: Dorothy L. Ellis, CEO

In Business: Two years

Clients: Hitachi Automotive, Lexington Community College, ISSC—Lexington, KY, Altex Electronics, Furness—Newburg.

Qualifications/Certifications: RAB Quality Systems Auditor; Certified Quality Engineer (CQE); IBM Quarter Century Club (IBM Retiree); American Society for Quality Control—Education Chairman—Lexington, Kentucky Chapter.

Additional Information:

E-mail Address 2: Dorothyle.@AOL.COM

ENVIRONMENTAL SCIENCE SERVICES-RETTEW ASSOCIATES, INC.

119 Centerville Road
Lancaster, PA 17603, U.S.A.
Tel: 717-898-7971; Fax: 717-290-8653
E-mail Address: ep@Rettew.com

Contact: Edwin Pinero, Consultant

Contact 2: Ted A. Miller

In Business: Seven years

Clients: Predominantly private sector commercial and industrial; varying from Fortune 500 to small and medium sized enterprises. Specific lists available upon request. Geographic areas include domestic and international.

Qualifications/Certifications: Certified and Registered Professionals, geologists, registered engineers, EARA-approved, advanced lead assessor training for EMS.

ENVIRONMENTAL TECHNOLOGY & MANAGEMENT

2323 Clear Lake City Boulevard.
Suite 180-237
Houston, TX 77062-8032, U.S.A.
Tel: 713-480-8019; Fax: 713-480-8019

Contact: John B. McVaugh, P.E., R.E.M., Principal

In Business: Two years

Clients: Akzo Nobel, Allied Signal, Anheuser Busch, BASF Corp., Delco Electronic Systems, Eli Lilly and Company, Exxon, Fina Oil and Chemical, GE Plastics, General Motors de Mexico, Inland Container, La Roche Industries, Lawrence Livermore National Lab, Los Alamos National Lab, NASA, OxyChem, Reilly Industries, Reynolds Electrical and Engineering, Sandia National LaB.S., Texaco Inc., The Coastal Corp., The Dow Chemical Company, TRW Automotive, Union Carbide.

Qualifications/Certifications: Environmental Technology & Management consultants offer many decades of business, operations, environmental, and quality management expertise gained in the oil, chemical, and aerospace industries and in government. ET & M consultants offer practical insights into organizational structure and management system integration are focused to minimize implementation costs.

EtQ MANAGEMENT CONSULTANTS, INC.

399 Conklin Street
Suite 208
Farmingdale, NY 11735, U.S.A.
Tel: 800-354-4476/516-293-0949;
Fax: 516-293-0784
E-mail Address: 71763.2023@CompuServe.COM
Home Page Address:
http://ourworld.compuserve.com/homepages/EtQHome

Contact: Glenn McCarty, CEO

Contact 2: Nancy Bongiorno, Sales Manager

In Business: Five years

Clients: General Motors, Zenith, Westinghouse, Festo Corp., Rogers Corp., Carlisle Syntec, Interstate Steel, Arlington Metals, Williams and Company, Pass & Seymour/Legrand and over a hundred more.

Qualifications/Certifications: All consultants are either RAB/IQA Lead Assessors or ASQC

CQA/CQE and have a bachelor's in engineering. Software developers are certified Lotus Notes Developers and system administrators.

EuroQuest

8351 Roswell Road
Suite 171
Atlanta, GA 30350, U.S.A.
Tel: 770-395-0124/800-355-3876;
Fax: 770-395-0737
E-mail Address: tmg_eq@ix.netcom.com

Contact: Lynn Soylemez, Marketing Director

Contact 2: Sezer Soylemez, Principal

In Business: Four years

Clients: W.R. Grace, Mead Paper Products, USCan, AT&T, Rockwell, Prestolite, Wire, CR Bard, Coin Acceptors, Computer Image Technologies.

Qualifications/Certifications: EuroQuest consultants have broad-based industry experience covering electronics, automotive, chemical, medical, paper, packaging, information technologies, retail, and finance industries. EuroQuest Auditor General is a fellow of the Institute of Quality Assurance of London. All consultants are registered lead auditors of quality systems. Some are TickIT trained. The average consultant has over 20 years of industry experience.

EXCEL PARTNERSHIP INC.

75 Glen Road
Sandy Hook, CT 06482, U.S.A.
Tel: 203-426-3281 / 800-374-3818;
Fax: 203-426-7811
E-mail Address: xlp@xlp.com
Home Page Address: http://www.xlp.com/

Contact: Dana M. Hatfield, Director of Customer Relations

Contact 2: David Middleton, Vice President

In Business: Eight years

Clients: Over 1000 companies in the United States including: Compaq Computer Corporation, BP Chemicals, Inc., Corning Incorporated, Chevron Chemical Co., Hewlett-Packard, Chrysler Corporation, Ford Motor Company, GM Powertrain, GE, Raychem, and United Technologies.

Qualifications/Certifications: The EXCEL Partnership Inc. is committed to providing the highest quality training. Excel's lead auditor training course is accredited by both the RAB in the United States and the IRCA in the United Kingdom. As a demonstration of Excel's commitment to quality service, Excel is registered to ISO 9001 for the design, development, and delivery of training services.

Additional Information:

E-mail Address 2: compuserv 74521,1366
WWW Address: http://www.xlp.com/

EXTENDED MARKETING, INC.

P.O. Box 28276
Seattle, WA 98118, U.S.A.
Tel: 206-725-1770; Fax: 206-725-1770
E-mail Address: homepage@eeservices.com
Home Page Address: http://www.eeservices.com

Contact: Steve Habib Rose, President

In Business: Two years

Clients: Brownfield Remediation Services, Environmental Services Directory for Washington State, Integra Technology Solutions Center, Law Seminars International, TPS Technologies.

Qualifications/Certifications: Lotus Certified Notes Consultant; Publisher of the Environmental Industry Web Site (www.enviroindustry.com), the ISO 14000 InfoCenter (www.iso14000.com), the Brownfield InfoCenter (www.brownfield.com), the Environmental Cleanup InfoCenter (www.cleanup.com), and numerous other Internet resources for the environmental industry; and Host of The Environment on the Net Conference.

FED-PRO, INC.

2811 19th Avenue.
Rockford, IL 61108, U.S.A.
Tel: 800-833-3776; Fax: 815-282-4304

Contact: Kathy Attebery, Senior Consultant

In Business: 13 years

Clients: Eli Lilly & Co., James River Corp., E.I. DuPont, GAF Building Materials, Allied Signal, TRW, Raytheon Corp., Fairchild Aircraft Corp., AT&T, General Electric, Olin Chemicals, North American Salt Co., Hoke Rubber Products Co., Eastman Kodak, Martin Marietta.

Qualifications/Certifications: FED-PRO, Inc., is a multifaceted consulting and training firm headquartered in Rockford, Illinois. The firm specializes in the area of quality assurance and contract administration. Since 1983, FED-PRO consultants have written and published numerous training manuals for clients nationwide, conducted seminars, and provided consulting services on topics such MIL-I-45208A, MIL-Q-9858A, TQM and ISO 9000. In addition to consulting on quality issues, Fed-Pro consultants have provided services in the area of contract disputes for clients doing business with the U.S. Department of Defense. This includes legal representation of clients to the Board of Contract Appeals.

FOUNDRY QUALITY SYSTEMS

P.O. Box 15606
Loves Park, IL 61132, U.S.A.
Tel: 815-961-9972; Fax: 815-961-9956
E-mail Address: FQS@LL.NET
Home Page Address:
http://www.LL.net/fqs/quality.html

Contact: Ralph Teetor, III

In Business: Four years

Clients: Waupaca Foundry, Ft. Wayne Foundry, Neenah Foundry, Impact Industries, Meridian Magnesium, Chicago White Metals.

Qualifications/Certifications: CQA, CQE, ISO 9000 Auditor IRCA, QS-9000 Trained, Member AIAG

GEO S. OLIVE & CO.

201 N. Illinois Street
Suite 700
Indianapolis, IN 46204, USA
Tel: 317-383-4052; Fax: 317-383-4059
WWW Address: WWW@GSO.COM

Contact: Kirk Eggebrecht

In Business: 80 years

Clients: North American Van Lines, Tokheim Corporation, Mitchel & Scott Machine Co., Carr Metal Products and a number of small to medium-sized manufacturers, distributors and service companies.

Qualifications/Certifications: All clients have successfully gained registration on the first attempt and have generated tangible quality improvements and cost savings in the process. The staff comes primarily from companies who have been registered. Our staff is also subcontracting to registrars in performing Pre-assessments, Registration Audits and Surveillance Audits. Much of the company's work is in manufacturing, but we have a growing list of references in the Transportation, Distribution, and Calibration services industries.

G. R. TECHNOLOGIES

9011 Leslie Street
Suite 211
Richmond Hill, Ontario L4B 3B6, Canada
Tel: 905-886-1307; Fax: 905-886-6327

Contact: Mayer Bernstein

Contact 2: Dwight Cameron

Contact 3: Gitte Jorgensen

In Business: 19 years

Clients: Lever Brothers Ltd., Domtar, Magna, Motor Coach Industries (MCI), Johnson & Johnson, Clemmer Industries, Edscha of Canada, Woodbridge Inoac, Johnson Matthey Ltd., Malette Kraft Pulp & Power Division, Huls Canada, Divacco.

Qualifications/Certifications: 19 years of experience in quality and quality systems. Facilitated, trained, and consulted for over 400 companies. Successfully assisted more than 20 companies to ISO 9000 and QS-9000 registration. Registered (IRCA & RAB) Lead Assessors on staff.

GEORGIA INSTITUTE OF TECHNOLOGY

Center for International Standards & Quality

Economic Development Institute
151 6th Street
Room 143, O'Keefe Building
Atlanta, GA 30332-0640, U.S.A.
Tel: 800-859-0968/ 404-894-0968
Fax: 404-894-1192
E-mail Address: donna.ennis@edi.gatech.edu

Contact: Donna M. Ennis, Marketing Manager

Contact 2: David S. Clifton, Jr., Manager

In Business: Five years

Clients: CISQ is a national organization serving the southeast and U.S. industry with clients in Latin America.

Qualifications/Certifications: CISQ draws from the expertise of its Atlanta and regional office network staff which includes Georgia Tech engineers, researchers, and trainers with extensive industry experience. Staff are certified lead and quality systems auditors and registered professional engineers. They hold a number of certifications in specific industries.

GLADHILL ASSOCIATES INTERNATIONAL

4846 West Cochise Drive
Glendale, AZ 85302, U.S.A.
Tel: 602-435-9114; Fax: 602-939-4412

Contact: Burt Gold

In Business: Five years

Clients: All sizes of manufacturing and service businesses.

Qualifications/Certifications: Gladhill Associates International has assisted over 50 companies to achieve ISO 9000 Registration.

Gladhill has trained over 1,500 people, many of whom went on to become lead auditors. Gladhill's employees are Registered Lead Auditors with over 30 years' quality assurance experience.

GRAND RAPIDS COMMUNITY COLLEGE

Applied Technology Center
151 Fountain NE
Grand Rapids, MI 49503-3263, U.S.A.
Tel: 616-771-3600; Fax: 616-771-3605

Contact: Robert Suchy, Quality Systems Coordinator/Instructor

Contact 2: Jane Neil-Chappell, Director of Business Services

Contact 3: John Gebhart, Statistical Methods Practitioner

In Business: (GRCC)—80 years, (B&TT)—15 years

Clients: GRCC and B&TT serve manufacturing, service, and public organizations in the deployment of technology, skill, and business practices on a local, regional, national, and international basis. The range includes the very small to Fortune 500 companies. References list available on request.

Qualifications/Certifications: B&TT trainers and consultants are drawn from the applied world of business and industry and have formal qualifications where applicable. IRCA or RAB certified assessors provide ISO/QS-9000 services. Academic and practically qualified trainers (with various certifications) and consultants have permanent staff status to provide ongoing stability for our clients.

GRANT THORNTON LLP

One Cleveland Center
1375 E. Ninth Street
Suite 970
Cleveland, OH 44114-1724, U.S.A.
Tel: 216-623-1462; Fax: 216-771-1409

Contact: Victor Murawa, Senior Manager

In Business: 72 years

Clients: All sizes of manufacturing and service businesses within every industry.

Qualifications/Certifications: Grant Thornton LLP has led hundreds of companies through to registration.

H.J. STEUDEL AND ASSOCIATES, INC.

Management Consultants
6410 Enterprise Lane
Suite 200
Madison, WI 53719, U.S.A.
Tel: 608-271-3121; Fax: 608-271-4755

Contact: Sherry Soehnlein, Vice President Operations

Contact 2: Harold J. Steudel

In Business: 22 years

Clients: Many companies in the auto industry; some produce robotics, gear drives, heating and air conditioning systems, instrument panels, and other decorative products.

The company's clients manufacture materials for the apparel industry, medical fields, oil industry, food applications, foundries, and the paper industry. Recent additions to our client list include manufacturers in the semiconductor industry.

Qualifications/Certifications: Harold Steudel, Ph.D., P.E. is an RAB Certified Lead Auditor and has met the requirements for qualification to conduct audits or assessments to QS-9000. His 20 years of consulting and management experience in implementing leading edge techniques for quality improvement. Mr Streudel is an international instructor and speaker on quality and engineering related topics. All of our consultants are experts in their fields and are RAB Certified Quality Systems Lead Auditors.

HEDMAN CONSULTING SERVICES

8809 49th Avenue North
Suite One
Minneapolis, Minnesota 55428, U.S.A.
Tel.: 612-535-5616; Fax: 612-535-0908

Contact: Stephen Hedman
Tel 2: 612-722-6571

Contact 2: Linda Simon, Trainer

In Business: 5 years

Clients: Small and medium sized Fortune 500 companies

Qualifications/Certifications: Mr. Hedman has over 30 years of Quality and Operations experience and is an ASQC Certified Quality Auditor, a RAB Certified Lead Auditor, a former Minnesota Quality Award Examiner and a contract assessor for the Registrar, SGS International Certification Services, Inc. He is a member of the American Association of Cereal Chemists, Institute of Food Technologist, American Production and Inventory Control Society and is a senior member of the American Society of Control.

HESTER ASSOCIATES, INC.

P.O. Box 669
Chatham, MA 02633, U.S.A.
Tel: 508-945-4860; Fax: 508-945-4862

Contact: William F. Hester, President

In Business: Eight years consulting, plus 25 years in industry

Corporation/Clients: General Signal, Stanley Tools, Harrow Industries, TAKATA Restraint Systems, Highland Industries Inc., Trafalgar House, AT&T, The New Can Corporation, PresMet, and others.

Qualifications/Certifications: Hester's clients have been successfully certified to ISO 9000 and QS-9000. In addition, the company offers the capability to document a system with or without management writing procedures. The company looks at ISO 9000/QS-9000 from a business process improvement viewpoint, not as note takers.

Hester has also authored a book entitled The Statistical Problem Solving Handbook the company recently received a national award from the ASQC for our Activity Based Costing Methods used to determine "True Quality Cost."

HUNT QUALITY SERVICES

2833 N.E. Everett Street
Portland, OR 97232-3246, U.S.A.
Tel: 503-235-4668; Fax: 503-235-4668

E-mail Address: dhunt@teleport.com
Home Page Address:
http://www.teleport.com/~dhunt

Contact: Diane M. Hunt, Quality Systems Auditor

In Business: Three years

Clients: Manufacturers of medical device equipment, audio test equipment, computers, ion beam workstations, presentation projection systems, printed circuit boards, industrial pumps, precision metal parts, sawmill equipment, and cables. Distributors of electronic equipment, seals and o-rings, and industrial tooling. Service providers of temporary help and training design and delivery.

Qualifications/Certifications: Diane Hunt is an RAB Certified Auditor of ISO 9000-based quality systems. She is a certified trainer for the EPiC ISO 9000 Training SystemTM, a series of ISO 9000 classes which has been favorably reviewed by the ISO Central Secretariat in Geneva. She is also certified to present the ISO 9000 classes developed by the Oregon-based Partnerships for Quality program. She has served as an active member of the Portland ISO 9000 Users Group Steering Team since 1992 and also serves on the Oregon Quality Initiative's ISO 9000 Technical Advisory Team.

I:Q;9000

P.O. Box 34066
Winnipeg, Manitoba R3T 5T5, Canada
Tel: 204-269-5098; Fax: 204-269-5098

Contact: Patrick Illet

In Business: Three years

Clients: References available on request

Qualifications/Certifications: Personnel registered with the IRCA, ASQC, and members of IQA.

IBM CANADA/ISO 9000 CONSULTING SERVICES

3600 Steeles Avenue. E.
E4/300
Markham, Ontario L3R 9Z7, Canada
Tel: 905-316-5917; Fax: 905-316-4665

E-mail Address: ISO_9000@vnet.ibm.com
Home Page Address: http://www.can.ibm.com (under this, click on "what we offer"; then under this click on "service")

Contact: Patti Dawson, Consultant

Contact 2: John Birke, Manager

In Business: Four years

Clients: Kelly Temporary Services, Moloney Electric, Goodyear Inc., Lantic Sugar, Lafarge Canada, Corfin Inc., Automatech Industrielle Inc., Les Plastiques Cy-Bo Inc., Industries Kancorp, Construction C+G Beaulieu Inc., Purolator Courier, Reynolds Extrusion, Valvoline Canada, BASF Canada.

Qualifications/Certifications: RAB auditors training on QS-9000 through AIAG in Detroit.

Additional Information:

E-mail Address 2: PDAWSON@vnet.ibm.com

WWW Address: www.can.ibm.com

INFORMATION MAPPING, INC.

300 Third Avenue
Waltham, MA 02154, U.S.A.
Tel: 800-MAP-4544/617-890-7003
Fax: 617-890-1339
Home Page Address: http://www.infomap.com

Contact: Jerry Paradis, Director of ISO/QS-9000 Services

In Business: 27 years

Clients: AT&T, Bell Atlantic, Dow Chemical Co., Eastman Kodak, Intel Corp., Shell Oil Co., Monsanto Chemical, Allergan, ACT Manufacturing, Exxon, Bausch & Lomb.

Qualifications/Certifications: IMI experts have made presentations at many ISO 9000/quality conferences, including: ASQC regional conferences, BOSCON, and Northeast Quality Control Conference. IMI staff maintains membership in many ISO/quality organizations, including

ASQC. IMI consultants include RAB Accredited Lead Assessors.

INNOVATIVE QUALITY SERVICES

5703 Willow Elm Drive
Arlington, TX 76017-4022, U.S.A.
Tel: 817-561-4319; Fax: 817-561-1642
E-mail Address: iqserv@iqserv.com
Home Page Address: http://www.iqserv.com

Contact: Greg Westall, President

In Business: Eight years

Clients: Aerospace and Commercial Fastener Industry

Qualifications/Certifications: ASQC Certified Quality Auditor; Staff writer for the *American Fastener Journal*

INTERNATIONAL ISO GROUP

11405 Sebring Drive
Cincinnati, OH 45240, U.S.A.
Tel: 800-ISO-3066/513-742-7500;
Fax: 513-851-5881

Contact: Terri Parker-Halpin, Principal

Contact 2: Victor C. Halpin, Principal
OMIC-ISO Group
15-6 Nihonbashi Kabuto-Cho
Chuo-Ku, Tokyo, Japan
Tel 3: 011-81-3-3669-5281;
Fax 3: 011-81-3-3669-5190

Contact 3: Jun Sato, Senior Consultant

In Business: Seven years

Clients: The International ISO Group specializes in the automotive industry, manufacturers, and service companies. Many of our clients have international markets and IIG services them in many countries.

Qualifications/Certifications: The International ISO Group's consultants are also instructors. They have extensive manufacturing, industry, quality systems, and business experience as well as years of ISO 9000 experience. Some 116 consultants work and train in Japanese; many clients are Japanese transplant companies. IIG has over 25 years' experience in documentation systems, training, and consulting, and is on the ASQC approved list of consultants/trainers.

INTERNATIONAL QUALITY ASSOCIATES, INC.

15455 NW Greenbrier Parkway
Suite 210
Beaverton, OR 97006, U.S.A.
Tel: 503-531-0302; Fax: 503-531-7259
E-mail Address: intlqual@aol.com

Contact: Jeff Omelchuck, ISO 14000 Contact

Contact 2: Ron Schwartz, ISO 9000 Contact

Contact 3: Roy Shima, CE Mark

In Business: 4 years

Clients: Hewlett Packard, Intel, Tektronix, Mentor Graphics, Freightliner Corp., Consolidated Freightways, Inc., Xilinx, Megatest, Domtar, Hanna Andersson, Danner Shoe, David Evans and Associates and many small businesses.

Qualifications/Certifications: IQA's consultants include ISO 9000 Lead Assessors, Certified Quality Engineers, State Quality Award examiners and EARA registered Principal Environmental Auditors.

INTERNATIONAL SYSTEMS REGISTRARS LTD.

18980 Told Woods Drive
Unit 29
Brookfield, WI 53045, U.S.A.
Tel: 800-785-1867/414-785-1867
Fax: 414-785-1867
E-mail Address: isrglt@execpc.com
Home Page Address:
http://www.execpc.com:bo/~isrglt

Contact: Dennis L. Gorectke, President

In Business: International Systems Registrars Ltd. has been incorporated since July 21, 1992.

Clients: ACE Machine & Stamping Inc., Bio-Systems Corporation, Quest Technologies Inc., Quality Calibration Services, Eaton Corporation—Cutler Hammer Products Division, Ford Fairlane Training & Development Worldwide, SpacesAvenuer Corporation, Plexus Corporation—Electronics Assembly Division, Nelson Container Corporation, Monopanel Technologies, Johnson Controls - Battery Division, Northwestern Industrial Distributors, Trostel Corporation, Xymox Corporation.

Qualifications/Certifications: Dennis Gorectke is president of International Systems Registrars Ltd. He is an IRCA Registered Internal Auditor, Certified Quality Auditor by the ASQC, a Certified Quality Engineer, and a Certified Quality Technician. Mr. Gorectke has more than 20 years of quality engineering and quality auditing experience in many industries including automotive with Ford Fairlane Training, industrial, and electromechanical controls, motorcycles, transformers, robotics, and others. In addition, he has consulted with more than 30 Wisconsin area corporations, helping twelve achieve ISO 9001/9002 registration.

INTERNATIONAL TECHNOLOGY TRANSFER INC. (ITTI)

P.O. Box 1043
Waukesha, WI 53187, U.S.A.
Tel: 414-547-5303/414-547-8086;
Fax: 414-547-5351

Contact: Ashok M. Thakkar, CEO

Contact 2: Bharati A. Thakkar, Vice President

In Business: 11 years

Clients: California MicrowAvenue, Redwing Shoes, McMillan Electric Co., Cold Spring Granite Co., Telsmith Inc., Gemtron Corp., AFG Industries, Marsh Electronics, Meehan Seaway, WS Tyler Inc., Downtown Radio, Eaton Corp., Milwaukee Machine & Engineering, Northwestern Industrial, Weimer Bearings, Boggis Johnson, Process Technology Holdings.

Qualifications/Certifications: Mr. Thakkar has traveled to over 20 countries and is a frequent speaker at various national and international conferences. He has also worked with the late Dr. Kaoru Ishikawa of Japan. In November, 1992, the

Governor of Wisconsin gave an award to Ashok
Thakkar (CEO) for being a "state leader in the
practice of TQM."

INTERQUAL

6168 Verdura Avenue
Goleta, CA 93117-2004, U.S.A.
Tel: 805-967-9958; Fax: 805-964-0767
E-mail: InterQua@silcom.com
Home Page Address: InterQual@aol.com

Contact: Eliana Borges, CEO

Contact 2: George Hudak, Management Associate

In Business: Two years

Clients: Air Liquide, Brylen Laboratories, Buena
Biosystems, Condor DC Power Supplies, CUI Cor-
poration, Discus Dental, Gingi-Pak, Industrias SL,
SA de CV, Karl Storz Imaging, QAD Inc., RIFOCS
Corporation, SL Waber, St.Jon Laboratories, Waber
de Mexico, Westside Packaging Inc.

Qualifications/Certifications: Eliana Borges is a
Certified Quality Systems Lead Auditor, a UCSB
Instructor, and a Consultant with Intertek Services
(Registrar). Ms. Borges has participated in numerous national and international certification audits.
Ellie is fluent in Spanish, Portuguese, and English,
and she is an Effective and Motivational Trainer,
with expertise in ISO 9000 Systems implementation and assessments. Ellie Borges is an ISO 9000
Quality Systems lead auditor certified by the Registrar Accreditation Board. She has 18 years of experience in high tech manufacturing and quality
control.

INTERTEK INC.
Inchcape Testing Services

9900 Main Street
Suite 500
Fairfax, VA 22031, U.S.A.
Tel: 703-591-1320; Fax: 703-273-4124

Contact: Dennis A. Taylor, Director, Consultancy Services and International Business Development

In Business: 22 years

Clients: Over 400 customers in the aerospace, aircraft,
automotive, computer, communications, electronics, medical, chemical, utilities, and many other
industries.

Qualifications/Certifications: INTERTEK's quality
system is certified to ISO 9002 by an independent
registrar with accreditation traceability to the
NACCB and the RAB. This made INTERTEK the
first U.S.-based contract labor organization
certified to ISO 9000. INTERTEK's consulting services have been evaluated and registered to the
same ISO 9000 series standards that it prepares its
clients for. INTERTEK's trained consultants have a
proven track record of preparing clients for a successful registration assessment.

IQS, INC.

19706 Center Ridge Road
Cleveland, OH 44116, U.S.A.
Tel: 800-635-5901 / 216-333-1344
Fax: 216-333-3752
E-mail Address: Iqs@ix.netcom.com

Contact: John M. Cachat, President

Contact 2: Craig L.Young, Vice President

Contact 3: Rick Cross, Sales and Business
Development

In Business: Since 1986

Clients: Azko Nobel Chemicals, Beckman Instruments,
Parker Hannifin, The Timeken Company, TRW Automotive.

Qualifications/Certifications: IQS has years of hands-
on experience, backed by extensive academic credentials. IQS has helped organizations prepare for
and pass ISO 9000 and QS-9000 assessments. IQS
offers organizations direct access to the fundamental strategies behind the IQS Business System with
its field tested team of quality assurance specialists
and planners. With a consortium of customers and
industry experts over several years, IQS has developed a proven methodology and a unique business
model for quality measuring and managing
business operations and processes.

ISO SYSTEMS, INC.

1234 Clematis Drive
Streamwood, IL 60107, U.S.A.
Tel: 708-830-5814; Fax: 708-830-5814
E-mail Address: Stanorm@aol.com

Contact: Stanley G.Ormbrek, CQA, President

In Business: Since 1992.

Clients: DoALL Corporation, Plitek, VisionTek, APAC, ACME Industries, Rival Company, Greenlee, Columbus Packaging, General Converting.

J. P. RUSSELL & ASSOCIATES

5980 East Bay Boulevard
Gulf Breeze, FL 32561, U.S.A.
Tel: 904-916-9496; Fax: 904-916-9497
E-mail Address: jprussell@mcimail.com

Contact: J. P. Russell, President

In Business: 10 years

Clients: Allied-Signal American Cyanamid, Internal Revenue Service, Owens/Corning Fiberglass, John Deere, Inc., Cargill, Inc., Connaught Laboratories, AEG Transportation Systems, Tecumseh Products Company.

Qualifications/Certifications: JPR's staff consists of certified auditors.

JAMES LAMPRECHT, Ph.D.

1420 NW Gilman Boulevard
Suite 2576
Issaquah, WA 98027-7001, U.S.A.
Tel: 206-644-9504; Fax: 206-644-9524
E-Mail Address: jiml@wolfenet.com

Contact: James Lamprecht

In Business: 8 years

Clients: ADEC, Applied Microsystems, AMTECH, Avtech, Pentz, Safety Supply, Ederer, PACCAR, Electroimpact, Computing Devices International, AMI, ILE, CDS, ECT, Audio Precision, Microdisk Services, Antipodes, NetExpress, Vickers, Hoechst

Celanese, Petrolite, Diamond Shamrock, FINA, GE Medical, Ethyl Corporation, Brown & Root, Red Dot, Manhatan Cable Company, LaRoche Chemicals, Miniature Precision Bearings, Intalco, DATA I/O, Amperif, BIO-RAD, Minco, Telxon Corporation.

Qualifications/Certifications:

- ISO 9000 Lead Assessor; ISO 9000 Certified Assessor IQA Registration AFAQ (France)

- ISO 14000 Auditor

- Full member of U.S. TAG committee on ISO 9000 TC/176, ISO 14000, TC/207 and Statistical committee TC 69

- Chair of the October 1994 conference: "Straight Talk on ISO 9000"

Author of:

> *ISO 9000: Preparing for Registration,*
>
> *Implementing the ISO 9000 Series,*
>
> *ISO 9000 and the Service Sector,*
>
> *ISO 9000 Implementation for Small Business,* and
>
> *ISO 14000: Issues and Implementation Guidelines for Responsible Environmental Management.*

J-E-T-S, INC.

2843 Edwin Jones Drive
Charlotte, NC 28269, U.S.A.
Tel: 800-944-1994; Fax: 704-547-9178

Contact: John H. Johnson, Vice President

In Business: Since 1984.

Clients: J-E-T-S specialty is working with highly regulated environments, including nuclear power plants, insurance, government, government contractors, and banking. A partial list of clients includes:

U.S. Department of Energy, Commonwealth Edison Company, Pacific Gas & Electric, Apex Environmental Consultants, Duke Power, Florida Power and Light, EG&G, Brookhaven National Laboratories, Rockwell International, Northeast Utilities Group.

Qualifications/Certifications: J-E-T-S consultants are qualified personnel serving a variety of needs, including administrative, operations, advanced manufacturing, processing, and customer service environments. Personnel are available with almost any certification that may be required, including Level III per NQA-1, lead auditor, certified quality auditor or engineer, professional engineer, and others. J-E-T-S also has personnel with a variety of security and access clearances, including nuclear power plant and DOE "Q" clearances.

JOHN A. KEANE
AND ASSOCIATES, INC.

575 Ewing Street
Princeton, NJ 08540, U.S.A.
Tel: 609-924-7904; Fax: 609-924-1078
E-mail Address: Keane@cnj.digex.net
Home Page Address:
HTTP://WWW.CNJ.DIGEX.NET/~Keane

Contact: Michael Kane, Marketing Director

Contact 2: Leonard F. Newton, Senior Associate

Contact 3: Seymour Altucher, Senior Associate

In Business: 26 years

Clients: American National Can, BASF, Caterpillar, Digital Equipment, Dupont, General Electric, Hoffman La Roche, IBM, Lockheed, Monsanto, Siemens/ Stromberg Carlson, Schlumberger, Vickers.

Qualifications/Certifications: John A. Keane and Associates, Inc., has been a leader in computer-based, integrated quality information systems for more than 25 years. The company has helped over 100 mid- and large-sized manufacturers plan, design, select, implement, and operate industrial strength quality execution solutions for ISO 9000/QS-9000, MIL-Q-9858A, FDA-GMP/GLP, and other regulated/audited environments.

Additional Information:

Home Page Address 2:
http://www.cnj.digex/~keane

JOSEPH TIRATTO
AND ASSOCIATES, INC.

5 North Longview Road
Howell, NJ 07731-1701, U.S.A.
Tel: 908-367-0837; Fax: 908-367-8898

Contact: Joseph Tiratto, President

In Business: Five years

Clients: Manufacturers of a wide variety of products including machinery, electrical, electronics, plastics, medical devices, chemicals, autos, automotive parts; steel mills, service industries, mining, shipping, and shipbuilding in the United States and worldwide.

Qualifications/Certifications: Joseph Tiratto has more than 35 years of engineering and quality management service. He has conducted quality system audits in 17 countries worldwide. He has B.S. degrees in naval architecture, marine engineering, and mechanical engineering. He has an M.S. degree in quality management. His qualifications also include Professional Engineer's license, Marine Engineer's license, Registered European Engineer (EU and EFTA Countries), Chartered Engineer (England), Registered Lead Auditor IQA/IRCA and RAB, Approved Lead Auditor Instructor, past member and secretary of the board of directors of the Registrar Accreditation Board (RAB). Member of ASQC Z-1 Committee on QA, and member of TAG to ISO 176 Committee on QA.

KELLY CONSULTING GROUP

7314 Sunshine Circle
Tampa , FL 33634, U.S.A.
Tel: 813-884-2195; Fax: 813-884-2195

Contact: A.M.Kelly, President

In Business: Three years

Clients: Small to medium size manufacturing and service businesses. Specializing in chemicals, plastics, metal fabrication, telecommunications, electronics, medical supplies, distribution systems, and professional services. Specific references available upon request.

Qualifications/Certifications: Trained and extensively experienced hands-on professionals with ASQC,

CQE, and/or CQA certification and RAB RLA and assessors.

KEMA-INTERNATIONAL QUALITY CONSULTANTS U.S.A.

4379 County Line Road
Chalfont, PA 18914, U.S.A.
Tel: 215-822-4227; Fax: 215-822-4267

Contact: Ron Cerzosimo, Vice President

In Business: 15 years

Clients: Includes companies (large and small) in research & development, engineering, manufacturing, and services. Industries cover communications, transportation, aerospace, pulp and paper, medical products and pharmaceuticals, chemical/process, petrochemical, energy and environmental, utilities, regulatory agencies, and national laboratories

Qualifications/Certifications: KEMA-International Quality Consultants (IQC)-U.S.A. is a division of KEMA-Powertest, Inc., the U.S. susidiary of KEMA, the Netherlands, a world-class leader in independent electrical testing, certification, engineering, and consulting services.

KEM-IQC has assisted numerous organizations (including Fortune 500) establish measures and document provisions to achieve quality systems maturity and certification (ISO 9001 and 9002).

THE KENNEDY COMPANY

5 Nesenkeag Drive
Litchfield, NH 03052-2420, U.S.A.
Tel: 603-429-3210; Fax: 603-424-7021

Contact: Robert Kennedy

In Business: 3 years

Clients: ABB, Digital Equipment Corporation, Mercury Computer, Dialogic Corporation, Honeywell, Wyman Gordon Investment Castings, Rotondo Pre-Cast, Lockheed Martin, Gerber Scientific, Boston Technology, BBN, Biometrix, Magnetometric Devices Inc., Nickerson Assembly, New Venture

Technology, New Hampshire Technical Institute, Rock Valley College, Center for Quality Management, Government of China—Academy of Science.

Qualifications/Certifications: Robert Kennedy is a member of the U.S. delegation to ISO Technical Committee 176.

Mr. Kennedy has extensive, practical ISO 9000 quality system implementation experience. While employed by Digital Equipment Corporation, he was responsible for the planning, justification and implementation of the Corporate ISO 9000 program. He is an RAB certified Lead Assessor. Robert Kennedy holds degrees in Electrical Engineering Management, Logistics and Repair Depot operations. He is a frequent speaker at National and International forums.

KENNEDY and DONKIN QUALITY INC.

25 Corporate Drive
Suite 230
Burlington, MA 01803, U.S.A.
Tel: 800-500-4ISO/617-229-2302
Fax: 617-229-2035
E-mail Address: PGMAC@TIAC.NET

Contact: Patricia Macdonald, Marketing & Administration Coordinator

Contact 2: Ed Haynes, Vice President

In Business: Kennedy & Donkin has over 105 years' experience in consulting and has provided quality related services since 1928.

Clients: Kennedy and Donkin has a list of clients from all over the world including: (United Kingdom) Bechtel, British Rail Infrastructure Services, Davis & Metcalf, The Highways Agency, Eve Construction Ltd.; (Germany) ABB, Siemens; (Japan) Fuji Electric Co., Inteco, Toshiba; (Hong Kong) China Light & Power, Hong Kong Government; (United States) McDonnell Douglas Corporation, Westinghouse Electric.

Qualifications/Certifications: Kennedy and Donkin operates its own in-house quality assurance program and has been certified under the UKAS (formally known as NACCB) in conformance with ISO 9001. They are members of: Association of

Consulting Engineers, British Consultants Bureau, British Quality Association, Institute of Quality Assurance, as a Participating Organization, and Trainers of Assessors of Quality Management Systems, F.D.I.C., Department of Industrial Safety, British Standards Institute, Manchester Chamber of Commerce and Industry, NAFLIC. RAB/IQA Accredited Lead Assessor Training Course.

KOLKA and ASSOCIATES
International Legal Consultants

2193 Spear Point Drive
Marietta, GA 30062, U.S.A.
Tel: 770-977-4049; Fax: 404-894-0485

Contact: James W. Kolka, President

In Business: 30 years

Clients: Georgia Tech Research Institute, Kaiser Aluminum & Chemical Corporation, Society for Automotive Engineers, International, MEDMARC Insurance Company, National Electrical Manufacturers Association/Diagnostic Imaging Division, American Petroleum Institute, Cymer Laser Technologies, Inc., Schlumberger Technologies ATE Division, Bayer Diagnostics, Johnson & Johnson, ASQC.

Qualifications/Certifications: Dr. Kolka has a Ph.D. in political science and international affairs from the University of Kansas, a J.D. in product liability, environmental law from the University of Wisconsin-Madison and a B.S. in political science (Economics/Chemistry) from the University of Wisconsin-Eau Claire. He served as a full professor, served as vice president at Drake University and in the University System of Georgia. He was instrumental in creating the Center for International Standards and Quality at Georgia Tech and has conducted over 200 seminars and written 100 articles and six books concerning product liability, environmental liability, services liability, product safety, European Union Technical Standards, ISO 9000, QS-9000 and ISO 14000, exporting and competitiveness, and has lectured throughout the United States, Canada, Mexico, Europe, and Latin America.

L. MARVIN JOHNSON AND ASSOCIATES, INC.

822 Montezuma Way
West Covina, CA 91791, U.S.A.
Tel: 818-919-1728; Fax: 818-919-7128

Contact: L. Marvin Johnson, CEO

In Business: 25 years

Clients: Partial listing of sponsors and attendees to L. Marvin Johnson and Associates' audit seminars includes the following: AT&T, BF Goodrich, General Electric, General Motors, Boeing, Hughes, Boston Edison, U.S. Army, U.S. Department of Energy, NASA, U.S. Navy, and the French, Italian, Swedish, and British governments.

Qualifications/Certifications: CEO L. Marvin Johnson graduated from the University of Southern California in 1949 with a degree in industrial engineering. He is a registered professional industrial engineer and professional quality engineer. He has 45 years of experience in quality assurance and related fields. Mr. Johnson lectures and conducts quality audit courses for industrial contractors and government agencies. He is author of *Quality Assurance Program Evaluation* and *Quality Assurance Evaluator's Workbook*. He is certified by the IQA/IRCA of the United Kingdom as a quality system lead assessor.

LEADS CORPORATION

230 North Elm Street
Greensboro, NC 27401, U.S.A.
Tel: 800-315-3237/910-275-9989
Fax: 910-275-9952
E-mail Address: leadscorp@aol.com
Home Page Address: www.leadscorp.com

Contact: J. Michael Crouch, Chairman and CEO

In Business: 10 years

Clients: Broadband Technologies, Washburn Graphics, U.S. Dept. of Agriculture, NISH, U.S. Dept. of Commerce, (Patent and Trademark office), U.S. Dept. of Justice (Immigration & Naturalization Service), U.S. Dept. of Defence (Joint STARS, Ft.

George S. Meade, TARDEC, USM.C. Log. Command, SWFLANT, etc.).

Qualifications/Certifications: The experience of the Leads' consultants includes senior management expertise in the service, manufacturing, human resources, aerospace, and engineering areas. Leads Corporation is certified by the U.S. federal government and the Arizona Department of Transportation to provide ISO 9000 consulting and training services. All consultants bring hands-on experience to every task and client.

Additional Information:

E-mail Address 2: leadsnc@aol.com
E-mail Address 3: richmall@aol.com
E-mail Address 4: mchapman@ossinc.net
BB.S. Address: mikrouch@aol.com

LEXINGTON COMMUNITY COLLEGE—EAST

The Center for Community Partnerships

817 E. Third Street
Lexington, KY 40505, U.S.A.
Tel: 606-257-3463; Fax: 606-323-1915

Contact: Winston Dodson, ISO/QS-9000 Training Manager

In Business: Five years

Clients: Lexington Safety Products, LexMark Inc., White Hydraulic, Rexroth Corp., Buckhorn Inc., Corning Inc., Fosroc Inc., Central Manufacturing Co., Westvaco Corporation, Glen Tech Co., Toyota-Tsusho America.

Qualifications/Certifications: The Center for Community Partnerships, an extension of the University of Kentucky, is a nonprofit entrepreneurial group which has been proactive to the needs of manufacturing companies in regard to quality assurance training. For five years, the Center has provided and continues to provide quality services at a reasonable cost. Training is provided by an RAB Auditor.

LITVAN ENTERPRISES

Route 4, Box 62-A
Winona, MN 55987, U.S.A.
Tel: 507-643-6057; Fax: 507-643-6057
E-mail Address: LITVAN@dakota.polaristel.net

Contact: Len Litvan, President

In Business: Four years

Clients: Fastenal, IBM-Rochester, Electrochemical Products Inc., Peerless Chain Company, Red Wing Shoe, Equality Die Cast, Dayco PTI, DB Industries, Pepin Manufacturing Company, CANAMER International, Miller FELPAX, RTP Company, A &L Machine, Polymer Composites Inc., Fiberite, ALTEC International, and others.

Qualifications/Certifications: Litvan Enterprises is an affiliate of QNET, a Twin Cities quality assurance and international trade training firm which provides ISO 9000 and QS-9000 training, consulting, and funding information. QNET's network of experienced affiliates has assisted over 300 firms achieve registered status through their offices in Arizona, California, Iowa, Michigan, Minnesota, New Jersey, New Mexico, and Texas. Specialists of Litvan Enterprises have passed the IRCA/RAB certified lead assessor course, conduct baseline audits and are adjunct faculty at Minnesota technical colleges and St. Mary's University of Minnesota.

MANAGEMENT CONTROL SYSTEMS, INTERNATIONAL

17843 North 17th Place
Suite 100
Phoenix, AZ 85022, U.S.A.
Tel: 602-992-5197; Fax: 602-992-5297
E-mail Address: MCSINTL@Primenet.com

Contact: Patrick J. McDermott, President

In Business: Management Control Systems, International has been in business for five years; President, Patrick J. McDermott has been in the quality/environment business for 21 years.

Clients: Management Control Systems, International's clients consist of government, manufacturing, service, and process companies of all sizes concerned with quality and environmental management systems.

Qualifications/Certifications: Registered Lead Auditor, IRCA/IQA (International Register of Certified Auditors). International Offices, London, England.

MANAGEMENT STANDARDS INTERNATIONAL, LTD.

636 South Central Avenue
Suite 201
Atlanta, GA 30354-1988, U.S.A.
Tel: 404-766-9000; Fax: 404-767-3217
E-mail Address: 76053.165@Compuserve.com

Contact: Steve Holladay, President

Contact 2: Larry Bissell, Vice President

Contact 3: Bryan Robertson, Vice President

In Business: MSI has been incorporated since November, 1993.

Clients: ABB Power T&D, Coleman Powermate, Coleman Outdoor Products, Concert Telecommunications, Gaylord Container Corporation, Hitachi Telecom, International Paper, Kawneer Aluminum Fabrication, PPG Industries, Rhone Poulenc, University of Missouri, Union Camp

Qualifications/Certifications: Entire staff has RAB Quality Systems Lead Auditor Status. The staff includes two ASQC Certified Quality Engineers, two MBNQA Examiners, a voting member on TC176 and international committee on ISO 10011, two reviewers of RAB auditor applications, and two RAB auditors of ISO 9000 and QS-9000 Registrar. The company has a 100 percent success rate in over 50 ISO 9000 Registration Audits. The staff has conducted over 700 ISO 9000 site audits both domestic and international.

Additional Information:

E-mail Address 2:
103214.3705@Compuserve.com

E-mail Address 3:
102433.1762@Compuserve.com

MANAGEMENT SYSTEMS ANALYSIS, INC.

P.O. Box 136
Royersford, PA 19468, U.S.A.
Tel: 610-409-0168; Fax: 610-409-0167
E-mail Address: 74776.1527.compuserve.com

Contact: James Highlands, President

In Business: MSA, Inc. was founded in 1986.

Qualifications/Certifications: Mr. Highlands has 25 years of experience. He is a registered lead assessor with both RAB in the U.S. and IQA in the UK. He is the vice chair of the U.S. SubTag to ISO TC 207 on Environmental management and a delegate to both ISO TC 176 on Quality Management (ISO 9000) and ISO TC 207 on Environment (ISO 14001) and one of the original delegates to ISO's Strategic Advisory Group on Environment. He is a certified lead assessor (IQA-London), a certified lead auditor (RAB-US), a certified quality auditor (ASQC) and a certified nuclear lead auditor (NQA-1/ANSI N45.2.23).

MGMT ALLIANCES INC.

513-1755 Robson Street
Vancouver, BC V6G 3B7, Canada
Tel: 604-669-6490; Fax: 604-669-6496
E-mail Address: mgmt@mgmt14k.com
Home Page Address: http://mgmt14k.com/

Contact: Isis Fredericks, Director

In Business: Four years

Clients: Oil/Gas, Chemical, Service, Manufacturing, Fabrication, Education, Pulp/Paper, Energy, Government, Distribution.

Qualifications/Certifications: B.Sc., CQE

MOORHILL INTERNATIONAL GROUP, INC.

2015 N. Dobson Road
Suite #4-B56
Chandler, AZ 85224, U.S.A.

Tel: 800-ISO-1987/602-491-2007;
Fax: 602-491-2101
E-mail Address: 80375@EF.GC.MARICOPA.EDU

Contact: Erik V. Myhrberg, Director, CEO

In Business: Began operations as Myhrberg & Associates in January 1992, incorporated as Moorhill International Group, Inc., in July 1993.

Clients: Arizona Department of Commerce, Japan External Trade Organization, CDI, Inc., MicroAge Computer, IBM (Phoenix and Tuscon, AZ), Innova, Inc., Geraghty & Miller, Inc., Unifiber, Inc.

Qualifications/Certifications: All Moorhill International Group, Inc., directors have completed a recognized assessor/lead assessor training course and have passed the written examination. All senior trainers are recognized as Certified Lead Assessors (including training time and assessments). The company also implements courses in several languages and cultural content. Each director holds higher education degrees, including: M.I.M., M.S.E., M.B.A., J.D., B.S. Memberships include the ASQC and the World Trade Center.

MOYE COMPANY LIMITED

59 Breadner Drive, Suite 102
Toronto, Ontario M9R 3M5, Canada
Tel: 416-248-9187; Fax:416-248-9187

Contact: Bohdan Dyczkowsky, Principal Engineer, Director

In Business: Six years

Clients: Honeywell Canada, Centennial College, Ontario Provincial Government , Quebecor Printing, Fuller-F.L. Smidth, Ford Electronics, Imperial Oil, Saft Nife.

Qualifications/Certifications: Bohdan Dyczkowsky, director of IC Company Limited, is a native of Toronto, Ontario. He graduated from the University of Toronto with a BASC in mechanical engineering. He has broad manufacturing experience in construction, steel fabricating, nuclear piping, and electronics. He was secretary of the Canadian Delegation to the ISO Technical Committee 176 during the development of the ISO 9000 series. Mr. Dyczkowsky is a registered professional engineer, a

registered lead auditor with the IQA in England and the RAB in the United States and an active member of the ASQC.

N. C. KIST and ASSOCIATES, INC.

900 East Porter Avenue
Naperville, IL 60540, U.S.A.
Tel: 630-357-1180; Fax: 630-357-3349

Contact: Nicolaas C. Kist, President

In Business: 24 years

Clients: Siebe Environmental Controls, Barber Coleman, International Computers, Ltd., Rhinelander Paper, Sonoco Fibre Drum, Lincoln-Smitweld, Milwaukee Gear, Stainless Foundry & Engineering, Hoechst Celanese, AEC Engineering, Stepan, Caterpillar.

Qualifications/Certifications: N. C. Kist and Associates, Inc., specializes in developing cost-effective quality systems based on national and international standards. Since 1972, the company has assisted more than 250 clients worldwide. In 1987, N. C. Kist and Associates began to offer ISO 9000 consulting with 30 ISO 9000 customers in the United States, Europe, and the Pacific Rim. Twenty customers have obtained ISO 9001 or 9002 certification. N. C. Kist also has RAB certified lead auditors on staff.

NATIONAL INSTITUTE FOR QUALITY IMPROVEMENT

17 Forest Avenue
Fond du Lac, WI 54935, U.S.A.
Tel: 414-923-9600/800-840-5095
Fax: 414-921-8228
E-mail Address: NIQI@aol.com

Contact: Henry J. Lindborg, Ph.D., Executive Director

In Business: Eight years

Clients: John Deere, Firstar Bank, Giddings and Lewis, Midstates Aluminum, Southwest Metal, Thomson Newspapers, hospitals, schools, professional organizations, state, and private colleges and universities, education and transportation departments in state and city government.

Qualifications/Certifications: Experienced U.S. and international consultants in organizational development and quality. Principals hold graduate degrees from the University of Wisconsin and the University of Chicago. Dr. Lindborg and his associates teach organizational development at the graduate level. Experts in organizational values, learning, and team development techniques. Developers of the Quality Opportunity Index, (QUOIN(TM), survey instrument. Dr. Lindborg is chair-elect of ASQC's Education Division.

NORMAC-WILL and ASSOCIATES
Medical Device Quality and Regulatory Consultants

20 Canary Court
Guilford, CT 06437-1428, U.S.A.
Tel: 203-457-9250; Fax: 203-457-9250
E-mail Address: normacwill@aol.com

Contact: William G. McMahon, Senior Partner

In Business: Three years

Clients: Bio-Imaging Research Inc., Elscint Inc., Hitachi Medical Systems, IGS, Lorad Corporation, Oldelft Corporation of America, Pantak Inc., Philips Medical Systems, Siemens Medical Systems, Varian-Canada, Wallach Surgical Corporation.

Qualifications/Certifications: Normac-Will personnel have over 25 years' background and experience in both an engineering and quality/regulatory environment, and are RAB Certified Quality Systems Auditors, IRCA Certified Quality Systems Auditors, and ASQC Certified Quality Auditors.

Additional Information:

E-mail Address 2: HHBP74A@PRODIGY.COM

E-mail Address 3: HHBP74@MSN

OMNEX TRAINING SEMINARS

777 Eisenhower
Suite 315
Ann Arbor, MI 48108, U.S.A.
Tel: 313-761-4940; Fax: 313-761-4966
E-mail Address: MHILLEGO@OMNEX.COM

Contact: Sherrie Smith, Training Coordinator

Contact 2: Mark Hillegonds, Aqua Sales and Technical Support (Software)

In Business: Five years

Clients: 3M, A.C. Delco, Aeroquip, Alcoa, Associated Spring, Caterpillar, Chrysler, CPI, Dana, Digitron, Duracell, Eaton Steel, Exxon, Ford, GM, Hercules Steel, ITT, Johnson Controls, Mascotech, Motorola, Motor Wheel, National Tech Team, Rockwell, Rouge Steel, Teleflex, Texas Instruments, Valvoline, Walbro, Westinghouse, Whirlpool.

Qualifications/Certifications: Certified auditors/RAB

OMNI TECH INTERNATIONAL, LTD.
Consulting and Staffing Services

2715 Ashman Street
Suite 100
Midland, MI 48640, U.S.A.
Tel: 517-631-3377; Fax: 517-631-7360
E-mail Address: INFO@OMNITECHINTEL.COM

Contact: Douglas B. Todd, Vice President, Quality, Environmental & Safety

In Business: 10 years

Clients: Akzo Chemicals, BASF, Ciba-Geigy, Ford, Hewlett-Packard, Monsanto, Occidental Chemical, Rhone-Poulenc, Texaco, Union Carbide, Chevron Chemical, SC Johnson, Kerr-McGee, Pioneer Chlor Alkali, A.P. Parts, Johnson Controls, Cabot, Amoco, American Axle.

Qualifications/Certifications: Omni Tech's qualifications include provisional IQA Lead Assessors, RAB Lead Auditors, member of ASQC-CQA Chemical Interest Committee and TAG to ISO/TC 207.

P. D. A. INTERNATIONAL, INC.

182 Forbes Road, Suite 224
Braintree, MA 02184, U.S.A.
Tel: 617-356-2282; Fax: 617-380-3516

Contact: Peter Kay, President

Contact 2: Geoff Miller, Vice President

In Business: 14 years

Clients: Adcole Corporation, Alpha Analytical LaB.S., Bermer Tool & Die Inc., Haartz Corporation, Trans Commodities (Switzerland), DHL International Ltd. (UK), DHL Worldwide, Belgium, GKN Defense, Commercial Sheetmetal, Chase & Sons, Lever Brothers Limited, Lindberg Heat Treating Honematic Machine Corporation, Bradford Industries Inc., Metal Logic Inc., American Engineered Components Inc.

Qualifications/Certifications: PDA is registered to ISO 9001. Senior consultants are registered or trained lead assessors and ASQC certified with over 10 years' experience in the quality field. All have strong backgrounds and proven track records in quality management and successful records in consultancy and training.

P-E HANDLEY-WALKER, INC.

6000 Freedom Square Drive
Suite 140
Cleveland, OH 44131, U.S.A.
Tel: 800-644-9004/216-524-2200
Fax: 216-524-1488

Contact: Valerie Baker, Marketing Manager

Contact 2: Michael V. Mathews, CEO

Contact 3: James Nance, Manager of Quality Systems

In Business: P-E Handley-Walker, Inc.—30 years. P-E International—60 years.

Clients: Baxter Healthcare Corp., Apple Computer, Bausch & Lomb, Honeywell, Pennzoil, Du Pont, Georgia Pacific, Western Digital, Shell Oil Company, Morton International, Kraft Foods.

Qualifications/Certifications: P-E Handley Walker, Inc. is part of the largest ISO 9000 consulting and training company in the world and has assisted over 4,000 clients through certification. P-E Handley-Walker operates in the following countries: United Kingdom, Ireland, Netherlands, France, Hungary, Hong Kong, Malaysia, India, Indonesia, Taiwan, Singapore, Australia, Mexico, the United States, and Canada. Several divisions of P-E International are ISO 9000 certified and many others are in the process with a certification target of late 1996.

PAPA & ASSOCIATES INC.
Management Associates

298 Sheppard Avenue West
Suite 200
Willowdale, Ontario M2N 1N5, Canada
Tel: 416-512-PAPA; Fax: 416-590-1729
E-mail Address: papa@hookup.net

Contact: Peter Papakostantinu, C.E.O., Partner
Tel 2: 416-756-8535; Fax 2: 416-756-8555

Contact 2: Barry Butcher, C.O.O., Partner

In Business: Seven years

Clients: Small, medium, and large multinationals. In order to protect confidentiality, letters of reference and a comprehensive list of clients will be provided on a project by project basis upon request.

Qualifications/Certifications: Papa principals have held executive positions in Canada and United States in Quality Management, General Management, Marketing and Sales in technical, food, consumer product, telecommunications, nuclear, and process industries. Papa is a member of ASQC, company principals serve on the CSA subcommittee for software quality control and are currently active members of Canadian Advisory Committee (CAC) TC 176/SC2, the Canadian working group for ISO 9000.

PERRY JOHNSON, INC.

3000 Town Center
Suite 2960
Southfield, MI 48075, U.S.A.
Tel: 800-800-0450/810-356-4410;
Fax: 810-356-4230
E-mail Address: PJI@wwnet.com
Home Page Address: http://www.pji.com

Contact: Carrie Hayden, ISO Services Manager

Contact 2: Alicia Garrison

In Business: 14 years

Clients: PJI has a clientele base of 18,000+ companies, including 85 percent of the Fortune 500. Clients include: AT & T, Amoco, Johnson Controls, Borg Warner Automotive, Goodyear Tire & Rubber, Rubbermaid, Lear Seating, U.S. Steel, and Allied Signal Automotive.

Qualifications/Certifications: Since 1983, PJI has implemented TQM in more than 500 facilities; presented seminars to more than 2,000,000 people worldwide; and trained in excess of 1,300,000 others through its extensive catalog of training products.

PJI's lead assessor training course is registered by both the International Register of Certified Auditors (IRCA) of Great Britain and the Registrar Accreditation Board of the United States.

Additional Information:

E-mail Address 2: pji@oeonline.com

P. F. J. EVERETT and ASSOCIATES

4017 Carling Avenue
Suite 305
Kanata, Ontario K2K 2A3, Canada
Tel: 613-592-3963; Fax: 613-592-2682
E-mail Address: peverett@igs.net
Home Page Address:
http://www.quality.co.uk/quality/isocons/pfje.htm

Contact: Peter Everett

In Business: Six years

Clients: NASA, Canadian Space Agency (CSA), Spar Aerospace, AIT Corporation, Neptec, Republic, Hughes Aircraft, TRW.

Qualifications/Certifications: All P. F. J. personnel are Certified Lead Assessors or Assessors. P. F. J. also provides specialists to CSA for project management.

PHILIP CROSBY ASSOCIATES, INC.

3260 University Boulevard
P.O. Box 6006
Winter Park, FL 32793-6006, U.S.A.
Tel: 800-722-1474/407-677-3084;
Fax: 407-677-3050

Contact: Becky White, Client Services

In Business: 17 years

Clients: AMOCO Petroleum Additives Company, Bristol Aerospace, Chevron, Inc., DSI Transports, Heinz Europe, Kone Elevators, Marconi Communications, Inc., McDermott International, Perle Systems Ltd., Rheem Australia, Riverwood International, Sicartsa Mexico, Trustee Savings Bank, Western Atlas, Wilkerson Corporation, Zimmer.

Qualifications/Certifications: Philip Crosby Associates, Inc., has helped over 1,500 companies (many Fortune 500) reach their quality goals in the last 17 years and has also helped many businesses register and achieve ISO 9000 certification.

PROCESS MANAGEMENT INTERNATIONAL

7801 E. Bush Lake Road
Suite 360
Minneapolis, MN 55439-3115, U.S.A.
Tel: 612-893-0313; Fax: 612-893-0502
E-mail Address: PMIRJRBER@AOL.COM

Contact: Robert J. Reber, Vice President

In Business: 12 years

Clients: Alusuisse Flexible Packaging, Inc., BHP Engineering and Construction Company, DATALINK, E-C Apparatus Corporation, Fleischmann Kurth Malting, Hopeman Brothers, IMED Corporation, Leeds Precision Instruments, Inc., Liquid Carbonic Corporation, Monsanto Company, Star Tex Corporation, Weather-Rite, Inc., Whale Scientific, Inc., Zytec Corporation, ICL, Phillips.

Qualifications/Certifications: PMI's teaming with OSQ has: a proven ISO 9000 and QS-9000 model, based on over seven years of practice and research; registration experience with more than

250 organizations; a network of experienced quality consultants nationwide who are able to integrate TQM methods with ISO 9000 and QS-9000 quality standards; and certified lead auditor training by consultants who know first-hand how internal auditing fits with the ISO 9000 and QS-9000 effort.

PRODUCTIVITY IMPROVEMENT CENTER OF DURHAM COLLEGE

1610 Champlain Avenue
Whitby, Ontario L1N 6A7, Canada
Tel: 800-263-3735/905-721-3320
Fax: 905-721-3339
Home Page Address:
HTTP://DURHAM.DURHAMC.ON.CA

Contact: Jacqui Sharpe, Marketing Manager
1425 West Pioneer Drive
Suite 128
Irving, TX 75061, U.S.A.
Tel 2: 800-727-6222/214-254-2815
Fax 2: 214-254-2344

Contact 2: Elena Johnson, Manager

Contact 3: Alan Duffy, Director

In Business: 12 years

Clients: All sizes of manufacturing and service businesses and individuals. Clients include: Honda, Gulf States Paper, Magna, A.G. Simpson, Rockwell International, Siemens, Scott Aviation, Cable Tech, Stackpole, Rothmans Benson & Hedges. (A more extensive list is available on request.)

Qualifications/Certifications: PIC is an ISO 9001 Registered Company; Boeing Advanced Quality System (AQS) Certified Training Center; Lead Auditors on staff (RAB QMI). PIC has assisted many companies in implementing quality management systems to achieve ISO 9000 registration as well as automotive, aerospace, and government standards.

PRODUCTIVITY MANAGEMENT CONSULTANTS

849 Harbor Island
Clearwater, FL 34630, U.S.A.
Tel: 813-447-6409; Fax: 813-447-6409

Contact: Peter J. Hunt, President

In Business: 21 years

Clients: ACF Industries, Akzo Coatings and Chemicals, American Cyanamid, American Paint and Coatings Journal, Ameron-Enmar Finishes, Aries Software, Ashland Chemicals, Avenuery, B.F. Goodrich , BAPCO Corporation, BASF, Bee Chemical, Camet, Cargill, Ciba-Geigy, Collaborative Testing Services, Cook Paint and Varnish, Coors Brewing, Crown Metro Aerospace Coatings, DeSoto, Devoe and Raynolds Coating, Dexter-Midland, Drew Chemicals, DuPont, EMB Corporation, The Glidden Company, Engelhard Corporation, Federation of Societies for Coatings Tech., Ferro, Ford Motor Company, Gaco Western Corporation, General Electric.

Qualifications/Certifications:

- B.S., M.S., Ph.D. in mechanical engineering and industrial/organizational psychology.

- Special training in ISO standards.

- Directly involved in the financial turn around of Ford Motor Co. and the auto industry during the 1980s. Lead consultant to ISO 9000 in Europe beginning 1987.

PRODUCTIVITY-QUALITY SYSTEMS INC. (PQ Systems)

10468 Miamisburg Springboro Rd.
Miamisburg, OH 45342, U.S.A.
Tel: 800-777-3020; Fax: 513-885-2252
E-mail Address:
sales@pqsys-hq.mhs.compuserv.com

Contact: Kurt Stueve

Contact 2: Steve Kreitzer

In Business: 16 years

Clients: Alaska Department of Education, American Cast Iron, Arcadian Corporation, Arthur Anderson, Becton-Dickinson, Child Care Group of Dallas, Cincinnati Milacron, Community Mutual/Blue Cross Blue Shield, ConAgra Frozen Foods, Conoco, Continental Can, Dayton (Ohio) City Schools, Evanston Hospital, Fort Dearborn Lithograph, Foth & Van Dyke, GenCorp Automotive, Intalco Aluminum, Jackson Community College, Kaneka Texas Corporation, Katema Heat Transfer Division, Kurt J. Lesker Company, Lancaster (Ohio) City Schools, Lexington (Kentucky) Community College, Lima (Ohio) City Schools, Maxwell Air Force Base, Merten Printing, Midwest Optical Laboratories, Navistar International.

Qualifications/Certifications: The company's 16-year history has focused on helping companies improve their organizations. This focus has enabled PQ Systems to develop an international reputation in consulting services and software products.

QMR CONSULTING, INC.

10,000 Richmond Avenue
Suite 680
Houston, TX 77042, U.S.A.
Tel: 713-974-1872; Fax: 713-974-6336

Contact: Lanny Gookin, President

Contact 2: Duane Wuestner, Vice President of Operations
428 Kyle Drive
Dallas Operations
Coppell, TX 75019, U.S.A.
Tel 3: 214-393-1444; Fax 3: 214-471-1172

Contact 3: Ken Peurifoy, Dallas Operations Manager

In Business: QMR Consulting, Inc. was incorporated in 1995 as a successor corporation to Lanny Gookin & Associates, Specialized Technical Services and Quality Consultants International. The combined businesses have been established since 1985.

Clients: American Petroleum Institute; Arrow Oil Tools; Baroid Drilling Fluids; Bestline Liner Systems; Coflexip Stena Offshore; Conoco Oil; Curley's Fishing Tools; DuraQuest Elastomer Products; Enpro Systems; Fluoro-Seal; Ford, Bacon and Davis; General Pipe Services; GMR International; H.M.T.; Ingram Cactus Company; Jim Ray Company; Kalyani Steels Limited; Marathon Oil; Mark Products; MASX Energy Services; Numerical Precision.

Qualifications/Certifications: The five principal partners of QMRC have over 100 years' experience in providing quality assistance. QMRC and STS staff

credentials are as follows: Registered IQA Lead Auditors; RAB Registered Lead Auditors; ASQC Certified Quality Engineer (CQE); ASQC Certified Auditors (CQA); Texas Quality Award Examiner since 1993; Past and Present Chairmen of American Petroleum Institute Quality Committees; Past ASQC Petroleum Committee Chairman; Member of the U.S. Technical Advisory Group to the ISO/TC176, ASNT Level III in RT, UT, MT and PT; American Welding Society Certified Welding Inspectors; Expert Witness for Quality Systems and Quality Assurance.

QNET TRAINING AND CONSULTING

P.O. Box 527
Elk River, MN 55330, U.S.A.
Tel: 612-441-0899; Fax: 612-441-0898
E-mail Address: Q9000@aol.com

Contact: Yvonne Halpaus, Director Sales & Marketing
QNET Technical Services
P.O. Box 527
Elk River, MN 55330, U.S.A.
Tel 2: 612-441-0187; Fax 2: 612-441-2462

Contact 2: Sandra Reitsma

Clients: All sizes of manufacturing and service businesses, including the first medical service company to be registered in North America. A comprehensive list available on request.

Qualifications/Certifications: QNET has assisted over 100 companies to ISO 9000 registration through step by step training and consulting that reduces the time and cost required. QNET delivers their public ISO 9000/QS-9000 courses through colleges and universities in Minnesota, Iowa, Wisconsin, and Arizona.

QUALIFIED SPECIALISTS, INC. (QSI)

363 North Belt
Suite 630
Houston, TX 77069, U.S.A.
Tel: 800-856-5366/713-448-5622
Fax: 713-448-6015
E-mail Address: qsi@sccsi.com
Home Page Address:
http://www.sccsi.com/qsi/index.htm

Contact: R. T. "Bud" Weightman, President

Contact 2: Andrew Bergman, Vice President

In Business: Seven years

Clients: International Fortune 500 to small/medium-sized firms. Industries served are: manufacturing, Petrochemical, Electronics, Software, Transportation, Oil and Gas, Automotive, Chemical, Mining, Utility, Aerospace, Telecommunications, Distribution and Process. Partial list of Clients: Shell Oil, BJ services, Grant, Prideco, National Semiconductor, Varel Manufacturing, EIMCO Process Equipment, ARCO, American Axle & Manufacturing, Lafarge, JM Huber Group, Criterion Catalyst Co., LP, and Eastman Christensen

Qualifications/Certifications: The president, R. T. "Bud" Weightman, has over 23 years quality experience. He is an IRCA (UK) Lead Assessor, RAB Lead Auditor, and an ASQC CQA. He is a member of the ISO/TC 176 and TC 67 WG2 (Oilfield Certification Principles). He is an international speaker and author. Mr. Weightman's well-known "How to Select a Registrar" article has been published and used internationally. The vice president, Andrew Bergman is an IRCA Lead Assessor, ASQC CQE and has a B.S. in computer integrated manufacturing technology. Mr. Bergman is the former manager of certification at OTS Quality Registrar. QSI staff members have strong academic credentials and business experience.

QUALITY APPLICATIONS, INC.

9699 N. Hayden Road
Suite 108
Scottsdale, AZ 85258, U.S.A.
Tel: 602-951-4555; Fax: 602-951-4255

Contact 1: Michelle Young

Contact 2: Al Kitlica, President

In Business: Seven years

Clients: Whirlpool Corp., AGFA, Hoechst/Celanese, Hyclone Laboratories, American Tank & Fabricating, Sumitomo Corporation, Seymour Tubing, Eigen, Precision Printers, Inc., IBM, Westinghouse Airbrake Company, Lockheed, Honeywell Inc., McDonnell Douglas.

Qualifications/Certifications: Certified Quality Engineer, Certified Reliability Engineer, Registered Lead Auditor (IRCA, RAB).

QUALITY ASSURANCE SERVICES OF NORTH AMERICA (QASNA)

5251 McFadden Avenue
Huntington Beach, CA 92649, U.S.A.
Tel: 800-781-2065/714-898-0893
Fax: 800-781-2067/714-373-3767
E-mail Address: thaney1@ix.netcom.com

Contact: Thomas F. Haney, Director of Operations

In Business: Five years

Clients: Our clients include The Aerospace Corporation, Allied Signal Aerospace, Baxter Healthcare Corporation, Beckman Instruments Corporation, BioRad Laboratories, Chevron U.S.A., Chiron Vision, Coast/Cooper Vision, Collagen Corporation, The Edison Company, Fluor Daniel, GM Hughes (various divisions), Grupo Maseca, Jet Propulsion Laboratory, Kennedy Space Center, The Lockheed Martin Corporation (various divisions), The Loral Corporation (various divisions), McDonnell Douglas Aerospace (various divisions), Mobil Oil U.S.A.

Qualifications/Certifications: QASNA has an in-house staff with extensive ISO 9000 registration experience extending to over 500 ISO 9000 registrations worldwide. Training and consulting services are offered in English as well as Spanish to assist in transitioning the multicultural workforce.

THE QUALITY IMPROVEMENT NETWORK, INC.,

119 Russell Street
Littleton, MA 01460-1274, U.S.A.
Tel: 508-486-0010; Fax: 508-486-9482
E-mail Address: ISOQUAL@AOL.COM

Contact: Rick Barbieri, Marketing Director

Contact 2: Loretta Summers, Sales Manager

In Business: Five years

Clients: All sizes and types of manufacturing and service businesses. A comprehensive list available on request.

Qualifications/Certifications: RAB Certified ISO 9000 Quality Systems Lead Auditors, ASQC Certified Quality Engineers, ASQC Certified Quality Auditors, AIAG Certified QS-9000 Quality Systems Lead Auditors, ASQC Certified Quality Managers.

QUALITY INSTITUE OF AMERICA, INC.

355 E. NASA Suite B Road 1
Webster, TX 77598, U.S.A.
Tel: 713-332-6174 Fax: 713-554-5220

Contact: Wali Alam

In Business: Two years

Clients: In excess of 30, including international clients. Client size ranges from small organization to major Fortune 500 companies spanning numerous industries including the manufacting and service sectors.

Qualifications/Certifications: Qualifications for full-time and associate staff include lead auditor RAB; IRCA registered lead auditors approved to conduct QS-9000 and TickIT auditor status; PE, CQE, advanced degrees; typically in engineering, business administration.

QUALITY INTEGRATION SERVICES (QIS)

6320 Hunting Ridge Lane
McLean, VA 22101, U.S.A.
Tel: 703-448-3310; Fax: 703-448-3310
E-mail Address: QISGROUP@aol.com

Contact: Stephen T. Gaw, Owner

In Business: Two years

Clients: Medium and small businesses. QIS provides government agencies the same services. Emphasis is in the area of service and information technology organizations.

Qualifications/Certifications: QIS has trained lead assessors and consultants available to support ISO 9000 quality management system development and implementation.

QUALITY MANAGEMENT ASSISTANCE GROUP (QMAG)

1528 Ballard Road
Appleton, WI 54911, U.S.A.
Tel: 800-236-7802; Fax: 414-738-7802
E-mail Address: QMAG@AOL.COM

Contact: Roy E. Rodgers, Principal

In Business: 10 years

Clients: QMAG has serviced numerous clients in the automotive, nuclear, electronic, distribution, service, machinery, and continuous process industries. Specific references are provided with each proposal.

Qualifications/Certifications: Principals have IQA - RAB lead assessor training, and over 10 years' experience as managers in their field. Quality Consultants have ASQC certification as Certified Quality Engineers.

Additional Information:

E-mail Address 2:
RODGER.R.28086@ASQCNET.ORG

QUALITY MANAGEMENT ASSOCIATES, INC.

420 Scottsdale Drive
Salisbury, NC 28146, U.S.A.
Tel: 704-637-2299; Fax: 704-637-6181

Contact: Mark L. Crossley, CMC, President

In Business: 15 years

Clients: Ritz-Carlton, U.S. Air Force, Hewlett-Packard, Motorola, Sandoz Chemicals, Lonza Chemicals, Bristol-Myers, Heartland Hospital, Abbot Laboratories, First Union National Bank, Mennens Corporation, Dukane Corporation, U.S. Coast Guard,

EDS Corporation, San Juan Regional Medical Center, Sacramento Regional Blood Bank, Channel Master, Glaxo, Sara Lee Corporation, Ingersoll-Rand and others.

Qualifications/Certifications: QMA, Inc. is an international consulting firm offering professional assistance in all aspects of the quality sciences. Mark L. Crossley, president, has over 20 years' experience in quality and holds a bachelor's degree in chemistry. He is certified by the American Society for Quality Control (ASQC), and is a recognized Certified Management Consultant (CMC), a Certified Quality Engineer, a Certified Reliability Engineer, a Certified Quality Auditor and is a certified lead assessor for ISO quality systems, LA-QS. Mark McComb is an associate and is certified by the ASQC as a Certified Quality Engineer and a senior member of the ASQC.

QUALITY MANAGEMENT CONSULTING SERVICES, INC.

62 Murray Drive
Oceanside, NY 11572-5722, U.S.A.
Tel: 516-536-1859; Fax: 516-536-1859
E-mail Address: dane@pbfreenet.seflin.lib.fl.us

Contact: Dan Epstein, Senior Advisor
6148 Brightwater Terrace
Boynton Beach, FL 33437-4136, U.S.A.
Tel 2: 407-499-6804 ; Fax 2: 407-499-6804

Contact 2: Jeff Epstein, Vice President Training and Development

Services: Quality Management Consulting Services' ISO 9000, QS-9000, and ISO 14000

In Business: Five years

Clients: Koehler Instrument Company, Davis Vision Systems (Excelsior Winner), Technology Systems Corporation, ADEMCO division of Pittway Corporation, Renco Electronics, Stony Point Electronics, AIL Systems, Inc. (Excelsior Winner), Cardion, Inc., Mini-Circuits division of Scientific Components, Mini-Circuits Laboratory, Inc.

Qualifications/Certifications: Dan Epstein has more than 45 years' experience in standards, reliability,

quality control, quality assurance, safety, hazardous materials, and motivational training. He also had 16 years' experience as a senior vice president of a major east coast electronics manufacturer. Dan Epstein is an author and lecturer and has provided testimony to Congress. He was an examiner for the Malcolm Baldrige National Quality Award and an examiner for the New York State Excelsior Award. Dan Epstein has been certified as an ISO 9000 lead assessor/auditor by QMI Ltd. (UK).

QUALITY MANAGEMENT INTERNATIONAL, INC.

Box 271, Exton Square Parkway
Exton, PA 19341, U.S.A.
Tel: 800-666-9001/ 800-971-4001
Fax: 800-FAX-9004/ 800-611-4004
E-mail Address: helenmcc@ix.netcom.com
Home Page Address:
http://www.exitlog.com/~leebee/actplan.htm

Contact: John R. Broomfield, President/CEO

Contact 2: Gordon Gray McPhail

Contact 3: Helen McCallum

In Business: Since 1986.

Clients: Connelly Containers, Connelly Paper Mill, EDS, Fuller Company, Gaylord Containers, The Knoll Group, Northern Telecom, Worthington Steel, Westcode.

Qualifications/Certifications: Accredited by the RAB as a course provider accredited by IRCA to train Quality Systems Lead Auditors and Internal Auditors.

All instructors are Registered Lead Auditors and most have a M.A. in business/management.

Additional Information:

Home Page Address 2:
http://www.exitlog.com/~leebee/training.htm

Home Page Address 3:
www.quality.management.industry.net

Home Page Address 4 :
http://www.stroller.com/isofiles/qmiimc.htm

QUALITY MANAGEMENT SOLUTIONS

P.O. Box 349
Uncasville, CT 06382, U.S.A.
Tel: 800-628-6424/860-442-9393
Fax: 860-442-3393
E-mail Address: ConsultQMS@AOL.COM

Contact: Barry Schnell, Business Development
Manager

Contact 2: Craig R. Mesler, President

Contact 3: James Malone, Vice President

In Business: 7 years

Clients: QMS services clients in 38 states, Canada,
Mexico, Puerto Rico, Chile, Italy, France, and
Japan. QMS's client base of over 300 companies
includes: Ensign Bickford, Pratt & Whitney, Stana-
dyne Automotive, MascoTech, Thermo Electron,
Enthone-OMI, Mobil Oil Corporation, ASARCO,
Wardwell Braiding Machine, Echlin Automotive,
Delphi Energy and Engine (Division of GMC),
Ashland Chemical, General Cable Corporation,
Seagate Technology, Advanced Chemical, and
C&M Corporation.

Qualifications/Certifications: Since 1989, QMS has
assisted over 300 clients in ISO 9000, QS-9000,
Malcolm Baldrige, and other quality and management improvement initiatives. Through 1995, over
75 of QMS's full-service clients have achieved
successful implementation and registration of their
quality management systems. All of QMS's consultants and trainers have significant and relevant experience in actual development, implementation,
maintenance, and improvement of effective quality
systems in operating business environments.

QUALITY PRACTITIONERS U.S., INC.

350-2 Gulf Boulevard
Indian Rocks Beach, FL 34635, U.S.A.
Tel: 813-596-2296; Fax: 813-595-4054
E-mail Address: mcrobert@interlog.com
Home Page Address:
http://www.interlog.com/~mcrobert

Contact: Charles McRobert, President
3212 Magwood Road
Unit 2
Mississauga, Ontario L5L 5K6, Canada
Tel 2: 905-569-6431; Fax 2: 905-569-7651

In Business: Eight years.

Clients: Bombardier, Imperial Oil, Novacor Chemicals,
Tektronix, Woodhead Aeromotive, Dufferin Construction. QPI has taken 75 companies to registration to ISO 9000.

Qualifications/Certifications: QPI has two RAB
accredited training courses for lead auditor
training. The training and consulting staff are all
RAB accredited lead auditors who conduct audits
for registrars and who have practical experience in
taking companies to registration.

QUALITY SCIENCES CONSULTANTS, INC. (QSCI)

22531 S.E. 42nd Court
Issaquah, WA 98029, U.S.A.
Tel: 206-392-4006; Fax: 206-392-2621
E-mail Address: Caplan.F.7729@ASQCNET.ORG

Contact: Frank Caplan, President

In Business: QSCI was incorporated in 1989 as a suc-
cessor corporation to Quality Services, Inc.
Altogether, the business has been in existence since
July 1983.

Clients: Bellevue Community College, Rock Valley
Community College, St. Louis Community College,
South Seattle Community College. AT&T Microelectronics—Hi Performance IC's, AT&T Microelectronics—Lightwave, AT&T Microelectronics—MOS,
AT&T Microelectronics—Submarine Systems, Au-
gat Communications, Seamed, Claircom.

Qualifications/Certifications: QSCI personnel are drawn
from a large pool of ISO 9000 Registered Lead Assessors and other professionals. Most of them are
CQEs or CQAs, as well. The President of the company is a Fellow and CQE of ASQC and holds a P.E.
in quality. He is also the recipient of ASQC's Eugene
L. Grant Award for lifelong efforts and achievements
in the educational field vis-a-vis quality.

QUALITY SYSTEMS ENHANCEMENT, INC. (QSE, Inc.)

1005 Anston Drive
Roswell (Atlanta), GA 30075-2979, U.S.A.
Tel: 770-518-9967; Fax: 770-518-9968

E-mail Address: AQCNET address is KOTTE.U.18962

Contact: Baskar Kotte, ISO Certified Lead Auditor

In Business: Four years

Clients: Georgia Pacific, Jefferson Smurfit, Georgia Tech, American Quality Assessors, Satellite Technology Management, Caterpillar, Buckeye Container, CIBA Vision, Liuski International, United Nations, Excel Machine and Foundry, Willamette Industries.

Qualifications/Certifications: ISO/QS-9000 Certified Auditors, Active Membership in TC 176, TC 69 and TC 207; International Activities; 30 Facilities Achieved Registrations (First Attempt); Certified Quality Engineers; Certified Quality Auditors; Exceptional Training Talents.

QUALITY SYSTEMS INTEGRATORS

P.O. Box 91
60 Pottstown Pike
Eagle, PA 19480, U.S.A.
Tel: 800-458-0539; Fax: 610-458-7555
E-mail Address: turocy@chestnet.com
Home Page Address: WWW.webimpact.com\QSI

Contact: Marti Turocy, President

In Business: 10 years

Clients: All sizes of manufacturing and service businesses. References and demonstrations are available upon request.

Qualifications/Certifications: QSI's 10 year history has helped quality managers cut costs, improve efficiency, and increase customer satisfaction, which ultimately improves the company's market share.

QUALITY TECHNOLOGY COMPANY

1161 Tower Road
Schaumburg, IL 60173, U.S.A.
Tel: 847-884-1900; Fax: 847-884-7280
E-mail Address: qtc1239@startnetinc.com
Home Page Address: qtcom.com

Contact: Praveen Gupta, President

Contact 2: Stanley Hon, Vice President

In Business: Eight years

Clients: Small to large sized companies in electronics manufacturing, auto industry, and service corporations.

Qualifications/Certifications: Quality Technology Company (QTC), an ISO 9001 registered firm. QTC instructors are internationally recognized. They have taught more than 7000 quality professionals through training courses held at Motorola University, College of DuPage, and QTC.

QUALITY ALERT ISO AND QS-9000 GROUP

A Division of Tompkins Associates, Inc.

2809 Millbrook Road
Suite 200
Raleigh, NC 27604, U.S.A.
Tel: 800-789-1257; Fax: 610-264-2205

Contact: John O. Brown, General Manager, Quality Practice-Corporate
944 Marcon Boulevard.
Suite 110
Allentown, PA 18103 U.S.A.
Tel 2: 610-264-1041; Fax 2: 610-264-2205

Contact 2: Christopher B. Gregg, Business Development Manager—Quality Practice

In Business:

- Warehousing, distribution, material handling and organizational excellence training and consulting—Founded 1974.
- ISO 9000 & QS-9000 training and consulting—Founded 1990.

Clients: United Parcel Service of America Airline, Ground, District, and Logistics Operations, Coca-Cola U.S.A., Bell of Canada, Allied-Signal, Sun Company, BP Oil, New York City Transit Authority, Otis Elevator, Eastman-Kodak, Bridgestone-Firestone, Delphi Harrison Thermal Systems, Delphi Saginaw Steering Systems, Allison Transmission, Delco Electronics, Delphi Energy and Engine Management Systems, PACCAR-Kenworth & Peterbilt Trucks, The Budd

Company, Abbott Laboratories, Abbott Diagnostics, Hoffmann-LaRoche.

Qualifications/Certifications: The company helps organizations with systems training, consulting, and implementation services they need to improve and control the quality of their products and services. QualityAlert's ISO 9000 and QS-9000 group started over six years ago. Our staff is made up of individuals who were prior executives with Fortune 500 companies and have helped large, mid-size, and small companies through the entire ISO 9000 and QS-9000 preparation and registration process.

QUALITYQUEST

1541 W. Richmond Street
Arlington Heights, IL 60004, U.S.A.
Tel: 847-870-0822; Fax: 847-870-0872
E-mail Address: MikeMick@ix.netcom.com

Contact: Michael J. Micklewright, President

In Business: Two years

Clients: All sizes of manufacturing and service businesses including: Action Technology, Advance Dial, Baxter Healthcare, Blistex, CFC International, Coilplus-Illinois, Dow Brands, FelPro Inc., Fort Transfer, Hoffer Plastics, Hydraforce Inc., Jensen International, Johnson and Higgins, Kelsey-Hayes, Mastercoil Spring, McDonald's Corporation, National Recovery Systems, Saturn Corporation, Seaquist Perfect, Tenneco Packaging, Teroson, Thatcher Tubing, and Videojet International.

Qualifications/Certifications: ISO 9000 Certified Auditor; QS-9000 Certified Auditor; ASQC CQA (Certified Quality Auditor); ASQC CQE (Certified Quality Engineer); Certified Instructor of Six Thinking Hats and Lateral Thinking.

QUALTEC QUALITY SERVICES, INC.

1400 Centrepark Boulevard.
Suite 600
West Palm Beach, FL 33401, U.S.A.
Tel: 407-775-8300; Fax: 407-775-8301
E-mail Address: isoinfo@qualtec.com
Home Page Address: http://www.qualtec.com

Contact: Reilly Sierra, Marketing

In Business: Nine years

Clients: CompuCom Systems, Inc., AT&T Microelectronics, Ameritech, BMW, Sensormatic, Mercedes-Benz.

Qualifications/Certifications: Qualtech is accredited by the Southern Association of Colleges and Schools to award Continuing Education Units and accredited by the American Council on Education (ACE) for college credits.

Qualtech has assisted organizations in 18 countries worldwide, in industries ranging from banking and insurance to healthcare, manufacturing, and government to successfully implement quality management systems.

Additional Information:

E-mail Address 2: marketing@qualtec.com

WWW Address: www.qualtec.com

QUEST USA INC.

27941 Harper
St. Clair Shores, MI 48081-1580, U.S.A.
Tel: 800-878-1669/ 810-774-9480
Fax: 810-774-2709
E-mail Address: info@qfe.com

Contact: Brian Cooke, Director of Operations North America

Contact 2: Brian Cherry, Sales Executive

In Business: Quest International Group of Companies England since (May 1990).

   • Quest U.S.A., Inc. (May 1992)

Clients: Basic Vegetable Products, Bell South Telecommunications, Cincinnati Milacron, Delta-Faucet, GTE Government Systems, Motorola, Plastipak Packaging, Snyder General, Sumitomo Electric, U.S. West Communications, Valentine Inc., and others.

Qualifications/Certifications: Quest U.S.A., Inc., is registered to ISO 9001: 1987 Quality Systems with SGS ICS. Quest has guided many U.S. companies to successfully implement ISO 9001, ISO 9002, ISO 9000 3, and QS-9000.

r. bowen international, inc. (rbi)

149 W Market Street
York, PA 17401-1314, U.S.A.
Tel: 717-843-4880; Fax: 717-854-5345
E-mail Address: rbi@aol.com

Contact: Robert D. Bowen, President

Contact 2: Michael D. Lohenitz, Business Manager

In Business: Five years

Clients: AT&T Microelectronics, Storage Technology, Hershey Foods, Carbide Graphite, Donsco Incorporated, Dentsply International, The BYRNES Group, Campbell's Soup, School District of Lancaster. r. bowen international, inc., also serves clients in the following industries: electronics, staffing, public education, computer manufacturers, micrographics, metallurgy, insurance, and food processing.

Qualifications/Certifications: Robert D. Bowen is ASQC certified as a Quality Engineer and Quality Auditor with 25 years' experience.

RICHARD JONES-MANAGEMENT CONSULTANTS INC. (RJMCI)

413 Briarwood Drive
Kingston, Ontario K7M 7V2, Canada
Tel: 613-389-3475; Fax: 613-389-7559
E-mail Address: rjmci@shadow.kingston.net

Contact: Richard J. Jones, President

In Business: 11 years

Clients: The Canadian Standards Association, The Quality Management Institute, General Electric (GECTS), The Ontario Ministry of Health, The Ontario Ministry of Transportation, The Ontario Transportation Capital Corporation, ETM Industries Inc., Agriculture and Agri-Food Canada, Ontario Hydro, Phillips Cables, Alcatel Canada Ltd., Queen's University.

Qualifications/Certifications: Registered to ISO 9001:1994 through QMI; Certified Quality Auditors through QMI on staff; Corporate psychologist (PhD) on staff; Highly trained organizational development consultants with degrees.

Additional information:

BB.S. Address: Shadow Communications System: 613-384-8772

ROBERT PEACH AND ASSOCIATES, INC.

200 W. Cornwall Road, #126
Cary, NC 27511-3802, U.S.A.
Tel: 919-319-1982; Fax: 919-319-1984
E-mail Address: 74654.1747@compuserve.com

Contact: Robert W. Peach, Principal

In Business: 14 years

Clients: North American Philips, Sematech, Institute of Industrial Launderers, Georgia Institute of Technology, University of Wisconsin, World Bank, APA (Engineered Wood Assn.), Instrument Society of America.

Qualifications/Certifications: As a member of the U.S. delegation to the ISO TC176 Committee on Quality Assurance, Mr. Peach served as convener of the working group that developed ISO 9004-1, quality management and quality system elements guidelines. He is immediate past chair of the ASQC's Registrar Accreditation Board, where he continues as a board member. He established and managed the quality assurance activity at Sears, Roebuck, and Company for over 25 years.

ROCHESTER INSTITUTE OF TECHNOLOGY

The John D. Hromi Center for Quality and Applied Statistics (CQAS)

Hugh L. Carey Building
98 Lomb Memorial Drive
Rochester, NY 14623-5604, U.S.A.
Tel: 716-475-6990; Fax: 716-475-5959
E-mail Address: CQAS@RIT.EDU

Contact: Donald D. Baker, Director

In Business: 12 years

Clients: During the past decade, CQAS has conducted courses and programs on quality standards, quality management, statistical analysis, experimental design and reliability at RIT or on site for more than 400 industry, business, government, and service agencies and for thousands of individuals. A growing client base includes small and medium-sized companies.

Qualifications/Certifications: The staff at CQAS are consultants, respected speakers, prominent authors, and active members of professional societies with industrial or business experience. The staff serves on the boards of registrars as well as on the RAB and Technical Committee 176.

RPI QUALITY MANAGEMENT

P.O. Box 111
Drexel Hill, PA 19026, U.S.A.
Tel: 610-284-9900; Fax: 610-284-9900

Contact: Robin Plummer

In Business: Eight years

Clients: All sizes of manufacturing, service companies and government agencies; and over 2000 people trained on lead auditor courses.

Qualifications/Certifications: Lead auditor training course (IRCA), Lead Auditors (RAB & IRCA), 15 years' experience auditing and consultancy in quality management systems that meet recognized international standards (ISO 9000). Contributing author of the *Gower Handbook of Quality Management,* published by Gower, 2nd edition 1995; and video on Reliability in the Micro-Electronics Industry.

SANDERS & ASSOCIATES

2470 Gray Falls
Suite 220
Houston, TX 77077, U.S.A.
Tel: 800-856-8772/ 713-531-9184
Fax: 713-531-5690
E-mail Address: sanders@accesscom.net

Contact: Judith A. Sanders, CEO

Contact 2: Marianne Gooch, Director of Marketing

Contact 3: Dr. Bill McNeese, Senior Consultant

In Business: 15 years

Clients: Motorola, VISTA Chemical, General Dynamics, Quanex Tube Group, Groth Corporation, Hollywood Marine, Mayer Electric, Industrial Distribution Association, Ethyl Corporation, Lyondell, PDVSA Services, Inc., PDVSA Services BV, Dixie Industrial Supply, Boring-Smith Industrial Supply, Stolt Parcel Tankers, and GLNX Railway Repair.

Qualifications/Certifications: Sanders & Associates has a 15-year track record of success helping Fortune 500, mid-size, and small companies integrate continuous quality improvement into operations and sales in both service and manufacturing environments.

Sanders' principals, Dr. Judith A. Sanders and Dr. William H. McNeese, are recognized authors in the field of quality. The company counts two Malcolm Baldrige winners among its clients, and has been featured in Bob Waterman's *Quest for Quality.*

SCOTT TECHNICAL SERVICES

137 Boston Post Road
Wayland, MA 01778, USA
Tel: 508-358-3470; Fax: 508-358-6119

Contact: Otis Russell, Senior Consultant

In Business: 11 years

Clients: Abiomed, Inc., Allied-Signal Aerospace Corporation (Bendix Communications Division), Cobra Industries Inc., ELBIT Computer Ltd., GTE-Government Systems, Inductotherm Corp., Lavolin Corporation, LEPEL Corp., MBTA-Mass. Bay Transportation Authority; Pacesetter Systems, Polymer Technology Corporation; Schwartzkopf Technologies corporation; Thermatool Corp., Bausch & Lomb

Qualifications/Certifications: Qualifications include more than 100 years of combined experience in all phases of quality assurance and operations management. Initially focused on the aerospace and defense industries, STS services now include the computer, telecommunications, medical devices, electronics, capital equipment, transportation and service industries.

SCS ENGINEERS

2405 140th Avenue, NE, Suite 107
Bellevue, WA 98005-1877, U.S.A.
Tel: 206-746-4600; Fax: 206-746-6747
E-mail Address: 1120JVK@be.scseng.com
Home Page Address: http:\www.iso 14000.com

Contact: John Kinsella, Vice President

Contact 2: Kris Baldyga

In Business: 25 years

Clients: Manufacturing, aerospace, oil and chemicals, wood products, state, provincial and federal governments, energy, and military facilities

Qualifications/Certifications: ISO 9000/14000 auditors; Environmental auditors

Additional Information:

E-mail Address 2: jkinsella@scseng.com

SERVICE-GROWTH CONSULTANTS, INC.

880-1140 West Pender Street
Box 30
Vancouver, BC V6E 4G1, Canada
Tel: 800-337-5787; Fax: 604-684-8283
E-mail: 72772.3262@compuserve.com
Home Page Address:
http://www.service-growth.com

Contact: Dorothy Riddle, Principal

In Business: Six years

Clients: A list is available upon request; not publicly available.

Qualifications/Certifications: The firm's principal, Dr. Dorothy Riddle, CMC, is a Registered ISO 9000 Specialist and author of *Quality Assurance in Services: An ISO 9000 Workbook for Small Professional Service Firms* (1996). Canada Communications Group-Publishing Ottawa, Canada K1A 0S9. The firm is registered to ISO 9001 (with SGS).

SERVICE PROCESS CONSULTING, INC.

76 George Avenue
Edison, NJ 08820, U.S.A.
Tel: 908-321-0045; Fax: 908-549-9117
E-mail Address: IDurand@aol.com

Contact: Ian Durand, President

Contact 2: April Cormaci, Executive Vice President

In Business: Seven years

Clients: AT&T, Dorman-Roth Foods, The Gillette Company, Hackettstown Community Hospital, Hexcel Corp., Krueger International, New Zealand Milk Products, Pitmann, W.R. Grace & Co., Westinghouse.

Qualifications/Certifications: The President and Principal Consultant has been actively involved in ISO/TC 176 since 1986, and holds major leadership positions at the national and international level. Mr. Durand has served as a senior examiner for the Malcolm Baldrige National Quality Award, specializing in service organizations. Prior to forming Service Process Consulting, Mr. Durand held a variety of technical and management positions during 32 years at AT&T Bell Laboratories, including managing departments responsible for training in quality management and product realization, quality of long-distance services, and field engineering.

SOFTWARE SYSTEMS QUALITY CONSULTING

2269 Sunny Vista Drive
San Jose, CA 95128, U.S.A.
Tel: 408-985-4476; Fax: 408-985-4476
E-mail Address: deibs@shell.portal.com
Home Page Address:
http://www.billboard.com/ssqc/index.html

Contact: William J. Deibler, II, Partner

Contact 2: Robert C. Bamford, Partner
Tel 2: 408-866-4792

In Business: Six years

Clients: Actel, Advantest, Amdahl, Antaries Alliance, Atria Software, Auspex, Autodesk, Becton Dickin-

son, Cadence, Centigram, CTB McGraw-Hill,
Digital MicrowAvenue, EDS, Gallo Winery,
Hewlett-Packard, Hitachi Data Systems, Hitachi Microsystems Inc., ICOT, Informix, Lam Research,
MDL Information Systems, Mentor Graphics, Motorola, Netframe Systems, Northern Telecom, Objectivity Inc., Octel, Oracle Corporation, Pacific Bell.

Qualifications/Certifications: Bill Deibler has an
M.Sc. in computer science and 16 years' experience in the software and computer industry in the
areas of software development and software quality
assurance.

Bob Bamford has a B.S. and MAT in mathematics.
Since 1975, he has managed training development,
technical publications, professional services, and
third-party software development for a variety of
high technology firms in California's Silicon Valley.

They are active U.S. TAG members in the ISO/IEC
JTC 1/SC7—Software Engineering Standards subcommittee. They were principal authors and editors
of *A Guide to Software Quality System Registration
under ISO 9001.*

Additional Information:

E-mail Address 2: ssqc@cris.com

WWW Address: http://www.cris.com/~ssqc

STAT-A-MATRIX GROUP/THE SAM GROUP

One Quality Place
2124 Oak Tree Road
Edison, NJ 08820-1059, U.S.A.
Tel: 908-548-0600/908-906-6105/6101
Fax: 908-548-0409
E-mail Address: Inc@theSAMgroup.com

Contact: Paul Berman, Vice President, New Business
Development
2711 Jefferson Davis Highway
Suite 200
Arlington, VA 22202, U.S.A.
Tel 2: 703-415-2591; Fax 2: 703-415-1684

Contact 2: Ira Epstein, Vice President, Government
Services
200 E. Big Beaver Road
Americenter, Suite 167
Troy, MI 48083, U.S.A.
Tel 3: 810-680-4673; Fax 3: 810-680-4664

Contact 3: Frederick W. Love, Automotive Industry
Services

In Business: 28 years

Clients: Fortune 500 companies, middle-market companies, government agencies, and small firms.
Industries served include medical devices and diagnostics, pharmaceutical, manufacturing, services,
software, aerospace/defense, automotive, chemicals, fibers, graphics, telecommunications, and
others.

Qualifications/Certifications: SAM's worldwide staff
of more than 100 staff members includes approximately 75 management consultants, certified auditors, and certified lead auditors. The average staff
member has over 25 years of industry experience,
plus an advanced degree and/or professional credentials, such as PE, CQA, CQE, IRCA and/or
RAB Lead Auditor certification. The SAM Group
staff includes TC 176 members, ASQC
fellows, and Baldrige Award Examiners. SAM was
the first U.S.-based organization registered by the
IRCA to provide lead auditor certification training
and the first in the world to receive
RAB accreditation for lead auditor training.

Additional Information:

E-mail Address 2: institute@thesamgroup.com

STATISTICAL QUALITY CONTROL AND CIRCLE INSTITUTE

P.O. Box 812
318 Pheasant Court
Fond Du Lac, WI 54936, U.S.A.
Tel: 414-923-7600; Fax: 414-923-7154
E-mail Address: stats@uscybar.com

Contact: Nima Ingle, Vice President, Marketing

Contact 2: Sud Ingle, President

In Business: 14 years

Clients: Allen Bradley, AT&T, Boeing, Ecanoprint,
FAA, Hughes Aircraft, Prince Macaroni, Old Fash-
ioned Cheese, St. Agnes Hospital, J.C. Penney,
Ziegler Manufacturing.

Qualifications/Certifications: Mr. Sud Ingle,
President, Statistical Quality Control & Circle In-

stitute, has authored more than 15 books and many video-programs in the field of quality control, teamwork and ISO 9000 standards. He has conducted training courses in manufacturing as well as service organizations in Canada, Brazil, Mexico, India, United Kingdom, Singapore, and Malaysia). He has a M.S. in industrial engineering from Purdue University and a M.B.A. from the University of Wisconsin.

STRATEGIC QUALITY ALLIANCE

1 Fern Road
Andover, MA 01810-6210, U.S.A.
Tel: 508-475-9686; Fax: 508-475-9686
E-mail Address: StratQual@aol.com

Contact: Richard J. Desjardins, President

In Business: Two years

Clients: Strategic Quality Alliance has performed process improvement and/or ISO 9000 certification activities for all types of software organizations (Commercial, MIS, System, etc.). Key clients include Viewlogic Systems, Computervision, Stratus Computer, BBN Software Products, ISG Technologies, Interleaf Inc., Delta Air Lines, Cigna, and CSC Healthcare Systems.

Qualifications/Certifications: The founder of Strategic Quality Alliance has nearly 20 years of software development, management, and process improvement experience. He is an IRCA certified provisional auditor for ISO 9000 and has helped numerous software organizations achieve their ISO 9000 goals. Strategic Quality Alliance partners with The David Consulting Group, an organization of IFPUG certified function point specialists, for all of its function point analysis and training needs.

SUCCESSFULL RESOURCES GROUP, INC.

6055 NE Glisan
Suite A
Portland, OR 97213, U.S.A.
Tel: 800-308-5113/503-231-1920
Fax: 503-238-4713
E-mail Address: success@teleport.com
Home Page Address:
http://www.teleport.com/~success

Contact: Scott Jenkins, President

In Business: Six years

Clients: Manufacturing, government

Qualifications/Certifications: Member ASTD (American Society of Training and Development); Certified Presenters of D.D.I. (Development Dimensions International); Certified Facilitators of L.M.I. (Leadership Management Institute).

SUNCOAST MANUFACTURING TECHNOLOGY CENTER

7431 114th Avenue, North, Suite 104
Largo, FL, 33773, U.S.A.
Tel: 813-545-2438; Fax: 813-544-8537
E-mail: enterpri@packet.net
Home Page Address:
http:\\www.enterprisecorp.com\sbdc

Contact: Arny Bereson, Director

In Business: Two years

Clients: Small to mid-sized manufacturing companies in Florida. The mission is the promotion of growth, competitiveness, and profitability to improve job opportunities in the state of Florida for Charlotte, Collier, DeSoto, Glades, Hardee, Hendry, Hernando, Highlands, Hillsborough, Lee, Manatee, Pasco, Pinellas, Polk, and Sarasota counties.

Qualifications/Certifications: Ten field engineers possess the following skills and expertise in various manufacturing fields including: metal fabrication, plastics molding, chemical manufacture, ship building, missiles, film and paper, and small business management. SMTC personnel include APICS certifications; Certified Quality Engineer, RAB & IRCA Registered QS-LA, Professional Engineers, CAD/CAM expertise, and experienced trainers.

TEAM 14000, INC.

P.O. Box 325
New Milford, NJ 07646, U.S.A.
Tel: 201-837-5934; Fax: 201-837-8893
E-mail Address: Team 14000@aol.com
Home Page Address: http://www.team14000.com

Contact: Alan Schoffman, Principal
8 Strafford Circle Road
Medford, NJ 08055 U.S.A.
Tel 2: 609-953-9163 ; Fax 2: 609-953-3877

Contact 2: Allan Tordini, Principal

In Business: One year

Clients: A client listing is available on request.

Qualifications/Certifications: The Team 14000 principals include a Ph.D. in chemistry, an M.S. in environmental science and a Certified Quality Auditor
with ISO 9000 experience. Associates include
technical and legal specialists. Personnel have published articles, presented papers, conducted seminars for industry, and have provided consulting
services in their respective areas of expertise
including (quality systems, environmental law, and
engineering, government relations, product testing
and certification, conformity assessment, technical
operations and business management).

Additional Information:

E-mail Address 2: aschoffm@ios.com
E-mail Address 3: AlTordini@aol.com

TEAM QUALITY INTERNATIONAL (U.S.A.)

70 High Howe Lane
Bournemouth, Dorset BH11 9QX, England
Tel: 44 1 202 590377; Fax: 44 1 202 590388
E-mail Address: tqi@bournemouth-net.co.uk
Home Page Address: http://www.angel.co.uk/tqi

Contact: Gary Vaissiére, Senior Consultant

In Business: Vaissiére Industries has been trading in
quality since 1990 as Team Quality International.

Clients: Vaisala Oy Group—Meteorological
Equipment/Services (H.O. Finland), Knürr A.G.
Group—Electronic Enclosures and Technical Fur-
niture (H.O. Germany).

Qualifications/Certifications: The Quality System
used is compliant with B.S. EN ISO 9001: 1994.
International Register of Certified Auditors. Registered head auditor/auditors used as appropriate.

Additional Information:

E-mail Address 2: tqi@angel.co.uk (Team Quality
International USA)

TECHNOLOGY INTERNATIONAL INC.

609 Twin Ridge Lane
Richmond, VA 23235, U.S.A.
Tel: 800-242-8399/804-560-5334
Fax: 804-560-5342
E-mail Address: Mktg@TechIntl.com

Contact: Kristin Eckhardt

Contact 2: Wayland Stephenson

In Business: Four years

Clients: General Motors, Convex Computers, Filtroil,
Homelite, AVO International, Hamilton
Beach/Proctor-Silex.

Qualifications/Certifications: Mr. Mather, quality
division director of Technology International, Inc.,
is a recognized expert with over 15 years' experience in Quality Management Consultancy, ISO
9000 auditing, training, and implementation in a
wide range of industrial and commercial sectors.

Mr. Mather is an accredited lead assessor and has
been accredited for many years under the United
Kingdom's National Assessment Approvals and
Certifications Board (NAACB) system.

THORNHILL USA

P.O. Box 3643
Wilmington, DE 19807, U.S.A.
Tel: 610-444-3998; Fax: 610-444-1365
E-Mail Address: THORN100@AOL.COM

Contact: Ralph Schmidt, Director

In Business: Six years

Clients: Thornhill USA's clients are primarily in the following industries: Service providers, automotive
suppliers, plastic resin compounders, major chemical and petroleum companies in North and South
America.

Qualifications/Certifications: Each consultant has lead assessor training and over five years of hands-on quality systems experience with Fortune 500 companies prior to joining the firm. Members include RAB Certified Lead Auditors as well as Certified Members. Thornhill USA is affiliated with Alberto Levy & Company of Buenos Aires.

TQM CONSULTING

9718 Braesmont Drive
Houston, TX 77096, U.S.A.
Tel: 713-723-6390; Fax: 713-721-5401
E-mail Address: 75362.2746@compuserve.com

Contact: William E. Cox

In Business: Five years

Clients: Exxon Chemical Co., Lyondell Petrochemical, Bayer Corporation, Lyondell Citgo Refining, Haltermann Ltd., Loral Corporation, NASA, Anheuser-Busch, Delco Electronic Systems, General Motors, Los Alamos National Lab, Montell USA, Lagoven and PDVSA (Venezuela), Arancia Ingredientes Especiales (Mexico), Intercor Operador (Colombia). Over 100 clients have achieved ISO 9000 certification.

Qualifications/Certifications: TQM Consulting associates have a wide variety of experience in petrochemical, refining, food, aerospace, service and manufacturing industries. All are certified as quality auditors and/or environmental managers. Two associates served on the Board of Examiners for the Texas Quality Award in 1993 and 1994, and two coauthored the ASQC book, *ISO 9000 Guidelines for the Chemical and Process Industries,* 2nd Edition, 1996. Memberships are maintained in ISO Technical Committees 176 and 207 which are responsible for development of quality and environmental standards.

TRI-TECH SERVICES INC.

55 Old Clairton Road
Pittsburgh, PA 15236, U.S.A.
Tel: 412-655-8970; Fax: 412-655-8973
Home Page Address:
http://www.TTS9000@aol.com

Contact: David Kudlock, Vice President;
Ross Marino, Sales Manager
4100 Edison Lakes Parkway
Mishawaka, IN 46545, U.S.A.
Tel 2: 2190273-7301; Fax 2: 219-273-7608

2301 W. Meadowview Road
Suite 101
Greensboro, NC 27407, U.S.A.
Tel 3: 910-294-9833; Fax 3: 910-294-9683

In Business: 10 years

Clients: Over 200 clients in all industrial fields.

Qualifications/Certifications: ASQC; RAB Lead Assessors; CQA, CQE.

Additional Information:

Home Page Address 2:
http://www.industry.met/c/pto-memdir/company/597

U.S. MANAGEMENT CONSULTING GROUP
Corporate Headquarters

32 Edgewood Drive
Wallingford, CT 06492, U.S.A.
Tel: 203-265-1771;/Fax: 203-265-1771

Contact: Gary L. Powell, Executive Vice President, New Business
225 Ann Place
Suite 100
Chicago , IL 60614, U.S.A.
Tel 2: 708-920-9075

Contact 2: Stephanie A. Cuningham, Program Director

In Business: Two years

Clients: Various size Aerospace, Plastic, Chemical, Medical, Metrology, Manufacturing, Computer, Container, Packaging, government, Automotive, and Service Industries. A qualified list is available upon request.

Qualifications/Certifications: The staff of U.S. Management Consulting Group has provided professional services to hundreds of Fortune 500 companies as well as smaller corporations. All consultants are degreed, Certified Quality Auditors/Assessors and professionals, with many years' experience in their fields of expertise and ISO 9000.

THE VICTORIA GROUP INCORPORATED

10340 Democracy Lane
Suite 204
Fairfax, VA 22030, U.S.A.
Tel: 800-845-0567/ 703-691-8484
Fax: 800-845-0767/703-691-2542
E-mail Address:
jennifer_cook@victoria.global.ibmmail.com

Contact: Jennifer Cook, Marketing Manager

In Business: Five years

Clients: Ford Motor Company, Hewlett Packard, Pacific Bell, AT&T, Georgia Pacific, Jaguar Cars North America, Kodak, Johnson & Johnson, U.S. Army Corps of Engineers.

Qualifications/Certifications: The Victoria Group has experience in implementing quality systems based on ISO 9000 or similar standards; the principals of the company have performed several hundred successful registrations. All have been functionally responsible for the successful implementation and management of ISO 9001 or ISO 9002 based quality systems. They have extensive experience with many industry and national standards, including MIL-Q-9858A, AQAP-1, Ford Q1, IBM, and others.

W. A. GOLOMSKI AND ASSOCIATES

N 9690 County U
Algoma, WI 54201, U.S.A.
Tel: 414-487-7248; Fax: 414-487-7249

Contact: William A. Golomski, President

In Business: 47 years

Clients: Food industry, metals industries, banking, insurance, hospitality and transportation industries, governments, auto and truck industries and their

suppliers, personal safety products industries, plastic fabrication, and textiles.

Qualifications/Certifications: Consultants have CQEs, Professional Engineer, membership in National Academy of Engineers; holder of American Deming Medal; Lead Auditors.

W. R. WAYMAN AND ASSOCIATES

Management Consultants

3722 Twin Oak Court
Flower Mound, TX 75028, U.S.A.
Tel: 972-539-0335; Fax: 972-539-0335

Contact: W. R. Wayman, Principal Consultant

In Business: Six years

Clients: All sizes of manufacturing and service companies throughout North America. References are available upon request.

Qualifications/Certifications: W.R. Wayman and Associates has successfully assisted many companies in achieving ISO 9000 Registration. Mr. Wayman has over 30 years of experience with both commercial and military quality systems including implementing, managing, auditing, training, and consulting. He is an ASQC Certified Quality Engineer and Quality Auditor and an RAB certified Quality Systems Auditor. All Associates are experienced, trained, and certified.

WHITSELL AND COMPANY, P. C.

1250 East Copeland Road
Suite 600
Arlington, TX 76011, U.S.A.
Tel: 817-261-4454; Fax: 817-861-5944

Contact: Daniel G. Whitsell, President

Contact 2: Michael Natishyn, Director of Marketing

In Business: Nine years

Clients: Privately owned manufacturing, distribution, and service businesses.

Qualifications/Certifications: AICPA; Peer Review.

WICKHAM INTERNATIONAL MANAGEMENT SERVICES INC.

1313 North Market Street
Suite 3410 NE, Hercules Plaza
Wilmington, DE 19801-1151, U.S.A.
Tel: 800-599-2338; Fax: 800-599-2338
E-mail Address: 100443.1633@compuserve.com

Contact: Bill Ferguson, Director and Principal Consultant

In Business: 10 years

Clients: Wickham International Management Services has guided over 50 companies worldwide to successful ISO 9000 accreditation. Client organizations range from Blue Chip multinationals to small businesses. Expertise is primarily in the packaging, chemicals, petrochemicals, oil, printing, textiles, general manufacturing, warehousing, and distribution industries.

Qualifications/Certifications: Bill Ferguson is a Chartered Engineer, Fellow of the Institute of Energy, Fellow of the Institute of Quality Assurance. In his capacity as a lead assessor, registered with both the RAB and IRCA, he is retained by a number of assessing bodies. He has been employed in quality assurance for 25 years and during this time has defined quality requirements for chemicals, papers and textiles on behalf of the U.K. Ministry of Defense. His responsibilities included evaluating existing and potential suppliers' quality management systems and the subsequent approval and monitoring of their performance.

WILMINGTON QUALITY ASSOCIATES

303 Water Street
Wrightsville Beach, NC 28480, U.S.A.
Tel: 910-256-8149; Fax: 910-256-8149

Contact: Thomas Hudgin, President

In Business: Four years

Clients: Information on clients in the United States and Europe available upon request.

Qualifications/Certifications: Thomas Hudgin has been employed in the chemical and pharmaceutical industries for 27 years with Warren-Teed, Hoechst-Roussel, Adria (Montedison/Farmitalia), Glaxo, and Applied Analytical Industries. During this period, he managed the disciplines of quality assurance, Total Quality Management, regulatory affairs, project management, document control, production planning, new business development in manufacturing, and services industries. He has a B.S. in chemistry and an M.B.A.

WINONA TECHNICAL COLLEGE
Custom Services Division

1250 Homer Road
P.O. Box 409
Winona, MN 55987-0409, U.S.A.
Tel: 800-372-8164/507-454-4600
Fax: 507-452-1564
E-mail Address: tomt@win.tec.mn.us
Home Page Address:
http://wwwsmp.smumn.edu/rwwtc/

Contact: Tom Tourville, Dean of Custom Services

In Business: 25 years

Clients: IBM-Rochester, Fastenal, Equality Die Cast, Peerless Chain Company, Pepin Manufacturing Company, CANAMER International, RTP Company, A &L Machine, Polymer Composites Inc., Fiberite, ALTEC International, Miller FELPAX, and others

Qualifications/Certifications: ISO 9000 training and assistance are provided by a local affiliate of the QNET network of ISO 9000 and QS-9000 quality professionals. QS-9000, software quality systems, and other areas are addressed by appropriate QNET specialists. Additionally, Winona Technical College has experienced, certified training professionals in a variety of related quality fields including Total Quality Transformation (TQT), Zenger-Miller, and others.

WMD AND ASSOCIATES

Quality Services Group

86 Frisbee Road
Cassadaga, NY 14718, U.S.A.
Tel: 800-819-0811; Fax: 800-819-0811

Contact: William M. Dorman, President

Contact 2: Tammy Montgomery, Vice President of
Operations

In Business: Seven years

Clients: Anderson Screw Products, Inc., Cummins En-
gine Company, DSM Chemicals, DSM Resins,
Praxair, Rand Winters, ASQC, NQA-US
(Registrar), AQA (Registrar), VCA (Registrar),
Graybar Electric, TCIE.

Qualifications/Certifications: WMD and Associates
has provided ISO-9000 related consultancy
services for seven years. Extensive auditing experience (in excess of 200 third-party audits) for several internationally accredited registrars enhances
their ability to (1) guide companies seeking registration to ISO/QS-9000 and (2) train internal auditors. Over 500 students have been trained directly
by WMD and Associates as internal auditors or as
lead auditors (through accredited lead auditor
courses).

ISO 9000/QS-9000/ISO 14000 Registrars and Area of Expertise (As of July 1996)

Registrar	ISO 9000	QS-9000	ISO 14000
A.G.A. Quality, A Service of International Approval Services 8501 E Pleasant Valley Road Cleveland, OH 44131 Tel: 216-524-4990 Fax: 216-642-3463 Contact: Charles Russo	Yes	Yes	
ABS Quality Evaluations, Inc. 16855 Northchase Drive Houston, TX 77060, USA Tel: 713-873-9400 Fax: 713-874-9564 Contact: Patti L. Wigginton	Yes	Yes	Yes
AFAQ Woodfield Executive Center 1101 Perimeter Drive, Suite 450 Schaumburg, IL 60173 Tel: 847-330-0606 or 800-241-3412 Fax: 847-330-0707 Contact: Maritza Robbennolt	Yes		Yes

Registrar	ISO 9000	QS-9000	ISO 14000
AIB Registration Services 1213 Bakers Way Manhattan, KS 66502 Tel: 913-537-4750 Fax:913-537-1493 Contact: William Pursley	Yes		
AIB-Vincotte (AV Qualité) 2900 Wilcrest, Suite 300 Houston, TX 77042 Tel: 713-465-2850 Fax: 713-465-1182 Contact: Alice Koblenz	Yes	Yes	Yes
The American Association for Laboratory Accreditation 656 Quince Orchard Road, #620 Gaithersburg, MD 20878-1409 Tel: 301-670-1377 Fax: 301-869-1495 Contact: Peter Unger	Yes		
American Quality Assessors 1200 Main Street, Suite 107M Columbia, SC 29201 Tel: 803-779-8150 Fax: 803-779-8109 Contact: Frank Degar	Yes	Yes	Yes
The American Society of Mechanical Engineers, United Engineering Center 345 E 47th Street New York, NY 10017 Tel: 212-705-8590 Fax: 212-705-8599 Contact: David Wizda E-mail Address: accreditation@asme.org Home Page Address: http/www.asme.org	Yes		
ASCERT USA, Inc. 1054 31st Street, NW Suite 330 Washington, DC 20007 Tel: 202-337-3214 Fax: 202-337-3709 Contact: William Coles	Yes		
Asociacion Española de Normalizacion y Certificacion C/ Fernandez de la Hoz, 52 Madrid, 28010 Spain Tel: 34-1-432-6000; Fax: 34-1-310-4518 Contact: Mr. Eliseo Gutiérrez	Yes		

Registrar	ISO 9000	QS-9000	ISO 14000
Associated Offices Quality Certification 650 N Sam Houston Parkway E, Suite 228 Houston, TX 77060, USA Tel: 713-591-7882 Fax: 713-448-1401 Contact: Todd Fleckenstein	Yes		
AT&T Quality Registrar 650 Liberty Avenue Union, NJ 07083, USA Tel: 908-851-3058 Fax: 908-851-3158 Contact: John Malinauskas	Yes	Yes	
Bellcore Quality Registration 6 Corporate Place, Room 1A230 Piscataway, NJ 08854, USA Tel: 908-699-3739 Fax: 908-336-2244 Contact: Susan Schatzman	Yes		
British Standards Institution Quality Assurance British Standards Institution, Inc. 8000 Towers Crescent Drive, Suite 1350 Vienna, VA 22182, USA Tel: 703-760-7828 Fax: 703-761-2770 Contact: Reg Blake	Yes	Yes	Yes
The Bureau de normalisation du Quebec, Quality System Registration 70, rue Dalhousie, Bureau 220 Quebec, Quebec G1K 4B2, Canada Tel: 418-643-5813 Fax: 418-646-3315 Conact: Francois Lambert	Yes		
Bureau Veritas Quality International (North America), Inc. North American Central Offices 509 N Main Street Jamestown, NY 14701, USA Tel: 716-484-9002 Fax: 716-484-9003 Contact: Michael Danielson	Yes	Yes	Yes
Canadian General Standards Board 222 Queen Street, Suite 1402 Ottawa, Ontario K1A 1G6, Canada Tel: 613-941-8657 Fax: 613-941-8706 Contact: Claudette Tremblay	Yes		

Registrar	ISO 9000	QS-9000	ISO 14000
Centerior Registration Services Corporate Office 300 Madison Avenue, Edison Plaza Toledo, OH 43652, USA Tel: 419-249-5268 Fax: 419-249-5450 Contact: William J. Niedzwiecki	Yes	Yes	Yes
Ceramic Industry Certification Scheme Ltd. Queens Road, Penkhull Stoke-On-Trent, ST4 7LQ, England Tel: 44-1782-411-008 Fax: 44-1782-412-331 Contact: Myriam Brundson	Yes		
International Approval Services Canada, Inc. 55 Scarsdale Road Toronto, Ontario M3B 2R3, Canada Tel: 416-447-6468 Fax: 416-447-7067 Contact: Allan Nicholson	Yes	Yes	Yes
Davy Registrar Services, Inc. One Oliver Plaza Pittsburgh, PA 15222-2604, USA Tel: 412-566-3086 Fax: 416-566-5290 Contact: Leroy W. Pfennigwerth	Yes	Yes	
Defense Supply Center, Columbus 3990 E Brad Street DSCC-VQ Columbus, OH 43126-5000, USA Tel: 614-692-7604 Contact: Darrell Hill	Yes	Yes	
Det Norske Veritas DNV Certification, Inc. 16340 Park Ten Place, Suite 100 Houston, TX 77084, USA Tel: 281-721-6600 Fax: 281-721-6903 Contact: Steve Cumings	Yes	Yes	Yes
DLS Quality Technology Associates, Inc. 108 Hallmore Drive Camillus NY 13031, USA Tel: 315-468-5811 Fax: 315-468-5811 Contact: Rocco Lupo	Yes	Yes	Yes

Registrar	ISO 9000	QS-9000	ISO 14000
EAGLE Registrations, Inc. 3212 Winding Way Dayton, OH 45419, USA Tel: 513-293-3377 Fax: 513-293-0220 Contact: Chris Shillito	Yes	Yes	
Electronic Industries Quality Registry 2500 Wilson Boulevard Arlington, VA 22201-3834, USA Tel: 703-907-7563 or 800-222-9001 Fax: 703-907-7966 E-mail Address: EQR@EIA.ORG Miguel Gaitan	Yes	Yes	Yes
Entela, Inc., Quality System Registration Division 2890 Madison Avenue SE Grand Rapids, MI 49548-1206 Tel: 616-247-0515 or 800-888-3787 Fax: 616-247-7527 E-mail Address: info@entela.com Contact: Brandon Kerkstra	Yes	Yes	Yes
Factory Mutual Research Corporation 1151 Boston-Providence Turnpike PO Box 9102 Norwood, MA 02062, USA Tel: 617-255-4883 Fax: 617-762-9375 Contact: Tom Broderick	Yes		
GBJD Registrars Limited 4950 Yonge Street, Suite 2200 North York, Ontario M2N6K1 Tel: 416-218-5594 or 416-222-0286 Fax: 416-222-1146 Contact: Ray Grayston	Yes	Yes	Yes
DQS—German American Registrar for Management Systems, Inc. 804 Harvard Street Wilmette, IL 60091, USA Tel: 847-256-0523 Fax: 847-0572 E-mail Address: DQSofUSA@aol.com Contact: Stefan Heinloth	Yes	Yes	Yes

Registrar	ISO 9000	QS-9000	ISO 14000
Global Registrars, Inc. 4700 Clairton Boulevard Pittsburgh, PA 15236, USA Tel: 412-884-2290 Fax: 412-884-2268 Contact: Joseph Fabian	Yes	Yes	Yes
HSB Registration Services One State Street PO Box 5024 Hartford, CT 06102-5024, USA Tel: 800-472-1866 or 860-722-5294 Fax: 860-722-5530 Contact: Jill Bellino	Yes	Yes	
Inchcape Systems Registration/Intertek Services 313 Speen Street, Suite 200 Natick, MA 01760, USA Tel: 508-647-5147 Fax: 508-647-6714 Contact: James P. O'Neil	Yes	Yes	Yes
Instituto Mexicano de Normalización y Certificación A.C. Manuel Maria Contreras No. 133, 1 er. Piso Col. Cuauhtémoc Mexico, Distrito Federal 06470, Mexico Tel: 525-566-4750 Fax: 525-546-4546 Contact: Joel Narváez Nieto	Yes		
International Certifications Ltd. 7127 E Becker Lane, Suite 83 Scottsdale AZ 85254, USA Tel: 800-315-4486 Fax: 602-998-8288 Contact: Richard Gilkcson	Yes		
ISOQAR PO Box 347 East Longmeadow, MA 01028-0347 Tel: 413-567-1297 Fax: 413-567-4514 Contact: H. Joseph Murphy	Yes		
KEMA Registered Quality, Inc. 4379 County Line Road Chalfont, PA 18914, USA Tel: 215-822-4258 Fax: 215-822-4285 Contact: H. Pierre Sallé	Yes	Yes	Yes

Registrar	ISO 9000	QS-9000	ISO 14000
Kemper Registrar Services, Inc. Plaza One Building, Suite 305 1 State Highway 12 Flemington, NJ 08822-1731, USA Tel: 908-806-7498 or 800-555-2928 Fax: 908-806-6937 Contact: Albert Egreczky	Yes		
KPMG Quality Registrar 150 John F. Kennedy Parkway Short Hills, NJ 07078, USA Tel: 800-716-5595 Fax: 201-912-6050	Yes	Yes	
Litton Systems Canada Limited, Quality System Registrars 25 City View Drive Etobicoke, Ontario M9W 5A7, Canada Tel: 800-267-0861 or 416-249-1231 x2308 Fax: 416-246-2049 Contact: Stephen J. Barfoot	Yes	Yes	
Lloyd's Register Quality Assurance Limited 33-41 Newark Street Hoboken, NJ 07030, USA Tel: 201-963-1111 Fax: 201-963-3299 Contact: Kevin Mullaney, Sales Manager	Yes	Yes	Yes
Loss Prevention Certification Board Ltd., The Melrose Avenue Borehamwood, Hertfordshire WD6 2BJ, England Tel: 44-081-207-2345 Fax: 44-081-207-6305 Contact: Lulie Sasin	Yes		
National Quality Assurance USA 4 Post Office Square Road Acton, MA 01720 Tel: 800-649-5289 or 508-635-9256 Fax: 508-263-0785 Contact: Robert S. Parsons	Yes	Yes	Yes
National Standards Authority of Ireland Worldwide Certification Services 5 Medallion Center (Greeley Street) Merrimack, NH 03054 Tel: 603-424-7070 Fax: 603-429-1427 richb@nsaieast.com or nsaiwest@cerfnet.com Contact: Richard Bernier	Yes	Yes	Yes

Registrar	ISO 9000	QS-9000	ISO 14000
NSF International Strategic Registrations Ltd 2100 Commonwealth Boulevard Suite 100 Ann Arbor, MI 48105, USA Tel: 313-669-0098 or 888-NSF-9000 Fax: 313-669-0196 Contact: David Fitzwilliam	Yes	Yes	Yes
OMNEX-Automotive Quality Systems Registrar, Inc. 3025 Boardwalk Drive Suite 190 Ann Arbor, MI 48108, USA Tel: 313-913-8055 Fax: 313-913-8152 Contact: Dennis Hughey	Yes	Yes	
Orion Registrar, Inc. PO Box 5070 Arvada, CO 80060, USA Tel: 303-456-6010 Fax: 303-456-6681	Yes	Yes	Yes
OTS Quality Registrars, Inc. 10700 Northwest Freeway, Suite 455 Houston, TX 77092, USA Tel: 713-688-9494; Fax: 713-688-9590 Contact: Ron Platt	Yes	Yes	
Performance Review Institute 163 Thornhill Road Warrendale, PA 15086-7527 Tel: 412-772-1616 or 800-352-7293 Fax: 412-772-1699 Contact: James Borczyk	Yes	Yes	Yes
Perry Johnson Registrars, Inc. 3000 Town Center, Suite 630 Southfield, MI 48075 Tel: 800-800-7910; Fax: 810-358-0882 E-mail Address: pji@wwnet.com Home Page Address: http:\\www.pji.com Contact: Terry Boboige	Yes	Yes	Yes
NTS-CS 1146 Massachusetts Borborough, MA 01719 Tel: 508-263-4811 Fax: 508-635-1037 Contact: Richard Dunne	Yes	Yes	Yes

Registrar	ISO 9000	QS-9000	ISO 14000
Quality Certification Bureau, Inc. 9650 - 20 Avenue Suite 103, Advanced Technology Centre Edmonton, Alberta T6N 1G1, Canada Tel: 800-268-7321 or 403-496-2463 Fax: 403-496-2464 Contact: Julie Press, Director of Registration	Yes	Yes	Yes
Quality Management Institute 90 Burnhamthorpe Road W, Suite 800 Mississauga, Ontario L5B 3C3, Canada Tel: 905-272-3920 or 800-465-3717 Fax: 905-272-8503 Contact: Michael J. Haycock	Yes	Yes	Yes
Quality Systems Assessment Registrar (Head Office) 7250 W Credit Avenue Mississauga, Ontario L5N 5N1, Canada Tel: 800-461-9001 or 905-542-0547 Fax: 905-542-1318 Contact: Richard Kitney, General Manager of Market Development	Yes	Yes	
Quality Systems Registrars, Inc. 13873 Park Center Road, Suite 217 Herndon, VA 22071-3279, USA Tel: 703-478-0241; Fax: 703-478-0645 E-mail Address: qsrdr@msn.com Home Page Address: http://home.navisoft.com/qsr Contact: Marshall Courtois	Yes	Yes	Yes
Scott Quality Systems Registrars, Inc. 8 Grove Street, Suite 200 Wellesley, MA 02181 Tel: 617-239-1110 Fax: 617-239-0433 Contact: Warren Riddle	Yes	Yes	

Registrar	ISO 9000	QS-9000	ISO 14000
SGS International Certification Services Canada, Inc. 90 Gough Road, Unit 4 Markham, Ontario L3R 5V5, Canada Tel: 905-479-1160 or 800-636-0847 Fax: 905-479-9452 Contact: Client Services Department	Yes	Yes	Yes
SGS International Certification Services, Inc. Meadows Office Complex 301 Route 17 N Rutherford, NJ 07070 Tel: 800-747-9047 Fax: 201-935-4555 Contact: Lois O'Brien	Yes	Yes	Yes
Sira Certification Service/Sira Test and Certification Ltd. South Hill Chislehurst, Kent BR7 5EH Tel: 44-181-467-2636 Fax: 44-181-295-1990 Contact: Peter Blackwell	Yes		
Smithers Quality Assessments, Inc. 425 W Market Street Akron, OH 44303-2099, USA Tel: 330-762-4231 Fax: 330-762-7447 Contact: John Sedlak	Yes	Yes	
Steel Related Industries Quality System Registrars SRI Quality System Registrar 2000 Corporate Drive, Suite 330 Wexford, PA 15090-7605, USA Tel: 412-934-9000 Fax: 412-935-6825 Contact: Peter Lake	Yes	Yes	Yes
TRA Certification 700 E Beardsley Avenue PO Box 1081 Elkhart, IN 46515-1081 Tel: 800-398-9282 or 219-264-0745 Fax: 219-264-0740 Contact: Dean Hupp	Yes	Yes	

Registrar	ISO 9000	QS-9000	ISO 14000
TRADA Certification, Inc. Lakewood Executive Center Suite 190 1233 Shelbourne Road Burlington, VT 05403, USA Tel: 802-865-4764 or 800-865-4540 Fax: 802-865-4849	Yes	Yes	Yes
TUV America and TUV Product Service TUV America, Inc. and TUV Product Service, Inc. (Headquarters) 5 Cherry Hill Drive Danvers, MA 01923, USA Tel: 508-777-7999 Fax: 508-762-8414 E-mail Address: CSTOCKWELL@TUVAM.COM or QUALITY@TUVPS.COM Contact: Chris Stockwell	Yes	Yes	Yes (TUV America)
TÜV ESSEN 2099 Gateway Plaza, Suite 200 San Jose, CA 95110, USA Tel: 408-441-7888 or 800-TUV-4630 Fax: 408-441-7111 E-mail Address: 76035.2136@compuserve.com Home Page Address: http://www.tuvessen.com Contact: Olga Rada	Yes	Yes	
TUV Rheinland of North America, Inc. TUV Rheinland, North American Headquarters 12 Commerce Road Newtown, CT 06470, USA Tel: 203-426-0888 Fax: 203-270-8883 Contact: Martin Langer	Yes	Yes	
Underwriters Laboratories, Inc. 1285 Walt Whitman Road Melville, NY 11747-3081, USA Tel: 516-271-6200 Fax: 516-271-3356 Contact: Dan Keck	Yes	Yes	Yes
Underwriters Laboratories of Canada 7 Crouse Road Scarborough, Ontario M1R 3A9, Canada Tel: 416-757-3611 Fax: 416-757-8915 Contact: Vijay K Aggarwal	Yes	Yes	

Registrar	ISO 9000	QS-9000	ISO 14000
United Registrar of Systems Ltd. 11A Rossall Road Thorton-Cleveleys Blackpool FY5 1AP, England Tel: 125-382-0060 Fax: 125-386-3310 Contact: David Riggs	Yes	Yes	Yes
Vehicle Certification Agency VCA North America, Colonial House Office Park, Suite 140 42000 W Six Mile Road Northville, MI 48167, USA Tel: 810-344-2190 Fax: 810-344-2191 Contact: Trevor Davies	Yes		
Inchcape Testing Service NA Ltd. 8810 Elmslie Street LaSalle, Quebec H8R 1V8, Canada Tel: 514-366-3100 Fax: 514-366-5350 Contact: Daniel Désilets	Yes	Yes	Yes

Additional Resources

This appendix contains additional ISO 9000 and European Community resources, including the following:

- Economic development centers: MTCs and TAACs
- Hotlines
- Networks and support groups
- Electronic information
- Publications, including

    Accreditation/certification

    European community

    International trade

    ISO 9000/QS-9000

    ISO 14000

    Standards and directives

    Quality in general

- Software
- Audio/Video tapes
- CD interactive
- Other

ECONOMIC DEVELOPMENT CENTERS

Economic Development Centers offer help to companies struggling to maintain their status in a technology-driven world. Included are the National Institute for Standards and Manufacturing Technology Centers (MTCs) and the U.S. Department of Commerce's Trade Adjustment Assistance Centers (TAACs)

The MTCs, designed to bridge the "technology gap" between sources of improved manufacturing technology and the companies that need it, offer assistance in helping smaller companies institute new high-technology practices and encourage the establishment of continuous improvement programs.

The TAACs offer financial assistance to companies that require consulting services, inluding costs related to ISO 9000 certification. Companies that have experienced recent declines in sales and employment, due at least in part to increasing imports of competitive products, are eligible for the program. The government pays up to 75 percent of the cost of consulting services.

MANUFACTURING TECHNOLOGY CENTERS

A.L. Philpott Manufacturing Center

Contact: Gerry Ward
Location: P.O. Box 5311
645 Patriot Avenue
Martinsville, VA 24115
Tel: 540-666-8890 Fax: 540-666-8892

Alabama Technology Network

Contact: Sara Dennis
Location: 1500 Resource Drive
Birmingham, AL 35242
Tel: 205-250-4747

Arkansas Manufacturing Extension Network

Contact: Julie S. Welch
Location: 100 Main Street
Suite 450
Little Rock, AK 72201
Tel: 501-324-9006

California Manufacturing Technology Center (CMTC)

Contact: David Braunstein
Location: 13430 Hawthorne Boulevard
Hawthorne, CA 90250
Tel: 310-355-3060

Chicago Manufacturing Center

Contact: Rheal Turcotte
Location: Homan Square
3333 W. Arthington
Chicago, IL 60624
Tel: 312-265-2020

Connecticut State Technology Extension Program (CONN/STEP)

Contact: Peter LaPlace
Location: 179 Middle Turnpike, U-52
Storrs, CT 06269-5052
Tel: 860-486-2585

Delaware Manufacturing Alliance

Contact: John J. Shwed
Location: Delaware Technology Park,
One Innovation Way
Suite 301
Newark, DE 19711
Tel: 302-452-2520

Delaware Valley Industrial Resource Center (DVIRC)

Contact: Joseph Houldin
Location: 12265 Townsend Road
Suite 500
Philadelphia, PA 19154-1286
Tel: 215-464-8550

Florida Manufacturing Extension Partnership

Contact: William G. Brundage
Location: 200 S. Orange Avenue
Suite 1200
Orlando, FL 32801
Tel: 407-425-5313

Georgia Manufacturing Extension Alliance

Contact: Charles Estes
Location: Georgia Institute of Technology
223 O'Keefe Building
Atlanta, GA 30332-0640
Tel: 404-894-8989 Fax: 404-894-8194

Great Lakes Manufacturing Technology Center (GLMTC)

Contact: Ed Kwiatkowski
Location: Prospect Park Building
4600 Prospect Avenue
Cleveland, OH
44103-4314
Tel: 216-432-5322

Hudson Valley Technology Development Center

Contact: Douglas Koop
Location: Hudson Valley Technology Development
Center
300 Westage Business Center
Suite 225
Fishkill, NY 12524
Tel: 914-896-6934 Fax: 914-896-7006

Idaho Manufacturing Alliance

Contact: James Steinfort
Location: 1021 Manitou Avenue
Boise, ID 83706
E-mail: jsteinf@bsu.idbsu.edu.
Tel: 208-385-3767 Fax: 208-385-3877

Industrial Modernization Center (IMC)

Contact: James K. Shillenn
Location: Industrial Modernization Center, Inc.
Farm Complex
R.R. #5, Box 220-62A
Montoursville, PA 17754
Tel: 717-368-8361

Industry Network Corporation (INC)

Contact: Randy Grissom
Location: 1155 University Boulevard S.E.
Albuquerque, NM 87106
Tel: 505-843-4250

Industry Technology Assistance Corporation

Contact: Sara Garretson
Location: 253 Broadway
Room 302
New York, NY 10007
Tel: 212-240-6920 Fax: 212-240-6879

Iowa Manufacturing Technology Center (Iowa MTC)

Contact: Del Shepard
Location: Des Moines Area Community College

ATC Building, 3E
2006 South Ankeny Boulevard
Ankeny, IA 50021
Tel: 515-965-7040 Fax: 515-965-7050

Kentucky Technology Service

Contact: Don Smith
Location: P.O. Box 1125
Lexington, KY 40589
Tel: 606-252-7801

Lake Erie Manufacturing Extension Partnership

Contact: Charles P. Alter
Location: 1700 N. Westwood Street
Toledo, OH 43607
Tel: 419-534-3709 Fax: 419-531-8465

Maine Manufacturing Extension Partnership

Contact: Diane Wescott
Location: 87 Winthrop Street
Augusta, ME 04330
Tel: 207-621-6350

MAMTC Missouri Rolla Regional Office

Contact: Ken Wells
Location: 800 W. 14th Street
Suite 111
Rolla, MO 65401
Tel: 573-364-8570 or 800-956-2682 (95-MAMTC)

Manufacturing Technology Industrial Resource Center (MANTC)

Contact: Jack E. Minnich
Location: MANTEC, Inc.
227 W. Market Street
P.O. Box 5046
York, PA 17405
Tel: 717-843-5054

Maryland Manufacturing Modernization Network

Contact: Joe McGrath
Location: Maryland Department of Economic

Development
Division of Business
217 East Redwood Street
Baltimore, MD 21202
Tel: 410-767-6476

Massachusetts Manufacturing Partnership (MMP)

Contact: Jan Pounds
Location: Bay State Skills Corp.
101 Summer Street
4th Floor
Boston, MA 02110
Tel: 617-292-5100, ext. 271

Miami Valley Manufacturing Extension Center

Contact: David L. Chalk
Location: 3155 Research Boulevard
Suite 205
Kettering, OH 45420
Tel: 513-258-6190 Fax: 513-258-6189

Michigan Manufacturing Technology Center (MMTC)

Contact: W.C. (Butch) Dyer
Location: P.O. Box 1485
2901 Hubbard Road
Ann Arbor, MI 48106
Tel: 800-292-4484

Mid-America Manufacturing Technology Center (MAMTC)

Contact: Paul Clay
Location: 10561 Barkley
Suite 602
Overland Park, KS 66212
Tel: 913-649-4333 or 800-653-4333 if calling from
CO, KS, MO, WY

Minnesota Manufacturing Technology Center (MnMTC)

Contact: Todd Loudenslager
Location: 111 Third Avenue S.
Suite 400
Minneapolis, MN 55401
Tel: 612-338-7722

Mississippi Polymer Institute and Pilot Manufacturing Extension Center

Contact: Robert K. Schlatzer
Location: P.O. Box 10003
Hattiesburg, MS 39406
Tel: 601-266-4607 Fax: 601-266-5635

Montana Manufacturing Extension Center

Contact: William R. Taylor
Location: Montana State University
315 Roberts Hall
Bozeman, MT 59717
Tel: 406-994-3812 Fax: 406-994-6098

Nebraska Industrial Competitiveness Service (NICS)

Contact: Tommy Thorne
Location: 8800 O Street
Lincoln, NE 68520
Tel: 402-437-2535 Fax: 402-437-2404

New Hampshire Regional Manufacturing Technology Center

Contact: Keith Bird
Location: 505 Amherst Street
Nashua, NH 03061-2052
Tel: 603-882-6923

New Jersey Manufacturing Extension Partnership

Contact: Don Sebastian
Location: Center for Manufacturing Systems
New Jersey Institute of Technology
218 Central Avenue
Suite 350 ITC
Newark, NJ 07102
Tel: 201-642-4869 Fax: 201-596-6056

New York Manufacturing Extension Parnership (New York MEP)

Contact: John Crews
Location: New York State Science and Technology Foundation
99 Washington Avenue
Suite 1730

Albany, NY 12210-2875
Tel: 518-474-4349 Fax: 518-473-6876

North Carolina Manufacturing Extension Partnership

Contact: Robert L. Edwards
Location: IES Technical Services
Box 7902
North Carolina State University
Raleigh, NC 27695-7902
Tel: 919-515-5408 Fax: 919-515-6519

North Dakota Maufacturing Extension Center

Contact: Warren Enyart
Location: 1833 E. Bismarck Expressway
Bismarck, ND 58504
Tel: 701-328-5300

North/East Pennsylvania Manufacturing Extension Partnership

Contact: Edith Ritter
Location: 125 Goodman Drive
Bethlehem, PA 18015
Tel: 610-758-5599

Northern California Manufacturing Extension Center

Contact: Phil Nanzetta
Location: 39550 Liberty Street
2nd Floor
Fremont, CA 94537
Tel: 510-249-1480

Northwest Wisconsin Manufacturing Outreach Center

Contact: Naidu Katuri
Location: University of Wisconsin
Stout Technology Transfer Institute
103 First Avenue West
Menomonie, WI 54751
Tel: 715-232-2397

Oklahoma Alliance for Manufacturing Excellence

Contact: Randy Goldsmith
Location: 525 South Main Street

Suite 500
Tulsa, OK 74103
Tel: 918-592-0722 Fax: 212-592-1417

Plastics Technology Deployment Center

Contact: William Roche
Location: c/o Penn State-Erie
Behrend College Station Road
Erie, PA 16563
Tel: 814-898-6122

Plastics Technology Deployment Center (PTDC)

Contact: David Thomas-Graves
Location: GLMTC Manufacturing Outreach
Program
Prospect Park Building
4600 Prospect Avenue
Cleveland, OH 44103
Tel: 216-432-5300

Puerto Rico Manufacturing Extension, Inc.

Contact: Miguel Burset
Location: Economic Development Administration
355 F.D. Roosevelt Avenue
Hato Rey, PR 00918
Tel: 809-766-0616 or 809-764-1415

Rhode Island Manufacturing Extension Services

Contact: William Ferrante
Location: 70 Lower College Road
Kingston, RI 02881
Tel: 401-874-5516 Fax: 401-874-2355

Southeast Manufacturing Technology Center (SMTC)

Contact: Belford E. Cross
Location: 1136 Washington Street
Suite 300
Columbia, SC 29201
Tel: 803-252-6976 or 803-254-8512

Southwestern Pennsylvania Manufacturing Extension Partnership

Contact: Ray Christman
Location: 4516 Henry Street
Pittsburgh, PA 15213
Tel: 412-687-0200 ext. 234

Tennessee Manufacturing Extension Partnership

Contact: Henry Tupis
Location: Department of Economic
and Community Development
Rachel Jackson Building
7th Floor
320 6th Avenue, North
Nashville, TN 37243-0405
Tel: 615-741-2826 Fax:615-741-5829

Texas Manufacturing Assistance Center

Contact: Mike Klonsinski
Location: 1700 Congress Avenue
Suite 200
Austin, TX 78701
Tel: 512-936-0235

University Industry Public Partnership for Economic Growth (UnIPEG)

Contact: Kay Adams
Location: 61 Court Street
6th Floor
Binghamton, NY 13901
Tel: 607-774-0022 Fax: 607-774-0026

Utah Manufacturing Extension Partnership

Contact: David Sorensen
Location: Western Coalition for Advanced
Manufacturing Processes
435 Crabtree Technology Building
Brigham Young University
Provo, UT 84602-4211
Tel: 801-378-9000

Vermont Manufacturing Extension Center

Contact: Muriel Durgin
Location: P.O. Box 500
Randolph Center, VT 05061
Tel: 802-728-1432 Fax: 802-728-3026
email: rzider@night.vtc.vsc.edu

Virginia Alliance for Manufacturing Competitiveness

Contact: Bill Dickinson
Location: P.O. Box 1163
Richmond, VA 23218
Tel: 804-786-3501 Fax: 804-371-2945

West Virginia Partnership for Industrial Modernization

Contact: Virginia Tucker
Location: 634 8th Street
Huntington, WV 25701
Tel: 304-525-1916

Western New York Technology Development Center

Contact: Melissa Kelly-McCabe
Location: 1576 Sweet Home Road
Amherst, NY 14228
Tel: 716-636-3626

Wisconsin Manufacturing Extension Partnership

Contact: Chris Thompson
Location: Wisconsin Center for Manufacturing
and Productivity
432 N. Lake Street
Room B121-B
Madison, WI 53706-1498
Tel: 608-262-2069 Fax: 608-262-0872

TRADER ADJUSTMENT ASSISTANCE CENTERS

Great Lakes TAAC

Contact: Maureen Burns, Director
Location: University of Michigan

School of Business Administration
506 E. Liberty Street
3rd Floor
Carver Building
Ann Arbor, MI 48104-2210
Tel: 313-998-6213 Fax: 313-998-6202

Mid-America TAAC

Contact: Paul Schmid, Director
Location: University of Missouri at Columbia
University Place
Suite 1700
Columbia, MO 65211
Tel: 314-882-6162 Fax: 314-882-6156

Mid-Atlantic TAAC

Contact: Raymond G. Hufnagel, Director
Location: 486 Norristown Road
Suite 130
Blue Bell, PA 19422-2353
Tel: 610-825-7819

Mid-West TAAC

Contact: Howard Yefsky, Director
Location: Applied Strategies
150 N. Wacker Drive
Suite 2240
Chicago, IL 60606
Tel: 312-368-4600 Fax: 312-368-9043

New England TAAC

Contact: Richard McLaughlin, Director
Location: 120 Boylston Street
Boston, MA 02116
Tel: 617-542-2395 Fax: 617-542-8457

New Jersey TAAC

Contact: John Walsh, Director
Location: 200 S. Warren Street
Trenton, NJ 08625
Tel: 609-292-0360 Fax: 609-984-4301

New York State TAAC

Contact: John Lacey, Director
Location: 117 Hawley Street
Suite 102

Binghamton, NY 13901
Tel: 607-771-0875/212-921-1662 (NYC)
Fax: 607-724-2404

Northwest TAAC

Contact: Gary Kuhar, Director
Location: Bank of California Center
900 4th Avenue
Suite 2430
Seattle, WA 98164
Tel: 206-622-2730 Fax: 206-622-1105

Rocky Mountain TAAC

Contact: Edvard M. Hag, Acting Director
Location: 5353 Manhattan Circle, Suite 200
Boulder, CO 80303
Tel: 303-499-8222 Fax: 303-499-8298

Southeastern TAAC

Contact: Paul Lewis, Director
Location: Georgia Institute of Technology
Research Institute
O'Keefe Building,151 6th Street, Room 224
Atlanta, GA 30332
Tel: 404-894-6789 Fax: 404-853-9172

Southwest TAAC

Contact: Robert Velasquez, Director
Location: 1222 N. Main, Suite 740
San Antonio, TX 78212-4414
Tel: 210-558-2490 Fax: 210-558-2491

Western TAAC

Contact: Daniel Jiminez, Director
Location: University of Southern California
3716 S Hope Street, Suite 200
Los Angeles, CA 90007
Tel: 213-743-2732 Fax: 213-746-9043

NETWORKS & SUPPORT GROUPS

Cleveland Consortium

Contact: Ray DePuy
Tel: 216-432-5300 Fax: 216-361-2900

The ISO 9000/QS-9000 Support Group

Contact: Rick Clements
Tel: 616-891-9114 Fax: 616-891-9114
Home Page Address:
http://www.cris.com/~isogroup

Washington ISO 9000 Initiative (WAISO)

Contact: K.C. Ayers
Tel: 206-392-7610 Fax: 206-392-7630

Portland ISO 9000 Users Group

Contact: Diane Hunt
Tel: 503-235-4668; Fax: 503-235-4668
Home Page Address:
http://www.teleport.com/~dhunt
E-mail Address: dhunt@teleport.com

Minnesota Technology Inc.

Contact: David Hepenstal
Tel: 800-803-6446 Fax: 612-339-5214

TRADE HOTLINES

EU Hotline

301-921-4164

GATT Hotline

301-975-4041

The Export Hotline

800-USA-XPORT (800-872-9767)

The Trade Information Center

800-USA-TRADE (800-872-8723)

PUBLICATIONS

Accreditation/Certification

Books

ISO/IEC Compendium of Conformity Assessment Documents

Author: ISO and IEC

Publisher: ANSI
Tel: 212-642-4900
Price Information: $95
Directories

Directory of Federal Government Certification Programs (NBS SP 739)

Author: NIST

Publisher: National Technical Information Service (NTIS)
Tel: 703-487-4650
Price Information: $51.50
ISBN or Reference Number: PB 88-201512/LL

Directory of Federal Government Laboratory Accreditation/Designation Programs (NIST SP 808)

Author: NIST

Publisher: National Technical Information Service (NTIS)
Tel: 703-487-4650
Price Information: $31.50
ISBN or Reference Number: PB 91-167379/LL

Directory of International and Regional Organizations Conducting Standards-Related Activities (NIST SP 767)

Author: NIST

Publisher: National Technical Information Service (NTIS)
Tel: 703-487-4650
Price Information: $71.50
ISBN or Reference Number: PB 89-221147/LL
Publisher: Global Engineering Documents

Tel: 800-854-7179
Order #: Cat. SP767

Directory of Private Sector Product Certification Programs (NIST SP 774)

Author: NIST

Publisher: National Technical Information Service (NTIS)
Tel: 703-487-4650
Price Information: $51.50
ISBN or Reference Number: PB 90-161712/LL

Directory of State and Local Government Laboratory Accreditation/Designation Programs (NIST SP 815)

Author: NIST

Publisher: National Technical Information Service (NTIS)
Tel: 703-487-4650
Price Information: $27.00
ISBN or Reference Number: PB 92-108968/LL

Directory of Professional/Trade Organization Laboratory Accreditation/Designation Programs (NIST SP 831)

Author: NIST

Publisher: National Technical Information Service (NTIS)
Tel: 703-487-4650
Price Information: $35.00
ISBN or Reference Number: PB 92-181940/LL

Reports

The ABC's of Standards-Related Activities in the United States (NBSIR 87-3576)

Author: NIST

Publisher: National Technical Information Service (NTIS)
Tel: 703-487-4650

Price Information: $12.50
ISBN or Reference Number: PB 87-224309/LL

The ABC's of Certification Activities in the United States (NBSIR 88-3821)

Author: NIST

Publisher: National Technical Information Service (NTIS)
Tel: 703-487-4650
Price Information: $24.50
ISBN or Reference Number: PB 88-239793/LL

Laboratory Accreditation in the United States (NISTIR 4576)

Author: NIST

Publisher: National Technical Information Service (NTIS)
Tel: 703-487-4650
Price Information: $24.50
ISBN or Reference Number: PB 91-194415

Survey on the Implementation of ISO/IEC Guide 25 by National Laboratory Accreditation Programs (NISTIR 5473)

Author: NIST

Publisher: National Technical Information Service (NTIS)
Tel: 703-487-4650
Price Information: $21.50
ISBN or Reference Number: PB94-210150/LL

EUROPEAN UNION

Books

Guide to Product Liability in Europe

Author: William C. Hoffman and Susanne Hill-Arning

Publisher: Kluwer Law & Taxation Publishers
Tel: 800-577-8118
Price: $100.00

The European Union Machinery Safety Directive: Compliance Manual for Trade

Authors: Bruce McIntosh, B.S.M.E. and James W. Kolka, Ph.D., J.D.

Publisher: Simcom, Inc.
Tel: (404)303-7799 Fax: 404-303-0192
Home page: http://www.eurocom.com
E-mail: simcom@mindspring.com
Price: Call for pricing.

The European Union Electromagnetic Compatibility and Low Voltage Directives: Compliance Manual for Trade

Authors: Nissen Isakov and Peter Perkins, P.E.

Publisher: Simcom, Inc.
Tel: (404)303-7799 Fax: 404-303-0192
Home page: http://www.eurocom.com
E-mail: simcom@mindspring.com
Price: Call for pricing.

A Summary of the New European Community Approach to Standards Development (NBSIR 88-3793-1)

Author: NIST

Publisher: National Technical Information Service (NTIS)
Tel: 703-487-4650
Price Information: $24.50
ISBN or Reference Number: PB 88-229489/AS

EU Electrical and Electronics Equipment Compendium, 2nd edition, 1996

Author: International Quality Press

Publisher: International Quality Press
Tel: 716-685-0534
Price: $595.00

EU Medical Device Compendium, 3rd edition, 1996

Author: International Quality Press

Publisher: International Quality Press
Tel: 716-685-0534
Price: $595.00

EU Medicinal Products Compendium, 3rd edition, 1996

Author: International Quality Press

Publisher: International Quality Press
Tel: 716-685-0534
Price: $595.00
Individual titles are available from International Quality Press. Call for a book list.

Loose-Leaf Binders

The CE Mark: A Handbook on Product Certification and Testing in the European Union

Author: European Document Research

Publisher: European Document Research/LeDroit Publishing
Tel: 202-785-8594
Price: $695. (one update)

The CE Mark for Medical Devices

Author: European Document Research

Publisher: European Document Research/LeDroit Publishing
Price: $995.00 (one update)
Tel: 202-785-8594

EC Legislation on Health and Safety at Work

Author: European Document Research

Publisher: European Document Research/LeDroit Publishing
Tel: 202-785-8594
Price: $495.00 (one update)

EU Magazines

EUROPE Magazine

Author: Delegation of the European Commission (Washington Office)

Publisher: Delegation of the European Commission (Washington Office)
Tel: 202-862-9555
Price: Available by Subscription, but sample copies are sent on request

The Complete European Trade Digest (CETD)TM

Author: Simcom, Inc.

Publisher: Simcom, Inc.
Tel: (404)303-7799 Fax: 404-303-0192
Home page: http://www.eurocom.com
E-mail: simcom@mindspring.com
Price: Call for pricing.

EU Newsletters

EC Bulletin

Author: Price Waterhouse—EC Services

Publisher: Price Waterhouse—EC Services
Tel: 32-2-773-4911 (Belgium); Fax: 32-2-762-5100

Eurowatch: Economics, Policy, and Law in the New Europe

Author: LRP Publications

Publisher: LRP Publications
Tel: 800-333-1291 Fax: 215-784-9639
Price: $797.00 per year

EURECOM

Author: Delegation of the European Commission (New York Office)

Publisher: Delegation of the European Commission
Tel: 212-371-3804 Fax: 212-688-1013
Price: Free (call to be put on the mailing list)

European Document Research— Current Report

Author: European Document Research

Publisher: European Document Research/LeDroit Publishing
Price: $767.00 per year (50 issues)
Tel: 202-785-8594

ISI Update

Publisher: International Standards Initiative
Tel: 206-392-7610

EU Reports

European Community and Environmental Policy and Regulations

Author: Single Internal Market Information Service

Publisher: U.S. Department of Commerce
Tel: 202-482-5823
Price Information: Free

The EC Liberalizes Transportation Rules and Upgrades Its Infrastructure

Author: Single Internal Market Information Service

Publisher: U.S. Department of Commerce
Tel: 202-482-5823
Price Information: Free

The European Community's Policy and Regulations on Food and Beverages

Author: Single Internal Market Information Service

Publisher: U.S. Department of Commerce
Tel: 202-482-5823
Price Information: Free

EC Single Market Law Affecting Exporting and Distribution: Agents, Distributors, Franchises

Author: Single Internal Market Information Service

Publisher: U.S. Department of Commerce
Tel: 202-482-5823
Price Information: Free

EC Telecommunications in the Single Market

Author: Single Internal Market Information Service

Publisher: U.S. Department of Commerce
Tel: 202-482-5823
Price Information: Free

EC Testing and Certification Procedures Under the Internal Market Program

Author: Single Internal Market Information Service

Publisher: U.S. Department of Commerce
Tel: 202-482-5823
Price Information: Free

Single Market for Intellectual Property Protection

Author: Single Internal Market Information Service

Publisher: U.S. Department of Commerce
Tel: 202-482-5823
Price Information: Free

The EFTA Industrial Development Fund for Portugal

Author: EFTA Press and Information Service

Publisher: EFTA Press and Information Service
Tel: 41-22-749-1111 (Switzerland)

EFTA: What It Is, What It Does

Author: EFTA

Publisher: EFTA Press and Information Service
Tel: 41-22-749-1111 (Switzerland)

The European Free Trade Association

Author: EFTA Press and Information Service

Publisher: EFTA Press and Information Service
Tel: 41-22-749-1111 (Switzerland)

The Free Trade Agreements of the EFTA Countries With Third Partner Countries

Author: EFTA Press and Information Service

Publisher: EFTA Press and Information Service
Tel: 41-22-749-1111 (Switzerland)

Medical Device Research Report

Author: AAMI

Publisher: AAMI
Tel: 703-525-4890

INTERNATIONAL TRADE

Books

EuroMarketing A Strategic Planner for Selling into the New Europe

Author: Rick Arons

Publisher: Irwin Professional Publishing
Tel: 800-773-4607
Price Information: $37.50
ISBN or Reference Number: 1557382018

Trade Directory

Faulkner and Gray's European Business Directory

Author: Faulkner and Gray

Publisher: Faulkner and Gray
Tel: 800-535-8403
Price Information: $295

Trade Magazines

Business America

Publisher: U.S. Department of Commerce

Tel: 202-512-1800 Fax: 202-512-2250
Price Information: $53.00 (US), $66.25
(International)

CE: Compliance Engineering Magazine

Publisher: Compliance Engineering

Tel: 508-681-6600
Price: Free

The Exporter

Publisher: Trade Data Reports

Tel: 212-587-1340
Price: $168.00 per year
ISBN or Reference Number: 0736-9239

Trade Newsletters

TMO Survey Week

Publisher: The Marley Organization, Inc.

Tel: 203-438-3801
Price: $465 annually which includes 11 monthly
issues of TMO Update which is also available as a
separate publication for $90.00

TMO Update

Publisher: The Marley Organization, Inc.

Tel: 203-438-3801
Price: $90.00

ISO 9000 Checklists

Author: Canadian Standards Association

Publisher: Canadian Standards Association
Tel: 414-747-4044

PLUS 9001—The ISO 9000 Essentials

Author: Canadian Standards Association

Publisher: Canadian Standards Association
Tel: 414-747-4044
Price: $38

The ISO 9000 Essentials: A Handbook for Medical Device Manufacturers

Author: Canadian Standards Association

Publisher: Canadian Standards Association
Tel: 414-747-4044
Price: $38

ISO 9000 & QS-9000 Books

ISO 9000 Compendium: International Standards for Quality Management

Author: ISO Central Secretariat

Publisher: ANSI
Tel: 212-642-4900
Price: $250
ISBN or Reference Number: 92-67-101 72-2

Documenting Quality for ISO 9000 and Other Industry Standards

Author: Gary E. MacLean

Publisher: ASQC Quality Press
Tel: 800-248-1946
Price: (List Price) $28.00, (ASQC Member Price) $25.00
ISBN or Reference Number: 0-87389-212-7, Item # H0761

ISO 9000: Preparing for Registration

Author: James L. Lamprecht

Publisher: ASQC Quality Press
Tel: 800-248-1946
Price: (List Price) $49.75, (ASQC Member Price) $44.75
ISBN or Reference Number: 0-8247-8741-2, Item # H0776

ISO 9000 Implementation for Small Business

Author: James L. Lamprecht

Publisher: ASQC Quality Press
Tel: 800-248-1946
Price: (List Price) $35.00, (ASQC Member Price) $31.50
ISBN or Reference Number: 0-87389-350-6, Item # H0907

Eight-Step Process to Successful ISO 9000 Implementation: A Quality Management System Approach

Author: Lawrence A. Wilson

Publisher: ASQC Quality Press
Tel: 800-248-1946
Price: (List Price) $35.00, (ASQC Member Price) $31.50
ISBN or Reference Number: 0-87389-327-1, Item # H0878

Integrating QS-9000 with Your Automotive Quality System, 2nd Edition

Author: D.H. Stamatis

Publisher: ASQC Quality Press
Tel: 800-248-1946
Price: (List Price) $38.00, (ASQC Member Price) $34.00
ISBN or Reference Number: 0-87389-408-1, Item #H0951

ISO 9000 and the Service Sector: A Critical Interpretation of the 1994 Revisions

Author: James L. Lamprecht

Publisher: ASQC Quality Press
Tel: 800-248-1946
Price: (List Price) $35.00, (ASQC Member Price) $31.50
ISBN or Reference Number: 0-87389-313-1, Item # H0871

How to Qualify for ISO 9000

Author: Sanders & Associates

Publisher: AMA
Tel: 800-262-9699
Price: $130, (AMA Member Price) $117
ISBN or Reference Number: Stock # 95003AG1

ISO 9000 Documentation: A Quality Manual and 27 Procedures

Author: Jack Kanholm

Publisher: AQA Co.
Tel: 213-222-3600
Price: $290

ISO 9000 Explained: 65 Requirements Checklist and Compliance Guide

Author: Jack Kanholm

Publisher: AQA Co.
Tel: 213-222-3600
Price: $49

ISO 9000 Quality System: Implementation Guide

Author: Jack Kanholm

Publisher: AQA Co.
Tel: 213-222-3600
Price: $69

ISO 9000 In Our Company: Self-Study Course for Personnel

Author: Jack Kanholm

Publisher: AQA Co.
Tel: 213-222-3600
Price: $9

QS-9000 Procedures: A Quality Manual and 40 Procedures

Author: Jack Kanholm

Publisher: AQA Co.
Tel: 213-222-3600
Price: $390

QS-9000 Requirements: 107 Requirements Checklist and Compliance Guide

Author: Jack Kanhom

Publisher: AQA Co.
Tel: 213-222-3600
Price: $59

QS-9000 In Our Company: Self-Study Course for Personnel

Author: Jack Kanholm

Publisher: AQA Co.
Tel: 213-222-3600
Price: $9

The ISO 9000 Book: A Global Competitor's Guide to Compliance & Certification (Second Edition)

Authors: John T. Rabbitt and Peter A. Bergh

Publisher: Quality Resources
Tel: 800-247-8519
Price Information: $26.95
ISBN or Reference Number: ISBN 0-527-76258-X, Item # 76258X

The Miniguide to ISO 9000

Authors: John T. Rabbitt and Peter A. Bergh

Publisher: Quality Resources
Tel: 800-247-8519
Price: $4.50
ISBN or Reference Number: ISBN 0-527-76302-0, Item # 763020

The Miniguide to QS-9000

Authors: John T. Rabbitt and Peter A. Bergh

Publisher: Quality Resources
Tel: 800-247-8519

Price: $4.50
ISBN or Reference Number: 0-527-76323-3

Quality Management Benchmark Assessment (Second Edition)

Author: J. P. Russell

Publisher: Quality Resources
Tel: 800-247-8519
Price: $21.00
ISBN or Reference Number: 0-527-76295-4, Item # 762954

Assessing Your Company for ISO 9001: 1994

Author: Ian Durand

Publisher: International Quality Systems
Tel: 508-480-9249
Price: $19.95

Implementing an ISO 9000: 1994 Quality System

Author: Ian Durand

Publisher: International Quality Systems
Tel: 508-480-9249
Price: $49.95

A Manager's View of ISO 9000: 1994 Quality System Registration

Author: Ian Durand

Publisher: International Quality Systems
Tel: 508-480-9249
Price: $24.95

Selecting an ISO 9000 Quality System Registrar

Author: Ian Durand

Publisher: International Quality Systems
Tel: 508-480-9249
Price: $24.95

Understanding the ISO 9000: 1994 Quality Assurance Standards

Author: Ian Durand

Publisher: International Quality Systems
Tel: 508-480-9249
Price: $49.95

Using Support Tools

Author: Ian Durand

Publisher: International Quality Systems
Tel: 508-480-9249
Price: $19.95
Or you may buy the entire IQS Documentation Package for the low price of $149.00

Documenting and Auditing for ISO 9000 and QS-9000

Tools for Ensuring Certification or Registration
Author: D.H. Stamatis

Publisher: Irwin Professional Publishing
Tel: 800-773-4607
Price: $50.00
ISBN or Reference Number: 0-7863-0862-1

Integrating QS-9000 with Your Automotive Quality System (distributed for ASQC Quality Press)

Author: D.H. Stamatis

Publisher: Irwin Professional Publishing
Tel: 800-773-4607
Price: $35.00
ISBN or Reference Number: 0-87389-338-7

ISO 9000 Survey Comprehensive Data and Analysis of U.S. Registered Companies, 1996

Author: Irwin Professional Publishing

Publisher: Irwin Professional Publishing and Dun & Bradstreet Information Services
Tel: 800-773-4607

Price: $150.00
ISBN or Reference Number: 0-7863-0901-6

QS-9000 Pioneers: Registered Companies Share Their Strategies for Success

Author: Subir Chowdhury and Ken Zimmer

Publisher: Irwin Professional Publishing
Tel: 800-773-4607
Price: $50.00
ISBN or Reference Number: 0-7863-0865-6

ISO 9000: A Comprehensive Guide to Registration, Audit Guidelines, and Successful Certification

Author: Greg Hutchins

Publisher: John Wiley & Sons
Tel: 800-225-5945
Price: $42.50
ISBN or Reference Number: 0471132055

The ISO 9000 Implementation Manual: Ten Steps to ISO 9000 Registration

Author: Greg Hutchins

Publisher: John Wiley & Sons, Inc.
Tel: 800-225-5945
Price: $60.00
ISBN or Reference Number: 0471131849

The ISO 9000 Workbook: A Comprehensive Guide to Developing Quality Manuals and Procedures

Author: Greg Hutchins

Publisher: John Wiley & Sons, Inc.
Tel: 800-225-5945
Price Information: $59.95
ISBN or Reference Number: 047114245x

Taking Care of Business: How to Become More Efficient and Effective Using ISO 9000

Author: Greg Hutchins

Publisher: John Wiley & Sons, Inc.
Tel: 800-225-5945
Price: $27.50
ISBN or Reference Number: 0471132047

ISO 9000: Meeting the New International Standards

Author: Perry Johnson

Publisher: McGraw-Hill, Inc.
Tel: 800-262-4729
Price: $40
ISBN or Reference Number: 0-07-032691-6

The Road to Quality: An Orientation Guide to ISO 9000

Author: Irwin Professional Publishing

Publisher: Irwin Professional Publishing
Tel: 800-773-4607
Price: $11.95
ISBN or Reference Number: 1-883337-32-1

ISO 9000: Motivating the People, Mastering the Process, Achieving Registration

Authors: David Stevenson and Craig Westover

Publisher: Irwin Professional Publishing and the National Center for Manufacturing Sciences
Tel: 800-773-4607
Price: $45.00
ISBN or Reference Number: 07863-0115-5

GMP/ISO 9000® Quality Audit Manual for Healthcare Manufacturers and Their Suppliers, Fourth Edition

Author: Leonard Steinborn

Publisher: Interpharm Press
Tel: 847-459-8480
Price: $298.00

ISO 9000 Standards on Quality Management and Quality Systems

Author: ASTM

Publisher: Interpharm Press
Tel: 847-459-8480
Price: $181.50

Preparing Your Company for QS-9000: A Guide for the Automotive Industry, (Second Edition)

Author: Richard Clements, Stanley M. Sidor, and Rand E. Winters Jr.

Publisher: ASQC Quality Press
Tel: 800-248-1946
Price: (List Price) $22.00; (ASQC Member Price) $19.00
ISBN or Reference Number: 0-87389-344-1, Item # H0928

ISO 9000 for Software Developers, Revised Edition

Author: Charles H. Schmauch

Publisher: ASQC Quality Press
Tel: 800-248-1946
Price: (List Price) $45.00, (ASQC Member Price) $40.00
ISBN or Reference Number: 0-87389-348-4, Item # H0901

Aviation Industry Quality Systems: ISO 9000 and the Federal Aviation Regulations

Author: Michael Dreikorn

Publisher: ASQC Quality Press
Tel: 800-248-1946
Price: (List Price) $60.00, (ASQC Member Price) $54.00
ISBN or Reference Number: 0-87389-331-X, Item # H0888

Managing Records for ISO 9000 Compliance

Author: Eugenia K. Brumm

Publisher: ASQC Quality Press
Tel: 800-248-1946
Price: (List Price) $49.00,

(ASQC Member Price) $44.00
ISBN or Reference Number: 0-87389-312-3, Item # H0870

ISO 9000 Audit Questionnaire and Registration Guidelines

Author: Praful (Paul) C. Mehta

Publisher: ASQC Quality Press
Tel: 800-248-1946
Price: (List Price) $13.00, (ASQC Member Price) $10.50
ISBN or Reference Number: 0-87389-299-2, Item # MB102

The ISO 9000 Auditor's Companion

Author: Kent A. Keeney with Foreword by Joseph J. Tsiakals

Publisher: ASQC Quality Press
Tel: 800-248-1946
Price: (List Price) $27.00, (ASQC Member Price) $24.00
ISBN or Reference Number: 0-87389-324-7, Item # H0880

The Audit Kit

Author: Kent A. Keeney

Publisher: ASQC Quality Press
Tel: 800-248-1946
Price: (List Price) $44.00, (ASQC Member Price) $39.50
ISBN or Reference Number: 0-87389-328-X, Item # H0880A
Complete Set contains *The ISO 9000 Auditor's Companion* and *The Audit Kit.*
Publisher: ASQC Quality Press
Tel: 800-248-1946
List Price: $68.00, (ASQC Member Price) $61.00
Item# H0880B

QS-9000 Handbook: A Guide to Registration and Audit

Author: Jayanta K. Bandyopadhyay

Publisher: St. Lucie Press
Tel: 407-274-9906

Price: $59.95
ISBN or Reference Number: 1-57444011-X

The 90-Day ISO 9000 Manual: Basics Manual and Implementation Guide

Authors: James Stewart, Peter Mauch and Frank Straka

Publisher: St. Lucie Press
Tel: 407-274-9906
Price: $189.00
ISBN or Reference Number: 1-884015-11-5

ISO 9000: An Implementation Guide for Small to Mid-sized Businesses

Authors: Frank Voehl, Peter Jackson, and David Ashton

Publisher: St. Lucie Press
Tel: 407-274-9906
Price: $47.95
ISBN or Reference Number: 1-884015-10-7

The QS-9000 Answer Book

Author: Radley M. Smith

Publisher: Paton Press
Price: $24.95 + $5.00 shipping and handling
Tel: 916-529-4125
ISBN or Reference Number: 0-9650445-0-5

The Memory Jogger™ 9000

Authors: Robert W. Peach and Diane S. Ritter

Publisher: GOAL/QPC
Tel: 800-207-5813
Price: Call for pricing (discount quantities available)
ISBN or Reference Number: 1-879364-82-4

Understanding the Value of ISO 9000

Authors: Bryn Owen and Peter Malkovich

Publisher: SPC Press, Inc.
Tel: 800-545-8602
Price: $25.00

Achieving ISO 9000 Registration

Authors: Bryn Owen, Tom Cothran and Peter Malkovich

Publisher: SPC Press, Inc.
Tel: 800-545-8602
Price: $75.00

ISO 9000 Directories
Chemical Week

Publisher: Chemical Week Associates
Tel: 212-621-4802; For advertising inquiries please contact Betsy Punsalan at 212-621-4945; To obatin copies of the directory, contact Ari Roman at 212-621-4802.
Price: Call for pricing.

ISO 9000 Registered Company Directory

Author: Irwin Professional Publishing

Publisher: Irwin Professional Publishing
Tel: 800-773-4607
Price: $295
ISBN or Reference Number: 0-7863-0777-3

ISO 9000 Newsletters
ISO Bulletin (1996)

Author: ISO Central Secretariat

Publisher: ANSI
Tel: 212-642-4900
Price Information: $85.00

Quality Systems Update (QSU)

Author: Irwin Professional Publishing

Publisher: Irwin Professional Publishing
Tel: 800-773-4607
Price: (Executive Service subscription) $595; Associate Service (newsletter only): $395
ISBN or Reference Number: 1060 1821

Molten Metal ISO 9000 Journal

Author: Ralph J. Teetor III

Publisher: Foundry Quality Systems
Tel: 815-961-9972
Price: $12.00 per year

ISO 9000 News Service

Author: ISO Central Secretariat

Publisher: ANSI
Tel: 212-642-4900
Price Information: $360.

IS0 9000 Reports

More Questions and Answers: the ISO 9000 Standard Series and Related Issues

Author: NIST

Publisher: U.S. Department of Commerce
Tel: 703-487-4650
Price: Free
ISBN or Reference Number: PB-93-140689

Questions and Answers on Quality, the ISO 9000 Standard Series, Quality System Registration, and Related Issues

Author: NIST

Publisher: U.S. Department of Commerce
Tel: 703-487-4650
Price: Free
ISO 14000

Environment and ISO 14000 Books

The Road to ISO 14000: An Orientation Guide to the Environmental Management Standards

Authors: Glenn K. Nestel and Roy F.Weston, Inc.

Publisher: Irwin Professional Publishing
Tel: 800-773-4607
Price: $13.95
ISBN or Reference Number: 0-7863-0866-4

ISO 14000: A Guide to the New Environmental Management Standards

Authors: Tom Tibor with Ira Feldman

Publisher: Irwin Professional Publishing
Tel: 800-773-4607
Price: $35.00
ISBN or Reference Number: 0-7863-0523-1

Implementing An Environmental Audit: How to Gain a Competitive Advantage Using Quality and Environmental Responsibility

Authors: Grant Ledgerwood, Elizabeth Street, and Riki Therivel

Publisher: Irwin Professional Publishing
Tel: 800-773-4607
Price: $45.00
ISBN or Reference Number: 0-7863-0142-2

The Handbook of Environmentally Conscious Manufacturing: From Design and Production to Labeling and Recycling

Authors: Robert E. Cattanach, Jake M. Holdreith, Daniel P. Reinke, and Larry K. Sibik

Publisher: Irwin Professional Publishing
Tel: 800-773-4607
Price: $75.00
ISBN or Reference Number: 0-7863-0147-3

Going Green: How to Communicate Your Company's Environmental Commitment

Author: E. Bruce Harrison

Publisher: Irwin Professional Publishing
Tel: 800-773-4607
Price: $32.50
ISBN or Reference Number: 1-55623-945-9

Inside ISO 14000: The Competitive Advantage of Environmental Management

Author: Don Sayre

Publisher: St. Lucie Press
Tel: 407-274-9906
Price: $32.95
ISBN or Reference Number: 1-57444-028-4

Magazines

Environmental Protection

Publisher: Stevens Publishing Corp.
Tel: 817-776-9000
Price: $113 per year

STANDARDS AND DIRECTIVES

Books

CEN—Standards for Access to the European Market (formerly known as The CEN Technical Program: Standards for Europe)

Author: CEN

Publisher: CEN; Distributor: ON-CEN SALES POINT
Fax: 43 1 21 00 818 (Vienna, Austria)
Price: ATS612 (excluding postage and VAT (10%) where applicable)
ISBN or Reference: 92-9097-384-6l, Item #: AC-CESS95
This Information is also available from ANSI (American National Standard Institute)
Tel: 212-642-4900
Price: $100.00

The New Approach: Legislation and Standards on the Free Movement of Goods in Europe

Author: CEN

Publisher: CEN; Distributor: ON-CEN SALES POINT
Fax: 43 1 213 00 818 (Vienna, Austria)

Price: ATS 560 (excluding postage and VAT (10%) where applicable)
ISBN or Reference Number: 92-9097-336-6; Item #: NOUVEL-E

CEN Machinery Program, Third Edition

Author: CEN

Publisher: CEN; Distributor: ON-CEN SALES POINT
Fax: 43 1 213 00 818 (Vienna, Austria)
Price: ATS 856 (excluding portage and VAT (10%) where applicable)
Item #: MACH 3

Standards and Directives Catalogs

CEN Catalogue

Author: CEN

Publisher: ANSI
Tel: 212-642-4900
Price: $135.00

CENELEC Catalogue

Author: CENELEC

Publisher: ANSI
Tel: 212-642-4900
Price: $110.00

Standards for Access to the European Market

Author: CEN

Publisher: ANSI
Tel: 212-642-4900
Price: $100.00

Standards and Directives Directories

Standards Activities of Organizations in the United States (NIST SP 806)

Author: NIST

Publisher: National Technical Information Service (NTIS)
Tel: 703-487-4650
Price: $115.00
ISBN or Reference Number: PB 91-177774/LL
Publisher: Global Engineering Documents
Tel: 800-854-7179
Order #: Cat. SP806

Directory of European Regional Standards-Related Organizations (NIST SP 795)

Author: NIST

Publisher: National Technical Information Service (NTIS)
Tel: 703-487-4650
Price: $44.50
ISBN or Reference Number: PB 91-107599/LL or
Publisher: Global Engineering Documents
Tel: 800-854-7179
Order #: Cat. 0258-3

Standards and Directives Magazines

BSI News

Author: BSI

Publisher: BSI
Tel: 703-760-7828
Price: Free for BSI Members

Consensus

Author: National Standards System

Publisher: Standards Council of Canada
Tel: 613-238-3222
Price: $24/year (Canada), $36/year (outside Canada)

Standards and Directives Newsletters

Compliance Alert

Author: International Quality Press

Publisher: International Quality Press
Tel: 716-685-0534
Price: Call for Pricing

ANSI Reporter

Publisher: American National Standards Institute
Tel: 212-642-4900
Price: $250.00 (includes ANSI Standards Action Newsletter)

ANSI Standards Action

Author: ANSI

Publisher: ANSI
Tel: 212-642-9000
Price: Free for members, $250/year for non-members (includes ANSI Reporter)

TBT News

Author: NIST

Publisher: National Center for Standards and Certification Information, Office of Standards Services
Tel: 301-975-4040
Price: Subscription is free upon request
Reports

Standards and Directives Reports

ANSI Global Standardization Reports

Author: ANSI

Publisher: ANSI
Tel: 212-642-4900
Price: Prices vary between $24 and $75 for each report

CENELEC Technical Report (1995)

Author: CENELEC

Publisher: ANSI
Tel: 212-642-4900
Price: $400.00

A Review of U.S. Participation in International Standards Activities (NBSIR 88-3698)

Author: NIST

Publisher: National Technical Information Service
(NTIS)
Tel: 703-487-4650
Price: $27.00
ISBN or Reference Number: PB 88-164165/LL

Chemicals and the European Union

Author: Single Internal Market Information
Service

Publisher: U.S. Department of Commerce
Tel: 202-482-5823
Price: Free

EC Product Standards Under the Internal Market Program

Author: Single Internal Market Information
Service

Publisher: U.S. Department of Commerce
Tel: 202-482-5823
Price: Free

Single Market for Medical Devices

Author: Single Internal Market Information
Service

Publisher: U.S. Department of Commerce
Tel: 202-482-5823
Price: Free

Quality Magazines
Quality Digest Magazine

Publisher: QCI International
Tel: 916-893-4095
Price: $75.00
ISBN or Reference Number (ISSN) 1049-8699

Quality Magazine

Publisher: Hitchcock Publishing Co.
Tel: 708-665-1000
Price: $70.00 per year

P. I. Quality Magazine

Publisher: Hithcock Publishing Co.
Tel: 708-665-1000
Price: $40.00 per year

Quality Progress Magazine

Publisher: ASQC Quality Press
Tel: 800-248-1946
Price: (U.S.) $50.00/year; (Canada) $85.00

EU Books Catalog
The Interpharm 1996 Catalog

Author: Interpharm Press

Publisher: Interpharm Press
Tel: 847-459-8480
Price: Free

SOFTWARE

This is a partial list of ISO/Quality related software that is available to the marketplace. For a more extensive list of software companies, please consult Quality Progress' all inclusive 13th Annual QA/QC Software Directory found in the April 1996 issue. Quality Progress lists nearly 300 software companies in an easy to read format. Readers can contact the companies directly for detailed software information. If interested, please call:

Quality Progress Magazine

Publisher: ASQC Quality Press
Price: (U.S.) $50.00/year; (Canada) $85.00
Tel: 800-248-1946

ISO 9000, QS-9000, ISO 14000 Software

allCLEAR III for Windows—Intelligent Flowcharting, Fast and Easy

Author: CLEAR Software, Inc.

Tel: 617-965-6755
Price: $299.00

Audit Management

Author: JK Technologies, Inc.

Tel: 954-984-9311
Price: $495 single user, $200 each additional user
E-mail Address: 103262.324@CompuServe.com
Home Page Address:
http://WWW.Gate/~HAMMERHD

Audit Master

Author: The Harrington Group, Inc.

Tel: 407-382-7005
Price: $199
E-mail Address: hgi@harrington-group.com
Home Page Address: http://www.harrington-group.com

Audit Pro 9000

Author: ISO Software Solutions, Inc.

Tel: 619-275-1726
Price: $699.00
E-mail Address: 72242.2621@compuserve.com

Calibration Recall

Author: The Harrington Group, Inc.

Tel: 407-382-7005
Price: $199

E-mail Address: hgi@harrington-group.com
Home Page Address:
http://www.harrington-group.com

Cal PM 9000

Author: ISO Software Solutions

Tel: 619-275-1726
Price: $699.99

Cal PRO 9000

Author: ISO Software Solutions

Tel: 619-275-1726
Price: $499.99

CAR-Pro 9000

Author: ISO Software Solutions

Tel: 619-275-1726
Price: $499.99

CAS9000: Corrective Action

Author: JK Technologies, Inc.

Tel: 954-984-9311
Price: $495 single user, $200 each additional user
E-mail Address: 103262.324@CompuServe.com
Home Page Address:
http://WWW.Gate/~HAMMERHD

Chek-Mate 9000™

Publisher: International Quality Technologies, Inc.

Tel: 800-245-9722
Price: $395 (discounts for volume and site license)
E-mail Address: Sales@IQT.COM

CLEAR Process for Windows v1.0

Author: CLEAR Software, Inc.

Tel: 617-965-6755
(Introductory) Price: $495.95

COQ-Pro 9000

Author: ISO Software Solutions, Inc.

Tel: 619-275-1726
Price: $499
E-mail Address: 72242.2621@compuserve.com

Corrective Action

Author: The Harrington Group, Inc.

Tel: 407-382-7005
Price: $299
E-mail Address: hgi@harrington-group.com
Home Page Address: http://www.harrington-group.com

Cost of Quality

Author: The Harrington Group, Inc.

Tel: 407-382-7005
Price: $299
E-mail Address: hgi@harrington-group.com
Home Page Address: http://www.harrington-group.com

DBQ®: Database for Quality

Author: Murphy Software

Tel: 800-892-3328
Price: Call for pricing.
E-mail Address: salesinfo@Murphyco.com
Home Page Address: http://www.murphyco.com/
DBQ is a registered trademark of Murphy Software

DOC-Pro 9000

Author: ISO Software Solutions, Inc.

Tel: 619-275-1726
Price: $399
E-mail Address: 72242.2621@compuserve.com

Document Control

Author: JK Technologies, Inc.

Tel: 954-984-9311
Price: $495 single user, $200 each additional user
E-mail Address: 103262.324@CompuServe.com
Home Page Address:
http://WWW.Gate/~HAMMERHD

Document Control System

Author: The Harrington Group, Inc.

Tel: 407-382-7005
Price: $199
E-mail Address: hgi@harrington-group.com
Home Page Address: http://www.harrington-group.com

The Essential Electronic Collection: A New ISO 9000 Diskette Offering

Author: Canadian Standards Association

Tel: 416-747-4044
Price: $295
The Electronic Collection (CSA/ISO 9000 Disk-94)

EtQ 9000 Maps®

Author: EtQ Management Consultants, Inc.

Tel: 800-354-4ISO
Price: $499 per user. Site licenses available.
E-mail Address: 71763.2023@CompuServe.COM
Home Page Address:
http://ourworld.compuserve.com/homepages/EtQHome

EtQ Solutions®

Author: EtQ Management Consultants, Inc.

Tel: 800-354-4ISO

Price: Call for site license price.
E-mail: 71763.2023@CompuServe.COM
Home Page Address:
http://ourworld.compuserve.com/homepages/EtQHome

Impact Software (Integrated Manufacturing Planning and Control Tool)

Author: Integral Solutions, Inc.

Tel: 810-543-6040
Price: Call for more information and to receive a
free information package.

Indoctrination to Quality Training Course

Author: FED-PRO, Inc.

Publisher: FED-PRO, Inc.
Telephone: 800-833-3776
Price: $195
(Additional student manuals available. Call for
pricing structure.)

Inspection and Test System

Author: The Harrington Group, Inc.

Tel: 407-382-7005
Price: $199
E-mail Address: hgi@harrington-group.com
Home Page Address: http://www.harrington-group.com

IQS Calibrate™ (Calibration Management)

Author: IQS, Inc.

Tel: 800-635-5901
Price: Call IQS for pricing
E-mail Address: sales@iqs.com
Home Page Address: http://www.iqs.com

IQS COLLECT™ (Data Collection)

Author: IQS, Inc.

Tel: 800-635-5901

Price: Call for pricing
E-mail Address: sales@iqs.com
Home Page Address: http://www.iqs.com

IQS Correct™ (Correct Action)

Author: IQS, Inc.

Tel: 800-635-5901
Price: Call for pricing
E-mail Address: sales@iqs.com
Home Page Address: http://www.iqs.com

IQS Customer™ (Customer Management)

Author: IQS, Inc.

Tel: 800-635-5901
Price: Call IQS for current pricing
E-mail Address: sales@iqs.com
Home Page Address: http://www.iqs.com

IQS Involve™ (Employee Involvement)

Author: IQS, Inc.

Tel: 800-635-5901
Price: Call for pricing
E-mail Address: sales@iqs.com
Home Page Address: http://www.iqs.com

IQS NCM™ (Nonconformance Management)

Author: IQS, Inc.

Tel: 800-635-5901
Price: Call for pricing
E-mail Address: sales@iqs.com
Home Page Address: http://www.iqs.com

IQS QCOST™ (Quality Costs)

Author: IQS, Inc.

Tel: 800-635-5901
Price: Call for pricing

E-mail Address: sales@iqs.com
Home Page Address: http://www.iqs.com

IQS PPAP MANAGER™

Author: IQS, Inc.

Tel: 800-635-5901
Price: Call for pricing
E-mail Address: sales@iqs.com
Home Page Address: http://www.iqs.com

IQS Prevent™ (Preventative Maintenance)

Author: IQS, Inc.

Tel: 800-635-5901
Price: Call IQS for current pricing
E-mail Address: sales@iqs.com
Home Page Address: http://www.iqs.com

IQS Process™ (Process Documentation)

Author: IQS, Inc.

Tel: 800-635-5901
Price: Call IQS for current pricing
E-mail Address: sales@iqs.com
Home Page Address: http://www.iqs.com

IQS Product™ (Product Documentation)

Author: IQS, Inc.

Tel: 800-635-5901
Price: Call IQS for current pricing
E-mail Address: sales@iqs.com
Home Page Address: http://www.iqs.com

IQS SPC™ (Statistical Process Control)

Author: IQS, Inc.

Tel: 800-635-5901

Price: Call IQS for current pricing
E-mail Address: sales@iqs.com
Home Page Address: http://www.iqs.com

IQS Supplier™ (Supplier Management)

Author: IQS, Inc.

Tel: 800-635-5901
Price: Call for pricing
E-mail Address: sales@iqs.com
Home Page Address: http://www.iqs.com

IQS SYSDOC™ (System Documentation)

Author: IQS, Inc.

Tel: 800-635-5901
Price: Call for pricing
E-mail Address: sales@iqs.com
Home Page Address: http://www.iqs.com

ISO 9000 Checklists

Publisher: G.R. Technologies Ltd.
Tel: 416-886-1307
Price: $105.00, plus $7.50 shipping and handling

ISO 9001, 9002, or 9003 Quality Manual Software

Author: FED-PRO, Inc.

Tel: 800-833-3776
Price: $345.00, $295.00, and $245.00

ISO 9000 Quality Management System

Publisher: ISO Systems, Inc.
Tel: 708-830-5814
Price: Call for pricing.
E-mail Address: stanorm@aol.com

ISO-Internal Auditor 9000

Author: ISO Software Solutions, Inc.

Tel: 619-275-1726
Price: $299
E-mail Address: 72242.2621@compuserve.com

ISO-Supplier Auditor 9000

Author: ISO Software Solutions, Inc.

Tel: 619-275-1726
Price: $299
E-mail Address: 72242.2621@compuserve.com

ISOxPERT and QSxPERT

Author: Management Software International, Inc.

Price: Call for pricing.
Tel: 800-476-3279
E-mail Address: custserv@isoxpert.com
Home Page Address: http://www.isoeasy.com

Machine Maintenance

Author: JK Technologies, Inc.

Tel: 954-984-9311
Price: $1,495 single user, $200 each additional user
E-mail Address: 103262.324@compuserve.com
Home Page Address:
http://WWW.Gate/~HAMMERHD

PPAP"Express"" 9000

Author: ISO Software Solutions

Price: $699.99
Tel: 619-275-1726

QMS/9000+™

Author: John A. Keane and Associates, Inc.

Tel: 609-924-7904
Price: Call for pricing.
E-mail Address: keane@cnj.digex.net
Home Page Address:
http://www.cnj.digex.net/~keane

Q-Pulse ISO 9000 Record Management Software

Author: W.C. Terry and Associates

Tel: 219-622-7950
Prices: Call for pricing.
E-mail Address: wcterry@cris.com
Home Page Address:
http://www.cris.com/~wcterry/

Quality Control

Author: JK Technologies, Inc.

Tel: 954-984-9311
Price: $495 single user, $200 each additional user
E-mail Address: 103262.324@compuserve.com
Home Page Address: http://www.Gate/~hammerhd

QualityTracker

Author: Cyber Quality Inc.

Tel: 713 589-5113 or 800 296-7406.
Price: Call for pricing. For a free demo, visit our world-wide web site at: http://www.cqinc.com/
E-mail Address: cqi@cqinc.com
Home Page Address: http://www.cqinc.com/

Self-Assessment Utility for Q9000

Author: The Harrington Group, Inc.

Tel: 407-382-7005
Price: $199
E-mail Address: hgi@harrington-group.com
Home Page Address: http://www.harrington-group.com

Software Systems Quality Consulting

Author: Software Systems Quality Consulting

Tel: 408-985-4476
Price: Call for information or visit home page.
E-mail Address: deibs@shell.portal.com
E-mail Address 2: ssqc@cris.com
Home Page Address:
http://www.billboard.com/ssqc/index.html
Home Page Address 2: http://www.cris.com/~ssqc

Supplier Rating Manager

Author: The Harrington Group, Inc.

Tel: 407-382-7005
Price: $99
E-mail Address: hgi@harrington-group.com
Home Page Address:
http://www.harrington-group.com

TeamFlow

Author: CFM, Inc.

Tel: 617-275-5258
Price: Single User: $295
E-mail Address: info@teamflow.com
Home Page Address: http://www.teamflow.com
Quantity discounts available.
Network version starts at $1350 for five concurrent
users.

Toolkit for ISO 9000/QS-9000/ISO 14000

Author: IQS, Inc.

Tel: 800-635-5901
Price: Call IQS for current pricing
E-mail Address: sales@iqs.com
Home Page Address: http://www.iqs.com

TQS-9000

Author: Advanced Systems and Design, Inc.

Tel: 810-616-9818

Price: $495
E-mail Address: asdpm@aol.com

TQS-Audit

Author: Advanced Systems and Design, Inc.

Tel: 810-616-9818
Price: $295
E-mail Address: asdpm@aol.com

TQS-Gage

Author: Advanced Systems and Design, Inc.

Tel: 810-616-9818
Price: $295
E-mail Address: asdpm@alo.com

Track and Flow 9000: Quality Task Management Software

Author: SystemCorp

Track & Flow Step by Step 9000-Lotus Notes
Tel: 514-339-1067
E-mail Address: sys@systemcorp.com
Price: $635.00 (stand alone)*
Track and Flow Step by Step 9000-Symantec®
Tel: 514-339-1067
E-mail Address: sys@systemcorp.com
Price: $635 (stand alone)*
*LAN Versions available

Training Accelerator

Author: ISO Software Solutions, Inc.

Tel: 619-275-1726
Price: $399
E-mail Address: 72242.2621@compuserve.com

Training Management and Document Control System

Knowledge Management and Transfer

Author: Quality Systems Integrators

Price: $5995 and up
Contact: Marti Turocy 800 458-0539
E-mail Address: turocy@chestnet.com
Home Page Address: www.webimpact.com/qsi

Training Manager

Author: The Harrington Group, Inc.

Tel: 407-382-7005
Price: $199
E-mail Address: hgi@harrington-group.com
Home Page Address: http://www.harrington-group.com

AUDIO/VIDEO TAPES

A Guide to ISO 9000: A Video Series (contains the titles listed below)

Author: The Media Group

Price: (Video Series Only) $975.00; (Video Set with Workbooks) $1075.00

- *ISO 9000 Making Your Company Competitive: A Management Overview*

  (Video Only) $139.00; (Video with Study Guide) $165.00

  (Video Only) $129.00; (Video with Study Guide) $149.00; (Available in Spanish) $129.00

- *A Practical Guide to Documenting and Implementing ISO 9000*

  (Video Only) $379.00; (Video with Study Guide) $409.00

- *Internal Auditing for ISO 9000*

  (Video Only) $379.00; (Video with Study Guide) $409.00

Tel: 800-678-1003

Employee Intro To QS-9000: The Automotive Perspective

Author: The Media Group

- (Video) $199.00

- (Video with Study Guide) $229.00

Tel: 800-678-1003

Implementing ISO 9000 or ANSI/ASQC Q90 Standards

Author: Statistical Quality Control and Circle Institute

Tel: 414-922-7938
Price Information: $199 each, or $525 for all three Video Programs written by Sud Ingle.

Price: $595.00

Indoctrination to Quality Training Course

Author: FED-PRO, Inc.

Tel: 800-833-3776
Price: $195
(Additional student manuals available—call for pricing structure.)

ISO 9000: 1994 Implementation Guide featuring Ian Durand

Author: Irwin Professional Publishing

Tel: 800-773-4607
Price: $595
ISBN or Reference Number: 1-883495-08-3

ISO 9000: Executive Briefing

Author: International Quality Systems

Tel: 508-480-9249
Price: $195
E-mail Address: RBLISS9000@AOL.COM

ISO 9000: 1994 Implementation Guide

Author: International Quality Systems

Tel: 508-480-9249
Price: $595
E-mail Address: rbliss9000@aol.com

ISO 9000: Executive Briefing featuring Ian Durand

Author: Irwin Professional Publishing

Tel: 800-773-4607
Price: $195
ISBN or Reference Number: 1-883495-02-4

ISO 9000: Opportunity Within Confusion

Author: Richard B. Clements

Tel: 800-248-1946
Price: $19.95
E-mail Address: isogroup@cris.com
Home Page Address:
http://www.cris.com./~isogroup/

ISO 9000: Tips and Techniques

Author: Richard B. Clements

Tel: 800-248-1946
Price: $19.95
E-mail Address: isogroup@cris.com
Home Page Address:
http://www.cris.com./~isogroup/

ISO 9000 Video Seminar includes:

Author: International Quality Systems

Tel: 508-480-9249
Price: $ 595
E-mail Address: rbliss9000@aol.com

ISO 9000 Video Seminar featuring Ian Durand

Author: Irwin Professional Publishing

Tel: 800-773-4607
Price: $595
ISBN or Reference Number: 1-883495-07-5

ISO 14000 In Focus: A Business Perspective For Sound Environmental Management

Author: The Media Group

Tel: 800-678-1003
Price: $199.95

Product Safety and Liability/CE Marking Video Training Program

Author: FED-PRO, Inc.

Tel: 800-833-3776
Price: $245

Selecting an ISO 9000 Quality System Registrar featuring Ian Durand

Author: Irwin Professional Publishing

Tel: 800-773-4607
Price: $99.95
ISBN or Reference Number: 1-883495-09-1

Simplifying ISO 9000: A Documentation Kit

Author: The Media Group

Tel: 800-678-1003
Price: $479

Supplier Certification/Quality in Purchasing Video Training Program

Author: FED-PRO, Inc.

Telephone: 800-833-3776
Price Information: $195

CD INTERACTIVE

H. James Harrington's Step by Step Multimedia

Author: SystemCorp

H. James Harrington's ISO 9000 Step-by-Step Multimedia Implementation

Tel: 514-339-1067
E-mail Address: sys@systemcorp.com
Price: $995 (stand alone)*
H. James Harrington's QS-9000 Step by Step Multimedia Implementation
Publisher: SystemCorp
Tel: 514-339-1067
E-mail Address: sys@systemcorp.com
Price: $1,295.00 (stand alone)*
*LAN versions available

Standards Infodisk—CD-ROM

Author: Infodisk Inc.

Tel: 201-703-8418
Price: $2195

Eurolaw: The Official Full Text Legal Database of the European Union

Author: Infodisk Inc.

Tel: 201-703-8418
Price: $2495

ISO 9000 Registration Series
QS-9000 Compliance Series
ISO 14000 EMS Conformance Series

Author: Reality Interactive, Inc.

Tel: 612-996-6777
Price: Average is $995 per CD. Varies depending on title.
E-mail Address: webmaster@ry.com
Home Page Address: http://www.realtools.com

OTHER

CE pro™

Author: Simcom International, LLC

Tel: 404-303-7799
Price: Call for pricing
E-mail Address: simcom@mindspring.com
Home Page Address:
http://www.eurocom.com

The ISO 9000/ QS-9000 Support Group

Author: National ISO 9000 Support Group

Tel: 800-248-1946
Price: $250.
E-mail Address: isogroup@cris.com
Home Page Address:
http://www.cris.com./~isogroup/

Vision 2000: The Strategy for the ISO 9000 Series Standards in the '90s

by Donald Marquardt, Jacques Chove, K.E. Jensen, Klaus Petrick, James Pyle, and Donald Strahle

Editor's Note: *The Vision 2000 document sets out the development strategy adopted by ISO/TC176 for the ISO 9000 standards into and beyond the Phase 1 revisions published by ISO in 1994. Further planning has been and is being undertaken within ISO/TC176. Those plans are consistent with and will not replace this Vision 2000 document but are intended to complement and extend the Vision 2000 strategy. Vision 2000 was first printed in Quality Progress, May 1991.[1] It has been reprinted by ISO[2] and in a number of individual countries.*

INTRODUCTION

This article is adapted from the report of the Ad Hoc Task Force of the International Organization for Standardization (ISO) Technical Committee 176. The task force was commissioned to prepare a strategic plan for ISO 9000 series architecture, numbering, and implementation. The task force report, which has become known as *Vision 2000,* was prepared by the authors of this article. TC 176 unanimously adopted the strategic principles of the report at its meeting in Interlaken, Switzerland, in October 1990. This article is part of a worldwide communication effort to gain broad acceptance of these principles and to globally influence standardization activities in the quality arena.

This article contains four major sections:

1. The Stake
   - The ISO 9000 Series Standards
   - Global Trends
   - The Critical Issues
2. Basic Concepts
   - Generic Product Categories
   - Industry/Economic Sectors
3. Analysis of the Marketplace
   - Preventing Proliferation
   - Segmenting the Markets
4. Vision 2000
   - Migration to Product Offerings Involving Several Generic Product Categories
   - Implications for Standards Development
   - Recommendations on Implementation

1. THE STAKE

The ISO 9000 Series Standards

In the years just prior to 1979, when TC 176 was formed, quality was rapidly emerging as a new emphasis in commerce and industry. Various national and multinational standards had been developed in the quality systems arena for commercial and industrial use and for military and nuclear power industry needs. Some standards were guidance documents. Other standards were for contractual use between customer and supplier organizations.

Despite some historical commonalities, these various standards were not sufficiently consistent for widespread use in international trade. Terminology in these standards and in commercial and industrial practice was also inconsistent and confusing.

The publication of the ISO 9000 series in 1987, together with the accompanying terminology standard (ISO 8402), has brought harmonization on an international scale and has supported the growing impact of quality as a factor in international trade. The ISO 9000 series has quickly been adopted by many nations and regional bodies and is rapidly supplanting prior national and industry-based standards. This initial marketplace success of the ISO 9000 series is testimony to the following two important achievements of TC176:

- The ISO 9000 series embodies comprehensive quality management concepts and guidance, together with several models for external quality-assurance requirements. Using an integrated systems architecture, the standards are packaged under a harmonized, easily memorized numbering system. These features have high value in meeting the commercial and industrial needs of current international trade.

- The ISO 9000 series was published in time to meet the growing need for international standardization in the quality arena and the wide adoption of third-party quality systems certification schemes.

More recently, TC 176 has been preparing additional international standards in quality management, quality assurance, and quality technology. Some of these will become part numbers to the 9000 series, while others will be in a 10000 series that has been reserved by ISO for use by TC 176.

Global Trends

Global Competition

Globalization has become a reality in the few years since the ISO 9000 series was published. Today, all but the smallest or most local commercial and industrial enterprises are finding that their principal marketplace competitors include companies headquartered in other countries.

Consequently, product development and marketing strategies must be global to reckon with global competition. Quality continues to grow in importance as a factor in marketplace success.

The European Union

The rapid implementation of the European Union (EU) single-market arrangement, targeted for full operation in 1992, has become a major driving force. EU 92 has global significance in quality because it places new marketplace pressures on all producers worldwide that wish to trade with European companies or even compete with European companies in other markets.

The EU 92 plan rests on the use of TC 176-produced standards as the requirement documents for its third-party certification scheme for quality systems registration, and for auditing compliance to the requirements.

Under such certification schemes, a company arranges to be audited by a single accredited independent (third-party) registrar organization. If the company's quality systems documentation and implementation are found to meet the requirements of the applicable ISO 9000 series international standard, the registrar grants certification and lists the company in its register of companies with certified quality systems. All purchasers of the company's products can then accept the third-party certification as evidence that the company's quality systems meet the applicable ISO 9000 series requirements.

Such a third-party certification scheme provides a number of benefits. Certification demonstrates that a company has implemented an adequate quality system for the products or services it offers. By this, better internal commitment as well as enhanced customer confidence can be achieved.

In the EU 92 scheme, quality system certification often will be a prerequisite for product certification or product conformity statements. In addition, from a broader national viewpoint, the scheme will result in improvements in the quality capability of a large fraction of commercial and industrial organizations.

An important corollary benefit for any organization is reduction of the cost of multiple assessments by multiple trading partners. In practice, customer organizations often audit portions of quality systems, but because of supplier quality system certification, the customer does not have to duplicate the, say, 80 percent that has already been audited by the third-party auditor.

Quality as a Competitive Weapon

Quality assurance continues to be a competitive weapon for companies, even in markets where third-party certification has become widespread. The competitive advantage can be achieved by means of second-party (customer) quality system requirements and audits that supplement (i.e., go beyond) the requirements of the ISO 9000 series contractual standards. This approach can be carried another step by setting up mutually advantageous partnership arrangements between customer and supplier, supplementing third-party audits.

Such partnerships focus on mutual efforts toward continuous quality improvement and the use of innovative quality technology. In instances where customer–supplier partnerships are fully developed, third-party certification often plays an important early role but might become relatively less important as the partnership develops and progresses beyond the requirements of the ISO 9001, ISO 9002, or ISO 9003 contractual standards. Various quality awards conferred at the company, national, or multinational level also provide further motivation for excellence in quality.

Critical Issues

Proliferation of Standards

The ISO 9000 series standards—in particular, those for contractual use (ISO 9001, ISO 9002, and ISO 9003)—are being employed in many industries for many different kinds of products and services. Some groups are evaluating or implementing local or industry-specific adaptations of the ISO 9000 series for guidance, certification, auditing, and documentation.

These include national or regional bodies (such as CEN/CENELEC, the European regional standardization organization) and international standards committees for industry sectors (such as other technical committees of ISO and IEC). These developments are indicators of success of the ISO 9000 series and indicators of concerns that TC 176 must address.

If the ISO 9000 series were to become only the nucleus of a proliferation of localized standards derived from, but varying in content and architecture, from the ISO 9000 series, then there would be little worldwide standardization. The growth of many localized certification schemes would present further complications. Once again, there could be worldwide restraint of trade because of proliferation of standards and inconsistent requirements.

Inadequacies of the ISO 9000 Series Standards

Careful study of the ISO 9000 series standards by certain major groups of users or potential users has identified a number of needs that are not easily met with the ISO 9000 series contractual standards in their present form. One example of such users or potential users is large companies, such as electric power providers or military organizations, that purchase complex products to specific functional design.

These users request, for example, a requirement for a quality plan to document how the generic requirements of the ISO 9000 series standards will be adapted to the specific needs of a particular contract. The requirement for a quality plan can improve the consistency of audits.

The position of such purchasers in the supply chain and their size enable their actions to expedite or hinder the worldwide implementation of harmonized external quality assurance standards. Moreover, there appears to be a large number of other users that would optionally want some of the same changes to the standards. At the same time, it is important to preserve simplicity of ISO 9000 series application for smaller companies.

At its meeting in Interlaken, Switzerland, in October 1990, TC 176 took actions to reckon with these critical issues in formulating future policy for international standards. To understand the basis for the TC 176 actions, some terminology and concepts must be introduced.

2. BASIC CONCEPTS

A Terminology Distinction

A product can be classified in a generic sense in two separate ways. The task force has introduced two terms to describe this important distinction. The first term is *generic product category*. The second term is *industry/economic* sector in which the product is present.

Generic Product Categories.   The task force has identified the following four generic product categories:

- Hardware

- Software

- Processed materials

- Services

These four generic product categories are described in Table F–1. Presently, Subcommittee 1 of TC 176 is developing formal definitions based on the descriptions of the four generic product categories in this table.

Subcommittee 1 has developed and submitted for international comment definitions for *product* and the generic product category *service. Product* is defined as the result of activities or processes. Notes to the definition point out that a product can be tangible or intangible or a combination thereof and that, for practical reasons, products can be classified in the four generic product categories introduced in Table F–1.

Table F–1   Generic Product Categories

Hardware	Products consisting of manufactured pieces, parts, or assemblies thereof.
Software	Products, such as computer software, consisting of written or otherwise recordable information, concepts, transactions, or procedures.
Processed Materials	Products (final or intermediate) consisting of solids, liquids, gases, or combinations thereof, including particulate materials, ingots, filaments, or sheet structures. **NOTE:** *Processed materials typically are delivered (packaged) in containers such as drums, bags, tanks, cans, pipelines, or rolls.*
Services	Intangible products, which may be the entire or principal offering or incorporated features of the offering, relating to activities such as planning, selling, directing, delivering, improving, evaluating, training, operating, or servicing a tangible product. **NOTE:** *All generic product categories provide value to the customer only at the times and places the customer interfaces with and perceives benefits from the product. However, the value from a service often is provided primarily by activities at a particular time and place of interface with the customer.*

Service is defined as the results generated by activities at the interface between the supplier and the customer, and by supplier internal activities to meet customer needs. Notes to the definition point out that the supplier or customer may be represented at the interface by personnel or equipment, that customer activities at the interface may be essential to the service delivery, that delivery or use of tangible products may form part of the service delivery, and that a service may be linked with the manufacture and supply of tangible products.

We believe the four generic product categories are all the kinds of product that need explicit attention in quality management and quality assurance standardization.

Industry/Economic Sectors

The term *industry/economic* sector applies to all sectors of the economy, including service sectors. The dual use of *industry sector* and *economic sector* recognizes that each term is used for the intended meaning in specific countries or languages. Such sectors include administration, aerospace, banking, chemicals, construction, education, food, healthcare, insurance, medical, retailing, telecommunications, textiles, tourism, and so forth. The number of industry/economic sectors and potential subsectors is extremely large. An industry/economic sector can be described as a grouping of suppliers whose offerings meet similar customer needs or whose customers are closely interrelated in the marketplace.

Required Combinations of Generic Product Categories

Two or more of the generic product categories have to be present in the marketplace offerings of any organization, whatever the industry/economic sector in which the organization operates.

An electric power utility is an example where the offering combines many characteristics of a service with delivery of a form of processed material (electric current) via a conducting cable.

Project management is another example where the offering typically combines many characteristics of a service with production and/or delivery of a hardware and/or software product.

Analytical instruments are examples where hardware, software, processed materials (such as titrating solutions or reference standard materials), and services (such as training) might all be important features of the offering.

Confusion Due to Intermixing Terms

It has been common practice to intermix the terms for generic product categories and the terms for industry/economic sectors. The result has been confusion and misunderstanding.

The Ad Hoc Task Force of the International Organization for Standardization Technical Committee 176 (TC 176) set forth four strategic goals for the ISO 9000 series standards and their related ISO 10000 series standards developed by TC 176:

- Universal acceptance
- Current compatibility
- Forward compatibility
- Forward flexibility.

In the following table, informal illustrative tests are described for each strategic goal. These tests are not meant to be strict requirements but only examples of indicators as to whether a strategic goal has been satisfied adequately.

These goals and tests are intended to apply particularly to standards used for external quality assurance. They are important but less critical for quality management guidance documents.

These four strategic goals for TC 176-developed standards will require constant managerial attention by the participants in TC 176 as well as affected user communities. Proposals that are beneficial to one of the goals might be detrimental to another goal. As in all standardization, compromises and paradoxes might be needed in specific situations. Experience continues to show that, when all viewpoints are put forth objectively, a harmonized standard can result, providing benefits to all parties.

Strategic Goals and Illustrative Tests for TC 176 Standards

Goal	Tests
Universal Acceptance	• The standards are widely adopted and used worldwide.
	• There are few complaints from users in proportion to the volum of use.
	• Few sector-specific supplementary or derivative standards are being used or developed.
Current Compatibility	• Part-number supplements to existing standards do not change or conflict with requirements in the existing parent document.
	• The numbering and the clause structure of a supplemental facilitate combined use of the parent document and the supplement.
	• Supplements are not stand-alone documents but are to be used with their parent document.
Forward Compatibility	• Revisions affecting requirements in existing standards are few in number and minor or narrow in scope.
	• Revisions are accepted for existing and new contracts.

(continued)

Strategic Goals and Illustrative Tests for TC 176 Standards	
Goal	**Tests**
Forward Flexibility	• Supplements are few but can be combined as needed to meet the needs of virtually any industry/economic sector or generic category of products.
	• Supplement or addendum architecture allows new features or requirements to be consolidated into the parent document at a subsequent revision if the supplement's provisions are found to be used (almost) universally.

Market Need
In managing the ISO 9000 series standards during the 1990s, market need also must be kept in focus. There is no value in a standard that is not wanted or not used in the marketplace. We encourage TC 176 to expedite the process of identifying and balloting internationally proposed new work items for new standards documents to meet new market needs. TC 176 must always apply the test of market need before embarking on a new standards project, however.

A prime example involves the terms *process industries* and *hardware industries*. In retrospect, these terms are seen to be collective names for the industry/economic sectors where processed materials and hardware are the primary kinds of product. All products are produced by processes.

People in the process industries know from experience that the classic quality management approaches, especially the quality control techniques and quality technology from the hardware industries, are not adequate to deal with the complexities they regularly encounter in the production processes and measurement processes for processed materials. In the language of ISO 9001 and ISO 9004, almost all processes in the process industries are special processes. These differences are crucial matters of degree and emphasis in a quality systems sense.

Terms such as *process industries* and *hardware industries* will continue to be useful when discussing collections of industry/economic sectors. However, for purposes of standardization in the quality arena, we recommend precise use of the four generic product category terms in written documents and oral communication.

3. ANALYSIS OF THE MARKETPLACE FOR THE ISO 9000 SERIES
Preventing Proliferation

The strategy regarding proliferation of supplementary or derivative standards must face up to a political and marketplace reality: TC 176 cannot legislate that industry groups, regional bodies, or other standards organizations will not produce supplementary or derivative documents.

This implies that to continue to influence the marketplace it serves, TC 176 must design its products and provide them in a timely way to prevent (or at least minimize) unhealthy proliferation of industry/economic sector schemes based on supplementary or derivative documents.

Fortunately, the current global trends are driving many standard users toward strategic recognition that they need and should conform to international standards. This suggests that the marketplace will resist proliferation if TC 176 meets the marketplace needs in a timely way.

It is easy to see why people in various industry/economic sectors would be motivated to create supplementary or derivative documents when confronted with implementing the initial ISO 9000 series standards published in 1987. The initial series is truly generic in scope and is applicable to all four generic product categories.

Nevertheless, the language and many details refer mainly to products in the hardware generic category. People in industry/economic sectors that involve products in the other generic product categories could easily conclude that what they need is a supplementary or derivative document for their specific industry/economic sector. This turns out, in our opinion, to be the wrong solution for the right problem.

We believe the proliferation of supplementary or derivative documents and industry-sector-specific schemes can be prevented or minimized by recognizing criteria that segment the markets for standardization in quality management and external quality assurance.

We recommend a development path for ISO 9000 series architecture that mirrors these market segmenting criteria. This path will allow sufficient flexibility in the ISO 9000 series to meet the needs of users in all generic product categories and all industry/economic sectors. There are, in our view, three criteria that are important to segment the markets.

Market-Segmenting Criterion 1: Generic Product Categories

Having distinguished generic product categories and industry/economic sectors as ways of classifying, we found an unanticipated commonality of our viewpoints on standardization policy. The members of the *ad hoc* task force represent a variety of national standards systems, generic product categories, industry/economic sectors, areas of personal experience, and current functional roles within our own organizations.

From all these points of view, we conclude the following:

- Guidance standards written by TC 176 should explicitly deal with the special needs of each generic product category. In fact, at the October 1990 meeting of TC 176, supplementary documents (guidelines) for services and software were advanced to international standard status, and work continued on a supplementary document for processed materials. Only these four generic product categories are represented by such formal work items. This reflects, we believe, a heretofore unarticulated international consensus on generic product categories.

- Neither TC 176 nor other groups should write supplementary or derivative standards (whether guidance standards or quality assurance requirements standards) for specific industry/economic sectors. In fact, experience in several nations has been that their industry-specific schedules or guidance documents have soon fallen into disuse because they had only transient tutorial value.

Market-Segmenting Criterion 2: Complexity of Purchaser Need and Product and Process Characteristics

A second dimension that segments the market for quality management and quality assurance standards has to do with the differing complexities of customer need and product characteristics as well as the differing complexities of designing and operating the process for producing and delivering the product or service. These differences are most obvious in external quality assurance situations.

The existing ISO 9000 series deals with this market-segmenting criterion by having three levels, or models, for external quality assurance requirements (ISO 9001, ISO 9002, and ISO 9003), by providing guidance for selecting the appropriate model (ISO 9000, Clause 8.2), and by tailoring the model within a particular contract (ISO 9000, Subclause 8.5.1).

In some industry/economic sectors, the options available in the existing ISO 9000 series are felt to be insufficient to meet all the critical needs. These deficiencies, for example, underlie the unmet needs of certain large organizations that purchase complex products to specific functional design, as discussed previously.

We believe that modest additions and revisions of the existing ISO 9000 series contractual standards can resolve these unmet near-term needs while preserving the necessary compatibility and flexibility.

Market-Segmenting Criterion 3: Contractual versus Noncontractual

For completeness, we remark that the distinction between contractual and noncontractual situations is a fundamental market segmenting criterion. The distinction is built into the existing ISO 9000 series architecture: ISO 9001, ISO 9002, and ISO 9003 are contractual requirements for quality assurance, and ISO 9004 provides guidance to a producer for implementing and managing a quality system. We believe the existing architecture meets the needs of this market-segmenting criterion.

Combined Use of the Market-Segmenting Criteria

Market-segmenting criteria 1, 2, and 3 currently provide an opportunity for many options of requirements to meet the needs of the various industry/economic sectors. When the recommendations of the *ad hoc* task force report are implemented, additional options will be provided. We believe that the need for sector-specific supplementary or derived standards and schemes will disappear with this added flexibility of implementation.

4. VISION 2000

The background, analyses, and goals discussed in the previous sections provide the basis for anticipating some critical features of standardization in the quality arena by the year 2000. We call this *Vision 2000*. We believe our recommendations will enhance progress toward *Vision 2000*.

Migration to Product Offerings Involving Several Generic Product Categories

Underlying *Vision 2000* is recognition of where current marketplace trends are leading. In all industry/economic sectors, there is an across-the-board migration toward product offerings that are combinations of two or more of the generic product categories (hardware, software, processed materials, and services).

For example, many products today involve production of processed materials that are then incorporated into manufactured parts and into hardware assemblies in which computer software also is an incorporated feature and the service aspects of selling, delivering, and servicing are important features of the total offering. A related example is firmware, where computing software is an integral part of the hardware of a product.

During the 1990s, this migration will continue. This means that organizations will have to learn about and implement the quality management and quality assurance terminology, skills, and emphases for all four generic product categories.

Global competition will be a powerful driving force in this process. Various requirements of society will also be a driving force, including laws, statutes, rules and regulations, codes, environmental considerations, health and safety factors, and conservation of energy and materials.

The boundaries for standardization program responsibilities will have to be worked out as a strategic issue in the 1990s. In this context, government agencies are one group representing the society as a customer whose needs and requirements must be satisfied and for whom appropriate quality management and quality assurance tools must be available.

Today, there is the impression that quality management and quality assurance for hardware, software, processed materials, and services are substantially different from each other. It is true that the relative emphases of quality system elements might differ and the sophistication of the quality technology being applied might differ, but the underlying generic quality system elements and needs are the same.

Today, there is the impression that quality systems and quality technology for hardware products are most mature because that has been the predominant kind of product in earlier standards and quality technology literature. However, we believe that the most rapid development of new quality technologies will occur in the other three generic product categories during the 1990s.

Implications for Standards Development

The standards now under development by TC 176 include a generic guidance document for implementing ISO 9001, ISO 9002, ISO 9003, and guidance documents for software, processed materials, and services. All of these reflect, in their detailed content, the migration we have described. All of these are expected to be published within a couple of years. It can be expected that the 1990s will be a transition period in standards development. During the early 1990s, we will need these separate documents for the various generic product categories.

As product offerings continue to migrate toward combinations of two or more generic product types, suppliers will find it increasingly necessary for employees in different functional activities to refer to different (but interrelated) supplementary standards, each written in a somewhat different style and format. Organizations have the managerial task of achieving compatibility in their quality systems for their offerings containing combinations of several generic product categories. This will be state-of-the-art and acceptable for the early 1990s, but a more strategic approach is needed for the longer term.

We envision that, by the year 2000, there will be an intermingling, a growing together, of the terminology, concepts, and technology used in all four generic product categories. This vision implies that, by the year 2000, the need for separate documents for the four generic product categories will have diminished. Terminology and procedures for all generic product categories will be widely understood and used by practitioners, whatever industry/economic sector they might be operating in.

Consequently, our *Vision 2000* for TC 176 is to develop a single, quality management standard (an updated ISO 9004 that includes new topics as appropriate) and an external quality assurance requirements standard (an updated ISO 9001) tied together by a roadmap standard (an updated ISO 9000). There would be a high degree of commonality in the concepts and architecture of ISO 9004 and ISO 9001. The requirements in ISO 9001 would continue to be based on a selection of the guidance elements in ISO 9004. Supplementary standards that provide expanded guidance could be provided by TC 176 as needed.

Multiple models of external quality assurance (now exemplified by ISO 9001, ISO 9002, and ISO 9003 in accord with the complexity criterion) might still be needed.

ISO standards must be reaffirmed or revised at approximately five-year intervals. At its October 1990 meeting, TC 176 adopted a two-phase strategy to meet the needs for revision of the ISO 9000 series. The first phase is to meet near-term needs for the (nominal) 1992 revision, with no major changes in architecture or numbering. Work on the second phase will begin during 1991, with the intent of implementing *Vision 2000*.

The target date is 1996—intentionally earlier than the nominal five years for the second cycle of revision. Working groups were set up within TC 176 to do these revisions. Formal comments have already been received from many nations as a basis for the revisions, and additional comments are being received.

Vision 2000 emphatically discourages the production of industry/economic-sector-specific generic quality standards supplemental to, or derived from, the ISO 9000 series. We believe such proliferation would constrain international trade and impede progress in quality achievements. A primary purpose of the widespread publication of this article is to prevent the proliferation of supplemental or derivative standards.

It is, however, well understood that product-specific standards containing technical requirements for specific products or processes or describing specific product test methods are necessary and have to be developed within the industry/economic sector.

A Visual Portrayal of International Quality Standardization

Figure F–2 shows graphically how the evolving system of standards meets the combined needs for quality management and quality assurance from both the producer's and the purchaser's viewpoints during the early 1990s.

Recommendations on Implementation

This section discusses some recommendations that are not entirely within the responsibility of TC 176. Nevertheless, reasonable conformance to these recommendations will be critical to success of the ISO 9000 series in the 1990s international environment. The recommendations include the following:

- TC 176 is encouraged to prepare standards that might be needed for an intermediate period for the four generic product categories. At the same time, TC 176 and all other standards-writing bodies are discouraged from writing standards for specific industry/economic sectors. The rationale for this recommendation was given earlier in this article.

  Comment: This recommendation is not intended for standards dealing with technical requirements for specific products, processes, or measurements that are outside the scope of TC 176.

- Quality system certification schemes worldwide should register suppliers only to ISO 9001, ISO 9002, ISO 9003, and other ISO 9000 series requirements documents that might hereafter be published and to their national equivalents translated exactly from the ISO 9000 series.

- There should be no industry/economic-sector-specific external quality system standards used as the assessment documents for such certification schemes. This recommendation applies to both third-party and second-party accredited assessment organizations.

Figure F–2

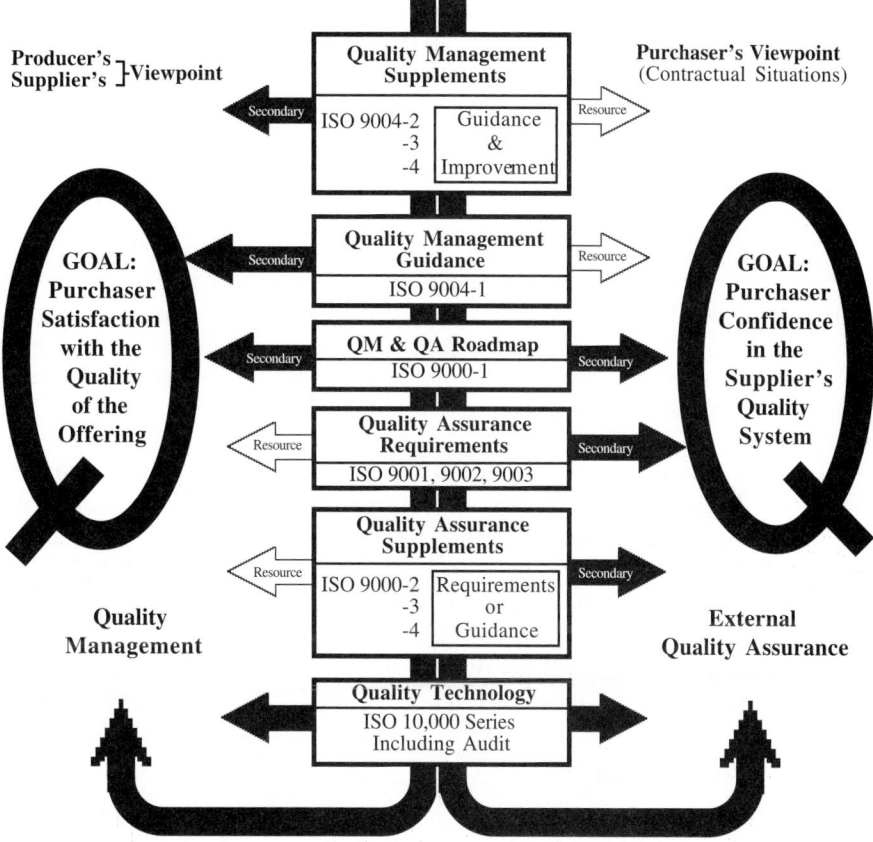

- Auditor accreditation (certification) schemes worldwide should be based on the ISO 10011 series audit standards.

- Auditors should be accredited (certified) generically, not on an industry/economic-sector basis. Each audit team should include at least one person knowledgeable in the industry/economic sector(s) involved in a particular audit. This knowledge might reside in the accredited auditors on the team or in technical experts on the audit team.

- TC 176 should help promote the development of mutual recognition arrangements among national quality system certification schemes worldwide. Fairness and consistency must be satisfied adequately, but failure to accomplish mutual recognition could severely restrain trade.

- The European Union has adopted a series of European standards (EN 45000 series) dealing with general criteria for operation, assessment, and accreditation of laboratories; certification bodies relating to certification of products, quality systems, and personnel; and suppliers' declaration of conformity.

Preexisting ISO/IEC guides also deal with these topics. Implementation of the ISO 9000 series and ISO 10000 series standards interfaces with such criteria. At the October 1990 meeting, TC 176 requested its chairman and secretary to investigate the desirability and options to harmonize the European standards in the EN 45000 series and the relevant ISO/IEC guides and to transform the results into international standards.

ACKNOWLEDGEMENTS

We acknowledge our indebtedness to R.N. Shaughnessy, TC 176 chairman, for his early appreciation of the strategic issues and his leadership in causing this task force to be formed, and to K.C. Ford, TC 176 secretariat, for his continuing and incisive support for our work. We also acknowledge the suggestions made by H. Kume and K. A. Rutter. Many other people have contributed indirectly through participating in other TC 176 activities or through contacts with task force members.

Donald Marquardt, Du Pont, United States; Jacques Chove, Conseil, France; K.E. Jensen, Alkatel Kirk A/S, Denmark; Klaus Petrick, DIN, Germany; James Pyle, British Telecom, United Kingdom; and Donald Strahle, Ontario Hydro, Canada, are members of the Ad Hoc Task Force of the International Organization for Standardization Technical Committee 176. Marquardt is the chairman of the task force.

References

1. Donald Marquardt, Jacques Chové, K.E. Jensen, Klaus Petrick, James Pyle, and Donald Strahle, "Vision 2000: The Strategy for the ISO 9000 Series Standards in the 90s," *Quality Progress,* May 1991, pp. 25–31.

2. Donald Marquardt, Jacques Chové, K.E. Jensen, Klaus Petrick, James Pyle, and Donald Strahle, "Vision 2000: The Strategy for the ISO 9000 Series Standards in the '90s," *ISO 9000 Quality Management Standards Compendium,* Sixth Edition, 1996, International Organization for Standardization, Geneva, Switerland, pp. 1–14.

E-mail and Home Page Addresses

Accademia Qualitas

E-mail Address: assist@ACCADEMIA.COM
E-mail Address 2: denis@accademia.com

Advanced Systems and Design, Inc.

E-mail Address: ASDPM@AOL.COM

Advent Management International, Ltd.

E-mail Address: 102463.430@Compuserv.com

Alamo Learning Systems

E-mail Address: alamols@aol.com

American Association for Laboratory Accreditation

E-mail Address: a2la@aol.com
Home Page Address:
http://users.aol.com/a2la/a2la.htm

American Electronics Association

E-mail Address: Jennifer_O'Donnell@aeanet.org

American National Standards Institute Online

Home Page Address: http://www.ansi.org/

American Productivity and Quality Center

Home Page Address: http://www.apqc.org

American Society for Quality Control

Home Page Address: http://www.asqc.org/

American Society for Testing and Materials

Home Page Address: http://www.astm.org

Anoka-Hennepin Technical College

E-mail Address: C.Skotterud@ank.tec.mn.us

Applied Quality Systems

E-mail Address: AQSYSTEMS@AOL.COM

ASQC

E-mail Address: asqc@asqc.org
Home Page Address: http://www.asqc.org
QS-9000 Web Site: http://www.asqc.org/9000

ASQC Quality Press

E-mail Address: molson@asqc.org
E-mail Address 2: bstratton@asqc.org
Home Page Address: http://www.asqc.org

Associated Quality Consultants

E-mail Address: help@quality.org
E-mail Address 2: bill@casti.com
E-mail Address 3: e_media@cais.com
Home Page Address: http://www.quality.org/qc
Home Page Address 2:
http://www.casti.com/casti/Bill.html

B-K Education Services

E-mail Address: BAIQUALITY@AOL.COM
Home Page Address: http://www.bai-bk.com

Babcock and Wilcox

Naval Nuclear Fuel Division
E-mail Address:
taz.daughtrey@bw.mcdermott.com

The Benchmarking Exchange

Home Page Address: http://www.benchnet.com

Besterfield and Associates Continuous Quality Improvement Consultants

Home Page Address: DBSTRFIELD@AOL.COM

Brewer & Associates, Inc.

E-mail Address: BAIQUALITY@AOL.COM
Home Page Address: http://www.BAI-BK.com

Center for Manufacturing Systems New Jersey Institute of Technology

E-mail Address: gold@admin.njit.edu

CFM Inc.

E-mail Address: info@teamflow.com
Home Page Address:
http://WWW.TEAMFLOW.COM

Charro Publishers, Inc. The Trainers' Workshop

E-mail Address: CHARROPubs@AOL.COM

Chemical Week Associates

E-mail Address: bpunsalan@chemweek.com
Home Page Address: http://www.chemweek.com

Chilton/Hitchcock Publishing Company

E-mail Address: gparr@chiltonco.com
Home Page Address: http://qualitymag.com

Clear Software, Inc.

E-mail Address: sales@clearsoft.com
colleen@clearsoft.com
Home Page Address: http://www.clearsoft.com

Compliance Engineering

E-mail Address: ce@ix.netcom.com
Home Page Address: www.ce-mag.com

Continuous Quality Improvement Server

Home Page Address:
http://deming.eng.clemson.edu/

Cordis

Home Page Address://www.cordis.lu

Cyber Quality Inc.

E-mail Address: cqi@cqinc.com
Home Page Address: http://www.cqinc.com/

D.A. Martin Software

Home Page Address: http://www.spcorch.com

DBS Quality Management International

E-mail Address: MIKEDOCRT@AOL.COM

Delaware County Community College Center for Quality and Productivity

E-mail Address:
PLMCQUAY@DCCCNET.DCCC.EDU
E-mail Address 2:
DEntner@DCCCNET.DCCC.EDU
E-mail Address 3:
AKitson@DCCC.NET.DCCC.EDU

Delegation of the European Commission

E-mail Address: nick16@ix.netcom.com
Home Page Address: http://www.cec.lu

Department of Energy

E-mail Address: bud.danielson@hq.doe.gov
Home Page Address: http://www.explorer.doe.gov

Devon Hunter Consulting

E-mail Address: mhunter@fox.nstn.ca
E-mail Address 2: da390@freenet.carleton.ca

Distribution Solutions, Inc.

E-mail Address: ROGERGARVER@MSN.COM

Eastern Michigan University Center for Quality

E-mail Address: Kathy.Trent@emich.edu
E-mail Address 2: Terry.Carew@emich.edu
Home Page Address:
http://www.emich.edu/public/cq/cq.html

EBA/NYS

E-mail Address: info@eba-nys.org
Home Page Address: http://www.eba-nys.org

Ellis and Associates, Ltd. Quality Engineering Consultants

E-mail Address: Jimikellis.@AOL.COM
E-mail Address 2: Dorothyle.@AOL.COM

Entela, Inc. Quality System Registration Division

E-mail Address: sdemarco@entela.com
E-mail Address 2: bvosberg@entela.com
E-mail Address 3: bkerkstra@entela.com

Environmental Business Opportunities

Home Page Address:
http://www.enviroindustry.com/opportunities.html

Environmental Cleanup InfoCenter

Home Page Address: http://www.cleanup.com/

The Environmental Industry Web Site

Home Page Address:
http;//www.enviroindustry.com/

The Environment on the Net Conference

Home Page Address:
http://www.enviroindustry.com/conference/

Environmental Science Services-Rettew Associates, Inc.

E-mail Address: ep@Rettew.com

E. Roberts Alley and Associates, Inc.

E-mail Address: alley@edge.net
Home Page Address: http://www.edge.net/alley/

ETQ Management Consultants, Inc.

E-mail Address: 71763.2023@CompuServe.COM
Home Page Address:
http://ourworld.compuserve.com/homepages/EtQHome

Europa

Home Page Address: http://www.cec.lu

EuroQuest

E-mail Address: tmg_eq@ix.netcom.com

Excel Partnership Inc.

E-mail Address: xlp@xlp.com
E-mail Address 2: compuserv 74521,1366
Home Page Address: http://www.xlp.com

Extended Marketing, Inc.

E-mail Address: homepage@eeservices.com
Home Page Address: http://www.eeservices.com

The Farrell Group

E-mail Address: farrellgrp@aol.com
Home Page Address:
http://www.exit.109.com/~leebee

FEDNet

E-mail Address: dlucarelli@fed.org
Home Page Address: http://www.saic.com:80/fed/

Foundry Quality Systems

E-mail Address: FQS@LL.NET
Home Page Address:
http://www.LL.net/fqs/quality.html

George Mason University, Executive Programs Corporate and International Training CPD

E-mail Address: cgibbs1@gmu.edu
E-mail Address 2: jniblock@gmu.edu

The George Washington University Continuing Engineering Education

E-mail Address: ceepinfo@ceep.gwu.edu
Home Page Address: http://www.gwu.edu/~ceep

Georgia Institute of Technology Center for International Standards and Quality

E-mail Address: donna.ennis@edi.gatech.edu

Gil Friend and Associates

E-mail Address: gfriend@eco-ops.com
Home Page Address:
http://www.igc.apc.org/eco-ops/

Global Environmental Management Systems

E-mail Address: Dburd52011@aol.com

GOAL/QPC

E-mail Address: service@GOAL.com
Home Page Address: http://www.goalqpc.com

GTW Associates

E-mail Address: Gwillingmy@aol.com

The Harrington Group, Inc.

E-mail Address: hgi@harrington-group.com
Home Page Address:
http://www.harrington-group.com

Hewlett-Packard SAC

Home Page Address:
http://www.corp.hp.com/Publish/iso9000-sac

Heurikon

E-mail Address: BOBG@HEURIKON.COM
Home Page Address: http://www.heurikon.com

Hunt Quality Services

E-mail Address: dhunt@teleport.com
Home Page Address:
http://www.teleport.com/~dhunt

IBM Canada/ISO 9000 Consulting Services

E-mail Address: ISO_9000@vnet.ibm.com
E-mail Address 2: PDAWSON@vnet.ibm.com
Home Page Address: http://www.can.ibm.com
(under this click on what we offer; then under this click on service)

I'meurope

Home Page Address: http://www.echo.lu

Inchcape Systems Registration Intertek Services

E-mail Address:
ITSUSBOS3!NATICK"CLEMAY"@inchqsw.attmal.com

Infodisk Inc.

E-mail Address: 103002,2360@Compuserve.COM

Information Mapping, Inc.

E-mail Address: http://www.infomap.com
Home Page Address: http://www.infomap.com

Innovative Quality Services

E-mail Address: iqserv@iqserv.com
Home Page Address: http://www.iqserv.com

INQC Consulting

E-mail Address: pernix01@AOL.COM
E-mail Address 2: 73367.1725@compuserv.com

Institute of Certified Management Consultants of Canada

E-mail Address: Compuserv-102774,1361

Integrated Quality Dynamics, Inc.

Home Page Address: http://www.iqd.com/quality/

International Organization for Standardization

Home Page Address: http:// iso.ch.com

International Quality Systems

E-mail Address: RBLISS9000@AOL.COM

International Quality Technologies, Inc.

E-mail Address: Sales@IQT.COM

International Systems Registrars, Ltd.

E-mail Address: isrglt@execpc.com
Home Page Address:
http://www.execpc.com:bo/~isrglt

Internet Resources for the Environmental Industry

Home Page Address:
http://www.enviroindustry.com/resources.html

InterQual

E-mail Address: InterQua@silcom.com
E-mail Address 2: eborges@aol.com
Home Page Address: InterQual@aol.com

IQS, Inc.

E-mail Address: Iqs@ix.netcom.com
E-mail Address 2: sales@iqs.com
Home Page Address: http://www.iqs.com

Irwin Professional Publishing

E-mail Address: Ipro@Irwin.com

ISO 9000 Network

E-mail Address: iso9000@best.com
Home Page Address:
http://www.best.com/~iso9000

ISO 9000 Web Directory

E-mail Address: info@iso9000directory.com
Home Page Address:
http://www.iso9000directory.com

ISO 14000 Info Center

Home Page Address: http://www.iso14000.com/

ISO Communications

E-mail Address: ISOCOM@AOL.COM

ISO Easy

E-mail Address: leebee@exit109.com
Home Page Address:
http://www.exit.109.com/~leebee

ISO Software Solutions, Inc.

E-mail Address: 72242.2621@compuserve.com

ISO Systems, Inc.

E-mail Address: Stanorm@aol.com

ISPO

Home Page Address: http://www.ispo.cec.be

JK Technologies, Inc.

E-mail Address: 103262.324@CompuServe.com
Home Page Address:
http://WWW.Gate/~HAMMERHD

J. P. Russell and Associates

E-mail Address: jprussell@mcimail.com

John A. Keane and Associates, Inc.

E-mail Address: Keane@cnj.digex.net
Home Page Address:
http://WWW.CNJ.DIGEX.NET/~Keane

Kennedy and Donkin Quality Inc.

E-mail Address: PGMAC@TIAC.NET

Kluwer Law International

E-mail Address: kli@world.std.com
Home Page Address: http://www.kli.com

Lanchester Press Inc.

Home Page Address: http://www.lanchester.com

Leads Corporation

E-mail Address: MIKROUCH@AOL.COM
E-mail Address 2: leadscorp@aol.com
E-mail Address 3: richmall@aol.com
E-mail Address 4: mchapman@ossinc.net
Home Page Address: www.leadscorp.com

Litvan Enterprises

E-mail Address: LITVAN@dakota.polaristel.net

Management Control Systems, International

E-mail Address: MCSINTL@Primenet.com

Management Software International, Inc.

E-mail Address: custserv@isoxpert.com
Home Page Address: http://www.isoeasy.com

Management Alliances Inc.

E-mail Address: mgmt@mgmt14k.com
Home Page Address: http://mgmt14k.com/

Moorhill International Group, Inc.

E-mail Address: 80375@EF.GC.MARICOPA.EDU

Murphy Software

E-mail Address: salesinfo@Murphyco.com
Home Page Address: http://www.murphyco.com/

National Institute for Quality Improvement

E-mail Address: NIQI@aol.com

National ISO Support Group

E-mail Address: isogroup@cris.com
Home Page Address:
http://www.cris.com./~isogroup/

NIST
Office of Standards Services

E-mail address: rlglad@micf.nist.gov

NIST Quality Program

E-mail Address: cap@micf.nist.gov
Home Page Address:
http://www.nist.gov/quality_program/

Normac-Will Associates
Medical Device Quality and
Regulatory Consultants

E-mail Address: normacwill@aol.com
E-mail Address 2: HHBP74A@PRODIGY.COM
E-mail Address 3: HHBP74@MSN

Office of Standards Services' Weights and Measures Program

E-mail Address: gharris@nist.gov

Omnex Training Seminars

E-mail Address: MHILLEGO@OMNEX.COM

Omni Data Sciences Inc.

Home Page Address:
http://www.iserv.net/omnidata/home.htm

Omni Tech International, Ltd.
Consulting and Staffing Services

E-mail Address: INFO@OMNITECHINTEL.COM

Online Quality Resource List

Home Page Address:
http://pages.prodigy.com/J/O/N/john/onlineqlist.html

Papa and Associates Inc.
Management Associates

E-mail Address: papa@hookup.net

Paton Press

E-mail Address: paton@aol.com

Perry Johnson, Inc.

E-mail Address: PJI@wwnet.com
E-mail Address 2: pji@oeonline.com
Home Page Address: http://www.PJI.com

P. F. J. Everett and Associates

E-mail Address: peverett@igs.net
Home Page Address:
http://www.quality.co.uk/quality/isocons/pfje.htm

Process Management International

E-mail Address: PMIRJRBER@AOL.COM

Productivity Improvement Center of Durham College

Home Page Address:
http://DURHAM.DURHAMC.ON.CA

Productivity-Quality Systems Inc. (PQ Systems)

E-mail Address: sales@pqsys-hq.mhs.compuserv.com

QEMA

Home Page Address: http://www.tqm.com/

QFD Institute

E-mail Address: qfdi@qfdi.org
Home Page Address:
http://www.nauticom.net/www/qfdi/

QNET Training and Consulting

E-mail Address: Q9000@aol.com

Qualified Specialists, Inc. (QSI)

E-mail Address: qsi@scci.com
Home Page Address:
http://www.sccsi.com/qsi/index.htm

QualiNET

E-mail Address: mksview@netview.com
Home Page Address: http://www.netview.com/

Qualitran Professional Services, Inc.

Home Page Address: http://www.barint.on.ca/qps/

Quality Assurance Services of North America (QASNA)

E-mail Address: thaney1@ix.netcom.com

Quality Auditor

E-mail Address: rj143@earthlink.net
Home Page Address:
http://www.lookup.com/Homepages/56694/qa-home.htm

Quality Digest Magazine

E-mail Address: QualityDig@aol.com
Home Page Address: http://www.tqm.com

The Quality Improvement Network, Inc.

E-mail Address: ISOQUAL@AOL.COM

Quality Integration Services (QIS)

E-mail Address: QISGROUP@aol.com

The Quality Junction Web Site

Home Page Address:
http://www.cris.com/~wcterry/

Quality Magazine

E-mail Address: GParr@chiltonco.com

Quality Management Assistance Group (QMAG)

E-mail Address: QMAG@AOL.COM
E-mail Address 2:
RODGER.R.28086@ASQCNET.ORG

Quality Management Consulting Services, Inc.

E-mail Address: dane@pbfreenet.seflin.lib.fl.us

Quality Management Institute

E-mail Address: qmi@inforamp.net
Home Page Address:
http://www.pic.nt/qmi/index.html

Quality Management International, Inc.

E-mail Address: helenmcc@ix.netcom.com
Home Page Address: http://www.exitlog.com/~lee-bee/actplan.htm
Home Page Address 2:
http://www.exitlog.com/~leebee/training.htm
Web Location:
www.quality.management.industry.net

Quality Management Principles

E-mail Address: k.forsberg@wineasy.se
Home Page Address: http://www.wineasy.se/qmp/

Quality Management Solutions

E-mail Address: ConsultQMS@AOL.COM

The Quality Network

E-mail Address: help@quality.co.uk
Home Page Address:
http://www.quality.co.uk/quality/home.htm

Quality Practitioners U.S., Inc.

E-mail Address: mcrobert@interlog.com
Home Page Address:
http://www.interlog.com/~mcrobert

Quality Publishing Inc.

Home Page Address:
http://www.qualitypublishing.com/QualityPublishing

Quality Resources International

E-mail Address: qri@quality-qri.com
Home Page Address: http://www.quality-qri.com

Quality Resources Online

Home Page Address: http://www.quality.org.qc

Quality Sciences Consultants, Inc. (QSCI)

E-mail Address: Caplan.F.7729@ASQCNET.ORG

Quality Systems Enhancement, Inc. (QSE, Inc.)

E-mail Address: AQCNET address is
KOTTE.U.18962

Quality Systems Integrators

E-mail Address: turocy@chestnet.com
Home Page Address: WWW.webimpact.com\QSI

Quality Technology Company

E-mail Address: qtc1239@startnetinc.com
Home Page Address: qtcom.com

QualityQuest

E-mail Address: MikeMick@ix.netcom.com

Qualtec Quality Services, Inc.

E-mail Address: isoinfo@qualtec.com
E-mail Address 2: marketing@qualtec.com
Home Page Address: http://www.qualtec.com

Quality Wave

E-mail Address: e.dawis@creacon.com
Home Page Address: http://www.xnet.com/~crea-
con/Q4Q

Quest USA, Inc.

E-mail Address: info @qfe.com

r. bowen international, inc. (rbi)

E-mail Address: rbi@aol.com

Reality Interactive, Inc.

E-mail Address: webmaster@ry.com
Home Page Address: htp://www.realtools.com

Richard Jones-Management Consultants Inc. (RJMCI)

E-mail Address: rjmci@shadow.kingston.net

Robert Peach and Associates, Inc.

E-mail Address:
74654.1747@COMPUSERVCOM

Rochester Institute of Technology

The John D. Hromi Center for Quality and Applied
Statistics (CQAS)
E-mail Address: CQAS@RIT.EDU

Sanders and Associates

E-mail Address: sanders@accesscom.net

Sapling Corporation

Home Page Address: http://www.sapling.com

Scitor Corp.

Home Page Address: http://www.scitor.com

SCS Engineers

E-mail Address: 1120JVK@be.scseng.com
E-mail Address 2: jkinsella@scseng.com
Home Page Address: http:\www.iso 14000.com

Service-Growth Consultants Inc.

E-mail Address: 72772.3262@compuserve.com
Home Page Address:
http://www.service-growth.com

Service Process Consulting, Inc.

E-mail Address: IDurand@aol.com

SIMCOM, Inc.

E-mail Address: simcom@mindspring.com
Home Page Address: http://www.eurocom.com

Software Engineering Process Technology

E-mail Address:
73211.2144@COMPUSERV.COM

Software Systems Quality Consulting

E-mail Address: deibs@shell.portal.com
E-mail Address 2: ssqc@cris.com
Home Page Address:
http://www.billboard.com/ssqc/index.html

SPSS Inc.

Home Page Address: http://www.spss.com

Stat-A-Matrix Group/The SAM Group

E-mail Address: Inc@theSAMgroup.com
E-mail Address 2: Inc@theSAMgroup.com
E-mail Address 3: institute@thesamgroup.com

Statistical Quality Control and Circle Institute

E-mail Address: stats@uscybar.com

Strategic Quality Alliance

E-mail Address: StratQual@aol.com

StatSoft Inc.

Home Page Address: http://www.statsoftinc.com

SuccessFull Resources Group, Inc.

E-mail Address: success@teleport.com
Home Page Address:
http://www.teleport.com/~success

SystemCorp
Ernst and Young LLP

E-mail Address: sys@systemcorp.com

Team 14000, Inc.

E-mail Address: Team 14000@aol.com
E-mail Address 2: aschoffm@ios.com
E-mail Address 3: AlTordini@aol.com
Home Page Address: http://www.team14000.com

Team Quality International (U.S.A.)

E-mail Address: tqi@bournemouth-net.co.uk
E-mail Address 2: tqi@angel.co.uk
Home Page Address: http://www.angel.co.uk/tqi

Technology Economics International

Home Page Address:
http://www.teintl.com/pub/teintl/homepage.htm

Technology International Inc.

E-mail Address: Mktg@TechIntl.com

Thomas F. Brandt Associates
Innovative Technology Management

E-mail Address:
71441.2051@COMPUSERV.COM

3C Technologies, Inc.

E-mail Address: 3c@3ctech.com
Home Page Address: http://www.3ctech.com

TMO Survey Week

E-mail Address: chyertmo@usa.nai.net

TQM Consulting

E-mail Address: 75362.2746@compuserve.com

Trade Data Reports

Home Page Address: 1-800-952-0122 (newsNet)

Tri-Tech Services Inc.

Home Page Address:
http://www.TTS9000@aol.com
Home Page Addres 2:
http://www.industry.met/c/pto-
memdir/company/597

The Victoria Group Incorporated Inc.

E-mail Address:
jennifer_cook@victoria.global.ibmmail.com

W.C. Terry and Associates

E-mail Address: wcterry@cris.com
Home Page Address:
http://www.cris.com/~wcterry/
Home Page Address 2: WCTerry-Quality-
Junction@postoffice.worldnet.att.net

Wickham International Management Services Inc.

E-mail Address: 100443.1633@compuserve.com

Winona Technical College Custom Services Division

E-mail Address: tomt@win.tec.mn.us
Home Page Address:
http://wwwsmp.smumn.edu/rwwtc/

Appendix

Acronyms and Glossary

ACRONYMS

A2LA American Association for Laboratory Accreditation

AAMI Association for the Advancement of Medical Instrumentation

ABCB Association of British Certification Bodies

AENOR Asociación Española de Normalización y Certificación (Spanish Association for Normalization and Certification)

AFAQ Association Française pour l'Assurance de la Qualité (French accreditation body and quality system registrar)

AFNOR Association française de normalisation (French standards association)

AIAG Automotive Industry Action Group

ASME American Society of Mechanical Engineers

ANSI American National Standards Institute

AQAP-1 Allied Quality Assurance Publication 1

ASC Accredited Standards Committee

ASQC American Society for Quality Control

BEC British Electrotechnical Committee

BSI British Standards Institution

CD Committee Draft

CDRH Center for Devices and Radiological Health, FDA

CE Mark Conformité Européene (official French name for the EC mark)

CEB/BEC Comité électrotechnique Belge/Belgisch Elektrotechnisch Comité (Belgian Electrotechnical Committee)

CEI Comitato Elettrotecnico Italiano (Italian Electrotechnical Committee)

CEN Comité Européen de Normalisation (European Committee for Standardization)

CENELEC Comité Européen de Normalisation Électrotechnique (European Committee for Electrotechnical Standardization)

CGA Canadian Gas Association

CMA Canadian Manufacturers' Association

CQA Certified Quality Auditor

DEK Dansk Elektroteknisk Komité (Danish Electrotechnical Committee)

DFARS	DoD Federal Acquisition Regulation Supplement	**EQNET**	European Network for Quality System Assessment and Certification
DHHS	U.S. Department of Health and Human Services	**EQS**	European Committee for Quality System Assessment and Certification
DIN	Deutsches Institut für Normung (German Standards Institute)	**ETCI**	Electro-Technical Council of Ireland
DIS	Draft International Standard	**ETSI**	European Telecommunications Standards Institute
DITI	UK's Department of Trade and Industry	**EU**	European Union
DKE	Deutsche Elektrotechnische Kommission im DIN und VDE (German Electrotechnical Commission)	**FAA**	Federal Aviation Administration, DOT
		FAR	Federal Acquisition Regulation
DOC	U.S. Department of Commerce	**FDA**	U.S. Food and Drug Administration, DHHS
DoD	U.S. Department of Defense	**GATT**	General Agreements on Tariffs & Trade
DOE	U.S. Department of Energy	**GMP**	Good Manufacturing Practices (US Food and Drug Administration)
DOT	U.S. Department of Transportation	**GSA**	General Services Administration
DS	Dansk Standardiseringsrad (Danish standards body)	**HD**	Harmonization Documents
EAC	European Accreditation of Certification	**HIMA**	Health Industry Manufacturers Association
EC	European Community	**IAAR**	Independent Association of Accredited Registrars
EEA	European Economic Area		
EFTA	European Free Trade Association	**IBN/BIN**	Institut belge de normalisation/Belgisch Instituut voor Normalisatie (Belgian Institute for Standardization)
ELOT	Hellenic Organization for Standardization		
EMAS	Ecomanagement and Audit Scheme	**IEC**	International Electrotechnical Commission
EMS	Environmental Management System		
EN	European Norm	**INTERTEK**	Intertek Services Corporation
ENV	European Prestandards	**IPQ**	Instituto Português da Qualidade (Portuguese Quality Institute)
EOQ	European Organization for Quality	**IQA**	Institute for Quality Assurance
EOTA	European Organization for Technical Approvals	**IRCA**	International Registrar of Certified Auditors
EOTC	European Organization for Testing and Certification	**ISO**	International Organization for Standardization
EPE	Environmental Performance Evaluation	**ITA**	U.S. International Trade Administration
EPI	Envirnomental Performance Indications	**ITM**	Inspection du travail et des mines (Luxembourg inspecting body)

ITQS	Recognition Arrangement for Assessment and Certification of Quality Systems in the Information Technology Sector
JAS-ANZ	Australia/New Zealand Accreditation Body for Quality System Registrars
MBNQA	Malcolm Baldrige National Quality Award
MOU	Memorandum of Understanding
MRA	Mutual Recognition Agreement
MTC	Manufacturing Technology Center, NIST
NAC-QS	Comité National pour L'Accreditation Des Organismes de Certification (Belgian Organization responsible for the accreditation of quality system registrars)
NACE	Nomenclature Générale des Activités Économique dans les Communautés Européennes
NATO	North Atlantic Treaty Organization
NCCLS	National Committee for Clinical Laboratory Standards
NCSCI	National Center for Standards and Certification Information, NIST
NEC	Nederlands Elektrotechnisch Comité (Dutch Electrotechnical Committee)
NEMA	National Electrical Manufacturers Association
NIST	National Institute of Standards and Technology, DOC
NNI	Nederlands Normalisatie Instituut (Dutch Normalization Institute)
NRC	Nuclear Regulatory Commission
NTIS	National Technical Information Service
NVCASE	National Voluntary Conformity Assessment System Evaluation, formerly the CASE Program
NVLAP	National Voluntary Laboratory Accreditation Program, NIST

OEM	Original Equipment Manufacturers
prEN	Proposed European Norm
QA	Quality Assurance
QM	Quality Manual
QMI	Quality Management Institute
QS	Quality System
QSR	Quality Systems Registrars, Inc.
RAB	Registrar Accreditation Board
RvA	Raad voor de Certificatie (Dutch Council for Certification)
SC	Subcommittee
SCC	Standards Council of Canada
SCI	Standards Code and Information Program
SIC	Standard Industrial Classification code
SPC	Statistical Process Control
Sub-TAG	Subgroup of a TAG
TAAC	Trade Adjustment Assistance Centers
TAG	Technical Advisory Group
TC	Technical Committee, as in TC 176
TickIT	UK Quality System Registration Scheme for Software Companies
TQM	Total Quality Management
UKAS	United Kingdom Accreditation Service
UNI	Ente Nazionale italiano di unificazione (Italian national association for standardization)
UTE	Union technique de l'électricité (French electrotechnical union)
WD	Working Draft
WG	Working Group

GLOSSARY

Accreditation   Procedure by which an authoritative body formally recognizes that a body or person is competent to carry out specific tasks. (ISO/IEC Guide 2)

Accreditation Mark   An insignia that indicates accreditation. Only accredited certification bodies and the companies they certify are allowed to use an accreditation mark. Non-accredited certification bodies and the companies they certify may not.

ANSI   American National Standards Institute. (Adopts but does not write American standards.) Assures that member organizations that do write standards follow rules of consensus and broad participation by interested parties. ANSI is the U.S. member of ISO.

ASQC   American Society for Quality Control. A technical society of over 140,000 quality professionals. Individual members from throughout the world, but primarily from the United States. Publishes quality-related American national standards.

Assessment   An estimate or determination of the significance, importance, or value of something. (ASQC Quality Auditing Technical Committee)

Assessment Body   Third party that assesses products and registers the quality systems of suppliers.

Assessment System   Procedural and managerial rules for conducting an assessment leading to issue of a registration document and its maintenance.

Audit   (See Quality Audit)

Audit Program   The organizational structure, commitment, and documented methods used to plan and perform audits. (ASQC Quality Auditing Technical Committee)

Auditee   An organization being audited. (ISO 8402, Clause 4.12)

Auditor (quality)   A person qualified to perform quality audits. (ISO 8402, Clause 4.11)

BSI   British Standards Institution. This is the U.K.'s standards-writing body.

BSI QA   BSI Quality Assurance. One of 15 accredited certification bodies (registrars) in the U.K. Assesses suppliers for conformance to the appropriate ISO 9000 series standard and registers those that meet the requirements. Organizationally separate from BSI.

CE Mark   Conformité Européenne. The mark of approval used by the European Union. This mark signifies that the equipment complies with all applicable directives and product standards.

CEN   European Committee for Standardization. Publishes regional standards (for EU and EFTA) covering nonelectrical, nonelectronic subject fields. (See also CENELEC.)

CENELEC   European Committee for Electrotechnical Standardization. Publishes regional standards (for EU and EFTA) covering electrical/ electronic subject fields. (See also CEN.)

Certification   Procedure by which a third party gives written assurance that a product, process, or service conforms to specified requirements. (ISO/IEC Guide 2)

Certified   The quality system of a company, location, or plant is certified for compliance to ISO 9000 after it has demonstrated such compliance through the audit process. When used to indicate quality system certification, it means the same thing as registration.

Compliance   An affirmative indication or judgment that the supplier of a product or service has met the requirements of the relevant specifications, contract, or regulation; also the state of meeting the requirements. (ANSI/ISO/ASQC A3) (See also conformance.)

Conformance   An affirmative indication or judgment that a product or service has met the requirements of the relevant specifications, contract or regulation; also the state of meeting the requirements. (ANSI/ISO/ASQC A3) (See also compliance.)

Conformity Assessment   Conformity assessment includes all activities that are intended to assure the conformity of products to a set of standards. This can include testing, inspection, certification, quality system assessment, and other activities.

Contractor   The organization that provides a product to the customer in a contractual situation. (ISO 8402, Clause 1.12)

Convention   A customary practice, rule, or method. (ASQC Quality Auditing Technical Committee)

Corrective Action   An action taken to eliminate the causes of an existing nonconformity, defect, or other undesirable situation in order to prevent recurrence. (ISO 8402, Clause 4.14)

Customer   Recipient of a product provided by the supplier. (ISO 8402 Clause 1.9)

Design Review   A formal, documented, comprehensive, and systematic examination of a design to evaluate the design requirements and the capability of the design to meet these requirements and to identify problems and propose solutions. (ISO 8402, Clause 3.11)

EEC   The European Economic Community. This comprises the EU and EFTA countries.

EFQM   European Federation for Quality Management. An organization of upper-level managers concerned with quality.

EFTA   European Free Trade Association. A group of nations whose goal is to remove import duties, quotas, and other obstacles to trade and to uphold nondiscriminatory practices in world trade.

EN 45000   A series of standards set up by the EU to regulate and harmonize certification, accreditation, and testing activities. National accreditation structures and inspection body standards are still being developed.

EOQ   European Organization for Quality, formerly EOQC (European Organization for Quality Control.) An independent organization whose mission is to improve quality and reliability of goods and services primarily through publications, conferences, and seminars. Members are quality-related organizations from countries throughout Europe, including Eastern-European countries. ASQC is an affiliate society member.

EOTC   European Organization for Testing and Certification. Set up by the EU and EFTA to focus on conformity assessment issues in the nonregulated spheres.

EQS   European Committee for Quality System Assessment and Certification. The function of EQS is to harmonize rules for quality system assessment and certification (registration), facilitate mutual recognition of registrations, and provide advice and counsel to other committees in the EOTC framework on matters related to quality system assessment and certification.

EU   European Union. The EU is a framework within which member states have agreed to integrate their economies and eventually form a political union. Current members are Austria, Belgium, Denmark, Finland, France, Germany, Greece, Ireland, Italy, Luxembourg, Netherlands, Portugal, Spain, Sweden, and the United Kingdom.

Finding   A conclusion of importance based on observation(s). (ASQC Quality Auditing Technical Committee)

Follow-Up Audit   An audit whose purpose and scope are limited to verifying that corrective action has been accomplished as scheduled and to determining that the action effectively prevented recurrence. (ASQC Quality Auditing Technical Committee)

Grade   A category or rank given to entities which have the same functional use but different requirements for quality. (ISO 8402, Clause 2.2)

IEC   International Electrotechnical Commission. A worldwide organization that produces standards in the electrical and electronic fields. Members are the national committees, composed of representatives of the various organizations which deal with electrical/electronic standardization in each country. Formed in 1906.

Inspection   Activities such as measuring, examining, testing, and gauging one or more characteristics of a product or service and comparing these with specified requirements to determine conformity. (ISO 8402, Clause 2.15)

IQA   Institute for Quality Assurance. A British organization of quality professionals; operates a widely-recognized system of certification of auditors for quality systems.

ISO   International Organization for Standardization. A worldwide federation of national standards bodies (92 at present). ISO produces standards in all fields, except electrical and electronic (which are covered by IEC). Formed in 1947.

Modules   The EU has devised a conformity assessment system, consisting of modules, to handle the diversity of testing, inspection, and certification activities. Modules in the "modular approach" range from manufacturer declaration—through a variety of routes involving design and type approval—to full third-party certification.

MOU   Memorandum of understanding; a written agreement among a number of organizations covering specific activities of common interest. There are a number of MOUs covering mutual recognition of quality system registrations in which one of the signatories is a non-European registrar.

MRA   Mutual Recognition Agreement. A company holds the title of a certificate issued by a signer of an MRA. The other signer recognizes the certification performed, and an attestation of equivalence signed by both certification bodies will be delivered to the company upon request.

Nonconformity   The nonfulfillment of a specified requirement. (ISO 8402, Clause 2.20)

Notified Body   A notified body is a testing organization that has been selected to perform assessment activities for (a) particular directive(s). It is approved by the competent authority of its member state and notified to the European Commission and all other member states.

Organization   A company, corporation, firm, enterprise, or association, or part thereof, whether incorporated or not, public or private, that has its own functions and administration. (ISO 8402, Clause 1.7)

Organizational Structure   The responsibilities, authorities, and relationships, arranged in a pattern, through which an organization performs its functions. (ISO 8402, Clause 1.8)

Procedure   A specified way to perform an activity. (ISO 8402, Clause 1.3)

Process   A set of interrelated resources and activities which transform inputs into outputs. (ISO 8402, Clause 1.2)

Process Quality Audit   An analysis of elements of a process and appraisal of completeness, correctness

of conditions, and probable effectiveness. (ISO 8402, Clause 4.9)

Product   The result of activities or processes. (ISO 8402, Clause 1.4)

Product Quality Audit   A quantitative assessment of conformance to required product characteristics. (ISO 8402, Clause 4.9)

Protocol Agreement   An agreement signed between two organizations that operate in different but complementary fields of activity and that commit themselves to take into account their respective assessment results according to conditions specified in advance.

Purchaser   The customer in a contractual situation (ISO 8402, Clause 1.11)

Qualification Process   The process of demonstrating whether an entity is capable of fulfilling specified requirements. (ISO 8402, Clause 2.13)

Quality   The totality of features and characteristics of an entity that bear on its ability to satisfy stated or implied needs. (ISO 8402, Clause 2.1)

Quality Assurance   All the planned and systematic activities implemented within the quality system and demonstrated as needed, to provide adequate confidence that an entity will fulfill requirements for quality. (ISO 8402, Clause 3.5)

Quality Audit   A systematic and independent examination to determine whether quality activities and related results comply with planned arrangements and whether these arrangements are implemented effectively and are suitable to achieve objectives. (ISO 8402 Clause 4.9)

Quality Control   The operational techniques and activities that are used to fulfill requirements for quality. (ISO 8402, Clause 3.4)

Quality Management   All activities of the overall management function that determine the quality policy, objectives, and responsibilities, and implement them by means such as quality planning, quality control, quality assurance, and quality improvement within the quality system. (ISO 8402, Clause 3.2)

Quality Manual   A document stating the quality policy and describing the quality system of an organization. (ISO 8402, Clause 3.12)

Quality Plan   A document setting out the specific quality practices, resources, and sequence of activities relevant to a particular product, project, or contract. (ISO 8402, Clause 3.13)

Quality Planning   The activities that establish the objectives and requirements for quality and for the application of quality system elements. (ISO 8402, Clause 3.3)

Quality Policy   The overall quality intentions and direction of an organization with regard to quality, as formally expressed by top management. (ISO 8402, Clause 3.1)

Quality Surveillance   The continuing monitoring and verification of the status of procedures, methods, conditions, products, processes, and services and the analysis of records in relation to stated references to ensure that requirements for quality are being met. (ISO 8402, Clause 4.7)

Quality System   The organizational structure, procedures, processes, and resources needed to implement quality management. (ISO 8402, Clause 3.6)

Quality System Audit   A documented activity to verify, by examination and evaluation of objective evidence, that applicable elements of the quality system are suitable and have been developed, documented, and effectively implemented in accordance with specified requirements. (ANSI/ISO/ASQC A3)

Quality System Review   A formal evaluation by management of the status and adequacy of the quality system in relation to quality policy and/or new objectives resulting from changing circumstances. (ANSI/ISO/ASQC A3)

RAB   Registrar Accreditation Board. A U.S. organization whose mission is to recognize the competence and reliability of registrars of quality systems, and to achieve international recognition of registrations issued by accredited registrars.

Registered   A procedure by which a body indicates the relevant characteristics of a product, process, or service, or the particulars of a body or person, in a published list. ISO 9000 registration is the evaluation of a company's quality system against the requirements of ISO 9001, 9002, or 9003.

Registration   Procedure by which a body indicates relevant characteristics of a product, process, or service, or particulars of a body or person, and then includes or registers the product, process, or service in an appropriate publicly available list. (ISO/IEC Guide 2)

Registration Document   Documentation that a supplier's quality system conforms to specified standards. Issued by an assessment body.

Requirements of Society   Requirements including laws, statutes, rules, and regulations, codes, environmental considerations, health and safety factors, and conservation of energy and materials. (9004-1, Clause 3.3)

Root Cause   A fundamental deficiency that results in a nonconformance and must be corrected to prevent recurrence of the same or similar nonconformance. (ASQC Quality Auditing Technical Committee)

RvA   Raad voor de Accreditatie (Dutch Council for Accreditation). The Dutch authority for recognizing the competence and reliability of organizations that perform third-party certification of products, accreditation of laboratories, and registration of quality systems. The first such organization, formed in 1980.

Service   The result generated, by activities at the interface between the supplier and the customer and by supplier internal activities to meet the customer needs. (ISO 8402, Clause 1.5)

Service Delivery   Those supplier activities necessary to provide the service. (ISO 8402, Clause 1.6)

Software   An intellectual creation consisting of information, instructions, concepts, transactions, or procedures. (ISO 9000-1, Clause 3.2)

Software Product   Complete set of computer programs, procedures, and associated documentation, and data designated for delivery to a user. (ISO 9000-3, Clause 3.7)

Specification   The document that prescribes the requirements with which the product or service must conform. (ISO 8402, Clause 3.14)

Subcontractor   An organization that provides a product to the supplier. (ISO 8402, Clause 1.13)

Supplier   An organization that provides a product to the customer. (ISO 8402, Clause 1.10)

Survey   An examination for some specific purpose such as, to inspect or consider carefully or to review in detail. (ASQC Quality Auditing Technical Committee)

TAG   Technical advisory group. A term used specifically in the United States for groups that are responsible for input on international standards within their respective scopes; other countries may use other terms.

Testing   A means of determining an item's capability to meet specified requirements by subjecting them to a set of physical, chemical, environmental, or operating actions and conditions. (ANSI/ISO/ASQC A3)

Total Quality Management   A management approach of an organization, centered on quality, based on the participation of all its members and aiming at long-term success through customer satisfaction and benefits to the members of the organization and society. (ISO 8402, Clause 3.7)

Traceability   The ability to trace the history, application, or location of an entity, by means of recorded identifications. (ISO 8402, Clause 3.16)

UKAS   United Kingdom Accredition Service. This is the British authority for recognizing the competence and reliability of organizations that perform third-party certification of products and registration of quality systems. Formed in 1984, it is the world's second such organization.

Validation   Confirmation by examination and provision of objective evidence (2.19) that the particular requirements for a specific intended use are fulfilled. (ISO 8402, Clause 2.17)

Verification   The act of reviewing, inspecting, testing, checking, auditing, or otherwise establishing and documenting whether items, processes, services, or documents conform to specified requirements. (ISO 8402, Clause 2.17)

Appendix

ANSI/ISO/ASQC Q9000 Series Standards

American National Standard

ANSI/ISO/ASQC Q9000-1-1994
QUALITY MANAGEMENT AND QUALITY ASSURANCE STANDARDS—
GUIDELINES FOR SELECTION AND USE

Prepared by
American Society for Quality Control
Standards Committee
for
American National Standards Committee
Z-1 on Quality Assurance

Descriptors: Quality management, quality assurance, quality audit, quality systems, selection, use, general conditions.

American National Standards: An American National Standard implies a consensus of those substantially concerned with its scope and provisions. An American National Standard is intended as a guide to aid the manufacturer, the consumer, and the general public.

The existence of an American National Standard does not in any respect preclude anyone, whether he or she has approved the standard or not, from manufacturing, purchasing, or using products, processes, or procedures not conforming to the standard. American National Standards are subject to periodic review and users are cautioned to obtain the latest edition.

Caution Notice: This American National Standard may be revised or withdrawn at any time. The procedures of the American National Standards Institute require that action be taken to reaffirm, revise, or withdraw this standard no later than five years from the date of publication. Purchasers of American National Standards may receive current information on all standards by calling or writing the American National Standards Institute.

ASQC Mission: To facilitate continuous improvement and increase customer satisfaction by identifying, communicating, and promoting the use of quality principles, concepts, and technologies; and thereby be recognized throughout the world as the leading authority on, and champion for, quality.

10 9 8 7 6 5 4 3 2 1

Printed in the United States of America

Printed on acid-free recycled paper

Published by:
ASQC
611 E. Wisconsin Avenue
Milwaukee, WI 53202

CONTENTS

1	Scope	7.7	Quality assurance: production, installation, and servicing
2	Normative reference	7.8	Quality assurance: final inspection
3	Definitions	7.9	Quality management
4	Principal concepts	7.10	Services
4.1	Key objectives and responsibilities for quality	7.11	Processed materials
4.2	Stakeholders and their expectations	7.12	Quality improvement
4.3	Distinguishing between quality-system requirements and product requirements	7.13	Audits
4.4	Generic product categories	7.14	Auditors
4.5	Facets of quality	7.15	Managing audits
4.6	Concept of a process	7.16	Quality assurance for measurement
4.7	Network of processes in an organization	8	Selection and use of International Standards and American National Standards for external quality assurance
4.8	Quality system in relation to the network of processes	8.1	General guidance
4.9	Evaluating quality systems	8.2	Selection of model
5	Roles of documentation	8.3	Demonstration of conformance to the selected model
5.1	The value of documentation	8.4	Additional considerations in contractual situations
5.2	Documentation and evaluation of quality systems		
5.3	Documentation as a support for quality improvement		**Annexes**
5.4	Documentation and training	A	Terms and definitions taken from ISO 8402
6	Quality-system situations	B	Product and process factors
7	Selection and use of International Standards and American National Standards on quality	B.1	Purpose
7.1	General	B.2	Factors
7.2	Selection and use	C	Proliferation of standards
7.3	Application guidelines	D	Cross-reference list of clause numbers for corresponding topics
7.4	Software	E	Bibliography
7.5	Dependability		
7.6	Quality assurance: design, development, production, installation, and servicing		

FOREWORD

(This Foreword is not a part of American National Standard *Quality Management and Quality Assurance Standards—Guidelines for Selection and Use.*)

This American National Standard corresponds to the International Standard ISO 9000-1:1994. This standard is the road map document for use of all other International Standards in the entire ISO 9000 family; that is, all the standards published by ISO Technical Committee 176, Quality Management and Quality Assurance. The initial five ISO 9000 series standards, ISO 9000, ISO 9001, ISO 9002, ISO 9003, and ISO 9004, when published in the United States as American National Standards in 1987, were designated as ANSI/ASQC Q90 through ANSI/ASQC Q94 respectively. The five 1987 standards in their 1994 international revisions are now designated ISO 9000-1, ISO 9001, ISO 9002, ISO 9003, and ISO 9004-1 respectively. Their publication as American National Standards are now designated ANSI/ISO/ASQC Q9000-1-1994, ANSI/ISO/ASQC Q9001-1994, ANSI/ISO/ASQC Q9002-1994, ANSI/ISO/ASQC Q9003-1994, and ANSI/ISO/ASQC Q9004-1-1994 respectively. This new numbering system is intended to emphasize the word-for-word correspondence of the International and American National Standards.

ISO (the International Organization for Standardization) is a worldwide federation of national standards bodies (ISO member bodies). The work of preparing International Standards is normally carried out through ISO technical committees. Each member body interested in a subject for which a technical committee has been established has the right to be represented on that committee. International organizations, governmental and nongovernmental, in liaison with ISO, also take part in the work. ISO collaborates closely with the International Electrotechnical Commission (IEC) on all matters of electrotechnical standardization. The American National Standards Institute (ANSI) is the U.S. member body of ISO. ASQC is the U.S. member of ANSI responsible for quality management and related standards.

This standard provides guidelines for selection and use of all standards in the ISO 9000 family. At this writing not all of the ISO 9000 family standards have been adopted as American National Standards, but this standard provides guidelines for the entire family as currently published internationally. See Clause 7 in the standard for detailed discussion of the designations for all standards in the ISO 9000 family.

Users should note that all ANSI/ISO/ASQC standards undergo revision from time to time. In the case of International Standards adopted as American National Standards, the revision timing is influenced by the international revision timing. Reference herein to any other standard implies the latest American National Standard revision unless otherwise stated.

Comments concerning this standard are welcome. They should be sent to the sponsor of the standard, American Society for Quality Control, 611 E. Wisconsin Avenue, P.O. Box 3005, Milwaukee, WI 53201-3005, c/o Standards Administrator.

INTRODUCTION

Organizations—industrial, commercial, or governmental—supply products intended to satisfy customers' needs and/or requirements. Increased global competition has led to increasingly more stringent customer expectations with regard to quality. To be competitive and to maintain good economic performance, organizations/suppliers need to employ increasingly effective and efficient systems. Such systems should result in continual improvements in quality and increased satisfaction of the organization's customers and other stakeholders (employees, owners, subsuppliers, society).

Customer requirements often are incorporated in "specifications." However, specifications may not in themselves guarantee that a customer's requirements will be met consistently if there are any deficiencies in the organizational system to supply and support the product. Consequently, these concerns have led to the development of quality system standards and guidelines that complement relevant product requirements given in the technical specifications. The International Standards in the ISO 9000 family are intended to provide a generic core of quality system standards applicable to a broad range of industry and economic sectors (Clause 7).

The management system of an organization is influenced by the objectives of the organization, by its products, and by the practices specific to the organization and, therefore, quality systems also vary from one organization to another. A major purpose of quality management is to improve the systems and processes so that continual improvement of quality can be achieved.

ANSI/ISO/ASQC Q9000-1-1994, which has the role of road map for the ISO 9000 family, has been expanded substantially. In particular, it contains guidance concepts not included in the 1987 version. These additional concepts

- Are needed for effective understanding and current application of the ISO 9000 family.

- Are planned for complete integration into the architecture and content of future revisions of the ISO 9000 family.

In revision of the ISO 9000 family, there are no major changes in the architectures of ANSI/ISO/ASQC Q9001-1994, ANSI/ISO/ASQC Q9002-1994, ANSI/ISO/ASQC Q9003-1994, and ANSI/ISO/ASQC Q9004-1-1994. (However, ANSI/ISO/ASQC Q9003-1994 does contain additional clauses compared to the 1987 version.) Each of these American National Standards has had small-scale changes. These changes move toward future revisions to meet better the needs of users.

ANSI/ISO/ASQC Q9000-1-1994 and all International Standards in the ISO 9000 family are independent of any specific industry or economic sector. Collectively they provide guidance for quality management and general requirements for quality assurance.

The International Standards in the ISO 9000 family describe what elements quality systems should encompass but not how a specific organization implements these elements. It is not the purpose of these International Standards to enforce uniformity of quality systems. Needs of organizations vary. The design and implementation of a quality system must necessarily be influenced by the particular objectives, products, and processes, and by specific practices of the organization.

ANSI/ISO/ASQC Q9000-1-1994 clarifies the principal quality-related concepts contained within the quality management and quality assurance International Standards generated by ISO/TC 176 and provides guidance on their selection and use.

Quality Management and Quality Assurance Standards— Guidelines for Selection and Use

1 Scope

ANSI/ISO/ASQC Q9000-1-1994

a) Clarifies principal quality-related concepts and the distinctions and interrelationships among them.

b) Provides guidance for the selection and use of the ISO 9000 family of International Standards and the ANSI/ISO/ASQC Q9000 series of American National Standards on quality management and quality assurance.

2 Normative Reference

The following standard contains provisions which, through reference in this text, constitute provisions of this part of ISO 9000. At the time of publication, the edition indicated was valid. All standards are subject to revision, and parties to agreements based on ANSI/ISO/ASQC Q9000-1-1994 are encouraged to investigate the possibility of applying the most recent edition of the standard indicated below. Members of IEC and ISO maintain registers of currently valid International Standards.

ISO 8402:1994, *Quality management and quality assurance—Vocabulary*

3 Definitions

This revision of ANSI/ISO/ASQC Q9000-1-1994, ANSI/ISO/ASQC Q9001-1994, ANSI/ISO/ASQC Q9002-1994, ANSI/ISO/ASQC Q9003-1994, and ANSI/ISO/ASQC Q9004-1994 has improved the

Table 1 Relationships of organizations in the supply chain.

ANSI/ISO/ASQC Q9000-1-1994	Subsupplier	→	supplier or organization	→	customer
ANSI/ISO/ASQC Q9001-1994 ANSI/ISO/ASQC Q9002-1994 ANSI/ISO/ASQC Q9003-1994	Subcontractor	→	supplier	→	customer
ANSI/ISO/ASQC Q9004-1-1994	Subcontractor	→	organization	→	customer

harmonization of terminology for organizations in the supply chain. Table 1 shows the supply-chain terminology used in these American National Standards.

The usage of all of these terms conforms with their formal definitions in ISO 8402. The remaining differences in terminology in Table 1 reflect, in part, a desire to maintain historical continuity with usage in the 1987 edition of these American National Standards.

NOTES

1 In all these American National Standards, the grammatical format of the guidance or requirements text is addressed to the organization in its role as a supplier of products (the third column of Table 1).

2 In the ANSI/ISO/ASQC Q9000-1-1994 row of Table 1, the use of "subsupplier" emphasizes the supply-chain relationship of the three organizational units, using the self-defining term in relation to "supplier." Where appropriate, especially in discussing quality-management situations, the term "organization" is used rather than "supplier."

3 In the ANSI/ISO/ASQC Q9001-1994, ANSI/ISO/ASQC Q9002-1994, and ANSI/ISO/ASQC Q9003-1994 rows of Table 1, the use of "subcontractor" reflects the fact that, in an external quality-assurance context, the relevant relationship often is (explicitly or implicitly) contractual.

4 In the ANSI/ISO/ASQC Q9004-1-1994 row of Table 1, the use of "organization" reflects the fact that quality-management guidance is applicable to any organizational unit, irrespective of the categories of products it may supply, or whether it is a free-standing unit or part of a larger organization.

For the purposes of ANSI/ISO/ASQC Q9000-1-1994, the definitions given in ISO 8402, together with the following definitions, apply.

NOTE

5 For the convenience of users of ANSI/ISO/ASQC Q9000-1-1994, some relevant definitions from ISO 8402 are contained in Annex A.

3.1 Hardware

Tangible, discrete product with distinctive form.

NOTE

6 Hardware normally consists of manufactured, constructed, or fabricated pieces, parts, and/or assemblies.

3.2 Software

An intellectual creation consisting of information expressed through supporting medium.

NOTES

7 Software can be in the form of concepts, transactions, or procedures.

8 A computer program is a specific example of software.

3.3 Processed material

Tangible product generated by transforming raw material into a desired state.

NOTES

9 The state of processed material can be liquid, gas, particulate material, ingot, filament, or sheet.

10 Processed material is typically delivered in drums, bags, tanks, cylinders, cans, pipelines, or rolls.

3.4 Industry/economic sector

A grouping of suppliers whose offerings meet similar customer needs and/or whose customers are closely interrelated in the marketplace.

NOTES

11 Dual use of "industry sector" and "economic sector" recognizes that each term is used for the intended meaning in specific countries or languages.

12 Industry/economic sectors include administration, aerospace, banking, chemicals, construction, education, food, healthcare, leisure, insurance, mining, retailing, telecommunications, textiles, tourism, and so forth.

13 Industry/economic sectors apply to the global economy or a national economy.

3.5 Stakeholder

An individual or group of individuals with a common interest in the performance of the supplier organization and the environment in which it operates.

3.6 ISO 9000 family

All those International Standards produced by the technical committee ISO/TC 176.

NOTE

14 At present, the family comprises

a) all the International Standards numbered ISO 9000 through to ISO 9004, including all parts of ISO 9000 and ISO 9004;

b) all the International Standards numbered ISO 10001 through 10020, including all parts; and

c) ISO 8402.

4 Principal concepts

4.1 Key objectives and responsibilities for quality

An organization should do the following:

a) Achieve, maintain, and seek to improve continuously the quality of its products in relationship to the requirements for quality.

b) Improve the quality of its own operations, so as to meet continually all customers' and other stakeholders' stated and implied needs.

c) Provide confidence to its internal management and other employees that the requirements for quality are being fulfilled and maintained, and that quality improvement is taking place.

d) Provide confidence to the customers and other stakeholders that the requirements for quality are being, or will be, achieved in the delivered product.

e) Provide confidence that quality-system requirements are fulfilled.

4.2 Stakeholders and their expectations

Every organization as a supplier has five principal groups of stakeholders: its customers, its employees, its owners, its subsuppliers, and society.

The supplier should address the expectations and needs of all its stakeholders.

Supplier's stakeholders	Typical expectations/ needs
Customers	Product quality
Employees	Career/work
Owners	Investment performance
Subsuppliers	Continuing business opportunity
Society	Responsible stewardship

The International Standards in the ISO 9000 family focus their guidance and requirements on satisfying the customer.

The requirements of society, as one of the five stakeholders, are becoming more stringent worldwide. In addition, expectations and needs are becoming more explicit for considerations such as: workplace health and safety; protection of the environment (including conservation of energy and natural resources); and security. Recognizing that the ISO 9000 family of International Standards provides a widely used approach for management systems that can meet requirements for quality, these management principles can be useful for other concerns of society. Compatibility of the management-system approach in these several areas can enhance the effectiveness of an organization. In the same manner that product and process technical specifications are separate from management-systems requirements, the technical specifications in these other areas should be separately developed.

4.3 Distinguishing between quality-system requirements and product requirements

The ISO 9000 family of International Standards makes a distinction between quality-system requirements and product requirements. By means of this distinction, the ISO 9000 family applies to organizations providing products of all generic product categories, and to all product quality characteristics. The quality-system requirements are complementary to the technical requirements of the product. The applicable technical specifications of the product (e.g., as set out in product standards) and technical specifications of the process are separate and distinct from the applicable ISO 9000 family requirements or guidance.

International Standards in the ISO 9000 family, both guidance and requirements, are written in terms of the quality-system objectives to be satisfied. These International Standards do not prescribe how to achieve the objectives but leave that choice to the management of the organization.

4.4 Generic product categories

It is useful to identify four generic product categories (see Clause 3 and Annex A), as follows:

a) Hardware

b) Software

c) Processed materials

d) Services

These four generic product categories encompass all the kinds of product supplied by organizations. International Standards in the ISO 9000 family are applicable to all four generic product categories. The quality-system requirements are essentially the same for all generic product categories, but the terminology and management-system details and emphases may differ.

Two or more of the generic product categories usually are present in the marketplace offerings of any organization, whatever the industry/economic sector (see clause 3) in which the organization operates. For example, most organizations that supply hardware, software, or processed materials have a service component to their offering. Customers (and other stakeholders) will look for value in each generic product category that is present in the offering.

Analytical instruments are examples where hardware (i.e., the instrument), software (for computing tasks within the instrument), processed materials (such as titrating solutions or reference standard materials), and services (such as training or maintenance servicing) might all be important features of the offering. A service organization such as a restaurant will have hardware, software, and processed materials as well as service components.

4.5 Facets of quality

Four facets that are key contributors to product quality may be identified as follows.

a) Quality due to definition of needs for the product

The first facet is quality due to defining and updating the product, to meet marketplace requirements and opportunities.

b) Quality due to product design

The second facet is quality due to designing into the product the characteristics that enable it to meet marketplace requirements and opportunities, and to provide value to customers and other stakeholders. More precisely, quality due to product design is the product design features that influence intended performance within a given grade, plus product design features that influence the robustness of product performance under variable conditions of production and use.

c) Quality due to conformance to product design

The third facet is quality due to maintaining day-to-day consistency in conforming to product design and in providing the designed characteristics and values for customers and other stakeholders.

d) Quality due to product support

The fourth facet is quality due to furnishing support throughout the product life-cycle, as needed, to provide the designed characteristics and values for customers and other stakeholders.

For some products, the important quality characteristics include dependability characteristics. Dependability (i.e., reliability, maintainability, and availability) may be influenced by all four facets of product quality.

A goal of the guidance and requirements of the International Standards in the ISO 9000 family is to meet the needs for all four facets of product quality. Some facets of quality may be specifically important, for example, in contractual situations, but, in general, all facets contribute to the quality of the product. The ISO 9000 family explicitly provides generic quality-management guidance and external quality-assurance requirements on facets a, b, c, and d.

When considering the complete product offering, the customer will bear in mind additional factors. These include the following.

- The supplier's market status and strategy—if the supplier has an established and reputable marketplace status and/or a strategy that is achieving a satisfactory market share, the customer is likely to place higher value on the supplier's offering.

- The supplier's financial status and strategy—if the supplier has an established and reputable financial status and/or a strategy that is improving financial performance, the customer is likely to place higher value on the supplier's offering.

- The supplier's human resources status and strategy—if the supplier has an established and reputable human resources status and/or a strategy that is developing improved skills, diversity, and commitment in its human resources, the customer is likely to place higher value on the supplier's offering.

These additional factors are of vital importance in managing a supplier organization as a total enterprise.

NOTE

15 Product value involves both quality and price and, as such, price is not a facet of quality.

4.6 Concept of a process

The International Standards in the ISO 9000 family are founded upon the understanding that all work is accomplished by a process (see Figure 1 below). Every process has inputs. The outputs are the results of the process. The outputs are products, tangible or intangible. The process itself is (or should be) a transformation that adds value. Every process involves people and/or other resources in some way. An output may be, for example, an invoice, computing software, liquid fuel, a clinical device, a banking service, or a final or intermediate product of any generic category. There are opportunities to make measurements on the

inputs, at various places in the process, as well as on the outputs. As shown in Figure 2 (page 244), inputs and outputs are of several types.

Type	Examples
Product-related	Raw materials
(solid lines in Fig. 2)	Intermediate product
	Final Product
	Sampled product
Information-related	Product requirements
(dashed lines in Fig. 2)	Product characteristics and status information
	Support-function communications
	Feedback on product performance needs
	Measurement data from sampled product

Figure 2 shows the supplier in a supply-chain relationship to a subsupplier and a customer. In this supply-chain structure, the various inputs and outputs need to flow in different directions, as illustrated in Figure 2. It is emphasized that in this context "product" includes all four generic product categories.

Quality management is accomplished by managing the processes in the organization. It is necessary to manage a process in two senses:

- The structure and operation of the process itself within which the product or information flows.

- The quality of the product or information flowing within the structure.

Figure 1 All work is accomplished by a process.

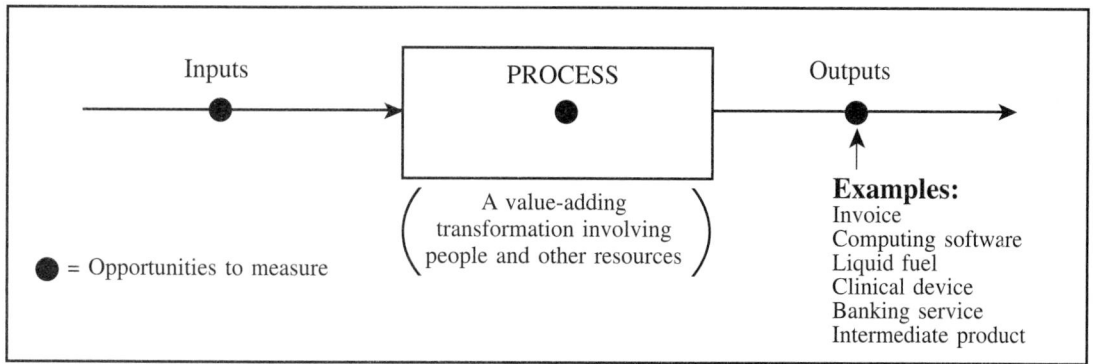

4.7 Network of processes in an organization

Every organization exists to accomplish value-adding work. The work is accomplished through a network of processes. The structure of the network is not usually a simple sequential structure, but typically is quite complex.

In an organization there are many functions to be performed. They include production, product design, technology management, marketing, training, human resources management, strategic planning, delivery, invoicing, and maintenance. Given the complexity of most organizations, it is important to highlight the main processes and to simplify and prioritize processes for quality-management purposes.

An organization needs to identify, organize, and manage its network of processes and interfaces. The organization creates, improves, and provides consistent quality in its offerings through the network of processes. This is a fundamental conceptual basis for the ISO 9000 family. Processes and their interfaces should be subject to analysis and continuous improvement.

Problems tend to arise where people have to manage several processes and their interrelationships, particularly for large processes that may span several functions. To clarify interfaces, responsibilities, and authorities, a process should have an owner as the person responsible. The quality of executive management's own processes, such as strategic planning, is especially important.

4.8 Quality system in relation to the network of processes

It is conventional to speak of a quality system as consisting of a number of elements. The quality system is carried out by means of processes, which exist both within and across functions. For a quality system to be effective, these processes and the associated responsibilities, authorities, procedures, and resources should be defined and deployed in a consistent manner. A system is more than a sum of processes. To be effective, the quality system needs coordination and compatibility of its component processes, and definition of their interfaces.

4.9 Evaluating quality systems

4.9.1 General
When evaluating quality systems, there are three essential questions that have to be asked in relation to every process being evaluated, as follows.

a) Are the processes defined and their procedures appropriately documented?

b) Are the processes fully deployed and implemented as documented?

c) Are the processes effective in providing the expected results?

The collective answers to these questions relating respectively to the approach, deployment, and results, will determine the outcome of the evaluation. An evaluation of a quality system may vary in scope, and encompass a wide range of activities, some of which are discussed in 4.9.2 and 4.9.3.

4.9.2 Management review
One of the important activities that executive management of the supplier organization needs to carry out systematically is an evaluation of the status and adequacy of the quality system, including the quality policy, in relation to the expectations of the stakeholders.

Figure 2 Supply-chain relationship of processes, with product-related and information-related flow.

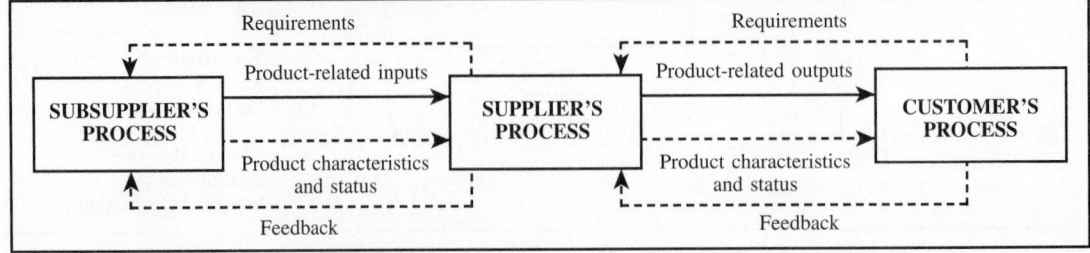

Management reviews usually take into account many additional factors beyond the requirements found in ANSI/ISO/ASQC Q9001-1994, ANSI/ISO/ASQC Q9002-1994, and ANSI/ISO/ASQC Q9003-1994. The results of internal audits and external audits are an important source of information. It is important that the outcome of the management review should lead to increased effectiveness and efficiency of the quality system.

4.9.3 Quality-system audits

In evaluating the effectiveness of a quality system, audits are an important element. Audits may be conducted by, or on behalf of, the organization itself (first party), its customers (second parties), or independent bodies (third parties). The second- or third-party audit may provide an enhanced degree of objectivity from the customer's perspective.

First-party internal quality audits may be conducted by members of the organization or by other persons on behalf of the organization. These provide information for effective management review and corrective, preventive, or improvement action.

Second-party quality audits may be conducted by customers of the organization, or by other persons on behalf of the customer where there is a contract or a series of contracts under consideration. These provide confidence in the supplier.

Third-party quality audits may be carried out by competent certification bodies to gain certification/registration, thereby providing confidence to a range of potential customers.

The basic requirements for quality systems are contained in ANSI/ISO/ASQC Q9001-1994, ANSI/ISO/ASQC Q9002-1994, and ANSI/ISO/ASQC Q9003-1994. ANSI/ISO/ASQC Q10011-1-1994, ANSI/ISO/ASQC Q10011-2-1994, and ANSI/ISO/ASQC Q10011-3-1994 give guidance on auditing.

NOTE

16 A first-party audit is often called an "internal" audit, whereas second-party and third-party quality audits are often called "external" quality audits.

5 Roles of documentation

5.1 The value of documentation

In the context of the ISO 9000 family, the preparation and use of documentation is intended to be a dynamic high-value-adding activity. Appropriate documentation is essential for several critical roles, including:

- Achieving required (product) quality.
- Evaluating quality systems.
- Quality improvement.
- Maintaining the improvements.

5.2 Documentation and evaluation of quality systems

For auditing purposes, documentation of procedures is objective evidence that

- A process has been defined.
- The procedures are approved.
- The procedures are under change control.

Only under these circumstances can internal or external audits provide a meaningful evaluation of the adequacy of both deployment and implementation.

5.3 Documentation as a support for quality improvement

Documentation is important for quality improvement. When procedures are documented, deployed, and implemented, it is possible to determine with confidence how things are done currently and to measure current performance. Then reliable measurement of the effect of a change is enhanced. Moreover, documented standard operating procedures are essential for maintaining the gains from quality-improvement activities.

5.4 Documentation and training

Maintaining consistency of the procedures that are deployed and implemented results from a combination of the documentation and the skills and training of personnel. In each situation an appropriate balance between the extent of documentation and the extent of skills and training should be sought, so as to keep documentation to a reasonable level that can be maintained at appropriate intervals. Quality-system audits should be performed with this necessary balancing in mind.

6 Quality-system situations

The ISO 9000 family is intended to be used in four situations:

    a) Guidance for quality management.

    b) Contractual, between first and second parties.

    c) Second-party approval or registration.

    d) Third-party certification/registration.

The supplier's organization should install and maintain a quality system designed to cover all the situations (among those listed under a, b, c, and d) that the organization meets.

For situation a, this system will strengthen its own competitiveness to fulfill the requirements for product quality in a cost-effective way.

In situation b, the customer may be interested in certain elements of the supplier's quality system which affect the supplier's ability consistently to produce product to requirements, and the associated risks. The customer, therefore, contractually requires that certain quality-system elements and processes, as appropriate, be part of the supplier's quality system, by specifying a particular quality-assurance model.

In situation c, the supplier's quality system is assessed by the customer. The supplier may be given formal recognition of conformance with the standard.

In situation d, the supplier's quality system is evaluated by the certification body, and the supplier agrees to maintain the quality system for all customers unless otherwise specified in an individual contract. This type of quality-system certification or registration often reduces the number and/or extent of quality-system assessments by customers.

A single supplier often will be involved in situations of all types. The supplier may purchase some materials or components from standard inventory without contractual quality-system requirements, and purchase others with contractual quality-system requirements. The same supplier may sell some products in noncontractual situations, with or without customers expecting quality-system certification, and may sell other products in contractual situations.

A supplier can elect to use the ISO 9000 family in either of two ways, which may be called "management motivated" and "stakeholder motivated," respectively. In either case, the supplier should initially consult ANSI/ISO/ASQC Q9000-1-1994, the road map

for the ISO 9000 family, to understand basic concepts and the types of standards available in the family.

The stakeholder-motivated approach is the predominant practice in many nations and industry/economic sectors. The increasing use of quality-system certification/registration is a factor in the spread of this approach.

In the stakeholder-motivated approach, the supplier initially implements a quality system in response to immediate demands by customers or other stakeholders. The selected quality system conforms to the requirements of ANSI/ISO/ASQC Q9001-1994, ANSI/ISO/ASQC Q9002-1994, or ANSI/ISO/ASQC Q9003-1994, as applicable. The supplier's management must play a significant leadership role in this approach, but the effort is driven by external stakeholders. Typically, the supplier finds that significant improvements in product quality, costs, and internal operating results are obtained. At the same time, or later, the supplier may initiate a quality-management effort to gain further improvements, building a more comprehensive quality system from the selected quality-assurance model as a core building block.

In the management-motivated approach, the supplier's own management initiates the effort in anticipation of emerging marketplace needs and trends. In this route, ANSI/ISO/ASQC Q9004-1-1994 (and other applicable parts of ISO 9004) are used first, to guide a quality-management approach to installing a quality system that will enhance the supplier's quality achievement. Subsequently, the supplier can use the applicable requirement standard, ANSI/ISO/ASQC Q9001-1994, ANSI/ISO/ASQC Q9002-1994, or ANSI/ISO/ASQC Q9003-1994, as the quality-assurance model for demonstrating quality-system adequacy, possibly seeking certification/registration in advance of any customer requirement as a preparatory measure.

The quality system implemented in this management-motivated approach will normally be more comprehensive and fruitful than the model used for demonstrating quality-system adequacy.

7 Selection and Use of International Standards and American National Standards on Quality

7.1 General

For quality-management purposes, organizations should use the ISO 9000 family of International Standards in order to develop, implement, and improve their quality system in both the management-motivated and stakeholder-motivated situations.

The ISO 9000 family contains two types of guidance standards. Application guidance for quality-assurance purposes is provided by several parts of ISO 9000. Specialized application guidance for quality-management purposes is provided by the parts of ISO 9004. These parts of ISO 9004 are not intended to be used to interpret the requirements of the quality-assurance standards, however, they can provide useful references. Likewise, International Standards with numbers in the 10000 sequence may be used for reference.

Throughout the ISO 9000 family, emphasis is placed on the satisfaction of customers' needs, the establishment of functional responsibilities, and the importance of assessing (as far as possible) the potential risk and benefits. All these aspects should be considered in establishing and maintaining an effective quality system, and its continuous improvement.

Special attention should be paid to ANSI/ISO/ASQC Q9004-1-1994 which deals with quality management of any product (see 7.9) and applies to all generic product categories and all industry/economic sectors.

Using ANSI/ISO/ASQC Q9004-1-1994, the supplier should determine, according to a specific situation, the extent to which each quality-system element is applicable and which specific methods and technologies are to be applied; appropriate parts of the ISO 9000 family give further guidance.

Subclauses 7.2 to 7.16 give guidance to enable organizations to select appropriate International Standards from the ISO 9000 family and American National Standards from the ANSI/ISO/ASQC Q9000 series or Q10011 series that would provide useful information for implementing and operating quality systems.

7.2 Selection and use

ANSI/ISO/ASQC Q9000-1-1994, *Quality Management and Quality Assurance Standards—Guidelines for Selection and Use*

Reference should be made to ANSI/ISO/ASQC Q9000-1-1994 by any organization which is contemplating the development and implementation of a quality system.

Increased global competition has led to increasingly more stringent customer expectations with regard to quality. To be competitive and sustain good economic performance, organizations/suppliers need to employ increasingly effective and efficient systems.

ANSI/ISO/ASQC Q9000-1-1994 clarifies the principal quality-related concepts and provides guidance for the selection and use of the ISO 9000 family for this purpose.

7.3 Application guidelines

ISO 9000-2:1993, *Quality Management and Quality Assurance Standards—Part 2: Generic Guidelines for the Application of ISO 9001, ISO 9002, and ISO 9003*

ISO 9000-2 should be selected when assistance is needed in the implementation and application of ANSI/ISO/ASQC Q9001-1994, ANSI/ISO/ASQC Q9002-1994, and ANSI/ISO/ASQC Q9003-1994 (see clause 8).

It provides guidance on the implementation of the clauses in the quality assurance standards and is particularly useful during the initial implementation.

7.4 Software

ISO 9000-3:1991, *Quality Management and Quality Assurance Standards—Part 3: Guidelines for the Application of ISO 9001 to the Development, Supply, and Maintenance of Software*

(ISO 9000-3 deals exclusively with computer software.)

Reference should be made to ISO 9000-3 by supplier organizations implementing a quality system in accordance with ANSI/ISO/ASQC Q9001-1994 for a software product or a product which includes a software element.

The process of development, supply, and maintenance of software is different from that of most other types of industrial products in that there is no distinct manufacturing phase. Software does not "wear out" and, consequently, quality activities during the design phase are of paramount importance to the final quality of the product.

ISO 9000-3 sets out guidelines to facilitate the application of ISO 9001 in organizations developing, supplying, and maintaining software, by suggesting appropriate controls and methods for this purpose.

7.5 Dependability

ISO 9000-4:1993, *Quality Management and Quality Assurance Standards—Part 4: Guide to Dependability Programme Management*

ISO 9000-4 should be selected when the supplier needs to provide assurance of the dependability (i.e., reliability, maintainability, and availability) characteristics of a product.

Society's increasing reliance upon services such as transportation, electricity, telecommunications, and information services leads to higher customer requirements and expectations with regard to quality of service. The dependability of products used for such services is a major contributing factor to their quality of service.

ISO 9000-4 provides guidance on dependability-program management. It covers the essential features of a comprehensive dependability programme for the planning, organization, direction, and control of resources to produce products that will be reliable and maintainable.

7.6 Quality assurance: design, development, production, installation, and servicing

ANSI/ISO/ASQC Q9001-1994, *Quality Systems—Model for Quality Assurance in Design, Development, Production, Installation, and Servicing*

ANSI/ISO/ASQC Q9001-1994 should be selected and used when the need is to demonstrate the supplier's capability to control the processes for design as well as production of conforming product. The requirements specified are aimed primarily at achieving customer satisfaction by preventing nonconformity at all stages from design through to servicing. ANSI/ISO/ASQC Q9001-1994 specifies a quality-assurance model for this purpose.

7.7 Quality assurance: production, installation, and servicing

ANSI/ISO/ASQC Q9002-1994, *Quality systems—Model for Quality Assurance in Production, Installation, and Servicing*

ANSI/ISO/ASQC Q9002-1994 should be selected and used when the need is to demonstrate the supplier's capability to control the processes for production of conforming product. ANSI/ISO/ASQC Q9002-1994 specifies a quality-assurance model for this purpose.

7.8 Quality assurance: final inspection

ANSI/ISO/ASQC Q9003-1994, *Quality systems—Model for Quality Assurance in Final Inspection and Test*

ANSI/ISO/ASQC Q9003-1994 should be selected and used when conformance to specified requirements is to be assured by the supplier solely at final inspection and test. ANSI/ISO/ASQC Q9003-1994 specifies a quality-assurance model for this purpose.

7.9 Quality management

ANSI/ISO/ASQC Q9004-1-1994, *Quality Management and Quality System Elements—Guidelines*

Reference should be made to ANSI/ISO/ASQC Q9004-1-1994 by any organization intending to develop and implement a quality system.

In order to meet its objectives, the organization should ensure that the technical, administrative, and human factors affecting the quality of its products will be under control, whether hardware, software, processed materials, or services.

ANSI/ISO/ASQC Q9004-1-1994 describes an extensive list of quality-system elements pertinent to all phases and activities in the life-cycle of a product to assist an organization to select and apply elements appropriate to its needs.

7.10 Services

ISO 9004-2:1991, *Quality Management and Quality System Elements—Part 2: Guidelines for Services*

Reference should be made to ISO 9004-2 by organizations that provide services or whose products include a service component.

The characteristics of a service can differ from those of other products and can include such aspects as personnel, waiting time, delivery time, hygiene, credibility, and communication delivered directly to the final customer. Customer assessment, often very subjective, is the ultimate measure of the quality of a service.

ISO 9004-2 supplements the guidance of ANSI/ISO/ASQC Q9004-1-1994 with respect to products in the services category. It describes the concepts, principles, and quality-system elements which are applicable to all forms of service offerings.

7.11 Processed materials

ISO 9004-3:1993, *Quality Management and Quality System Elements—Part 3: Guidelines for Processed Materials*

Reference should be made to ISO 9004-3 by organizations whose products (final or intermediate) are prepared by transformations, and which consist of solids, liquids, gases, or combinations thereof (including particulate materials, ingots, filaments, or sheet structures). Such products are typically delivered in bulk systems such as pipelines, drums, bags, tanks, cans, or rolls.

By their nature, processed (bulk) materials present unique difficulties with regard to the verification of the product at important points in the production process. This increases the importance of the use of statistical sampling and evaluation procedures and their application to in-process controls and final product specifications.

ISO 9004-3 supplements the guidance of ANSI/ISO/ASQC Q9004-1-1994 with respect to products in the processed-materials category.

7.12 Quality improvement

ISO 9004-4:1993, *Quality Management and Quality System Elements—Part 4: Guidelines for Quality Improvement*

Reference should be made to ISO 9004-4 by any organization wishing to improve its effectiveness (whether or not it has implemented a formal quality system).

A constant goal of management of all functions and at all levels of an organization should be to strive for customer satisfaction and continuous quality improvement.

ISO 9004-4 describes fundamental concepts and principles, management guidelines, and methodology (tools and techniques) for quality improvements.

7.13 Audits

ANSI/ISO/ASQC Q10011-1-1994, *Guidelines for Auditing Quality Systems—Auditing*

ANSI/ISO/ASQC Q10011-1-1994 should be selected when establishing, planning, carrying out, and documenting audits of quality systems. It provides guidelines for verifying the existence and implementation of elements of a quality system, and for verifying the system's ability to achieve defined quality objectives.

7.14 Auditors

ANSI/ISO/ASQC Q10011-2-1994, *Guidelines for Auditing Quality Systems—Qualification Criteria for Quality Systems Auditors*

ANSI/ISO/ASQC Q10011-2-1994 should be selected when staff selection and training for quality-systems auditors is needed.

It provides guidance on the qualification criteria for quality-systems auditors. It contains guidance on the education, training, experience, personal attributes, and management capabilities needed to carry out an audit.

7.15 Managing audits

ANSI/ISO/ASQC Q10011-3-1994, *Guidelines for Auditing Quality Systems—Management of Audit Programs*

ANSI/ISO/ASQC Q10011-3-1994 should be selected when planning the management of an audit program. It provides basic guidelines for managing quality-systems audit programs. It is consistent with ANSI/ISO/ASQC Q10011-1-1994 and ANSI/ISO/ASQC Q10011-2-1994.

7.16 Quality assurance for measurement

ISO 10012-1:1992, *Quality Assurance Requirements for Measuring Equipment—Part 1: Metrological Confirmation System for Measuring Equipment*

ISO 10012-1 should be selected when the product or process quality depends heavily on the ability to measure accurately. It specifies the main features of the confirmation system to be used for a supplier's measuring equipment. It contains the quality-assurance requirements for a supplier's measurement equipment to ensure that measurements are made with the intended accuracy and consistency. It contains more detailed requirements than those found in ANSI/ISO/ASQC Q9001-1994, ANSI/ISO/ASQC Q9002-1994, and ANSI/ISO/ASQC Q9003-1994, and is presented with guidance for implementation.

8 Selection and Use of International Standards and American National Standards for External Quality Assurance

8.1 General guidance

In second-party approval or registration (situations b and c in Clause 6), the supplier and the other party should agree on which International Standard or American National Standard will be used as the basis for approval. The selection and application of a model for quality assurance appropriate to a given situation should provide benefits to both customer and supplier.

Examining the benefits, risks, and costs for both parties will determine the extent and nature of reciprocal information and the measures each party should take to provide adequate confidence that the intended quality will be achieved. The supplier has the responsibility to select the model for subcontracts unless otherwise agreed on with the customer.

In third-party certification/registration, the supplier and the certification body should agree on which International Standard or American National Standard will be used as the basis for certification/registration. The selected model should be adequate and not misleading from the point of view of the supplier's customers. For example, the role and character of design activities, if any, is especially important in selecting between ANSI/ISO/ASQC Q9001-1994 and ANSI/ISO/ASQC Q9002-1994. The selection and application of a model for quality assurance appropriate to a given situation should also support the supplier's objectives. Examining the scope of the supplier's activities which will be encompassed by the certificate will determine the extent and nature of reciprocal information and the measures each party should take to provide confidence that the certifica-

tion/registration is maintained in accordance with the requirements of the selected model.

8.2 Selection of model

8.2.1 Three models for quality assurance
As indicated in 7.6 to 7.8, in the three relevant American National Standards, certain quality-system elements have been grouped to form three distinct models suitable for the purpose of suppliers demonstrating their capabilities and for assessment of such supplier capability by external parties.

  a) ANSI/ISO/ASQC Q9001-1994: for use when conformance to specified requirements is to be assured by the supplier during design, development, production, installation, and servicing.

  b) ANSI/ISO/ASQC Q9002-1994: for use when conformance to specified requirements is to be assured by the supplier during production, installation, and servicing.

NOTE
17 ANSI/ISO/ASQC Q9002-1994 is identical to ANSI/ISO/ASQC Q9001-1994 except for the deletion of all quality-system requirements for design control.

  c) ANSI/ISO/ASQC Q9003-1994: for use when conformance to specified requirements is to be assured by the supplier at final inspection and test.

In 4.6 to 4.8 and elsewhere, a process perspective is emphasized. The goal of the quality system is to fulfill the requirements for quality in the results from the supplier's processes. But the quality-system requirements are directed toward the procedures for these processes. Therefore, specific quality-system requirements in ANSI/ISO/ASQC Q9001-1994, ANSI/ISO/ASQC Q9002-1994, and ANSI/ISO/ASQC Q9003-1994 usually are phrased: "The supplier shall establish and maintain documented procedures...."

8.2.2 Selection
The scopes of the American National Standards as summarized in 8.2.1 indicate how the choice should be made among ANSI/ISO/ASQC Q9001-1994, ANSI/ISO/ASQC Q9002-1994, or ANSI/ISO/ASQC Q9003-1994 consistent with the situations a, b, c, and d in Clause 6.

8.3 Demonstration of conformance to the selected model

The quality-system elements should be documented and demonstrable in a manner consistent with the requirements of the selected model.

Demonstration of the quality-system elements and their associated processes provides confidence on:

a) adequacy of the quality system; and

b) capability to achieve product conformity with the specified requirements.

The responsibility for demonstrating the adequacy and effectiveness of the quality system lies with the supplier. However, the supplier may need to consider the expectations for demonstration of the relevant interested parties as described in clause 6 b, c, and d. These considerations may determine the means adopted to demonstrate conformance to the selected model. Methods may include:

- Supplier's declaration of conformity.
- Providing basic documented evidence.
- Providing evidence of approvals or registrations by other customers.
- Audit by the customer.
- Audit by a third party.
- Providing evidence of competent third-party certificates.

Any of these means or a combination of them may apply in situations b and c of clause 6. In the 6d situation, the last two means are applicable.

The nature and degree of demonstration may vary from one situation to another in accordance with such criteria as:

a) The economics, uses, and conditions of use of the product.

b) The complexity and innovation required to design the product.

c) The complexity and difficulty of producing the product.

d) The ability to judge product quality on the basis of final inspection and test alone.

e) The requirements of society regarding the product.

f) The past performance of the supplier.

g) The degree of partnership in the relationship with the customer.

8.4 Additional considerations in contractual situations

8.4.1 Tailoring and contractual elements

Experience has shown that with a small fixed number of International Standards and American National Standards available, it is possible in almost every given contractual situation to select one that will meet needs adequately. However, on occasions, certain quality-system elements or subelements called for in the selected International Standard or American National Standard may be deleted and, on other occasions, elements or subelements may be added. Tailoring may also concern the degree of demonstration of quality-system elements. If tailoring should prove necessary, it should be agreed between the customer and the supplier, and should be specified in the contract.

8.4.2 Review of contractual quality-system elements

Both parties should review the proposed contract to be sure that they understand the quality-system requirements and that the requirements are mutually acceptable considering the economics and risks in their respective situations.

8.4.3 Supplementary quality-assurance requirements

There may be a need to specify supplementary requirements in the contract, such as statistical process control or systems requirements for safety-critical items.

8.4.4 Precontract assessment

Assessments of a supplier's quality system according to ANSI/ISO/ASQC Q9001-1994, ANSI/ISO/ASQC Q9002-1994, or ANSI/ISO/ASQC Q9003-1994 and, when appropriate, supplementary requirements often are utilized prior to a contract to determine the supplier's ability to satisfy the requirements. In many cases, assessments are performed directly by the customer.

8.4.5 Audits after award of the contract

Continuing demonstration of the supplier's quality system after award of the contract may be achieved by a series of quality audits conducted by the customer, the customer's agent, or an agreed on third party.

ANNEX A (NORMATIVE)

Terms and Definitions Taken from ISO 8402

A.1 Quality

Totality of characteristics of an entity that bear on its ability to satisfy stated and implied needs.

NOTES

18 In a contractual environment, or in a regulated environment, such as the nuclear safety field, needs are specified, whereas in other environments, implied needs should be identified and defined.

19 In many instances, needs can change with time; this implies periodic review of requirements for quality.

20 Needs are usually translated into characteristics with specified criteria. Needs may include, for example, aspects of performance, usability, dependability (availability, reliability, maintainability), safety, environment, economics, and aesthetics.

21 The term "quality" is not used as a single term to express a degree of excellence in a comparative sense nor is it used in a quantitative sense for technical evaluations. To express these meanings, a qualifying adjective should be used. For example, use can be made of the following terms:

   a) "relative quality" where entities are ranked on a relative basis in the "degree of excellence" or "comparative" sense (not to be confused with "grade");

   b) "quality level" in a "quantitative" sense (as used in acceptance sampling) and "quality measure" where precise technical evaluations are carried out.

22 The achievement of satisfactory quality involves all stages of the quality loop as a whole. The contributions to quality of these various stages are sometimes identified separately for emphasis, for example, quality due to definition of needs, quality due to product design, quality due to conformance, quality due to product support throughout its lifetime.

23 In some references, quality is referred to as "fitness for use" or "fitness for purpose" or "cus-

tomer satisfaction" or "conformance to the requirements." These represent only certain facets of quality, *as defined above.*

A.2 Quality policy

Overall intentions and direction of an organization with regard to quality, as formally expressed by top management.

NOTE

24 The quality policy forms one element of the corporate policy and is authorized by top management.

A.3 Quality management

All activities of the overall management function that determine the quality policy, objectives and responsibilities, and implement them by means such as quality planning, quality control, quality assurance and quality improvement within the quality system.

NOTES

25 Quality management is the responsibility of all levels of management but must be led by top management. Its implementation involves all members of the organization.

26 In quality management, consideration is given to economic aspects.

A.4 Quality system

Organizational structure, procedures, processes, and resources needed to implement quality management.

NOTES

27 The quality system should only be as comprehensive as needed to meet the quality objectives.

28 The quality system of an organization is designed primarily to meet the internal managerial needs of the organization. It is broader than the requirements of a particular customer, who evaluates only the relevant part of the quality system.

29 For contractual or mandatory quality assessment purposes, demonstration of the implementation

of the identified quality system elements may be required.

A.5 Quality control

Operational techniques and activities that are used to fulfill the requirements for quality.

NOTES
30 Quality control involves operational techniques and activities aimed both at monitoring a process and at eliminating causes of unsatisfactory performance at all stages of the quality loop in order to result in economic effectiveness.

31 Some quality control and quality assurance actions are interrelated.

A.6 Quality assurance

All the planned and systematic activities implemented within the quality system, and demonstrated as needed, to provide adequate confidence that an entity will fulfill requirements for quality.

NOTES
32 There are both internal and external purposes for quality assurance:

   a) Internal quality assurance: within an organization quality assurance provides confidence to the management.

   b) External quality assurance—in contractual or other situations, quality assurance provides confidence to the customers or others.

33 Some quality control and quality assurance actions are interrelated.

34 Unless requirements for quality fully reflect the needs of the user, quality assurance may not provide adequate confidence.

A.7 Quality improvement

Actions taken throughout the organization, to increase the effectiveness and efficiency of activities and processes to provide added benefits to both the organization and its customers.

A.8 Product

Result of activities or processes.

NOTES
35 A product may include service, hardware, processed materials, software, or a combination thereof.

36 A product can be tangible (e.g., assemblies or processed materials) or intangible (e.g., knowledge or concepts), or a combination thereof.

37 Product can be intended (e.g., offering to customers) or unintended (e.g., pollutant or unwanted effects).

A.9 Service

Result generated by activities at the interface between the supplier and the customer and by supplier internal activities, to meet customer needs.

NOTES
38 The supplier or the customer may be represented at the interface by personnel or equipment.

39 Customer activities at the interface with the supplier may be essential to the service delivery.

40 Delivery or use of tangible products may form part of the service delivery.

41 A service may be linked with the manufacture and supply of tangible product.

A.10 Customer

Recipient of a product provided by the supplier.

NOTES
42 In a contractual situation, the customer is called the "purchaser."[1]

43 The customer may be, for example, the ultimate consumer, user, beneficiary, or purchaser.

44 The customer can be either external or internal to the organization.

A.11 Supplier

Organization that provides a product to the customer.

NOTES

45 In a contractual situation, the supplier may be called the "contractor."

46 The supplier may be, for example, the producer, distributor, importer, assembler, or service organization.

47 The supplier can be either external or internal to the organization.

A.12 Process

Set of interrelated resources and activities which transform inputs into outputs.

NOTE

48 Resources may include personnel, finance, facilities, equipment, techniques, and methods.

ANNEX B (INFORMATIVE)

Product and process factors

B.1 Purpose

Product and process characteristics are important in the application of the ISO 9000 family. This annex highlights a number of product and process factors that should be considered, for example:

a) By a supplier's management for quality-management purposes, when planning the approach and extent of implementing a quality-system element (see 7.1).

b) By auditors, when planning first-, second- or third-party audits (see 4.9.3).

c) By supplier and customer jointly when selecting and/or tailoring quality-system requirements for a two-party contract (see 8.4).

NOTE

49 In ANSI/ISO/ASQC Q90-1987, these factors were given as guidance only for purposes.

B.2 Factors

a) Complexity of designing

This factor deals with difficulty of designing the product and designing the production and support processes if they have to be designed, or if the design needs periodic change.

b) Maturity and stability of product designs

This factor deals with the extent to which the total product design is known and proven, either by performance testing or field experience.

c) Production process complexity

This factor deals with

1) The availability of proven production processes.

2) The need for development of new processes.

3) The number and variety of processes required.

4) The impact of the process(es) on the performance of the product.

5) The need for process control.

d) Product characteristics

This factor deals with the complexity of the product, the number of interrelated characteristics, and whether each characteristic is critical to the performance.

e) Product safety

This factor deals with the risk of the occurrence of failure and the consequences of such failure.

f) Economics

This factor deals with the economic costs, to both supplier and customer, of the preceding factors weighed against the risk of costs due to nonconformities in the product.

ANNEX C (INFORMATIVE)

Proliferation of Standards

This ISO 9000 family—in particular, the International Standards for contractual, assessment or certification/registration use (ISO 9001, ISO 9002 and ISO 9003)—are being employed worldwide in many industry/economic sectors for products in all four generic product categories. Various schemes have been developed specific to particular industry/economic sectors.

It is important to distinguish schemes which implement, without change, the ISO 9000 family, from schemes which involve localized versions of these

International Standards. If the ISO 9000 family were to become only the nucleus of a proliferation of localized standards derived from, but varying in content and architecture from, the ISO 9000 family, then there would be little worldwide standardization. Once again, there could be worldwide restraint of trade because of the proliferation of standards and inconsistent requirements.

Fortunately, current global marketplace trends are driving many standards users toward strategic recognition that they need and should conform to International Standards. The International Standards in the ISO 9000 family, and the plans for continuing revision, are

intended to provide the needed scope, content, and flexibility to meet current and emerging marketplace needs in a timely way.

Figure C.1 (below) shows in matrix form which standards-implementation activities are recommended in each of four implementation domains, within the quality management and quality assurance arena. Any third-party assessment and certification scheme should operate under procedures that conform fully to all the International Standards, guides, and practices as required for mutual international recognition of quality-system certification.

Figure C.1   Activity matrix for quality-assurance standards.

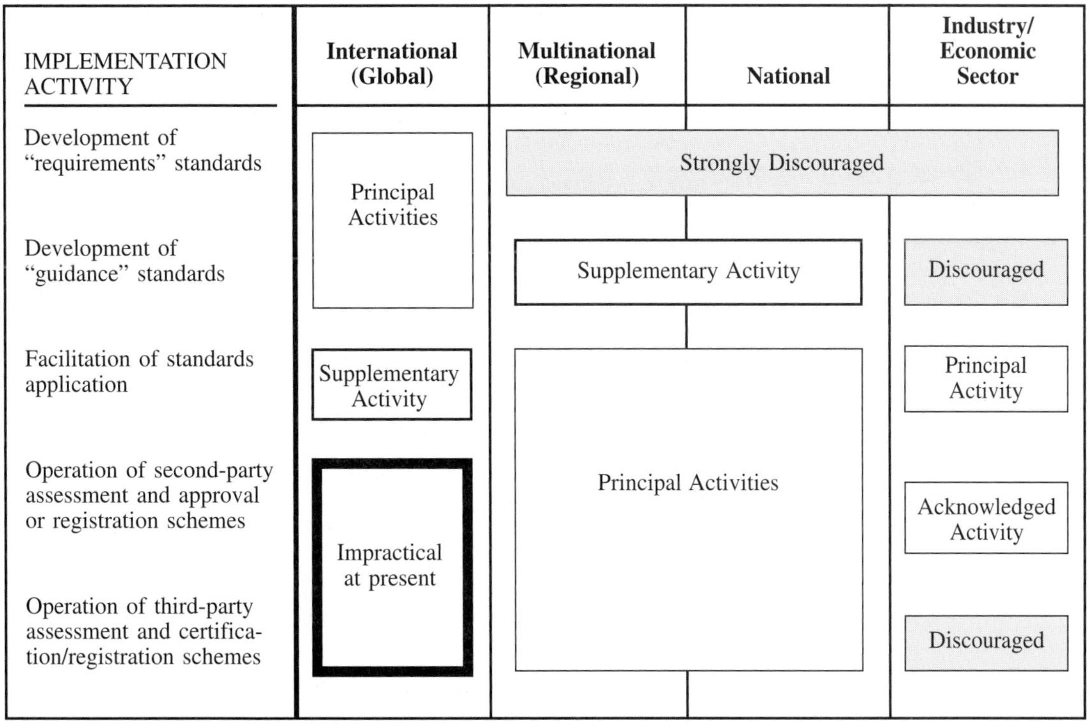

IMPLEMENTATION ACTIVITY	International (Global)	Multinational (Regional)	National	Industry/ Economic Sector
Development of "requirements" standards	Principal Activities	Strongly Discouraged		
Development of "guidance" standards		Supplementary Activity		Discouraged
Facilitation of standards application	Supplementary Activity	Principal Activities		Principal Activity
Operation of second-party assessment and approval or registration schemes	Impractical at present			Acknowledged Activity
Operation of third-party assessment and certification/registration schemes				Discouraged

ANNEX D (INFORMATIVE)

Table D.1 Cross-reference list of clause numbers for corresponding topics.

ANSI/ISO/ASQC Q9001-1994	ANSI/ISO/ASQC Q9002-1994	ANSI/ISO/ASQC Q9003-1994	Clause Title in ANSI/ISO/ASQC Q9001-1994	QM Guidance ANSI/ISO/ASQC Q9004-1 1994	Road Map ANSI/ISO/ASQC Q9001-1 1994

Key:
• = Comprehensive requirement
ø = Less-comprehensive requirement than ANSI/ISO/ASQC Q9001-1994 and ANSI/ISO/ASQC Q9002-1994
§ = Element not present

ANNEX E (INFORMATIVE)

Bibliography

[1] ANSI/ISO/ASQC Q9001-1994, *Quality Systems—Model for Quality Assurance in Design, Development, Production, Installation, and Servicing.*

[2] ANSI/ISO/ASQC Q9002-1994, *Quality Systems—Model for Quality Assurance in Production, Installation, and Servicing.*

[3] ANSI/ISO/ASQC Q9003-1994, *Quality Systems—Model for Quality Assurance in Final Inspection and Test.*

[4] ANSI/ISO/ASQC Q9004-1-1994, *Quality Management and Quality Systems—Guidelines.*

[5] ANSI/ISO/ASQC Q10011-1-1994, *Guidelines for Auditing Quality Systems—Auditing.*

[6] ANSI/ISO/ASQC Q10011-2-1994, *Guidelines for Auditing Quality Systems—Qualification Criteria for Quality Systems Auditors.*

[7] ANSI/ISO/ASQC Q10011-3-1994, *Guidelines for Auditing Quality Systems—Management of Audit Programs.*

[8] ISO 9000-2:1993, *Quality Management and Quality Assurance Standards—Part 2: Generic Guidelines for the Application of ISO 9001, ISO 9002, and ISO 9003.*

[9] ISO 9000-3:1991, *Quality Management and Quality Assurance Standards—Part 3: Guidelines for the Application of ISO 9001 to the Development, Supply and Maintenance of Software.*

[10] ISO 9000-4:1993, *Quality Management and Quality Assurance Standards—Part 4: Guide to Dependability Programme Management.*

[11] ISO 9004-2:1991, *Quality Management and Quality System Elements—Part 2: Guidelines for Services.*

[12] ISO 9004-3:1993, *Quality Management and Quality System Elements—Part 3: Guidelines for Processed Materials.*

[13] ISO 9004-4:1993, *Quality Management and Quality System Elements—Part 4: Guidelines for Quality Improvement.*

[14] ISO 10012-1:1992, *Quality Assurance Requirements for Measuring Equipment—Part 1: Metrological Confirmation System for Measuring Equipment.*

[15] ISO 10013:—[2]*Guidelines for Developing Quality Manuals.*

[16] ISO Handbook 3:1989, *Statistical Methods.*

Endnotes

[1] The recommended harmonized term is "customer" as shown in Table 1 of ANSI/ISO/ASQC Q9000-1-1994. The term "purchaser" was used in ANSI/ASQC Q91-1987, ANSI/ASQC Q92-1987, and ANSI/ASQC Q93-1987.

[2] To be published.

American National Standard

ANSI/ISO/ASQC Q9001-1994

QUALITY SYSTEMS—MODEL FOR QUALITY ASSURANCE IN DESIGN, DEVELOPMENT, PRODUCTION, INSTALLATION, AND SERVICING

[Revision of first edition (ANSI/ASQC Q91-1987)]

Prepared by

American Society for Quality Control
Standards Committee

for

American National Standards Committee
Z-1 on Quality Assurance

Descriptors: Quality assurance, quality assurance program, quality systems, design, development (work), production, installation, after-sales services, reference models.

American National Standards: An American National Standard implies a consensus of those substantially concerned with its scope and provisions. An American National Standard is intended as a guide to aid the manufacturer, the consumer, and the general public. The existence of an American National Standard does not in any respect preclude anyone, whether he or she has approved the standard or not, from manufacturing, purchasing, or using products, processes, or procedures not conforming to the standard. American National Standards are subject to periodic review and users are cautioned to obtain the latest edition.

Caution Notice: This American National Standard may be revised or withdrawn at any time. The procedures

of the American National Standards Institute require that action be taken to reaffirm, revise, or withdraw this standard no later than five years from the date of publication. Purchasers of American National Standards may receive current information on all standards by calling or writing the American National Standards Institute.

ASQC Mission: To facilitate continuous improvement and increase customer satisfaction by identifying, communicating, and promoting the use of quality principles, concepts, and technologies; and thereby be recognized throughout the world as the leading authority on, and champion for, quality.

10 9 8 7 6 5 4 3 2 1

Printed in the United States of America

Printed on acid-free recycled paper

Published by:
ASQC
611 E. Wisconsin Avenue
Milwaukee, WI 53202

CONTENTS

1   Scope
2   Normative reference
3   Definitions
4   Quality-system requirements
4.1   Management responsibility
4.2   Quality system
4.3   Contract review
4.4   Design control

4.5   Document and data control
4.6   Purchasing
4.7   Control of customer-supplied product
4.8   Product identification and traceability
4.9   Process control
4.10   Inspection and testing
4.11   Control of inspection, measuring, and test equipment

4.12   Inspection and test status

4.13   Control of nonconforming product

4.14   Corrective and preventive action

4.15   Handling, storage, packaging, preservation, and delivery

4.16   Control of quality records

4.17   Internal quality audits

4.18   Training

4.19   Servicing

4.20   Statistical techniques

Annex

A   Bibliography

FOREWORD

(This Foreword is not a part of American National Standard *Quality Systems—Model for Quality Assurance in Design, Development, Production, Installation, and Servicing.*)

This American National Standard corresponds to the International Standard ISO 9001:1994. The initial five ISO 9000 series standards, ISO 9000, ISO 9001, ISO 9002, ISO 9003, and ISO 9004, when published in the United States as American National Standards in 1987, were designated as ANSI/ASQC Q90 through ANSI/ASQC Q94 respectively. The five 1987 standards in their 1994 international revisions are now designated ISO 9000-1, ISO 9001, ISO 9002, ISO 9003, and ISO 9004-1 respectively. Their publication as American National Standards are now designated ANSI/ISO/ASQC Q9000-1-1994, ANSI/ISO/ASQC Q9001-1994, ANSI/ISO/ASQC Q9002-1994, ANSI/ISO/ASQC Q9003-1994, and ANSI/ISO/ASQC Q9004-1-1994 respectively. This new numbering system is intended to emphasize the word-for-word correspondence of the International and American National Standards.

ISO (the International Organization for Standardization) is a worldwide federation of national standards bodies (ISO member bodies). The work of preparing International Standards is normally carried out through ISO technical committees. Each member body interested in a subject for which a technical committee has been established has the right to be represented on that committee. International organizations, governmental and nongovernmental, in liaison with ISO, also take part in the work. ISO collaborates closely with the International Electrotechnical Commission (IEC) on all matters of electrotechnical standardization. The American National Standards Institute (ANSI) is the U.S. member body of ISO. ASQC is the U.S. member of ANSI responsible for quality management and related standards.

Users should note that all ANSI/ISO/ASQC standards undergo revision from time to time. In the case of

International Standards adopted as American National Standards, the revision timing is influenced by the international revision timing. Reference herein to any other standard implies the latest American National Standard revision unless otherwise stated.

Comments concerning this standard are welcome. They should be sent to the sponsor of the standard, American Society for Quality Control, 611 E. Wisconsin Avenue, P.O. Box 3005, Milwaukee, WI 53201-3005, c/o Standards Administrator.

INTRODUCTION

This American National Standard is one of three American National Standards dealing with quality-system requirements that can be used for external quality-assurance purposes. The quality-assurance models, set out in the three American National Standards listed below, represent three distinct forms of quality-system requirements suitable for the purpose of a supplier demonstrating its capability, and for the assessment of the capability of a supplier by external parties.

a) *ANSI/ISO/ASQC Q9001-1994, Quality Systems—Model for Quality Assurance in Design, Development, Production, Installation, and Servicing*

For use when conformance to specified requirements is to be assured by the supplier during design, development, production, installation, and servicing.

b) ANSI/ISO/ASQC Q9002-1994, *Quality Systems—Model for Quality Assurance in Production, Installation, and Servicing*

For use when conformance to specified requirements is to be assured by the supplier during production, installation, and servicing.

c) ANSI/ISO/ASQC Q9003-1994, *Quality Systems—Model for Quality Assurance in Final Inspection and Test*

For use when conformance to specified requirements is to be assured by the supplier solely at final inspection and test.

It is emphasized that the quality-system requirements specified in this American National Standard, ANSI/ISO/ASQC Q9002-1994, and ANSI/ISO/ASQC Q9003-1994 are complementary (not alternative) to the technical (product) specified requirements. They specify requirements which determine what elements quality systems have to encompass, but it is not the purpose of these American National Standards to enforce uniformity of quality systems. They are generic and independent of any specific industry or economic sector. The design and implementation of a quality system will be influenced by the varying needs of an organization, its particular objectives, the products and services supplied, and the processes and specific practices employed.

It is intended that these American National Standards will be adopted in their present form, but on occasions they may need to be tailored by adding or deleting certain quality-system requirements for specific contractual situations. ANSI/ISO/ASQC Q9000-1-1994 provides guidance on such tailoring as well as on selection of the appropriate quality-assurance model, *viz.* ANSI/ISO/ASQC Q9001-1994, ANSI/ISO/ASQC Q9002-1994, or ANSI/ISO/ASQC Q9003-1994.

Quality Systems—Model for Quality Assurance in Design, Development, Production, Installation, and Servicing

1 Scope

This American National Standard specifies quality-system requirements for use where a supplier's capability to design and supply conforming product needs to be demonstrated.

The requirements specified are aimed primarily at achieving customer satisfaction by preventing nonconformity at all stages from design through to servicing.

This American National Standard is applicable in situations when:

a) design is required and the product requirements are stated principally in performance terms, or they need to be established, and

b) confidence in product conformance can be attained by adequate demonstration of a

supplier's capabilities in design, development, production, installation, and servicing.

NOTE
1 For informative references, see Annex A.

2 Normative reference

The following standard contains provisions which, through reference in this text, constitute provisions of this American National Standard. At the time of publication, the edition indicated was valid. All standards are subject to revision, and parties to agreements based on this American National Standard are encouraged to investigate the possibility of applying the most recent edition of the standard indicated below. The American National Standards Institute and members of IEC and ISO maintain registers of currently valid American National Standards and International Standards.

ISO 8402:1994, *Quality Management and Quality Assurance—Vocabulary.*

3 Definitions

For the purposes of this American National Standard, the definitions given in ISO 8402 and the following definitions apply.

3.1 Product

Result of activities or processes.

NOTES

2 *A product may include service, hardware, processed materials, software, or a combination thereof.*

3 *A product can be tangible (e.g., assemblies or processed materials) or intangible (e.g., knowledge or concepts), or a combination thereof.*

4 *For the purposes of this American National Standard, the term "product" applies to the intended product offering only and not to unintended "by-products" affecting the environment. This differs from the definition given in ISO 8402.*

3.2 Tender

Offer made by a supplier in response to an invitation to satisfy a contract award to provide product.

3.3 Contract; accepted order

Agreed requirements between a supplier and customer transmitted by any means.

4 Quality-system requirements

4.1 Management responsibility

4.1.1 Quality policy

The supplier's management with executive responsibility shall define and document its policy for quality, including objectives for quality and its commitment to quality. The quality policy shall be relevant to the supplier's organizational goals and the expectations and needs of its customers. The supplier shall ensure that this policy is understood, implemented, and maintained at all levels of the organization.

4.1.2 Organization
4.1.2.1 Responsibility and authority

The responsibility, authority, and the interrelation of personnel who manage, perform, and verify work affecting quality shall be defined and documented, particularly for personnel who need the organizational freedom and authority to:

a) initiate action to prevent the occurrence of any nonconformities relating to product, process, and quality system;

b) identify and record any problems relating to the product, process, and quality system;

c) initiate, recommend, or provide solutions through designated channels;

d) verify the implementation of solutions;

e) control further processing, delivery, or installation of nonconforming product until the deficiency or unsatisfactory condition has been corrected.

4.1.2.2 Resources

The supplier shall identify resource requirements and provide adequate resources, including the assignment of trained personnel (see 4.18), for management, performance of work, and verification activities including internal quality audits.

4.1.2.3 Management representative

The supplier's management with executive responsibility shall appoint a member of the supplier's own management who, irrespective of other responsibilities, shall have defined authority for

a) Ensuring that a quality system is established, implemented, and maintained in accordance with this American National Standard.

b) Reporting on the performance of the quality system to the supplier's management for review and as a basis for improvement of the quality system.

NOTE

  5 The responsibility of a management representative may also include liaison with external parties on matters relating to the supplier's quality system.

4.1.3 Management review

The supplier's management with executive responsibility shall review the quality system at defined intervals sufficient to ensure its continuing suitability and effectiveness in satisfying the requirements of this American National Standard and the supplier's stated quality policy and objectives (see 4.1.1). Records of such reviews shall be maintained (see 4.16).

4.2 Quality system

4.2.1 General

The supplier shall establish, document, and maintain a quality system as a means of ensuring that product conforms to specified requirements. The supplier shall prepare a quality manual covering the requirements of this American National Standard. The quality manual shall include or make reference to the quality-system procedures and outline the structure of the documentation used in the quality system.

NOTE

  6 Guidance on quality manuals is given in ISO 10013.

4.2.2 *Quality-system procedures*

The supplier shall

a) Prepare documented procedures consistent with the requirements of this American National Standard and the supplier's stated quality policy.

b) Effectively implement the quality system and its documented procedures.

For the purposes of this American National Standard, the range and detail of the procedures that form part of the quality system depend on the complexity of the work, the methods used, and the skills and training needed by personnel involved in carrying out the activity.

NOTE

7   Documented procedures may make reference to work instructions that define how an activity is performed.

4.2.3 *Quality planning*

The supplier shall define and document how the requirements for quality will be met. Quality planning shall be consistent with all other requirements of a supplier's quality system and shall be documented in a format to suit the supplier's method of operation. The supplier shall give consideration to the following activities, as appropriate, in meeting the specified requirements for products, projects, or contracts:

a) The preparation of quality plans.

b) The identification and acquisition of any controls, processes, equipment (including inspection and test equipment), fixtures, resources, and skills that may be needed to achieve the required quality.

c) Ensuring the compatibility of the design, the production process, installation, servicing, inspection, and test procedures, and the applicable documentation.

d) The updating, as necessary, of quality control, inspection, and testing techniques, including the development of new instrumentation.

e) The identification of any measurement requirement involving capability that exceeds the known state of the art, in sufficient time for the needed capability to be developed.

f) The identification of suitable verification at appropriate stages in the realization of product.

g) The clarification of standards of acceptability for all features and requirements, including those which contain a subjective element.

h) The identification and preparation of quality records (see 4.16).

NOTE

8   The quality plans referred to (see 4.2.3a) may be in the form of a reference to the appropriate documented procedures that form an integral part of the supplier's quality system.

4.3 Contract review

4.3.1 *General*

The supplier shall establish and maintain documented procedures for contract review and for the coordination of these activities.

4.3.2 *Review*

Before submission of a tender, or at the acceptance of a contract or order (statement of requirement), the tender, contract, or order shall be reviewed by the supplier to ensure the following:

a) The requirements are adequately defined and documented; where no written statement of requirement is available for an order received by verbal means, the supplier shall ensure that the order requirements are agreed before their acceptance.

b) Any differences between the contract or accepted order requirements and those in the tender are resolved.

c) The supplier has the capability to meet the contract or accepted order requirements.

4.3.3 *Amendment to contract*

The supplier shall identify how an amendment to a contract is made and correctly transferred to the functions concerned within the supplier's organization.

4.3.4 *Records*

Records of contract reviews shall be maintained (see 4.16).

NOTE
9 Channels for communication and interfaces with the customer's organization in these contract matters should be established.

4.4 Design control

4.4.1 General
The supplier shall establish and maintain documented procedures to control and verify the design of the product in order to ensure that the specified requirements are met.

4.4.2 Design and development planning
The supplier shall prepare plans for each design and development activity. The plans shall describe or reference these activities, and define responsibility for their implementation. The design and development activities shall be assigned to qualified personnel equipped with adequate resources. The plans shall be updated, as the design evolves.

4.4.3 Organizational and technical interfaces
Organizational and technical interfaces between different groups which input into the design process shall be defined and the necessary information documented, transmitted, and regularly reviewed.

4.4.4 Design input
Design-input requirements relating to the product, including applicable statutory and regulatory requirements, shall be identified, documented, and their selection reviewed by the supplier for adequacy. Incomplete, ambiguous, or conflicting requirements shall be resolved with those responsible for imposing these requirements.

Design input shall take into consideration the results of any contract-review activities.

4.4.5 Design output
Design output shall be documented and expressed in terms that can be verified against design-input requirements and validated (see 4.4.8).

Design output shall do the following:
a) Meet the design-input requirements.
b) Contain or make reference to acceptance criteria.

c) Identify those characteristics of the design that are crucial to the safe and proper functioning of the product (e.g., operating, storage, handling, maintenance, and disposal requirements).

Design-output documents shall be reviewed before release.

4.4.6 Design review
At appropriate stages of design, formal documented reviews of the design results shall be planned and conducted. Participants at each design review shall include representatives of all functions concerned with the design stage being reviewed, as well as other specialist personnel, as required. Records of such reviews shall be maintained (see 4.16).

4.4.7 Design verification
At appropriate stages of design, design verification shall be performed to ensure that the design-stage output meets the design-stage input requirements. The design-verification measures shall be recorded (see 4.16).

NOTE
10 In addition to conducting design reviews (see 4.4.6), design verification may include activities such as

- Performing alternative calculations.
- Comparing the new design with a similar proven design, if available.
- Undertaking tests and demonstrations.
- Reviewing the design-stage documents before release.

4.4.8 Design validation
Design validation shall be performed to ensure that product conforms to defined user needs and/or requirements.

NOTES
11 Design validation follows successful design verification (see 4.4.7).
12 Validation is normally performed under defined operating conditions.

13 Validation is normally performed on the final product, but may be necessary in earlier stages prior to product completion.

14 Multiple validations may be performed if there are different intended uses.

4.4.9 Design changes

All design changes and modifications shall be identified, documented, reviewed, and approved by authorized personnel before their implementation.

4.5 Document and data control

4.5.1 General

The supplier shall establish and maintain documented procedures to control all documents and data that relate to the requirements of this American National Standard including, to the extent applicable, documents of external origin such as standards and customer drawings.

NOTE

15 Documents and data can be in the form of any type of media, such as hard copy or electronic media.

4.5.2 Document and data approval and issue

The documents and data shall be reviewed and approved for adequacy by authorized personnel prior to issue. A master list or equivalent document-control procedure identifying the current revision status of documents shall be established and be readily available to preclude the use of invalid and/or obsolete documents.

This control shall ensure that:

a) The pertinent issues of appropriate documents are available at all locations where operations essential to the effective functioning of the quality system are performed.

b) Invalid and/or obsolete documents are promptly removed from all points of issue or use, or otherwise assured against unintended use.

c) Any obsolete documents retained for legal and/or knowledge-preservation purposes are suitably identified.

4.5.3 Document and data changes

Changes to documents and data shall be reviewed and approved by the same functions/organizations that performed the original review and approval, unless specifically designated otherwise. The designated functions/organizations shall have access to pertinent background information upon which to base their review and approval.

Where practicable, the nature of the change shall be identified in the document or the appropriate attachments.

4.6 Purchasing

4.6.1 General

The supplier shall establish and maintain documented procedures to ensure that purchased product (see 3.1) conforms to specified requirements.

4.6.2 Evaluation of subcontractors

The supplier shall do the following:

a) Evaluate and select subcontractors on the basis of their ability to meet subcontract requirements including the quality system and any specific quality-assurance requirements.

b) Define the type and extent of control exercised by the supplier over subcontractors. This shall be dependent upon the type of product, the impact of subcontracted product on the quality of final product, and, where applicable, on the quality audit reports and/or quality records of the previously demonstrated capability and performance of subcontractors.

c) Establish and maintain quality records of acceptable subcontractors (see 4.16).

4.6.3 Purchasing data

Purchasing documents shall contain data clearly describing the product ordered, including (where applicable):

a) The type, class, grade, or other precise identification.

b) The title or other positive identification, and applicable issues of specifications, drawings, process requirements, inspection instructions, and other relevant technical data, including requirements for approval or qualification of product, procedures, process equipment, and personnel.

c) The title, number, and issue of the quality-system standard to be applied.

The supplier shall review and approve purchasing documents for adequacy of the specified requirements prior to release.

4.6.4 Verification of purchased product
4.6.4.1 Supplier verification at subcontractor's premises
Where the supplier proposes to verify purchased product at the subcontractor's premises, the supplier shall specify verification arrangements and the method of product release in the purchasing documents.

4.6.4.2 Customer verification of subcontracted product
Where specified in the contract, the supplier's customer or the customer's representative shall be afforded the right to verify at the subcontractor's premises and the supplier's premises that subcontracted product conforms to specified requirements. Such verification shall not be used by the supplier as evidence of effective control of quality by the subcontractor.

Verification by the customer shall not absolve the supplier of the responsibility to provide acceptable product, nor shall it preclude subsequent rejection by the customer.

4.7 Control of customer-supplied product

The supplier shall establish and maintain documented procedures for the control of verification, storage, and maintenance of customer-supplied product provided for incorporation into the supplies or for related activities. Any such product that is lost, damaged, or is otherwise unsuitable for use shall be recorded and reported to the customer (see 4.16).

Verification by the supplier does not absolve the customer of the responsibility to provide acceptable product.

4.8 Product identification and traceability

Where appropriate, the supplier shall establish and maintain documented procedures for identifying the product by suitable means from receipt and during all stages of production, delivery, and installation.

Where and to the extent that traceability is a specified requirement, the supplier shall establish and maintain documented procedures for unique identification of individual product or batches. This identification shall be recorded (see 4.16).

4.9 Process control

The supplier shall identify and plan the production, installation, and servicing processes which directly affect quality and shall ensure that these processes are carried out under controlled conditions. Controlled conditions shall include the following:

a) Documented procedures defining the manner of production, installation, and servicing, where the absence of such procedures could adversely affect quality.

b) Use of suitable production, installation, and servicing equipment, and a suitable working environment.

c) Compliance with reference standards/codes, quality plans, and/or documented procedures.

d) Monitoring and control of suitable process parameters and product characteristics.

e) The approval of processes and equipment, as appropriate.

f) Criteria for workmanship, which shall be stipulated in the clearest practical manner (e.g., written standards, representative samples, or illustrations).

g) Suitable maintenance of equipment to ensure continuing process capability.

Where the results of processes cannot be fully verified by subsequent inspection and testing of the product and where, for example, processing deficiencies may become apparent only after the product is in use, the processes shall be carried out by qualified operators and/or shall require continuous monitoring and control of process parameters to ensure that the specified requirements are met.

The requirements for any qualification of process operations, including associated equipment and personnel (see 4.18), shall be specified.

NOTE

16 Such processes requiring prequalification of their process capability are frequently referred to as special processes.

Records shall be maintained for qualified processes, equipment, and personnel, as appropriate (see 4.16).

4.10 Inspection and testing

4.10.1 General
The supplier shall establish and maintain documented procedures for inspection and testing activities in order to verify that the specified requirements for the product are met. The required inspection and testing, and the records to be established, shall be detailed in the quality plan or documented procedures.

4.10.2 Receiving inspection and testing
4.10.2.1
The supplier shall ensure that incoming product is not used or processed (except in the circumstances described in 4.10.2.3) until it has been inspected or otherwise verified as conforming to specified requirements. Verification of the specified requirements shall be in accordance with the quality plan and/or documented procedures.

4.10.2.2
In determining the amount and nature of receiving inspection, consideration shall be given to the amount of control exercised at the subcontractor's premises and the recorded evidence of conformance provided.

4.10.2.3
Where incoming product is released for urgent production purposes prior to verification, it shall be positively identified and recorded (see 4.16) in order to permit immediate recall and replacement in the event of nonconformity to specified requirements.

4.10.3 In-process inspection and testing
The supplier shall do the following:

    a) Inspect and test the product as required by the quality plan and/or documented procedures.

    b) Hold the product until the required inspection and tests have been completed or necessary reports have been received and verified, except when product is released under positive-recall procedures (see 4.10.2.3). Release under positive-recall procedures shall not preclude the activities outlined in 4.10.3a.

4.10.4 Final inspection and testing
The supplier shall carry out all final inspection and testing in accordance with the quality plan and/or documented procedures to complete the evidence of conformance of the finished product to the specified requirements.

The quality plan and/or documented procedures for final inspection and testing shall require that all specified inspection and tests, including those specified either on receipt of product or in-process, have been carried out and that the results meet specified requirements.

No product shall be dispatched until all the activities specified in the quality plan and/or documented procedures have been satisfactorily completed and the associated data and documentation are available and authorized.

4.10.5 Inspection and test records
The supplier shall establish and maintain records which provide evidence that the product has been inspected and/or tested. These records shall show clearly whether the product has passed or failed the inspections and/or tests according to defined acceptance criteria. Where the product fails to pass any inspection and/or test, the procedures for control of nonconforming product shall apply (see 4.13).

Records shall identify the inspection authority responsible for the release of product (see 4.16).

4.11 Control of inspection, measuring, and test equipment

4.11.1 General
The supplier shall establish and maintain documented procedures to control, calibrate, and maintain inspection, measuring, and test equipment (including test software) used by the supplier to demonstrate the conformance of product to the specified requirements. Inspection, measuring, and test equipment shall be

used in a manner which ensures that the measurement uncertainty is known and is consistent with the required measurement capability.

Where test software or comparative references such as test hardware are used as suitable forms of inspection, they shall be checked to prove that they are capable of verifying the acceptability of product, prior to release for use during production, installation, or servicing, and shall be rechecked at prescribed intervals. The supplier shall establish the extent and frequency of such checks and shall maintain records as evidence of control (see 4.16).

Where the availability of technical data pertaining to the measurement equipment is a specified requirement, such data shall be made available, when required by the customer or customer's representative, for verification that the measuring equipment is functionally adequate.

NOTE

17 For the purposes of this American National Standard, the term "measuring equipment" includes measurement devices.

4.11.2 Control procedure
The supplier shall do the following:

a) Determine the measurements to be made and the accuracy required, and select the appropriate inspection, measuring, and test equipment that is capable of the necessary accuracy and precision.

b) Identify all inspection, measuring, and test equipment that can affect product quality, and calibrate and adjust them at prescribed intervals, or prior to use, against certified equipment having a known valid relationship to internationally or nationally recognized standards. Where no such standards exist, the basis used for calibration shall be documented;

c) Define the process employed for the calibration of inspection, measuring, and test equipment, including details of equipment type, unique identification, location, frequency of checks, check method, acceptance criteria, and the action to be taken when results are unsatisfactory.

d) Identify inspection, measuring, and test equipment with a suitable indicator or approved identification record to show the calibration status.

e) Maintain calibration records for inspection, measuring, and test equipment (see 4.16).

f) Assess and document the validity of previous inspection and test results when inspection, measuring, and test equipment is found to be out of calibration.

g) Ensure that the environmental conditions are suitable for the calibrations, inspections, measurements, and tests being carried out.

h) Ensure that the handling, preservation, and storage of inspection, measuring, and test equipment is such that the accuracy and fitness for use are maintained.

i) Safeguard inspection, measuring, and test facilities, including both test hardware and test software, from adjustments which would invalidate the calibration setting.

NOTE

18 The metrological confirmation system for measuring equipment given in ISO 10012 may be used for guidance.

4.12 Inspection and test status

The inspection and test status of product shall be identified by suitable means, which indicate the conformance or nonconformance of product with regard to inspection and tests performed. The identification of inspection and test status shall be maintained, as defined in the quality plan and/or documented procedures, throughout production, installation, and servicing of the product to ensure that only product that has passed the required inspections and tests [or released under an authorized concession (see 4.13.2)] is dispatched, used, or installed.

4.13 Control of nonconforming product

4.13.1 General
The supplier shall establish and maintain documented procedures to ensure that product that does not conform to specified requirements is prevented from unintended use or installation. This control shall provide for identification, documentation, evaluation, segregation (when practical), disposition of nonconforming product, and for notification to the functions concerned.

4.13.2 Review and disposition of nonconforming product

The responsibility for review and authority for the disposition of nonconforming product shall be defined.

Nonconforming product shall be reviewed in accordance with documented procedures. It may be

a) reworked to meet the specified requirements,

b) accepted with or without repair by concession,

c) regraded for alternative applications, or

d) rejected or scrapped.

Where required by the contract, the proposed use or repair of product (see 4.13.2b) which does not conform to specified requirements shall be reported for concession to the customer or customer's representative. The description of the nonconformity that has been accepted, and of repairs, shall be recorded to denote the actual condition (see 4.16).

Repaired and/or reworked product shall be reinspected in accordance with the quality plan and/or documented procedures.

4.14 Corrective and preventive action

4.14.1 General

The supplier shall establish and maintain documented procedures for implementing corrective and preventive action.

Any corrective or preventive action taken to eliminate the causes of actual or potential nonconformities shall be to a degree appropriate to the magnitude of problems and commensurate with the risks encountered.

The supplier shall implement and record any changes to the documented procedures resulting from corrective and preventive action.

4.14.2 Corrective action

The procedures for corrective action shall include:

a) The effective handling of customer complaints and reports of product nonconformities.

b) Investigation of the cause of nonconformities relating to product, process, and quality system, and recording the results of the investigation (see 4.16).

c) Determination of the corrective action needed to eliminate the cause of nonconformities.

d) Application of controls to ensure that corrective action is taken and that it is effective.

4.14.3 Preventive action

The procedures for preventive action shall include:

a) The use of appropriate sources of information such as processes and work operations which affect product quality, concessions, audit results, quality records, service reports, and customer complaints to detect, analyze, and eliminate potential causes of nonconformities.

b) Determination of the steps needed to deal with any problems requiring preventive action.

c) Initiation of preventive action and application of controls to ensure that it is effective.

d) Confirmation that relevant information on actions taken is submitted for management review (see 4.1.3).

4.15 Handling, storage, packaging, preservation, and delivery

4.15.1 General

The supplier shall establish and maintain documented procedures for handling, storage, packaging, preservation, and delivery of product.

4.15.2 Handling

The supplier shall provide methods of handling product that prevent damage or deterioration.

4.15.3 Storage

The supplier shall use designated storage areas or stock rooms to prevent damage or deterioration of product, pending use or delivery. Appropriate methods for authorizing receipt to and dispatch from such areas shall be stipulated.

In order to detect deterioration, the condition of product in stock shall be assessed at appropriate intervals.

4.15.4 Packaging

The supplier shall control packing, packaging, and marking processes (including materials used) to the extent necessary to ensure conformance to specified requirements.

4.15.5 Preservation

The supplier shall apply appropriate methods for preservation and segregation of product when the product is under the supplier's control.

4.15.6 Delivery

The supplier shall arrange for the protection of the quality of product after final inspection and test. Where contractually specified, this protection shall be extended to include delivery to destination.

4.16 Control of quality records

The supplier shall establish and maintain documented procedures for identification, collection, indexing, access, filing, storage, maintenance, and disposition of quality records.

Quality records shall be maintained to demonstrate conformance to specified requirements and the effective operation of the quality system. Pertinent quality records from the subcontractor shall be an element of these data.

All quality records shall be legible and shall be stored and retained in such a way that they are readily retrievable in facilities that provide a suitable environment to prevent damage or deterioration and to prevent loss. Retention times of quality records shall be established and recorded. Where agreed contractually, quality records shall be made available for evaluation by the customer or the customer's representative for an agreed period.

NOTE
19 Records may be in the form of any type of media, such as hard copy or electronic media.

4.17 Internal quality audits

The supplier shall establish and maintain documented procedures for planning and implementing internal quality audits to verify whether quality activities and related results comply with planned arrangements and to determine the effectiveness of the quality system.

Internal quality audits shall be scheduled on the basis of the status and importance of the activity to be audited and shall be carried out by personnel independent of those having direct responsibility for the activity being audited.

The results of the audits shall be recorded (see 4.16) and brought to the attention of the personnel having responsibility in the area audited. The management personnel responsible for the area shall take timely corrective action on deficiencies found during the audit.

Follow-up audit activities shall verify and record the implementation and effectiveness of the corrective action taken (see 4.16).

NOTES
20 The results of internal quality audits form an integral part of the input to management review activities (see 4.1.3).

21 Guidance on quality-system audits is given in ANSI/ISO/ASQC Q10011-1-1994, ANSI/ISO/ASQC Q10011-2-1994, and ANSI/ISO/ASQC Q10011-3-1994.

4.18 Training

The supplier shall establish and maintain documented procedures for identifying training needs and provide for the training of all personnel performing activities affecting quality. Personnel performing specific assigned tasks shall be qualified on the basis of appropriate education, training, and/or experience, as required. Appropriate records of training shall be maintained (see 4.16).

4.19 Servicing

Where servicing is a specified requirement, the supplier shall establish and maintain documented procedures for performing, verifying, and reporting that the servicing meets the specified requirements.

4.20 Statistical techniques

4.20.1 Identification of need

The supplier shall identify the need for statistical techniques required for establishing, controlling, and verifying process capability and product characteristics.

4.20.2 Procedures

The supplier shall establish and maintain documented procedures to implement and control the application of the statistical techniques identified in 4.20.1.

ANNEX A (INFORMATIVE)

Bibliography

[1] ANSI/ISO/ASQC Q9000-1-1994, *Quality Management and Quality Assurance Standards—Guidelines for Selection and Use.*

[2] ANSI/ISO/ASQC Q9002-1994, *Quality Systems—Model for Quality Assurance in Production, Installation, and Servicing.*

[3] ANSI/ISO/ASQC Q9003-1994, *Quality Systems—Model for Quality Assurance in Final Inspection and Test.*

[4] ANSI/ISO/ASQC Q10011-1-1994, *Guidelines for Auditing Quality Systems—Auditing.*

[5] ANSI/ISO/ASQC Q10011-2-1994, *Guidelines for Auditing Quality Systems—Qualification Criteria for Quality Systems Auditors.*

[6] ANSI/ISO/ASQC Q10011-3-1994, *Guidelines for Auditing Quality Systems—Management of Audit Programs.*

[7] ISO 9000-2:1993, *Quality Management and Quality Assurance Standards—Part 2: Generic Guidelines for the Application of ISO 9001, ISO 9002 and ISO 9003.*

[8] ISO 9000-3:1991, *Quality Management and Quality Assurance Standards—Part 3: Guidelines for the Application of ISO 9001 to the Development, Supply and Maintenance of Software.*

[9] ISO 10012-1:1992, *Quality Assurance Requirements for Measuring Equipment—Part 1: Metrological Confirmation System for Measuring Equipment.*

[10] ISO 10013:—[1] *Guidelines for Developing Quality Manuals.*

Endnote

[1] To be published.

American National Standard

ANSI/ISO/ASQC Q9002-1994

QUALITY SYSTEMS—MODEL FOR QUALITY ASSURANCE IN PRODUCTION, INSTALLATION, AND SERVICING

[Revision of first edition (ANSI/ASQC Q92-1987)]

Prepared by

American Society for Quality Control
Standards Committee

for

American National Standards Committee
Z-1 on Quality Assurance

Descriptors: Quality assurance, quality assurance program, quality systems, production, installation, after-sale services, reference models.

American National Standards: An American National Standard implies a consensus of those substantially concerned with its scope and provisions. An American National Standard is intended as a guide to aid the manufacturer, the consumer, and the general public. The existence of an American National Standard does not in any respect preclude anyone, whether he or she has approved the standard or not, from manufacturing, purchasing, or using products, processes, or procedures not conforming to the standard. American National Standards are subject to periodic review and users are cautioned to obtain the latest edition.

Caution Notice: This American National Standard may be revised or withdrawn at any time. The procedures of the American National Standards Institute require

that action be taken to reaffirm, revise, or withdraw this standard no later than five years from the date of publication. Purchasers of American National Standards may receive current information on all standards by calling or writing the American National Standards Institute.

ASQC Mission: To facilitate continuous improvement and increase customer satisfaction by identifying, communicating, and promoting the use of quality principles, concepts, and technologies; and thereby be recognized throughout the world as the leading authority on, and champion for, quality.

10 9 8 7 6 5 4 3 2 1

Printed in the United States of America

Printed on acid-free recycled paper

Published by:
ASQC
611 E. Wisconsin Avenue
Milwaukee, WI 53202

CONTENTS

1   Scope
2   Normative reference
3   Definitions
4   Quality-system requirements
4.1   Management responsibility
4.2   Quality system
4.3   Contract review
4.4   Design control
4.5   Document and data control
4.6   Purchasing

4.7   Control of customer-supplied product
4.8   Product identification and traceability
4.9   Process control
4.10   Inspection and testing
4.11   Control of inspection, measuring, and test equipment
4.12   Inspection and test status
4.13   Control of nonconforming product
4.14   Corrective and preventive action
4.15   Handling, storage, packaging, preservation, and delivery
4.16   Control of quality records

4.17 Internal quality audits

4.18 Training

4.19 Servicing

4.20 Statistical techniques

FOREWORD

(This Foreword is not a part of American National Standard *Quality Systems—Model for Quality Assurance in Production, Installation, and Servicing*.)

This American National Standard corresponds to the International Standard ISO 9002:1994. The initial five ISO 9000 series standards, ISO 9000, ISO 9001, ISO 9002, ISO 9003, and ISO 9004, when published in the United States as American National Standards in 1987, were designated as ANSI/ASQC Q90 through ANSI/ASQC Q94 respectively. The five 1987 standards in their 1994 international revisions are now designated ISO 9000-1, ISO 9001, ISO 9002, ISO 9003, and ISO 9004-1 respectively. Their publication as American National Standards are now designated ANSI/ISO/ASQC Q9000-1-1994, ANSI/ISO/ASQC Q9001-1994, ANSI/ISO/ASQC Q9002-1994, ANSI/ISO/ASQC Q9003-1994, and ANSI/ISO/ASQC Q9004-1-1994 respectively. This new numbering system is intended to emphasize the word-for-word correspondence of the International and American National Standards.

ISO (the International Organization for Standardization) is a worldwide federation of national standards bodies (ISO member bodies). The work of preparing International Standards is normally carried out through ISO technical committees. Each member body interested in a subject for which a technical committee has been established has the right to be represented on that committee. International organizations, governmental and nongovernmental, in liaison with ISO, also take part in the work. ISO collaborates closely with the International Electrotechnical Commission (IEC) on all matters of electrotechnical standardization. The American National Standards Institute (ANSI) is the U.S. member body of ISO. ASQC is the U.S. member of ANSI responsible for quality management and related standards.

Users should note that all ANSI/ISO/ASQC standards undergo revision from time to time. In the case of International Standards adopted as American National Standards, the revision timing is influenced by the international revision timing. Reference herein to any other standard implies the latest American National Standard revision unless otherwise stated.

Comments concerning this standard are welcome. They should be sent to the sponsor of the standard, American Society for Quality Control, 611 E. Wisconsin Avenue, P.O. Box 3005, Milwaukee, WI 53201-3005, c/o Standards Administrator.

Annex

A   Bibliography

INTRODUCTION

This American National Standard is one of three American National Standards dealing with quality-system requirements that can be used for external quality-assurance purposes. The quality-assurance models, set out in the three American National Standards listed below, represent three distinct forms of quality-system requirements suitable for the purpose of a supplier demonstrating its capability, and for the assessment of the capability of a supplier by external parties.

a) ANSI/ISO/ASQC Q9001-1994, *Quality Systems—Model for Quality Assurance in Design, Development, Production, Installation, and Servicing*

For use when conformance to specified requirements is to be assured by the supplier during design, development, production, installation, and servicing.

b) ANSI/ISO/ASQC Q9002-1994, *Quality Systems—Model for Quality Assurance in Production, Installation, and Servicing*

For use when conformance to specified requirements is to be assured by the supplier during production, installation, and servicing.

c) ANSI/ISO/ASQC Q9003-1994, *Quality Systems—Model for Quality Assurance in Final Inspection and Test*

For use when conformance to specified requirements is to be assured by the supplier solely at final inspection and test.

It is emphasized that the quality-system requirements specified in this American National Standard, ANSI/ISO/ASQC Q9001-1994 and ANSI/ISO/ASQC Q9003-1994 are complementary (not alternative) to the technical (product) specified requirements. They specify requirements which determine what elements quality systems have to encompass, but it is not the purpose of these American National Standards to enforce uniformity of quality systems. They are generic and independent of any specific industry or economic sector. The design and implementation of a quality system will be influenced by the varying needs of an organization, its particular objectives, the products and services supplied, and the processes and specific practices employed.

It is intended that these American National Standards will be adopted in their present form, but on occasions they may need to be tailored by adding or deleting certain quality-system requirements for specific contractual situations. ANSI/ISO/ASQC Q9000-1-1994 provides guidance on such tailoring as well as on selection of the appropriate quality-assurance model, viz. ANSI/ISO/ASQC Q9001-1994, ANSI/ISO/ASQC Q9002-1994, or ANSI/ISO/ASQC Q9003-1994.

Quality Systems—Model for Quality Assurance in Production, Installation, and Servicing

1 Scope

This American National Standard specifies quality-system requirements for use where a supplier's capability to supply conforming product to an established design needs to be demonstrated.

The requirements specified are aimed primarily at achieving customer satisfaction by preventing nonconformity at all stages from production through to servicing.

This American National Standard is applicable in situations when

  a) the specified requirements for product are stated in terms of an established design or specification, and

  b) confidence in product conformance can be attained by adequate demonstration of a supplier's capabilities in production, installation, and servicing.

NOTE
  1 For informative references, see Annex A.

2 Normative reference

The following standard contains provisions which, through reference in this text, constitute provisions of this American National Standard. At the time of publication, the edition indicated was valid. All standards are subject to revision, and parties to agreements based on this American National Standard are encouraged to investigate the possibility of applying the most recent edition of the standard indicated below. The American National Standards Institute and members of IEC and ISO maintain registers of currently valid American National Standards and International Standards.

ISO 8402:1994, *Quality Management and Quality Assurance—Vocabulary*

3 Definitions

For the purposes of this American National Standard, the definitions given in ISO 8402 and the following definitions apply.

3.1 Product

Result of activities or processes.

NOTES
  2 A product may include service, hardware, processed materials, software, or a combination thereof.

  3 A product can be tangible (e.g., assemblies or processed materials) or intangible (e.g., knowledge or concepts), or a combination thereof.

  4 For the purposes of this American National Standard, the term "product" applies to the intended product offering only and not to unintended "by-products" affecting the environment. This differs from the definition given in ISO 8402.

3.2 Tender

Offer made by a supplier in response to an invitation to satisfy a contract award to provide product.

3.3 Contract; accepted order

Agreed requirements between a supplier and customer transmitted by any means.

4 Quality-system requirements

4.1 Management responsibility

4.1.1 Quality policy

The supplier's management with executive responsibility shall define and document its policy for quality, including objectives for quality and its commitment to quality. The quality policy shall be relevant to the supplier's organizational goals and the expectations and needs of its customers. The supplier shall ensure that this policy is understood, implemented, and maintained at all levels of the organization.

4.1.2 Organization
4.1.2.1 Responsibility and authority

The responsibility, authority, and the interrelation of personnel who manage, perform, and verify work affecting quality shall be defined and documented, particularly for personnel who need the organizational freedom and authority to do the following:

a) Initiate action to prevent the occurrence of any nonconformities relating to product, process, and quality system.

b) Identify and record any problems relating to the product, process, and quality system.

c) Initiate, recommend, or provide solutions through designated channels.

d) Verify the implementation of solutions.

e) Control further processing, delivery, or installation of nonconforming product until the deficiency or unsatisfactory condition has been corrected.

4.1.2.2 Resources

The supplier shall identify resource requirements and provide adequate resources, including the assignment of trained personnel (see 4.18), for management, performance of work, and verification activities including internal quality audits.

4.1.2.3 Management representative

The supplier's management with executive responsibility shall appoint a member of the supplier's own management who, irrespective of other responsibilities, shall have defined authority for

a) Ensuring that a quality system is established, implemented, and maintained in accordance with this American National Standard.

b) Reporting on the performance of the quality system to the supplier's management for review and as a basis for improvement of the quality system.

NOTE

5 The responsibility of a management representative may also include liaison with external parties on matters relating to the supplier's quality system.

4.1.3 Management review

The supplier's management with executive responsibility shall review the quality system at defined intervals sufficient to ensure its continuing suitability and effectiveness in satisfying the requirements of this American National Standard and the supplier's stated quality policy and objectives (see 4.1.1). Records of such reviews shall be maintained (see 4.16).

4.2 Quality system

4.2.1 General

The supplier shall establish, document, and maintain a quality system as a means of ensuring that product conforms to specified requirements. The supplier shall prepare a quality manual covering the requirements of this American National Standard. The quality manual shall include or make reference to the quality-system procedures and outline the structure of the documentation used in the quality system.

NOTE

6 Guidance on quality manuals is given in ISO 10013.

4.2.2 Quality-system procedures

The supplier shall do the following:

a) Prepare documented procedures consistent with the requirements of this American National Standard and the supplier's stated quality policy.

b) Effectively implement the quality system and its documented procedures.

For the purposes of this American National Standard, the range and detail of the procedures that form part of the quality system depend on the complexity of the work, the methods used, and the skills and training needed by personnel involved in carrying out the activity.

NOTE

7 Documented procedures may make reference to work instructions that define how an activity is performed.

4.2.3 Quality planning

The supplier shall define and document how the requirements for quality will be met. Quality planning shall be consistent with all other requirements of a supplier's quality system and shall be documented in a format to suit the supplier's method of operation. The supplier shall give consideration to the following activities, as appropriate, in meeting the specified requirements for products, projects or contracts:

a) The preparation of quality plans.

b) The identification and acquisition of any controls, processes, equipment (including inspection and test equipment), fixtures, resources, and skills that may be needed to achieve the required quality.

c) Ensuring the compatibility of the production process, installation, servicing, inspection, and test procedures and the applicable documentation.

d) The updating, as necessary, of quality control, inspection, and testing techniques, including the development of new instrumentation.

e) The identification of any measurement requirement involving capability that exceeds the known state of the art, in sufficient time for the needed capability to be developed.

f) The identification of suitable verification at appropriate stages in the realization of product.

g) The clarification of standards of acceptability for all features and requirements, including those which contain a subjective element.

h) The identification and preparation of quality records (see 4.16).

NOTE

8 The quality plans referred to (see 4.2.3a) may be in the form of a reference to the appropriate documented procedures that form an integral part of the supplier's quality system.

4.3 Contract review

4.3.1 General

The supplier shall establish and maintain documented procedures for contract review and for the coordination of these activities.

4.3.2 Review

Before submission of a tender, or the acceptance of a contract or order (statement of requirement), the tender, contract, or order shall be reviewed by the supplier to ensure the following:

a) The requirements are adequately defined and documented; where no written statement of requirement is available for an order received by verbal means, the supplier shall ensure that the order requirements are agreed before their acceptance.

b) Any differences between the contract or accepted order requirements and those in the tender are resolved.

c) The supplier has the capability to meet the contract or accepted order requirements.

4.3.3 Amendment to a contract

The supplier shall identify how an amendment to a contract is made and correctly transferred to the functions concerned within the supplier's organization.

4.3.4 Records

Records of contract reviews shall be maintained (see 4.16).

NOTE

9 Channels for communication and interfaces with the customer's organization in these contract matters should be established.

4.4 Design control

The scope of this American National Standard does not include quality-system requirements for design control. This subclause is included to align the clause numbering with ANSI/ISO/ASQC Q9001-1994.

4.5 Document and data control

4.5.1 General

The supplier shall establish and maintain documented procedures to control all documents and data that relate to the requirements of this American National Standard including, to the extent applicable, documents of external origin such as standards and customer drawings.

NOTE

10 Documents and data can be in the form of any type of media, such as hard copy or electronic media.

4.5.2 Document and data approval and issue

The documents and data shall be reviewed and approved for adequacy by authorized personnel prior to issue. A master list or equivalent document-control procedure identifying the current revision status of documents shall be established and be readily available to preclude the use of invalid and/or obsolete documents.

This control shall ensure the following:

a) The pertinent issues of appropriate documents are available at all locations where operations essential to the effective functioning of the quality system are performed.

b) Invalid and/or obsolete documents are promptly removed from all points of issue or use, or otherwise assured against unintended use.

c) Any obsolete documents retained for legal and/or knowledge-preservation purposes are suitably identified.

4.5.3 Document and data changes

Changes to documents and data shall be reviewed and approved by the same functions/organizations that performed the original review and approval, unless specifically designated otherwise. The designated functions/organizations shall have access to pertinent background information upon which to base their review and approval.

Where practicable, the nature of the change shall be identified in the document or the appropriate attachments.

4.6 Purchasing

4.6.1 General

The supplier shall establish and maintain documented procedures to ensure that purchased product (see 3.1) conforms to specified requirements.

4.6.2 Evaluation of subcontractors

The supplier shall do the following:

a) Evaluate and select subcontractors on the basis of their ability to meet subcontract requirements including the quality system and any specific quality-assurance requirements.

b) Define the type and extent of control exercised by the supplier over subcontractors. This shall be dependent upon the type of product, the impact of subcontracted product on the quality of final product, and, where applicable, on the quality audit reports and/or quality records of the previously demonstrated capability and performance of subcontractors.

c) Establish and maintain quality records of acceptable subcontractors (see 4.16).

4.6.3 Purchasing data

Purchasing documents shall contain data clearly describing the product ordered, including (where applicable):

a) The type, class, grade, or other precise identification.

b) The title or other positive identification, and applicable issues of specifications, drawings, process requirements, inspection instructions, and other relevant technical data, including requirements for approval or qualification of product, procedures, process equipment, and personnel.

c) The title, number, and issue of the quality-system standard to be applied.

The supplier shall review and approve purchasing documents for adequacy of the specified requirements prior to release.

4.6.4 Verification of purchased product
4.6.4.1 Supplier verification at subcontractor's premises

Where the supplier proposes to verify purchased product at the subcontractor's premises, the supplier shall specify verification arrangements and the method of product release in the purchasing documents.

4.6.4.2 Customer verification of subcontracted product

Where specified in the contract, the supplier's customer or the customer's representative shall be afforded the right to verify at the subcontractor's premises and the supplier's premises that subcontracted product conforms to specified requirements. Such verification shall not be used by the supplier as evidence of effective control of quality by the subcontractor.

Verification by the customer shall not absolve the supplier of the responsibility to provide acceptable product, nor shall it preclude subsequent rejection by the customer.

4.7 Control of customer-supplied product

The supplier shall establish and maintain documented procedures for the control of verification, storage, and maintenance of customer-supplied product provided for incorporation into the supplies or for related activities. Any such product that is lost, damaged, or is otherwise unsuitable for use shall be recorded and reported to the customer (see 4.16).

Verification by the supplier does not absolve the customer of the responsibility to provide acceptable product.

4.8 Product identification and traceability

Where appropriate, the supplier shall establish and maintain documented procedures for identifying the product by suitable means from receipt and during all stages of production, delivery, and installation. Where and to the extent that traceability is a specified requirement, the supplier shall establish and maintain documented procedures for unique identification of individual product or batches. This identification shall be recorded (see 4.16).

4.9 Process control

The supplier shall identify and plan the production, installation, and servicing processes which directly affect quality and shall ensure that these processes are carried out under controlled conditions. Controlled conditions shall include the following:

a) Documented procedures defining the manner of production, installation, and servicing, where the absence of such procedures could adversely affect quality.

b) Use of suitable production, installation, and servicing equipment, and a suitable working environment.

c) Compliance with reference standards/codes, quality plans, and/or documented procedures.

d) Monitoring and control of suitable process parameters and product characteristics.

e) The approval of processes and equipment, as appropriate.

f) Criteria for workmanship, which shall be stipulated in the clearest practical manner (e.g., written standards, representatives samples, or illustrations).

g) Suitable maintenance of equipment to ensure continuing process capability.

Where the results of processes cannot be fully verified by subsequent inspection and testing of the product and where, for example, processing deficiencies may become apparent only after the product is in use, the processes shall be carried out by qualified operators and/or shall require continuous monitoring and control of process parameters to ensure that the specified requirements are met.

The requirements for any qualification of process operations, including associated equipment and personnel (see 4.18), shall be specified.

NOTE
11 Such processes requiring prequalification of their process capability are frequently referred to as special processes.

Records shall be maintained for qualified processes, equipment, and personnel, as appropriate (see 4.16).

4.10 Inspection and testing

4.10.1 General

The supplier shall establish and maintain documented procedures for inspection and testing activities in order to verify that the specified requirements for product are met. The required inspection and testing, and the records to be established, shall be detailed in the quality plan or documented procedures.

4.10.2 Receiving inspection and testing
4.10.2.1

The supplier shall ensure that incoming product is not used or processed (except in the circumstances described in 4.10.2.3) until it has been inspected or otherwise verified as conforming to specified requirements. Verification of the specified requirements shall be in accordance with the quality plan and/or documented procedures.

4.10.2.2

In determining the amount and nature of receiving inspection, consideration shall be given to the amount of control exercised at the subcontractor's premises and the recorded evidence of conformance provided.

4.10.2.3

Where incoming product is released for urgent production purposes prior to verification, it shall be positively identified and recorded (see 4.16) in order to permit immediate recall and replacement in the event of nonconformity to specified requirements.

4.10.3 In-process inspection and testing

The supplier shall do the following:

a) Inspect and test the product as required by the quality plan and/or documented procedures.

b) Hold the product until the required inspection and tests have been completed or necessary reports have been received and verified, except when product is released under positive-recall procedures (see 4.10.2.3). Release under positive-recall procedures shall not preclude the activities outlined in 4.10.3a.

4.10.4 Final inspection and testing

The supplier shall carry out all final inspection and testing in accordance with the quality plan and/or documented procedures to complete the evidence of conformance of the finished product to the specified requirements.

The quality plan and/or documented procedures for final inspection and testing shall require that all specified inspection and tests, including those specified either on receipt of product or in-process, have been carried out, and that the results meet specified requirements.

No product shall be dispatched until all the activities specified in the quality plan and/or documented procedures have been satisfactorily completed and the associated data and documentation are available and authorized.

4.10.5 Inspection and test records

The supplier shall establish and maintain records which provide evidence that the product has been inspected and/or tested. These records shall show clearly whether the product has passed or failed the inspections and/or tests according to defined acceptance criteria. Where the product fails to pass any inspection and/or test, the procedures for control of nonconforming product shall apply (see 4.13).

Records shall identify the inspection authority responsible for the release of product (see 4.16).

4.11 Control of inspection, measuring, and test equipment

4.11.1 General

The supplier shall establish and maintain documented procedures to control, calibrate, and maintain inspection, measuring, and test equipment (including test software) used by the supplier to demonstrate the conformance of product to the specified requirements. Inspection, measuring, and test equipment shall be used in a manner which ensures that the measurement uncertainty is known and is consistent with the required measurement capability.

Where test software or comparative references such as test hardware are used as suitable forms of inspection, they shall be checked to prove that they are capable of verifying the acceptability of product, prior to release for use during production, installation, or servicing, and shall be rechecked at prescribed intervals. The supplier shall establish the extent and frequency of such checks and shall maintain records as evidence of control (see 4.16).

Where the availability of technical data pertaining to the measurement equipment is a specified requirement, such data shall be made available, when required by the customer or customer's representative, for verification that the measuring equipment is functionally adequate.

NOTE

12   For the purposes of this American National Standard, the term "measuring equipment" includes measurement devices.

4.11.2 *Control procedure*
The supplier shall do the following:

a) Determine the measurements to be made and the accuracy required, and select the appropriate inspection, measuring, and test equipment that is capable of the necessary accuracy and precision.

b) Identify all inspection, measuring, and test equipment that can affect product quality, and calibrate and adjust them at prescribed intervals, or prior to use, against certified equipment having a known valid relationship to internationally or nationally recognized standards. Where no such standards exist, the basis used for calibration shall be documented.

c) Define the process employed for the calibration of inspection, measuring, and test equipment, including details of equipment type, unique identification, location, frequency of checks, check method, acceptance criteria, and the action to be taken when results are unsatisfactory.

d) Identify inspection, measuring, and test equipment with a suitable indicator or approved identification record to show the calibration status.

e) Maintain calibration records for inspection, measuring, and test equipment (see 4.16).

f) Assess and document the validity of previous inspection and test results when inspection, measuring, or test equipment is found to be out of calibration.

g) Ensure that the environmental conditions are suitable for the calibration, inspections, measurements, and tests being carried out.

h) Ensure that the handling, preservation, and storage of inspection, measuring, and test equipment is such that the accuracy and fitness for use are maintained.

i) Safeguard inspection, measuring, and test facilities, including both test hardware and test software, from adjustments which would invalidate the calibration setting.

NOTE

13   The metrological confirmation system for measuring equipment given in ISO 10012 may be used for guidance.

4.12 Inspection and test status

The inspection and test status of product shall be identified by suitable means, which indicate the conformance or nonconformance of product with regard to inspection and tests performed. The identification of inspection and test status shall be maintained, as defined in the quality plan and/or documented procedures, throughout production, installation, and servicing of the product to ensure that only product that has passed the required inspections and tests [or released under an authorized concession (see 4.13.2)] is dispatched, used, or installed.

4.13 Control of nonconforming product

4.13.1 *General*
The supplier shall establish and maintain documented procedures to ensure that product that does not conform to specified requirements is prevented from unintended use or installation. This control shall provide for identification, documentation, evaluation, segregation (when practical), disposition of nonconforming product, and for notification to the functions concerned.

4.13.2 *Review and disposition of nonconforming product*
The responsibility for review and authority for the disposition of nonconforming product shall be defined.

Nonconforming product shall be reviewed in accordance with documented procedures. It may be

 a) reworked to meet the specified requirements,

 b) accepted with or without repair by concession,

 c) regraded for alternative applications, or

 d) rejected or scrapped.

Where required by the contract, the proposed use or repair of product (see 4.13.2b) which does not conform to specified requirements shall be reported for concession to the customer or customer's representative. The description of the nonconformity that has been accepted, and of repairs, shall be recorded to denote the actual condition (see 4.16).

Repaired and/or reworked product shall be reinspected in accordance with the quality plan and/or documented procedures.

4.14 Corrective and preventive action

4.14.1 General
The supplier shall establish and maintain documented procedures for implementing corrective and preventive action.

Any corrective or preventive action taken to eliminate the causes of actual or potential nonconformities shall be to a degree appropriate to the magnitude of problems and commensurate with the risks encountered.

The supplier shall implement and record any changes to the documented procedures resulting from corrective and preventive action.

4.14.2 Corrective action
The procedures for corrective action shall include:

 a) The effective handling of customer complaints and reports of product nonconformities.

 b) Investigation of the cause of nonconformities relating to product, process, and quality system, and recording the results of the investigation (see 4.16).

 c) Determination of the corrective action needed to eliminate the cause of nonconformities.

 d) Application of controls to ensure that corrective action is taken and that it is effective.

4.14.3 Preventive action
The procedures for preventive action shall include:

 a) The use of appropriate sources of information such as processes and work operations which affect product quality, concessions, audit results, quality records, service reports, and customer complaints to detect, analyze, and eliminate potential causes of nonconformities.

 b) Determination of the steps needed to deal with any problems requiring preventive action.

 c) Initiation of preventive action and application of controls to ensure that it is effective.

 d) Confirmation that relevant information on actions taken is submitted for management review (see 4.1.3).

4.15 Handling, storage, packaging, preservation, and delivery

4.15.1 General
The supplier shall establish and maintain documented procedures for handling, storage, packaging, preservation, and delivery of product.

4.15.2 Handling
The supplier shall provide methods of handling product that prevent damage or deterioration.

4.15.3 Storage
The supplier shall use designated storage areas or stock rooms to prevent damage or deterioration of product, pending use or delivery. Appropriate methods of authorizing receipt to and dispatch from such areas shall be stipulated.

In order to detect deterioration, the condition of product in stock shall be assessed at appropriate intervals.

4.15.4 Packaging
The supplier shall control packing, packaging, and marking processes (including materials used) to the extent necessary to ensure conformance to specified requirements.

4.15.5 Preservation
The supplier shall apply appropriate methods for preservation and segregation of product when the product is under the supplier's control.

4.15.6 Delivery

The supplier shall arrange for the protection of the quality of product after final inspection and test. Where contractually specified, this protection shall be extended to include delivery to destination.

4.16 Control of quality records

The supplier shall establish and maintain documented procedures for identification, collection, indexing, access, filing, storage, maintenance, and disposition of quality records.

Quality records shall be maintained to demonstrate conformance to specified requirements and the effective operation of the quality system. Pertinent quality records from the subcontractor shall be an element of these data.

All quality records shall be legible and shall be stored and retained in such a way that they are readily retrievable in facilities that provide a suitable environment to prevent damage or deterioration and to prevent loss. Retention times of quality records shall be established and recorded. Where agreed contractually, quality records shall be made available for evaluation by the customer or the customer's representative for an agreed period.

NOTE

14 Records may be in the form of any type of media, such as hard copy or electronic media.

4.17 Internal quality audits

The supplier shall establish and maintain documented procedures for planning and implementing internal quality audits to verify whether quality activities and related results comply with planned arrangements and to determine the effectiveness of the quality system.

Internal quality audits shall be scheduled on the basis of the status and importance of the activity to be audited and shall be carried out by personnel independent of those having direct responsibility for the activity being audited.

The results of the audits shall be recorded (see 4.16) and brought to the attention of the personnel having responsibility in the area audited. The management personnel responsible for the area shall take timely corrective action on deficiencies found during the audit.

Follow-up audit activities shall verify and record the implementation and effectiveness of the corrective action taken (see 4.16).

NOTES

15 The results of internal quality audits form an integral part of the input to management review activities (see 4.1.3).

16 Guidance on quality-system audits is given in ANSI/ISO/ASQC Q10011-1-1994, ANSI/ISO/ASQC Q10011-2-1994, and ANSI/ISO/ASQC Q10011-3-1994.

4.18 Training

The supplier shall establish and maintain documented procedures for identifying training needs and provide for the training of all personnel performing activities affecting quality. Personnel performing specific assigned tasks shall be qualified on the basis of appropriate education, training, and/or experience, as required. Appropriate records of training shall be maintained (see 4.16).

4.19 Servicing

Where servicing is a specified requirement, the supplier shall establish and maintain documented procedures for performing, verifying, and reporting that the servicing meets the specified requirements.

4.20 Statistical techniques

4.20.1 Identification of need
The supplier shall identify the need for statistical techniques required for establishing, controlling, and verifying process capability and product characteristics.

4.20.2 Procedures
The supplier shall establish and maintain documented procedures to implement and control the application of the statistical techniques identified in 4.20.1.

ANNEX A (INFORMATIVE)

Bibliography

[1] ANSI/ISO/ASQC Q9000-1-1994, *Quality Management and Quality Assurance Standards— Guidelines for Selection and Use.*

[2] ANSI/ISO/ASQC Q9001-1994, *Quality Systems— Model for Quality Assurance in Design, Development, Production, Installation, and Servicing.*

[3] ANSI/ISO/ASQC Q9003-1994, *Quality Systems— Model for Quality Assurance in Final Inspection and Test.*

[4] ANSI/ISO/ASQC Q10011-1-1994, *Guidelines for Auditing Quality Systems—Auditing.*

[5] ANSI/ISO/ASQC Q10011-2-1994, *Guidelines for Auditing Quality Systems—Qualification Criteria for Quality Systems Auditors.*

[6] ANSI/ISO/ASQC Q10011-3-1994, *Guidelines for Auditing Quality Systems—Management of Audit Programs.*

[7] ISO 9000-2:1993, *Quality Management and Quality Assurance Standards—Part 2: Generic Guidelines for the Application of ISO 9001, ISO 9002, and ISO 9003.*

[8] ISO 9000-3:1991, *Quality Management and Quality Assurance Standards—Part 3: Guidelines for the Application of ISO 9001 to the Development, Supply, and Maintenance of Software.*

[9] ISO 10012-1:1992, *Quality Assurance Requirements for Measuring Equipment—Part 1: Metrological Confirmation System for Measuring Equipment.*

[10] ISO 10013:—[1] *Guidelines for Developing Quality Manuals.*

Endnote

[1] To be published.

American National Standard

ANSI/ISO/ASQC Q9003-1994

QUALITY SYSTEMS—MODEL FOR QUALITY ASSURANCE IN FINAL INSPECTION AND TEST

[Revision of first edition (ANSI/ASQC Q93-1987)]

Prepared by

American Society for Quality Control
Standards Committee

for

American National Standards Committee
Z-1 on Quality Assurance

Descriptors: Quality assurance, quality assurance program, quality systems, tests, inspection, reference models.

American National Standards: An American National Standard implies a consensus of those substantially concerned with its scope and provisions. An American National Standard is intended as a guide to aid the manufacturer, the consumer, and the general public. The existence of an American National Standard does not in any respect preclude anyone, whether he or she has approved the standard or not, from manufacturing, purchasing, or using products, processes, or procedures not conforming to the standard. American National Standards are subject to periodic review and users are cautioned to obtain the latest edition.

Caution Notice: This American National Standard may be revised or withdrawn at any time. The procedures of the American National Standards Institute require that action be taken to reaffirm, revise, or withdraw this standard no later than five years from the date of publication. Purchasers of American National Standards may receive current information on all standards by calling or writing the American National Standards Institute.

ASQC Mission: To facilitate continuous improvement and increase customer satisfaction by identifying, communicating, and promoting the use of quality principles, concepts, and technologies; and thereby be recognized throughout the world as the leading authority on, and champion for, quality.

10 9 8 7 6 5 4 3 2 1

Printed in the United States of America

Printed on acid-free recycled paper

Published by:
ASQC
611 E. Wisconsin Avenue
Milwaukee, WI 53202

CONTENTS

1   Scope
2   Normative reference
3   Definitions
4   Quality-system requirements
4.1   Management responsibility
4.2   Quality system
4.3   Contract review
4.4   Design control
4.5   Document and data control
4.6   Purchasing
4.7   Control of customer-supplied product
4.8   Product identification and traceability
4.9   Process control
4.10   Inspection and testing
4.11   Control of inspection, measuring, and test equipment
4.12   Inspection and test status
4.13   Control of nonconforming product
4.14   Corrective action
4.15   Handling, storage, packaging, preservation, and delivery
4.16   Control of quality records

4.17   Internal quality audits

4.18   Training

4.19   Servicing

4.20   Statistical techniques

FOREWORD

(This Foreword is not a part of American National Standard *Quality Systems—Model for Quality Assurance in Final Inspection and Test*.)

This American National Standard corresponds to the International Standard ISO 9003:1994. The initial five ISO 9000 series standards, ISO 9000, ISO 9001, ISO 9002, ISO 9003, and ISO 9004, when published in the United States as American National Standards in 1987, were designated as ANSI/ASQC Q90 through ANSI/ASQC Q94 respectively. The five 1987 standards in their 1994 international revisions are now designated ISO 9000-1, ISO 9001, ISO 9002, ISO 9003, and ISO 9004-1 respectively. Their publication as American National Standards are now designated ANSI/ISO/ASQC Q9000-1-1994, ANSI/ISO/ASQC Q9001-1994, ANSI/ISO/ASQC Q9002-1994, ANSI/ISO/ASQC Q9003-1994, and ANSI/ISO/ASQC Q9004-1-1994 respectively. This new numbering system is intended to emphasize the word-for-word correspondence of the International and American National Standards.

ISO (the International Organization for Standardization) is a worldwide federation of national standards bodies (ISO member bodies). The work of preparing International Standards is normally carried out through ISO technical committees. Each member body interested in a subject for which a technical committee has been established has the right to be represented on that committee. International organizations, governmental and nongovernmental, in liaison with ISO, also take part in the work. ISO collaborates closely with the International Electrotechnical Commission (IEC) on all matters of electrotechnical standardization. The American National Standards Institute (ANSI) is the U.S. member body of ISO. ASQC is the U.S. member of ANSI responsible for quality management and related standards.

Users should note that all ANSI/ISO/ASQC standards undergo revision from time to time. In the case of International Standards adopted as American National Standards, the revision timing is influenced by the international revision timing. Reference herein to any other standard implies the latest American National Standard revision unless otherwise stated.

Comments concerning this standard are welcome. They should be sent to the sponsor of the standard, American Society for Quality Control, 611 East Wisconsin Avenue, P.O. Box 3005, Milwaukee, WI 53201-3005, c/o Standards Administrator.

Annex

A   Bibliography

INTRODUCTION

This American National Standard is one of three American National Standards dealing with quality-system requirements that can be used for external quality-assurance purposes. The quality-assurance models, set out in the three American National Standards listed below, represent three distinct forms of quality-system requirements suitable for the purpose of a supplier demonstrating its capability, and for the assessment of the capability of a supplier by external parties.

a) ANSI/ISO/ASQC Q9001-1994, Quality Systems—Model for Quality Assurance in Design, Development, Production, Installation, and Servicing

For use when conformance to specified requirements is to be assured by the supplier during design, development, production, installation, and servicing.

b) ANSI/ISO/ASQC 9002-1994, Quality Systems—Model for Quality Assurance in Production, Installation, and Servicing

For use when conformance to specified requirements is to be assured by the supplier during production, installation, and servicing.

c) ANSI/ISO/ASQC Q9003-1994, Quality Systems—Model for Quality Assurance in Final Inspection and Test

For use when conformance to specified requirements is to be assured by the supplier solely at final inspection and test.

It is emphasized that the quality-system requirements specified in this American National Standard, ANSI/ISO/ASQC Q9001-1994, and ANSI/ISO/ASQC Q9002-1994 are complementary (not alternative) to the technical (product) specified requirements. They specify requirements which determine what elements quality systems have to encompass, but it is not the purpose of these American National Standards to enforce uniformity of quality systems. They are generic and independent of any specific industry or economic sector. The design and implementation of a quality system will be influenced by the varying needs of an organization, its particular objectives, the products and services supplied, and the processes and specific practices employed.

It is intended that these American National Standards will be adopted in their present form, but on occasions they may need to be tailored by adding or deleting certain quality-system requirements for specific contractual situations. ANSI/ISO/ASQC Q9000-1-1994 provides guidance on such tailoring as well as on selection of the appropriate quality-assurance model, viz. ANSI/ISO/ASQC Q9001-1994, ANSI/ISO/ASQC Q9002-1994, or ANSI/ISO/ASQC Q9003-1994.

Quality Systems—Model for Quality Assurance in Final Inspection and Test

1 Scope

This American National Standard specifies quality-system requirements for use where a supplier's capability to detect and control the disposition of any product nonconformity during final inspection and test needs to be demonstrated.

It is applicable in situations when the conformance of product to specified requirements can be shown with adequate confidence providing that certain suppliers' capabilities for inspection and tests conducted on finished product can be satisfactorily demonstrated.

NOTE
  1 For informative references, see Annex A.

2 Normative reference

The following standard contains provisions which, through reference in this text, constitute provisions of

this American National Standard. At the time of publication, the edition indicated was valid. All standards are subject to revision, and parties to agreements based on this American National Standard are encouraged to investigate the possibility of applying the most recent edition of the standard indicated below. The American National Standards Institute and members of IEC and ISO maintain registers of currently valid American National Standards and International Standards.

ISO 8402:1994, *Quality management and quality assurance—Vocabulary*

3 Definitions

For the purposes of this American National Standard, the definitions given in ISO 8402 and the following definitions apply.

3.1 Product

Result of activities or processes.

NOTES
  2 A product may include service, hardware, processed materials, software, or a combination thereof.

  3 A product can be tangible (e.g., assemblies or processed materials) or intangible (e.g., knowledge or concepts), or a combination thereof.

  4 For the purposes of this American National Standard, the term "product" applies to the intended product offering only and not to unintended "by-products" affecting the environment. This differs from the definition given in ISO 8402.

3.2 Tender

Offer made by a supplier in response to an invitation to satisfy a contract award to provide product.

3.3 Contract; accepted order

Agreed requirements between a supplier and customer transmitted by any means.

4 Quality-system requirements

4.1 Management responsibility

4.1.1 Quality policy

The supplier's management with executive responsibility shall define and document its policy for quality, including objectives for quality and its commitment to quality. The quality policy shall be relevant to the supplier's organizational goals and the expectations and needs of its customers. The supplier shall ensure that this policy is understood, implemented, and maintained at all levels of the organization.

4.1.2 Organization

4.1.2.1 Responsibility and authority

The responsibility, authority, and the interrelation of personnel who manage, perform, and verify work that is subject to the requirements of this American National Standard shall be defined and documented, particularly for personnel who need the organizational freedom and authority to do the following:

a) Conduct final inspection and tests.

b) Ensure that finished product that does not conform to specified requirements is prevented from being used or delivered.

4.1.2.2 Resources

The supplier shall identify resource requirements and provide adequate resources, including the assignment of trained personnel (see 4.18), for management, performance of work, and verification activities including internal quality audits.

4.1.2.3 Management representative

The supplier's management with executive responsibility shall appoint a member of the supplier's own management who, irrespective of other responsibilities, shall have defined authority for the following:

a) Ensuring that a quality system is established, implemented, and maintained in accordance with this American National Standard.

b) Reporting on the performance of the quality system to the supplier's management for review and as a basis for improvement of the quality system.

NOTE

5 The responsibility of a management representative may also include liaison with external par-

ties on matters relating to the supplier's quality system.

4.1.3 Management review

The supplier's management with executive responsibility shall review the quality system at defined intervals sufficient to ensure its continuing suitability and effectiveness in satisfying the requirements of this American National Standard and the supplier's stated quality policy and objectives (see 4.1.1). Records of such reviews shall be maintained (see 4.16).

4.2 Quality system

4.2.1 General

The supplier shall establish, document, and maintain a quality system as a means of ensuring that product on completion conforms to specified requirements. The supplier shall prepare a quality manual covering the requirements of this American National Standard. The quality manual shall include or make reference to the quality-system procedures and outline the structure of the documentation used in the quality system.

NOTE

6 Guidance on quality manuals is given in ISO 10013.

4.2.2 Quality-system procedures

The supplier shall do the following:

a) Prepare documented procedures consistent with the requirements of this American National Standard and the supplier's stated quality policy.

b) Effectively implement the quality system and its documented procedures.

For the purposes of this American National Standard, the range and detail of the procedures that form part of the quality system depend on the complexity of the work, the methods used, and the skills and training needed by personnel involved in carrying out the activity.

NOTE

7 Documented procedures may make reference to work instructions that define how an activity is performed.

4.2.3 Quality planning

The supplier shall define and document how the requirements for quality of the finished product will be met. Quality planning shall be consistent with all other requirements of a supplier's quality system and shall be documented in a format to suit the supplier's method of operation. The supplier shall give consideration to the following activities (as appropriate):

a) The preparation of a quality plan for final inspection and tests.

b) The identification and acquisition of any final inspection and test equipment, resources, and skills that may be needed to achieve the required quality.

c) The updating, as necessary, of final inspection and testing techniques.

d) The identification of any final inspection and test measurement requirement involving capability that exceeds the known state of the art in sufficient time for the needed capability to be developed.

e) The identification of suitable verification at the finished product state.

f) The clarification of standards of acceptability for all features and requirements, including those which contain a subjective element.

g) The identification and preparation of quality records (see 4.16).

NOTE

  8   The quality plan referred to (see 4.2.3a) may be in the form of a reference to the appropriate documented procedures that form an integral part of the supplier's quality system.

4.3 Contract review

4.3.1 General

The supplier shall establish and maintain documented procedures for contract review and for the coordination of these activities.

4.3.2 Review

Before submission of a tender, or the acceptance of a contract or order (statement of requirement), the tender, contract, or order shall be reviewed by the supplier to ensure the following:

a) The requirements are adequately defined and documented; where no written statement of requirement is available for an order received by verbal means, the supplier shall ensure that the order requirements are agreed before their acceptance.

b) Any difference between the contract or accepted order requirements and those in the tender are resolved.

c) The supplier has the capability to meet the contract or accepted order requirements for finished product.

4.3.3 Amendment to a contract

The supplier shall identify how an amendment to a contract is made and correctly transferred to the functions concerned within the supplier's organization.

4.3.4 Records

Records of contract reviews shall be maintained (see 4.16).

NOTE

  9   Channels for communication and interfaces with the customer's organization in these contract matters should be established.

4.4 Design control

The scope of this American National Standard does not include quality-system requirements for design control. This subclause is included to align the clause numbering with ANSI/ISO/ASQC Q9001-1994.

4.5 Document and data control

4.5.1 General

The supplier shall establish and maintain documented procedures to control all documents and data that relate to the requirements of this American National Standard including, to the extent applicable, documents of external origin such as standards and customer drawings.

NOTE

  10   Documents and data may be in the form of any type of media, such as hard copy or electronic media.

4.5.2 Document and data approval and issue

The documents and data shall be reviewed and approved for adequacy by authorized personnel prior to issue. A master list or equivalent document-control procedure identifying the current revision status of documents shall be established and be readily available to preclude the use of invalid and/or obsolete documents.

This control shall ensure the following:

a) The pertinent issues of appropriate documents are available at all locations where operations essential to the effective functioning of the quality system are performed.

b) Invalid and/or obsolete documents are promptly removed from all points of issue or use, or otherwise assured against unintended use.

c) Any obsolete documents retained for legal and/or knowledge-preservation purposes are suitably identified.

4.5.3 Document and data changes

Changes to documents and data shall be reviewed and approved by the same functions/organizations that performed the original review and approval, unless specifically designated otherwise. The designated functions/organizations shall have access to pertinent background information upon which to base their review and approval.

Where practicable, the nature of the change shall be identified in the document or the appropriate attachments.

4.6 Purchasing

The scope of this American National Standard does not include quality-system requirements for purchasing. This subclause is included to align the clause numbering with ANSI/ISO/ASQC Q9001-1994.

4.7 Control of customer-supplied product

The supplier shall establish and maintain documented procedures for the control of verification, storage, and maintenance of customer-supplied product provided for incorporation into the finished product or for related activities. Any such product that is lost, damaged, or is otherwise unsuitable for use shall be recorded and reported to the customer (see 4.16).

Verification by the supplier does not absolve the customer of the responsibility to provide acceptable product.

4.8 Product identification and traceability

Where and to the extent that traceability is a specified requirement, the supplier shall establish and maintain documented procedures for unique identification of individual product or batches. This identification shall be recorded (see 4.16).

4.9 Process control

The scope of this American National Standard does not include quality-system requirements for process control. This subclause is included to align the clause numbering with ANSI/ISO/ASQC Q9001-1994.

4.10 Inspection and testing

4.10.1 General

The supplier shall establish and maintain documented procedures for final inspection and testing activities in order to verify that the specified requirements for finished product are met. The required final inspection and testing, and the records to be established, shall be detailed in the quality plan or documented procedures.

4.10.2 Final inspection and testing

The supplier shall carry out all final inspection and testing in accordance with the quality plan and/or documented procedures and maintain appropriate records to complete the evidence of conformance of product to the specified requirements. When the specified requirements cannot be fully verified on the finished product, then a verification of acceptable results of other necessary inspection and tests performed previously shall be included for the purpose of verifying product requirements at final inspection and test.

Records shall identify the inspection authority responsible for the release of conforming product (see 4.16).

4.11 Control of inspection, measuring, and test equipment

4.11.1 General

The supplier shall establish and maintain documented procedures to control, calibrate, and maintain final inspection, measuring, and test equipment (including test software) used by the supplier to demonstrate the conformance of product to the specified requirements. Inspection, measuring, and test equipment shall be used in a manner which ensures that the measurement uncertainty is known and is consistent with the required measurement capability.

Where test software or comparative references such as test hardware are used as suitable forms of inspection, they shall be checked to prove that they are capable of verifying the acceptability of product, prior to release for use during final inspection and testing, and shall be rechecked at prescribed intervals. The supplier shall establish the extent and frequency of such checks and shall maintain records as evidence of control (see 4.16).

Where the availability of technical data pertaining to the measurement equipment is a specified requirement, such data shall be made available, when required by the customer or customer's representative, for verification that the measuring equipment is functionally adequate.

NOTE
11 For the purposes of this American National Standard, the term "measuring equipment" includes measurement devices.

4.11.2 Control procedure
The supplier shall do the following:

a) Determine the measurements to be made and the accuracy required, and select the appropriate inspection, measuring, and test equipment that is capable of the necessary accuracy and precision.

b) Identify all inspection, measuring, and test equipment that can affect product quality, and calibrate and adjust them at prescribed intervals, or prior to use, against certified equipment having a known valid relationship to internationally or nationally recognized standards. Where no such standards exist, the basis used for calibration shall be documented.

c) Define the process employed for the calibration of inspection, measuring, and test equipment, including details of equipment type, unique identification, location, frequency of checks, check method, acceptance criteria, and the action to be taken when results are unsatisfactory.

d) Identify inspection, measuring, and test equipment with a suitable indicator or approved identification record to show the calibration status.

e) Maintain calibration records for inspection, measuring, and test equipment (see 4.16).

f) Assess and document the validity of previous inspection and test results when inspection, measuring, or test equipment is found to be out of calibration.

g) Ensure that the environmental conditions are suitable for the calibrations, inspections, measurements, and tests being carried out.

h) Ensure that the handling, preservation, and storage of inspection, measuring, and test equipment is such that the accuracy and fitness for use are maintained.

i) Safeguard inspection, measuring, and test facilities, including both test hardware and test software, from adjustments which would invalidate the calibration setting.

NOTE
12 The metrological confirmation system for measuring equipment given in ISO 10012 may be used for guidance.

4.12 Inspection and test status

The inspection and test status of product shall be identified by suitable means, which indicate the conformance or nonconformance of product with regard to inspection and tests performed. The identification of inspection and test status shall be maintained, as defined in the quality plan and/or documented procedures, to ensure that only product that has passed the required final inspection and test [or released under an authorized concession (see 4.13)] is dispatched.

4.13 Control of nonconforming product

The supplier shall establish and maintain control of product that does not conform to specified requirements to ensure that unintended use or delivery is avoided.

Control shall provide for identification, documentation, evaluation, segregation (when practical), disposition of nonconforming product, and for notification to the functions concerned.

The description of repairs, and of any nonconformity that has been accepted under authorized concession, shall be recorded to denote the actual condition (see 4.16).

Repaired and/or reworked product shall be reinspected in accordance with the quality plan and/or documented procedures.

4.14 Corrective action

The supplier shall do the following:

a) Investigate nonconformities that have been identified from the analysis of final inspection and test reports and customer complaints of product.

b) Determine and implement appropriate corrective action on the nonconformities.

c) Ensure that relevant information on the actions taken is submitted for management review (see 4.1.3).

4.15 Handling, storage, packaging, preservation, and delivery

4.15.1 General
The supplier shall establish and maintain documented procedures for handling, storage, packaging, preservation, and delivery of completed product after final inspection and test.

4.15.2 Handling
The supplier shall provide methods of handling product that prevent damage or deterioration.

4.15.3 Storage
The supplier shall use designated storage areas or stock rooms to prevent damage or deterioration of product, pending delivery. Appropriate methods for authorizing receipt to and dispatch from such areas shall be stipulated.

In order to detect deterioration, the condition of product in stock shall be assessed at appropriate intervals.

4.15.4 Packaging
The supplier shall control packing, packaging, and marking processes (including materials used) to the extent necessary to ensure conformance to specified requirements.

4.15.5 Preservation
The supplier shall apply appropriate methods for preservation and segregation of product when the product is under the supplier's control.

4.15.6 Delivery
The supplier shall arrange for the protection of the quality of product after final inspection and test. Where contractually specified, this protection shall be extended to include delivery to destination.

4.16 Control of quality records

The supplier shall establish and maintain control of appropriate quality records to demonstrate conformance of the finished product to specified requirements and the effective operation of the quality system.

Quality records shall be legible and identifiable to the product involved. Quality records that substantiate conformance of the finished product with the specified requirements and the effective operation of the quality system shall be retained for an agreed period and made available on request.

NOTE
13 Records may be in the form of any type of media, such as hard copy or electronic media.

4.17 Internal quality audits

The supplier shall carry out internal quality audits to verify whether quality activities and related results comply with planned arrangements covering the requirements of this American National Standard and to determine the effectiveness of the quality system.

Internal quality audits shall be scheduled on the basis of the status and importance of the activity to be audited and shall be carried out by personnel independent of those having direct responsibility for the activity being audited.

The results of the audits shall be recorded (see 4.16) and brought to the attention of the personnel having responsibility in the area audited. The management personnel responsible for the area shall take timely corrective action on deficiencies found during the audit.

Follow-up audit activities shall verify and record the implementation and effectiveness of the corrective action taken (see 4.16).

NOTES

14 The results of internal quality audits form an integral part of the input to management review activities (see 4.1.3).

15 Guidance on quality-system audits is given in ANSI/ISO/ASQC Q10011-1-1994, ANSI/ISO/ASQC Q10011-2-1994, and ANSI/ISO/ASQC Q10011-3-1994.

4.18 Training

Personnel performing final inspection and test activities covering the requirements of this American National Standard shall have appropriate experience and/or training, including any necessary qualification for specific assigned tasks. Appropriate records of training shall be maintained (see 4.16).

4.19 Servicing

The scope of this American National Standard does not include quality-system requirements for servicing. This subclause is included to align the clause numbering with ANSI/ISO/ASQC Q9001-1994.

4.20 Statistical techniques

The supplier shall

a) identify the need for statistical techniques required for the acceptability of product characteristics; and

b) implement and control the application of the statistical techniques.

ANNEX A (INFORMATIVE)

Bibliography

[1] ANSI/ISO/ASQC Q9000-1-1994, *Quality Management and Quality Assurance Standards—Guidelines for Selection and Use.*

[2] ANSI/ISO/ASQC Q9001-1994, *Quality Systems—Model for Quality Assurance in Design, Development, Production, Installation, and Servicing.*

[3] ANSI/ISO/ASQC Q9002-1994, *Quality Systems—Model for Quality Assurance in Production, Installation, and Servicing.*

[4] ANSI/ISO/ASQC Q10011-1-1994, *Guidelines for Auditing Quality Systems—Auditing.*

[5] ANSI/ISO/ASQC Q10011-2-1994, *Guidelines for Auditing Quality Systems—Qualification Criteria for Quality Systems Auditors.*

[6] ANSI/ISO/ASQC Q10011-3-1994, *Guidelines for Auditing Quality Systems—Management of Audit Programs.*

[7] ISO 9000-2:1993, *Quality Management and Quality Assurance Standards—Part 2: Generic Guidelines for the Application of ISO 9001, ISO 9002 and ISO 9003.*

[8] ISO 9000-3:1991, *Quality Management and Quality Assurance Standards—Part 3: Guidelines for the Application of ISO 9001 to the Development, Supply and Maintenance of Software.*

[9] ISO 10012-1:1992, *Quality Assurance Requirements for Measuring Equipment—Part 1: Metrological Confirmation System for Measuring Equipment.*

[10] ISO 10013:—[1] *Guidelines for Developing Quality Manuals.*

Endnotes

[1] To be published.

American National Standard

ANSI/ISO/ASQC Q9004-1-1994

QUALITY MANAGEMENT AND QUALITY SYSTEM ELEMENTS—GUIDELINES

Prepared by

American Society for Quality Control
Standards Committee

for

American National Standards Committee
Z-1 on Quality Assurance

Descriptors: Quality management, quality systems, components, general conditions.

American National Standards: An American National Standard implies a consensus of those substantially concerned with its scope and provisions. An American National Standard is intended as a guide to aid the manufacturer, the consumer, and the general public. The existence of an American National Standard does not in any respect preclude anyone, whether he or she has approved the standard or not, from manufacturing, purchasing, or using products, processes, or procedures not conforming to the standard. American National Standards are subject to periodic review and users are cautioned to obtain the latest edition.

Caution Notice: This American National Standard may be revised or withdrawn at any time. The procedures of the American National Standards Institute require that action be taken to reaffirm, revise, or withdraw this standard no later than five years from the date of publication. Purchasers of American National Standards may receive current information on all standards by calling or writing the American National Standards Institute.

ASQC Mission: To facilitate continuous improvement and increase customer satisfaction by identifying, communicating, and promoting the use of quality principles, concepts, and technologies; and thereby be recognized throughout the world as the leading authority on, and champion for, quality.

10 9 8 7 6 5 4 3 2 1

Printed in the United States of America

Printed on acid-free recycled paper

Published by:
ASQC
611 E. Wisconsin Avenue
Milwaukee, WI 53202

CONTENTS

1   Scope
2   Normative references
3   Definitions
4   Management responsibility
5   Quality-system elements
6   Financial considerations of quality systems
7   Quality in marketing
8   Quality in specification and design
9   Quality in purchasing
10   Quality of processes
11   Control of processes
12   Product verification
13   Control of inspection, measuring, and test equipment
14   Control of nonconforming product
15   Corrective action
16   Postproduction activities
17   Quality records
18   Personnel
19   Product safety
20   Use of statistical methods

Annex

A   Bibliography

FOREWORD

(This Foreword is not a part of American National Standard *Quality Management and Quality System Elements—Guidelines*.)

This American National Standard corresponds to the International Standard ISO 9004-1:1994. The initial five ISO 9000 series standards, ISO 9000, ISO 9001, ISO 9002, ISO 9003, and ISO 9004, when published in the United States as American National Standards in 1987, were designated as ANSI/ASQC Q90 through ANSI/ASQC Q94 respectively. The five 1987 standards in their 1994 international revisions are now designated ISO 9000-1, ISO 9001, ISO 9002, ISO 9003, and ISO 9004-1 respectively. Their publication as American National Standards are now designated ANSI/ISO/ASQC Q9000-1-1994, ANSI/ISO/ASQC Q9001-1994, ANSI/ISO/ASQC Q9002-1994, ANSI/ISO/ASQC Q9003-1994, and ANSI/ISO/ASQC Q9004-1-1994 respectively. This new numbering system is intended to emphasize the word-for-word correspondence of the International and American National Standards.

ISO (the International Organization for Standardization) is a worldwide federation of national standards bodies (ISO member bodies). The work of preparing International Standards is normally carried out through ISO technical committees. Each member body interested in a subject for which a technical committee has been established has the right to be represented on that committee. International organizations, governmental and nongovernmental, in liaison with ISO, also take part in the work. ISO collaborates closely with the International Electrotechnical Commission (IEC) on all matters of electrotechnical standardization. The American National Standards Institute (ANSI) is the U.S. member body of ISO. ASQC is the U.S. member of ANSI responsible for quality management and related standards.

Users should note that all ANSI/ISO/ASQC standards undergo revision from time to time. In the case of International Standards adopted as American National Standards, the revision timing is influenced by the international revision timing. Reference herein to any other standard implies the latest American National Standard revision unless otherwise stated.

Comments concerning this standard are welcome. They should be sent to the sponsor of the standard, American Society for Quality Control, 611 E. Wisconsin Avenue, P.O. Box 3005, Milwaukee, WI 53201-3005, c/o Standards Administrator.

INTRODUCTION

0.1 General

ANSI/ISO/ASQC Q9004-1-1994 and all the International Standards in the ISO 9000 family are generic and independent of any specific industry or economic sector. Collectively they provide guidance for quality management and models for quality assurance.

The International Standards in the ISO 9000 family describe what elements quality systems should encompass, but not how a specific organization should implement these elements. Because the needs of organizations vary, it is not the purpose of these International Standards or the corresponding American National Standards to enforce uniformity of quality systems. The design and implementation of a quality system will be influenced by the particular objectives, products, processes, and individual practices of the organization.

A primary concern of any organization should be the quality of its products. (See 3.5 for the definition of "product," which includes service.)

In order to be successful, an organization should offer products that.

a) meet a well-defined need, use, or purpose,

b) satisfy customers' expectations,

c) comply with applicable standards and specifications,

d) comply with requirements of society (see 3.3),

e) reflect environmental needs,

f) are made available at competitive prices, and

g) are provided economically.

0.2 Organizational goals

In order to meet its objectives, the organization should ensure that the technical, administrative, and human factors affecting the quality of its products will be under control, whether hardware, software, processed materials, or services. All such control should be oriented towards the reduction, elimination, and, most importantly, prevention of quality nonconformities.

A quality system should be developed and implemented for the purpose of accomplishing the objectives set out in the organization's quality policy.

Each element (or requirement) in a quality system varies in importance from one type of activity to another and from one product to another.

In order to achieve maximum effectiveness and to satisfy customer expectations, it is essential that the quality system be appropriate to the type of activity and to the product being offered.

0.3 Meeting customer/organization needs and expectations

A quality system has two interrelated aspects, as follows:

a) The customer's needs and expectations

For the customer, there is a need for confidence in the ability of the organization to deliver the desired quality as well as the consistent maintenance of that quality.

b) The organization's needs and interests

For the organization, there is a business need to attain and to maintain the desired quality at an optimum cost; the fulfillment of this aspect is related to the planned and efficient utilization of the technological, human, and material resources available to the organization.

Each of the above aspects of a quality system requires objective evidence in the form of information and data concerning the quality of the system and the quality of the organization's products.

0.4 Benefits, costs, and risks

Benefit, cost, and risk considerations have great importance for both the organization and customer. These considerations are inherent aspects of most products. The possible effects and ramifications of these considerations are given in a to c.

a) Benefit considerations

For the customer, consideration has to be given to reduced costs, improved fitness for use, increased satisfaction, and growth in confidence.

For the organization, consideration has to be given to increased profitability and market share.

b) Cost considerations

For the customer, consideration has to be given to safety, acquisition cost, operating, mainte-

nance, downtime and repair costs, and possible disposal costs.

For the organization, consideration has to be given to costs due to marketing and design deficiencies, including unsatisfactory product, rework, repair, replacement, reprocessing, loss of production, warranties, and field repair.

c) Risk considerations

For the customer, consideration has to be given to risks such as those pertaining to the health and safety of people, dissatisfaction with product, availability, marketing claims, and loss of confidence.

For the organization, consideration has to be given to risks related to deficient products which lead to loss of image or reputation, loss of market, complaints, claims, liability, and waste of human and financial resources.

0.5 Conclusions

An effective quality system should be designed to satisfy customer needs and expectations while serving to protect the organization's interests. A well-structured quality system is a valuable management resource in the optimization and control of quality in relation to benefit, cost, and risk considerations.

Quality Management and Quality System Elements—Guidelines

1 Scope

ANSI/ISO/ASQC Q9004-1-1994 provides guidance on quality management and quality-system elements.

The quality-system elements are suitable for use in the development and implementation of a comprehensive and effective in-house quality system, with a view to ensuring customer satisfaction.

ANSI/ISO/ASQC Q9004-1-1994 is not intended for contractual, regulatory, or certification use. Consequently, it is not a guideline for the implementing of ANSI/ISO/ASQC Q9001, ANSI/ISO/ASQC Q9002-1994, and ANSI/ISO/ASQC Q9003-1994. ISO 9000-2 should be used for that purpose.

The selection of appropriate elements contained in this part of ANSI/ISO/ASQC Q9004-1-1994 and the extent to which these elements are adopted and applied by an organization depends upon factors such as the market being served, nature of the product, production processes, and customer and consumer needs.

References in ANSI/ISO/ASQC Q9004-1-1994 to a "product" should be interpreted as applicable to the generic product categories of hardware, software, processed materials or service (in accordance with the definition of "product" in ISO 8402).

NOTES
1 For further guidance, see ISO 9004-2 and ISO 9004-3.

2 For informative references, see Annex A.

2 Normative references

The following standards contain provisions which, through reference in this text, constitute provisions of ANSI/ISO/ASQC Q9004-1-1994. At the time of publication, the editions indicated were valid. All standards are subject to revision, and parties to agreements based on ANSI/ISO/ASQC Q9004-1-1994 are encouraged to investigate the possibility of applying the most recent editions of the standards indicated below. Members of IEC and ISO maintain registers of currently valid International Standards.

ANSI/ISO/ASQC Q9000-1-1994, Quality Management and Quality Assurance Standards—Guidelines for Selection and Use.

ISO 8402:1994, *Quality Management and Quality Assurance—Vocabulary*.

3 Definitions

This revision of ANSI/ISO/ASQC Q94-1987 has improved the harmonization of terminology with other American National Standards in the ANSI/ISO/ASQC Q9000 series and with other International Standards in the ISO 9000 family. Table 1 shows the supply-chain terminology used in these American National Standards.

Thus, the term "subcontractor" is used rather than the term "supplier" in ANSI/ISO/ASQC Q9004-1-1994 to avoid confusion with the meaning of the term "supplier" in ANSI/ISO/ASQC Q9000-1-1994 and ANSI/ISO/ASQC Q9001-1994. (See ANSI/ISO/ASQC Q9000-1-1994 for a fuller explanation of the basis for usage of these terms.)

For the purposes of ANSI/ISO/ASQC Q9004-1-1994, the definitions given in ISO 8402 apply.

For the convenience of users of ANSI/ISO/ASQC Q9004-1-1994, the following definitions are quoted from ISO 8402.

3.1 Organization

Company, corporation, firm, enterprise, or institution, or part thereof, whether incorporated or not, public or private, that has its own functions and administration.

3.2 Customer

Recipient of a product provided by the supplier.

NOTES
3 In a contractual situation, the customer is called the "purchaser."[1]

4 The customer may be, for example, the ultimate consumer, user, beneficiary, or purchaser.

5 The customer can be either external or internal to the organization.

Table 1 Relationships of organizations in the supply chain.

ANSI/ISO/ASQC Q9000-1-1994	Subsupplier	→	supplier or organization	→	customer
ANSI/ISO/ASQC Q9001-1994 ANSI/ISO/ASQC Q9002-1994 ANSI/ISO/ASQC Q9003-1994	Subcontractor	→	supplier	→	customer
ANSI/ISO/ASQC Q9004-1-1994	Subcontractor	→	organization	→	customer

3.3 Requirements of society

Obligations resulting from laws, regulations, rules, codes, statutes, and other considerations.

NOTES

6 "Other considerations" include protection of the environment, health, safety, security, and conservation of energy and natural resources.

7 All requirements of society should be taken into account when defining the requirements for quality.

8 Requirements of society include jurisdictional and regulatory requirements. These may vary from one jurisdiction to another.

3.4 Quality plan

Document setting out the specific quality practices, resources, and sequence of activities relevant to a particular product, project, or contract.

NOTES

9 A quality plan usually makes reference to the parts of the quality manual applicable to the specific case.

10 Depending on the scope of the plan, a qualifier may be used, for example, quality assurance plan, quality management plan.

3.5 Product

Result of activities or processes.

NOTES

11 A product may include service, hardware, processed materials, software, or a combination thereof.

12 A product can be tangible (e.g., assemblies or processed materials) or intangible (e.g., knowledge or concepts), or a combination thereof.

13 A product can be intended (e.g., offering to customers) or unintended (e.g., pollutant or unwanted effects).

3.6 Service

Result generated by activities at the interface between the supplier and the customer and by supplier internal activities to meet the customer needs.

NOTES

14 The supplier or the customer may be represented at the interface by personnel or equipment.

15 Customer activities at the interface with the supplier may be essential to the service delivery.

16 Delivery or use of tangible products may form part of the service delivery.

17 A service may be linked with the manufacture and supply of tangible product.

4 Management responsibility

4.1 General

The responsibility for and commitment to a quality policy belongs to the highest level of management. Quality management encompasses all activities of the overall management function that determine the quality policy, objectives, and responsibilities, and implement them by means such as quality planning, quality control, quality assurance, and quality improvement within the quality system.

4.2 Quality policy

The management of an organization should define and document its quality policy. This policy should be consistent with other policies within the organization. Management should take all necessary measures to ensure that its quality policy is understood, implemented, and reviewed at all levels of the organization.

4.3 Quality objectives

4.3.1

Management should document objectives and commitments pertaining to key elements of quality, such as fitness for use, performance, safety, and dependability.

4.3.2

The calculation and evaluation of costs associated with all quality elements and objectives should always be an important consideration, with the objective of minimizing quality losses.

4.3.3

Appropriate levels of management should document specific quality objectives consistent with quality policy as well as other objectives of the organization.

4.4 Quality system

4.4.1

A quality system is the organizational structure, procedures, processes, and resources needed to implement quality management.

4.4.2

The organization's management should develop, establish, and implement a quality system to accomplish the stated policies and objectives.

4.4.3

The quality system should be structured and adapted to the organization's particular type of business and should take into account the appropriate elements outlined in ANSI/ISO/ASQC Q9004-1-1994.

4.4.4

The quality system should function in such a manner as to provide confidence that

    a) the system is understood, implemented, maintained, and effective,

    b) the products actually do satisfy customer needs and expectations,

    c) the needs of both society and the environment have been addressed, and

    d) emphasis is placed on problem prevention rather than dependence on detection after occurrence.

5 Quality-system elements

5.1 Extent of application

5.1.1

The quality system typically applies to, and interacts with, all activities pertinent to the quality of a product. It will involve all phases in the life-cycle of a product and processes, from initial identification of market needs to final satisfaction of requirements. Typical phases are

    a) marketing and market research,

    b) product design and development,

    c) process planning and development,

    d) purchasing,

    e) production, or provision of services,

    f) verification,

    g) packaging and storage,

    h) sales and distribution,

    i) installation and commissioning,

    j) technical assistance and servicing,

    k) after sales, and

    l) disposal or recycling at the end of useful life.

NOTE

18 Figure 1 gives a schematic representation of the typical life-cycle phases of a product.

5.1.2

In the context of interacting activities within an organization, marketing and design should be emphasized as especially important for the following:

- Determining and defining customer needs, expectations, and other product requirements.

- Providing the concepts (including supporting data) for producing a product to documented specifications at optimum cost.

5.2 Structure of the quality system

5.2.1 *General*

Input from the market should be used to improve new and existing products and to improve the quality system.

Management is ultimately responsible for establishing the quality policy and for decisions concerning the initiation, development, implementation, and maintenance of the quality system.

5.2.2 *Responsibility and authority*

Activities contributing to quality, whether directly or indirectly, should be defined and documented, and the following actions taken:

    a) General and specific quality-related responsibilities should be explicitly defined.

    b) Responsibility and authority delegated to each activity contributing to quality should be clearly established. Responsibility, organizational freedom, and authority to act should be sufficient to attain the assigned quality objectives with the desired efficiency.

c) Interface control and coordination measures between different activities should be defined.

d) In organizing a well-structured and effective quality system, emphasis should be placed on the identification of potential or actual quality problems and the implementation of preventive or corrective action (see Clauses 14 and 15).

5.2.3 *Organizational structure*
Functions related to the quality system should be clearly established within the overall organizational structure. The lines of authority and communication should be defined.

5.2.4 *Resources and personnel*
Management should identify resource requirements, and provide sufficient and appropriate resources essential to the implementation of the quality policy and the achievement of quality objectives. For example, these resources can include:

a) Human resources and specialized skills.

b) Design and development equipment.

c) Manufacturing equipment.

d) Inspection, test, and examination equipment.

e) Instrumentation and computer software.

Management should determine the level of competence, experience, and training necessary to ensure the capability of personnel (see clause 18).

Management should identify quality-related factors affecting market position and objectives relative to products, processes, or associated services, in order to allocate organization resources on a planned and timely basis.

Programs and schedules covering these resources and skills should be consistent with the organization's overall objectives.

5.2.5 *Operational procedures*
The quality system should be organized in such a way that adequate and continuous control is exercised over all activities affecting quality.

The quality system should emphasize preventive actions that avoid occurrence of problems, while maintaining the ability to respond to and correct failures, should they occur.

Documented operational procedures coordinating different activities with respect to an effective quality system should be developed, issued, and maintained to implement the quality policy and objectives. These documented procedures should specify the objectives and performance of the various activities having an impact on quality (see Figure 1).

All documented procedures should be stated simply, unambiguously, and understandably, and should indicate methods to be used and criteria to be satisfied.

5.2.6 *Configuration management*
The quality system should include documented procedures for configuration management to the extent appropriate. This discipline is initiated early in the design phase and continues through the whole life-cycle of a product. It assists in the operation and control of design, development, production, and use of a product, and gives management visibility of the state of documentation and product during its life-time.

Configuration management can include: configuration identification, configuration control, configuration status accounting, and configuration audit. It relates to several of the activities described in this part of ANSI/ISO/ASQC Q9004-1-1994.

5.3 Documentation of the quality system

5.3.1 *Quality policies and procedures*
All the elements, requirements, and provisions adopted by an organization for its quality system should be documented in a systematic, orderly, and understandable manner in the form of policies and procedures. However, care should be taken to limit documentation to the extent pertinent to the application.

The quality system should include adequate provision for the proper identification, distribution, collection, and maintenance of all quality documents.

5.3.2 *Quality-system documentation*
5.3.2.1
The typical form of the main document used to demonstrate or describe a documented quality system

Figure 1 Main activities having an impact on quality.

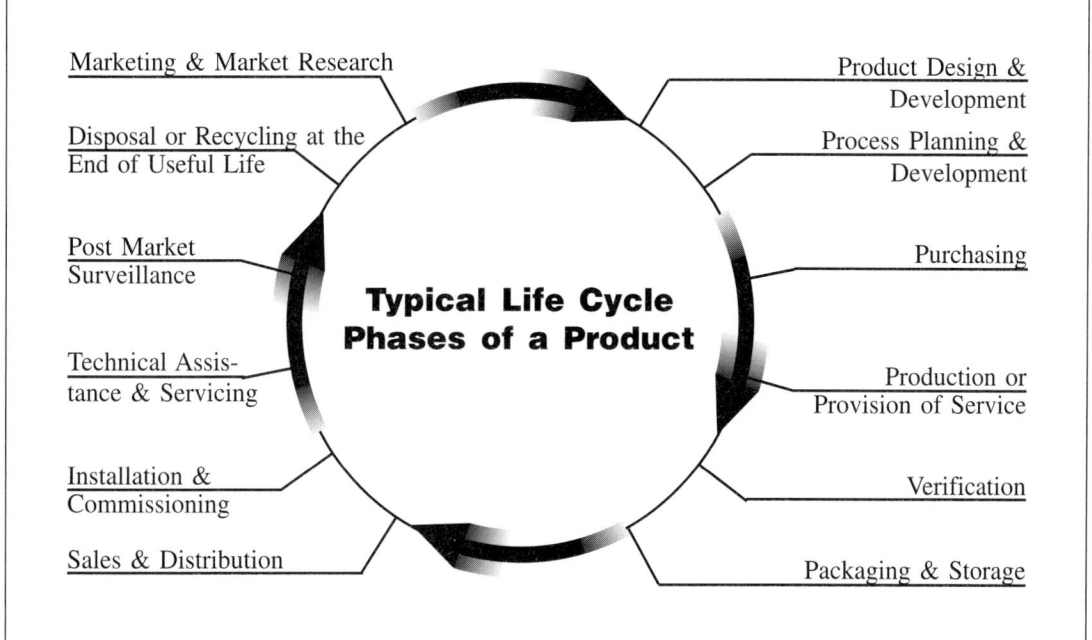

is a "quality manual." (For further guidance, see ISO 10013.)

5.3.2.2
The primary purpose of a quality manual is to define an outline structure of the quality system while serving as a permanent reference in the implementation and maintenance of that system.

5.3.2.3
Documented procedures should be established for making changes, modifications, revisions, or additions to the contents of a quality manual.

5.3.2.4
Supporting the quality manual are documented quality-system procedures (e.g., design, purchasing, and process work instructions). These documented procedures can take various forms, depending on the following:

- The size of the organization.
- The specific nature of the activity.
- The intended scope and structure of the quality manual.

Documented procedures may apply to one or more parts of the organization.

5.3.3 *Quality plans*
For any product or process, management should ensure that documented quality plans are prepared and maintained. These should be consistent with all other requirements of the organization's quality system, and should ensure that specified requirements for a product, project, or contract are met. A quality plan may be a part of a larger overall plan. A quality plan is particularly necessary for a new product or process, or when there is significant change to an existing product or process.

Quality plans should define the following:

a) The quality objectives to be attained (e.g., characteristics or specifications, uniformity, effectiveness, aesthetics, cycle time, cost, natural resources, utilization, yield, and dependability).

b) The steps in the processes that constitute the operating practice of the organization (a flowchart or similar diagram can be used to demonstrate the elements of the process).

c) The specific allocation of responsibilities, authority, and resources during the different phases of the project.

d) The specific documented procedures and instructions to be applied.

e) Suitable testing, inspection, examination, and audit programs at appropriate stages (e.g., design and development).

f) A documented procedure for changes and modifications in a quality plan as projects proceed.

g) A method for measuring the achievement of the quality objectives.

h) Other actions necessary to meet the objectives.

Quality plans may be included or referenced in the quality manual, as appropriate.

To facilitate achievement of the objectives of a quality plan, documented operational control as described in this part of ANSI/ISO/ASQC Q9004-1-1994 should be used.

5.3.4 Quality records

Quality records, including charts pertaining to design, inspection, testing, survey, audit, review, or related results, should be maintained as important evidence to demonstrate conformance to specified requirements and the effective operation of the quality system (see Clause 17).

5.4 Auditing the quality system

5.4.1 General

Audits should be planned and carried out to determine if the activities and related results of the organization's quality system comply with planned arrangements, and to determine the effectiveness of the quality system. All elements should be internally audited and evaluated on a regular basis, considering the status and importance of the activity to be audited. For this purpose, an appropriate audit program should be established and implemented by the organization's management.

5.4.2 Audit program

The audit program should cover the following:

a) Planning and scheduling the specific activities and areas to be audited.

b) Assignment of personnel with appropriate qualifications to conduct audits.

c) Documented procedures for carrying out audits, including recording and reporting the results of the quality audit and reaching agreement on timely corrective actions on the deficiencies found during the audit.

Apart from planned and systematic audits, other factors necessitating audits can be organizational changes, market feedback, nonconformity reports, and surveys.

5.4.3 Extent of audits

Objective evaluations of quality-system activities by competent personnel should include the following activities or areas:

a) Organizational structures.

b) Administrative, operational, and quality-system procedures.

c) Personnel, equipment, and material resources.

d) Work areas, operations, and processes.

e) Products being produced (to establish the degree of conformance to standards and specifications).

f) Documentation, reports, and record keeping.

Personnel conducting audits of quality-system elements should be independent of those having direct responsibilities for the specific activities or areas being audited. An audit plan should be prepared and documented to include the items listed in a to f.

5.4.4 Audit reporting

Audit observations, conclusions, and agreements on timely corrective action should be recorded and submitted for appropriate action by the management responsible for the area audited, and communicated for the information of management with executive responsibility for quality.

The following items should be covered in the audit report:

a) All examples of nonconformities or deficiencies.

b) Appropriate and timely corrective action.

5.4.5 Follow-up action

Implementation and effectiveness of corrective actions resulting from previous audits should be assessed and documented.

NOTE

19 For further guidance on quality auditing, qualifications of auditors and management of audit programs, see ANSI/ISO/ASQC Q10011-1-1994, ANSI/ISO/ASQC Q10011-2-1994, and ANSI/ISO/ASQC Q10011-3-1994.

5.5 Review and evaluation of the quality system

The organization's management should provide for independent review and evaluation of the quality system at defined intervals. The reviews of the quality policy and objectives should be carried out by top management, and the review of supporting activities should be carried out by management with executive responsibilities for quality and other appropriate members of management, utilizing competent independent personnel as decided on by the management.

Reviews should consist of well-structured and comprehensive evaluations which include:

a) Results from internal audits centred on various elements of the quality system (see 5.4.3).

b) The overall effectiveness in satisfying the guidance of ANSI/ISO/ASQC Q9004-1-1994 and the organization's stated quality policy and objectives.

c) Considerations for updating the quality system in relation to changes brought about by new technologies, quality concepts, market strategies, and social or environmental conditions.

Observations, conclusions, and recommendations reached as a result of review and evaluation should be documented for necessary action.

5.6 Quality improvement

When implementing a quality system, the management of an organization should ensure that the system will facilitate and promote continuous quality improvement.

Quality improvement refers to the actions taken throughout the organization to increase the effectiveness and efficiency of activities and processes to provide added benefits to both the organization and its customers.

In creating an environment for quality improvement, consideration should be given to the following:

a) Encouraging and sustaining a supportive style of management.

b) Promoting values, attitudes, and behavior that foster improvement.

c) Setting clear quality-improvement goals.

d) Encouraging effective communication and teamwork.

e) Recognizing successes and achievements.

f) Training and educating for improvement.

NOTE

20 Further guidance is given in ISO 9004-4.

6 Financial considerations of quality systems

6.1 General

It is important that the effectiveness of a quality system be measured in financial terms. The impact of an effective quality system upon the organization's profit and loss statement can be highly significant, particularly by improvement of operations, resulting in reduced losses due to error and by making a contribution to customer satisfaction.

Such measurement and reporting can provide a means for identifying inefficient activities, and initiating internal improvement activities.

By reporting quality-system activities and effectiveness in financial terms, management will receive the results in a common business language from all departments.

6.2 Approaches to financial reporting of quality-system activities

6.2.1 General

Some organizations find it useful to report the financial benefits using systematic quality financial reporting procedures.

The approach(es) to financial reporting selected and used by particular organizations will be dependent upon their individual structures, their activities, and the maturity of their quality systems.

6.2.2 *Approaches*

There are various approaches to gathering, presenting, and analyzing the elements of financial data. The approaches given in a to c have been found to be useful, but do not exclude others, or adaptations or combinations of them.

a) Quality-costing approach

This approach addresses quality-related costs, which are broadly divided into those arising from internal operations and external activities.

Cost elements for internal operations are analyzed according to the PAF (prevention, appraisal, failure) costing model.

Prevention and appraisal costs are considered as investments, while failure costs are considered as losses. The components of the costs are:

1) Prevention: efforts to prevent failures.

2) Appraisal: testing, inspection, and examination to assess whether requirements for quality are being fulfilled.

3) Internal failure: costs resulting from a product failing to meet the quality requirements prior to delivery (e.g., re-performing a service, reprocessing, rework, retest, scrap).

4) External failure: costs resulting from a product failing to meet the quality requirements after delivery (e.g., product maintenance and repair, warranties and returns, direct costs and allowances, product recall costs, liability costs).

b) Process-cost approach

This approach analyzes the costs of conformity and the costs of nonconformity for any process, both of which can be the source of savings. These are defined as:

1) Cost of conformity—cost to fulfill all of the stated and implied needs of customers in the absence of failure of the existing process.

2) Cost of nonconformity—cost incurred due to failure of the existing process.

c) Quality-loss approach

This approach focuses on internal and external losses due to poor quality and identifies tangible and intangible loss types. Typical external intangible losses are loss of future sales due to customer dissatisfaction. Typical internal intangible losses arise from lower work efficiency due to rework, poor ergonomics, missed opportunities, etc. Tangible losses are internal and external failure costs.

6.3 *Reporting*

The financial reporting of quality activities should be regularly provided to and monitored by management, and be related to other business measures such as "sales," "turnover," or "added value" in order to provide for a realistic, entrepreneurial

- evaluation of the adequacy and effectiveness of the quality system,

- identification of additional areas requiring attention and improvement, and

- establishment of quality and cost objectives for the following period.

The elements of financial quality reports are in many cases already available in the organization, but in other forms. Their reporting as a financial quality report can require regrouping of individual elements from other reports.

7 Quality in marketing

7.1 Marketing requirements

The marketing function should establish adequately defined and documented requirements for the quality of the product. Particularly at this early stage in the product life-cycle, it is important to consider the requirements for all the elements of the total product, whether hardware, software, processed materials, or services. In fact, all products involve some element of service, and many products involve several generic product categories. The marketing function should do the following:

a) Determine the need for a product.

b) Define the market demand and sector, so that product grade, quantity, price, and timing can be determined.

c) Determine specific customer requirements, or review general market needs; actions include assessment of any unstated expectations or biases held by customers.

d) Communicate all customer requirements within the organization.

e) Ensure that all relevant organizational functions agree that they have the capability to meet customer requirements.

7.2 Defining product specification

The marketing function should provide the organization with a formal statement or outline of product requirements. Specific customer and general market requirements and expectations should be translated into a preliminary set of specifications as the basis for subsequent design work. Among the elements that may be included are the following requirements:

a) Performance characteristics (e.g., environmental and usage conditions and dependability).

b) Sensory characteristics (e.g., style, color, taste, smell).

c) Installation, arrangement layout or fit.

d) Applicable standards and statutory regulations.

e) Packaging.

f) Quality verification and/or assurance.

7.3 Customer feedback information

The marketing function should establish an information-monitoring and feedback system on a continuous basis. All information pertinent to the customers' use of and satisfaction with the quality of a product should be analyzed, collated, interpreted, verified, and reported in accordance with documented procedures. Such information will help to determine the nature and extent of product problems in relation to customer experience and expectations. In addition, the feedback information can lead to management action resulting in product improvement or to new product offerings (see also 8.8, 8.9, Clause 15, and 16.6).

8 Quality in specification and design

8.1 Contribution of specification and design to quality

The specification and design function should provide for the translation of customer needs into technical specifications for materials, products, and processes. This should result in a product that provides customer satisfaction at an acceptable price that gives a satisfactory financial return for the organization. The specification and design should be such that the product is producible, verifiable, and controllable under the proposed production, installation, commissioning, or operational conditions.

8.2 Design planning and objectives (defining the project)

8.2.1

Management should prepare plans that define the responsibility for each design and development activity inside and/or outside the organization, and ensure that all those who contribute to design are aware of their responsibilities in relation to the full scope of the project.

8.2.2

In its delegation of responsibilities and authority for quality, management should ensure that design functions provide clear and definitive technical data for procurement, the execution of work, and verification of conformance of products and processes to specification requirements.

8.2.3

Management should establish time-phased design programs with holdpoints appropriate to the nature of the product and process. The extent of each phase, and the position of the holdpoints at which evaluations of the product or the process will take place, can depend upon several elements, such as the following:

• The product's application.

• Its design complexity.

• The extent of innovation and technology being introduced.

• The degree of standardization and similarity with past proven designs.

8.2.4

In addition to customer needs, consideration should be given to the requirements relating to safety, environmental, and other regulations, including items in the organization's quality policy which may go beyond existing statutory requirements (see also 3.3).

8.2.5

The design should unambiguously and adequately define characteristics important to quality, such as the acceptance criteria. Both fitness for purpose and safeguards against misuse should be considered. Product definition can also include dependability and serviceability through a reasonable life expectancy, including benign failure and safe disposability, as appropriate.

8.3 Product testing and measurement

The methods of measurement and test, and the acceptance criteria applied to evaluate the product and processes during both the design and production phases, should be specified. These should include the following:

a) Performance target values, tolerances, and attribute features.

b) Acceptance criteria.

c) Test and measurement methods, equipment, and computer software (see clause 13).

8.4 Design review

8.4.1 General

At the conclusion of each phase of design development, a formal, documented, systematic, and critical review of the design results should be planned and conducted. This should be distinguished from a project progress meeting. Participants at each design review should include representatives of all functions affecting quality, as appropriate to the phase being reviewed. The design review should identify and anticipate problem areas and inadequacies, and initiate corrective actions to ensure that the final design and supporting data meet customer requirements.

8.4.2 Elements of design reviews

As appropriate to the design phase and product, the elements outlined in a to c should be considered.

a) Items pertaining to customer needs and satisfaction include:

1) Comparison of customer needs expressed in the product specification with technical specifications for materials, products, and processes.

2) Validation of the design through prototype tests.

3) Ability to perform under expected conditions of use and environment.

4) Unintended uses and misuses.

5) Safety and environmental compatibility.

6) Compliance with regulatory requirements, national and International Standards, and organization practices.

7) Comparisons with competitive designs.

8) Comparison with similar designs, especially analysis of the history of internal and external problems to avoid repeating problems.

b) Items pertaining to product specification include:

1) Dependability and serviceability requirements.

2) Permissible tolerances and comparison with process capabilities.

3) Product acceptance criteria.

4) Installability, ease of assembly, storage needs, shelf-life, and disposability.

5) Benign failure and fail-safe characteristics.

6) Aesthetic specifications and acceptance criteria.

7) Failure mode and effect analysis, and fault tree analysis.

8) Ability to diagnose and correct problems.

9) Labeling, warnings, identification, traceability requirements, and user instructions.

10) Review and use of standard parts.

c) Items pertaining to process specification include:

1) Ability to produce product conforming to the design, including special process needs, mechanization, automation, assembly, and installation of components.

2) Capability to inspect and test the design, including special inspection and test requirements.

3) Specification of materials, components, and subassemblies, including approved supplies and subcontractors as well as availability.

4) Packaging, handling, storage, and shelf-life requirements, especially safety factors relating to incoming and outgoing items.

8.4.3 Design verification

All designs should be verified to ensure that product specifications are fulfilled (see 7.2). In addition to design review, design verification should include one or more of the following methods:

a) Performing alternative calculations, made to verify the correctness of the original calculations and analyses.

b) Testing and demonstrations (e.g., by model or prototype tests); if this method is adopted, the test programs should be clearly defined and the results documented.

c) Independent verification, to verify the correctness of the original calculations and/or other design activities.

8.5 Design qualification and validation

The design process should provide periodic evaluation of the design at significant stages. Such evaluation can take the form of analytical methods, such as FMEA (failure mode and effect analysis), fault tree analysis, or risk assessment, as well as inspection and test of prototype models and/or actual production samples. The amount and degree of testing (see 8.3) should be related to the identified risks. Independent evaluation can be used, as appropriate, to verify original calculations, provide alternative calculations, or perform tests. A number of samples should be examined by tests and/or inspection to provide adequate statistical confidence in the results. The tests should include the following activities:

a) Evaluation of performance, durability, safety, reliability, and maintainability under expected storage and operational conditions.

b) Inspections to verify that all design features conform to defined user needs and that all authorized design changes have been accomplished and recorded.

c) Validation of computer systems and software.

The results of all tests and evaluations should be documented regularly throughout the qualification test cycle. Review of test results should include nonconformity and failure analysis.

8.6 Final design review and production release

The final design should be reviewed and the results appropriately documented in specifications and drawings, which then form the design baseline. Where appropriate, this should include a description of initial test units and any modifications made to correct deficiencies identified during the qualification test programmes. The total document package that defines the design baseline (output) should require approval at appropriate levels of management affected by or con-

tributing to the product. This approval constitutes the production release and signifies that the design can be realized.

8.7 Market-readiness review

A determination should be made as to whether the organization has the capability to deliver the new or redesigned product. Depending upon the type of product, the review can cover the following points:

a) Availability and adequacy of installation, operation, maintenance, and repair manuals.

b) Existence of adequate distribution and customer after-sales service.

c) Training of field personnel.

d) Availability of spare parts.

e) Field trials.

f) Satisfactory completion of qualification tests.

g) Physical inspection of early production units and their packaging and labeling;

h) Evidence of process capability to meet specification on production equipment.

8.8   Design-change control

The quality system should include documented procedures for controlling the release, change, and use of documents that define the design input and the design baseline (output), and for authorizing the necessary work to be performed to implement changes and modifications that can affect product during its entire lifecycle, including changes in software and service instructions. The procedures should provide for various necessary approvals, specified points and times for implementing changes, removing obsolete drawings and specifications from work areas, and verification that changes are made at the appointed times and places. These procedures should handle emergency changes necessary to prevent production or delivery of nonconforming product. Consideration should be given to instituting formal design reviews and validation testing when the magnitude, complexity, or risk associated with the change warrant such actions.

8.9 Design requalification

Periodic evaluation of product should be performed in order to ensure that the design is still valid. This should include a review of customer needs and technical specifications in the light of field experiences, field performance surveys, or new technology and techniques. The evaluation should also consider process modifications. The quality system should ensure that any production and field experience indicating the need for design change is fed back for analysis. Care should be taken that design changes do not cause degradation of product quality for example, and that proposed changes are evaluated for their impact on all product characteristics in the original product specification.

8.10 Configuration management in design

This discipline may be initiated once the requirements have been defined, but is most useful during the design phase. It continues through the whole life-cycle of a product (see 5.2.6).

9 Quality in purchasing

9.1 General

Purchases become part of the organization's product and directly affect the quality of its product. All purchasing activities should be planned and controlled by documented procedures. Purchased services such as testing, calibration, and subcontracted processing should also be included. A close working relationship and feedback system should be established with each subcontractor. In this way, continual quality improvements can be maintained and disputes avoided or settled quickly. This close working relationship and feedback system will benefit both parties.

The quality system for purchasing should include the following elements as a minimum:

a) The applicable issue of specifications, drawings, purchase documents and other technical data (see 9.2).

b) Selection of qualified subcontractors (see 9.3).

c) Agreement on quality assurance (see 9.4).

d) Agreement on verification methods (see 9.5).

e) Provisions for settlement of disputes (see 9.6).

f) Receiving inspection procedures (see 9.7).

g) Receiving controls (see 9.7).

h) Receiving quality records (see 9.8).

9.2 Requirements for specifications, drawings, and purchase documents

The successful purchase of supplies begins with a clear definition of the requirements. Usually these requirements are contained in contract specifications, drawings, and purchase documents which are provided to the subcontractor.

The purchasing activity should develop documented procedures to ensure that the requirements for the supplies are clearly defined, communicated, and, most importantly, are completely understood by the subcontractor. These methods may include documented procedures for the preparation of specifications, drawings, and purchase documents, meetings with subcontractors prior to the release of the purchase document, and other activities appropriate for the supplies being procured.

Purchasing documents should contain data clearly describing the product ordered. Typical elements are as follows:

a) Precise identification of type, class, and grade.

b) Inspection instructions and applicable issue of specifications;

c) Quality-system standard to be applied.

Purchasing documents should be reviewed and approved for accuracy and completeness prior to release.

9.3 Selection of acceptable subcontractors

Each subcontractor should have a demonstrated capability to furnish product which meets all the requirements of the specifications, drawings, and purchase documents.

The methods of establishing this capability can include, but are not limited to, any combination of the following:

a) On-site evaluation of subcontractor's capability and/or quality system.

b) Evaluation of product samples.

c) Past history with similar products.

d) Test results of similar products.

e) Published experience of other users.

9.4 Agreement on quality assurance

The organization should develop a clear agreement with subcontractors for the assurance of product supplied. This can be achieved by one or more of the following:

a) Reliance on subcontractor's quality system.

b) Submission of specified inspection/test data and process-control records with shipments.

c) 100% inspection/testing by the subcontractor.

d) Lot acceptance inspection/testing by sampling by the subcontractor.

e) Implementation of a formal quality system by the subcontractor as specified by the organization; in certain cases, a formal quality-assurance model may be involved (see ANSI/ISO/ASQC Q9001-1994, ANSI/ISO/ASQC Q9002-1994, and ANSI/ISO/ASQC Q9003-1994 for further information).

f) Periodic evaluation of subcontractor quality practices by the organization or by a third party.

g) In-house receiving inspection or sorting.

9.5 Agreement on verification methods

A clear agreement should be developed with the subcontractor on the methods by which conformance to requirements will be verified. Such agreements may also include the exchange of inspection and test data with the aim of furthering quality improvements. Reaching agreement can minimize difficulties in the interpretation of requirements as well as inspection, test, or sampling methods.

9.6 Provisions for settlement of disputes

Systems and procedures should be established by which settlement of disputes regarding quality can be reached with subcontractors. Provisions should exist for dealing with routine and nonroutine matters.

A very important aspect of these systems and procedures is the provision of improved communication channels between the organization and the subcontractor on matters affecting quality.

9.7 Receiving inspection planning and control

Appropriate measures should be established to ensure that received materials are properly controlled. These procedures should include quarantine areas or other appropriate methods to prevent unintended use or installation of nonconforming materials (see 14.3).

The extent to which receiving inspection will be performed should be carefully planned. The characteristics to be inspected should be based on the cruciality of the product. The capability of the subcontractor should also be considered, taking into account the factors listed in 9.3. The level of inspection should be selected so as to balance the costs of inspection against the consequences of inadequate inspection.

It is also necessary to ensure, before the incoming product arrives, that all the necessary tools, gauges, meters, instruments, and equipment are available and properly calibrated. Personnel should be adequately trained.

9.8 Quality records related to purchasing

Appropriate quality records related to product received should be maintained. This will ensure the availability of historical data to assess subcontractor performance and quality trends.

In addition, it may be useful, and in certain instances essential, to maintain records of lot identification for the purposes of traceability.

10 Quality of processes

10.1 Planning for process control

10.1.1

Planning of processes should ensure that these proceed under controlled conditions in the specified manner and sequence. Controlled conditions include appropriate controls for materials, approved production, installation, and servicing equipment, documented procedures or quality plans, computer software, reference standards/codes, suitable approval of processes and personnel, as well as associated supplies, utilities, and environments.

The operation of processes should be specified to the necessary extent by documented work instructions.

Process-capability studies should be conducted to determine the potential effectiveness of a process (see 10.2).

Common practices that can be beneficially applied throughout the organization should be documented and referenced in all appropriate procedures and instructions. These should describe the criteria for determining satisfactory work completion and conformity to specification and standards of good workmanship. Workmanship criteria should be stipulated in the clearest practical manner by written standards, photographs, illustrations, and/or representative samples.

10.1.2
Verification of the quality status of a hardware product, process, software, processed material, service, or environment should be considered at important points in the production sequence to minimize effects of errors and to maximize yields. The use of control charts and statistical sampling procedures and plans are examples of techniques employed to facilitate process control (see also 12.2).

10.1.3
Monitoring and control of processes should relate directly to finished product specifications or to an internal requirement, as appropriate. If verification of the process variables through some measurement procedure is not physically or economically practical or feasible, then verification will have to depend primarily on verification of final product characteristics. In all cases, relationships between in-process controls, their specifications, and final product specifications should be developed, communicated to the personnel concerned, and then documented.

10.1.4
All in-process and final verifications should be planned and specified. Documented test and inspection procedures should be maintained for each quality characteristic to be checked. These should include the specific equipment to perform such checks and tests, and the specified requirements and workmanship criteria.

10.1.5
The appropriate methods of cleaning and preserving, and the details of packing, including moisture elimination, cushioning, blocking, and crating, should be established and maintained in documented procedures.

10.1.6 Efforts to develop new methods for improving process quality should be encouraged.

10.2 Process capability

Processes should be verified as being capable of producing product in accordance with specifications. Operations associated with product or process characteristics that can have a significant effect on product quality should be identified. Appropriate control should be established to ensure that these characteristics remain within the specification, or that appropriate modifications or changes are made.

Verification of processes should include material, equipment, computer system and software, procedures, and personnel.

10.3 Supplies, utilities, and environment

Where important to product quality characteristics, auxiliary materials and utilities, such as water, compressed air, electrical power, and chemicals used for processing, should be controlled and verified periodically to ensure uniformity of effect on the process. Where environmental conditions, such as temperature, humidity and cleanliness, are important to product quality, appropriate limits should be specified, controlled, and verified.

10.4 Handling

The handling of product requires proper planning, control, and a documented system for incoming, in-process, and final product; this applies not only during delivery but up to the time of being put into use.

The methods of handling of product should provide for the correct selection and use of suitable pallets, containers, conveyors, and vehicles to prevent damage or deterioration due to vibration, shock abrasion, corrosion, temperature, or any other conditions occurring during the production or delivery processes.

11 Control of processes

11.1 General

Product quality should be addressed in each phase of the life-cycle (see 5.1.1).

11.2 Material control, traceability, and identification

11.2.1 Material control

All materials and parts should conform to specified requirements before being introduced into a process. However, in determining the amount and nature of receiving inspection necessary, consideration should be given to cost impact and the effect that substandard material quality will have on production flow.

In-process product, including that in in-process inventory stockrooms, should be appropriately stored, segregated, handled, and preserved to maintain its suitability. Special consideration should be given to shelf-life and deterioration control, including assessment of product in stock at appropriate intervals. (For final product storage, see 16.1.)

11.2.2 Traceability

Where traceability of product is important, appropriate identification should be maintained throughout the process, from receipt and during all stages of production, delivery, and installation, to ensure traceability to original material identification and verification status (see 11.7 and 14.2).

11.2.3 Identification

The marking and labeling of materials should be legible, durable and in accordance with specifications. Materials should be uniquely identified from the time of initial receipt, to delivery and installation at the final destination. The identification should be in accordance with documented procedures, and should be recorded. This should enable a particular product to be identified in the event that a recall or special inspection becomes necessary.

11.3 Equipment control and maintenance

All equipment, including fixed machinery, jigs, fixtures, tooling, templates, patterns, and gauges, should be proved for accuracy prior to use. Special attention should be paid to computers used in controlling processes, and especially the maintenance of the related software (see 13.1).

Equipment should be appropriately stored and adequately protected between use, and verified or recalibrated at appropriate intervals to ensure that the requirements concerning accuracy (trueness and precision) are fulfilled.

A program of preventive maintenance should be established to ensure continued process capability. Special attention should be given to equipment characteristics that contribute to product quality.

11.4 Process-control management

Processes which are important to product quality should be planned, approved, monitored, and controlled. Particular consideration should be given to product characteristics which cannot be easily or economically measured, and those requiring special skills.

Process variables should be monitored, controlled, and verified at appropriate frequencies to assure the following:

a) The accuracy and variability of equipment used.

b) The skill, capability, and knowledge of operators.

c) The accuracy of measurement results and data used to control the process.

d) Process environment and other factors affecting quality, such as time, temperature, and pressure.

e) Appropriate documentation of process variables, equipment, and personnel.

In some cases, for example where process deficiencies may become apparent only after the product is in use, the results of processes cannot be directly verified by subsequent inspection or test of the product itself. Such processes require prequalification (validation) to ensure process capability and control of all critical variables during process operation.

11.5 Documentation

Documentation should be controlled as specified by the quality system (see 5.3 and 17.3).

11.6 Process-change control

Those responsible for authorization of process changes should be clearly designated and, where necessary, customer approval should be sought. As with design changes, all changes to production tooling or equipment, materials, or processes should be documented.

The implementation should be covered by defined procedures.

A product should be evaluated after any change to verify that the change instituted had the desired effect upon product quality. Any changes in the relationship between process and product characteristics resulting from the change should be documented and appropriately communicated.

11.7 Control of verification status

Verification status of product output should be identified. Such identification should be suitable means, such as stamps, tags, notations, or inspection records that accompany the product, or by computer entries or physical location. The identification should distinguish among unverified, conforming, or nonconforming product. It should also identify the organizational unit responsible for verification.

11.8 Control of nonconforming product

Provision should be made for the identification and control of all nonconforming products and materials (see clause 14).

12 Product verification

12.1 Incoming materials and parts

The method used to ensure quality of purchased materials, component parts and assemblies that are received into the production facility will depend on the importance of the item to quality, the state of control and information available from the subcontractor, and impact on costs (see clause 9, in particular 9.7 and 9.8).

12.2 In-process verification

Verification, typically by inspections or tests, should be considered at appropriate points in the process to verify conformity. Location and frequency will depend on the importance of the characteristics and ease of verification during processing. In general, verification should be made as close as possible to the point of realization of the characteristic.

Verifications for hardware products may include the following:

a) Set-up and first-piece inspection.

b) Inspections or tests by machine operator.

c) Automatic inspection or test.

d) Fixed inspection stations at intervals throughout the process;

e) Monitoring specific operations by patrolling inspectors.

Product should not be released for further use until it has been verified in accordance with the quality plan, except under positive recall procedures.

12.3 Finished product verification

To augment inspections and tests made during processing, two forms of verification of finished product are available. Either or both of the following may be used, as appropriate.

a) Acceptance inspections or tests may be used to ensure that finished product conforms to the specified requirements. Reference may be made to the purchase order to verify that the product to be shipped agrees in type and quantity. Examples include 100 percent inspection of items, lot sampling and continuous sampling.

b) Product-quality auditing of sample units selected as representative of completed lots may be either continuous or periodic.

Acceptance inspection and product-quality auditing may be used to provide rapid feedback for corrective action of product, process, or the quality system. Nonconforming product should be reported and reviewed, removed, or segregated, and repaired, accepted with or without concession, reworked, regraded, or scrapped (see clause 14). Repaired and/or reworked products should be reinspected or retested.

No product should be dispatched until all the activities specified in the quality plan or documented procedures have been satisfactorily completed and the associated data and documentation are available and authorized.

13 Control of inspection, measuring, and test equipment

13.1 Measurement control

Control should be maintained over all measuring systems used in the development, production, installation, and servicing of product to provide confidence in decisions or actions based on measurement data. Control should be exercised over gauges, instruments, sensors, special test equipment, and related test software. In addition, manufacturing jigs, fixtures such as test hardware, comparative references, and process instrumentation that can affect the specified characteristics of a product or process should be suitably controlled (see 11.3).

Documented procedures should be established to monitor and maintain the measurement process itself in a state of statistical control, including equipment, procedures, and operator skills. Inspection, measuring, and test equipment, including test software, should be used in conjunction with documented procedures to ensure that measurement uncertainty is known and is consistent with the required measurement capability. Appropriate action should be taken when accuracy is not adequate to measure properly the process and product.

13.2 Elements of control

The procedures for control of inspection, measuring, and test equipment and test methods should include, (as appropriate):

a) Suitable specification and selection, including range, accuracy, and robustness, under specified environmental conditions.

b) Initial calibration prior to first use in order to validate the required accuracy (accuracy and precision); the software and procedures controlling automatic test equipment should also be tested.

c) Periodic recall for adjustment, repair, and recalibration, considering the manufacturer's specification, the results of prior calibration, and the method and extent of use, to maintain the required accuracy in use.

d) Documentary evidence covering unique identification of instruments, frequency of recalibration, calibration status, and procedures for recall, handling, preservation, and storage, adjustment, repair, calibration, installation, and use.

e) Traceability to reference standards of known accuracy and stability, preferably to nationally or internationally recognized standards; where such standards do not exist, the basis used for calibration should be documented.

13.3 Subcontractor measurement controls

The control of measuring and test equipment and test methods may be extended to all subcontractors.

13.4 Corrective action

Where measuring processes are found to be out of control, or where inspection, measuring, and test equipment are found to be out of calibration, appropriate action is necessary. Evaluation should be made to determine the effects on completed work and to what extent reprocessing, retesting, recalibration, or complete rejection may be necessary. In addition, investigation of cause is important in order to avoid recurrence. This can include review of calibration methods and frequency, training, and adequacy of test equipment.

13.5 Outside testing

The facilities of outside organizations may be used for inspection, measurement, testing, or calibration to avoid costly duplication or additional investment, provided that the conditions given in 13.2 and 13.4 are satisfied. (For further information, see ISO 10012-1.)

14 Control of nonconforming product

14.1 General

The steps for dealing with nonconforming product should be established and maintained in documented procedures. The objectives of procedures for nonconformity control are to prevent the customer from inadvertently receiving nonconforming product and to avoid the unnecessary costs of further processing nonconforming product. The steps outlined in 14.2 to 14.7 should be taken as soon as indications occur that materials, components, or completed product do not, or may not, conform to the specified requirements.

14.2 Identification

Suspected nonconforming items or lots should be immediately identified and the occurrence(s) recorded. Provision should be made as necessary to examine or reexamine previous lots.

14.3 Segregation

The nonconforming items should be segregated, when practical, from the conforming items and adequately identified to prevent further unintended use of them until the appropriate disposition is decided.

14.4 Review

Nonconforming product should be subjected to review by designated persons to determine whether it can be accepted with or without repair by concession, repaired, reworked, regraded, or scrapped. Persons carrying out the review should be competent to evaluate the effects of the decision on interchangeability, further processing, performance, dependability, safety, and aesthetics (see 9.7 and 11.8).

14.5 Disposition

Disposition of nonconforming product should be taken as soon as practicable. A decision to accept such product should be documented, together with the reason for doing so, in authorized waivers, with appropriate precautions.

14.6 Action

Action should be taken as soon as possible to prevent unintended use or installation of nonconforming product. This action can include review of other product designed or processed following the same procedures as the product found to be nonconforming, and/or previous lots of the same product.

For work in progress, corrective action should be instituted as soon as practical in order to limit the costs of repair, reworking, or scrapping. Repaired, reworked, and/or modified product should be reinspected or retested to verify conformance with specified requirements.

In addition, it may be necessary to recall completed product, whether in a finished product warehouse, in transit to distributors, in their stores, or already in use (see 11.2). Recall decisions are affected by considerations of safety, product liability, and customer satisfaction.

14.7 Avoidance of recurrence

Appropriate steps should be taken to avoid the recurrence of nonconformity (see 15.5 and 15.6).

15 Corrective action

15.1 General

The implementation of corrective action begins with the detection of a quality-related problem and involves taking measures to eliminate or minimize the recurrence of the problem. Corrective action also presupposes the repair, reworking, recall, or scrapping of unsatisfactory product. The need for action to eliminate the cause of nonconformities can originate from sources such as the following:

a) Audits (internal and/or external).

b) Process-nonconformity reports.

c) Management reviews.

d) Market feedback.

e) Customer complaints.

Specific actions to eliminate the causes of either an existing nonconformity or a potential nonconformity are given in steps 15.2 to 15.8.

15.2 Assignment of responsibility

The responsibility and authority for instituting corrective action should be defined as part of the quality system. The coordination, recording, and monitoring of corrective action related to all aspects of the quality system should be assigned within the organization. The analysis and implementation may involve a variety of functions, such as design, purchasing, engineering, processing, and quality control.

15.3 Evaluation of importance

The significance of a problem affecting quality should be evaluated in terms of its potential impact on such aspects as processing costs, quality-related costs, performance, dependability, safety, and customer satisfaction.

15.4 Investigation of possible causes

Important variables affecting the capability of the process to meet specified requirements should be identified. The relationship of cause and effect should be determined, with all potential causes considered. The results of the investigation should be recorded.

15.5 Analysis of problem

In the analysis of a quality-related problem, the root cause or causes should be determined before corrective action is planned. Often the root cause is not obvious, thus requiring careful analysis of the product specifications and of all related processes, operations, quality records, servicing reports, and customer complaints. Statistical methods can be useful in problem analysis (see clause 20).

Consideration should be given to establishing a file listing nonconformities to help identify those problems having a common source, contrasted with those that are unique occurrences.

15.6 Elimination of causes

Appropriate steps should be taken to eliminate causes of actual or potential nonconformities. Identification of the cause or potential causes may result in changes to production, packing, service, transit or storage processes, a product specification, and/or revision of the quality system. Action should be initiated to a degree appropriate to the magnitude of the problem and to avoid the recurrence of nonconformities.

15.7 Process controls

Sufficient controls of processes and procedures should be implemented to avoid recurrence of the problem. When the corrective action is implemented, its effect should be monitored in order to ensure that desired goals are met.

15.8 Permanent changes

Permanent changes resulting from corrective action should be recorded in work instructions, production-process documentation, product specifications, and/or the quality-system documentation. It may also be necessary to revise the procedures used to detect and eliminate potential problems.

16 Postproduction activities

16.1 Storage

Appropriate storage methods should be specified to ensure shelf-life and to avoid deterioration. Storage conditions and the condition of product in stock should be checked at appropriate intervals for compliance with specified requirements and to detect any loss, damage, or deterioration of product (see also 10.1.5 and 10.4).

16.2 Delivery

Provision for protection of the quality of product is important during all phases of delivery. All product, in particular product with limited shelf-life or requiring special protection during transport or storage, should be identified and procedures established, documented, and maintained to ensure that deteriorated product is not shipped and put into use.

16.3 Installation

Installation procedures, including warning notices, should contribute to proper installations and should be documented. They should include provisions which preclude improper installation or factors degrading the quality, reliability, safety, and performance of any product or material.

16.4 Servicing

16.4.1
Special-purpose tools or equipment for handling and servicing products during or after installation should have their design and function validated, as for any new product (see 8.5).

16.4.2
Inspection, measuring, and test equipment used in the field should be controlled (see clause 13).

16.4.3
Documented procedures and associated instructions for field assembly and installation, commissioning, operation, administration of spares or parts lists, and servicing of any product should be comprehensive and be established and supplied in a timely manner. The suitability of instructions for the intended reader should be verified.

16.4.4
Adequate logistic back-up, to include technical advice, spares or parts supply, and competent servicing, should be assured. Responsibility should be clearly assigned and agreed among subcontractors, distributors and customers.

16.5 After sales

Consideration should be given to the establishment of an early warning system for reporting instances of product failure of shortcomings, to ensure rapid corrective action.

Information on complaints, the occurrence and modes of failure, or any problem encountered in use should be made available for review and corrective action in the design, processing, and/or use of the product.

16.6 Market feedback

A feedback system regarding performance in use should exist to monitor the quality characteristics of products throughout the life-cycle. This system can permit the analysis, on a continuing basis, of the degree to which the product satisfies customer requirements or expectations on quality, including safety and dependability.

17 Quality records

17.1 General

The organization should establish and maintain documented procedures as a means for identification, collection, indexing, access, filing, storage, maintenance, retrieval, and disposition of pertinent quality records. Policies should be established concerning availability and access of records to customers and subcontractors. Policies concerning documented procedures should also be established for changes and modifications in various types of documents.

17.2 Quality records

The quality system should require that sufficient records be maintained to demonstrate conformance to specified requirements and verify effective operation of the quality system. Analysis of quality records provides an important input for corrective action and improvement. The following are examples of the types of quality records, including charts, requiring control:

- Inspection reports.
- Test data.
- Qualification reports.
- Validation reports.
- Survey and audit reports.
- Material review reports.
- Calibration data.
- Quality-related cost reports.

Quality records should be retained for a specified time, in such a manner as to be readily retrievable for analysis, in order to identify trends in quality measures and the need for, and the effectiveness of, corrective action.

While in storage, quality records should be protected in suitable facilities from damage, loss and deterioration (e.g., due to environmental conditions).

17.3 Quality-records control

The quality system should require that sufficient documentation be available to follow and demonstrate conformance to specified requirements and the effective operation of the quality system. Pertinent subcontractor documentation should be included. All documentation should be legible, dated (including revision dates), clean, readily identifiable, retrievable, and maintained in facilities that provide a suitable environment to minimize deterioration or damage and to prevent loss. Records may be in the form of any type of media such as hard copy, electronic media, etc.

In addition, the quality system should provide a method for defining retention times, removing and/or disposing of documentation when that documentation has become outdated.

The following are examples of the types of documents requiring control:

- Drawings.
- Specifications.
- Inspection procedures and instructions.
- Test procedures.
- Work instructions.
- Operation sheets.
- Quality manual (see 5.3.2).
- Quality plans.
- Operational procedures.
- Quality-system procedures.

18 Personnel

18.1 Training

18.1.1 General
The need for training of personnel should be identified, and documented procedures for providing that training should be established and maintained. Appro-

priate training should be provided to all levels of personnel within the organization performing activities affecting quality. Particular attention should be given to the qualifications, selection, and training of newly recruited personnel and personnel transferred to new assignments. Appropriate records of training should be maintained.

18.1.2 Executive and management personnel

Training should be given which will provide executive management with an understanding of the quality system, together with the tools and techniques needed for full executive management participation in the operation of the system. Executive management should also be aware of the criteria available to evaluate the effectiveness of the system.

18.1.3 Technical personnel

Training should be given to the technical personnel to enhance their contribution to the success of the quality system. Training should not be restricted to personnel with primary quality assignments, but should include assignments such as marketing, purchasing, and process and product engineering. Particular attention should be given to training in statistical techniques, such as those listed in 20.2.

18.1.4 Process supervisors and operating personnel

All process supervisors and operating personnel should be trained in the procedures and skills required to perform their tasks, such as the following:

- The proper operation of instruments, tools and machinery they have to use,

- Reading and understanding the documentation provided,

- The relationship of their duties to quality, and

- Safety in the workplace.

As appropriate, personnel should be certified in their skills, such as welding. Training in basic statistical techniques should also be considered.

18.2 Qualification

The need to require and document qualifications of personnel performing certain specialized operations, processes, tests, or inspections should be evaluated and implemented where necessary, in particular for

safety-related work. The need to assess periodically and/or require demonstrations of skills and/or capability should be addressed. Considerations should also be given to appropriate education, training, and experience.

18.3 Motivation

18.3.1 General

Motivation of personnel begins with their understanding of the tasks they are expected to perform and how those tasks support the overall activities. Personnel should be made aware of the advantages of proper job performance at all levels, and of the effects of poor job performance on other people, customer satisfaction, operating costs, and the economic well-being of the organization.

18.3.2 Applicability

Efforts to encourage personnel toward quality of performance should be directed not only at production workers, but also at personnel in marketing, design, documentation, purchasing, inspection, test, packing and shipping, and servicing. Management, professional, and staff personnel should be included.

18.3.3 Quality awareness

The need for quality should be emphasized through an awareness program which can include introduction and elementary programs for new personnel, periodic refresher programs for long-standing personnel, provision for personnel to initiate preventive and corrective actions, and other procedures.

18.3.4 Measuring quality

Where appropriate, objective and accurate means of measuring quality achievements should be developed. These may be publicized to let personnel see for themselves what they, as a group or as individuals, are achieving. This can encourage them to improve quality. Recognition of performance should be provided.

19 Product safety

Consideration should be given to identifying safety aspects of products and processes with the aim of enhancing safety. Steps can include:

a) Identifying relevant safety standards in order to make the formulation of product specifications more effective.

b) Carrying out design evaluation tests and proto-type (or model) testing for safety and documenting the test results.

c) Analyzing instructions and warnings to the user, maintenance manuals, and labeling and promotional material in order to minimize misinterpretation, particularly regarding intended use and known hazards.

d) Developing a means of traceability to facilitate product recall (see 11.2, 14.2 and 14.6).

e) Considering development of an emergency plan in case recall of a product becomes necessary.

20 Use of statistical methods

20.1 Applications

Identification and correct application of modern statistical methods are important elements to control every phase of the organization's processes. Documented procedures should be established and maintained for selecting and applying statistical methods to the following:

a) Market analysis.

b) Product design.

c) Dependability specification, longevity, and durability prediction.

d) Process-control and process-capability studies.

e) Determination of quality levels in sampling plans.

f) Data analysis, performance assessment, and nonconformity analysis.

g) Process improvement.

h) Safety evaluation and risk analysis.

20.2 Statistical techniques

Specific statistical methods for establishing, controlling, and verifying activities include, but are not limited to, the following:

a) Design of experiments and factorial analysis.

b) Analysis of variance and regression analysis.

c) Tests of significance.

d) Quality-control charts and cusum techniques.

e) Statistical sampling.

NOTE
21 Guidance on the International Standards to be used for the statistical techniques that are identified may be found in ISO/TR 13425 and *ISO Handbook 3*. For guidance on dependability applications, reference should be made to ISO 9000-4 and to IEC publications.

ANNEX A (INFORMATIVE)

Bibliography

[1] ANSI/ISO/ASQC Q9000-1-1994, *Quality Management and Quality Assurance Standards—Guidelines for Selection and Use.*

[2] ANSI/ISO/ASQC Q9001-1994, *Quality Systems—Model for Quality Assurance in Design, Development, Production, Installation, and Servicing.*

[3] ANSI/ISO/ASQC Q9002-1994, *Quality Systems—Model for Quality Assurance in Production, Installation, and Servicing.*

[4] ANSI/ISO/ASQC Q9003-1994, *Quality Systems—Model for Quality Assurance in Final Inspection and Test.*

[5] ANSI/ISO/ASQC Q10011-1-1994, *Guidelines for Auditing Quality Systems—Auditing.*

[6] ANSI/ISO/ASQC Q10011-2-1994, *Guidelines for Auditing Quality Systems—Qualification Criteria for Quality Systems Auditors.*

[7] ANSI/ISO/ASQC Q10011-3-1994, *Guidelines for Auditing Quality Systems—Management of Audit Programs.*

[8] ISO 9000-2:1993, *Quality Management and Quality Assurance Standards—Part 2: Generic Guidelines for the Application of ISO 9001, ISO 9002, and ISO 9003.*

[9] ISO 9000-3:1991, *Quality Management and Quality Assurance Standards—Part 3: Guidelines for the Application of ISO 9001 to the Development, Supply, and Maintenance of Software.*

[10] ISO 9000-4:1993, *Quality Management and Quality Assurance Standards—Part 4: Guide to Dependability Programme Management.*

[11] ISO 9004-2:1991, *Quality Management and Quality System Elements—Part 2: Guidelines for Services.*

[12] ISO 9004-3:1993, *Quality Management and Quality System Elements—Part 3: Guidelines for Processed Materials.*

[13] ISO 9004-4:1993, *Quality Management and Quality System Elements—Part 4: Guidelines for Quality Improvement.*

[14] ISO 10012-1:1992, *Quality Assurance Requirements for Measuring Equipment—Part 1: Metrological Confirmation System for Measuring Equipment.*

[15] ISO 10013:—[2] *Guidelines for Developing Quality Manuals.*

[16] ISO Handbook 3:1989, *Statistical methods.*

Endnotes

[1] The recommended harmonized term is "customer" as shown in Table 1 of ANSI/ISO/ASQC Q9004-1-1994. The term "purchaser" was used in ANSI/ASQC Q91-1987, ANSI/ASQC Q92-1987, and ANSI/ASQC Q93-1987.

[2] To be published.

Index

A

Acceptable performance, 269
Access ramps, 221
Access to records, 379
Accreditation, 19, 26
    auditor certification, 645–648
    in Canada, 634–635
    in Europe, 630–631
    mutual recognition, 625
    RAB criteria, 631–632
    recognition of registration
        certificates, 639–643
    registrar criteria, 635
    of registrars, 167–170, 630–633
    in the United States, 631–634
Accreditation body, 167
    in Europe, 630
Accredited calibration laboratories,
    570
Accuracy, measurements of, 274
Achieving customer satisfaction by
    preventing nonconformance,
    61
Action plans, 281
Addressing nonconformances, 296,
    297
Adequacy audits, 288
Adjustment of plans, 223
Amendment records, 343
Amendment to contract, 76
American Industrial Hygiene
    Association (AIHA),
    545, 551
American National Standards Insti-
    tute, 543, 572, 661
American Petroleum Institute, 627
American Society for Quality
    Control, 545, 631
American Society of Safety
    Engineers (ASSE), 545
Analysis of options, 283
API STEP, 493
Appearance items, 112
Appraisal of records, 380

ASQC, 545, 631
Assessment, 176, 183
    of current status, 248
    formal registration assessment,
        296
    systemwide, 294
    validation, 295
Assessment instruments, OHSMS
    and, 547–550
Audit, 56, 192. *See also* Internal
    Audit
    auditee stress, 208
    body language, 210
    checklist, 197, 198
    communication skills, 208–212
    conflicts/difficult situations,
        210–211
    environmental, 482
    ethics, 211
    interview techniques, 209
    introductory and closing meet-
        ing, 183
    ISO 14001, 506
    listening skills, 209
    objectivity, 211
    Quality System Requirements
        9000, 578
    questions, 200
Auditor
    assessment by, 295
    certification programs, 643–647
    qualifications, 177
Audit report, 204
Audit team, 178
Automotive industry, 748

B

Baldrige. *See* Malcolm Baldrige
    National Quality Award
Bamford, Robert C., 700
Barriers to implementation, 224
Barriers to trade, 607
Belfit, Robert W., Jr., 724

Benefits of ISO 9000, 7–8
Bigelow, James S., 730
Block diagrams, 259
Bootstrap, 718–719
Bowen, Robert, 409
British Standards Institute (BSI),
    491
Brumm, Eugenia K., Ph.D., 371
Business plan, 70
Business processes, 258

C

Calibration control system, 228
Calibration laboratories, 570
CAN-P-10, 634
Capability evaluation, 713–714
Capability requirements, 111
Capur, Devan, 589
CASCO, 640, 658, 661
Casis de Dijon, 608
CE mark, 613
CE mark directive, 614
Center for Devices and Radiologi-
    cal Health (CDRH), 560
Certification, 19, 637. *See also*
    Registration
    EMS system, 490
    third-party, 25
Certification and Testing Bodies.
    See Conformity Assessment
Certified Quality Auditor (CQA),
    646–647
Champion, recognition of, 241
Change control, establishment of,
    261
Checklist, 391. *See also* Registra-
    tion
Chemical and process industry,
    724–745
    continuous improvement, 744
    contract review, 731
    corrective and preventive
        action, 735–736

Chemical and process industry (*continued*)
  customer-supplied product control, 733
  design control, 732, 741
  document and data control, 732
  handling, storage, packaging, preservation, and delivery, 736
  harmonization of process management initiatives, 724–729
  inspection, measuring, and test equipment control, 734
  inspection and testing, 733
  inspection and test status, 735
  internal quality audits, 736–737
  ISO 14000 and, 744
  management responsibility, 731
  measurement, 738
  measurement system analysis, 743
  nonconforming product control, 735
  process control, 733, 742
  product identification and traceability, 733
  production scheduling, 743
  purchasing, 732–733
  QS–9000 and, 740
  QS–9000 compared to ISO 9001, 738–740
  quality planning, 740
  quality records control, 736
  quality system, 731
  receiving inspection and testing, 742–743
  registrar selection, 737
  statistical techniques, 737, 744
Chemical Manufacturers Association (CMA), 493, 730
Chrysler Corporation, Quality System Requirements 9000 and, 576, 581
Clause, Quality System Requirements, 62
Client confidentiality, 188
Closeout, 206
Code of conduct, 682–685
Command-control standards, 540
Committee for Electrotechnical Standardization (CENELEC), 615, 616, 617

Committee for European Standardization (CEN), 615, 616
Common nonconformances, 297
Communication, Quality System Requirements 9000 and, 590, 596
Communication strategy, 238
Company-level data analysis and use, 70
Company management, responsibilities of, 300
Company profile, 342
Company-specific standards, 11
Companywide processes, charts of, 259
Competencies
  employee, 290
  ensuring, 290
Competency gaps, 291
Competent authority, 612
Competitive advantage, 216, 217
Competitor questions, in registrar choice, 171
Completeness, measurements of, 274
Compliance, demonstration of, 217
Compliance audit, 291
Computerized documentation, 336
Computer software, 668–670
Conditional approval, 184
Configuration management, 47, 56, 707
  beyond ISO 9000-3, 708–711
  ISO 9000-3 and, 707–708
Conflict of interest, 638, 679–681
Conformance standards, 40–41
"Conformite Europeene," 613
Conformity assessment, 17
  certification and testing bodies, 20, 167
  consistent procedures, 18–19, 617, 624
  definition, 606
  degree of complexity, 623
  versus industry standards, 627
  major components, 607
  modular approach. *See* Modular Approach
  quality assurance, 19–20
  US and EU, 626
Consistency and compatibility, ensuring of, 256

Construction industry
  application of ISO 9000 to, 457
  quality system model, 449
Continuous improvement, 580, 662–664
Continuous quality improvement, 44, 48
Contract, 181
Contractor, 35
Contract review, 75, 226–227, 346
  service organizations, 423
Contractual situation, 34–35, 44
Controlled documents. *See* Document and data control
Control limits, 273
Control of customer-supplied product, 349
  service organizations, 427
Control of inspection, measuring, and test equipment, 350
  service organizations, 432–433
Control of nonconforming product, 351
  service organizations, 434
Control of quality records, 144, 352
  service organizations, 437
Control of reworked product, 135
Core standards, 470
Cormaci, April, 215, 409, 465
Corporate strategy, ISO 9000 as part of, 221
Corrective action, 51, 184, 205, 220, 230, 488
Corrective action requests (CARs), 204
Corrective and preventive action, 137, 351
  service organizations, 435
Cost of registration. *See* ISO 9000
Costs of registration, 174
Council Committee on Conformity Assessment (CASCO), 640, 658, 661
Council of Ministers, 608
Country-specific standards, 11
Cox, William E., 730
Credibility. *See* Registration
Cross-linking of teams, 253
Cultural change, 218
Current performance
  acceptable, 269
  perceived, 269

Current quality system, use of, 222
Current status, assessment of, 248
Customer, 35
Customer audits, 217
Customer pressure, 216, 217
Customer requirements, 26, 269
Customer satisfaction, 70
Customer-supplied product, control
    of, 349
    service organizations, 427
Cycle of continuous improvement,
    662–664

D

Data analysis, 70
Data control. *See* Document and
    data control
Data gathering, 287
Daughtrey, Taz, 700
Deficiency, 201
Define business processes, 258
Definition of ISO, 15
Definition of project, 241
Definition of terms, 35
Definitions, 35, 39–41. *See also*
    Vocabulary
    service organizations, 416
Deibler, William J., 700
Deming, W. Edwards, 512–513
Department managers, 240
Department of Agriculture
    (USDA), 556
Department of Commerce (DOC),
    556, 633
Department of Defense (DoD),
    adoption by, 5
Department of Education (DOEd),
    556
Department of Energy (DOE), 559
Department of Health and Human
    Services (DHHS), 559
Department of Labor (DOL), 561
Department of State, 561
Dependability program
    management, 53–54
Derivative standards, 471
Descriptive requirements, 41
Design, 49
Design and development planning,
    79–80
Design changes, 83
Design control, 78–79, 346
    service organizations, 424

Design input, 80–81
Design output, 81
Design procedures, 283
Design review, 82
Design validation, 83
Design verification, 82–83
Determination of quality
    requirements, customer
    requirements, 269
Directives, 17–18, 608, 615
Disclosure, ISO 14001, 507
Discrepancy, 201
Document, terminology regarding,
    385
Document and data approval and
    issue, 90
Document and data change, 91
Document and data control, 85,
    227–228, 348
    service organizations, 426
Document and data control system,
    development of, 253
Documentation, 43–44, 47, 218,
    220, 223, 327, 338
    approval of, 288
    benefits, 304
    design of, 282
    levels, 315–318
    quality manual, 315
    records, 317
    trial, 287
    work instructions, 316
Documentation responsibilities, 286
Document control, 227
Documenting personnel qualifica-
    tions, 154–157
Document preparation guidelines,
    255
Document quality plans, 275
Document review, 182
Dual operation, 291
Durand, Ian, 5, 409, 465
Dutch Council for Certification
    (RvC), 630
Dyczkowsky, Bohdan, 696

E

EC 92, 17
Eco-Management and Audit
    Scheme (EMAS), 491
Education, 237
    of management, 247
    quality system, 451

Educational institution, 447
Electronic control of documents,
    147–148
Element design, validation of, 285
Elements of quality system, 246
Element teams, 255
    establishment of, 278
Employee competencies, 290
Employee involvement, 224
Employees
    feedback from, 292
    source of information, 282
Empowerment, 328. *See also*
    Responsibility
EMS system development
    certification issues, 490
    formulation strategy, 490
    implementation issues, 490
    initial review, 488
EN 45000 series, 613, 624, 626,
    636, 637
EN 45012, 167, 635–637,
    651–655
Eng, P., 696
Engineering approved product
    authorization, 136
Ensure competencies, 290
Environmental auditing, 482
Environmental labeling, 483
Environmental management
    standards
    advantages, 479
    auditing, 482
    corrective action, 488
    critical elements, 498
    development of, 480
    implementation, 487
    ISO 14001, 485
    labeling, 483
    Life Cycle Assessment (LCA),
        483
    management review, 488
    management systems, 482
    participation in development,
        495
    performance evaluation, 483
    planning, 487
    product standards, 485
    reasons for, 478
    relationship to ISO 9000, 493
    structure of, 481
Environmental management
    systems (EMS), 482, 670.
    See also Environment

Environmental performance evaluation, 483
Environmental policy, 486
Environmental Protection Agency (EPA), 539, 542
ESPRIT program, 718
Ethics, RAB code of conduct, 682–685
European Accreditation of Certification (EAC), 637, 643
European Commission, 718
European Committee for Information Technology Testing and Certification (ECITC), 644
European Committee for Quality System Assessment and Certification (EQS), 640
European Economic Area (EEA) Treaty, 20
European Free Trade Association (EFTA), 611, 616
European Network for Quality System Assessment and Certification (EQNET), 637, 639
European Organization for Technical Approvals (EOTA), 615
European Organization for Testing and Certification (EOTC), 624, 643–644
European Telecommunications Standards Institute (ETSI), 615, 618
European Union (EU), 643
    conformity assessment, 17
    directives, 17–18
    harmonized standards, 17–18
    regulatory hierarchy, 18
    role in implementation of standards, 16–17
Evaluation of subcontractors, 97
Expected benefits, 242
External resources, 223

F

Facilities, 246
Factors to be controlled
    identification of, 270
    prioritization of, 272
    selection of, 272

Federal agencies. *See* Government agencies
Federal Trade Commission (FTC), 562
Feedback, from employees, 292
Final inspection and testing, 120
Financial considerations, 48–49
Fleischli, Dave, 575
Flowcharts, 259, 332
Follow-up, 205
Ford Motor Company, Quality System Requirements 9000 and, 576, 582
Formal management involvement, 219
Formal registration assessment, 296
Frequency of measurement, 274
Future of ISO 9000, 6–7

G

Gap analysis, 251, 279, 447
General Motors North America, Quality System Requirements 9000 and, 576, 582
General statement, 320
Global economy
    international standards and, 9–10
    ISO 9000 standards and, 11–12
    technological basis of, 10
Global Environmental Management Initiative (GEMI), 491
Gold, Marc, 504
Good manufacturing practice (GMP), 560
Gore, Al (Vice President), 545
Goult, Roderick, 215, 303, 320, 328
Government, OHSMS and, 542, 544
Government agencies, 627
    ISO 9000 in, 556
Guide, 60, 68, 167, 637
Guide 25 (laboratory accreditation), 565–571
Guidelines
    for developing quality manuals, 368
    for implementation, 221
Guides, 640–641

H

Hale, Guy, 575
Handling, storage, packaging, preservation, and delivery, 138, 352
    service organizations, 436
Hardware, 36
Harmonized standards, 17–18, 607, 615–618
Healthcare, Quality System model, 453
Hooton, Bill, 575
Hunter, David, 724
Hutchins, Greg, 7

I

Identification. *See* Product Identification and Traceability
Identification of options, 283
IEEE Software Engineering Standards, 715, 716
Implementation
    case study: ISO 9002, 444
    EMS system, 490
    environmental management standards, 487
    guidelines for, 221
    of ISO 14001, 497
    Quality System Requirements 9000 (QS-9000), 583, 589–602
    roadblocks to, 224
    service organizations, 444, 446
    timeline for, 236
Implementation priorities, 247
Implementation process
    issues to be addressed, 237
    overview, 233
    phase 1: achieve commitment, 239
    phase 2: plan and organize, 248
    phase 3: define and analyze processes, 258
    phase 4: develop quality plans, 267
    phase 5: design quality system elements, 278
    phase 6: document quality system elements, 286
    phase 7: implement quality system elements, 289
    phase 8: validate implementation, 293

Implementation strategy, reevaluation of, 239
Inadequate records, 232
Incomplete corrective action, 230
Independent Association of Accredited Registrars (IAAR), 634
Industry Standards, 627
Information Technology (IT), 10, 645
Information Technology Sector (ITQS), 645
Initial formal audit, 299
In-process inspection and testing, 120
Inputs
  identification of suppliers for, 262
  validation of, 262
Inspection, control of, 228
Inspection, measuring, and test equipment, 51
  control of, 350
  service organizations, 432–433
Inspection and testing, 118, 350
  service organizations, 431
Inspection and test records, 121
Inspection and test status, 132, 351
  service organizations, 433
Institute for Quality Assurance (IQA), 646
Insurance industry, OHSMS and, 545
Interfaces, organizational and technical, 80
Internal audit
  auditing the internal audit system, 207
  audit report, 204
  closing meeting, 202
  collecting information, 199
  corrective action, 205–206
  information sources, 197
  nonconformities, 201
  objective and scope, 197
  opening meeting, 199
  selecting the team, 197
  verifying observations, 201
Internal auditing, 47
Internal audit system, 193
Internal operations, improvement in, 216, 218

Internal quality audits, 148, 192, 220, 353
  service organizations, 438
International Accreditation Forum (IAF), 642
International Chamber of Commerce (ICC), 491
International Electrotechnical Commission (IEC), 616, 717
International Loss Control Institute's International Safety Rating System (ILCI-ISRS), 547
International marketplace, influence of ISO 9000 on, 3–8
International Safety Rating System (ISRS), 547
International Trade Administration (ITA), 556
Inventory, of records, 377–378
ISO 9000
  construction industry, application to, 457
  ISO 9000-1, 41–44
  ISO 9000-2, 53
  ISO 9000-3, 53, 700–711
  ISO 9000-4, 53–54
  metals industry, 745–748
  origins of, 21
  in U.S. government agencies, 556
ISO 9000 industry, 657–658
  alternate routes to registration, 666
  computer software, 668–670
  continuous improvement, 662–664
  environmental management systems, 666
  industry-specific standards, 666–668
  registration credibility, 658–661
  responsibilities of organizations, 661–662
  standards interpretation, 665–666
  statistical techniques, 664–665
ISO 9000 series. See also Registration
  background of, 12–13
  compared to MBNQA, 526
  overview, 513

ISO 9001, 40–41
  4.1 management responsibility, 62
  4.2 quality system, 70–71
  4.3 contract review, 75
  4.4 design control, 78–79
  4.5 document and data control, 85
  4.6 purchasing, 91
  4.7 control of customer-supplied product, 106
  4.8 product identification and traceability, 107
  4.9 process control, 107
  4.10 inspection and testing, 118
  4.11 control of inspection, measuring, and test equipment, 125
  4.12 inspection and test status, 132
  4.13 control of nonconforming product, 133
  4.14 corrective and preventive action, 137
  4.15 handling, storage, packaging, preservation, and delivery, 138
  4.16 control of quality records, 144
  4.17 internal quality audits, 148
  4.18 training, 152
  4.19 servicing, 157
  4.20 statistical techniques, 158
  cycle of continuous improvement, 662–664
  differences from ISO 14001, 497
  integration with ISO 14001, 496
  interpretation, 649–650
  key points, 61–62
  language, 418
  service organization, use by, 416
  standard, 59
  TickIT and, 717
  underlying principles, 417
ISO 9002, 41
  case study: implementation, 444
  language, 418
  service organization, use by, 416
  underlying principles, 417

ISO 9003, 41
ISO 9004
    ISO 9004-1, 44
    ISO 9004-2, 55
    ISO 9004-3, 56
    ISO 9004-4, 56
    ISO 9004-7, 56
    service organizations, 441
ISO 10011, 56, 645
ISO 10012, 57
ISO 10013, 57
ISO 14000. *See* Environmental
    management standards
ISO 14000 series, 6, 648–649
ISO 14001
    auditing issues, 507
    compliance, reasons for, 505
    differences from ISO 9001, 497
    disclosure issues, 507
    elements of, 486
    implementation of, 497
    integration with ISO 9001, 496
    integration with responsible
      care, 499
    legal considerations, 504–509
    priority identification, 508
    standard of care, 509
    structure, 485
ISO 14010–12, 548
ISO defined, 15
ISO family of standards, 21–22
ISO/IEC Guides, 640–641

J

Job aids, development of, 286

K

KISS principle, 327
Kolka, James, 606

L

Labeling, environmental, 483
Labor, OHSMS support by, 544
Laboratories. *See* Conformity
    assessment
Laboratory accreditation, 565–571
Laboratory Accreditation Program
    (NVLAP), 557
Laboratory product, 568
Labor relations, effect of OHSMS
    on, 543

Lake, Peter B., 745
Language, service organizations,
    418
Legal issues
    ISO 14001, 504–509
    records requirements, 372
    statute of limitations, 376
Legal requirements, 26
Levine, Steven P., Ph.D., CIH,
    537
Liability, product liability and
    safety, 44
Liability concerns, 26–27
Life Cycle Assessment (LCA),
    483
Life cycle phases of a product, 38
Line departments, representatives
    of, 279
Line managers, 282
Long-term business strategy, 240

M

McVaugh, John (Jack), 499
Malcolm Baldrige National Quality
    Award (MBNQA), 442
    compared to ISO 9000, 526
    overview, 518
Management
    education of, 247
    responsibilities of, 300
Management commitment, 486
Management representative, 66
Management responsibility, 46–47,
    62, 343, 731
    for implementation, 235
    service organizations, 419
Management review, 43, 68–69,
    71, 206, 488
Management systems, 24
    environmental, 482
Managers, critical role in
    implementation, 239
Manuals, development of, 57
Manufacturing capabilities, 581
Marketing department quality, 49
Marketplace competition, 27
Marketplace forces, 473
Marketplace questions, in registrar
    choice, 170
Marquardt, Donald W., 4, 5, 6, 9,
    38, 657
MBNQA. *See* Malcolm Baldrige
    National Quality Award

Measurement, 57
    of accuracy, 274
    of completeness, 274
    control of, 228
    frequency of, 274
    of overall process performance,
      262
    performance tracking, 292
    of service, 274
    of timeliness, 274
    types of, 274
Measuring equipment. *See* Inspec-
    tion, measuring, and test
    equipment
Medical devices, 611
Memorandum of understanding
    (MOU), 173, 632
Metals industry, 745–748
    automotive industry and, 748
    change in, 745–746
    future of, 748
    internal improvements, 747
    registration, 746
    TQM, 747
Middleton, David, 192
Modification of processes, 264,
    278
Modified Preliminary or Ongoing
    Capability Requirements,
    111
Modular approach, 618–620, 643
    conformity to type, 622
    description of modules in, 620
    full quality assurance, 623
    internal control of production,
      620
    production quality assurance, 622
    product quality assurance, 622
    product verification, 622
    type-examination, 622
    unit-verification, 623
Morrow, Mark, 3, 689
MOU, 173, 632
MRA, 610, 625, 633
Mutual recognition, 608, 610, 625
Mutual Recognition Agreement
    (MRA), 610, 625, 633

N

NACCB, 625, 716
National Accreditation Council of
    Certification Bodies
    (NACCB), 625, 716

National Accreditation Program for Registration Organizations (NAPRO), 634

National Aeronautics and Space Administration (NASA), 562
adoption by, 5

National Committee for Quality Assurance (NCQA), 442

National Institute for Occupational Safety and Health, 559

National Institute of Standards and Technology (NIST), 557, 633

National Performance Review (NPR), 539, 545

National Sanitation Foundation (NSF), 491

National Voluntary Conformity Assessment System Evaluation (NVCASE), 557, 633

National Voluntary Laboratory Accreditation Program (NVLAP), 557

Nee, Paul A., 457

Nestel, Glen, 478

New-approach directives, 609
components, 609, 611–612

NIST, 557, 633

Nonaccredited registrars, 168

Nonconformance, 230
addressing, 296, 297
common, 297

Nonconforming product, 51

Nonconforming product, control of, 351
service organizations, 434

Nonconformities, 201

Nonconformity report, 203, 205

Noncontractual situation, 34

Non-regulated products, 608, 630, 643

Norrid, Warren, 587

Notified bodies, 612–613, 624, 625–626, 633–634

Nuclear Regulatory Commission (NRC), 563

NVCASE, 557, 633

O

Objective evidence, 199

Occupational health and safety management system (OHSMS)
AIHA guidance document, 551

Clinton administration's "reinvention" agenda and, 545

command-control standards, and, 540

development, current state of, 543

expected benefits of, 541

government, benefits to, 542

government, support by, 544

industry, benefits to, 542

insurance industry, and, 545

labor, support by, 544

labor relations, predicted effect on, 543

national/international benefits, 541

OSHA and, 550

overview, 538

public and private assessment instruments and, 547–550

rationale for, 538

relation to existing standards/policies, 546

standards-developing organizations, 545

support for, 543

Occupational Safety and Health Act, modifications to, 549

Occupational Safety and Health Administration (OSHA), 539, 542, 550

Official journal, 611

Old-approach directives, 609, 614, 643

Ongoing process performance requirements, 111

Operating procedures, 316, 322

Operations manual. See Quality Manual; Quality manual

Options
analysis of, 283
identification of, 283
revision of, 284
selection of, 283

Organization, 35, 64

Organizational and technical interfaces, 80

Organizational assessment, 248

Organizational chart, 321

Organization records, 385

Other useful standards in the ISO 9000 family, 53

Outputs
identification of customers for, 262
validation of, 262

Overall process performance, measurement of, 262

Overview of registration process, 298

P

Peach, Robert W., 33, 391

Peer review, 639

People-procedure analysis, 282

Perceived current performance, 269

Performance evaluation, environmental, 483

Performance tracking, 292

Personal records, 385

Personnel, 52

Personnel qualifications. See Training

Planning, 222, 487
adjustment of plans, 223

Policy statement, 320, 321

Potts, Elizabeth, 180

Pratt, Roger, 208

Pre-assessment, 182
for registration, 175

Preliminary process capability requirements, 111

Presumption of conformity, 609–610

Prevention. See Product liability and safety

Preventive action, 220. See also Corrective and preventive action

Principle of mutual recognition. See Mutual recognition

Priorities
identification and clarification of, 241
implementation, 247
position of implementation among, 247
refining of, 279
setting of, 256

Prioritization, 272

Procedure index, 317, 322

Procedures, 316, 322
administration and control, 333–334

Procedures (*continued*)
benefits, 322
common quality system
elements, 256
computerized documentation,
336
current practice, 323
development of, 286, 287
distribution, 334
documentation, 327
implementation and control,
337
need, 322–323
numbering system, 334
outline of, 284
planning and development, 322,
332
purpose, 326
putting into effect, 291
refinement of, 287
revision, 335–336
sample procedures, 355
scope, 325
service organizations, 446
structure and format, 332
templates, 332
Process, 43
Process changes, 112
Process charts
creation of, 259
second-level, 261
third-level, 261
Process control, 107, 220,
349
service organizations, 429
Process-cost approach, 49
Processed material, 36, 56
Processes
control, 50
modification of, 278
quality, 50
Process interfaces, characterization
of, 262
Process monitoring and operation
instructions, 110–111
Process performance, measurement
of, 262
Product, 35–36
hardware, 36
processed material, 36
services, 36
software, 36
Product authorization, 136
Product certification, 613

Product identification and
traceability, 105–107, 349
service organizations, 428
Production part approval process,
580
Productivity, 218
Product liability and safety, 44
introduction, 259
Product liability records, 376
Products, 245
Product safety, 52
Product standards, 485
Product type, 622
Project plan, development of,
253
Project planning, 448
Project structure
common, 252
establishment of, 251
Project team, 254, 279
Project team leader, 252
Publicity for registration, 187
Purchaser, 36
Purchaser-supplied product. *See*
Customer-supplied product
Purchasing, 50, 91, 348
service organizations, 427
Purchasing data, 98

Q

QS-9000. *See* Quality System
Requirements 9000
(QS-9000)
Qualitative needs and expectations,
269
Quality, 36–37
Quality assurance, 19–20, 38–39
quality management compared,
22–24
Quality assurance standard, 244
Quality concepts, 39
Quality cost approaches, 44
Quality-costing approach, 48
Quality economics, 44
Quality improvement, 48
Quality in marketing, 49
Quality loss approach, 49
Quality management, 37–38
quality assurance compared,
22–24
Quality management system, 24
Quality manual, 182, 255, 315,
319–321, 322

development of, 286, 289
guidelines for development, 368
sample, 341
Quality objectives, service
organizations, 422
Quality planning, 72, 219
Quality plans, 47, 256, 317
Quality policy, 63, 320–322
Quality policy statement, 343
Quality records, 47
Quality records, control of, 352
service organizations, 437
Quality requirements
determination of, 268
translation into control factors,
270
Quality standards. *See* Standards
Quality system, 24, 37, 70–72,
345, 659, 731
construction industry model,
449
education model, 451
healthcare model, 453
service organizations, 418, 421
staffing industry model, 455
Quality system assessment (QSA),
578
Quality system elements, 246
Quality system guidance, development of, 253
Quality system registrar, selection
of, 257
Quality system registration, service
organizations, 447
Quality System Requirements 9000
(QS-9000), 5–6
audit, 578
auto industry-specific standards,
583
benefits of, 587–588
Chrysler-specific requirements,
581
communication and, 592, 596
companies affected, 576
continuous improvement, 580
Ford-specific requirements,
582
General Motors-specific
requirements, 582
history of, 577
implementation, 583, 589–602
information sources, 579
interpretations, 577, 650
management process and, 589

Quality System Requirements 9000 (QS-9000) (*continued*)
manufacturing capabilities, 581
metals industry, 745–748
overview, 575
positioning for change, 590
production part approval process, 580
registrar accreditation, 649
registrars, 597–602
registration, 597
registration process, 576
requirements beyond ISO 9001 standard, 578
resources, 596
teams and, 592
training and, 596
truck manufacturers-specific requirements, 583

R

RAB. *See* Registrar Accreditation Board
Reason behind ISO 9000, 4–6
Reasons for implementing ISO 9000 systems, 216
Receiving inspection and testing, 118
Record maintenance, 291
Records, 52, 76, 317
access to, 379
appraisal, 380
inaccessible, 233
inadequate, 232
incomplete, training, 232
interview of key personnel regarding, 379–380
organization, 385
personal, 385
terminology regarding, 385
Records destruction, litigation imminent, 384
Records inventory, 377–378
Records management, 371, 376
statute of limitations, 376
Records requirements, 372
Records retention
distribution of schedule, 381–382
draft schedule, 380–381
implementation of schedule, 382
records on development of schedule, 383

schedule, 376
signatures, 381
Redinger, Charles, F., MPA, CIH, 537
Reevaluation of implementation strategy, 239
Refining of priorities, 279
Register of holders of controlled copies, 343
Registrar Accreditation Board (RAB), 572, 625, 631, 632, 658, 661
auditor certification, 646–647
code of conduct, 682–685
conflict of interest policy, principles and implementation, 679–681
MOU, 632
white paper defining scope, 671–678
Registrars
accreditation, 167. *See also* Accreditation
accreditation status, 170
audit team of, 178
background questions, 172
consulting versus registration, 638
financial security, 173
internal operations, 173
Quality System Requirements 9000, 597–602
selection of, 169
subcontracting by, 173
technical competence of assessors, 638
Registration, 19, 184, 543
alternate routes to, 666
application, 181
approval, 184
assessment, 183
benefits for individual companies, 691–696
Canadian system, status of, 696–697
checklist, 391
concerns about, 27
conditional approval, 184
costs, 186
disapproval, 184
document review, 182
globally, 691

growth of, 3, 689–697
pre-assessment, 182
publicizing, 187
Quality System Requirements 9000 and, 576, 597
reasons for, 3–4
scope of, 659
service organizations, 447
subcontractors, 27
surveillance, 185
third-party, 25, 467
time, 186
Registration assessment, 296
Registration costs, 174
Registration credibility, 658–661
Registration process. *See also* ISO 9000
overview of, 298
Regulated products, 608–609, 630
Reid, R. Dan, 587
Relationship of quality concepts, 39
Release for urgent production purposes, 119
Requirements, service organizations, 419
Research records, records retention, 380
Resources, 65
Quality System Requirements 9000, 596
Responsibilities of company management, 300
Responsibility, 321, 333, 731
and authority, 64
Responsible care, 493
history of, 503
integration with ISO 14001, 499
Review
management, 206
of quality system, 226
Review and disposition of nonconforming product, 135
Review inputs, 283
Revision of options, 284
Reworked product, 135
Roadblocks to implementation, 224

S

Safety of products. *See* Product liability and safety

Sample procedures, 355
Sample quality manual, 341
Scope
    laboratory accreditation, 569
    procedures, 325
    RAB white paper defining,
        671–678
Scope of quality system, 245
Second-level process charts, 261
Second-party registration, 35
SEI Capability Maturity Model,
    713–714
Selection of accredited registrar,
    169
Selection of options, 283
Selection of quality system regis-
    trar, 257
Senior managers, 240
Service, 36
    measurements of, 274
Service organizations
    characteristics, 411
    contract review, 423
    control of customer-supplied
        product, 427
    control of inspection,
        measuring, and test
        equipment, 432–433
    control of nonconforming
        product, 434
    control of quality records, 437
    corrective and preventive action,
        435
    definitions, 410, 416
    design control, 424
    document and data control, 426
    educational institution, 447
    examples of services, 411
    handling, storage, packaging,
        preservation, and delivery,
        436
    implementation, 444, 446
    inspection and testing, 431
    inspection and test status, 433
    internal quality audits, 438
    ISO 9004, 441
    language, 418
    Malcolm Baldrige National
        Quality Award (MBNQA),
        442
    management responsibility, 419
    National Committee for Quality
        Assurance (NCQA), 442
    procedures, 446

    process control, 429
    product identification and
        traceability, 428
    purchasing, 427
    quality objectives, 422
    quality system, 418, 421
    quality system registration, 447
    requirements, 419
    servicing, 440
    standards, 420
    statistical techniques, 441
    teamwork, 446
    training, 439
    use of ISO 9000, 413–415
    use of ISO 9001 or ISO 9002,
        416
Servicing, 157, 353
    service organizations, 440
Single internal market, 607
    objectives, 607
Society of Automotive Engineers,
    627
Software, 36, 668–670
    IEEE software engineering
        standards, 715, 716
    SEI capability maturity model,
        713–714
    TickIT, 716–717
Software capability evaluation,
    713–714
Software development, 700
    acceptance of ISO 9001, 700
    application of ISO 9001 to,
        700–701
    design, 703
    guidance in ISO 9000-3,
        701–706
    maintenance and servicing, 704
    mapping ISO 9001 to 9000-3,
        706–708
    production, 704
    terms and definitions, 703
    verification and validation,
        704–706
Software process assessment,
    713–714
Software processing assessment,
    SPICE, 717–718
Software producing unit (SPU),
    718
SPICE, 717–718
Staffing industry, quality system
    model, 455
Stakeholder, 43, 545

Standard, 244
Standardization, 14
Standards
    core standards, 470
    derivative standards, 471
    development and use of, 465
    future of, 465–475
    history of, 10–11
    international approaches, 14
    marketplace forces and, 473
    national adoption of, 13
    need for, 11
    proliferation of, 471
    relation of OHSMS to, 546
    service organizations, 420
Standards Council of Canada
    (SCC), 634
Standards-developing organizations,
    545
Standards interpretation, 665–666
Statement of purpose, 241
Statistical methods, 53
Statistical techniques, 158, 353,
    664–665
    service organizations, 441
Steering committee, 252, 254, 279
Stephens, Kenneth S., 511
Strategic Advisory Group (SAGE),
    480
Structure
    of ISO 9000, 41
    of ISO 14000, 481
    of ISO 14001, 485
Subcontracting, 625
    by registrar, 173
Subcontractors, 27, 36
Supplier, 36
Supplier Audit Confirmation (SAC)
    program, 572–573
Supplier-chain terminology, 35
Surveillance, 185
Systems outline, 322
Systemwide assessment, 294

                    T

Target date, 244
Team leader, 252
Teams, Quality System
    Requirements 9000 and, 592
Teamwork, service organizations,
    446
Technical Committee 207 (TC
    207), 538

Technical documents, 228
Technical interfaces, 80
Technical requirements, 609
Technical Standards, 611
Telecommunications products, 718
Templates, 332
Terms, 35, 39
Test data, 568
Test equipment. *See* Inspection, measuring, and test equipment
Test equipment, control of, 228
Testing, 350
    service organizations, 431
Testing and certification bodies. *See* Conformity Assessment
Test status, 351. *See also* Inspection and test status
    service organizations, 433
Third-level process charts, 261
Third-party inspectors, 542
Third-party registration, 25, 35, 467, 658
TickIT, 716–717
Time and costs of registration, 186

Timeliness, measurements of, 274
Time to register. *See* Registration
Timing of trials, 287
Tiratto, Joseph, 629
Total quality management. *See* TQM
TQM, 511–513, 530, 589, 747
Traceability. *See* Product identification and traceability
Trade barriers, 28
Training, 152, 220, 232, 237, 291, 353, 448
    Quality System Requirements 9000 and, 596
    service organizations, 439
Transition period, 611
Transportation technology, 10
Trial procedures, 284
Trials
    documentation, 287
    timing of, 287
Trillium, 718
Type examination, 611
Types of standards in the ISO 9000 series, 39

U

Unger, Peter S., 565
Uses of standards, 34–35
US International Trade Commission (USITC), 562
US Postal Service, 564

V

Validation, of element design, 285
Validation assessment, 295
Validation of implementation, 293
Verification, 51
    of purchased product, 99
Vocabulary, 35
Voluntary protections programs (VPP), 547
Voluntary standards, 610

W

Walsh, Steve, 587
Weightman, Bud, 166, 207
Weights and measures program, 558
Work instructions, 256, 316
    development of, 286